Optical Engineer's Desk Reference

About the Editor

William L. (Bill) Wolfe was born in Yonkers, NY, at a very early age. He attended Bucknell University in Lewisburg, PA, where he recently attended his 50th reunion. He graduated cum laude with a B.S. in physics. He went on to The University of Michigan, where he worked full time at The Institute of Science and Technology, taught at the university's Dearborn campus, and earned an M.S. in physics and an M.S.E. in electrical engineering. In 1966 he went to Honeywell's Radiation Center in Lexington, MA, where he was Chief Engineer and Department Manager of the Electro-optics Department; he also taught at Northeastern University. In 1969 he moved to Tucson, where he became Professor of Optical Science in the Optical Sciences Center of The University of Arizona. He remained in that position until 1996, when he became Professor Meritless (officially). He is a Fellow of both OSA and SPIE, served on many committees for both, and was President of SPIE in 1989. In 1999 he was awarded the gold medal of SPIE for his work on handbooks and optical material measurements and for mentoring "a remarkable number of students" in optics. Since his retirement he has written two books on infrared system design—one on radiometry and one on imaging spectrometers. He has also caught and released an incredible number of gorgeous trout and solved many, but not all, crossword puzzles. He has been wonderfully supported for 48 years by his trophy wife, Mary Lou. They raised three children and now are enjoying six grandchildren. The experiences of preparing this handbook have not completely discouraged him from undertaking other publications.

—Written by Bill Wolfe

Optical Engineer's Desk Reference

Edited by

William L. Wolfe

OSA®
Optical Society of America

The International Society for Optical Engineering

Optical Society of America
2010 Massachusetts Avenue, NW
Washington, DC 20036-1023
(202) 223-8130

SPIE—The International Society for Optical Engineering
P.O. Box 10
Bellingham, WA 98227-0010

When citing this publication, the following form for the citation is suggested in order to facilitate access to the original publication:

William L. Wolfe, *Optical Engineer's Desk Reference* (Optical Society of America, Washington, DC 2003).

ISBN Number 1-55752-757-1
LC Number 2003109710

Optical Society of America
2010 Massachusetts Ave., NW
Washington, DC 20036-1023
www.osa.org

SPIE Order Number PM 131
SPIE—The International Society for Optical Engineering
P.O. Box 10
Bellingham, WA 98227-0010
www.spie.org

Printed in the United States of America.

Contents

Contributors

Bjørn F. Andresen	ElOp Electrooptics Industries Ltd., Rehovot, Israel
Robert P. Breault	Breault Research Organization, Inc., Tucson, AZ
Eugene L. Church	Brookhaven National Laboratory, Morristown, NJ
Jasmin Côté	Side by Side Optical Engineering, Montreal, Canada
Michael Couture	Optics 1, Inc., Manchester, NH
Carolyn J. Sher DeCusatis	Freelance writer, Poughkeepsie, NY
Casimer DeCusatis	IBM Corporation, Poughkeepsie, NY
Robert E. Fischer	Optics 1, Inc., Westlake Village, CA
B. Roy Frieden	University of Arizona Optical Sciences Ctr., Tucson, AZ
J.D. Gaskill	University of Arizona Optical Sciences Ctr., Tucson, AZ
John E. Greivenkamp, Jr.	University of Arizona Optical Sciences Ctr., Tucson, AZ
Neelam Gupta	U.S. Army Research Laboratory, Adelphi, MD
Gerald C. Holst	JCD Publishing, Winter Park, FL
R. Barry Johnson	Optical ETC. Inc., Huntsville, AL
Shanalyn A. Kemme	Sandia National Laboratories, Albuquerque, NM
Vasudevan Lakshminarayanan	Dept. of Physics & College of Optometry, University of Missouri, St. Louis, MO
Emmett N. Leith	University of Michigan, Ann Arbor, MI
Angus Macleod	Thin Film Center, Inc., Tucson, AZ
Virendra N. Mahajan	The Aerospace Corp., El Segundo, CA
Arvind Marathay	University of Arizona Optical Sciences Ctr., Tucson, AZ
James F. McCalmont	Wright Patterson Air Force Base, Dayton, OH
J.M. Mooney	J.M. Mooney Imaging Technology, Irvine, CA
Kiebong Nahm	Hallym University, Chunchon, Kangwon, Korea
Aparna Raghuram	College of Optometry, University of Missouri, St. Louis, MO
Robert R. Shannon	University of Arizona Optical Sciences Ctr., Tucson, AZ
J. Shiefman	Shiefman Consulting, Tucson, AZ
Warren J. Smith	Kaiser Electro-Optics, Inc., Carlsbad, CA
William C. Sweatt	Sandia National Laboratories, Albuquerque, NM

P.Z. Takacs	Brookhaven National Laboratory, Upton, NY
Mary G. Turner	Zemax Development Corp., San Diego, CA
Mial E. Warren	Sandia National Laboratories, Albuquerque, NM
Douglas W. Wolfe	Raytheon, El Segundo, CA
William L. Wolfe	University of Arizona Optical Sciences Ctr., Tucson, AZ
James C. Wyant	University of Arizona Optical Sciences Ctr., Tucson, AZ

Preface

This handbook was initially the brainchild of John Greivenkamp and some of his students at the Optical Sciences Center at the University of Arizona. The baton was passed to me when they realized that they should get their education first! It was intended to be modeled after the *Physician's Desk Reference,* which is largely a compendium of medical formulas to which the doctor can refer. This handbook is meant to provide the formulas and procedures needed by the practicing optical engineer. Accordingly, it has no derivations and only enough description to make matters clear. I have learned that this kind of writing is very hard for most of my colleagues, who really like to teach about their respective subjects. So a little pedagogy has sneaked into these pages.

I understand the optical engineer to be someone who designs optical instruments or optical systems or the optical parts of larger systems. For these tasks there exist a wide variety of different disciplines of optics that need to be called on. I have tried to include all the pertinent ones. In addition, current practice often uses long, expensive and complicated computer codes to make some calculations. These codes have been referenced, but not described in detail. The citation of a particular program or a particular commercial vendor does not represent an endorsement or a guarantee that the code is just right and never wrong, or the best one for the reader to use. Competing programs usually have different advantages and disadvantages, and their arrangement may be better suited to some than others.

It is sometimes difficult to know when to stop. For instance, optical elements are mounted on optical benches in the lab or on a platform, and the characteristics of the mount in terms of stress, strain, vibration, and the like can be of the utmost importance. But these are mechanical considerations. The same is true of the designs for optical windows and mirror mounts. I have arbitrarily limited our scope in these and similar areas. The bounds of optical engineering are fuzzy, but the lines were drawn against quantum mechanics, relativity, nuclear physics, solid-state physics, electronics, mechanics, chemistry, the atmosphere, and vision and color (mostly). However, the vision aspects of display design have been included.

I have made every attempt at uniformity of symbols. But there are not enough letters in either the English or the Greek alphabets, and some authors have a very decided preference for some symbols. There are a few symbols that mean more than one thing and a few quantities that are described by more than one symbol, but there should be no confusion.

Although there is a rationale for the order of the chapters, most readers will simply open the book to their area of interest at the time—at least after an initial perusal. Accordingly, acronyms have been kept to an absolute minimum, and an extensive index has been prepared.

I would like to thank all of the chapter authors, who gave unstintingly of their time and knowledge. Thanks also to Publication Services, who prepared the comprehensive index. We all hope that this handbook will be of use to the practicing optical engineer.

William L. Wolfe
Tucson, Arizona
January, 2003

Glossary

Introduction

This section includes descriptions of the SI, Systeme Internationale, base and derived quantities, the values of atomic constants, numerical prefixes, and most of the symbols used in this reference book. There is now general agreement and almost universal use of both the base and derived quantities of the International System. There is also almost universal agreement on the symbols for the so-called atomic constants. There is surely not general agreement on many of the other symbols used in optics. The editor and the authors of this reference have strived to adhere to a single set of symbols with a minimum of redundancy.

Units

The Base SI units are presented first with names and units and the symbols for these units. They are consistent with those of the International Standards Organization, ISO,[1] and the International Union of Pure and Applied Physics, IUPAP.[2]

Prefixes

A table of numerical prefixes is given.[3] The table includes the scientific prefix name, the common name, where available, and the power of ten that it represents.

[1] Anonymous, *ISO Standards Handbook 2: Units of Measurement,* 2d ed. International Organization for Standardization, 1982.

[2] Anonymous, *Symbols, Units and Nomenclature in Physics,* Document U. I. P. 20, International Union of Pure and Applied Physics, 1978.

[3] Anonymous, Phys. Today, January 1998, and footnote 1.

Fundamental Constants

Much work has been done in establishing accurate values of the fundamental constants. They are given here in standard notation, indicating the value and the accuracy, along with their units.[4]

Mathematics

We have chosen i to represent the imaginary, $\sqrt{-1}$, arbitrarily because IUPAP lists both i and j and ISO lists neither. Scalars are represented by italicized Roman letters. Vectors are represented by boldfaced symbols. Matrices are represented by script or italic capital letters. The amplitudes of these vectors are represented by the same symbol but not in boldface. The vector operators are represented by standard symbols, ∇ for the gradient, $\nabla\cdot$ for the divergence and $\nabla\mathbf{x}$ for the curl, where, in Cartesian coordinates,

$$\nabla = \hat{x}\frac{\partial}{\partial x} + \hat{y}\frac{\partial}{\partial y} + \hat{z}\frac{\partial}{\partial z}.$$

The Laplacian Δ is used for the vector wave equation. A summation is indicated by Σ and a product function by Π. These usually have the running indexes indicated above and below the symbols. Average values are indicated either by bars over the functions or by enclosing the quantity inside "pointy brackets" $\langle\,\rangle$. In addition to the equals sign, =, the not-equals sign, $\neq$, the approximately equals sign, $\approx$, and the definition sign, $\equiv$, the greater-than and less-than signs, $> <$, and the proportionality $\propto$ and similar to $\sim$ signs are also used. The exponential function of x is represented either as $\exp(x)$ or e^x. The convolution operation is indicated by $\otimes$ and the correlation by $\oplus$.

Spectral Variables

These include symbols for frequency (both standard and circular or radian), wavelength and wave number. The frequency of radiation is represented by υ in cycles per second or hertz and ω (=$2\pi\upsilon$) in radians per second. We have reserved f for focal length, and in some cases an electronic frequency, but not a radiation frequency. The standard wave number, in waves per centimeter, is represented by σ. There should be no confusion with the Stefan–Boltzmann constant, and it saves considerable typography problems (compared with $\tilde{\nu}$) as well as being in agreement with ISO and IUPAP. The circular or radian wave number is k and the wave vector is **k.** Spatial frequencies are represented by ξ and η or f_x and f_y.

[4] E. Cohen and B. Taylor, *The Fundamental Physical Constants,* Phys. Today, 9, August 1990.

Radiometry

The most recent recommendations from ISO and IUPAP are followed: L is radiance; E, incidance, M, exitance; I, intensity; and Φ, radiant flux (power). We have dealt with the vexatious intensity by using field intensity and radiant intensity when there is a chance for ambiguity. These symbols are subscripted, when necessary, by u for energetic quantities, q for photonic quantities, and ν for luminous quantities. The processes of transmission, reflection, absorption and emission are indicated by these words, which end in *ion.* The properties of samples are indicated by transmittance, reflectance, absorptance, and emittance, whereas the properties of a legitimately representative sample are indicated by transmissivity, reflectivity, absorptivity, and emissivity, although often the differences between the sample and the ideal are not discernible.

Optical Design

For the most part ISO and IUPAP do not address symbols used for optical design. We have chosen f for focal length, and *efl* for effective focal length, NA for numerical aperture, and F for focal ratio. Aberrations are described in terms of the eikonal function, which is written as $W_{i,j,k}$.

Electromagnetic Fields

There is no argument that **E** and **H** are used for the electric- and magnetic-field vectors, q for quantity of charge, ρ for charge volume density, ρ_s for charge surface density, ε for permittivity, μ for permeability (with subscripts o and r for free space and relative values), **D** for displacement vector, **S** for the Poynting vector, and **J** for current spatial density.

Polarization

There are a number of conventions that must be understood in the treatment of polarization. First, the subscripts $\perp$ and $\|$ are used to indicate perpendicular and parallel polarization conditions. The convention for describing the handedness of circular polarization is that the vector rotates clockwise or counterclockwise based on the observer watching the light approach.

Base SI Quantities

Quantity	Unit Symbol	Unit Name
Length	m	meter
Time	s	second
Mass	kg	kilogram
Electric current	A	ampere

Temperature	K	Kelvin
Amount of substance	mol	mole
Luminous intensity	cd	candela

Derived SI Quantities

Electric charge	C	coulomb
Electric capacitance	F	farad
Electric conductance	siemens	
Electric potential	V	volt
Electric resistance	Ω	ohm
Energy	J	joule
Frequency	Hz	hertz
Force	N	newton
Magnetic flux	Wb	weber
Magnetic flux density	T	tesla
Illuminance	lx	lux
Inductance	H	henry
Luminance	lm	lumen
Plane angle	rad	radian
Power	W	watt
Pressure	Pa	pascal
Solid angle	sr	steradian

Metric Prefixes

Symbol	Name	Common Name	Exponent of ten
Y	yotta		24
Z	zetta		21
E	exa	quintillion[a]	18
P	peta	quadrillion[b]	15
T	tera	trillion[c]	12
G	giga	billion[d]	9
M	mega	million	6
k	kilo	thousand	3
h	hecto	hundred	2
da	deca	ten	1
d	deci	tenth	–1
c	centi	hundredth	–2
m	milli	thousandth	–3
μ	micro	millionth	–6
n	nano	billionth	–9
p	pico	trillionth	–12
f	femto		–15
a	atto		–18
z	zepto		–21
y	yocto		–24

[a]In some countries milliard; [b]in some countries billion; [c]in some countries billiard; [d]in some countries trillion.

Constants

c	Speed of light in vacuo	299792458	ms^{-1}
c_1	First radiation constant	$2\pi c^2h = 3.7417749 \times 10^{-16}$	Wm^{-2}
c_2	Second radiation constant	hc/k = 0.01438769	mK
e	Elementary charge	$1.60217733 \times 10^{-19}$	C
g_n	Free-fall constant	9.80665	ms^{-2}
h	Planck's constant	$6.6260755 \times 10^{-34}$	Ws
k_B	Boltzmann's constant	1.380658×10^{-23}	JK^{-1}
m_e	Mass of the electron	$9.1093897 \times 10^{-31}$	kg
N_A	Avogadro constant	6.0221367×10^{23}	mol
R_∞	Rydberg constant	10973731.534	m^{-1}
ε_0	Vacuum permittivity	$\mu_0^{-1}c^{-2} = 8.8541878 \times 10^{-12}$	$s^2N^{-1}A^{-1}m^2$
σ	Stefan–Boltzmann constant	5.67051×10^{-8}	$W\,m^{-2}\,K^{-4}$
μ_0	Vacuum permeability	$4\pi \times 10^{-7}$	NA^{-2}
μ_B	Bohr magneton	$9.2740154 \times 10^{-24}$	JT^{-1}

Symbols

Symbol	Meaning
A	Area
$\mathbf{B}$	Magnetic induction
B	Temporal bandwidth
c	Speed of light in vacuo
c_1	First radiation constant
c_2	Second radiation constant
C	Thermal capacitance
D	Electric displacement
D^*	Specific detectivity
E	Radiant incidance
$\mathbf{E}$	Electrical-field strength
e	Electronic charge
E_v	Illuminance
E_g	Bandgap energy
E_q	Photon incidance
f	Focal length, frequency
F	Focal ratio
G	Throughput
G	Thermal conductivity
$\mathbf{H}$	Magnetic-field strength
h	Height
h	Planck's constant
I	Radiant intensity
I_q	Photon intensity
I_v	Luminous intensity
i	$\sqrt{-1}$
I	Electric current
J	Electric current areal density
Im( )	Imaginary part
$\mathbf{k}$	Wave vector
k	Radian wave number
k	Boltzmann constant
L, M, N	Direction cosines
L	Radiance
L_q	Photon radiance, photance
L_v	Luminance
m	Linear magnification
M	Radiant exitance
MTF	Modulation transfer function
$\dot{N}$	Photon flux
N	Carrier number
n	(Real) refractive index
n	Number
$\tilde{n}$	Complex refractive index
NA	Numerical aperture
NEP	Noise equivalent power
NETD	Noise equivalent temperature difference
OPD	Optical path difference
p	Pyroelectric coefficient
Re( )	Real part
R	Resistance
$\mathbf{r}$	Position vector
r	Reflectivity amplitude
$\mathcal{R}$	Responsivity
S	Seebeck coefficient
s	Distance
T	Temperature
t	Time
u	Ray slope
V	Abbe number reciprocal dispersion
V	Voltage
x, y, z	Cartesian coordinates
Z	Atomic number

α	Absorptivity
α	Temperature coefficient of resistance
α	Detector angle
β	Detector angle
γ	Angle
δ	Phase difference
Δ	Difference
ε	Emissivity
ε	Permittivity
ζ	Spatial frequency
η	Efficiency
θ	Angle
κ	Extinction coefficient
λ	Wavelength
μ	Permeability
ν	Optical frequency
ξ	Spatial frequency
π	3.14159
ρ	Reflectivity
σ	Stefan–Boltzmann constant
σ	Wave number
σ	Conductivity
τ	Transmittance
τ	Time constant
φ	Angle
Φ	Flux
ω	Radian frequency
Ω	Solid angle
ψ	Wave function

Subscripts

a	atmosphere
b	bolometer
B	bandwidth
d	detector
D	dynamic
e	effective
e	electrical
i	current
L	load
n	noise
o	initial
o	optical
v	voltage

1

Probability and Statistics

Roy Frieden

1.1. Introduction

Probability theory has as its aim the construction of probability laws from a knowledge of inputs. This is a problem of analysis. Statistics attacks the reverse problem: the inference of inputs from an output histogram or probability law. This is an inverse mathematical problem. Both problems are addressed here.

1.2. Probability as a Frequency of Occurrence: Law of Large Numbers

In N repetitions of an experiment, if the event A occurs $n = n(A)$ times, the probability P of the event is

$$P(A) = \lim_{N \to \infty} \frac{n(A)}{N}.$$

This is the *law of large numbers.* The event A is usually one of many possible events A_m, $m = 1, \ldots, M$. If N is finite, the fractions $n(A_m)/N$ are called the *relative occurrences* of the events A_m and together they comprise the histogram curve of the events. The law of large numbers says that the histogram approaches the probability law in the limit $N \to \infty$. For finite N, see Section 1.44.

1.3. Kolmogorov's Three Axioms of Probability

All probability and statistics follow from these three axioms:

1. $P(A) \geq 0$. A probability is never negative.
2. $P(C) = 1$. The probability of the certain event is unity.
3. $P(A \text{ or } B) = P(A) + P(B)$ if A and B are disjoint, that is, the occurrence of A rules out the occurrence of B and vice versa.

1.4. Additive Property

If events $A_1, \ldots, A_n$ are disjoint, then

$$P(A_1 \text{ or } \cdots A_n) = P(A_1) + P(A_2) + \cdots + P(A_n).$$

1.5. Normalization Property

If events $A_1, \ldots, A_n$ are disjoint and together form the certain event C, then

$$P(A_1) + P(A_2) + \cdots + P(A_n) = 1.$$

1.6. Joint or Compound Events

The event AB denotes the occurrence of (A and B). Generally $A_1A_2 \cdots A_n$ denotes the joint event (A_1 and A_2 and $\cdots A_n$). The comma notation $A_1, A_2, \ldots, A_n$ is alternatively used to denote the joint event.

1.7. Marginal Probability

Given all the joint probabilities $P(A_mB_n)$, $m = 1, \ldots, M$, $n = 1,2, \ldots, N$, the elemental or marginal probability $P(A_m)$ obeys

$$P(A_m) = \sum_{n=1}^{N} P(A_mB_n), \quad m = 1,2, \ldots, M.$$

1.8. Laplace's Definition of Probability

For an event space A_1 or A_2 or $\cdots A_N$, where all A_m are disjoint and equally probable, $P(A_m) = 1/N$. Let event $B = A_1$ or A_2 or $\cdots A_n$, with $n \le N$; then $P(B) = n/N$. This is the number of ways the one event B can happen divided by the number of different ways all events A_m of the space can occur.

1.9. Optical Objects and Images as Probability Laws

A pixel event x_n may be defined as $x_n < x \le x_{n+1}$. By the law of large numbers (Section 1.2), if $i(x_n)$ denotes an image irradiance (or object irradiance distribution), then the probability $P(x_n)$ obeys

$$P(x_n) = \frac{i(x_n)}{\sum_m i(x_m)}.$$

This holds in the limit as the pixel size approaches zero and $P(x_n) \longrightarrow p(x)\mathrm{d}x$, where $p(x)$ is a probability density for the random position x.

1.10. Conditional Probability

The probability that event B will happen if A already has obeys by definition

$$P(B|A) = \frac{P(AB)}{P(A)}.$$

The vertical bar means "if." All variables to the right of the bar are assumed to be known constants.

1.11. Shannon Information

There are many forms of information (described in Section 1.51). The Shannon information in a message A about the possible occurrence of an event B is given by

$$I_S(A, B) = \log_b \left[\frac{P(B|A)}{P(B)} \right].$$

The Shannon information $I_S(A, B) > 0$ if B is more probable in the presence of A than in its absence. It is dimensionless and has the unit of bit if the base b is 2.

1.12. Statistical Independence

If $P(B|A) = P(B)$ then the events A and B are termed independent.

1.13. Multiply Independent Events

If the events $A_1, \ldots, A_n$ are all independent, then joint probability $P(A_1, \ldots, A_n) = P(A_1)P(A_2) \cdots P(A_n)$, the product of the individual probabilities.

1.14. Partition Law

If an event A_m occurs jointly with either B_1 or B_2 or $\cdots$ B_N, where the latter are disjoint, then

$$P(A_m) = \sum_{n=1}^{N} P(A_m|B_n)P(B_n).$$

1.15. Bayes' Rule

The order *AB* or *BA* chosen to name a joint event does not matter. Thus

$$P(AB) = P(BA),$$

or, equivalently,

$$P(B|A) = \frac{P(A|B)P(B)}{P(A)}.$$

1.16. Markov Events Sequence

A Markov events sequence is a sequence of repeated trial events (say, the consecutive purchases of brands *AABAAB*...) for which the probability of an event (say the purchase of brand *B*) on a given trial *n* depends on the preceding event $n - 1$ (that is, whether *A* or *B* was purchased). A "good" brand *A* has a high probability $P(A_n|A_{n-1})$ of repetition, whereas a "poor" brand has a low probability (for a smart shopper).

1.17. Complex-Number Events

If the event *A* is complex, $A = X + iY$, then $P(A) = P(XY)$, the joint probability.

1.18. Vector Events

If the event **A** is a vector, $\mathbf{A} = (A_1, A_2, \ldots, A_n)$, then $P(\mathbf{A}) = P(A_1, A_2, \ldots, A_n)$.

1.19. Continuous Random Variables and Probability Density Function

The event *A* that a number *x* lies anywhere on the fixed interval $a < x < b$ obeys

$$P(A) = \int_a^b p(x)\mathrm{d}x$$

by definition of $p(x)$, where $p(x)$ is the probability density function for the random variable *x*.

1.20. Rules for Forming Probability Density Functions

The equations for probability density functions can be easily obtained from the equations for discrete probabilities: One merely replaces the summations of the latter with continuous integrals. The following examples show this:

Marginal law (Section 1.7):

$$p(x) = \int p(x, y)\mathrm{d}y.$$

Independence (Section 1.13):

$$p(x_1, \ldots, x_n) = p_1(x_1)p_2(x_2)\cdots p_n(x_n).$$

Partition law (Section 1.14):

$$p(x) = \int p(x|y)p(y)\mathrm{d}y.$$

Bayes' rule (Section 1.15):

$$p(y|x) = \frac{p(x|y)p(y)}{p(x)}.$$

1.21. Imaging Equation as an Expression of the Partition Law

The probability that a random photon arrives at pixel position y in an image obeys

$$i(y) = \int o(x)s(y|x)\mathrm{d}x.$$

This is the imaging equation for incoherent light in which the probability density functions $i(y)$, $o(x)$, and $s(y|x)$ are the irradiances. The optical point-spread function $s(y|x)$ is the probability the image position is y for a photon that originated at object position x. The point-spread function is the probability density function.

1.22. Properties of the Dirac Delta Function

The perfect impulse or spike function is defined as

$$\delta(x) = 0 \text{ for } x \neq 0, \qquad \int \delta(x)\mathrm{d}x = 1.$$

This implies the "sifting property" of the delta function:

$$\int f(x)\delta(x - x_0)\mathrm{d}x = f(x_0).$$

A discrete event with possible values $A_1, A_2, \ldots, A_n$ may be expressed as a probability density as

$$p(x) = P(A_1)\delta(x - A_1) + \cdots + P(A_n)\delta(x - A_n),$$

where $\delta(x)$ is the Dirac delta function. This identity allows any expression containing the probability density function $p(x)$ to be converted to the case for which x is discrete.

1.23. Expected Value

The mean value of a function f of a random variable x is denoted as $\langle f \rangle$ and obeys

$$\langle f \rangle = \int f(x)p(x)\mathrm{d}x.$$

Other averages include the mode and the median: the most frequently occurring value and the middle value.

1.24. Moment

The nth moment of $p(x)$, $n = 1,2,3, \ldots$, is given by $m_n = \langle x^n \rangle = \int x^n p(x)\mathrm{d}x$.

1.25. Central Moment, Variance, and Standard Deviation

The moment about the mean is given by

$$\mu_n = \langle (x - \langle x \rangle)^n \rangle = \int (x - \langle x \rangle)^n p(x)\mathrm{d}x.$$

A special case is the second moment, called the variance; the variance can be written in terms of the first two moments:

$$\sigma^2 = \langle x^2 \rangle - \langle x \rangle^2.$$

The standard deviation is the square root of the variance.

1.26. Shannon Mutual Information or Transinformation

This is the average of the Shannon information (Section 1.11) over all events (x, y):

$$I_S(X, Y) = \iint p(x, y)\log\left[\frac{p(y|x)}{p(y)}\right]\mathrm{d}x\mathrm{d}y.$$

1.27. Entropy *H*

The Shannon information may be written as

$$I_{\mathrm{SH}}(X, Y) = \iint p(x, y)\log[p(y|x)]\mathrm{d}x\mathrm{d}y - \int p(y)\log[p(y)]\mathrm{d}y.$$

The second term $H = -\int p(y)\log[p(y)]\mathrm{d}y$ is the entropy of the random variable y.

1.28. Mean of a Weighted Sum

The mean of a weighted sum is the sum of the weighted means:

$$\left\langle \sum_{n=1}^{N} a_n x_n \right\rangle = \sum_{n=1}^{N} a_n \langle x_n \rangle,$$

where a_n are constants.

1.29. Distribution Function

The distribution function of the random variable x is the integral of the probability density function (Section 1.19):

$$P_F(-\infty \le x' \le x) = \int_{-\infty}^{x} p_F(x')\mathrm{d}x' = P_F(x).$$

The probability density function, accordingly, is the derivative of the distribution function, $p_F(x) = \mathrm{d}P_F/\mathrm{d}x$.

1.30. A Survey of Probability Laws

The most common probability laws that arise in optics are in the following 12 subsections, along with their first few moments and some optical applications.

1.30.1. Poisson Law

This describes photon statistics:

$$P_n = \frac{e^{-a}a^n}{n!}, \quad n = 0,1,2, \ldots, \quad \langle n \rangle = a, \quad \sigma^2 = a, \quad \mu_3 = a.$$

1.30.2. Uniform Law

This describes a uniform background, laser speckle, and the phase of a diffusor:

$$p(x) = \frac{1}{b}\,\mathrm{rect}\left(\frac{x-a}{b}\right), \quad \langle x\rangle = a, \quad \sigma^2 = \frac{b^2}{12}.$$

1.30.3. Exponential Law

This describes speckle intensity if it is represented as *x:*

$$p(x) = \frac{e^{-x/a}}{a}, \quad x \geq 0, \quad \langle x\rangle = a, \quad \sigma^2 = a^2.$$

1.30.4. Normal Law, One Dimension

$$p(x) = \frac{1}{\sqrt{2\pi\sigma^2}}\, e^{\frac{-(x-\langle x\rangle)^2}{2\sigma^2}}.$$

This describes phase or log intensity x in the presence of random turbulence. As a matter of notation, the expression $x = N(\langle \mathrm{x}\rangle, \sigma^2)$ means that the random variable x is normal with the indicated mean and variance. The *Gaussian* law is the normal law with the mean equal to zero.

1.30.5. Gaussian Law, Two Dimensions

This describes the probability of joint pupil phase values (x, y) in the presence of random turbulence. The corresponding normal law uses the replacements $x \rightarrow x - \langle x\rangle$, $y \rightarrow y - \langle y\rangle$ on the right-hand side of the following equation:

$$p(x, y) = \frac{1}{2\pi\sqrt{b_1 b_2(1-\rho^2)}}\, e^{-\{1/[2(1-\rho^2)]\}[(x^2/b_1)+(y^2/b_2)-(2\rho xy/\sqrt{b_1 b_2})]},$$

where

$$\langle x\rangle = \langle y\rangle = 0, \quad \sigma_x^2 = b_1, \quad \sigma_y^2 = b_2, \quad \rho = \frac{\langle xy\rangle}{\sigma_x\sigma_y}.$$

1.30.6. Gram–Charlier Law, a Skewed Gaussian

This describes glitter patterns on the surface of the ocean. The probability density function for intensity x is

$$p(x) = p_0(x)\left[1 + \sum_{n=1}^{N} \frac{c_n}{n!} H_n(x)\right],$$

where $p_0(x)$ is a Gaussian law with the same variance as that of $p(x)$, c_n are constants, and $H_n(x)$ are Hermite polynomials (see Chap. 2).

1.30.7. Geometric Law

This describes Bose–Einstein photon statistics and the law of reliability:

$$P_n = \frac{a_n}{(1+a)^{n+1}}, \quad \langle n \rangle = a, \quad \sigma^2 = a(a+1).$$

1.30.8. Binomial Law

This is a "checkerboard" model of photographic emulsion. It also describes Bernoulli or binary number sequences (Section 1.37):

$$P_n = \frac{N!}{n!(N-n)!} p^n q^{N-n}, \quad q = 1 - p, \quad \langle n \rangle = Np, \quad \sigma^2 = Npq.$$

1.30.9. Negative Binomial (Polya) Law

This describes the number n of laser speckles in an aperture of fixed size. It also describes the number of photons distributed among M degrees of freedom in a radiation field:

$$P_n = \frac{(n+M-1)!}{n!(M-1)!} p^n q^M, \quad p = \frac{\langle n \rangle}{\langle n \rangle + M}, \quad q = 1 - p.$$

The law generally arises when $n = \sum_{m=1}^{M} n_m$, where the random variables n_m are independent, identically distributed geometric random variables.

1.30.10. Cauchy Law

This is also called the Lorentzian law and describes the Doppler or natural line shape:

$$p(x) = \frac{1}{\pi}\left(\frac{a}{a^2 + x^2}\right), \quad \langle x \rangle = 0, \quad \sigma^2 = \infty.$$

1.30.11. Slit Diffraction Law

This describes the point-spread function for a diffraction-limited optical slit:

$$p(x) = \frac{w}{\lambda f}\,\mathrm{sinc}^2\left(\frac{\pi w}{\lambda f}\,x\right),$$

where w is the slit width, λ is the wavelength, f is the focal length, and $\mathrm{sinc}(x)$ is $\sin(x)/x$.

1.30.12. Chi-Squared (χ^2) Law

This describes laser-speckle averaging and mean-square wave-front phase error in turbulence. The probability density function $p(x)$ generally holds for a random variable x formed as

$$x = \sum_{n=1}^{N} y_n^2, \quad y_n = N(0, \sigma^2),$$

$$p(x) = \frac{x^{(N/2)-1}\, e^{-(x/2\sigma^2)}}{(\sqrt{2}\sigma)^N \Gamma(N/2)},$$

where y_n are identically and independently distributed Gaussian random variables. The parameter N is the *number of degrees of freedom* of the law.

1.31. Independent and Identically Distributed Random Variables

If the random numbers $x_1, x_2, \ldots, x_n$ are formed as independent samples from a single probability distribution function $p(x)$, then $x_1, x_2, \ldots, x_n$ are called identically and independently distributed (i.i.d.) random variables.

1.32. Characteristic Function

The Fourier transform of the probability density function is called the characteristic function:

$$\Phi(\omega) = \int p(x) e^{i\omega x} \mathrm{d}x, \quad i = \sqrt{-1}.$$

By the uniqueness of a Fourier transform and its inverse, it is equivalent to describe a random variable by either its probability density function or its characteristic function.

1.32.1. *Moment Generation*

The *n*th moment is generated from the characteristic function by

$$\Phi^{(n)}(0) = i^n \langle x^n \rangle.$$

1.32.2. *Shift Theorem*

The characteristic function for a random variable, with *a* and *b* as constants, has the following forms for the indicated probability laws:

Linear:

$$y = ax + b, \quad \Phi_y(\omega) = e^{i\omega b}\Phi_x(a\omega).$$

Poisson:

$$\Phi(\omega) = e^{a(e^{i\omega}-1)}.$$

One-dimensional normal:

$$\Phi(\omega) = e^{i\omega\langle x\rangle - \sigma^2\omega^2/2}.$$

Two-dimensional Gaussian:

$$\Phi(\omega) = e^{-(1/2)(\sigma_x^2\omega_1^2 + \sigma_y^2\omega_2^2 - 2\rho\sigma_x\sigma_y\omega_1\omega_2)}.$$

Sum of *n* identically and independently distributed random variables:

$$\Phi_x(\omega) = [\Phi_y(\omega)]^n.$$

1.32.3. *Application to the Random-Walk Process*

Use of the moment-generating property of the preceding characteristic function gives

$$\langle x \rangle = N\langle y \rangle, \quad \sigma_x^2 = N\sigma_y^2.$$

The mean and the variance are proportional to the number *N* of steps.

1.33. Central-Limit Theorem

In the random walk, as the number N of steps approaches infinity,

$$p(x) \longrightarrow \frac{e^{-(x-\langle x\rangle)^2/2\sigma_x^2}}{\sqrt{2\pi\,\sigma_x^2}}.$$

1.33.1. Error Function

The distribution function for a normal random variable requires the evaluation of integrals of the form

$$\frac{2}{\sqrt{\pi}}\int_0^y e^{-y^2}\mathrm{d}y = \mathrm{erf}(y).$$

This is called the *error function* because errors of measurement are usually distributed normally.

1.34. Transformation of a One-Dimensional Random Variable

If $p(x)$ is known and a new random variable $y = f(x)$ is formed, where $f(x)$ is a known function with a unique inverse $x = f^{-1}(y)$, then the probability density function for y obeys

$$p_y(y) = \frac{p_x(x)}{|\mathrm{d}y/\mathrm{d}x|_{x=f^{-1}(y)}}.$$

If, instead, there are multiple inverses, then the right-hand side is replaced with the sum over all the inverses.

1.35. Transformation of a Random Variable, Multiple Dimensions

If $p_x(x_1, x_2, \ldots, x_n)$ is known and a new vector of random variables $y_1, y_2, \ldots, y_n$ is formed in any deterministic way from the xs, then the joint probability density function on the ys obeys

$$p_y(y_1, \ldots, y_n) = p_x(x_1, \ldots, x_n)|J(x_1, \ldots, x_n/y_1, \ldots, y_n)|$$

as evaluated at $x_1, \ldots, x_n = f^{-1}(y_1, \ldots, y_n)$, and the Jacobian is defined as

$$J(x_1, \ldots, x_n/y_1, \ldots, y_n) = \det[\partial x_m/\partial y_n],$$

where the determinant is the determinant of all the partial derivatives of all the x variables with respect to all the y variables.

1.36. Bernoulli Trials Sequence

This is a sequence of N independent, binary events such as 0110001011000 1100010 (where $N = 20$). The probabilities $P(0)$ and $P(1)$ remain fixed over the sequence. The probability of exactly n events 1 in any order obeys the binomial law (Subsection 1.30.8).

In the rare-events limit $p \ll 1$, $N \gg 1$, such that $\langle n \rangle = Np = a$ is finite, the binomial law becomes the Poisson law. The Poisson law has the advantage of being defined by only one parameter, the mean, whereas the binomial law requires two parameters, N and p.

In the opposite limit of abundant events, where $\sigma^2 = Npq \gg 1$ and n is not too far from its mean $|n - \langle n \rangle| \leq \sigma$, the binomial law goes over to the normal law, which is easier to use in many practical problems requiring the probability of a number n lying on some finite interval (a, b). The normal representation reduces the problem to trivial table look-ups of the tabulated error function whereas use of the binomial representation can require the evaluation of a large sum of numbers that are so tiny as to be ill defined mathematically.

The probability of n events 1 to *consecutively* occur obeys the geometric law $P(1)^n[1 - P(1)]$. In the continuous limit the latter goes over to an exponential law, e.g., $p(t) = (1/\langle t \rangle)\, e^{-t/\langle t \rangle}$, $t \geq 0$, for the time t between collisions for a molecule in a gas.

1.37. Stochastic Processes

Stochastic processes describe randomly chosen curves as opposed to randomly chosen discrete numbers.

1.37.1. Definition

An ensemble is a collection of randomly chosen curves $f(x)$, where x is not random but merely the independent coordinate of the curve. The randomness is through a random variable λ that specifies a curve from the ensemble. The stochastic process is fully denoted by $f(x;\lambda)$. Any indicated average is over λ, often called the ensemble average.

1.37.2. Power Spectrum

The power spectrum of the process is

$$S_f(\omega) = \frac{1}{2L} \lim_{L\to\infty} \left\langle \left| \int_{-L}^{L} f(x;\lambda) e^{-i\omega x} \mathrm{d}x \right|^2 \right\rangle.$$

The variable ω is a general angular frequency with units of radians per unit of x.

1.37.3. Autocorrelation Function

The autocorrelation function is defined as

$$R_f(x;x_0) = \langle f(x_0;\lambda)f^*(x_0 + x;\lambda)\rangle,$$

where the asterisk indicates a complex conjugate.

1.37.4. Wide-Sense and Weak-Sense Stationarity

The absolute position x_0 is irrelevant to the mean and the autocorrelation:

$$R_f(x;x_0) = R_f(x), \quad \langle f(x;\lambda)\rangle = \text{constant}.$$

Equivalently,

$$R_f(x - x') = \langle f(x;\lambda)f^*(x';\lambda)\rangle.$$

1.37.5. Strict-Sense Stationarity

Consider N stochastic processes, $f_n(x_n;\lambda) = f_n$ for $n = 1,2, \ldots, N$, and their rigidly shifted versions $f_n(x_n + x;\lambda) = f'_n$.

If the joint probability density function $p(f_1,f_2, \ldots, f_N) = p(f'_1,f'_2, \ldots, f'_N)$, then the overall stochastic process obeys strict-sense stationarity. Optical phase in a pupil suffering from atmospheric turbulence obeys strict-sense stationarity of the order of $N = 2$. If a process obeys strict-sense stationarity, it also obeys wide-sense stationarity, but not vice versa.

1.37.6. Uncorrelated Noise Causes a White Power Spectrum

From Subsection 1.37.2 it can be shown that if $\langle f(x;\lambda)f^*(x';\lambda)\rangle = m_2(x)\delta(x - x')$, then $S_f(\omega)$ is a constant. The constant power spectrum is called white in analogy to equal-energy white light.

1.37.7. Additive Noise

This is defined to be noise that does not correlate with the signal. If a data curve $d(x)$, a signal curve $s(x)$, and an additive noise curve $n(x)$ are connected by a relation $d = s + n$, then $\langle s(x)n(x')\rangle = \langle s(x)\rangle\langle n(x')\rangle$ for all x and x', and in particular $\langle s(x)n(x)\rangle = 0$ if $\langle n(x')\rangle = 0$.

1.37.8. *Optical Transfer Function of Stochastic Turbulence*

The optical transfer function (OTF) for long-exposure imagery through atmospheric turbulence obeys

$$\mathrm{OTF}(\omega) = e^{-k^2\sigma^2[1-R_{\mathrm{OPD}}(\omega)]},$$

where $R_{\mathrm{OPD}}(\omega)$ is the autocorrelation function for the optical path difference between two points in the pupil separated by a distance $(\lambda f/2\pi)\omega$, where f is the image distance and λ is the average wavelength of light. For a range of seeing conditions and spatial frequencies, the autocorrelation is given by

$$R_{\mathrm{OPD}}(\omega) = 1 - K\omega^{5/3},$$

where K is a constant determined by the atmosphere.

1.37.9. *Transfer Theorem for Power Spectra*

If an input (object) signal of a power spectrum is filtered by a linear system with optical transfer function OTF(ω), then the output (image) signal from the system has a power spectrum of

$$S_i(\omega) = |\mathrm{OTF}(\omega)|^2 S_o(\omega) = \mathrm{MTF}(\omega)^2 S_o(\omega).$$

This permits the modulation transfer function (MTF), the absolute value of the optical transfer function, to be observed as the power spectral image of a white power spectral object.

1.37.10. *Mean-Squared Optical Transfer Function in the Presence of Turbulence*

$$\langle|\mathrm{OTF}(\omega)|^2\rangle \approx \left(\frac{\Delta D}{D}\right)^2 \mathrm{OTF}_{\mathrm{diff}}(\omega),$$

where D is the diameter of the optical pupil, ΔD is the average diameter of a "phase blob" (a region of constant phase) within the pupil, and $\mathrm{OTF}_{\mathrm{diff}}$ is the diffraction-limited optical transfer function of the pupil. The mean-squared optical transfer function exhibits diffraction-limited performance.

1.37.11. *Ergodicity*

For any (kth) member of the stochastic process $f_k(x)$, the fluctuations of f along x follow the same probability density function as do values f_k over all values of

k at any fixed x. As a corollary, any ensemble *average* equals its corresponding average along the curve $f_k(x)$.

1.37.12. Image Restoration

An image $i(x)$ is a stochastic process, obeying $i(x) = i_o(x) + n(x)$, $i_o(x) = \int o(x')\text{PSF}(x - x')\text{d}x'$. Here, there is a signal part $i_o(x)$, noise part $n(x)$, object $o(x)$, and point-spread function PSF(x). The objective of image restoration is to estimate the object $o(x)$, given $i(x)$ in the presence of noise $n(x)$. For this purpose an optimal filter function $Y(\omega)$ may be sought such that when it multiplies (modulates) the image data spectrum as $Y(\omega)I(\omega) = \hat{O}(\omega)$, the output $\hat{O}(\omega)$ differs from the true object spectrum minimally in a mean-square sense. If the noise is additive, then the filter is given by

$$Y(\omega) = \frac{\text{OTF}^*(\omega)S_o(\omega)}{|\text{OTF}|^2 S_o(\omega) + S_n(\omega)}.$$

1.37.13. Information Capacity of an Image

The maximum possible Shannon information over all possible objects is called the information capacity or channel capacity C of the imaging system. In the case of additive noise, this is

$$C = \int \ln\left[1 + \frac{|\text{OTF}(\omega)|^2 S_o(\omega)}{S_n(\omega)}\right] \text{d}\omega.$$

1.37.14. Superposition Process

The superposition process is a process described by $i(x) = \sum_{n=1}^{m} h(x - x_n)$, where x_n are identically and independently distributed according to a known probability law $p(x)$. Also, m is random according to a known law $P(m)$, and $h(x)$ is a fixed, known "disturbance" function.

1.37.15. Shot-Noise Process

The shot-noise process is a superposition process (Subsection 1.37.14) for the special case in which (1) coordinate $x = t$, the time; (2) $P(m)$ is Poisson (Subsection 1.30.1) with $\langle m \rangle = \langle \dot{n} \rangle \Delta t$, where $\langle \dot{n} \rangle$ is the average number of detected photons per unit time and Δt is the detection interval; and (3) $p(x)$ is uniform (Subsection 1.30.2). The statistics of $i(t)$ are as follows: The probability density function on $i(t_0)$, where t_0 is an arbitrary fixed point, is normal (Subsection 1.30.4). The moments are

$$\langle i(t_0)\rangle = \langle \dot{n}\rangle \int h(t)\mathrm{d}t, \quad \sigma^2(t_0) = \langle \dot{n}\rangle \int h^2(t)\mathrm{d}t,$$

$$R_i(t;t_0) = \langle \dot{n}\rangle \int h\,(t')h(t'+t)\mathrm{d}t'.$$

Because there are no t_0 dependencies on the right-hand sides, $i(t)$ is wide-sense stationary (Subsection 1.37.4).

1.38. Estimating a Parameter, General

Given a sample consisting of N measurements, y_n, $n = 1,2, \ldots, N$ of a parameter a, the aim is to estimate the parameter. Accordingly, we form an estimate $\hat{a}$ of a as an estimator function $\hat{a}(y_1,y_2, \ldots, y_N)$ of the data. Assume that the physics of the system is known in the form of a likelihood law $p(y_1,y_2, \ldots, y_N|a)$, i.e., the probability of the data in the presence of the parameter value. How good can an estimator be?

1.39. The Cramer–Rao Inequality

The mean-square error in an estimator is

$$e^2 = \langle (a-\hat{a})^2\rangle = \int p(y_1,y_1, \ldots, y_N|a)[a - \hat{a}(y_1,y_2, \ldots, y_N)]^2.$$

The error obeys the inequality

$$e^2 \geq \frac{1}{I_F} = \frac{1}{\int \left(\dfrac{\partial \log p}{\partial a}\right)^2 p\mathrm{d}y_1,\mathrm{d}y_2, \ldots, \mathrm{d}y_N},$$

where p is the likelihood law (Section 1.38). The integral is called the Fisher information, I_F. The higher it is, the lower is the mean square error, and vice versa.

1.40. Fisher Information for the Exponential Family

For a wide class of likelihood laws, including the exponential, normal, and Poisson cases, the Fisher information is equal to the reciprocal of the variance. These likelihood laws comprise the “exponential family.”

1.41. Efficient Estimator

An estimator $\hat{a}(y_1,y_2, \ldots, y_N)$ that attains the lower bound of $1/I_F$ on the error is called an efficient estimator. Efficient estimation can be attained for the exponential family of likelihood laws (Section 1.40).

The only kind of estimate that can attain efficiency is the maximum-likelihood estimate, i.e., the value of $a = \hat{a}_{ML}$ that makes the data maximum likely. This obeys

$$\log p(y_1, y_2, \dots, y_N|a)|_{a=\hat{a}_{ML}} = \text{maximum}.$$

The logarithm operation is not essential, but is often useful.

In the likelihood ratio test, one of two hypotheses, H_1 or H_2, is correct. If $p(y_1, y_2, \dots, y_N|H_1) > p(y_1, y_2, \dots, y_N|H_2)$, choose H_1; otherwise choose H_2.

1.42. Maximum-Likelihood Estimate of Mean *a*

If each datum independently obeys $y_n = a + x_n$, where each noise value x_n is identically and independently distributed Gaussian, with standard deviation σ, then the maximum-likelihood estimator of a is the *sample mean:*

$$\hat{a}_{ML}(y_1, y_2, \dots, y_N) = \frac{1}{N} \sum_{n=1}^{N} y_n = \bar{y}.$$

The mean-square error in the estimator is given by

$$e^2 = \frac{\sigma^2}{N}.$$

1.43. Maximum-Likelihood Estimate of a Variance σ^2

For the same type of data as those in Section 1.42, the maximum-likelihood estimator of the variance is the *sample variance:*

$$\sigma_{ML}^2(y_1, y_2, \dots, y_N) = \frac{1}{N-1} \sum_{n=1}^{N} (y - \bar{y})^2 = S^2.$$

The mean-squared error in the estimate is

$$e^2 = \left(\frac{2}{N-1}\right) \sigma^4.$$

Because the true value of the parameter is σ^2, the relative error (RE) is e/σ^2. Therefore

$$\text{RE} = \left(\frac{2}{N-1}\right)^{1/2}.$$

The relative error is purely a function of the *number of data.*

1.44. Maximum-Likelihood Estimate of a Probability

Many practical problems obey the following model: A coin of unknown bias x (x = probability of a "head") is flipped N times, with the result that n heads are observed (a single datum). What is the estimated bias? By Section 1.36, the coin flips comprise a Bernoulli trials sequence, so that the likelihood law is

$$P(n|x) = \frac{N!}{n!(N-n)!} x^n (1-x)^{N-n}.$$

By the maximum-likelihood principle (Section 1.41), the estimated value of bias x is

$$\hat{x}_{\mathrm{ML}} = \frac{n}{N}$$

(see Section 1.2). The mean-square error in this estimate is given by

$$e^2 = \frac{x(1-x)}{N}.$$

Taking the limit as N approaches infinity verifies the law of large numbers (Section 1.2).

1.45. Estimating a Probability Law by Use of Orthogonal Expansion

From a set of data $x_1, x_2, \ldots, x_n$ we sometimes must construct the continuous probability density function $p(x)$. By Section 1.44 a possible answer is $\hat{p}(x_i) = (n_i/N)$, where n_i is the number of times event x_i occurs. However, this may produce too lumpy or erratic a curve, $p(x_i)$. Instead, we represent $p(x)$ by a set of orthogonal functions

$$p(x) = \sum_{m=1}^{\infty} c_m f_m(x)$$

for a chosen set of orthonormal functions $f_n(x)$ that obey

$$\int_a^b f_m(x) f_n(x) \mathrm{d}x = \delta_{mn},$$

where δ_{mn} is the Kronecker delta (See Chap. 2). This allows the coefficients c_m to be found:

$$c_m = \int_a^b p(x) f_m(x) \mathrm{d}x = \langle f_m(x) \rangle$$

(Subsection 1.23). The maximum-likelihood estimate of the right-hand side is a sample mean, so

$$\hat{c}_m = \frac{1}{N} \sum_{n=1}^{N} f_m(x_n)$$

(see Section 1.42). The expansion for $p(x)$ should be cut off at a finite order M such that the standard deviation $\sigma_{\hat{c}_M}$ is

$$\sigma_{\hat{c}_M} = \hat{c}_M,$$

where

$$\sigma_{\hat{c}_M} = \frac{\hat{\sigma}_{f_M}}{\sqrt{N}}.$$

The quantity $\hat{\sigma}_{f_M}$ can be constructed as the sample variance (Section 1.43).

1.46. Estimating a Probability Law by Use of Maximum Entropy

Knowledge of a finite number of moments of $p(x)$ does not imply a unique probability density function $p(x)$. However, we can form a least-biased or maximally smooth estimate of $p(x)$ by requiring the estimate to have maximum entropy subject to these moments. The answer is the solution $p(x) = \hat{p}(x)$ to the variational problem

$$-\int p(x) \log p(x) dx + \sum_{n=0}^{N} \lambda_n \left[\int x^n p(x) dx - d_n \right] = \text{maximum}.$$

The Euler–Lagrange solution is

$$\hat{p}(x) = \exp\left(-1 + \sum_{n=0}^{N} \lambda_n x^n\right),$$

where λ_n obey

$$\int x^n \exp\left(-1 + \sum_{k} \lambda_k x^k\right) dx = m_n, \quad n = 0, 1, \ldots, N.$$

These are $N + 1$ equations in $N + 1$ unknowns λ_k, which can be solved by, e.g., Newton–Raphson iteration.

Examples: For $N = 0$, $\hat{p}(x)$ is a uniform law, for $N = 1$ it is an exponential law (Subsection 1.30.3), and for $N = 2$ it is a normal law (Subsection 1.30.4).

1.47. Estimating a Probability Law by Use of Minimum Fisher Information

For the same data case as that in Section 1.46, we attain an estimate $\hat{p}(x)$ that is maximally smooth in a different sense by requiring $\hat{p}(x)$ to have minimum Fisher information subject to the moment data. The answer is now the solution to the variational problem

$$4\int\left(\frac{dq}{dx}\right)^2 dx + \sum_{n=1}^{N} \lambda_n \left[\int x^n q^2(x)dx - m_n\right] = \text{minimum},$$

where $p(x) = q^2(x)$. The first term is the Fisher information I_F (defined in Section 1.39) for the case of a shift-invariant probability density function $p(y|a) = p(y - a) = p(x)$. The solution is the differential equation

$$q'' - \frac{1}{4} q \sum_n \lambda_n x^n = 0.$$

Comment: The requirement of solving a differential equation makes this approach slightly less practical than the maximum-entropy approach (Section 1.46). However, the results are "richer" because a differential equation admits of a family of solutions rather than a single one. Remarkably, these differential equations include the defining laws of physics: Schroedinger wave equation, Maxwell's equations, etc.

Examples: For $N = 0$, $\hat{p}(x)$ is either exponential (Subsection 1.30.3) or trigonometric, depending on the range of x; for $N = 1$ it is the square of an Airy function (quite close to the exponential maximum-entropy answer); and for $N = 2$ it is a general superposition of Hermitian–Gaussian functions, with a normal law case attaining the absolute-minimum Fisher information. The normal law is the corresponding maximum-entropy answer (Section 1.46).

1.48. The Chi-Square Test of Significance

In N trials, events x_k occur with corresponding frequencies m_k, $k = 1,2, \ldots, K$. The m_k are known data. We suspect that these violate the premise or hypothesis H_0 that a certain probability law P_k is present. How can we quantify our suspicion by testing the data in some way? We form a test statistic, the number

$$\chi^2 = \sum_{k=1}^{K} \frac{(m_k - NP_k)^2}{NP_k}.$$

The probability density function $p_{\chi^2}(x)$ for a general value χ^2 conditional on the premise H_0 is

$$p_{\chi^2}(x) = \frac{x^{(K-3)/2} e^{-x/2}}{2^{(K-1)/2}\Gamma[(K-1)/2]}, \quad x \geq 0,$$

and is called a chi-square law (also a gamma distribution). From this, the probability

$$P(\chi^2 \geq \chi_0^2) = \int_{\chi_0^2}^{\infty} p_{\chi}^2 (x)\mathrm{d}x$$

may be computed for the known number χ_0^2 and the number of degrees of freedom $(K - 1)$ of the problem. If χ_0^2 is so large that the probability is very small, say, less than 0.05, then the premise is rejected. The data x_k are said to be "significant at the 5% level." Stricter tests result from lower significance levels.

1.49. The Student *t* Test on the Mean

Measurements are made, $y_n = a + x_n$, $n = 1,2, \ldots, N$, to estimate a. The errors x_n are assumed to be Gaussian but of unknown variance. An effective estimate of a is the sample mean $\bar{y}$ (Section 1.42). However, this is liable to be in error. We want to bracket the estimate within a known interval $[\bar{y} - \Delta a, \bar{y} + \Delta a]$. How can Δa be found?

First, we guess at a largest reasonable value of Δa. This is equivalent to estimating a as $\hat{a} = \bar{y} + \Delta a$. From the data, we form the test statistic

$$t = \frac{(N-1)^{1/2}(\bar{y} - \hat{a})}{\left[\frac{1}{N}\sum_{n=1}^{N}(y_n - \hat{a})^2\right]^{1/2}}.$$

The probability density function for a general value of t in the presence of the hypothetical "true" value is

$$p_T(t) = \frac{1}{\sqrt{(N-1)\pi}} \frac{\Gamma(N/2)}{\Gamma(N/2 - 1/2)} \left(1 + \frac{t^2}{N-1}\right)^{-N/2}.$$

This is called the Student probability law. From this the probability

$$P(t \geq t_0) = \int_{t_0}^{\infty} p_T(t)\mathrm{d}t$$

can be computed for the known value t_0 of t and the given number of degrees of freedom, $N - 1$. If the resulting probability $P(t \geq t_0)$ is very small, say 0.05, this means that the test value is unreasonably far from $\bar{y}$ to be acceptable as an estimate. Thus the hypothetical value of Δa is "just rejected as a possibility." Accordingly, the sample mean is bracketed between $\bar{y} - \Delta a$ and $\bar{y} + \Delta a$.

The t test can also be used to test the hypothesis that two independent sets of data originate from phenomena that have the same parameter value a. If a is the mean, showing that the two data sets have a common mean is often used to imply that the two data sets arose from a common source. (Two pictures arose from the same roll of film, etc.)

1.50. The f Test on the Variance

A test can also be made to determine whether two sets of data arose from probability density functions that have the same variance value (cf. preceding section). Let H_0 be the hypothesis that they do. From their sample variances (Section 1.43), S_1^2 and S_2^2 define a statistic, the number

$$f_0 = \frac{S_1^2}{S_2^2}, \quad S_1^2 \geq S_2^2.$$

All the data are assumed to be normally distributed. Under the hypothesis H_0, the probability density function $p_F(x)$ for a general f obeys

$$p_F(x) = Kx^{(N_1-3)/2}(1 + rx)^{1-(N_1+N_2)/2},$$

$$K = \frac{\Gamma(N_1/2 + N_2/2 - 1)}{\Gamma(N_1/2 - 1/2)\Gamma(N_2/2 - 1/2)} r^{(N_1-1)/2}$$

where $x \geq 0$ and $r = [(N_1 - 1)/(N_2 - 1)]$.

$p_F(x)$ is called Snedecor's F distribution with $N_1 - 1$ and $N_2 - 1$ degrees of freedom. From this the probability

$$P(f \geq f_0) = \int_{f_0}^{\infty} p_F(x)\mathrm{d}x$$

can be computed. If f_0 is far from the ideal value of 1, i.e., is large, then this suggests that the hypothesis H_0 be rejected. Analytically the integral becomes very small, i.e., the event f_0 is very unlikely. If the integral is, say, 0.05, the hypothesis is said to be rejected on the level of 5%. A value of 1% would signify a more decisive level of rejection.

1.51. Various Information Measures

The information quantities in Subsections 1.51.2–1.51.6 are actually measures of distance between one probability density function $p(x)$ and another independent probability density function $q(x)$, called a *reference* function. An area of application lies in minimizing such an information quantity with respect to the choice of $p(x)$ and subject to input data constraints (see, e.g., Section 1.47). This gives a solution $\hat{p}(x)$ that is proportional to $q(x)$. This reference density function thereby acts as a "bias function," exhibiting the user's prior knowledge of expected answers for $p(x)$. The data constraints force departures from the bias function. Reality should be allowed to change the biases.

1.51.1. Fisher Information

See Sections 1.39 and 1.47.

1.51.2. Shannon Mutual Information

See Sections 1.11 and 1.26.

1.51.3. Kullback–Leibler Information

$$I_{\mathrm{KL}}(p, q) = -\int p(x) \log\left[\frac{p(x)}{q(x)}\right] \mathrm{d}x.$$

This is also called the *cross entropy* between the two general probability density functions $p(x)$ and $q(x)$.

1.51.4. Renyi Information

$$I_R(p, q, \alpha) = \log \int [p(x)]^{\alpha}\, q(x)^{1-\alpha}\, \mathrm{d}x, \quad \alpha \neq 1.$$

1.51.5. Wootters Information

$$I_W(p, q) = \arccos \int \sqrt{p(x)q(x)}\mathrm{d}x.$$

1.51.6. Hellinger Information

$$I_H(p, q, \alpha) = \int [(p(x)^{\alpha} - q(x)^{\alpha}]^{1/\alpha}\mathrm{d}x.$$

1.51.7. Tsallis Information

$$I_T(p, \alpha) = \frac{1}{\alpha - 1} \int [p(x) - p(x)^{\alpha}]\mathrm{d}x.$$

In the limit as α approaches 1,

$$I_T(p, \alpha) \rightarrow -\int p(x)\log p(x)\mathrm{d}x = H(x),$$

which is the entropy (Section 1.27).

1.52. Principal-Component Analysis

The aim of this technique is to represent a large set of data curves $f_j(x_n)$, $n = 1,2, \ldots, N$, $j = 1,2, \ldots, J$ by a much smaller set of principal-component curves $\hat{f}_k(x_n)$, $k = 1, 2, \ldots K$, with $K \ll J$. The process is to form the covariance matrix $\mathscr{C}$ between all possible pairs of data curves. This has the elements

$$c(i, j) = \frac{1}{N}\sum_{n=1}^{N} f_i(x_n)f_j(x_n) - \bar{f}_i\bar{f}_j,$$

where

$$\bar{f}_i = \frac{1}{N}\sum_{n=1}^{N} f_i(x_n).$$

An eigenvalue problem to be solved is

$$[\mathbf{C}]\mathbf{u}_k = \sigma_k^2\, \mathbf{u}_k$$

for the vectors $\mathbf{u}_k$ and eigenvalues σ_k^2, where $\mathbf{u}_k = [u_k(1)u_k(2)\cdots u_k(J)]^T$ and T denotes the transpose. These computed $\mathbf{u}_k(j)$ form the principal components of the problem:

$$\hat{f}_k(x_n) = \sum_{j=1}^{J} u_k(j)f_j(x_n), \quad k = 1,2, \ldots, K.$$

Because the matrix of elements $u_k(j)$ is unitary, the data can be represented in terms of the K principal components as

$$f_k(x_n) = \sum_{j=1}^{K} u_k(j)\hat{f}_j(x_n).$$

Typically the eigenvalues σ_k^2 fall off very rapidly in size once a critical value $k = K \ll J$ is reached. This identifies the upper limit in the preceding principal-component expansion. The result is that all J data curves f_k, $k = 1,2, \ldots, J$, are reconstructed from a much smaller set of principal-component curves $\hat{f}_k$, $k = 1,2, \ldots, K$, where $K \ll J$. This is a sizable reduction of the problem dimensionality.

1.53. Monte Carlo Calculation

This is a computer-based approach to solving the general problem of predicting a probability density function $p(x)$.

1.53.1. Aim

Individual events x_n are computed and "binned" to form a histogram $h(x_n)$, $n = 1,2, \ldots, N$. In the limit as $N \rightarrow \infty$, by the law of large numbers (Section 1.2), $h(x_n) \rightarrow p(x_n)\Delta x$, where Δx is the bin size. We compute the events x_n by randomly sampling from the probability density functions that in turn govern the transition from an input to an output value x_n. All probability density functions required of the calculation are presumed to be known as facets of a physical model of the system.

1.53.2. How to Randomly Sample from a Known Probability Density Function

First we generate a uniformly random number y between 0 and 1 by use of a computer random-number generator. The probability density function and therefore its distribution function $P_F(x)$ (Section 1.29) are assumed to be known. The equation $P_F(x) - y = 0$ is solved for its root x. This is the required random number x from $p(x)$.

Further Reading

B. R. Frieden, *Probability, Statistical Optics and Data Testing,* 3rd ed. (Springer-Verlag, 2001).

P. A. P. Moran, *An Introduction to Probability Theory* (Clarendon, 1968).

A. Papoulis, *Probability, Random Variables and Stochastic Processes* (McGraw-Hill, 1965).

E. Parzen, *Modern Probability Theory and Its Applications* (Wiley, 1966).

A. Stuart and J. K. Ord, *Kendall's Advanced Theory of Statistics,* 5th ed. (Arnold, 1991).

J. W. Goodman, *Statistical Optics* (Wiley, 1985).

E. L. O'Neill, *Introduction to Statistical Optics* (Addison-Wesley, 1963).

B. Saleh, *Photoelectron Statistics* (Springer-Verlag, 1978).

N. Wax, ed., *Selected Papers on Noise and Stochastic Processes* (Dover, 1954).

H. L. Van Trees, *Detection, Estimation and Modulation Theory, Part 1* (Wiley, 1971).

C. R. Smith and W. T. Grandy, eds., *Maximum Entropy and Bayesian Methods in Inverse Problems* (Reidel, 1985).

B. R. Frieden, *Physics from Fisher Information: A Unification* (Cambridge Univ. Press, 1998).

P. A. Jansson, ed., *Deconvolution of Images and Spectra,* 2nd ed. (Academic, 1997).

M. E. Newman and G. T. Barkema, *Monte Carlo Methods in Statistical Physics* (Clarendon, 1999).

The first five books are on probability and statistics; the next three are on statistical optics. In Wax's book, noise theory is described; Van Trees discusses parameter estimation. In the book by Smith and Grandy, maximum entropy is used to derive probability laws; Frieden (1998) describes the derivation of probability laws of physics from Fisher information; Jansson includes image restoration; and the final reference is to Monte Carlo simulation techniques.

2

Mathematical Functions and Relations

William L. Wolfe

2.1. Trigonometric Functions and Identities

$$\sin^2 x + \cos^2 x = 1, \quad \sec^2 x - \tan^2 x = 1,$$
$$\csc^2 x - \cot^2 x = 1, \quad \sin^2 x - \sin^2 y = \sin(x+y)\sin(x-y),$$
$$\cos^2 x - \cos^2 y = -\sin(x+y)\sin(x-y), \quad \cos^2 x - \sin^2 y = \cos(x+y)\cos(x-y),$$
$$\sin(x \pm y) = \sin x \cos y \pm \cos x \sin y \quad \cos(x \pm y) = \cos x \cos y \mp \sin x \sin y,$$
$$\tan(x \pm y) = \frac{\tan x \pm \tan y}{1 \mp \tan x \tan y}.$$

Functions of the double angles are obtained when y is set equal to x in the preceding equations:

$$\sin\frac{x}{2} = \pm\sqrt{\frac{1-\cos x}{2}}, \quad \cos\frac{x}{2} = \pm\sqrt{\frac{1+\cos x}{2}},$$

$$\tan\frac{x}{2} = \pm\sqrt{\frac{1-\cos x}{1+\cos x}} = \frac{1-\cos x}{\sin x} = \frac{\sin x}{1+\cos x},$$

$$\sin x \pm \sin y = 2\sin\frac{1}{2}(x \pm y)\cos\frac{1}{2}(x \mp y).$$

Some relations among inverse functions are

$$\arcsin x = \arccos\sqrt{1-x^2} = \arctan\frac{x}{\sqrt{1-x^2}},$$

$$\arccos x = \arcsin\sqrt{1-x^2} = \arctan\frac{\sqrt{1-x^2}}{x},$$

$$\arctan x = \arcsin\frac{x}{\sqrt{1+x^2}} = \arccos\frac{1}{\sqrt{1+x^2}}.$$

2.2. Hyperbolic Functions and Identities

$$\sinh(x) = \frac{e^x - e^{-x}}{2}, \quad \cosh(x) = \frac{e^x + e^{-x}}{2}, \quad \tanh(x) = \frac{\sinh(x)}{\cosh(x)} = \frac{e^x - e^{-x}}{e^x + e^{-x}},$$

$$\coth(x) = \frac{\cosh(x)}{\sinh(x)} = \frac{e^x + e^{-x}}{e^x - e^{-x}}.$$

2.3. Bessel Functions

There are several different Bessel functions, of different kinds and different orders. The zero-order Bessel function of the first kind is

$$J_0(x) = \frac{1}{\pi}\int_{-1}^{1} \frac{\cos(xt)}{\sqrt{1-t^2}}\,dt.$$

The nth-order Bessel function of the first kind is

$$J_n(x) = \frac{x^n}{1 \cdot 2 \cdot \ldots (2n-1)\pi}\int_{-1}^{1} \cos(xt)(1-t^2)^{n-(1/2)}\,dt$$

where n is a nonnegative integer.

For any parameter p, real or complex, it may be written as a series, with radial coordinate ρ:

$$J_p(\rho) = \sum_{m=0}^{\infty} \frac{(-1)^m}{m!\Gamma(p+m+1)}\left(\frac{\rho}{2}\right)^{p+2m}.$$

When p is an integer n, the relationship is

$$J_n(\rho) = \sum_{m=0}^{\infty} \frac{(-1)^m}{m!(n+m)!}\left(\frac{\rho}{2}\right)^{n+2m}.$$

The Bessel function of the second kind is

$$N_p(\rho) = \frac{1}{\sin p\pi}\,[J_p(\rho)\cos p\pi - J_{-p}(\rho)].$$

The Bessel functions of the third kind are

$$H_p^{(1)}(\rho) = J_p(\rho) + iN_p(\rho), \quad H_p^{(2)}(\rho) = J_p(\rho) - iN_p(\rho).$$

They are also known as Hankel functions.

Some properties of J are

$$J_0(0) = 1, \quad J_n(0) = 0, \quad J_{-n} = (-1)^n J_n, \quad \frac{\mathrm{d}}{\mathrm{d}x} J_0 = -J_1,$$

$$\frac{\mathrm{d}}{\mathrm{d}x} xJ_0 = xJ_0, \quad \frac{\mathrm{d}}{\mathrm{d}x} x^{\pm p} = \pm p^{\pm} J_{p\pm 1}, \quad \int_0^\infty J_p(ax)\mathrm{d}x = \frac{1}{a} \text{ when Re } (p) > 1,$$

$$\int_0^a J_1(x)\mathrm{d}x = 1 - J_0(a), \quad \int_a^\infty J_1(x)\mathrm{d}x = J_0(a).$$

2.4. Gamma Function

The gamma function is defined as

$$\Gamma(x) = \int_0^\infty e^{-t} t^{x-1} \mathrm{d}t.$$

If x is a positive integer n, then

$$\Gamma(n) = (n-1)!.$$

Some properties of Γ are

$$\Gamma(0) = \Gamma(-1) = \infty, \quad \Gamma(1) = \Gamma(2) = 1, \quad \Gamma(1/2) = \sqrt{\pi},$$

$$\Gamma(-1/2) = -2\sqrt{\pi}, \quad \Gamma(3/2) = \frac{\sqrt{\pi}}{2}, \quad \Gamma(-3/2) = \frac{4\sqrt{\pi}}{3},$$

$$\Gamma\left(n + \frac{1}{2}\right) = \frac{\sqrt{\pi}}{2^n} [1 \cdot 2 \cdot \ldots (2n-1)], \quad \Gamma(x+1) = x\Gamma(x),$$

$$\Gamma\left(\frac{1}{2} + x\right) \Gamma\left(\frac{1}{2} - x\right) = \frac{\pi}{\cos \pi x}, \quad \Gamma(1-x)\Gamma(x) = \frac{\pi}{\sin \pi x},$$

$$\Gamma\left(n - \frac{1}{2}\right) = \frac{(-1)^n \, 2^n \sqrt{\pi}}{1 \cdot 2 \cdot \ldots (2n-1)}.$$

2.5. The Correlation Function

The general correlation between two functions $f(x)$ and $g(x)$ is

$$R_{f,g}(x) = \int_{-\infty}^\infty f^*(u) g(u+x) \mathrm{d}u = f(x) \oplus g(x),$$

where $\oplus$ is the shorthand symbol for correlation. The autocorrelation is

$$R_{f,f}(x) = \int_{-\infty}^{\infty} f^*(u)f(u+x)\mathrm{d}u = f(x) \oplus f(x).$$

2.6. Convolution

The convolution between two functions f and g is given by

$$R_{f,g}(x) = \int_{-\infty}^{\infty} f(u)g(x-u)\mathrm{d}u = f \otimes g,$$

where $\otimes$ is the shorthand symbol for convolution. Convolutions are associative, distributive, commutative, and shift invariant:

$$f \otimes g = g \otimes f, \quad f \otimes (g \otimes h) = (f \otimes g) \otimes h, \quad f \otimes (g+h) = f \otimes g + f \otimes h,$$
$$f(x) \otimes g(x) = f(x - x_0) \otimes g(x).$$

A shift moves the position of the convolution, but not the shape or the magnitude. The infinite integral of the convolution of two functions is equal to the product of the integrals of the functions. The derivative of a convolution is equal to the convolution of the derivative of one function times the other:

$$h' = f' \otimes g = f \otimes g'.$$

The convolution of the mth and the nth derivatives of two functions is equal to the $(m + n)$th derivative of the convolution:

$$f^m \otimes g^n = (f \otimes g)^{m+n}.$$

2.7. Specially Defined Functions

A number of functions that have special definitions and special properties are useful. These are defined here, although care must be taken in that they are not regular in the sense of having all derivatives and continuity. They are defined such that the maximum is 1 and the area equals the magnitude of b—with the exception of the delta functions.

2.7.1. Delta Functions

$$\delta(x - x_0) = 0, \quad x \neq x_0, \quad \int_{-\infty}^{\infty} f(x)\delta(x-a)\mathrm{d}x = f(a).$$

The Dirac delta function may be defined as a limit of a Gaussian function:

$$\delta(x) = \lim_{b\to 0} \frac{1}{|b|} \text{Gaus}\left(\frac{x}{b}\right).$$

It may be thought of as having zero width, infinite height, and an area of 1.

2.7.1.1. Some Properties

$$\delta(-x) = \delta(x), \quad \delta(x - x_0) = \delta(-x + x_0),$$

$$\delta\left(\frac{x - x_0}{b}\right) = |b|\delta(x - x_0), \quad \delta(ax - x_0) = \frac{1}{|a|}\delta\left(x - \frac{x_0}{a}\right),$$

$$f(x)\delta(x - x_0) = f(x_0)\delta(x - x_0), \quad x\delta(x - x_0) = x_0\delta(x - x_0),$$

$$\delta(x)\delta(x - x_0) = 0, \quad x \neq 0, \quad \delta(x - x_0)\delta(x - x_0) \text{ is undefined},$$

$$\int_{-\infty}^{\infty} A\delta(x - x_0)\mathrm{d}x = A, \quad \int_{-\infty}^{\infty} \delta(z - x_0)\delta(x - z)\mathrm{d}z = \delta(x - x_0).$$

2.7.1.2. Double-Delta Functions

Even-impulse pair:

$$\delta\delta_e(x) = \delta(x + 1) + \delta(x - 1).$$

Odd-impulse pair:

$$\delta\delta_o(x) = \delta(x + 1) - \delta(x - 1).$$

2.7.1.3. Kronecker Delta

$$\delta_{ij} = 1 \text{ for } i = j, \quad \delta(i,j) = 0 \text{ for } i \neq j.$$

2.7.2. *Unit-Step Function*

$$\text{step}\left(\frac{x - x_0}{b}\right) = 0 \text{ for } \frac{x}{b} < \frac{x_0}{b}, \quad = \frac{1}{2} \text{ for } \frac{x}{b} = \frac{x_0}{b}, \quad = 0 \text{ for } \frac{x}{b} > \frac{x_0}{b}.$$

2.7.3. *Rectangle Function*

$$\text{rect}\left(\frac{x - x_0}{b}\right) = 0 \text{ for } \left|\frac{x - x_0}{b}\right| > \frac{1}{2}, \quad = \frac{1}{2} \text{ for } \frac{x - x_0}{b} = \frac{1}{2},$$

$$= 1 \text{ for } \frac{x - x_0}{b} < \frac{1}{2}.$$

2.7.4. *Ramp Function*

$$\text{ramp}\left(\frac{x-x_0}{b}\right) = 0 \text{ for } \frac{x}{b} \le \frac{x_0}{b}, \quad = \left|\frac{x-x_0}{b}\right| \text{ for } \frac{x}{b} > \frac{x_0}{b}.$$

A property of ramp functions: The integral goes from x_0 to x, that is,

$$\text{ramp}\left(\frac{x-x_0}{b}\right) = \frac{1}{b}\int_{x_0}^{x} \text{step}\left(\frac{z-x_0}{b}\right) dz.$$

2.7.5. *Triangle Function*

$$\text{tri}\left(\frac{x-x_0}{b}\right) = 0 \text{ for } \left|\frac{x-x_0}{b}\right| \ge 1, \quad = 1 - \left|\frac{x-x_0}{b}\right| \text{ for } \left|\frac{x-x_0}{b}\right| < 1.$$

2.7.6. *Gaussian Function*

$$\text{Gaus}\left(\frac{x-x_0}{b}\right) = e^{-\pi\left(\frac{x-x_0}{b}\right)^2}.$$

2.7.7. *Sign or Signum Function*

$$\text{sgn}\left(\frac{x-x_0}{b}\right) = -1 \text{ for } \frac{x}{b} < \frac{x_0}{b}, \quad = 0 \text{ for } \frac{x}{b} = \frac{x_0}{b}, \quad = 1 \text{ for } \frac{x}{b} > \frac{x_0}{b}.$$

Note that

$$\text{sgn}\left(\frac{x-x_0}{b}\right) = 2\,\text{step}\left(\frac{x-x_0}{b}\right) - 1.$$

2.7.8. *Comb or Sampling Function*

This is a series of displaced delta functions:

$$\text{comb}(x) = \sum_{n=-\infty}^{\infty} \delta(x-n).$$

2.7.9. *Sinc Function*

$$\mathrm{sinc}\left(\frac{x-x_0}{b}\right)=\frac{\sin\pi\left(\dfrac{x-x_0}{b}\right)}{\pi\left(\dfrac{x-x_0}{b}\right)}.$$

Note the use of π in this function. It is not used this way by everyone. A closely related function is sinc squared, the square of the sinc function.

2.7.10. *Sombrero Function*

This is a radial sinc function; it is also known as a bessinc function because it involves Bessel functions:

$$\mathrm{somb}\left(\frac{r}{d}\right)=\frac{2J_1\left(\dfrac{\pi r}{d}\right)}{\dfrac{\pi r}{d}},$$

where d is defined in the cylinder function.

2.7.11. *Cylinder Function*

$$\mathrm{cyl}\left(\frac{r}{d}\right)=1 \;\text{ for }\; 0\le r<\frac{d}{2}, \quad =\frac{1}{2}\;\text{ for }\; r=\frac{d}{2}, \quad =0 \;\text{ for }\; r>\frac{d}{2},$$

where r is the cylinder radius, d is its diameter, and J_1 is the first-order Bessel function of the first kind. The function has unit height and a volume of $\pi d^2/4$.

2.7.12. *Circular Gaussian Function*

$$\mathrm{Gaus}\left(\frac{r}{d}\right)=e^{-\pi(r/d)^2}.$$

This has unit height and a volume of d^2.

2.7.13. Polar Coordinate Translations

In general, a translation from the origin to a point (r_0,θ_0) is accomplished by

$$f(r - r_0) = g\left[\sqrt{r^2 + r_0^2 - 2rr_0 \cos(\theta - \theta_0)}\right].$$

2.7.14. Cartesian Two-Dimensional Functions

In general, the two-dimensional equivalent of a one-dimensional function is the product of the functions with appropriate variables, e.g.,

$$\text{rect}\left(\frac{x - x_0}{b}\frac{y - y_0}{c}\right) = \text{rect}\left(\frac{x - x_0}{b}\right)\text{rect}\left(\frac{y - y_0}{c}\right).$$

2.8. Fourier Transforms

Fourier transforms (see Tables 2.1 and 2.2) exist if the function $f(x)$ is piecewise continuous and the integral of the absolute value of $f(x)$ exists. At points of discontinuity the integral is equal to the average of the limits from the left and from the right. Both one- and two-dimensional transforms are used. The variables can be time and frequency, t and ν, or space and spatial frequency, x and ξ.

Table 2.1. Fourier Transform Properties

$f(t)$	$F(\omega)$	$f(t)$	$F(\omega)$
$f(at)$	$\frac{1}{\lvert a\rvert}F\left(\frac{\omega}{a}\right)$	$f(t/a)$	$\lvert a\rvert F(a\nu)$
$f(t) + g(t)$	$F(\omega) + G(\omega)$	$f(\pm t)$	$F(\pm\nu)$
$f(t - a)$	$e^{-i2\pi a}\nu F(\nu)$	$f^*(\pm t)$	$F^*(\mp\nu)$
$f(t \otimes g(t)$	$F(\omega)G(\omega)$	$F(\pm t)$	$f(\mp\nu)$
$f(t) \otimes f^*(-t)$	$\lvert F(\omega)\rvert^2$	$F^*(\pm t)$	$f^*(\pm\nu)$
$f'(t)$	$i2\pi\nu F(\omega)$	$f(t \pm t_0)$	$e^{-i\omega t_0}F(\nu)$
$\lvert a\rvert f(ax)$	$F(\nu/d)$	$f(x/a)$	$\lvert a\rvert F(a\nu)$

The one-dimensional Fourier exponential transform of $f(t)$ is given by

$$F(\omega) = \int_{-\infty}^{\infty} f(t)e^{i\omega t}\mathrm{d}t.$$

The inverse transform is given by

$$F(t) = \frac{1}{2\pi}\int_{-\infty}^{\infty} F(\omega)e^{i\omega t}\mathrm{d}\omega.$$

Some authors write this pair with $1/\sqrt{2\pi}$ multiplying both integrals. The sine and the cosine transforms are not treated here.

Table 2.2. Fourier Transform Pairs

$f(t)$	$F(\omega)$	$f(t)$	$F(\omega)$
1	$\delta(\omega)$	rect(t)	sinc(ν)
$\delta(t)$	1	sinc(t)	rect(ν)
$\delta(t \pm t_0)$	$e^{\pm i\omega t}$	tri(t)	$\mathrm{sinc}^2(\nu)$
$e^{\pm i\omega t}$	$\delta(\omega \mp \omega_0)$	$\mathrm{sinc}^2(t)$	tri(ν)
$\cos \omega t$	$\frac{1}{2\lvert\nu_0\rvert}\delta\delta_e\left(\frac{\nu}{\nu_0}\right)$	sgn(t)	$\frac{1}{i\pi\nu} = \frac{2}{i\omega}$
$\frac{1}{2\lvert t_0\rvert}\delta\delta_e\left(\frac{t}{t_0}\right)$	$\cos \omega t$	$\frac{1}{i\pi t}$	–sgn(ν)
$\sin \omega_0 t$	$\frac{1}{2\lvert\nu_0\rvert}\delta\delta_o\left(\frac{\nu}{\nu_0}\right)$	step(t)	$\frac{1}{2}\delta(\nu) + \frac{1}{i\omega}$
comb(t)	comb(ν)	Gaus(t)	Gaus(ν)
sech(t)	sech($\pi\nu$)	$\cos(\pi t^2)$	$\cos\pi(\nu^2 - 0.25)$
$\sin(\pi x^2)$	$-\sin \pi(\nu^2 - 0.25)$	$e^{\pm i\pi t^2}$	$e^{i\pi(\pm\nu^2 \pm 0.25)}$

2.9. Relations Between Space and Frequency Functions

Rayleigh's theorem states that

$$\int_{-\infty}^{\infty} |f(x)|^2 \mathrm{d}x = \int_{-\infty}^{\infty} |F(\xi)|^2 \mathrm{d}\xi.$$

The power theorem states that

$$\int_{-\infty}^{\infty} f(x)g^*(x)\mathrm{d}x - \int_{-\infty}^{\infty} F(\xi)G^*(\xi)\mathrm{d}\xi;$$

also,

$$\int_{-\infty}^{\infty} f(x)g(-x)\mathrm{d}x = \int_{-\infty}^{\infty} F(\xi)G(\xi)\mathrm{d}\xi.$$

The autocorrelation theorem states that

$$\int_{-\infty}^{\infty} f(x)f(x+u)\mathrm{d}x = |F(\xi)|^2.$$

The uncertainty relation is

$$\Delta x \Delta \xi \geq \frac{1}{4\pi}.$$

2.10. Hankel Transforms

The Hankel transform is a two-dimensional transform for circularly symmetric functions, those functions for which $f(x,y) = f(\mathbf{r})$, where $\mathbf{r}$ is the radius. The kernel of the Hankel transform is a Bessel function rather than an exponential. The Hankel transform is defined by

$$F(\xi) = 2\pi \int_0^\infty f(r) J_0(2\pi\xi r) r \mathrm{d}r;$$

$$f(\mathrm{r}) = 2\pi \int_0^\infty F(\xi) J_0(2\pi r\xi) r \mathrm{d}r;$$

Table 2.3 includes a selection of Hankel transforms and properties.

Table 2.3. Zero-Order Hankel Transforms and Properties

$f(r)$	$F(\xi)$	$f(r)$	$F(\xi)$
$\delta(r)$	πr	r^2 Gaus(r)	$(1/\pi - \xi^2)$Gaus(ξ)
$1/r$	$1/\xi$	$e^{-i\pi r^2}$	$e^{-i\pi\xi^2}$
cyl(r)	$(\pi/4)$somb(ξ)	Gaus(r)	Gaus(ξ)
e^{-r}	$2\pi/(4\pi^2\xi^2 + 1)^{3/2}$	cyl(r)	$(\pi/4)$somb(ξ)
$\cos(\pi r)$	$\sin(\pi\xi)$	$e^{\pm i\pi r^2}$	$\pm ie^{\mp i\pi\zeta^2}$
sinc(r)	$\dfrac{2\,\mathrm{cyl}(\xi)}{\pi\sqrt{1-4\xi^2}}$	$\cos(\pi r)/\pi r$	$\dfrac{2[1-\mathrm{cyl}(\xi)]}{\pi\sqrt{4\xi^2-1}}$

2.11. Laplace Transforms

The Laplace transform (see Table 2.4) is most useful for circuit considerations in which the signal is zero before the time of initiation. It also has the nice feature of accounting for initial conditions.

The Laplace transform of $f(t)$, a piecewise-regular exponentially bounded function that is zero for t less than zero, that is,

$$f(t) = 0 \text{ for } t < 0 \quad |f(t)| < e^{at} \text{ for } t > 0, \quad a > 0,$$

is given by

$$F(s) = \int_0^\infty f(t)e^{-st}\mathrm{d}t,$$

where s is the complex variable $\sigma + i\omega$.

The inverse Laplace transform is

$$f(t) = \int_0^\infty F(s)e^{st}\mathrm{d}s.$$

Table 2.5 is a compilation of some Laplace transform pairs and relations. In it, a is a constant and 0+ represents the limit of the function as the variable approaches zero from greater than zero. The function $\delta(t)$ is the delta function and step(t) is the unit-step function.

Table 2.4. Laplace Transform Properties

$f(t)$	$F(s)$	$f(t)$	$F(s)$
$f(t)$	$F(s)$	$e^{at}f(t)$	$F(s-a)$
$f'(t)$	$sF(s) - f(0)$	$(1/a)f(t/a)$	$F(as),\ a > 0$
$\int_0^t f(t)\mathrm{d}t$	$\frac{1}{s}F(s)$	$(1/t)f(t)$	$\int_s^\infty F(s)\mathrm{d}s$
$(-1)^n t^n f(t)$	$F^{(n)}(s)$	step$(t-a)$	$e^{-as}F(s),\ a > 0$
$e^{at}f(t)$	$F(s-a)$	$-tf(t)$	$F'(s)$

$f(t)$	$F(s)$
$f^{(n)}(t)$	$s^{(n)}F(s) - s^{(n-1)}F(0) - s^{(n-2)}F'(0) \ldots F^{(n-1)}(0)$

Table 2.5. Laplace Transforms

$f(t)$	$F(s)$	$f(t)$	$F(s)$
$c_1f(t) + c_2g(t)$	$c_1F(s) + c_2G(s)$	$t^n/n!$	$1/s^{n+1}$
$f'(t)$	$-f(0+) + sF(s)$	$\sin kt$	$k/(s^2 + k^2)$
$f^{(n)}(t)$	$-f(n-1)\cdots s^nF(s)$	$\cos kt$	$p/(s^2 + k^2)$
$e^{-at}f(t)$	$F(s+a)$	$\sinh kt$	$k/(s^2 - k^2)$
$\delta(t)$	$1/s$	$\cosh kt$	$p/(s^2 + k^2)$
$\int_0^t f(t)\mathrm{d}t$	$F(s)/s$	$t^ne^{-at}/n!$	$1/(s+a)^{n+1}$
e^{-at}	$1/(s+a)$	$e^{-at}\sin kt$	$k/[(s+a)^2 + k^2]$
t	$1/s^2$	$e^{-at}\sin kt$	$(s+a)/[(s+a)^2 + k^2]$

2.12. Z Transforms

These are used for sampled systems and are represented in terms of time t and sampled frequency z. If $f(t)$ is known at the sample times, 0, 1, 2, … , then the Z transform is

$$F(z) = f(0) + \frac{f(1)}{z} + \frac{f(2)}{z^2} + \cdots + \frac{f(n)}{z^n}.$$

Table 2.6 provides some Z transforms. For compactness of presentation, ξ is the reciprocal of z.

Table 2.6. Z Transforms

$f(n)$	$F(z)$	$f(n)$	$F(z)$
$a_n = a_0, a_1, a_2 \ldots$	$a_0 + a_1\xi + \cdots +$	$\sinh(an)$	$\dfrac{\xi \sinh(a)}{[1 - \xi^2 \cosh(a) + \xi^2]}$
1	$1/(1 - \xi)$	$\cosh(an)$	$\dfrac{1 - \xi \cosh(a)}{[1 - \xi^2 \cosh(a) + \xi^2]}$
n	$\xi/(1 - \xi)$	$f(n) + g(n)$	$F(z) + G(z)$
n^2	$\xi (1 + \xi)/(1 - \xi)^3$	$f(n - a)$	$\xi^a F(z), \quad a > 0$
e^{-an}	$1/(1 - e^{-a}\xi)$	$nf(n)$	$-zF'(z)$
$1 - e^{-an}$	$(1 - e^{-a})\xi/(1 - \xi)(1 - e^{-a}\xi)$	$(1 - n)f(n - 1)$	$F'(z)$
$\sin(an)$	$\xi \sin(a)/[1 - \xi^2 \cos(a) + \xi^2]$	$a^{-n}f(n)$	$F(az)$
$\cos(an)$	$\dfrac{[1 - \xi \cos(a)]}{[1 - \xi^2 \cos(a) + \xi^2]}$	$f(n) \otimes g(n)$	$F(z)G(z)$

2.13. Hilbert Transforms

The Hilbert transform is used for analytic signals, notably for the Kramers–Kronig method of obtaining optical constants. It is defined as

$$F(x) = \frac{1}{\pi}\int_{-\infty}^{\infty} \frac{f(u)}{u - x}\, du,$$

where one accounts for infinity at $u = x$ by taking the Cauchy principal value.

2.14. Series

Many functions can be represented by series by means of the Taylor expansion:

$$f(x) = f(a) + f'(a)\,\frac{x - a}{1}\;!\mathrm{cot}\cdots f^{(n)}(a)\,\frac{(x - a)^{n-1}}{(n - 1)!} + R_n,$$

$$R_n'(x) = -f^{(n)}(x)\,\frac{(a-x)^n}{n!}.$$

The Maclaurin series is the Taylor series with $a = 0$.

2.14.1. Convergence

Absolute convergence requires convergence for a series in which all the signs are positive. If a series that has negative signs converges, it is conditionally convergent.

The *Cauchy test* for convergence is

$$|s_m - s_n| = |a_{n+1} + a_{n+2} + \cdots + a_m| \rightarrow 0$$

for $m > n$ and both large. For example, the series $1/2^n$ has the partial sum

$$\frac{1}{2^n} + \frac{1}{2^{n+1}} + \cdots + \frac{1}{2^{m-1}} = \frac{1}{2^{n-1}}\left(\frac{1}{2} + \frac{1}{2^2} + \cdots + \frac{1}{2^{m-n}}\right) < \frac{1}{2^{n-1}}.$$

Then the difference between the sum of the infinite series and the finite series can be made smaller than some error ε if enough terms n, $\varepsilon = 1/2^{n-1}$, are taken.

The *comparison test* is if $|a_n| \leq b_n$, where $b_n > 0$ and the b series converges, then the a series converges.

The *comparison to the geometric series* is if from some term onward $|a_n| < cq^n$, where c is a positive number independent of n and q is any fixed positive number <1, then the a series converges.

The *ratio test* is if from some term onward $|a_{n+1}/a_n| < q$, then the a series converges.

The *root test* is if from some term onward $(|a_n|)^{1/n} < q$, then the a series converges.

2.14.2. Series Representations

$$e^x = 1 + x + \frac{x^2}{2!} + \cdots + \frac{x^n}{n!} + \frac{x^{n+1}}{(n+1)!}\,e^{qx},$$

$$\sin(x) = x - \frac{x^3}{3!} \cdots \frac{(-1)^n x^{(2n+1)}}{(2n+1)!} = \sum_{n=0}^{\infty} \frac{(-1)^n x^{(2n+1)}}{(2n+1)!},$$

$$\cos(x) = 1 - \frac{x^2}{2!} + \frac{x^3}{3!} - \frac{x^4}{4!} \cdots (-1)^n \frac{x^{2n}}{(2n)!} = \sum_{n=0}^{\infty} \frac{(-1)^n x^{2n}}{(2n)!},$$

$$\tan x = x + \frac{x^3}{3} + \frac{2}{15}x^5 + \frac{17}{315}x^7 + \frac{62}{2835}x^9 \ldots,$$

$$\arcsin x = x + \frac{x^3}{6} + \frac{1}{2}\frac{3}{4}\frac{x^5}{5} \ldots,$$

$$\arctan x = x - \frac{1}{3}x^3 + \frac{1}{5}x^5 - \frac{1}{7}x^7 \ldots,$$

$$\sinh(x) = x + \frac{x^3}{3!} \cdots \frac{x^{(2n+1)}}{(2n+1)!} = \sum_{n=0}^{\infty} \frac{x^{(2n+1)}}{(2n+1)!},$$

$$\cosh(x) = 1 + \frac{x^2}{2!} + \cdots + \frac{x^{2n}}{(2n)!} = \sum_{n=0}^{\infty} \frac{x^{2n}}{(2n)!}.$$

2.14.3. Series

Arithmetic series:

$$a + (a + d) + (a + 2d) \ldots a + (n - 1)\, d,$$

$$S = \frac{n}{2}(a + l) = \frac{n}{2}[2a + (n - 1)\, d],$$

where S is the sum, a is the first term, l is the last term, and d is the difference.

Geometric series:

$$a + ar + ar^2 + \cdots + ar^{n-1},$$

$$S = a\,\frac{1 - r^n}{1 - r} = \frac{a(r^n - 1)}{r - 1}.$$

The infinite sum is given by

$$S = \frac{a}{1 - r}.$$

Binomial series:

$$(1 + x)^a = 1 + ax + \frac{a(a - 1)}{2}!x^2 + \cdots + \frac{a(a - 1) \ldots (a - n + 1)}{n!}x^n + R_n,$$

$$R_n \leq \frac{2^b}{(b - 1)!}(n + b)^{b-1}x^{n+1},$$

where b is the largest integer less than or equal to a and $x > 0$.

$$R_n \leq \frac{K}{(b-1)!}(n+b)^{b-1}|x|^{n+1},$$

where $K = |(1+\theta x)^{a-1}|$, $0 \leq = \theta \leq 1$, and $-1 < x < 0$.

$$S(n) = 1 + 2 + \cdots + n = \frac{n(n+1)}{2},$$

$$S(n^2) = 1^2 + 2^2 + \cdots + = \frac{n(n+1)(2n+1)}{6},$$

$$S(n^3) = 1^3 + 2^3 + \cdots + = \frac{n^2(n+1)^2}{4}.$$

Stirling's formula for the approximation for factorials is

$$\sqrt{2n\pi}\left(\frac{n}{e}\right)^n < n! = \sqrt{2n\pi}\left(\frac{n}{e}\right)^n\left(1 + \frac{1}{12n-1}\right).$$

2.15. Matrices

The general matrix is represented here as a 3×3 matrix. They can be of other dimensions, not necessarily square. The one-dimensional versions are vectors:

$$\begin{bmatrix} a_{11} & a_{12} & a_{13} \\ a_{21} & a_{22} & a_{23} \\ a_{31} & a_{32} & a_{33} \end{bmatrix}.$$

Addition is

$$\begin{bmatrix} a_{11} & a_{12} & a_{13} \\ a_{21} & a_{22} & a_{23} \\ a_{31} & a_{32} & a_{33} \end{bmatrix} + \begin{bmatrix} b_{11} & b_{12} & b_{13} \\ b_{21} & b_{22} & b_{23} \\ b_{31} & b_{32} & b_{33} \end{bmatrix}$$

$$= \begin{bmatrix} a_{11}+b_{11} & a_{12}+b_{12} & a_{13}+b_{13} \\ a_{21}+b_{21} & a_{22}+b_{22} & a_{23}+b_{23} \\ a_{31}+b_{31} & a_{32}+b_{32} & a_{33}+b_{33} \end{bmatrix}.$$

Subtraction is the same, but the plus sign is replaced with the minus sign. Multiplication is the same, with multiplication substituted for addition or subtraction. Multiplication of three or more matrices proceeds from right to left. Matrix multiplication is generally noncommutative. When a matrix is multiplied by a constant, every element is multiplied by that constant.

The determinant of a matrix uses the same elements but uses them as a determinant:

$$\det\begin{bmatrix} a_{11} & a_{12} & a_{13} \\ a_{21} & a_{22} & a_{23} \\ a_{31} & a_{32} & a_{33} \end{bmatrix}$$

$$= a_{11}a_{22}a_{33} - a_{11}a_{23}a_{32} - a_{12}a_{21}a_{33} + a_{12}a_{21}a_{33} + a_{33}a_{21}a_{32} - a_{33}a_{22}a_{31}.$$

The minor of an element of a determinant is defined as the determinant when the row and the column in which the element resides are crossed out:

$$\min(a_{11}) = \begin{vmatrix} a_{11} & a_{12} & a_{13} \\ a_{21} & a_{22} & a_{23} \\ a_{31} & a_{32} & a_{33} \end{vmatrix} = a_{22}a_{33} - a_{23}a_{32}.$$

The *cofactor* is the minor times $(-1)^{-(i+j)}$.

The *identity* matrix has ones on the diagonal and zeros elsewhere. The *transpose* of a matrix is obtained by the interchange of rows and columns. An *adjoint* matrix is the transpose of the matrix in which each element is replaced with its cofactor.

The reciprocal of a matrix is found by calculation of the matrix of the minors, the matrix of the cofactors, the transposed matrix, and division by the determinant. The following examples illustrate them:

$$\mathscr{A} = \begin{bmatrix} a_{11} & a_{12} & a_{13} \\ a_{21} & a_{22} & a_{23} \\ a_{31} & a_{32} & a_{33} \end{bmatrix},$$

$$\mathscr{A}_{\text{minor}} = \begin{bmatrix} a_{22}a_{33} - a_{23}a_{32} & a_{11}a_{33} - a_{13}a_{31} & a_{21}a_{32} - a_{22}a_{31} \\ a_{12}a_{33} - a_{13}a_{32} & a_{11}a_{33} - a_{31}a_{13} & a_{21}a_{32} - a_{31}a_{22} \\ a_{12}a_{23} - a_{22}a_{13} & a_{11}a_{23} - a_{21}a_{13} & a_{11}a_{22} - a_{21}a_{12} \end{bmatrix},$$

$$\mathscr{A}_{\text{cofactor}} = \begin{bmatrix} a_{22}a_{33} - a_{23}a_{32} & -(a_{11}a_{33} - a_{13}a_{31}) & a_{21}a_{32} - a_{22}a_{31} \\ -(a_{12}a_{33} - a_{13}a_{32}) & a_{11}a_{33} - a_{31}a_{13} & -(a_{21}a_{32} - a_{31}a_{22}) \\ a_{12}a_{23} - a_{22}a_{13} & -(a_{11}a_{23} - a_{21}a_{13}) & a_{11}a_{22} - a_{21}a_{12} \end{bmatrix},$$

$$\mathscr{A}_{\text{transpose}} = \begin{bmatrix} a_{22}a_{33} - a_{23}a_{32} & a_{12}a_{33} - a_{13}a_{32} & a_{21}a_{23} - a_{22}a_{13} \\ a_{11}a_{33} - a_{13}a_{31} & a_{11}a_{33} - a_{31}a_{13} & a_{11}a_{23} - a_{21}a_{13} \\ a_{21}a_{32} - a_{22}a_{31} & a_{21}a_{32} - a_{31}a_{22} & a_{11}a_{22} - a_{21}a_{12} \end{bmatrix}.$$

The reciprocal matrix is then the transpose matrix divided by the determinant of the original matrix, as given above.

The *trace* of a matrix is the sum of its diagonal elements. The *determinant* of a matrix is a determinant of all the same elements in the same places.

A *Hermitian* matrix is one in which all elements satisfy $a_{ij} = a_{ji}^*$. It is Hermitian if the complex conjugate of its transpose is the same as the matrix itself.

2.16. Geometrical Relations

The properties of geometric figures are given below; a and b are the lengths of the sides, h is the height, and θ is the angle at which the sides intersect. The area is denoted by A, R is the radius of the circumscribed circle, r is that of the inscribed circle, D is the diameter, C is the circumference, and s_1 and s_2 are the major and the minor semiaxes of the ellipse, respectively:

Triangle,

$$A = \frac{1}{2} bh.$$

Rectangle,

$$A = ab.$$

Parallelogram,

$$A = ah = ab \sin \theta.$$

Trapezoid,

$$A = \frac{1}{2} h(a + b).$$

Regular polygon of n sides,

$$A = \frac{1}{4} na^2 \cot(\pi/2n), \quad R = \frac{a}{2} \csc(\pi/2n), \quad r = a/2 \cot(\pi/2n).$$

Circle,

$$A = \pi r^2, \quad A_{\text{sector}} = \frac{1}{2} r^2\theta, \quad A_{\text{segment}} = \frac{1}{2} r^2 (\theta - \sin \theta).$$

Ellipse,

$$A = \pi s_1 s_2.$$

Triangle relations (how to figure the last of the ménage à trois): For a triangle with sides *a, b,* and *c* and angles α, β, and γ (opposite sides *a, b,* and *c,* respectively), the sine relationship is

$$\frac{\sin\alpha}{\alpha} = \frac{\sin\beta}{\beta} = \frac{\sin\delta}{\delta}.$$

The cosine relationship is

$$a^2 = b^2 + c^2 + bc\cos\alpha.$$

3
Electromagnetism Foundations

William L. Wolfe

3.1. Maxwell's Equations

Maxwell's equations are

$$\nabla X \mathbf{E} = -\frac{\partial \mathbf{B}}{\partial t},$$

$$\nabla X \mathbf{H} = \mathbf{j} + \frac{\partial \mathbf{D}}{\partial t},$$

$$\nabla \cdot \mathbf{D} = \rho,$$

$$\nabla \cdot \mathbf{B} = 0,$$

$$\nabla \cdot \mathbf{j} = -\frac{\partial \rho}{\partial t},$$

where **E** is the electric field, **B** is the magnetic induction, **H** is the magnetic field, **D** is the electric displacement, **j** is the current (areal) density, ρ is the charge, ∇ is the gradient operator, and t is time. The material equations are

$$\mathbf{j} = \sigma \mathbf{E},$$

$$\mathbf{D} = \epsilon \mathbf{E},$$

$$\mathbf{B} = \mu \mathbf{H},$$

where σ is the conductivity, ϵ is the permitivity (dielectric constant), and μ is the permeability.

3.2. Poynting Vector

The Poynting vector **S** is

$$\mathbf{S} = \mathbf{E X H}.$$

For a plane wave traveling in the z direction, the Poynting vector is

$$\mathbf{S} = E_x H_y \hat{\mathbf{z}} = \hat{\mathbf{z}} \frac{E_0^2}{Z_0} \sin^2(kz - \omega t) = \hat{\mathbf{z}} \frac{E_0^2}{Z_0} \sin^2\left[2\pi\left(\frac{z}{\lambda} - \nu t\right)\right],$$

where $Z_0 = \sqrt{\mu_0/\epsilon_0}$ is the impedance of the vacuum and is equal to 377 Ω. The time-averaged electrical and magnetic energies are $(\epsilon/4)\mathbf{E} \cdot \mathbf{E}^*$ and $(\mu/4)\mathbf{H} \cdot \mathbf{H}^*$.

3.3. Wave Equations

The vector wave equation in the time domain is

$$\nabla^2 \mathbf{E} - \mu\epsilon \frac{\partial^2 \mathbf{E}}{\partial t^2} - \mu\sigma \frac{\partial \mathbf{E}}{\partial t} = 0,$$

where

$$\nabla^2 \mathbf{E} = \nabla\nabla \cdot \mathbf{E} - \nabla \mathbf{X} \nabla \mathbf{X}\mathbf{E},$$

and the equation applies to **E, B, D,** and **H.** In a vacuum the equation reduces to

$$\nabla^2 \mathbf{E} - \mu_0\epsilon_0 \frac{\partial^2 \mathbf{E}}{\partial t^2} = \nabla^2 \mathbf{E} - \frac{1}{c^2} \frac{\partial^2 \mathbf{E}}{\partial t^2} = 0.$$

The vector wave equation in the frequency domain, Helmholtz's equation, is

$$\nabla^2 \mathbf{E} + \omega^2 \mu\epsilon \mathbf{E} - i\omega\sigma \mathbf{E} = \nabla^2 \mathbf{E} + k^2 \mathbf{E} - i\omega\sigma \mathbf{E} = 0,$$

where ω is the radian frequency, k is the radian wave number, and the fields are the Fourier transforms of the time-varying fields. Solutions are subsequently given for rectangular coordinates, circular cylindrical coordinates, and spherical coordinates.

3.4. Boundary Conditions

The normal components of **D** and **B** are continuous across a dielectric boundary:

$$\hat{\mathbf{n}} \cdot \mathbf{D}_1 = \hat{\mathbf{n}} \cdot \mathbf{D}_2,$$
$$\hat{\mathbf{n}} \cdot \mathbf{B}_1 = \hat{\mathbf{n}} \cdot \mathbf{B}_2,$$

where $\hat{\mathbf{n}}$ is the normal unit vector. Across the boundary of a good conductor the tangential component of the **H** field is equal to the surface current density and the normal component of the **D** field is equal to the charge surface density:

$$\hat{\mathbf{n}} \mathbf{X} \mathbf{H} = \mathbf{j}_s,$$
$$\hat{\mathbf{n}} \cdot \mathbf{D} = \rho_s.$$

In a perfect conductor (and for very high-conductivity materials) the tangential electric field is continuous and therefore zero and the normal **B** field is also zero:

$$\hat{\mathbf{n}} \mathbf{X} \mathbf{E} = 0,$$
$$\hat{\mathbf{n}} \cdot \mathbf{B} = 0.$$

3.5. Plane Waves

Each Cartesian component of the field satisfies the scalar wave equation,

$$\nabla^2 E_x + k^2 E_x = \frac{\partial^2 E_x}{\partial x^2} + \frac{\partial^2 E_x}{\partial y^2} + \frac{\partial^2 E_x}{\partial z^2} + k_x^2 E_x = 0,$$

and the same for $\mathbf{E}_y$ and $\mathbf{E}_z$, where

$$k_x^2 + k_y^2 + k_z^2 = k_0^2.$$

Solutions can be written as sines and cosines or exponentials. For a field that is a function of z only the solution is

$$E_x(z) = Ae^{-ikz} + Be^{ikz}.$$

The physically real field is given by

$$E_x(z) = \text{Re}[E_x(z)] = \text{Re}(Ae^{-ikz} + Be^{ikz}) = A\cos(kz - \omega t) + B\sin(kz + \omega t),$$

where Re indicates "real part of."

Snell's laws for reflection and refraction in a nonabsorbing medium are

$$n_t \sin\theta_t = n_i \sin\theta_i, \quad \theta_r = \pi - \theta_i,$$

where n is the refractive index and the subscripts i, r, and t represent the incident, reflected, and transmitted light, respectively. The condition for total internal reflection, from a dense medium to a rare one, is

$$\sin\theta_t = \frac{n_t}{n_i},$$

where n_i is the refractive index of the denser medium and n_t is that of the rare medium. The reflectivity for total internal reflection is one.

The transmitted and the reflected amplitudes for light polarized parallel to the plane of incidence in a nonabsorbing medium are

$$\frac{E_t}{E_i} = \frac{2n_i\cos\theta_i}{n_t\cos\theta_i + n_i\cos\theta_t} = \frac{2\sin\theta_t\cos\theta_i}{\sin(\theta_i+\theta_t)\cos(\theta_i-\theta_t)},$$

$$\frac{E_r}{E_i} = \frac{n_t\cos\theta_i - n_i\cos\theta_t}{n_t\cos\theta_i + n_i\cos\theta_t} = \frac{\tan(\theta_i-\theta_t)}{\tan(\theta_i+\theta_t)}.$$

The corresponding equations for the perpendicular components are

$$\frac{E_t}{E_i} = \frac{2n_i\cos\theta_t}{n_i\cos\theta_i + n_t\cos\theta_t} = \frac{2n_t\sin\theta_t\cos(\theta_t)}{\sin(\theta_i+\theta_t)},$$

$$\frac{E_r}{E_i} = \frac{n_i\cos\theta_i - n_t\cos\theta_t}{n_i\cos\theta_i + n_t\cos\theta_t} = -\frac{\sin(\theta_i-\theta_t)}{\sin(\theta_i+\theta_t)}.$$

The corresponding intensities, absolute squares of the amplitude ratios, are

$$\rho_{\parallel} = r_{\parallel} = \left|\frac{E_{\parallel,r}}{E_{\parallel,i}}\right|^2 = \frac{\tan^2(\theta_i-\theta_t)}{\tan^2(\theta_i+\theta_t)},$$

$$\rho_{\perp} = r_{\perp} = \left|\frac{E_{\perp,r}}{E_{\perp,i}}\right|^2 = \frac{\sin^2(\theta_i-\theta_t)}{\sin^2(\theta_i+\theta_t)},$$

$$\tau_{\parallel} = \left|\frac{E_{\parallel,t}}{E_{\parallel,i}}\right|^2 = \frac{\sin^2\theta_i\sin\theta_t}{\sin^2(\theta_i+\theta_t)\cos^2(\theta_i-\theta_t)},$$

$$\tau_{\perp} = \left|\frac{E_{\perp,t}}{E_{\perp,i}}\right|^2 = \frac{\sin^2\theta_i\sin\theta_t}{\sin^2(\theta_i+\theta_t)}.$$

The special case of normal incidence yields the following expression for the reflectivity of both polarizations, which are indistinguishable:

$$\rho_{\perp} = \rho_{\parallel} = \rho = \left(\frac{n-1}{n+1}\right)^2,$$

where $n = n_t/n_i$.

The transmissivity is

$$\tau = \frac{4n}{(n+1)^2}.$$

At Brewster's angle, the polarizing angle, the transmitted parallel component of the field is zero. The reflected light is polarized perpendicularly. The angle is given by

$$\tan \theta_i = n.$$

3.6. Optical Properties of Metals

The wave equation for a metal is formally identical to that for a dielectric if the real refractive index is replaced with a complex refractive index:

$$\tilde{n} = n(1 + j\kappa),$$

where n is the real part of the complex refractive index and κ is the imaginary part, often called the extinction coefficient. Some authors use an alternative definition:

$$\tilde{n} = n - j\kappa.$$

In the first definition, the real and the imaginary parts of the refractive index are related to the other material properties of a metal as follows:

$$n^2 = \frac{1}{2}\left(\sqrt{\mu^2\epsilon^2 + \frac{4\mu^2\sigma^2}{\nu^2}} + \mu\epsilon\right),$$

$$n^2\kappa^2 = \frac{1}{2}\left(\sqrt{\mu^2\epsilon^2 + \frac{4\mu^2\sigma^2}{\nu^2}} - \mu\epsilon\right).$$

The electric field is represented by

$$\mathbf{E} = E_0 e^{-j[\tilde{k}\cdot r - \omega t]},$$

where

$$\tilde{k} = \frac{\omega n(1 + j\kappa)}{c}.$$

This is an attenuated plane wave, with an absorption coefficient of

$$\alpha = \frac{4\pi}{\lambda}\kappa$$

and a skin depth of

$$d = \frac{\lambda}{4\pi\kappa} = \frac{\lambda}{4\pi n\kappa}.$$

Snell's law for a metal is analogous to that for a dielectric:

$$n_i \sin\theta_i = \tilde{n}_t \sin\theta_t.$$

The wave in the metal is inhomogeneous; the surfaces of constant phase and constant amplitude do not generally coincide. The planes of constant amplitude are given by z = constant (for an xz plane of incidence); those of constant phase are given by

$$x\sin\theta_i + znq(\cos\gamma - \kappa\sin\gamma) = \text{constant},$$

where q and γ are given by

$$q^2\cos^2\gamma = 1 - \frac{1-\kappa^2}{n^2(1+\kappa^2)^2}\sin^2\theta_i,$$

$$q^2\sin^2\gamma = \frac{2\kappa}{n^2(1+\kappa^2)^2}\sin^2\theta_i.$$

Fresnel's equations for metals are

$$\rho = \frac{\rho_{12}e^{2v_2\eta} + \rho_{23}e^{-2v_2\eta} + 2\rho_{12}\rho_{23}\cos(\phi_{23} - \phi_{12} + 2u_2\eta)}{e^{2v_2\eta} + \rho_{12}^2\,\rho_{23}^2e^{-2v_2\eta} + 2\rho_{12}\rho_{23}\cos(\phi_{12} + \phi_{23} + 2u_2\eta)},$$

$$\tan\delta_r = \frac{\rho_{23}\,(1-\rho_{12}^2)\sin(2u_2\eta + \phi_{23}) + \rho_{12}(e^{2v_2\eta} - \rho_{23}^2e^{-2v_2\eta})\sin\phi_{12}}{\rho_{23}\,(1+\rho_{12}^2)\cos(2u_2\eta + \phi_{23}) + \rho_{12}(e^{2v_2\eta} + \rho_{23}^2e^{-2v_2\eta})\cos\phi_{12}},$$

$$\tau = \frac{n_3\cos\theta_3\,\tau_{12}^2\tau_{23}^2}{n_1\cos\theta_1[e^{2v_2\eta} + \rho_{12}^2\rho_{23}^2e^{-2v_2\eta} + 2\rho_{12}\rho_{23}\cos(\phi_{12} + \phi_{23} + 2u_2\eta)]},$$

$$\tan(\delta_t - \chi_{12} - \chi_{23} + u_2\eta) = \frac{e^{2v_2\eta}\sin^2 u_2\eta - \rho_{12}\rho_{23}\sin(\phi_{12} + \phi_{23})}{e^{2v_2\eta}\cos^2 u_2\eta + \rho_{12}\rho_{23}\cos(\phi_{12} + \phi_{23})},$$

where, for the TE wave,

$$\rho_{12}^2 = \frac{(n_1 \cos \theta_1 - u_2)^2 + \nu_2^2}{(n_1 \cos \theta_1 + u_2)^2 + \nu_2^2},$$

$$\tan \phi_{12} = \frac{2\nu_2 n_1 \cos \theta_1}{u_2^2 + \nu_2^2 - n_1^2 \cos \theta_1^2},$$

$$\rho_{23}^2 = \frac{(n_3 \cos \theta_3 - u_2)^2 + \nu_2^2}{(n_3 \cos \theta_3 + u_2)^2 + \nu_2^2},$$

$$\tan \phi_{23} = \frac{2\nu_2 n_3 \cos \theta_3}{u_2^2 + \nu_2^2 - n_3^3 \cos^2 \theta_3}.$$

For the TM wave,

$$\rho_{12}^2 = \frac{[n_2^2(1-\kappa^2)\cos \theta_1 - n_1 u_2^2]^2 + (2n_2^2\kappa^2 \cos \theta_1 - n_1\nu_2)^2}{[n_2^2(1-\kappa^2)\cos \theta_1 + n_1 u_2^2]^2 + (2n_2^2\kappa^2 \cos \theta_1 + n_1\nu_2]^2},$$

$$\tan \phi_{12} = \frac{2n_1 n_2^2 \cos \theta_1 \, [2\kappa_2 u_2 - (1-\kappa_2^2)\nu_2]}{n_2^4(1+\kappa^2)^2 \cos^2 \theta_1 - n_1^2(u_2^2 + \nu_2^2)},$$

$$\tau = \frac{n_1 \cos \theta_3 \tau_{12}^2 \tau_{23}^2}{n_3 \cos \theta_1 [e^{2\nu_2\eta} + \rho_{12}^2\rho_{23}^2 e^{-2\nu_2\eta} + 2\rho_{12}\rho_{23}\cos(\phi_{12} + \phi_{23} + 2u_2\eta)]},$$

$$\tan(\delta_t - \chi_{12} - \chi_{23} + u_2\eta) = \frac{e^{2\nu_2\eta} \sin^2 u_2\eta - \rho_{12}\rho_{23} \sin(\phi_{12} + \phi_{23})}{e^{2\nu_2\eta} \cos^2 u_2\eta + \rho_{12}\rho_{23} \cos(\phi_{12} + \phi_{23})}$$

(the same as for the TE wave).

For both,

$$2u_2^2 = n_2^2(1-\kappa^2) - n_1^2 \sin^2 \theta_1 + \sqrt{[n_2^2(1-\kappa^2) - n_1^2 \sin^2 \theta_1]^2 + 4n_2^4\kappa_2^2},$$

$$\nu_2^2 = -[n_2^2(1-\kappa^2) - n_1^2 \sin^2 \theta_1] + \sqrt{[n_2^2(1-\kappa^2) - n_1^2 \sin^2 \theta_1]^2 + 4n_2^4\kappa_2^2},$$

$$\eta = \frac{2\pi}{\lambda_0}.$$

3.7. Cylindrical Waves

In cylindrical coordinates (r, θ, z), the circular cylindrical scalar wave equations are

$$r\frac{\mathrm{d}}{\mathrm{d}r}\left[\frac{r\mathrm{d}R(r)}{\mathrm{d}r}\right] + [(k^2 - h^2)r^2 - p^2]R(r) = 0$$

for the radial function, $R(r)$, h, and p are separation constants determined by geometry, and

$$\frac{d^2\Theta(\theta)}{d\theta^2} + p^2\Theta(\theta) = 0$$

for the angular function $\Theta(\theta)$. The radial function is represented by a Bessel function and the angular function by an exponential. Thus particular solutions in the frequency domain can be constructed from waves of the form

$$\Psi_n = e^{jn\theta} Z_n\left(\sqrt{k^2 - h^2}\, r\right) e^{\pm ihz - i\omega t}.$$

The fields are

$$E_r = ih \sum a_n \frac{\partial \Psi_n}{\partial r} - \mu \frac{\omega}{r} \sum n b_n \Psi_n,$$

$$E_\theta = -\frac{h}{r} \sum n a_n \Psi_n - i\mu\omega \sum b_n \frac{\partial \Psi_n}{\partial r},$$

$$E_z = (k^2 - h^2) \sum a_n \Psi_n$$

$$H_r = \frac{k^2}{\mu\omega r} \sum n a_n \Psi_n + ih \sum b_n \Psi_n \frac{\partial \Psi_n}{\partial r},$$

$$H_\theta = \frac{ik^2}{\mu\omega} \sum a_n \frac{\partial \Psi_n}{\partial r} - \frac{h}{r} \sum n b_n \Psi_n,$$

$$H_z = (k^2 - h^2) \sum b_n \Psi_n,$$

where the summation goes from minus infinity to plus infinity, the sign of h determines the propagation direction, and the constants a_n and b_n are constants that are determined from the initial conditions.

3.8. Spherical Waves

The vector wave equation separates into three scalar equations. The radial scalar equation is

$$r^2 \frac{d^2R(r)}{dr^2} + 2r \frac{dR(r)}{dr} + (k^2r^2 - p^2)R(r) = 0.$$

The azimuthal equation is

$$\frac{1}{\sin\theta}\frac{\mathrm{d}}{\mathrm{d}\theta}\left[\sin\theta\,\frac{\mathrm{d}\Theta(\theta)}{\mathrm{d}\theta}\right]+\left(p^2-\frac{q^2}{\sin^2\theta}\right)\Theta(\theta)=0,$$

and the elevation equation is

$$\frac{\mathrm{d}^2\Phi(\phi)}{\mathrm{d}\phi^2}+q^2\Phi(\phi)=0,$$

where $R(r)$, $\Theta(\theta)$, and $\Phi(\phi)$ are the radial, azimuthal, and elevations functions, respectively. The solution is

$$R(r)=\frac{1}{\sqrt{kr}}Z_{n+(1/2)}(kr)P_n^m(\cos\theta)\mathrm{trig}(m\phi),$$

where Z is the spherical Bessel function, P is the Legendre polynomial, and trig is either sine or cosine.

3.9. Group and Phase Velocities

For a quasi-monochromatic plane wave, the velocity at which the planes of constant phase travel is given by

$$v_p=\frac{\bar{\omega}}{\bar{k}}.$$

The group velocity is given by

$$v_g=\frac{\mathrm{d}\omega}{\mathrm{d}k}.$$

3.10. Fermat's Principle

Also called the principle of least time, Fermat's principle can be used as a theoretical basis for many optical problems. It states that the optical length

$$\int_a^b n\cdot\mathrm{d}s$$

of a real ray from a to b is the shortest distance in a regular neighborhood.

3.11. Eikonal Equation

The eikonal, or image, equation is

$$\left(\frac{\partial S}{\partial x}\right)^2 + \left(\frac{\partial S}{\partial y}\right)^2 + \left(\frac{\partial S}{\partial z}\right)^2 = n^2(x, y, z),$$

where S, a function of x, y, and z, is the optical path and is often called the eikonal. The surfaces $S(x, y, z)$ = constant are geometrical wave fronts.

3.12. Dispersion

Dispersion theory is generally understood to mean the theory of the way in which the optical constants vary with frequency or wavelength. Based on the Lorentz–Lorenz theory of polarizability and the equations for a damped-harmonic oscillating electron, the expression for dispersion is

$$\frac{n^2 - 1}{n^2 + 2} = \frac{4\pi}{3} \frac{Ne^2}{m(\omega_0^2 - \omega^2)}.$$

For materials that have a refractive index of ~1, the expression can be written as

$$n^2 - 1 = \frac{4\pi Ne^2}{m(\omega_0^2 - \omega^2)}.$$

If the material has more than one resonance frequency, this can be written as a summation over them. Another useful form is the Cauchy equation:

$$n^2 - 1 = a + \sum_i \frac{a_i}{\lambda^2 - \lambda_i^2}.$$

A variant of this is the Sellmeier equation:

$$n^2 - 1 = \frac{A\lambda^2}{\lambda^2 - \lambda_0^2}.$$

3.13. Analytic Signal

The electric field of polychromatic light can be represented as

$$E^{(r)}(t) = \int_0^\infty a(\nu)\cos[\phi(\nu) - 2\pi\nu t]d\nu.$$

It is a real function. There are corresponding imaginary and complex functions:

$$E^{(i)}(t) = \int_0^{\infty} a(\nu)\sin[\phi(\nu) - 2\pi\nu t]\mathrm{d}\nu,$$

$$E(t) = \int_0^{\infty} a(\nu)e^{[\phi(\nu) - 2\pi\nu t]}\mathrm{d}\nu.$$

The complex function $E(t)$ is called the analytic signal associated with the real signal $E^{(r)}(t)$.

3.14. Hilbert Transforms

The real and the imaginary functions, $E^{(r)}(t)$ and $E^{(i)}(t)$, are related to each other by the following Hilbert transforms:

$$E^{(i)} = \frac{1}{\pi} P \int_{-\infty}^{\infty} \frac{E^{(r)}(t')}{t' - t}\,\mathrm{d}t', \quad E^{(r)} = \frac{1}{\pi} P \int_{-\infty}^{\infty} \frac{E^{(i)}(t')}{t' - t}\,\mathrm{d}t',$$

where P indicates the Cauchy principal value.

3.15. Kramers–Kronig Technique

The complex refractive index is $\tilde{n} = n - ik$. The real refractive index and the imaginary part (the extinction coefficient) are related by Hilbert transforms, as are the magnitude and the phase of reflection, $r = r_0 e^{i\phi}$. Thus we can measure the reflectivity over as wide a frequency range as possible and therefrom obtain both components. It is not necessary to measure to and from infinity, as the integrals are strongly weighted in the regions near the resonance.

The relationships are

$$\phi(\omega) = -\frac{2\omega}{\pi} \int_0^{\infty} \log \frac{r_0}{(\omega')^2 - \omega^2}\,\mathrm{d}\omega'.$$

The relationship between the refractive index n and the absorption coefficient is

$$n(\omega) - 1 = \frac{1}{2\pi^2} \int_0^{\infty} \frac{\alpha}{1 - \lambda^2/\lambda_0^2}\,\mathrm{d}\lambda.$$

3.16. Partial Coherence

The mutual coherence (correlation) of two light beams is

$$\Gamma_{12}(\tau) = \langle E_1(t + \tau)E_2(t)^*\rangle.$$

The normalized function, the complex degree of coherence (correlation), is

$$\gamma_{12}(\tau) = \frac{\Gamma_{12}(\tau)}{\sqrt{\Gamma_{11}(0)\Gamma_{2}(0)}}.$$

The mutual spectral density (cross-power spectrum) is

$$G_{12}(\nu) = \lim \left[\frac{e(x,\nu)e^{*}(x,\nu)}{2T} \right],$$

where the e's are the Fourier transforms of the fields and the limit is as T goes to infinity. The relationship between $\Gamma_{12}(\tau)$ and $G_{12}(\nu)$ is

$$\Gamma_{12}(\tau) = 4 \int_{0}^{\infty} G_{12}(\nu) e^{-i\omega\tau} d\nu.$$

The real and the imaginary parts of Γ_{12} are Hilbert transforms of each other. The coherency matrix is defined as

$$\begin{bmatrix} \langle E_x E^{*}{}_x \rangle & \langle E_x E^{*}{}_y \rangle \\ \langle E_y E^{*}{}_x \rangle & \langle E_y E^{*}{}_y \rangle \end{bmatrix}.$$

3.17. Further Reading

M. Born and E. Wolf, *Principles of Optics,* 7th ed., Pergamon, 1999.
J. A. Stratton, *Electromagnetic Theory,* McGraw-Hill, 1941.
R. Plonsey and R. Collins, *Principles and Applications of Electromagnetic Fields,* McGraw-Hill, 1961.
E. W. Hobson, *The Theory of Spherical and Ellipsoidal Harmonics,* Cambridge Univ. Press, 1931.
G. N. Watson, *Bessel Functions,* Cambridge Univ. Press, 1922.

Principles of Optics by Born and Wolf is the classic treatment of electromagnetic development of optics. Stratton's book is a detailed text of electromagnetic theory, whereas that of Plonsey and Collins is a more introductory text. The last two books give mathematical backgrounds for some of the wave-equation solutions.

4

Polarization

William L. Wolfe

4.1. Introduction

This chapter gives information about the four ways to describe polarized light, about Mueller and Jones calculi and generating the appropriate matrices and converting from one description to another, and some information on partially polarized light.

4.2. Some Concepts

A *linear polarizer* is an optical element that converts unpolarized light into linearly polarized light. A linear polarizer has two orthogonal axes, called the transmission axis and the extinction axis.

The *degree of polarization P* is the degree to which an optical beam is polarized. If $P < 1$, the light is partially polarized; if $P = 1$, the light is completely polarized; if $P = 0$, the light is unpolarized.

To measure the degree of polarization of an optical beam, the beam is first passed through a quarter-wave plate, then through a linear polarizer of high polarizance. If the intensity transmittance τ is not changed by any combinations of the orientations of the wave plate and the polarizer, then the light is unpolarized; $P = 0$. If the light is polarized, then the maximum transmittance τ_{max} is found by adjustment of the orientations of the two optical elements; then the minimum τ_{min} is found. The degree of polarization is then given by

$$P = \frac{\tau_{max} - \tau_{min}}{\tau_{max} + \tau_{min}}. \tag{4.1}$$

If τ_{min} is zero, then the light is completely polarized.

Polarizance is the degree of polarization produced by a polarizer when the incident light is unpolarized.

A *retarder* is an optical element that is often used to convert linearly polarized light into elliptically polarized light. A retarder has two optical axes, named the fast and slow axes.

Retardance is the change in relative phase produced by a retarder. A quarter-wave plate is identical with a 90° retarder.

Retardation is another name for retardance.

Partially polarized light can be described as consisting of one part that is polarized and a second that is unpolarized. If the flux densities of these are M_p and M_u, respectively, then the *degree of polarization* is given by the ratio of the polarized portion to the total:

$$P = \frac{M_p}{M_p + M_u}. \tag{4.2}$$

Partially polarized light may also be represented as the sum of orthogonal completely polarized parts, which parts are completely incoherent. The part with the greater flux density is the dominant part M_d, whereas the other part is the inferior part M_i. The degree of polarization is then written as

$$P = \frac{M_d - M_i}{M_d + M_i}. \tag{4.3}$$

4.3. Types of Polarization

Light can be fully polarized or partially polarized.

4.3.1. Fully Polarized Light

Polarization describes the motion of the electric vector of an optical wave. The wave is transverse, and the polarization state can be viewed as the pattern that would be projected onto a plane that the wave approaches, an end-on view. The most general pattern is an ellipse. Special, but very important, cases include circular polarization and linear polarization.

Circular polarization is further described by the direction it rotates—clockwise or counterclockwise. Clockwise is also described as right-circular polarization and counterclockwise as left. These are sometimes described as the handedness of circular polarization. Most authors describe clockwise and counterclockwise as if the light is approaching.

Linear polarization is further described by the direction of electric field—vertical, horizontal, or with respect to some reference plane (like the plane of incidence), e.g., perpendicular or parallel. There exists an infinite number of linear polarizations, each indicated by its angle with the vertical or horizontal. Most modern authors describe the linear polarization direction in terms of the electric vector; that is, horizontal, linear polarization occurs when the electric vector is in a horizontal plane, but this convention is not universal.

4.3.2. Partially Polarized Light

Light is said to be unpolarized if there is no preferred direction of vibration for the electric vector or if the degree of polarization is zero.

Elliptical polarization has the same handedness conventions as circular polarization. Elliptical polarization is described by its handedness, axial ratio, and azimuth, as shown in Fig. 4.1. If $2a$ and $2b$ are the axes of the ellipse, the ratio b/a is the (semi)axial ratio, where $a > b$. The azimuth is the angle α that the axis a makes with the horizontal. These are quantities that are readily determined with an ellipsometer.

There are essentially three forms of orthogonal polarization. Orthogonal linear-polarized components are 90° apart. Orthogonal circular components rotate in opposite directions. Orthogonal elliptical components have opposite handedness, azimuths at 90°, and equal axial ratios.

4.4. Phasor Representations

For a quasi-monochromatic plane wave traveling in the z direction, the polarization is in the x and y directions. The **E** vector in the x,y plane can be represented by two complex components, E_x and E_y, or by $A_x e^{i\phi_x}$, $A_y e^{i\phi_y}$, where the A's are real and the exponentials represent the phase. The absolute phase is hardly ever important, only that relative between the two components. In that case, one writes A_x and $A_y e^{i\gamma}$ or $A_x e^{-i\gamma}$ and A_y or symmetrically $A_x e^{-i\gamma/2}$ and

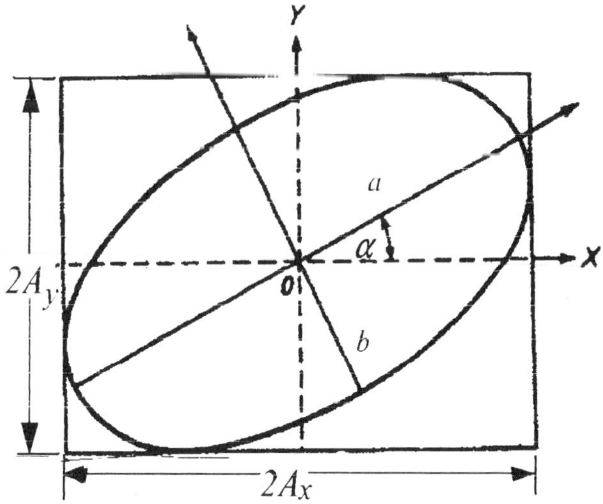

Figure 4.1. Polarization ellipse.

$A_y e^{i\gamma/2}$. When A_x is zero, the state of polarization is linear vertical. When A_y is zero, it is linear horizontal. If A_x and A_y are equal and the phase difference is +90°, the polarization is right circular. If the phase difference is –90°, it is left circular. If the amplitudes are not equal, the polarization is elliptical, right handed if the γ is between 0° and 180°. They are all normalized to unit intensity; therefore $s_0 = 1$.

4.5. The Poincaré Sphere[1]

This sphere, shown in Fig. 4.2, is a handy visual representation of polarization states that allows the formulation and solution of certain problems. It applies

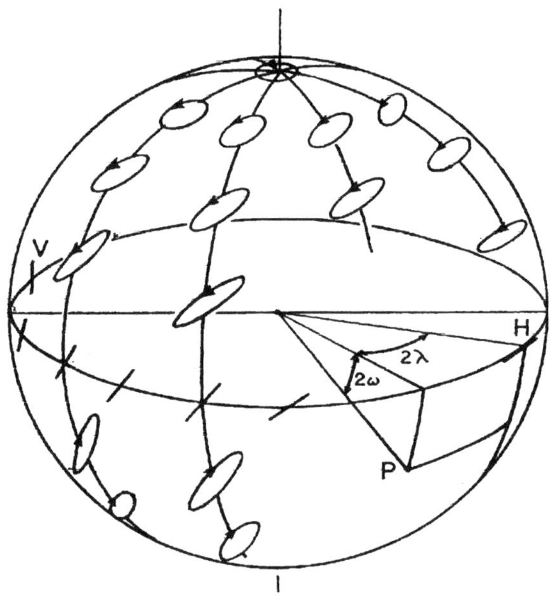

Figure 4.2. The Poincaré sphere.

to incident light that is completely polarized. The north pole of the sphere represents right-circular polarization; the south pole, left. The equator represents linear polarization, the zero degree of longitude, vertical polarization; the 180° of longitude, horizontal. Diametrically opposite points on the equator represent orthogonal polarization forms. The main use of the Poincaré sphere is to determine the change in polarization that is due to a given retardation. A general position on the sphere represents a completely polarized beam, the ellipse of which has azimuth λ, axial ratio tan ω, and handedness that is given by the sign of ω, plus for right and vice versa. The polarization is left if $2\omega < 0$ or right if $2\omega > 0$, and $p_1 = \sin 2\omega$, $p_2 = \cos 2\omega \cos 2\lambda$, and $p_3 = \cos 2\omega \sin 2\lambda$,

[1] H. Poincaré, *Thèorie mathèmatique de la lumiere,* Gauthiers-Villars, 1892.

where the p's are Cartesian coordinates of the Poincarè sphere of unit radius. The normal order of the p's is p_2, p_3, p_1; here they are arranged so that $p_1 = s_1$, $p_2 = s_2$, and $p_3 = s_3$.

The process of determining the effect of retardation on the Poincaré sphere is to find the point P that describes the original polarization state and then rotate the sphere (or move on spherical arcs) about the correct angle and through the correct angle. The axis is the radius vector from the center of the sphere to R that describes the fast eigenvector of the retarder. The angle is the value of the retardance.

An example is the effect of a 90° retardance with a fast axis at 22.5°, as shown in Fig. 4.3. The original point is P. The axis is at 22.5°; the rotation shown in Fig. 4.2. The Poincaré sphere is 90° about the axis R; P' is the result.

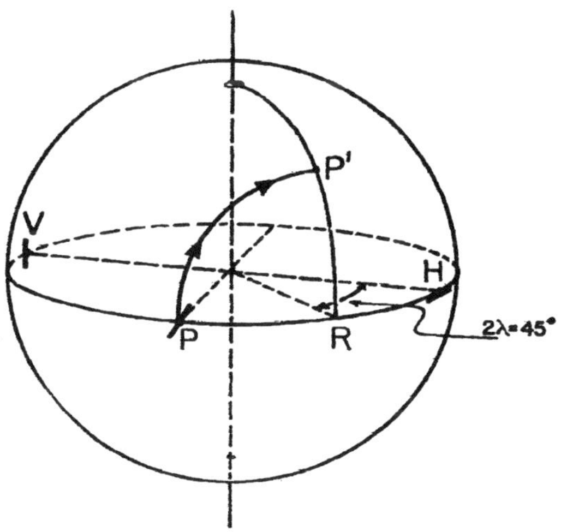

Figure 4.3. Using the Poincaré sphere.

4.6. Ellipsometer Description of a Vector

This method of describing the polarization vector is related to how an ellipsometer can be used to find the information. It uses the parameters b/a, the ratio of the semiaxes of the polarization ellipse, and the angle α that the major axis makes with the axis, as shown in Fig. 4.1.

4.7. Stokes Vectors[2]

The state of polarization of a beam of light may be described by the Stokes vector **S**, which in the row mode may be written as

[2] G. G. Stokes, "On the composition and resolution of streams of polarized light from different sources," Trans. Cambridge Philos. Soc. **9,** 399, 1852; in *Mathematical and Physical Papers,* Cambridge Univ. Press, 1904.

$$\mathbf{S} = \lfloor s_0\ s_1\ s_2\ s_3 \rfloor \tag{4.4}$$

or as a column vector

$$\mathbf{S} = \begin{bmatrix} s_0 \\ s_1 \\ s_2 \\ s_3 \end{bmatrix}. \tag{4.5}$$

The Stokes vector is not a vector in configuration space. The closely related Wiener coherency matrix is a 2×2 matrix in the x,y plane; both the Stokes vector and the Wiener matrix contain only observables, measurable quantities. The components of the Stokes vector may be obtained by the sequential insertion, in a steady beam of light, of the following ideal filters: nonpolarizing, linear horizontal, linear 45°, and right circular. Each should have 50% transmittance for unpolarized light. If the output signals are V_i, with proper constants, then

$$s_0 = 2V_1, \quad s_1 = 2(V_2 - V_1), \quad s_2 = 2(V_3 - V_1), \quad s_3 = 2(V_4 - V_1). \tag{4.6}$$

It is conventional to define the Stokes vector in terms of averages; thus

$$s_0 = \langle A_x^2 + A_y^2 \rangle, \quad s_1 = \langle A_x^2 - A_y^2 \rangle, \quad s_2 = \langle 2A_x A_y \cos\gamma \rangle,$$
$$s_3 = \langle 2A_x A_y \cos\gamma \rangle, \tag{4.7}$$

where the angle brackets signify a time average and γ is the phase angle between the electric-field components. The vectors may be normalized to unit intensity by dividing all terms by the first value s_0, which is then 1.

The Stokes vector can describe partially polarized light. The field intensity (flux density) of the polarized portion is

$$I_p = \sqrt{s_1^2 + s_2^2 + s_3^2}. \tag{4.8}$$

The degree of polarization is therefore I_p/s_0.

The normalized Stokes vector may be written in terms of ρ and γ as (defined in Sections 4.4 and 4.8)

$$S = \begin{bmatrix} 1 \\ \cos 2\rho \\ \sin 2\rho \cos\gamma \\ \sin 2\rho \sin\gamma \end{bmatrix}. \tag{4.9}$$

The normalized Stokes vector is closely related to the Cartesian components, p_1, p_2, and p_3, of the unit-radius Poincaré sphere in Section 4.5. If $\gamma = 2\lambda$ and $2\rho = 90° - \omega$, then $p_1 = s_1$, $p_2 = s_2$, and $p_3 = s_3$.

The Stokes vectors are given in Table 4.1 for some important states of polarization.

4.8. Jones Vectors[3]

Unlike the Stokes vector, which can describe partially polarized light, neither the Jones vector nor the ellipse in Fig. 4.1 can represent partially polarized light.

Unlike the Stokes vector, the four components of which are real and observables, the two components of the Jones vector **J** are complex and not fully observable. The components of the Jones vector are E_x and E_y and are assembled in the Jones vector as follows:

$$\mathbf{J} = \begin{bmatrix} E_x \\ E_y \end{bmatrix}. \tag{4.10}$$

The components of E_x and E_y carry the magnitude and the absolute phase of the x and y components of the electric vector. One is rarely interested in the absolute phase of a light beam, so little is lost if E_xE_y must be real. Then the **J** may be written as

$$\mathbf{J} = \begin{bmatrix} A_x e^{-i\gamma/2} \\ A_y e^{i\gamma/2} \end{bmatrix}, \tag{4.11}$$

where the components A_x and A_y are real and may have either sign, and where γ is the relative phase; **J** is described by three independent numbers. If the flux density is of no concern, then one may write

$$|E_x|^2 + |E_y|^2 = A_x^2 + A_y^2 - 1. \tag{4.12}$$

Then the Jones vector of unit intensity may be written as

$$\mathbf{J} = \begin{bmatrix} \cos \rho e^{-i\gamma/2} \\ \sin \rho e^{i\gamma/2} \end{bmatrix}, \tag{4.13}$$

where $\rho = \arctan A_y/A_x$. Table 4.1 summarizes the specifications of beams of

[3] R. C. Jones, "New calculus for the treatment of optical systems," I. J. Opt. Soc. Am. **31,** 488, 1941; II, **31,** 493, 1941; III, **31,** 500, 1941; IV, **32,** 486, 1942; V, **37,** 107, 1947; VI, **37,** 110, 1947; VII, **38,** 671, 1948; VIII, **46,** 126, 1956.

Table 4.1. Normalized Stokes and Jones Vectors and Ellipsometer Parameters[a]

Pattern	α	b/a	A_y/A_x	ρ	γ	$[s_0$	s_1	s_2	$s_3]$	Jones
Linear states: $\gamma = 0$, $\alpha = \rho$, $b/a = 0$										
Horizontal	0°	0	0	0°	0°	[1	1	0	0]	$\begin{bmatrix} 1 \\ 0 \end{bmatrix}$
Vertical	90°	0	∞	90°	0°	[1	–1	0	0]	$\begin{bmatrix} 0 \\ 1 \end{bmatrix}$
+45°	+45°	0	1	45°	0°	[1	0	1	0]	$\frac{1}{\sqrt{2}}\begin{bmatrix} 1 \\ 1 \end{bmatrix}$
–45°	–45°	0	–1	–45°	0°	[1	0	–1	0]	$\frac{1}{\sqrt{2}}\begin{bmatrix} 1 \\ -1 \end{bmatrix}$
General	ρ	0	$\tan \rho$	α	0°	[1	$\cos 2\rho$	$\sin 2\rho$	0]	$\begin{bmatrix} \cos \rho \\ \sin \rho \end{bmatrix}$
Circular states: $\gamma = \pm 90°$, $A_x = A_y$, $b/a = 1$										
Right	——	1	1	45°	90°	[1	0	0	1]	$\frac{1}{2}\begin{bmatrix} 1-i \\ 1+i \end{bmatrix}$
Left	——	–1	1	45°	–90°	[1	0	0	–1]	$\frac{1}{2}\begin{bmatrix} 1+i \\ 1-i \end{bmatrix}$
Ratio-two elliptical states										
Right horizontal	0°	$^1/_2$	$^1/_2$	26.57°	90°	[1	0.6	0	0.8]	$\frac{1}{\sqrt{10}}\begin{bmatrix} 2(1-i) \\ 1+i \end{bmatrix}$

	Right vertical	90°	$^1/_2$	2	63.43°	90°	[1	–0.6	0	0.8]	$\frac{1}{\sqrt{10}}\begin{bmatrix}1-i\\2(1+i)\end{bmatrix}$
	Left horizontal	0°	$-^1/_2$	$^1/_2$	26.57°	–90°	[1	0.6	0	–0.8]	$\frac{1}{\sqrt{10}}\begin{bmatrix}2(1+i)\\1-i\end{bmatrix}$
	Left vertical	90°	$-^1/_2$	2	63.43°	–90°	[1	–0.6	0	–0.8]	$\frac{1}{\sqrt{10}}\begin{bmatrix}1+i\\2(1-i)\end{bmatrix}$
	Slanted	22.5°	0.318	0.518	27°	45°	$\Big[1$	$\frac{1}{\sqrt{3}}$	$\frac{1}{\sqrt{3}}$	$\frac{1}{\sqrt{3}}\Big]$	$\begin{bmatrix}0.888e^{-\pi i/8}\\0.460e^{\pi i/8}\end{bmatrix}$

****

Most general polarized state

In terms of ρ and γ, the most general polarization state is represented by the Stokes or Jones column vectors:

$$\begin{bmatrix}1\\ \cos 2\rho\\ \sin 2\rho \cos \gamma\\ \sin 2\rho \sin \gamma\end{bmatrix} \qquad \begin{bmatrix}\cos \rho e^{-i\gamma/2}\\ \sin \rho e^{i\gamma/2}\end{bmatrix}$$

The corresponding values of α and *b*/*a* may be calculated by the equations in Subsections 4.9.2 and 4.9.3.

Unpolarized state

The unpolarized state cannot be represented by the Jones vector. The Stokes vector of unpolarized light is the column vector [1 0 0 0].

---

[a] In this table the handedness of the states is carried by the sign of $b/a, s_3$, and γ. The Jones vectors satisfy $E_x E_y$ = real and are normalized by $|E_x|^2 + |E_y|^2 = 1$. The Stokes vectors are normalized by $s_0 = 1$.

Note: In this table $\rho = \arctan A_y/A_x$, and the dashes in the α column mean that α is not defined.

different polarizations in the Stokes and Jones formats. The Jones vectors all have unit intensity so that $A_x^2 + A_y^2 = 1$, and they have symmetrical phases so that E_xE_y is real.

This concludes a description of the ways of describing polarized light. The remainder of the chapter includes ways to convert from one description to another and matrix representations of components that change the state of polarization of light that passes through them. Such flat, polarization-changing components include retarders, linear and circular polarizers, and rotators. One method is based on the Stokes vector; the other on the Jones vector.

4.9. Conversions

This section provides the information for converting from the three representations of polarization: the ellipsometer parameters *a, b,* and γ; the Stokes vector; and the Jones vector. The equations were derived for this handbook, and therefore they do not have the test of time, but have been verified by two independent sets of computer calculations. They are a guide and must be used in conjunction with physical interpretations.

The relations in Subsections 4.9.2 and 4.9.3 were derived from the following two equations:

$$s_2/s_1 = \tan 2\alpha = \tan 2\rho \cos \gamma, \tag{4.14}$$

$$s_3/s_0 = \sin 2\beta = \sin 2\rho \sin \gamma, \tag{4.15}$$

where $\rho = \arctan A_y/A_x$ and $\beta = \arctan b/a$.

4.9.1. *Stokes and Jones*

In this subsection the Stokes and Jones vectors need not have unit intensity. The intensities are related by $s_0 = A_x^2 + A_y^2$. The Stokes vector is quadratic in the electric vector, so the Stokes vector is to the Jones vector as watts are to volts.

The Stokes vector can be found from the Jones vector $\mathbf{J}(A_x, A_y, \gamma)$ by the following transformations:

$$s_0 = A_x^2 + A_y^2, \quad s_1 = A_x^2 - A_y^2, \quad s_2 = 2A_xA_y \cos \gamma,$$
$$s_2 = 2A_xA_y \sin \gamma. \tag{4.16}$$

The Jones vector can be found from the Stokes vector by the following transformations:

$$A_x = \sqrt{\frac{s_0 + s_1}{2}}, \quad A_y = \sqrt{\frac{s_0 - s_1}{2}}, \quad \gamma = \arctan\left(\frac{s_3}{s_2}\right). \tag{4.17}$$

The signs of the roots are determined as follows: with γ known, the product A_xA_y, including its sign, is given by

$$A_xA_y = \frac{1}{2}\frac{s_2}{\cos\gamma} = \frac{1}{2}\frac{s_3}{\sin\gamma}. \tag{4.18}$$

This is sufficient because if (A_x, A_y, γ) produces a given Stokes vector, so does $(-A_x, -A_y, \gamma)$.

4.9.2. Stokes and Ellipsometer

The ellipsometer parameters b/a and α can be found from the Stokes vector by

$$\frac{b}{a} = \tan[0.5\arcsin(s_3/s_0)], \quad \alpha = 0.5\arctan(s_2/s_1). \tag{4.19}$$

The principal value of the arcsine works well in the equation for b/a. The principal value of the arctangent in the equation for α puts α in the closed range from –45° to +45°. If it is to be in the range from 0° to 90°, then 90° must be added to the principal value of α when s_2/s_1 is negative. Then $s_2/s_1 = 0$ is on the edge; it can correspond either to $\alpha = 0°$ or $\alpha = 90°$.

The Stokes vector normalized to unit intensity can be found from the ellipsometer parameters by the following transformations:

$$s_0 = 1, \quad s_1 = \sqrt{\frac{1-s_3^2}{1+\tan^2(2\alpha)}}, \quad s_2 = s_1\tan(2\alpha),$$

$$s_3 = \sin[2\arctan(b/a)]. \tag{4.20}$$

It does not matter which of the signs is taken for s_1, because if $[1,s_1,s_2,s_3]$ produces a given ellipsometer state b/a, α, so does $[1,-s_1,-s_2,s_3]$.

4.9.3. Jones and Ellipsometer

The Jones vector can be found from the ellipsometer parameters by the following transformations:

$$\frac{A_y}{A_x} = \tan\left(0.5\arctan\left\{\sqrt{\frac{\tan^2(2\alpha)+\sin^2[2\arctan(b/a)]}{\cos^2[2\arctan(b/a)]}}\right\}\right) \tag{4.21}$$

$$\gamma = \arctan\left\{\frac{\sin(2\arctan b/a)}{\tan 2\alpha\cos[2\arctan(A_y/A_x)]}\right\}. \tag{4.22}$$

Care must be taken with the sign of the square root; a good guide is to take the negative root if α is negative.

The ellipsometer parameters can be found from the Jones vector by the following transformations:

$$\frac{b}{a} = \tan\left(\frac{1}{2}\arcsin\{\sin[2\arctan(A_y/A_x)]\sin(\gamma)\}\right),$$

$$\alpha = \frac{1}{2}\arctan\{\tan[2\arctan(A_y/A_x)]\cos(\gamma)\}. \tag{4.23}$$

The multiple values of arcsine and arctangent introduce complexities in the use of these equations, as does the sign of the square root. One way to avoid these troubles is to compute by means of the Stokes vector.

4.9.4. Stokes and Poincaré

If the parameters λ and ω of the Poincaré sphere are related to the ρ and γ of the Stokes vector as follows,

$$\gamma = 2\lambda, \quad \cos 2\rho = \sin 2\omega,$$

$$\text{alternatively,} \quad \sin 2\rho = \cos 2\omega, \quad 2\rho + 2\omega = \frac{\pi}{2}, \tag{4.24}$$

then the Poincaré p's are related to the normalized Stokes parameters by

$$p_1 = s_1, \quad p_2 = s_2, \quad p_3 = s_3. \tag{4.25}$$

A_x and A_y may take any real value when the Jones vectors are not normalized. The angle γ should lie in the range between –180° and +180°. The angle ρ should lie between –90° and +90°. The sign of γ indicates the handedness of the Jones polarization, and the sign of s_3 indicates the handedness of the Stokes polarization. The angle ω lies between 180° and 180°; the sign of ω indicates the handedness in the Poincaré sphere. In the ellipsometer parameters in this chapter, the angle α lies between 0° and 90°, and the axial ratio b/a lies between –1 and +1, the sign indicating the handedness. (It is counterintuitive that the handedness is carried by b/a, and not α). No matter which way the + and – signs are associated with right and left handedness, the equations in Section 4.9 are consistent. In Table 4.1, + is associated with right.

4.10. Description of the Calculi

Both the Jones and the Mueller calculi operate neatly with nearly monochromatic light that is incident normally upon a series of flat optical components, including retarders, polarizers, and rotators. They are most useful with more than a few components. They both use matrices that multiply *column* input vectors to obtain *row* output vectors by standard matrix multiplication (see Chap. 2).

4.10.1. Jones Calculus

The Jones calculus uses the Jones vector **J** as the input and output, and uses the Jones matrices $\mathcal{J}$ to obtain the output vector, which describes the output light, from the input vector. The Jones matrix contains four complex numbers arrayed in a 2×2 matrix:

$$\mathcal{J} = \begin{bmatrix} j_{11} & j_{12} \\ j_{21} & j_{22} \end{bmatrix}, \tag{4.26}$$

which, for simplicity, can be written as

$$\mathcal{J}_i = \begin{bmatrix} a_i & b_i \\ c_i & d_i \end{bmatrix}, \tag{4.27}$$

where all the elements can be complex numbers. Then, for example, if there are three components in a system in which the light enters component one and leaves component three, then

$$[E_x \quad E_y]_{\text{output}} = \begin{bmatrix} a_3 & b_3 \\ c_3 & d_3 \end{bmatrix} \begin{bmatrix} a_2 & b_2 \\ c_2 & d_2 \end{bmatrix} \begin{bmatrix} a_1 & b_1 \\ c_1 & d_1 \end{bmatrix} \begin{bmatrix} E_x \\ E_y \end{bmatrix}_{\text{input}}, \tag{4.28}$$

which may be written symbolically as

$$[E_x \quad E_y]_{\text{output}} = \mathcal{J}_1\mathcal{J}_2\mathcal{J}_3 \begin{bmatrix} E_x \\ E_y \end{bmatrix}_{\text{input}}, \tag{4.29}$$

or even more succinctly as

$$\mathbf{E}_{\text{output}} = \mathcal{J}_1\mathcal{J}_2\mathcal{J}_3\mathbf{E}_{\text{input}}, \tag{4.30}$$

and most succinctly as

$$\mathbf{E}_{\text{output}} = \mathcal{J}\mathbf{E}_{\text{input}}, \tag{4.31}$$

where $\mathcal{J} = \mathcal{J}_1\mathcal{J}_2\mathcal{J}_3$.

4.10.2. Illustration of the Jones Calculus

This simple example has only real numbers and only one optical component. The input light is completely linearly polarized at the orientation angle $\rho = \arctan 0.75 = 36.780$. Then

$$\mathcal{J}_{\text{input}} = \begin{bmatrix} \cos\rho \\ \sin\rho \end{bmatrix} = \begin{bmatrix} 0.8 \\ 0.6 \end{bmatrix}. \tag{4.32}$$

This light is incident upon a partial polarizer that is aligned with the coordinate axis, so that

$$\mathcal{J} = \begin{bmatrix} \tau_x & 0 \\ 0 & \tau_y \end{bmatrix}. \tag{4.33}$$

The output vector is

$$\mathcal{J}_{\text{output}} = [0.8\tau_x \quad 0.6\tau_y] = \begin{bmatrix} \tau_x & 0 \\ 0 & \tau_y \end{bmatrix} \begin{bmatrix} 0.8 \\ 0.6 \end{bmatrix}. \tag{4.34}$$

The output light is linearly polarized and oriented at the angle $\arctan(3\tau_y/4\tau_x)$. The retardation-free partial polarizer rotates the light, maintaining its linear polarization.

The flat, partial polarizers usually do have retardation, which can be ignored only if one of the τ's is zero.

4.10.3. Mueller Calculus

Mueller calculus uses the Stokes vector **S** to represent the input and output light and uses the Mueller matrix $\mathcal{M}$ to obtain the output light from the input light.

The Mueller matrix contains 16 real numbers in a 4×4 array:

$$\mathcal{M} = \begin{bmatrix} m_{00} & m_{01} & m_{02} & m_{03} \\ m_{10} & m_{11} & m_{12} & m_{13} \\ m_{20} & m_{21} & m_{22} & m_{23} \\ m_{30} & m_{31} & m_{32} & m_{33} \end{bmatrix}, \tag{4.35}$$

where the m's are all real.

For the three-component example, the symbolic process is

$$\mathbf{S}_{\text{output}} = \mathcal{M}_3\mathcal{M}_2\mathcal{M}_1\mathbf{S}_{\text{input}} \tag{4.36}$$

or

$$\mathbf{S}_{\text{output}} = \mathcal{M}\mathbf{S}_{\text{input}}, \tag{4.37}$$

where $\mathcal{M}$ is the triple product of the three matrices. This can be written out in more detail as

$$[s_0 \quad s_1 \quad s_2 \quad s_3]_{\text{output}} = \begin{bmatrix} m_{00} & m_{01} & m_{02} & m_{03} \\ m_{10} & m_{11} & m_{12} & m_{13} \\ m_{20} & m_{21} & m_{22} & m_{23} \\ m_{30} & m_{31} & m_{32} & m_{33} \end{bmatrix} \begin{bmatrix} s_0 \\ s_1 \\ s_2 \\ s_3 \end{bmatrix}_{\text{input}} . \tag{4.38}$$

4.10.5. Illustration of the Mueller Calculus

The example of Subsection 4.10.3 is used again, in which a linearly polarized input beam is shone on a linear polarizer with tan $\rho = 0.75$ and $\gamma = 0$. Then

$$\mathbf{S}_{\text{input}} = \begin{bmatrix} 1.0 \\ 0.28 \\ 0.96 \\ 0.0 \end{bmatrix}. \tag{4.39}$$

The Mueller matrix of a retardation-free partial polarizer with amplitude transmittances t_x and t_y is

$$\mathcal{M} = \frac{1}{2} \begin{bmatrix} t_x^2 + t_y^2 & t_x^2 - t_y^2 & 0 & 0 \\ t_x^2 - t_y^2 & t_x^2 + t_y^2 & 0 & 0 \\ 0 & 0 & 2t_x t_y & 0 \\ 0 & 0 & 0 & 2t_x t_y \end{bmatrix}. \tag{4.40}$$

The multiplication is

$$\mathbf{S}_{\text{output}} = \frac{1}{2}\,[t_x^2 + t_y^2 + 0.28(t_x^2 - t_y^2) \;\; t_x^2 - t_y^2 + 0.8(t_x^2 + t_y^2) \;\; 1.92 t_x t_y \;\; 0]$$

$$= \frac{1}{2} \begin{bmatrix} t_x^2 + t_y^2 & t_x^2 - t_y^2 & 0 & 0 \\ t_x^2 - t_y^2 & t_x^2 + t_y^2 & 0 & 0 \\ 0 & 0 & 2t_x t_y & 0 \\ 0 & 0 & 0 & 2t_x t_y \end{bmatrix} \begin{bmatrix} 1.0 \\ 0.28 \\ 0.96 \\ 0.0 \end{bmatrix}. \tag{4.41}$$

This can be simplified to

$$\mathbf{S}_{\text{output}} = [0.64t_x^2 + 0.36t_y^2 \quad 0.64t_x^2 - 0.36t_y^2 \quad 0.96t_x t_y \quad 0]. \tag{4.42}$$

The Mueller calculation is more ponderous than the Jones calculation, but it has several advantages. Both the Stokes vector and the Mueller matrix can handle partially polarized light, and the Mueller matrix can represent depolarization. All calculations are with real numbers. If only the output flux density is to be calculated, then only s_0 is necessary and only the top row needs to be evaluated.

4.10.6. *Finding the Jones Matrices*

A little experience with the Jones matrices allows one to write them almost by inspection. They are closely related to the physical effect of a component on the electric vector. All the Jones matrices that will be needed for almost any situation are in Section 4.11 and in Shurcliff's book (see Further Reading section).

4.10.7. *Finding the Mueller Matrices*

Mueller matrices are more difficult to generate. They can be obtained by trial and error, if the input and output Stokes vectors are known. Different arrangements of elements need to be tried. Some were worked out by Professor Hans Mueller of MIT in the late 1940s and were in his course notes. Shurcliff used these notes and trial and error to obtain others. His book and Section 4.11 give about all the Mueller matrices one will ever need.

Another way in which they may be generated is by calculation from the Jones matrices. R. W. Schmieder showed how the Mueller matrix can be calculated from *any given Jones matrix.*[4]

If the Jones matrix is represented by

$$\mathcal{J} = \begin{bmatrix} a & b \\ c & d \end{bmatrix}, \tag{4.43}$$

the Schmieder matrix that gives the Mueller matrix is

$$\mathcal{M} = \frac{1}{2}$$

$$\times \begin{bmatrix} a^2 + b^2 + c^2 + d^2 & a^2 - b^2 + c^2 - d^2 & 2\,\mathrm{Re}(a^*b + c^*d) & 2\,\mathrm{Im}(a^*b + c^*d) \\ a^2 + b^2 - c^2 - d^2 & a^2 - b^2 - c^2 + d^2 & 2\,\mathrm{Re}(a^*b - c^*d) & 2\,\mathrm{Im}(a^*b - c^*d) \\ 2\,\mathrm{Re}(a^*c + b^*d) & 2\,\mathrm{Re}(a^*c - b^*d) & 2\,\mathrm{Re}(a^*d + b^*c) & 2\,\mathrm{Im}(a^*d - b^*c) \\ -2\,\mathrm{Im}(a^*c + b^*d) & -2\,\mathrm{Im}(a^*c - b^*d) & -2\,\mathrm{Im}(a^*d + b^*c) & 2\,\mathrm{Re}(a^*d - b^*c) \end{bmatrix}. \tag{4.44}$$

[4] R. W. Schmeider, "Stokes-algebra formalism," J. Opt. Soc. Am. **59,** 297, 1969.

In this representation, a^*b means that the complex conjugate of a multiplies b. Re means the real part of the complex expression and Im is its imaginary part. That is, $\text{Re}(a + ib) = a$ and $\text{Im}(a + ib) = b$. Then the complex conjugate of $a + ib$ is $a - ib$. The terms that are squared, a^2, b^2, etc., are really the product of the term and its complex conjugate, $a^2 = aa^* = a^*a$.

The determination of m_{03} shows one way to make this evaluation:

$$\begin{aligned} m_{03} &= \text{Im}(a^*b + c^*d) = \left[\frac{1}{2i}(a^*b - b^*a + c^*d - d^*c)\right] \\ &= a_r b_i - a_i b_r + c_r d_i - c_i d_r, \end{aligned} \tag{4.45}$$

where a_r and a_i and b_r and b_i represent the real and imaginary parts of a and b in the Jones matrix. It can be seen by inspection that the Jones matrices for quarter-wave plates (Subsection 4.11.5) yield 0 for this calculation and the Mueller matrix element m_{03} is 0.

4.11. Polarization Matrices

Mueller and Jones matrices are presented here for a variety of useful flat optical components.

4.11.1. Rotation Matrices

Polarization matrices are usually presented in their simplest form, in which the principal axes of the polarizers and retarders are coincident with the x and y axes. Then, if the optical element is rotated through an angle θ in its own plane, the rotation operators are used.

The Jones rotation operation is

$$\mathcal{J}(\theta) = \mathcal{R}(\theta)\mathcal{J}(0)\mathcal{R}(-\theta), \tag{4.46}$$

where

$$\mathcal{R}(\theta) = \begin{bmatrix} \cos\theta & -\sin\theta \\ \sin\theta & \cos\theta \end{bmatrix}. \tag{4.47}$$

The Mueller rotation operation is

$$\mathcal{M}(\theta) = T(2\theta)M(0)T(-2\theta), \tag{4.48}$$

$$\mathcal{T}_{\text{rot}}(2\theta) = \begin{bmatrix} 1 & 0 & 0 & 0 \\ 0 & \cos 2\theta & -\sin 2\theta & \\ 0 & \sin 2\theta & \cos 2\theta & 0 \\ 0 & 0 & 0 & 0 \end{bmatrix}. \tag{4.49}$$

4.11.2. Standard Matrices

The following matrices are in standard form; the optical components are aligned with the coordinate axes.

Ideal plate of glass with 100% external transmittance:

$$\mathcal{M}_{\text{isotropic},1} = \begin{bmatrix} 1 & 0 & 0 & 0 \\ 0 & 1 & 0 & 0 \\ 0 & 0 & 1 & 0 \\ 0 & 0 & 0 & 1 \end{bmatrix}, \quad \mathcal{J}_{\text{isotropic},1} = \begin{bmatrix} 1 & 0 \\ 0 & 1 \end{bmatrix}. \tag{4.50}$$

Ideal isotropic plate with intensity transmittance τ and amplitude transmittance t:

$$\mathcal{M}_{\text{isotropic},\tau} = \tau \begin{bmatrix} 1 & 0 & 0 & 0 \\ 0 & 1 & 0 & 0 \\ 0 & 0 & 1 & 0 \\ 0 & 0 & 0 & 1 \end{bmatrix}, \quad \mathcal{J}_{\text{isotropic},t} = t \begin{bmatrix} 1 & 0 \\ 0 & 1 \end{bmatrix}. \tag{4.51}$$

Opaque plate:

$$\mathcal{M}_{\text{opaque}} = \begin{bmatrix} 0 & 0 & 0 & 0 \\ 0 & 0 & 0 & 0 \\ 0 & 0 & 0 & 0 \\ 0 & 0 & 0 & 0 \end{bmatrix}, \quad \mathcal{J}_{\text{opaque}} = \begin{bmatrix} 0 & 0 \\ 0 & 0 \end{bmatrix}. \tag{4.52}$$

Ideal depolarizer:

$$\mathcal{M}_{\text{depolariz}} = \begin{bmatrix} 1 & 0 & 0 & 0 \\ 0 & 0 & 0 & 0 \\ 0 & 0 & 0 & 0 \\ 0 & 0 & 0 & 0 \end{bmatrix}, \quad \text{no Jones matrix.} \tag{4.53}$$

4.11.3. Polarization Rotators

The most simple, nontrivial element used in polarization is a rotator; it causes the state of polarization of a polarization vector $\mathbf{V}$ to be rotated in the x–y plane through an angle θ. Simple examples are a crystal quartz plate with its optic axis normal to its surface and a cell containing a solution of sucrose.

A rotated Stokes vector is found from

$$\mathbf{S}(\theta) = \mathcal{T}(2\theta)\mathbf{S}_0. \tag{4.54}$$

A rotated Jones vector is found from

$$\mathcal{J}(\theta) = \mathcal{R}(\theta)\mathcal{J}_0. \tag{4.55}$$

The rotation matrices are given in Subsection 4.11.1.

4.11.4. Retarders

Retarders are also called wave plates and compensators. The most important are those with retardances δ of 90° and 180°, often called quarter- and half-wave plates. A retarder introduces a phase shift δ between the components of the electric vector along the two axes of the retarder.

With the fast axis horizontal, the retarder with general phase shift δ has the following matrices:

$$\mathcal{M}_{1/2,\delta} = \begin{bmatrix} 1 & 0 & 0 & 0 \\ 0 & 1 & 0 & 0 \\ 0 & 0 & \cos\delta & \sin\delta \\ 0 & 0 & -\sin\delta & \cos\delta \end{bmatrix}, \quad \mathcal{J}_{1/2,\delta} = \begin{bmatrix} e^{+i\delta/2} & 0 \\ 0 & e^{-i\delta/2} \end{bmatrix}, \tag{4.56}$$

where the angle δ is in radians.

4.11.5. Quarter-Wave Optical Retarders

The quarter-wave retarder ($\delta = 90°$) with its fast axis horizontal ($\theta = 0$) has the following Mueller and Jones matrices.

Fast axis horizontal:

$$\theta = 0°, \quad \mathcal{M}_{1/4,0} = \begin{bmatrix} 1 & 0 & 0 & 0 \\ 0 & 1 & 0 & 0 \\ 0 & 0 & 0 & 1 \\ 0 & 0 & -1 & 0 \end{bmatrix}, \quad \mathcal{J}_{1/4,0} = \begin{bmatrix} e^{i\pi/4} & 0 \\ 0 & e^{-i\pi/4} \end{bmatrix}. \tag{4.57}$$

Fast axis vertical:

$$\theta = 90°, \quad \mathcal{M}_{1/4,90} = \begin{bmatrix} 1 & 0 & 0 & 0 \\ 0 & 1 & 0 & 0 \\ 0 & 0 & 0 & -1 \\ 0 & 0 & 1 & 0 \end{bmatrix}, \quad \mathcal{J}_{1/4,90} = \begin{bmatrix} e^{-i\pi/4} & 0 \\ 0 & e^{i\pi/4} \end{bmatrix}. \tag{4.58}$$

The quarter-wave plates used at 45° have the following matrices:
Fast axis at +45°:

$$\mathcal{M}_{1/4,+45} = \begin{bmatrix} 1 & 0 & 0 & 0 \\ 0 & 0 & 0 & -1 \\ 0 & 0 & 1 & 0 \\ 0 & +1 & 0 & 0 \end{bmatrix}, \quad \mathcal{J}_{1/4,+45} = \frac{1}{\sqrt{2}} \begin{bmatrix} 1 & +1 \\ +1 & 1 \end{bmatrix}. \tag{4.59}$$

Fast axis at –45°:

$$\mathcal{M}_{1/4,-45} = \begin{bmatrix} 1 & 0 & 0 & 0 \\ 0 & 0 & 0 & +1 \\ 0 & 0 & 1 & 0 \\ 0 & -1 & 0 & 0 \end{bmatrix}, \quad \mathcal{J}_{1/4,-45} = \frac{1}{\sqrt{2}} \begin{bmatrix} 1 & 1 \\ -1 & 1 \end{bmatrix}. \tag{4.60}$$

The rotator matrices can be used to show that quarter-wave plates with fast axis at an angle θ are

$$\mathcal{M}_{1/4,0} = \begin{bmatrix} 1 & 0 & 0 & 0 \\ 0 & C_2^2 & C_2S_2 & -S_2 \\ 0 & C_2S_2 & S_2^2 & C_2 \\ 0 & S_2 & -C_2 & 0 \end{bmatrix},$$

$$\mathcal{J}_{1/4,\theta} = \begin{bmatrix} C_1^2 e^{i\pi/4} + S_1^2 e^{-i\pi/4} & \sqrt{2} i C_1 S_1 \\ \sqrt{2} i C_1 S_1 & C_1^2 e^{-i\pi/4} + S_1^2 e^{i\pi/4} \end{bmatrix}, \tag{4.61}$$

where $S_1 = \sin\theta$, $S_2 = \sin 2\theta$, $C_1 = \cos\theta$, and $C_2 = \cos 2\theta$.

A linear polarizer and a quarter-wave plate can be used to form any state of elliptical polarization and every elliptical polarizer.

4.11.6. *Half-Wave Optical Retarders*

A half-wave plate with $\delta = 180°$ has the same matrices whether the fast axis is aligned with the x or the y axis; that is, they are the same for $\theta = 0°$ or $\theta = 90°$:

$$\theta = 0° \text{ or } \theta = 90°, \quad \mathcal{M}_{1/2,0,90} = \begin{bmatrix} 1 & 0 & 0 & 0 \\ 0 & 1 & 0 & 0 \\ 0 & 0 & 1 & 0 \\ 0 & 0 & 0 & -1 \end{bmatrix}, \quad \mathcal{J}_{1/2,0,90} = \begin{bmatrix} 1 & 0 \\ 0 & -1 \end{bmatrix}. \tag{4.62}$$

For the special case of ±45°, the matrices are

$$\theta = \pm 45^\circ, \quad \mathcal{M}_{1/2,\pm 45} = \begin{bmatrix} 1 & 0 & 0 & 0 \\ 0 & -1 & 0 & 0 \\ 0 & 0 & 1 & 0 \\ 0 & 0 & 0 & -1 \end{bmatrix}, \quad \mathcal{J}_{1/2,\pm 45} = \begin{bmatrix} 0 & 1 \\ 1 & 0 \end{bmatrix}. \tag{4.63}$$

When the half-wave plate is rotated to the angle θ, the matrices become

$$\mathcal{M}_{1/2,\theta} = \begin{bmatrix} 1 & 0 & 0 & 0 \\ 0 & C_2^2 - S_2^2 & 2C_2S_2 & 0 \\ 0 & 2C_2S_2 & S_2^2 - C_2^2 & 0 \\ 0 & 0 & 0 & 1 \end{bmatrix}, \quad \mathcal{J}_{1/2,\theta} = \begin{bmatrix} C_2 & S_2 \\ S_2 & -C_2 \end{bmatrix}, \tag{4.64}$$

where the symbols are as previously defined.

4.11.7. *Ideal Linear Polarizers*

An ideal polarizer has 100% transmittance for linearly polarized light with its electric vector parallel to the transmission axis of the polarizer and has zero transmittance for the electric vector that is perpendicular to that direction. Nicol prisms and good sheet polarizers come close to this ideal. The ideal polarizer with its transmission axis (for the electric vector) horizontal has the matrices given by

$$\theta = 0^\circ, \quad \mathcal{M}_{\text{lin},0} = \frac{1}{2}\begin{bmatrix} 1 & 1 & 0 & 0 \\ 1 & 1 & 0 & 0 \\ 0 & 0 & 0 & 0 \\ 0 & 0 & 0 & 0 \end{bmatrix}, \quad \mathcal{J}_{\text{lin},0} = \begin{bmatrix} 1 & 0 \\ 0 & 0 \end{bmatrix}. \tag{4.65}$$

The matrix for the ideal polarizer with its transmission axis vertical is

$$\theta = 90^\circ, \quad \mathcal{M}_{\text{lin},90} = \begin{bmatrix} 1 & 1 & 0 & 0 \\ -1 & 1 & 0 & 0 \\ 0 & 0 & 0 & 0 \\ 0 & 0 & 0 & 0 \end{bmatrix}, \quad \mathcal{J}_{\text{lin},90} = \begin{bmatrix} 0 & 0 \\ 0 & 1 \end{bmatrix}. \tag{4.66}$$

With the ideal polarizer at ±45°, the matrices are

$$\theta = \pm 45^\circ, \quad \mathcal{M}_{\text{lin},\pm 45} = \frac{1}{2}\begin{bmatrix} 1 & 0 & \pm 1 & 0 \\ 0 & 1 & 0 & 0 \\ \pm 1 & 0 & 1 & 0 \\ 0 & 0 & 0 & 0 \end{bmatrix},$$

$$\mathcal{J}_{\text{lin},\pm 45} = \frac{1}{2}\begin{bmatrix} 1 & \pm 1 \\ \pm 1 & 1 \end{bmatrix}. \tag{4.67}$$

When the ideal polarizer is rotated to an angle θ, the matrices are

$$\mathcal{M}_{\text{lin},\theta} = \frac{1}{2}\begin{bmatrix} 1 & C_2 & S_2 & 0 \\ C_2 & C_2^2 & C_2S_2 & 0 \\ S_2 & C_2S_2 & S_2^2 & 0 \\ 0 & 0 & 0 & 0 \end{bmatrix}, \quad \mathcal{J}_{\text{lin},\theta} = \begin{bmatrix} C_1^2 & C_1S_1 \\ C_1S_1 & S_1^2 \end{bmatrix}. \tag{4.68}$$

where the symbols are as previously defined.

4.11.8. Partial Polarizers

The matrices for the partial polarizers aligned with the coordinate axes are

$$\text{partial}, \theta = 0, \quad \mathcal{M}_{\text{part},0} = \frac{1}{2}\begin{bmatrix} \tau_1 + \tau_2 & \tau_1 - \tau_2 & 0 & 0 \\ \tau_1 - \tau_2 & \tau_1 + \tau_2 & 0 & 0 \\ 0 & 0 & 2\sqrt{\tau_1\tau_2} & 0 \\ 0 & 0 & 0 & 2\sqrt{\tau_1\tau_2} \end{bmatrix},$$

$$\mathcal{J}_{\text{part},0} = \begin{bmatrix} t_1 & 0 \\ 0 & t_2 \end{bmatrix}, \tag{4.69}$$

where the τ's are the squares of the t's. The matrices assume zero retardation, whereas sheet polarizers do have linear retardation.

Both matrices seem to contain two independent parameters, but this is not so, because every matrix may have a scalar multiplier, e.g., the factor of 1/2. By multiplying outside by τ_1 and dividing inside by it, the matrix becomes

$$\mathcal{M}_{\text{part},0} = \frac{\tau_1}{2}\begin{bmatrix} 1+r & 1-r & 0 & 0 \\ 1-r & 1+r & 0 & 0 \\ 0 & 0 & 2r & 0 \\ 0 & 0 & 0 & 2r \end{bmatrix}, \quad \mathcal{J}_{\text{part},0} = \begin{bmatrix} 1 & 0 \\ 0 & s \end{bmatrix}, \tag{4.70}$$

where $r = \tau_2/\tau_1$ and $s = t_2/t_1$.

When retardance is taken into account, the matrices are

$$\mathcal{M}_{\text{part},0} = \frac{\tau_1}{2}\begin{bmatrix} 1+r & 1-r & 0 & 0 \\ 1-r & 1+r & 0 & 0 \\ 0 & 0 & 2\sqrt{r}\cos\delta & 2\sqrt{r}\sin\delta \\ 0 & 0 & -2\sqrt{r}\sin\delta & 2\sqrt{r}\cos\delta \end{bmatrix},$$

$$\mathcal{J}_{\text{part},\delta} = t_1\begin{bmatrix} e^{i\delta/2} & 0 \\ 0 & e^{-i\delta/2} \end{bmatrix}, \tag{4.71}$$

and the retardance δ is in radians.

The Jones matrix elements are often complex quantities but they are real for rotators, half-wave plates, ideal polarizers, and partial polarizers with zero retardance.

4.12. Determination of the Polarization of a Beam

The process to determine the Stokes vectors is defined by the operational definition of the Stokes vector. Use four filters that have equal transmittances for unpolarized light: one that has no polarization, one that is polarized linearly in the horizontal direction, one that is polarized linearly at +45°, and one that is right circularly polarized. Each is used independently. The first measurement gives s_0. The difference between the first and second measurements gives s_1. The difference between the third and first measurements gives s_2. The difference between the fourth and first measurements gives s_3.

4.13. Types of Polarizers

By any measure, the dominant form of polarizers used today is the sheet dichroic polarizers developed by Edwin H. Land and his colleagues. They are used in three-dimensional cinemas, sunglasses, photography, and many instruments. Before their development in the 1930s, the only polarizer suitable for scientific instruments was the Nicol prism and its relatives.

Reflections at the Brewster angle are fully polarized, and transmission polarizers made of piles of plane, parallel plates at the Brewster angle have been described. These are all described in Chaps. 4–6 of Shurcliff; retarders are described in Chap. 7, and practical uses of polarizers are described in Chaps. 9 and 10.

Sheet linear retarders suffer from chromatic dispersion. It is the path difference that tends to be constant, but the phase difference tends to be inversely proportional to the wavelength. The Fresnel rhomb is an achromatic quarter-wave retarder, but is awkward to use. In 1955 Pancharatnom ingeniously found a combination of three sheet retarders that is nearly achromatic; later improvements are in Shurcliff, p. 109.

Acknowledgment

I am pleased to acknowledge the considerable help I received from R. Clark Jones in preparing this chapter. He proofed and corrected the text many times and created the techniques for converting among the different ways for describing the states of polarization. He gets much credit for the good parts of the chapter; the ererrs are mine.

Further Reading

W. A. Shurcliff, *Polarized Light, Production and Use,* Harvard Univ. Press, 1962.

J. M. E. Collett, *Polarized Light, Fundamentals and Applications,* Marcel Dekker, 1992.

W. Swindell, *Polarized Light, Production and Use,* Harvard Univ. Press, 1962.

P. C. D. Hobbs, *Building Electro Optical Systems, Making It All Work,* Wiley, 2000.

J. B. Shumaker, *Self Study Manual on Optical Radiation Measurements, Part 1. Concepts,* National Institutes of Technology, 1977, Chap. 6.

J. Bennett, "Polarizers," in *Handbook of Optics,* M. Bass, E. W. Van Stryland, D. R. Williams, and W. L. Wolfe, eds., McGraw-Hill, 1995, Vol. II, Chap. 3.

B. H. Billings, ed., *Polarization,* Vol. MS23 of SPIE Milestone Series, SPIE, 1990 (how to deal with polarization in electro-optical systems).

W. Swindell, "Polarized light," in *Benchmark Papers in Optics,* Dowden, Hutchison, and Ross, Halsted.

Various authors, *Applications and Measurement of Polarized Radiation,* Proc. SPIE **891,** 1988; **1166,** 1990; **1746,** 1992; **2265,** 1994; **2873,** 1990; **3121,** 1997; and 3754, 1999.

The first three are good texts on the subject. The next two describe applications in optical systems and radiometry. Bennett describes the properties of many different polarizers. The volumes by Billings and Swindell are collections of reprints of the significant papers in the field. The last entry is a series of papers given at meetings on polarization applications, measurements, and analyses.

5

Interference

John E. Greivenkamp

5.1. Introduction

Interference results from the superposition of two or more electromagnetic waves. From a classical optics perspective, interference is the mechanism by which light interacts with light.

5.2. Waves and Wave Fronts

The electric-field vector that is due to an electromagnetic field at a point in space is composed of an amplitude and a phase:

$$\mathbf{E}(x,y,z,t) = \mathbf{A}(x,y,z,t)e^{i\phi(x,y,z,t)}, \tag{5.1}$$

or

$$\mathbf{E}(\mathbf{r},t) = \mathbf{A}(\mathbf{r},t)e^{i\phi(\mathbf{r},t)}, \tag{5.2}$$

where $\mathbf{r}$ is the position vector and both the amplitude $\mathbf{A}$ and the phase ϕ are functions of the spatial coordinate and time. The polarization state of the field is contained in the temporal variations of the amplitude vector. A linearly polarized monochromatic wave is represented by

$$\mathbf{E}(x,y,z,t) = \mathbf{A}(x,y,z)e^{i[\omega t-\phi(x,y,z)]}, \tag{5.3}$$

where ω is the angular frequency in radians per second and is related to the frequency ν by

$$\omega = 2\pi\nu. \tag{5.4}$$

5.2.1. Plane Wave

The plane wave is produced by a monochromatic point source at infinity and is approximated by a collimated light source:

$$\mathbf{E}(x,y,z,t) = \mathbf{E}(\mathbf{r},t) = \mathbf{A}e^{i(\omega t-\mathbf{k}\cdot\mathbf{r})}, \tag{5.5}$$

where $\mathbf{k}$ is the wave vector. The wave vector points in the direction of propagation, and its magnitude is the wave number $k = 2\pi/\lambda$, where λ is the wavelength. The wavelength is related to the temporal frequency by the speed of light v in the medium:

$$\lambda = \frac{v}{\nu} = 2\pi\frac{v}{\omega} = \frac{c}{n\nu} = 2\pi\frac{c}{n\omega}, \tag{5.6}$$

where n is the index of refraction and c is the speed of light in a vacuum. If the direction of propagation is parallel to the z axis, the plane wave simplifies to

$$\mathbf{E}(x,y,z,t) = \mathbf{A}e^{i(\omega t-kz)}. \tag{5.7}$$

5.2.2. Spherical Wave

The spherical wave that radiates from an isotropic point source is

$$E(r,t) = (A/r)e^{i(\omega t-kr)}, \tag{5.8}$$

where $r = (x^2 + y^2 + z^2)^{1/2}$ and the source is assumed to be located at the origin. At a large distance from the source, the spherical wave can be approximated by a plane wave.

5.2.3. Wave Fronts

Wave fronts represent surfaces of constant phase for the electromagnetic field. A given wave front also represents a surface of constant optical path length (OPL) from the source. The optical path length is defined by the following path integral:

$$\mathrm{OPL} = \int_S^P n(s)\mathrm{d}s, \tag{5.9}$$

where the integral goes from the source S to the observation point P and $n(s)$ is the index of refraction along the path.

5.3. Interference

The net complex amplitude is the sum of all of the component fields,

$$\mathbf{E}(x,y,z,t) = \sum_i \mathbf{E}_i(x,y,z,t), \tag{5.10}$$

and the resulting intensity (flux density) is the time average of the modulus squared of the total complex amplitude,

$$I(x,y,z,t) = \langle |\mathbf{E}(x,y,z,t)|^2 \rangle, \tag{5.11}$$

where the angle brackets denote a time average over a period much longer than $1/\nu$. When two interfering waves $\mathbf{E}_1$ and $\mathbf{E}_2$ are used, this equation simplifies to

$$I(x,y,z,t) = \langle |\mathbf{E}_1|^2 \rangle + \langle |\mathbf{E}_2|^2 \rangle + \langle \mathbf{E}_1 \cdot \mathbf{E}_2^* \rangle + \langle \mathbf{E}_1^* \cdot \mathbf{E}_2 \rangle \tag{5.12}$$

or

$$I(x,y,z,t) = I_1 + I_2 + \langle \mathbf{E}_1 \cdot \mathbf{E}_2^* \rangle + \langle \mathbf{E}_1^* \cdot \mathbf{E}_2 \rangle, \tag{5.13}$$

where I_1 and I_2 are the intensities that are due to the two beams individually. This general result is simpler for two linearly polarized monochromatic waves [Eq. (5.3)]:

$$I(x,y,z,t) = I_1 + I_2 + 2(\mathbf{A}_1 \cdot \mathbf{A}_2)\cos\{(\omega_1 - \omega_2)t - [\phi_1(x,y,z) - \phi_2(x,y,z)]\}. \tag{5.14}$$

5.3.1. Interference Fringes

The interference of two waves with the same optical frequency and with parallel linear polarizations results in the usual equation for interference fringes:

$$I(x,y,z) = I_1 + I_2 + 2\sqrt{I_1 I_2}\cos[\Delta\phi(x,y,z)], \tag{5.15}$$

where $\Delta\phi = \phi_1 - \phi_2$ is the phase difference. The detected intensity varies cosinusoidally with the phase difference between the two waves. These alternating bright and dark bands in the intensity pattern are referred to as interference fringes, and, along a particular fringe, the phase difference is constant.

The phase difference is related to the difference in the optical path lengths between the source and the observation point for the two waves. This is the optical path difference (OPD):

$$\text{OPD} = \text{OPL}_1 - \text{OPL}_2 = \left(\frac{\lambda}{2\pi}\right)\Delta\phi \tag{5.16}$$

or

$$\Delta\phi = \left(\frac{2\pi}{\lambda}\right)\text{OPD}. \tag{5.17}$$

The optical path difference is therefore constant along a fringe.

Constructive interference occurs when the two waves are in phase, and a bright fringe or maximum in the intensity pattern results. This corresponds to a phase difference of $2m\pi$ or an optical path difference of $m\lambda$, where m is an integer. A dark fringe or minimum in the intensity pattern results from destructive interference when the phase difference is $2(m + 1)\pi$ or the optical path difference is $(m + 1/2)\lambda$. The absolute value of m is called the order of interference. Between adjacent fringes, the phase difference changes by 2π and the optical path difference changes by λ.

5.3.2. *Fringe Visibility*

The fringe visibility or contrast can be defined if Eq. (5.15) is rewritten:

$$I(x,y) = I_0(x,y)\{1 + \gamma(x,y)\cos[2\pi\text{OPD}(x,y)/\lambda]\}, \tag{5.18}$$

where

$$\gamma(x,y) = \frac{2[I_1(x,y)I_2(x,y)]^{1/2}}{I_1(x,y) + I_2(x,y)}, \tag{5.19}$$

$I_0(x,y) = I_1(x,y) + I_2(x,y)$ is the average intensity, and $\gamma(x,y)$ is the local fringe contrast or visibility. The fringe visibility can also be equivalently calculated with the standard formula for modulation:

$$\gamma(x,y) = \frac{I_{\text{max}}(x,y) - I_{\text{min}}(x,y)}{I_{\text{max}}(x,y) + I_{\text{min}}(x,y)}, \tag{5.20}$$

where I_{max} and I_{min} are the maximum and the minimum intensities in the fringe pattern.

5.3.3. *Two Plane Waves*

The interference of two plane waves results in straight equispaced fringes. The two waves are incident upon the observation plane ($z = 0$) at θ_1 and θ_2 with respect to the z axis. The plane of incidence is the x–z plane (contains the two k vectors), and the fringe pattern is given by

$$I(x,y) = I_0\{1 + \cos[(2\pi x/\lambda)(\sin\theta_1 - \sin\theta_2)]\}, \tag{5.21}$$

where I_0 is twice the intensity of each of the individual waves. The fringes are parallel to the y axis, and the fringe period is

$$p = \frac{\lambda}{\sin\theta_1 - \sin\theta_2}. \tag{5.22}$$

A typical situation for interference is that the two angles of incidence are equal and opposite, $\theta_1 = -\theta_2 = \theta$. The angle between the two beams is 2θ. Under this condition, the period is

$$p = \frac{\lambda}{2\sin\theta} \approx \frac{\lambda}{2\theta}, \tag{5.23}$$

and the small-angle approximation is also given.

5.3.4. *Plane Wave and Spherical Wave*

The interference of a plane wave and a spherical wave is observed at the plane $z = R$. The plane wave is parallel to the z axis, and the spherical wave is due to a source at the origin. When the radial coordinate in the observation plane $\rho = (x^2 + y^2)^{1/2}$ is used, the resulting intensity pattern is

$$I(\rho) = I_0\left[1 + \cos\left(\frac{\pi\rho^2}{\lambda R}\right)\right]. \tag{5.24}$$

The fringe pattern comprises concentric circles, and the radial fringe spacing decreases as the radius ρ increases. The radius of the mth bright fringe is

$$\rho_m = \sqrt{2mR}, \tag{5.25}$$

and the order of interference m increases with radius. This fringe pattern is the Newton's ring pattern. The intensities of the two waves have been assumed to be equal at the observation plane, and this result is valid only when ρ is much smaller than R.

5.3.5. *Two Spherical Waves*

When the observation of the interference of two spherical waves is made on a plane perpendicular to a line connecting the two sources, a fringe pattern similar to that of Newton's rings is obtained. R is the average distance to the observation plane, and d is the separation between the two sources of the spherical waves. The observed phase difference is

$$\Delta\phi \approx \frac{2\pi d}{\lambda} - \left(\frac{\pi\rho^2}{\lambda}\right)\left(\frac{d}{R^2}\right). \tag{5.26}$$

The second term in ρ^2 produces the circular fringe pattern, and the first term is a constant phase shift related to the separation of the two sources. If this term is not a multiple of π, the center of the fringe pattern will not be a bright or dark fringe. As opposed to the Newton's ring pattern, the order of interference is a maximum at the center of the pattern.

The observation of the interference of two spherical waves on a plane parallel to the line connecting the sources results in straight equispaced fringes. L is the distance between the sources and the observation plane, and the resulting phase difference is

$$\Delta\phi(x,y) = \frac{2\pi xd}{\lambda L}. \tag{5.27}$$

The period of the fringes is $\lambda L/d$, and the approximations used require that L be much larger than ρ or d.

5.3.6. *Temporal Beats*

If light at different optical frequencies is interfered, a time-varying fringe pattern is produced:

$$I(x,y,t) = I_0\{1 + \cos[2\pi\Delta\nu t - \Delta\phi(x,y)]\}, \tag{5.28}$$

where $\Delta\nu = \nu_1 - \nu_2$. The spatial fringe pattern is temporally modulated at the beat frequency $\Delta\nu$.

5.4. Interference by Wave-Front Division

Interference by wave-front division occurs when two sections of a single wave front are selected and then redirected to a common volume in space. The common example is Young's double-slit experiment, in which monochromatic light from a single source illuminates an opaque screen with two pinholes or slits. The light diffracts from these pinholes and illuminates a viewing screen at a distance that is large compared with the pinhole separation. This geometry is shown in Fig. 5.1.

The diffracted light can be modeled as two spherical or cylindrical waves, and the interference pattern for two spherical waves results [Eq. (5.27)]. Equi-

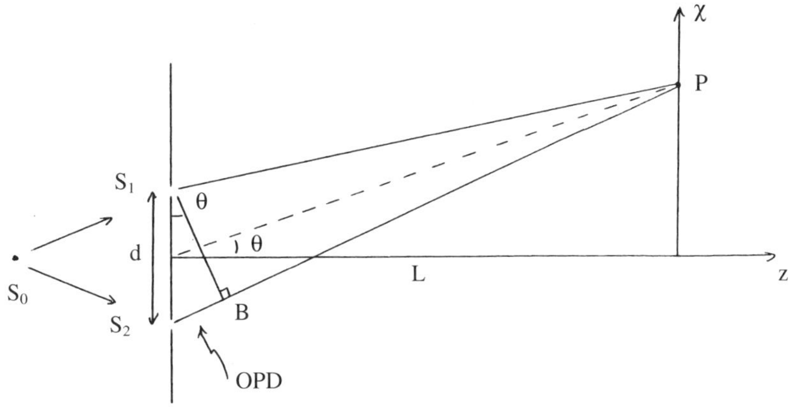

Fig. 5.1. Young's double-slit experiment.

spaced straight fringes are produced, and the fringes are oriented perpendicular to the line connecting the two pinholes or slits. The condition for the *m*th-order bright fringe is

$$\sin(\theta) \approx \theta = \frac{m\lambda}{d} \quad \text{or} \quad x = \frac{m\lambda L}{d}. \tag{5.29}$$

5.4.1. Slit Width

The overall interference intensity pattern from the two-slit interference pattern is modulated by the single-slit diffraction pattern (assuming slit apertures):

$$I(x) = I_0 \operatorname{sinc}^2\left(\frac{Dx}{\lambda L}\right)\left[1 + \gamma(x)\cos\left(\frac{2\pi xd}{\lambda L}\right)\right], \tag{5.30}$$

where D is the slit width and the definition of a sinc function is

$$\operatorname{sinc}(\alpha) = \frac{\sin(\pi\alpha)}{\pi\alpha}, \tag{5.31}$$

where the zeros of the function occur when the argument α is an integer. The term $\gamma(x)$ is included to account for other variations in the fringe visibility.

5.4.2. Source Spectrum

Each individual wavelength in a source spectrum produces an independent interference pattern that has a different scaling factor in λ. The net interference pattern is the sum of the individual fringe patterns:

$$I(x) = \int_0^\infty S(\lambda) I(x, \lambda)\, d\lambda = \int_0^\infty S(\nu) I(x, \nu) d\nu, \tag{5.32}$$

where $S(\lambda)$ or $S(\nu)$ is the source intensity spectrum that serves as a weighting function.

For a source with a uniform frequency distribution over a frequency range of $\Delta\nu$, the resulting intensity pattern is

$$I(x) = I_0 \left[1 + \text{sinc}\left(\frac{\text{OPD}\Delta\nu}{c} \right) \cos\left(\frac{2\pi\text{OPD}}{\lambda_0} \right) \right], \tag{5.33}$$

where λ_0 is the average wavelength and I_0 is the average intensity. A fringe pattern corresponding to the average wavelength results, but the fringe visibility is reduced by a sinc function that depends on $\Delta\nu$ and the optical path difference. Good fringe visibility is obtained only when either the spectral width is small (the source is quasi monochromatic) or the optical path difference is small. The first zero of the visibility function occurs when OPD = $c/\Delta\nu$. This distance is known as the coherence length as it is the path difference over which interference can be obtained.

5.5. Interference by Amplitude Division

Interference by amplitude division uses the same section of a wave front from a single source for the interfering wave fronts. Some sort of beam splitter divides the original wave-front amplitude into two or more parts, and each fraction is directed along a different optical path. These waves are then recombined to produce interference. There are a great many interferometer designs based on this method, including the Fizeau interferometer, the Michelson interferometer, the Twyman–Green interferometer, and thin-film interference filters.

5.5.1. *Extended Sources*

An extended source is modeled as a collection of independent point sources. If the source is quasi monochromatic, all of the point sources radiate at the same nominal frequency, but without a phase relationship. Each point source will produce its own interference pattern, and the net intensity pattern is the sum or integral of all the individual intensity patterns.

The fringes will be localized where the individual fringe position or spacing is not affected by the location of the point sources that comprise the extended source. If there is a location where the optical path difference is zero independent of source location, all of the individual interference patterns will be in phase and the net pattern will show good visibility. This usually occurs in a region where the optical path difference between the two interfering wave fronts is small.

5.5.2. *Fringes of Equal Inclination*

The Fresnel reflections from the two surfaces of a plane parallel plate can be used to produce interference fringes. The optical path difference between the two reflected wave fronts is constant, and no high-visibility fringes should result with an extended source. If, however, a lens is used to collect the light reflected from the plate, fringes are formed in the back focal plane of the lens. This situation is shown in Fig. 5.2, and any ray leaving the surface at a particular angle θ is focused to the same point *P*. Different points in the image plane correspond to different angles.

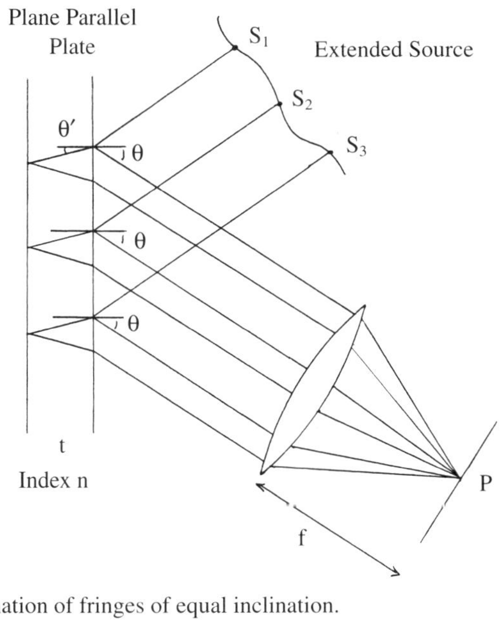

Fig. 5.2. Formation of fringes of equal inclination.

Taking into account the half-wave shift that is due to the phase change difference of π between an internal and an external reflection, a dark fringe will result for angles satisfying

$$2nt \cos \theta' = m\lambda \quad \text{or} \quad \cos \theta' = \frac{m\lambda}{2nt}, \tag{5.34}$$

where θ′ is the internal angle and *m* is an integer. These high-contrast fringes are called fringes of equal inclination and are localized at infinity.

When the axis of the lens is normal to the surfaces of the plate, a beam-splitter arrangement is required for allowing light from the extended source to be reflected into the lens. These symmetric fringes are called Haidinger fringes, and they are identical in appearance to Newton's rings.

5.5.3. *Fringes of Equal Thickness*

For plates of nonuniform thickness, the thickness changes are the primary source of optical path difference variations between the two reflected wave fronts. The resulting fringes are called fringes of equal thickness, and they are localized in regions where the optical path difference between the two reflections is small.

It is common to arrange the system so that the fringes are observed in a direction approximately normal to the surface. Taking into account the additional phase shift introduced at the reflection from one of the surfaces, the conditions for bright and dark fringes are then given by the following equations:

Bright:

$$2nt - \frac{\lambda}{2} = m\lambda. \tag{5.35}$$

Dark:

$$2nt = m\lambda, \tag{5.36}$$

where m is an integer greater than or equal to zero. If a wedged plate is used, the observed pattern will be straight equispaced fringes.

This same result holds for any plate for which the two surfaces are not parallel. If the surface angles are small and the plate is relatively thin, high-visibility fringes localized in the plate are observed. Along a given fringe the value of m is constant, and a fringe represents a contour of constant optical path length nt. The fringes provide a contour map of the plate thickness, and adjacent fringes correspond to a change of thickness of $\lambda/2n$.

5.6. Multiple-Beam Interference

There are many situations in which multiple-beam interference occurs. Two examples are the diffraction grating and a plane parallel plate. The net electric field is the sum of all of the component fields.

5.6.1. *Diffraction Grating*

A diffraction grating operates by division of wave front, and it can be modeled as a series of equispaced slits. The slit separation is d, and all of the diffracted beams are assumed to have equal intensity. The resulting interference pattern in terms of phase difference and diffraction angle is

$$I(\theta) = I_0 \left[\frac{\sin^2\left(\frac{N\Delta\phi}{2}\right)}{\sin^2\left(\frac{\Delta\phi}{2}\right)} \right] = I_0 \left\{ \frac{\sin^2\left[\frac{N\pi d \sin(\theta)}{\lambda}\right]}{\sin^2\left[\frac{\pi d \sin(\theta)}{\lambda}\right]} \right\}, \tag{5.37}$$

where I_0 is the intensity that is due to an individual slit.

This intensity interference pattern is plotted in Fig. 5.3 for $N = 5$. The result

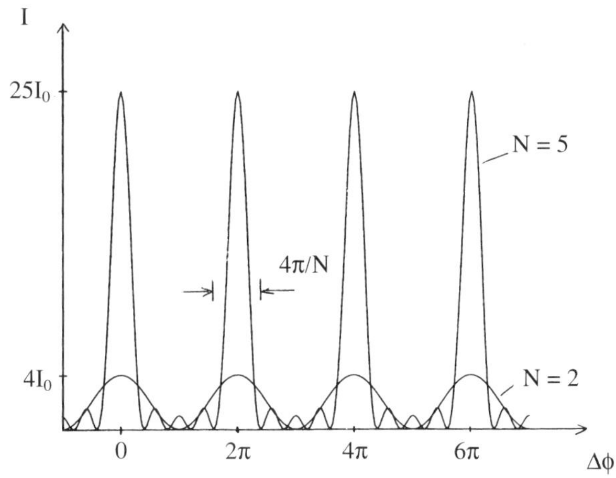

Fig. 5.3. Interference patterns produced by gratings with two and five slits.

for $N = 2$, which is the double-slit experiment, is also shown. The maxima occur at the diffraction angles given by

$$\sin(\theta) = \frac{m\lambda}{d}, \tag{5.38}$$

where m is an integer. The intensity at each maximum in the pattern increases to M^2 times that which is due to a single slit. The full width of a diffraction peak between intensity zeros corresponds to a phase difference $\Delta\phi$ of $4\pi/N$, and the number of intensity zeros between peaks is $N - 1$. As the number of slits increases, the angular resolution or resolving power of the grating greatly increases.

5.6.2. Plane Parallel Plate

The plane parallel plate serves as an example of the interference of multiple waves obtained by division of amplitude. When multiple reflections are included, many beams of differing intensity interfere. The phase difference between successive reflected or transmitted beams is the same as for fringes of equal inclination from a plane parallel plate:

$$\Delta\phi = \left[\frac{4\pi nt\cos(\theta')}{\lambda}\right], \tag{5.39}$$

where θ' is the angle inside the plate.

Under the assumption that there is no absorption in the plate, $I_0 = I_t + I_r$, and the expressions for the transmitted and the reflected intensities are

$$\frac{I_t}{I_0} = \frac{1}{1 + F\sin^2(\Delta\phi/2)}, \tag{5.40}$$

$$\frac{I_r}{I_0} = \frac{F\sin^2(\Delta\phi/2)}{1 + F\sin^2(\Delta\phi/2)}, \tag{5.41}$$

where the coefficient of finesse of the system F is defined as

$$F \equiv \left(\frac{2r}{1 - r^2}\right)^2. \tag{5.42}$$

The amplitude-reflection coefficient for each surface is r. With an extended source, fringes of equal inclination are formed, and they are localized at infinity. The locations of the maxima and the minima of the fringes are the same as those obtained with two-beam interference.

A plot of the transmitted fringe intensity as a function of $\Delta\phi$ is shown in Fig. 5.4 for several values of F. When the phase difference is a multiple of 2π, a bright fringe is obtained independent of F or r. When F is small, low-visibility fringes are produced.

5.6.3. *Fabry–Perot Interferometer*

A practical implementation of a multiple-beam interferometer based on division of amplitude is the Fabry–Perot interferometer. This interferometer serves as a high-resolution spectrometer and also an optical resonator. The system is shown in Fig. 5.5, and it consists of two highly reflective parallel surfaces separated by a distance t, much like a plane parallel plate. These two separated reflective plates are referred to as a Fabry–Perot etalon or cavity. The screen is located in the focal plane of the second lens so that fringes of equal inclination localized at infinity are viewed. The transmission is described by Eq. (5.40), and each wavelength in the spectrum produces a ring pattern with a different scale.

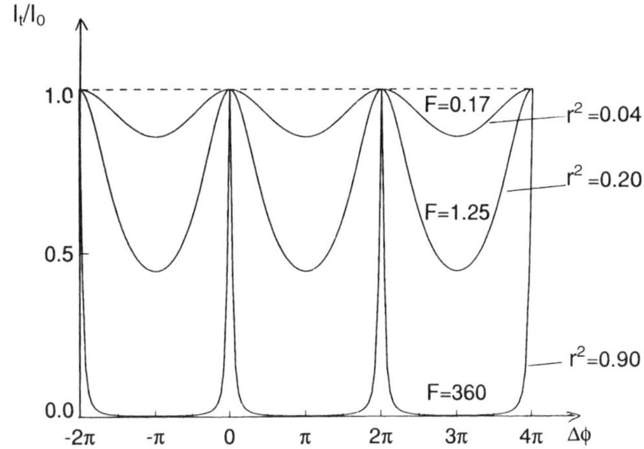

Fig. 5.4. Transmitted intensity of a multiple-beam interference pattern produced by a plane parallel plate.

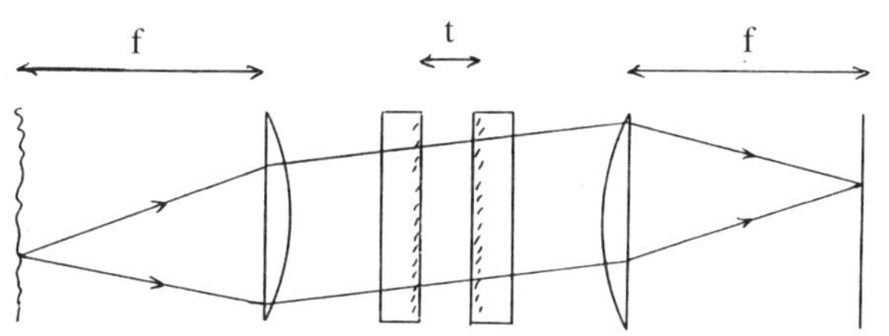

Fig. 5.5. Fabry–Perot interferometer.

A quantity that is often used to describe the performance of a Fabry–Perot cavity is the finesse $\mathcal{F}$. It is a measure of the number of resolvable spectral lines and is defined as the ratio of the phase difference between adjacent fringes to the FWHM of a single fringe:

$$\mathcal{F} \equiv \frac{2\pi}{\text{FWHM}} = \frac{\pi\sqrt{F}}{2} = \frac{\pi r}{1 - r^2} = \frac{\pi\sqrt{\rho}}{1 - \rho}, \tag{5.43}$$

where r is the amplitude reflectivity and ρ is the intensity reflectivity.

When the Fabry–Perot interferometer is used as a scanning spectrometer, it is useful to write its transmission in terms of frequency:

$$T = \frac{I_t}{I_0} = \frac{1}{1 + F\sin^2(2\pi nt\nu/c)}, \tag{5.44}$$

where normal incidence has been assumed. This function is plotted in Fig. 5.6, and a series of transmission spikes separated in frequency by $c/2nt$ occur. This separation is known as the free spectral range. If the separation of the mirrors is changed slightly, these transmission peaks will scan the frequency axis. To scan one free spectral range, the required change in separation is approximately t/m.

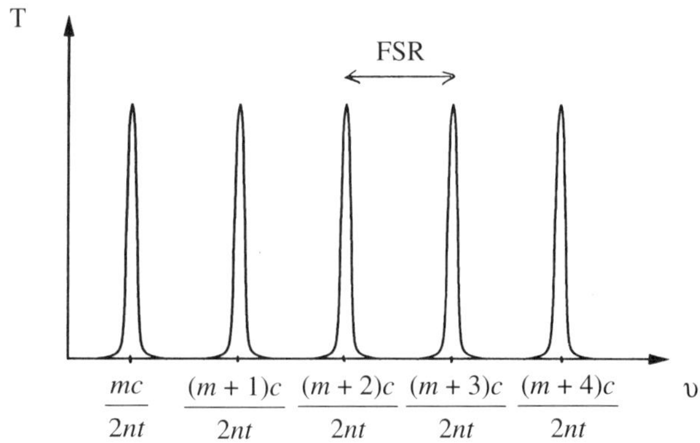

Fig. 5.6. Transmission of a Fabry–Perot cavity as a function of frequency.

5.7. Coherence and Interference

The observed fringe visibility and localization is a function of the spatial and the temporal coherence of the source. The assumption for the analysis is that every point on an extended source radiates independently and therefore produces its own interference pattern. The net intensity is the sum of all of the individual intensity patterns. In a similar manner, each wavelength or frequency of a nonmonochromatic source radiates independently, and the temporal average is the sum of the individual temporal averages. Coherence theory allows the interference between the light from two point sources to be analyzed and also accounts for the effects of the spectral bandwidth of the source.

Consider the interference of light from two point sources or pinholes, where the amplitudes of the light at the pinholes are $E_1(t)$ and $E_2(t)$. The mutual coherence function $\Gamma_{12}(\tau)$ is

$$\Gamma_{12}(\tau) = \langle E_1(t + \tau)E_2^*(t)\rangle, \tag{5.45}$$

which is the cross correlation of the two complex amplitudes. Normalizing the mutual coherence function by the square root of the product of the two self-coherence functions results in the complex degree of coherence:

$$\gamma_{12}(\tau) = \frac{\Gamma_{12}(\tau)}{\sqrt{\Gamma_{11}(0)\Gamma_{22}(0)}} = \frac{\Gamma_{12}(\tau)}{\sqrt{\langle|E_1(t)|^2\rangle\,\langle|E_2(t)|^2\rangle}} = \frac{\Gamma_{12}(\tau)}{\sqrt{I_1 I_2}}, \tag{5.46}$$

and interference pattern is given by

$$I = I_1 + I_2 + 2\sqrt{I_1 I_2}\ \mathrm{Re}\,[\gamma_{12}(\tau)]. \tag{5.47}$$

The variable τ is the temporal measure of the optical path difference (OPD) between the two sources and the observation point:

$$\tau = \frac{\mathrm{OPD}}{c}. \tag{5.48}$$

The complex degree of coherence contains both visibility and interference information. The resulting fringe visibility is

$$\gamma(\tau) = \frac{2\sqrt{I_1 I_2}}{I_1 + I_2}\,|\gamma_{12}(\tau)| \tag{5.49}$$

and is a function of the degree of coherence and τ.

If the two intensities are equal, the fringe visibility is simply the degree of coherence: $\gamma(\tau) = |\gamma_{12}(\tau)|$, and the degree of coherence will take on values between 0 and 1. The source is coherent when $|\gamma_{12}(\tau)| = 1$ and completely incoherent when $|\gamma_{12}(\tau)| = 0$. The source is said be partially coherent for other values. No fringes are observed with an incoherent source, and the visibility is reduced with a partially coherent source.

5.7.1. *Spatial Coherence*

Over the range of time delays between the interfering beams that are usually of interest, the degree of coherence is a slowly varying function and is approximately equal to the value at $\tau = 0$: $|\gamma_{12}(\tau)| = |\gamma_{12}(0)| = |\gamma_{12}|$. The van Cittert–Zernike theorem allows the degree of coherence in the geometry of Fig. 5.7 to be calculated. Let θ be the angular separation of the two pinholes as seen from the source. This theorem states that modulus of the degree of coherence between two points is the modulus of the scaled and normalized Fourier transform of the source intensity distribution:

$$|\gamma_{12}| = \left| \frac{\iint_S I(\xi, \eta) e^{i\frac{2\pi}{\lambda}(\xi\theta_x + \eta\theta_y)}\, d\xi d\eta}{\iint_S I(\xi, \eta)\, d\xi d\eta} \right|, \tag{5.50}$$

where θ_x and θ_y are the x and the y components of the pinhole separation θ, and the integrals are over the source.

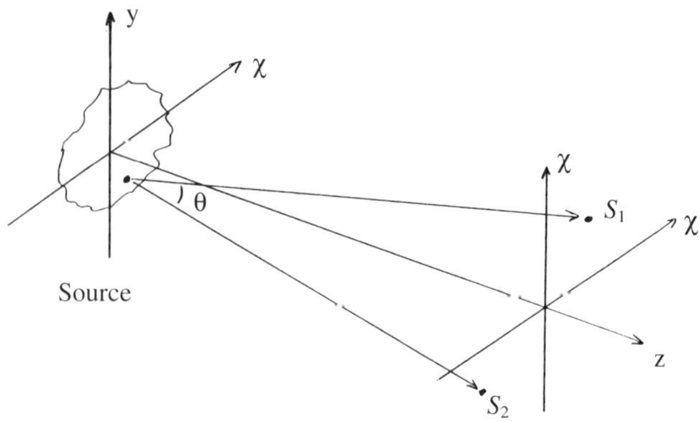

Fig. 5.7. Geometry for the van Cittert–Zernike theorem.

Two cases that are of particular interest are a slit source and a circular source. The application of the van Cittert–Zernike theorem yields the following two coherence functions.
Slit source of width *w*:

$$|\gamma_{12}| = \left|\operatorname{sinc}\left(\frac{w\theta_x}{\lambda}\right)\right| = \left|\operatorname{sinc}\left(\frac{wa}{\lambda z}\right)\right|. \tag{5.51}$$

Circular source of diameter *d*:

$$|\gamma_{12}| = \left|\frac{2J_1\left(\frac{\pi d\theta_x}{\lambda}\right)}{\frac{\pi d\theta_x}{\lambda}}\right| = \left|\frac{2J_1\left(\frac{\pi da}{\lambda z}\right)}{\frac{\pi da}{\lambda z}}\right|, \tag{5.52}$$

where a is the separation of the pinholes, z is the distance from the source to the pinholes, and J_1 is a first-order Bessel function.

5.7.2. *Temporal Coherence*

When temporal coherence effects are examined, light from a single source is implied, so the mutual coherence function becomes the self coherence function $\Gamma_{11}(\tau)$. The complex degree of temporal coherence $\gamma_{11}(\tau)$ is the normalized Fourier transform of the source intensity spectrum $S(\nu)$:

$$\gamma_{11}(\tau) = \frac{\mathrm{FT}[S(\nu)]}{\int_0^\infty S(\nu)d\nu} = \frac{\int_0^\infty S(\nu)e^{i2\pi\nu\tau}\,d\nu}{\int_0^\infty S(\nu)d\nu}. \tag{5.53}$$

Equal-intensity beams are assumed. The fringe visibility is the modulus of this result. Because $\gamma_{11}(\tau)$ has a maximum at $\tau = 0$, the maximum fringe visibility will occur when the time delay between the two beams is zero or the optical path difference is small.

As an example, consider a source with a uniform spectrum:

$$S(\nu) = \mathrm{rect}\left(\frac{\nu - \bar{\nu}}{\Delta\nu}\right), \tag{5.54}$$

where $\bar{\nu}$ is the average frequency and $\Delta\nu$ is the bandwidth. The resulting intensity pattern is

$$I = I_0\,\{1 + \mathrm{Re}[\gamma_{12}(\tau)]\} = I_0\,[1 + \mathrm{sinc}(\tau\,\Delta\nu)\cos(2\pi\tau\bar{\nu})], \tag{5.55}$$

where the sinc function is the Fourier transform of the rect function.

5.7.3. *Laser Sources*

The laser is an important source for interferometry as it is a bright source of coherent radiation. The laser is a Fabry–Perot cavity that contains a gain medium. Its output spectrum is therefore a series of discrete frequencies separated by $c/2nL$, where L is the cavity length. For gas lasers, the index is approximately equal to 1. If $G(\nu)$ is the gain bandwidth, the frequency spectrum is

$$S(\nu) = G(\nu)\mathrm{comb}\left(\frac{2L\nu}{c}\right), \tag{5.56}$$

where a comb function is a series of equally spaced delta functions. The resulting visibility function is

$$\gamma(\tau) = |\gamma_{11}(\tau)| = \left|\tilde{G}(\tau) * \mathrm{comb}\left(\frac{c\tau}{2L}\right)\right| = \left|\tilde{G}(\tau) * \mathrm{comb}\left(\frac{\mathrm{OPD}}{2L}\right)\right|, \tag{5.57}$$

where $\tilde{G}(\tau)$ is the normalized Fourier transform of the gain bandwidth and $*$ indicates convolution. $\tilde{G}(\tau)$ is duplicated at multiples of $2L$, indicating that high-visibility fringes are obtained if the optical path difference between the two optical paths is a multiple of twice the laser cavity length.

Further Reading

M. Born and E. Wolf, *Principles of Optics,* Pergamon, 1975.

R.W. Ditchburn, *Light,* Academic, 1976.

M. Francon, *Optical Interferometry,* Academic, 1966.

M.H. Freeman, *Optics,* Butterworths, 1990.

J.D. Gaskill, *Linear Systems, Fourier Transforms and Optics,* Wiley, 1978.

J.E. Greivenkamp, "Interference," in *Handbook of Optics,* M. Bass, E. Van Stryland, D. Williams and W. Wolfe, eds., McGraw-Hill, 1995, Chap. 2.

P. Hariharan, *Optical Interferometry,* Academic, Press, 1985.

P. Hariharan, ed., *Interferometry,* Vol. MS28 of SPIE Milestone Series, SPIE, 1990.

P. Hariharan, *Basics of Interferometry,* Academic, 1992.

E. Hecht, *Optics,* Addison-Wesley, 1989.

F.A. Jenkins and H.E. White, *Fundamentals of Optics,* McGraw-Hill, 1976.

R.A. Longhurst, *Geometrical and Physical Optics,* Longman, 1973.

D. Malacara, ed., *Optical Shop Metrology,* Vol. MS18 of SPIE Milestone Series, SPIE, 1990.

D. Malacara, Ed., *Optical Shop Testing,* 2nd ed., Wiley, 1992.

A.S. Marathay, *Elements of Optical Coherence Theory,* Wiley, 1982.

J.R. Meyer-Arendt, *Introduction to Classical and Modern Optics,* Prentice-Hall, 1989.

G.O. Reynolds, J.B. DeVelis, G.B. Parrent, Jr., and B.J. Thompson, *The New Physical Optics Notebook: Tutorials in Fourier Optics,* Vol. PM01 of SPIE Press Monographs Series, SPIE, 1989.

B.E.A. Saleh and M.C. Teich, *Fundamentals of Photonics,* Wiley, 1991.

W.H. Steel, *Interferometry,* Cambridge Univ. Press, 1967.

F.G. Smith and J.H. Thompson, *Optics,* Wiley, 1971.

R.W. Wood, *Physical Optics,* Optical Society of America, 1988.

6

Diffraction

A. S. Marathay, J. F. McCalmont, and J. Shiefman

6.1. Radiant Flux Density

Radiant flux density (in units of watts per square meter) is defined in terms of the time-averaged Poynting vector,

$$M = |\langle \mathbf{S} \rangle| = |\langle \mathbf{E} \times \mathbf{H} \rangle|, \tag{6.1}$$

where **E** and **H** are the total electric and magnetic fields, respectively, at the observation point. Radiant incidance or irradiance is the incident flux density. Both are related to the fields in the same way. The exitance is discussed here to avoid the use of E for incidance and E for electric field, as (strongly) suggested by the editor.

When the Poynting vector is not perpendicular to the plane of observation, the exitance is

$$M = |\langle \mathbf{S} \rangle \cdot \hat{\mathbf{n}}| \tag{6.2}$$

where $\hat{\mathbf{n}}$ is the unit vector normal to the plane of observation.

For the special case of a time and space harmonic plane wave $\mathbf{E} = \mathbf{E}_0 \cos(\omega t - kz)$ propagating in the z direction, the exitance is

$$M = \frac{1}{2}\sqrt{\frac{\varepsilon_0}{\mu_0}}\,|\mathbf{E}_0|^2 = \frac{1}{2}\sqrt{\frac{\mu_0}{\varepsilon_0}}\,|\mathbf{H}_0|^2\,\frac{\mathrm{W}}{\mathrm{m}^2}, \tag{6.3}$$

where $\mu_0 = 4\pi \times 10^{-7}$ H/m, $\varepsilon_0 = (\mu_0\, c^2)^{-1}$ F/m, and $c = 2.9979 \times 10^8$ m/s is the speed of light in a vacuum.

6.2. Huygens–Fresnel Diffraction Theory

A typical diffraction geometry is shown in Fig. 6.1. A dark screen is shown with a diffracting aperture A. The left-half space contains a monochromatic

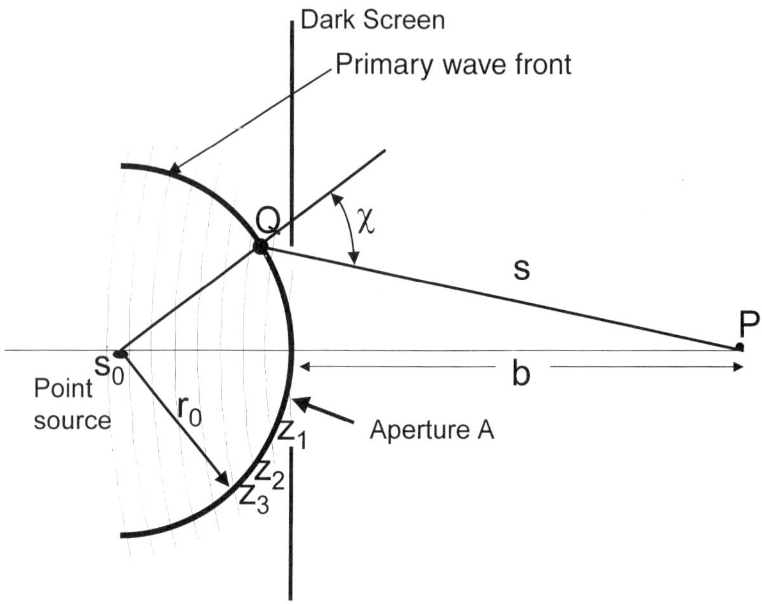

Figure 6.1. Setup for a diffraction experiment: Aperture A in a dark screen with point of observation P and Fresnel half-period zone construction.

point source at s_0 that produces a primary spherical wave incident upon the aperture A. The point of observation P is located in the right-half space.

6.3. Huygens–Fresnel Diffraction Formula

The Huygens–Fresnel diffraction formula is

$$\psi(P) = a\,\frac{\exp(ikr_0)}{r_0}\iint_A \frac{\exp(iks)}{s}K(\chi)\mathrm{d}A, \tag{6.4}$$

where $\psi(P)$ is the diffraction amplitude at P, r_0 is the radius of the primary wave front incident upon the aperture, whose amplitude is a, s is the distance from the secondary source at Q to the point of observation P, and $K(\chi)$ is the inclination factor assumed by Fresnel, with χ defined as the angle made by QP to S_0Q:

$$K(\chi) = \frac{1+\cos\chi}{2}. \tag{6.5}$$

The symbol A under the area integral in Eq. (6.4), with the differential element of area dA, indicates adding the contributions from all the secondary sources exposed by the open aperture A.

Evaluation of the diffraction integral is done when the incoming wave front is divided into half-period zones: Z_1, Z_2, Z_3 ... (see Fig. 6.1 again). Each zone is a distance $\lambda/2$ farther from P, and each zone exposed by the open aperture contributes to the amplitude at P. The magnitude of the contribution that is due to each successive zone has the following characteristics:

1. It is directly proportional to the area a_n of the zone.
2. It is inversely proportional to the average distance d_n of the zone to P; $(a_n/d_n) \approx \pi\lambda$.
3. It is proportional to $K(\chi)$, which varies slowly from one zone to the next.

6.4. Results of the Huygens–Fresnel Diffraction Theory

In summary, the Fresnel zone construction for the amplitude at P is

$$\psi = \psi_1 - \psi_2 + \psi_3 - \psi_4 + \psi_5 - \psi_6 + \psi_7 - \psi_8 + \cdots + (-1)^{n-1}\psi_n, \tag{6.6}$$

where ψ_n denotes the average contribution of the nth zone, $n = 1, 2, 3, \ldots$. The total contribution from all the zones when there is no diffracting aperture and the wave travels freely to P is found to a good approximation to be

$$\psi_{\mathrm{UN}} = \frac{1}{2}\psi_1. \tag{6.7}$$

The flux density at P for an unobstructed wave is

$$M_{\mathrm{UN}} = |\psi|^2 = \frac{1}{4}|\psi_1|^2 = \frac{1}{4}M_1, \tag{6.8}$$

where the subscript UN indicates an unobstructed wave and M_1 is the flux density of zone Z_1 alone.

6.4.1. *Special-Case Results of Huygens–Fresnel Diffraction Theory*

Typical results obtained from the preceding theory for different sized circular apertures are summarized in Table 6.1. It gives examples wherein the distance and the size of the aperture are such that, with respect to the point P, only one half, one, and two zones are uncovered in the aperture in turn.

Table 6.1. Flux Density at the Point of Observation P Predicted by the Huygens–Fresnel Theory for Special Cases

Number of Zones Uncovered in Aperture A	Amplitude at Point of Observation P on Axis	Flux Density at P on Axis
One half of Z_1	$\psi(P) = \frac{1}{2}\psi_1$	$M = \frac{1}{4}M_1' = M_{UN}$
One zone, Z_1	$\psi(P) = \psi_1$	$M = M_1' = 4M_{UN}$
Two zones, Z_1 and Z_2	$\psi(P) = \psi_1 - \psi_2 \approx 0$	$M = 0$
Opaque disk: ψ_1' is the contribution of the first zone drawn starting from the edge of the disk	$\psi(P) = \frac{1}{2}\psi_1'$	$M = \frac{1}{4}M_1' \approx M_{UN}$

6.4.2. *Fresnel Number*

The number of Fresnel zones in an aperture of diameter D as seen from an observation point P at a distance b from the aperture is

$$\text{Fresnel number} = N = \frac{D^2}{4b\lambda}. \tag{6.9}$$

6.4.3. *Far-Field Distance*

The dividing line between diffraction patterns of the Fresnel type and the diffraction patterns of the Fraunhofer type depends on the accuracy required. From *Standards on Antennas,*[1] this distance is taken to be

$$z_{\text{ff}} = \frac{2D^2}{\lambda}, \tag{6.10}$$

where the aperture diameter is D. This is the distance (and beyond) where Fraunhofer diffraction is observed.

6.5. On-Axis Diffraction by a Circular Aperture

A circular aperture of radius a in a dark screen is illuminated by a normally incident plane wave, $A\exp(ikz)$, where z is the axis perpendicular to the plane of the aperture. The on-axis diffracted field ψ_{cir} is

[1] IRE, *Standards on Antennas,* 1948.

$$\psi_{cir}(z) = A\exp(ikz) - \frac{Az}{\sqrt{a^2+z^2}}\exp(ik\sqrt{a^2+z^2}). \tag{6.11}$$

The corresponding exitance at z is

$$M_{cir}(z) = |\psi_{cir}|^2 = A^2\left\{1 + \frac{z^2}{a^2+z^2} - \frac{2z}{\sqrt{a^2+z^2}}\cos\left[k\left(\sqrt{a^2+z^2}-z\right)\right]\right\}. \tag{6.12}$$

See Fig. 6.2 for the on-axis flux density plotted against the distance z from the circular aperture.

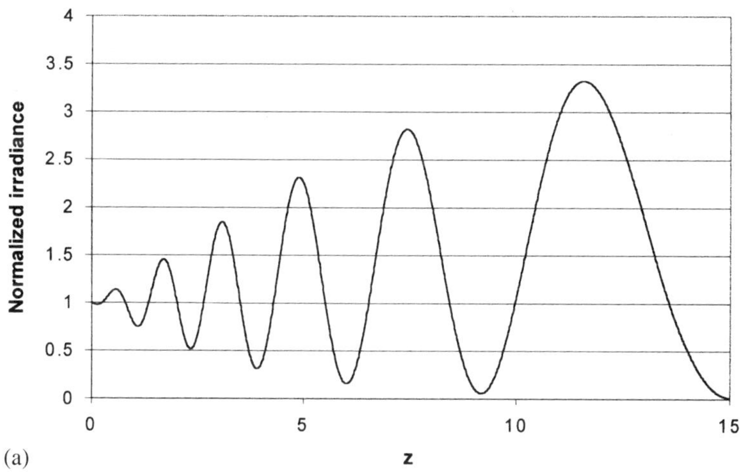

(a)

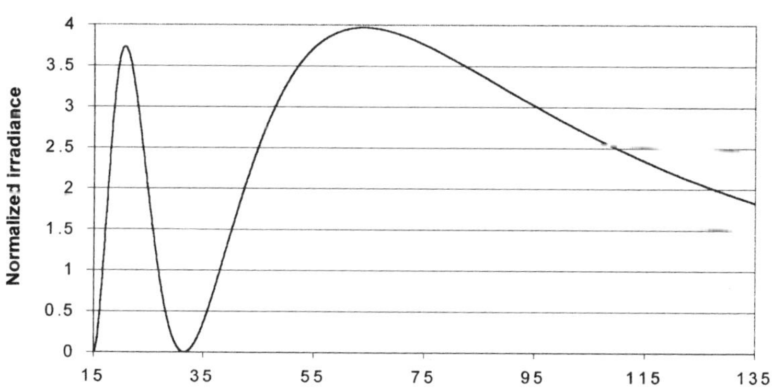

(b)

Figure 6.2. On-axis normalized flux density for the case of an open circular aperture of radius 8λ plotted against z, the distance from the aperture in terms of wavelengths. (a) $z = 0\lambda$–15λ. At $z = 0$, the assumed boundary condition of unity is obtained. (b) $z = 15\lambda$–135λ. The maxima of the oscillations increase monotonically to a maximum value of 4 at $z = D^2/4\lambda$ when one Fresnel zone fills the aperture. For $z > D^2/4\lambda$ the flux density decreases monotonically to zero.

6.6. On-Axis Diffraction by an Opaque Disk

The on-axis diffracted field for an opaque disk of radius a illuminated by a normally incident plane wave, $A\exp(ikz)$, is

$$\psi_{\text{disk}}(z) = \frac{Az}{\sqrt{a^2 + z^2}} \exp\left(ik\sqrt{a^2 + z^2}\right). \tag{6.13}$$

The corresponding flux density at z is

$$M_{\text{disk}}(z) = |\psi_{\text{disk}}|^2 = A^2 \frac{z^2}{a^2 + z^2}. \tag{6.14}$$

See Fig. 6.3 for the variation of the on-axis irradiance with distance z from the opaque disk of radius a.

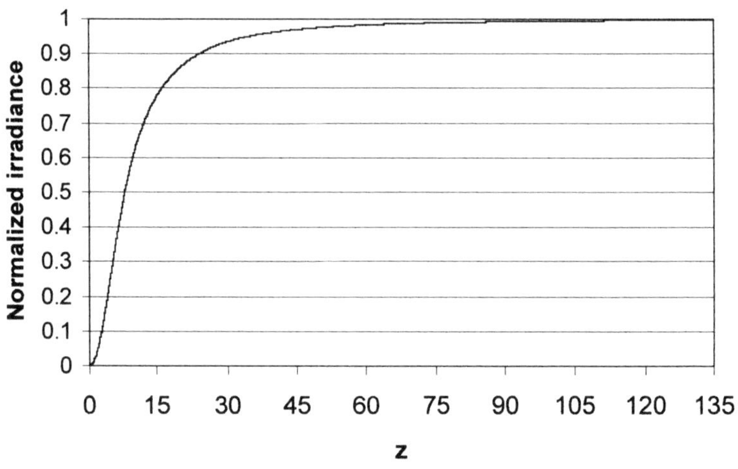

Figure 6.3. On-axis normalized flux density behind a disk of radius 8λ plotted against z, the distance from the aperture in terms of wavelengths. The flux density is zero at $z = 0$ and increases monotonically to unity for large z.

6.7. Babinet's Principle

The Babinet principle states that, if the wave disturbance $\psi_s(P)$ at point of observation P that is due to a diffracting screen S is added to the disturbance $\psi_{\text{CS}}(P)$ that is due to a complementary screen, the total disturbance at P equals the disturbance that is due to an unobstructed wave:

$$\psi_S + \psi_{\text{CS}} = \psi_{\text{UN}}. \tag{6.15}$$

See Fig. 6.4 for a diffracting aperture A and its corresponding complementary screen (Jackson[2] and Andrews[3]).

An immediate verification is offered by Eqs. (6.11) and (6.13):

$$\psi_{\text{cir}} + \psi_{\text{disk}} = \psi_{\text{UN}}, \tag{6.16}$$

where $\psi_{\text{UN}} = A\exp(ikz)$ is the freely propagating plane wave arriving at the on-axis point $(0, 0, z)$ without any obstructing screen.

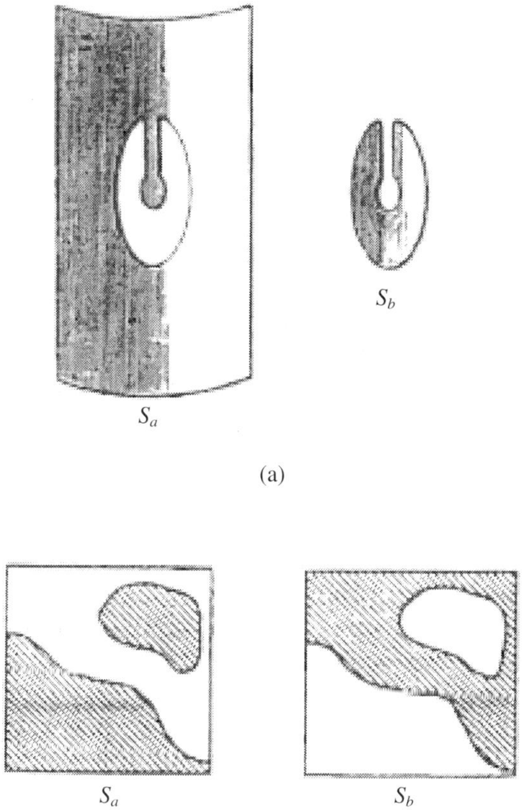

Figure 6.4. Examples of complementary diffraction screens.

6.8. Fresnel Zone Plate

Typical Fresnel zone plates are shown in Fig. 6.5. Both work just as effectively; the plate in Fig. 6.5(a) blocks all even-numbered Fresnel zones whereas the

[2] J. D. Jackson, *Classical Electrodynamics*, 3rd ed., Wiley, 1999.
[3] C. L. Andrews, *Optics of the Electromagnetic Spectrum*, Prentice-Hall, 1960.

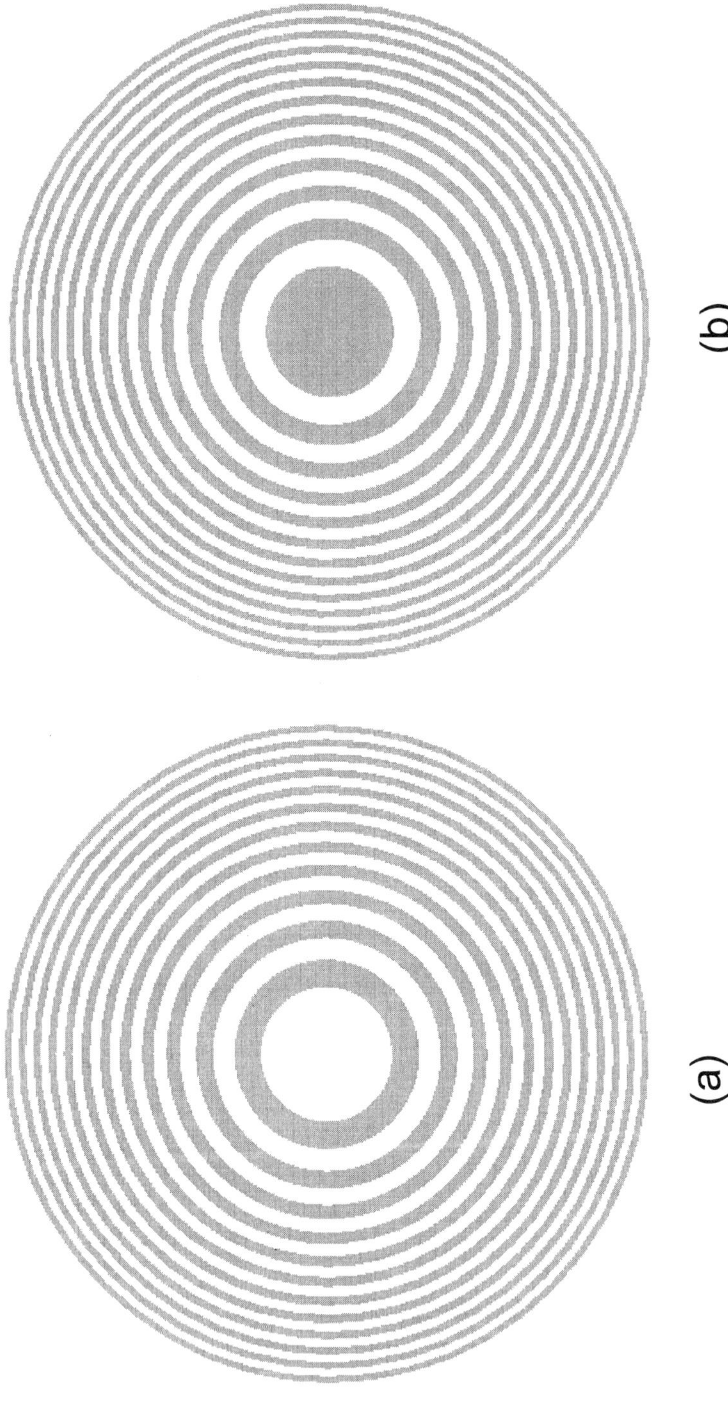

Figure 6.5. Two mutually complementary zone plates.

plate in Fig. 6.5(b) blocks all the odd. A zone plate has a primary focus and several subsidiary foci. Correspondingly, the zone plate produces a primary image and many subsidiary images. If a point source is placed on axis at a distance r_0 in front of the zone plate, then, to a good approximation, the primary image is found on axis at a distance b behind the zone plate, given by

$$\frac{1}{r_0} + \frac{1}{b} = \frac{1}{f_1}, \tag{6.17}$$

where f_1 is the primary focal length, given by

$$f_1 = \frac{R_m^2}{m\lambda}, \tag{6.18}$$

and R_m is the radius of the mth zone. Each zone of the plate matches one Fresnel zone. As a result, every alternate Fresnel zone is blocked and the same-sign amplitudes are added to produce a very bright spot. The on-axis flux density is approximately equal to m^2 times the flux density contributed by one zone alone.

The secondary foci are given by

$$f_{2n-1} = \frac{R_m^2}{(2n-1)m\lambda}, \quad n = 1,2,3, \ldots . \tag{6.19}$$

Images found at these foci are relatively less bright compared with the primary image. In these cases, each zone of the plate blocks $(2n - 1)$ Fresnel zones. The pair of zone plates of Fig. 6.5 are complementary diffracting screens.

6.9. Fresnel Zones and Cylindrical Wave Fronts

A line source or a slit source generates cylindrical wave fronts. Fresnel zones in this case are in the form of long rectangular strips, as shown in Fig. 6.6. Line

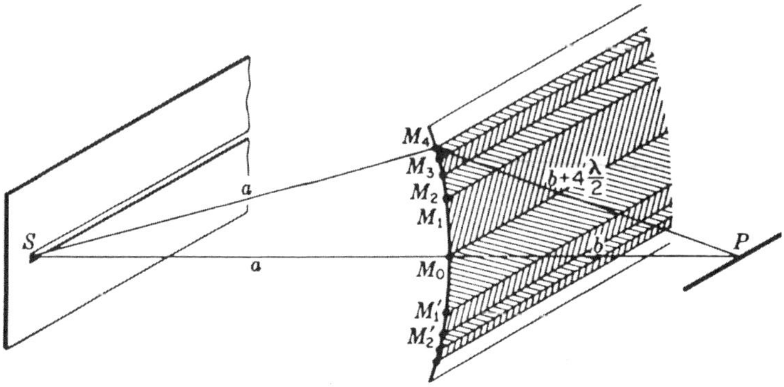

Figure 6.6. Cylindrical wave front with half-period strip zones. S is the slit source generating the wave front, and P is the point of observation.

M_0 on the wave front intersects at right angles to the line joining the source S and the point of observation P. Line M_0 is the axis line of the wave front with respect to the point P. The radius of the cylindrical wave front is a, and b is the distance that M_0 is from the point P. The Fresnel zones are strips parallel to the line marked M_0. The line pairs M_1M_1', M_2M_2', and so on, are $\lambda/2$ farther apart compared with M_0 from the point of observation P. Thus the Fresnel zones are now half-period strips. The arc length is

$$\text{arc length} = (M_m, M_{m+1}) \approx a\left(\sqrt{m+1} - \sqrt{m}\right)\sqrt{\frac{mb\lambda}{a(a+b)}}. \tag{6.20}$$

6.10. Fresnel Diffraction from Apertures with Rectangular Symmetry

Figure 6.7 shows the diffraction geometry for diffraction by a straight edge. The line or slit source is aligned parallel to the edge of the straight edge. The

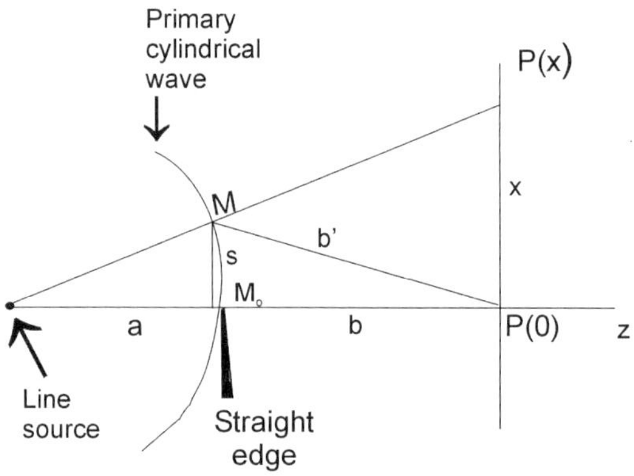

Figure 6.7. Geometry for straight-edge diffraction. The line source is along the y axis, which is perpendicular to the figure.

on-axis point of observation is $P(0)$; the off-axis point is $P(x)$ at x. The plane passing through $P(0)$ and $P(x)$ is perpendicular to the z axis. The radius of the cylindrical wave is a, and the distance between M_0 and $P(0)$ is b. The distance of a point above M_0, labeled M, to the point $P(0)$ is b'. The path difference $b' - b = \Delta$. To a good approximation,

$$\Delta = s^2 \frac{a+b}{2ab}, \tag{6.21}$$

where s is the arc length between M_0 and M but is approximated by the corresponding chord length. For computational purposes the arc length s is converted to a dimensionless parameter v if the phase difference is defined as

$$\delta = \frac{2\pi}{\lambda}\Delta = \frac{2\pi}{\lambda}s^2\frac{a+b}{2ab} \equiv \frac{\pi}{2}v^2 \Rightarrow v = s\sqrt{\frac{2(a+b)}{\lambda ab}}. \tag{6.22}$$

With Fig. 6.7 we can relate the arc length s with the coordinate x in the plane of observation:

$$x = s\left(\frac{a+b}{a}\right). \tag{6.23}$$

Therefore

$$v = s\sqrt{\frac{2(a+b)}{\lambda ab}} = x\sqrt{\frac{2a}{\lambda b(a+b)}}. \tag{6.24}$$

This equation relates the parameter v to the arc length s on the wave front and in turn relates v to the coordinate x in the plane of observation.

For cylindrical wave illumination of the straight edge, the double integral of Eq. (6.4) reduces to a single integral. The single integral describes integration along an arc on the cylindrical wave front. For $r_0 = a$ and b much larger than the region of integration on the cylindrical wave front, we use the following approximation:

$$\psi(P) = C\frac{\exp[ik(a+b)]}{ab}K(\chi)\int_{s_1}^{s_2}\exp(ik\Delta)\mathrm{d}s, \tag{6.25}$$

where Δ is given in Eq. (6.21), a is the radius of curvature of the cylindrical wave, and ds is the differential line element along s. The formal limits of integration, s_1 and s_2, designate the limits appropriate for a slit aperture oriented parallel to the line source. For the straight-edge problem, the upper limit is ∞ and the lower limit s_1 describes the location of the straight edge with respect to the position of the point of observation P along the x axis.

Thus the amplitude is proportional to

$$\psi(P) = C'\left[\int_{-\infty}^{s_1}\exp(ik\Delta)\mathrm{d}s\right] = C'\left[\int_{-\infty}^{0}\exp(ik\Delta)\mathrm{d}s\right.$$

$$\left.+\int_{0}^{s_1}\exp(ik\Delta)\mathrm{d}s\right], \tag{6.26}$$

$$\psi(P) = C'\left\{\left[\frac{1}{2}+C(v_1)\right]+i\left[\frac{1}{2}+S(v_1)\right]\right\}, \tag{6.27}$$

$$M(P) = |\psi(P)|^2 = \left\{\left[\frac{1}{2} + C(v_1)\right]^2 + \left[\frac{1}{2} + S(v_1)\right]^2\right\}, \tag{6.28}$$

where $C(v)$ and $S(v)$ are the standard Fresnel integrals, which are defined as

$$C(v) = \int_0^v \cos\left(\frac{\pi}{2} v'^2\right) dv', \quad S(v) = \int_0^v \sin\left(\frac{\pi}{2} v'^2\right) dv', \tag{6.29}$$

and the constant of proportionality has been omitted in Eq. (6.28).

6.11. Cornu's Spiral

Cornu's spiral is a two-dimensional geometrical depiction of the Fresnel integrals (see Fig. 6.8[4]) and has the following properties:

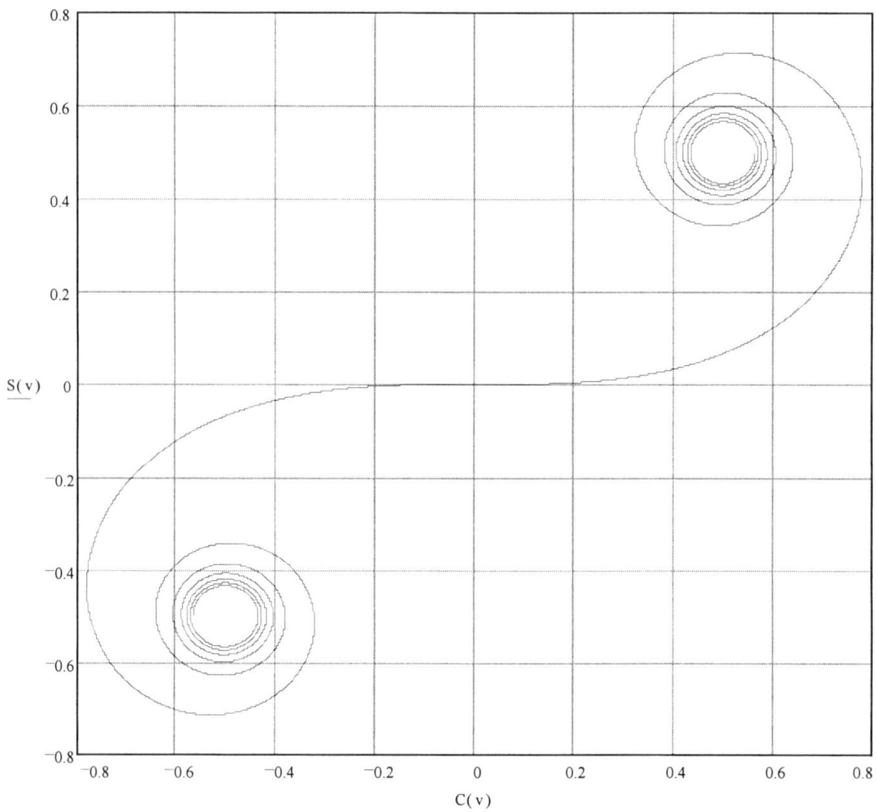

Figure 6.8. Vibration curve, "Cornu's spiral," for use with cylindrical waves.

[4] From F. A. Jenkins and H. E. White, *Fundamentals of Optics,* McGraw-Hill, 1957.

1. Cornu's spiral is a plot of $C(v)$ along the horizontal real axis and $S(v)$ along the vertical imaginary axis with v as a parameter: $-\infty \leq v \leq \infty$.
2. The arc length measured from the origin along the spiral is v; $\delta v = \sqrt{(\delta C)^2 + (\delta S)^2}$.
3. The vector length from the origin to any point v on the spiral is proportional to diffracted amplitude.
4. The angle ϕ made by the vector measured from the horizontal equals the phase of the diffracted light; $\tan \phi = (\delta S/\delta C) = \tan(\pi/2)v^2 \Rightarrow \phi = (\pi/2)v^2$.
5. The radius of curvature = $(\mathrm{d}v/\mathrm{d}\phi) = (\pi v)^{-1}$.
6. For large v, the spiral winds about two limit points:

$$v \to +\infty, \quad C(+\infty) = +\frac{1}{2}, \quad S(+\infty) = +\frac{1}{2},$$

$$v \to -\infty, \quad C(-\infty) = -\frac{1}{2}, \quad S(-\infty) = -\frac{1}{2}.$$

7. The magnitude of the diffraction integral has its maximum value when

$$\phi = \frac{3\pi}{4} \Rightarrow v = \sqrt{\frac{3}{2}}.$$

8. Subsidiary maxima at $\phi = (3\pi/4) + 2\pi n \Rightarrow v = \sqrt{(3/2) + 4n}$, $n = 1,2,3, \ldots$.
9. Minimum values at $\phi = (7\pi/4) + 2\pi m \Rightarrow v = \sqrt{(7/2) + 4m}$, $m = 0,1,2,3, \ldots$.

Cornu's spiral is tabulated in Table 6.2.

Plane-wave incidence is a special case in the limit in which the radius a of the cylindrical wave is ~100 times the distance b to the observation screen.

Figure 6.9 shows a diffraction pattern of light from a straight edge. The maxima and the minima follow the description given by properties 8 and 9 in the preceding list. Figure 6.10 is an example of Fresnel diffraction that is due to a single-slit aperture. Figure 6.11 is an example of Fresnel diffraction that is due to an opaque strip.

6.12. Mathematical Theory of Diffraction

These include the boundary conditions and the Fresnel–Kirchhoff and Rayleigh–Sommerfeld approaches.

Table 6.2. Tabulated Values of Fresnel Integrals

v	$C(v)$	$S(v)$	v	$C(v)$	$S(v)$
0.00	0.0000	0.0000	5.00	0.5636	0.4992
0.10	0.1000	0.0005	5.10	0.4998	0.5624
0.20	0.1990	0.0042	5.20	0.4389	0.4969
0.30	0.2994	0.0141	5.30	0.5078	0.4405
0.40	0.3975	0.0334	5.40	0.5572	0.5140
0.50	0.4923	0.0647	5.50	0.4784	0.5537
0.60	0.5811	0.1105	5.60	0.4517	0.4700
0.70	0.6597	0.1721	5.70	0.5385	0.4595
0.80	0.7228	0.2493	5.80	0.5298	0.5460
0.90	0.7648	0.3398	5.90	0.4486	0.5163
1.00	0.7799	0.4383	6.00	0.4995	0.4470
1.10	0.7628	0.5365	6.10	0.5495	0.5165
1.20	0.7154	0.6234	6.20	0.4676	0.5398
1.30	0.6386	0.6863	6.30	0.4760	0.4555
1.40	0.5431	0.7135	6.40	0.5496	0.4965
1.50	0.4453	0.6975	6.50	0.4816	0.5454
1.60	0.3655	0.6389	6.60	0.4690	0.4631
1.70	0.3238	0.5492	6.70	0.5467	0.4915
1.80	0.3336	0.4509	6.80	0.4831	0.5436
1.90	0.3945	0.3733	6.90	0.4732	0.4624
2.00	0.4883	0.3434	7.00	0.5455	0.4997
2.10	0.5816	0.3743	7.10	0.4733	0.5360
2.20	0.6363	0.4557	7.20	0.4887	0.4573
2.30	0.6266	0.5532	7.30	0.5393	0.5189
2.40	0.5550	0.6197	7.40	0.4601	0.5161
2.50	0.4574	0.6192	7.50	0.5160	0.4607
2.60	0.3889	0.5500	7.60	0.5156	0.5389
2.70	0.3925	0.4529	7.70	0.4628	0.4820
2.80	0.4675	0.3915	7.80	0.5395	0.4896
2.90	0.5624	0.4101	7.90	0.4760	0.5323
3.00	0.6057	0.4963	8.00	0.4998	0.4602
3.10	0.5616	0.5818	8.10	0.5228	0.5320
3.20	0.4663	0.5938	8.20	0.4638	0.4859
3.30	0.4057	0.5193	8.30	0.5377	0.4932
3.40	0.4385	0.4296	8.40	0.4709	0.5243
3.50	0.5326	0.4152	8.50	0.5142	0.4653
3.60	0.5880	0.4923	8.60	0.5025	0.5369
3.70	0.5419	0.5750	8.70	0.4827	0.4677
3.80	0.4481	0.5656	8.80	0.5280	0.5229
3.90	0.4223	0.4752	8.90	0.4661	0.4886
4.00	0.4984	0.4205	9.00	0.5354	0.4999
4.10	0.5737	0.4758	9.10	0.4666	0.5104
4.20	0.5417	0.5632	9.20	0.5291	0.4814
4.30	0.4494	0.5540	9.30	0.4763	0.5247
4.40	0.4383	0.4623	9.40	0.5180	0.4713
4.50	0.5260	0.4343	9.50	0.4873	0.5310
4.60	0.5672	0.5162	9.60	0.5081	0.4679
4.70	0.4914	0.5671	9.70	0.4955	0.5325
4.80	0.4338	0.4968	9.80	0.5019	0.4676
4.90	0.5002	0.4351	9.90	0.4996	0.5321

Note: This table is adapted from Jenkins and White; see footnote 4.

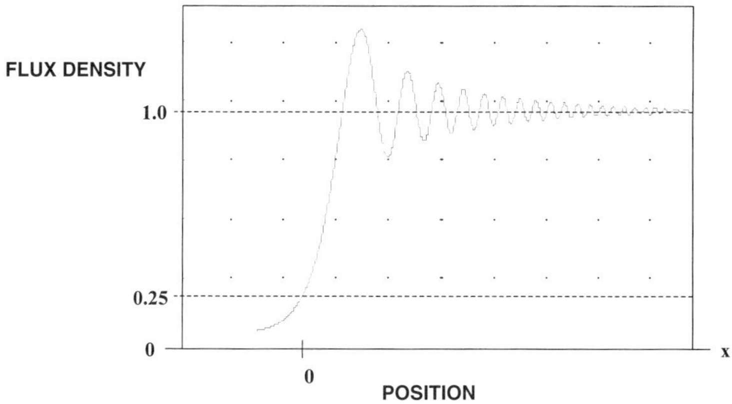

Figure 6.9. Diffraction pattern of light from a straight edge.

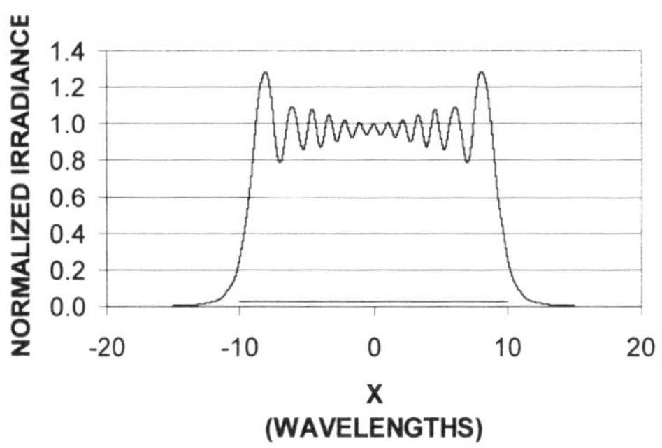

Figure 6.10. Fresnel diffraction pattern from a slit 20 λ wide.

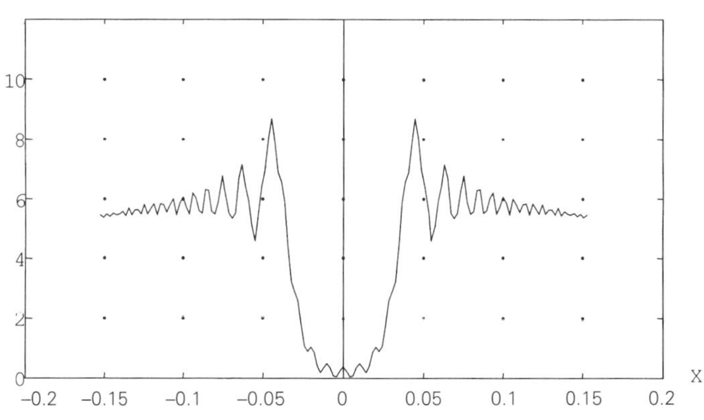

Figure 6.11. Fresnel diffraction pattern from an opaque strip in the strip region.

6.12.1. Kirchhoff Boundary Conditions

The diffraction geometry is shown in Fig. 6.12.[5] For the aperture plane, the following conditions are Kirchhoff's boundary conditions:

For the opaque part of the screen,

$$\psi = 0, \quad \frac{\partial \psi}{\partial n} = 0. \tag{6.30}$$

On the clear aperture A,

$$\psi = \psi_{\text{in}}, \quad \frac{\partial \psi}{\partial n} = \left.\frac{\partial \psi}{\partial n}\right|_{\text{in}}. \tag{6.31}$$

Thus the field and its normal derivative are assumed to be zero on the dark part of the screen, whereas the field and its normal derivative in the open aperture A are taken to be the same as the (unperturbed) incident field.

6.12.2. Fresnel–Kirchhoff Diffraction Formula

If the incident field on an aperture in an opaque screen is an expanding spherical wave, $(a/r)\exp(ikr)$ and $r \gg \lambda$ (a physically reasonable approximation), the diffracted field is

$$\psi(P) = -\frac{i}{\lambda}\iint_A \left[\frac{a}{r}\exp(ikr)\right]\left[\frac{\exp(iks)}{s}\right]\left[\frac{\cos(\hat{\mathbf{n}}, \mathbf{r}) - \cos(\hat{\mathbf{n}}, \mathbf{s})}{2}\right] \mathrm{d}A. \tag{6.32}$$

Owing to the assumed boundary conditions, the area integral is over only the open aperture A with a differential element of area dA. As shown in Fig. 6.12, $(\hat{\mathbf{n}}, \mathbf{r})$ and $(\hat{\mathbf{n}}, \mathbf{s})$ are the angles made by the vectors **r** and **s**, respectively, with the normal to the plane of the aperture.

6.12.3. Rayleigh–Sommerfeld Diffraction Formula

From a strictly mathematical point of view, the specification of the field and its normal derivative as stated in the Kirchhoff boundary conditions overspecifies the required boundary conditions. It is possible to modify the Green's function so that only the field ψ or its normal derivative $\partial\psi/\partial n$ needs to be specified. With this modification, the diffracted field becomes (see Goodman[6])

[5] Figure 6.12 is adapted from M. Born and E. Wolf, *Principles of Optics,* Pergamon, 1989, p. 379, Fig. 8.3.

[6] J. W. Goodman, *Introduction to Fourier Optics,* McGraw-Hill, 1968, pp. 30–76.

$$\psi(P) = -\frac{i}{\lambda}\iint_A \left[\frac{a}{r}\exp(ikr)\right]\left[\frac{\exp(iks)}{s}\right]\cos(\hat{\mathbf{n}}, \mathbf{s})\mathrm{d}A. \qquad (6.33)$$

More generally, the field (for a single temporal frequency) at the point of observation $P(x, y, z)$ may be expressed by

$$\psi(x, y, z) = \iint \psi(x_s, y_s, 0)\left[\frac{1}{2\pi}\frac{z}{s}(1 - iks)\frac{\exp(iks)}{s^2}\right]\mathrm{d}x_s\mathrm{d}y_s, \qquad (6.34)$$

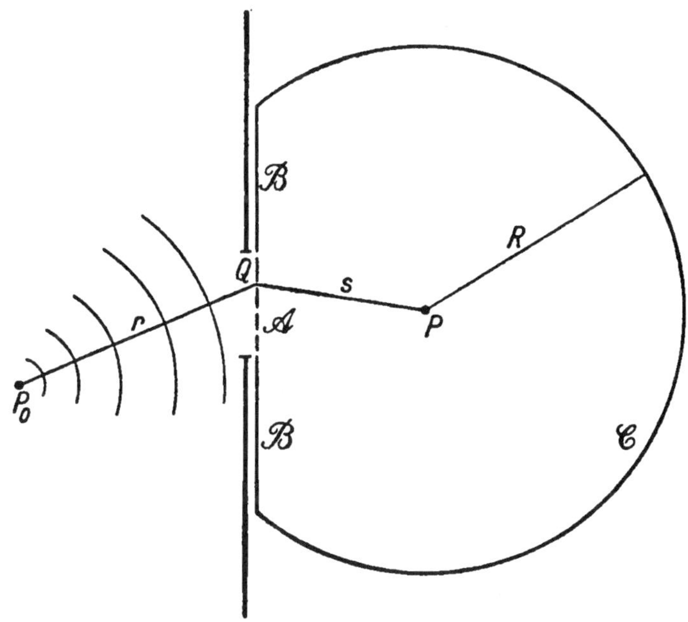

(a)

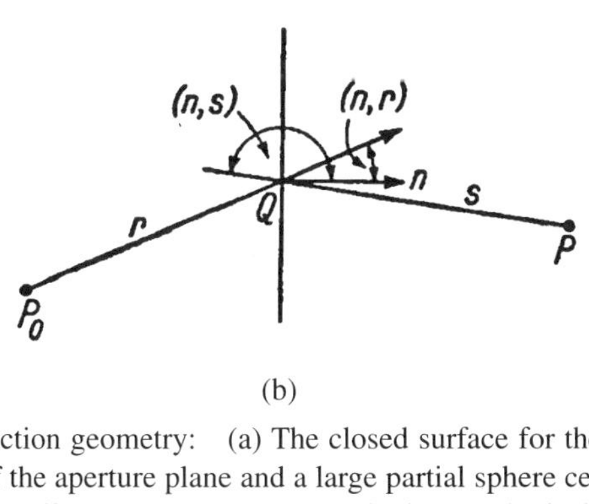

(b)

Figure 6.12. Diffraction geometry: (a) The closed surface for the diffraction problem. It is made up of the aperture plane and a large partial sphere centered at the point of observation P. (b) Diffracting aperture with an incident spherical wave to show the angles involved in the calculation with reference to Eq. (6.32).

where $\psi(x_s, y_s, 0)$ are the field values in the aperture A at $z = 0$. In this expression, the distance s from a point in the aperture to the point of observation is defined by

$$s = \sqrt{(x - x_s)^2 + (y - y_s)^2 + z^2}. \tag{6.35}$$

The ratio z/s is the direction cosine of the difference vector.

6.12.4. *Angular Spectrum*

Taking a two-dimensional spatial Fourier transform of both sides of Eq. (6.34) gives the product relation in terms of the angular spectrum,

$$\tilde{\psi}(L/\lambda, M/\lambda, z) = \tilde{\psi}(L/\lambda, M/\lambda, 0)\exp(ikNz), \tag{6.36}$$

where the transform is defined by

$$\psi(x, y, z) = \iint \tilde{\psi}(L/\lambda, M/\lambda, z)\exp\left[+i\frac{2\pi}{\lambda}(Lx + My)\right]\mathrm{d}(L/\lambda)\mathrm{d}(M/\lambda), \tag{6.37}$$

and the third direction cosine is defined by

$$N = +\sqrt{1 - L^2 - M^2} \quad \text{for } L^2 + M^2 \le 1,$$
$$N = +i\sqrt{L^2 + M^2 - 1} \quad \text{for } L^2 + M^2 > 1.$$

For $z = 0$ the self-consistent Rayleigh–Sommerfeld theory reproduces its own assumed boundary conditions, a property not shared by the Fresnel–Kirchhoff theory. See, for example, the $z = 0$ values in Fig. 6.2 for the circular aperture and in Fig. 6.3 for an opaque disk.

6.13. Fresnel and Fraunhofer Approximations

In instrumental optics, diffraction is studied mainly in the forward direction for small angles relative to the direction of propagation of the incident field. Also, the distances involved are much larger than the wavelength, $s \gg \lambda$. For such cases the distance s, as defined in Eq. (6.35), is approximated by

$$s \approx z\frac{x^2 + y^2}{2z^2} - \frac{xx_s + yy_s}{z} + \frac{x_s^2 + y_s^2}{2z}. \tag{6.38}$$

When the quadratic terms in the aperture variables are maintained, namely $x_s^2 + y_s^2$, we obtain the Fresnel diffraction approximation

$$\psi(x, y, z) = -\frac{i}{\lambda}\frac{\exp\left[ikz\left(1+\frac{x^2+y^2}{2z^2}\right)\right]}{z}\iint \psi(x_s, y_s, 0)$$

$$\times \exp\left(+ik\frac{x_s^2+y_s^2}{2z}\right)\exp\left(-ik\frac{xx_s+yy_s}{z}\right)\mathrm{d}x_s\mathrm{d}y_s. \quad (6.39)$$

This approximation is valid for both the Fresnel and the Fraunhofer regions. At far-field distances [see Eq. (6.10)], we obtain the Fraunhofer diffraction approximation

$$\psi(x, y, z) = -\frac{i}{\lambda}\frac{\exp\left[ikz\left(1+\frac{x^2+y^2}{2z^2}\right)\right]}{z}\iint \psi(x_s, y_s, 0)$$

$$\times \exp\left(-ik\frac{xx_s+yy_s}{z}\right)\mathrm{d}x_s\mathrm{d}y_s. \quad (6.40)$$

6.14. Image of a Point Object as an Example of Fraunhofer Diffraction

An image $U_{\mathrm{im}}(x, y)$ of a coherently illuminated object $U_{\mathrm{ob}}(\xi, \eta)$ that is due to a thin converging lens is described by

$$U_{\mathrm{im}}(x, y) = \exp\left[\frac{ik}{2s'}(x^2+y^2)\right]\iint \mathrm{d}\xi\mathrm{d}\eta\, U_{\mathrm{ob}}(\xi,\eta)\exp\left[\frac{ik}{2s}(\xi^2+\eta^2)\right]$$

$$\times A(x-m\xi, y-m\eta). \quad (6.41)$$

The function A is called the amplitude impulse response, and it is defined by

$$A(x-m\xi, y-m\eta) = \frac{1}{\lambda^2 ss'}\iint \mathrm{d}\alpha'\mathrm{d}\beta' L(\alpha',\beta')$$

$$\times \exp\left[-i2\pi\frac{(x-m\xi)\alpha'+(y-m\eta)\beta'}{\lambda s'}\right]. \quad (6.42)$$

The variables α' and β' are in the plane of the aperture of the thin lens. $L(\alpha', \beta')$ is the pupil function. The distances s and s' are measured from the lens to the object and from the lens to the image, respectively; m is the magnification, $m = -s'/s$. Fulfilling the geometrical imaging condition for a thin lens of focal length f, namely, $[(1/s) + (1/s')] = (1/f)$, leads to Eq. (6.41) for imaging of extended objects and to Eq. (6.42) for imaging a point impulse located at (ξ, η) in the object plane. It should be noted that Eq. (6.42) is in the form of a Fourier transform of the pupil function.

6.15. Fraunhofer Diffraction Pattern of a Double Slit

The Fraunhofer diffraction pattern of a double slit is given by

$$M = M_0 \left(\frac{\sin \alpha}{\alpha} \right)^2 (\cos \delta)^2, \tag{6.43}$$

where the flux density at the center of the pattern is $M_0 \propto 4b^2$. The width of the slit is b and the factor 4 comes from the square of the number of slits, in this case 2. The phase argument α is defined by $\alpha \equiv (1/2)kb \sin \theta$ and $\delta \equiv (1/2)kd \sin \theta$, where d is the slit separation and θ is the angular location of the point of observation measured from the optic axis. See Fig. 6.13 for the optical arrangement to observe the Fraunhofer diffraction pattern. In Eq. (6.43) the first factor with argument α describes the single-slit Fraunhofer diffraction and the second factor with argument δ gives the details of the interference pattern that is due to the two slits.

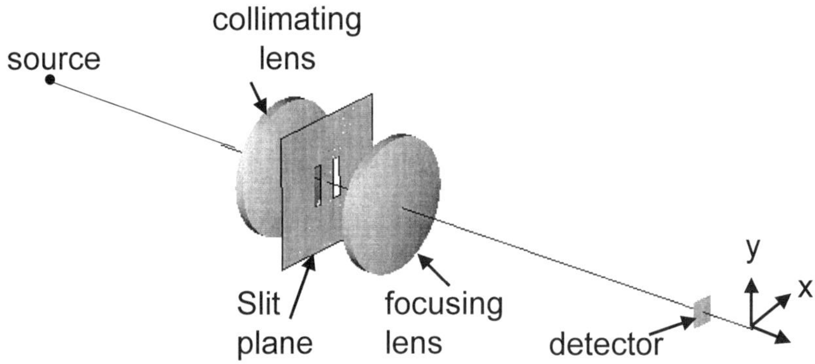

Figure 6.13. Optical arrangement for observing interference of light from two slits. The slit separation is d and the slit width is b.

6.16. Diffraction Grating

See Fig. 6.14 for the optical arrangement. The normalized irradiance distribution that is due to a diffraction grating illuminated by normally incident, collimated light is given by

$$M = M_0 \left(\frac{\sin \alpha}{\alpha} \right)^2 \left(\frac{\sin N\gamma}{N \sin \gamma} \right)^2. \tag{6.44}$$

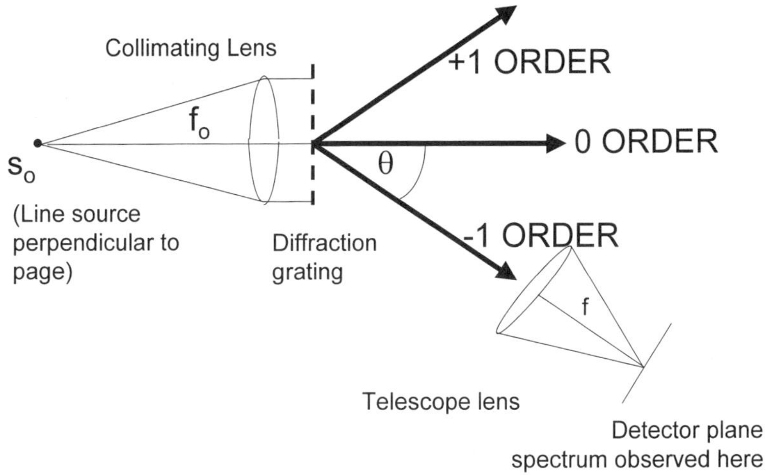

Figure 6.14. Optical arrangement for observing large-angle diffraction maxima from a diffraction grating. Collimated light from a line source is incident normally onto the grating plane whose slits are perpendicular to the page. An observing telescope arrangement is free to rotate along the arc of a circle centered at the midpoint of the grating.

The grating has N slits, each of width b and slit separation d. The flux density at the center of the pattern is $E_0 \propto N^2b^2$. The factor with the argument $\alpha \equiv (1/2)kb \sin\theta$ describes the single-slit diffraction pattern and the second factor with argument $\gamma \equiv (1/2)kd \sin\theta$ describes the interference of light from N slits. The interference term has principal maxima of value unity when both the numerator and denominator are zero, this happens for $\gamma = m\pi$, where m is an integer. In between the principal maxima there are $(N - 1)$ zeros of the pattern, and these are located at $\gamma = (\pi/N), (2\pi/N), \ldots, \{[(N-1)\pi]/N\}$.

6.17. Grating Equation

The grating equation that is valid for both transmission and reflection is

$$d(\sin\theta_m - \sin\theta_i) = m\lambda, \tag{6.45}$$

where θ_i is the angle of incidence relative to the normal and θ_m is the angle for the mth order. The zero order ($m = 0$, $\theta_i = \theta_m$) is dispersionless; all other orders show wavelength dependence. Real orders are observed for $m \leq (d/\lambda)$. For $m > (d/\lambda)$ the orders are evanescent or imaginary. The spectral resolving power is given by

$$\mathcal{R} = mN. \tag{6.46}$$

6.18. Stationary Phase Approximation

The diffracted field in the Rayleigh–Sommerfeld diffraction theory is given by

$$\psi(x, y, z) = \iint \psi(x_s, y_s, 0)\left[\frac{1}{2\pi}\frac{z}{s}(1 - iks)\frac{\exp(iks)}{s^2}\right]\mathrm{d}x_s\mathrm{d}y_s, \quad (6.47)$$

where $\psi(x_s, y_s, 0)$ is the field in the aperture. The diffracted field can also be represented by

$$\psi(x, y, z) = \int_{-\infty}^{\infty}\int_{-\infty}^{\infty} \psi(L/\lambda, M/\lambda, 0)\exp\left[i\frac{2\pi}{\lambda}(Lx + My + Nz)\right] \times \mathrm{d}(L/\lambda)\mathrm{d}(M/\lambda). \quad (6.48)$$

We now define the coefficient of $i2\pi/\lambda$ in the exponent as

$$\varphi(L, M) \equiv (Lx + My + Nz) = \left\{Lx + My + \sqrt{[1 - (L^2 + M^2)]}z\right\}. \quad (6.49)$$

The values of L and M that make the first partial derivatives of $\varphi(L, M)$ equal to zero, that is,

$$\frac{\partial\varphi}{\partial L} = \frac{\partial\varphi}{\partial M} = 0, \quad (6.50)$$

are

$$L_0 = \pm\frac{x}{r}, \quad M_0 = \pm\frac{y}{r}, \quad (6.51)$$

where $r = \sqrt{x^2 + y^2 + z^2}$. The negative sign is omitted for forward propagation, $z > 0$. The phase is then approximated by

$$\phi(L, M) \approx \phi(L_0, M_0) + \frac{1}{2}[\alpha(L - L_0)^2 + \beta(M - M_0)^2 + 2\gamma(L - L_0)(M - M_0)],$$

where the higher-order terms are neglected and α,β, and γ are the second partial derivatives evaluated at $L = L_0$ and $M = M_0$. These constant coefficients are given by

$$\varphi(L_0, M_0) = r; \quad (6.52)$$

$$\alpha \equiv \frac{\partial^2 \varphi}{\partial L^2} = -r\,\frac{x^2 + z^2}{z^2},$$

$$\beta \equiv \frac{\partial^2 \varphi}{\partial M^2} = -r\,\frac{y^2 + z^2}{z^2},$$

$$\gamma \equiv \frac{\partial^2 \varphi}{\partial L \partial M} = \frac{\partial^2 \varphi}{\partial M \partial L} = -r\,\frac{xy}{z^2}. \tag{6.53}$$

The resulting phase function is used in the double integral to obtain the diffracted field $\psi(x, y, z)$. The preceding procedure yields the stationary phase approximation for the Fraunhofer diffraction, given by

$$\psi(x, y, z) \approx -\frac{i}{\lambda}\frac{z}{r}\frac{\exp(ikr)}{r} \iint_A \psi(x_s, y_s, 0)$$
$$\times \exp\left[-i\,\frac{2\pi}{\lambda}\left(\frac{xx_s + yy_s}{r}\right)\right] \mathrm{d}x_s\,\mathrm{d}y_s. \tag{6.54}$$

6.19. Vector Diffraction

The following equations account for vector properties of the diffracted fields:

$$\mathbf{E}(\mathbf{r}) = \frac{-i}{2\lambda^2} \iiint_{\text{EPW}} \left\{ [\hat{\mathbf{R}}_s \times \mathbf{E}(\mathbf{r}_s)] \times \hat{\mathbf{R}}_s \right.$$
$$\left. - \sqrt{\frac{\mu_0}{\varepsilon_0}}\,[\hat{\mathbf{R}}_s \times \mathbf{H}(\mathbf{r}_s)] \right\} \frac{\exp(ikR_s)}{R_s}\,\mathrm{d}^{(3)}\mathbf{r}, \tag{6.55}$$

$$\mathbf{H}(\mathbf{r}) = \frac{-i}{2\lambda^2} \iiint_{\text{EPW}} \left\{ [\hat{\mathbf{R}}_s \times \mathbf{H}(\mathbf{r}_s)] \times \hat{\mathbf{R}}_s \right.$$
$$\left. + \sqrt{\frac{\varepsilon_0}{\mu_0}}\,[\hat{\mathbf{R}}_s \times \mathbf{E}(\mathbf{r}_s)] \right\} \frac{\exp(ikR_s)}{R_s}\,\mathrm{d}^{(3)}\mathbf{r}_s, \tag{6.56}$$

where $\mathbf{E}(\mathbf{r})$ and $\mathbf{H}(\mathbf{r})$ are the diffracted electric and magnetic fields at the point of observation $\mathbf{r}$, respectively. Under the integral, $\mathbf{E}(\mathbf{r}_s)$ and $\mathbf{H}(\mathbf{r}_s)$ are the fields of the wave front incident upon the aperture at point $\mathbf{r}_s$ and are related by

$$\mathbf{H}(\mathbf{r}_s) = \sqrt{\frac{\varepsilon_0}{\mu_0}}\,\hat{\mathbf{n}}(\mathbf{r}_s) \times \mathbf{E}(\mathbf{r}_s),$$

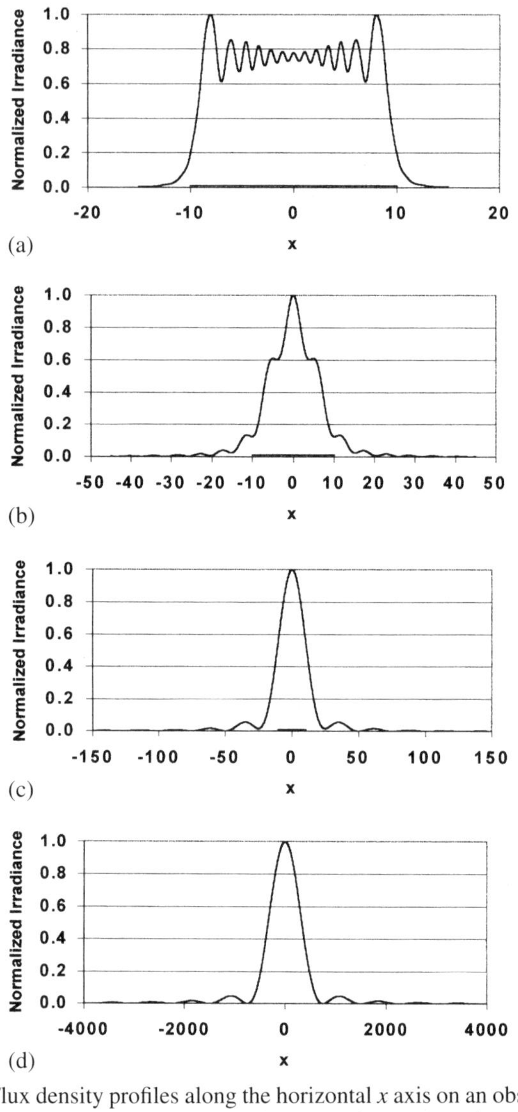

Figure 6.15. Flux density profiles along the horizontal x axis on an observation screen at the following distances: (a) 5λ, (b) 100λ, (c) 500λ, (d) 15000λ. These profiles are due to diffraction by a rectangular slit and are based on Eqs. (6.55) and (6.56). The slit is of width $W = 20\lambda$. The first zero of the Fraunhofer pattern is at 2.81°. The incident field is a plane wave of unit amplitude, falling normally on the plane of the aperture and polarized in the vertical y direction. The position on the x axis is in terms of wavelengths from the origin of the observation plane.

where $\hat{\mathbf{n}}(\mathbf{r}_s)$ is the normal to the wave front at that point. $\mathbf{R}_s \equiv \mathbf{r} - \mathbf{r}_s$; R_s is the magnitude and $\hat{\mathbf{R}}_s$ is the unit vector in the direction of $\mathbf{R}_s$. The letters EPW under the volume integral stand for "exposed parts of the primary wave front." The integration is restricted to the open areas of the aperture. (See McCalmont[7] and Marathay and McCalmont[8] for a more detailed description and additional references.) See Fig. 6.15 for a sequence of irradiance profiles that are due to the diffraction of light by a narrow slit based on Eqs. (6.55) and (6.56). The sequence describes diffraction from deep within the Fresnel region into the Fraunhofer zone.

Further Reading

E. Hecht, *Optics,* Addison, 1987.

J. D. Jackson, *Classical Electrodynamics,* 3rd ed., Wiley, 1999.

B. Rossi, *Optics,* Addison-Wesley, 1957.

A. E. Sommerfeld, *Electrodynamics,* Academic, 1964 (translation of *Über Theoretische Vorlesung Physik,* Vol. 3, *Elektrodynamik*).

[7] J. F. McCalmont, "A vector Huygens Fresnel model of the diffraction of electromagnetic waves," Ph.D. dissertation, University of Arizona, 1999.

[8] A. S. Marathay and J. F. McCalmont, "Vector diffraction theory for electromagnetic waves," J. Opt. Soc. Am. **18,** 2585 (2001).

7

Surface Scattering

E. L. Church and P. Z. Takacs

7.1. Introduction

In this chapter a number of analytic results are presented for the scattering of electromagnetic radiation from topographically rough surfaces. The discussion is limited to the scattering of linearly polarized plane waves from nominally flat surfaces made of perfect-electrical conducting material. Particular attention is paid to one- and two-dimensional problems, Gaussian and non-Gaussian statistics, statistical models, and fractallike surfaces.

In Section 7.2 the notation used is described. In Section 7.3 exact results for semicircular ridges and hemispherical bumps are discussed. In Sections 7.4 and 7.5 two more general but approximate methods—the small-perturbation method and the physical-optics method—are discussed. In Sections 7.7 and 7.8, statistical scattering and statistical models are discussed. The chapter ends with a Glossary and suggestions for further reading.

7.2. Notation

Surface scattering is described in terms of the standard spherical coordinate system shown in Fig. 7.1, where the subscripts i and f denote the initial (or incident) and the final states of scattering, respectively.

Note that θ_i is always positive, whereas θ_f can be positive or negative, and that φ_f is understood to be $\varphi_f - \varphi_i$, with $\varphi_i = 0$.

The wave vectors in the initial and the final directions are

$$\mathbf{k}_i = k\begin{bmatrix} +\sin(\theta_i) \\ 0 \\ -\cos(\theta_i) \end{bmatrix}, \quad \mathbf{k}_f = k\begin{bmatrix} +\sin(\theta_f)\cos(\varphi_f) \\ +\sin(\theta_f)\sin(\varphi_f) \\ +\cos(\theta_f) \end{bmatrix}, \tag{7.1}$$

where $k = 2\pi/\lambda$. The wave number or momentum change on scattering is then

$$\mathbf{k}_f - \mathbf{k}_i = 2\pi\xi = \frac{2\pi}{\lambda}\begin{bmatrix} \sin(\theta_f)\cos(\varphi_f) - \sin(\theta_i) \\ \sin(\theta_f)\sin(\varphi_f) \\ \cos(\theta_f) + \cos(\theta_i) \end{bmatrix}, \tag{7.2}$$

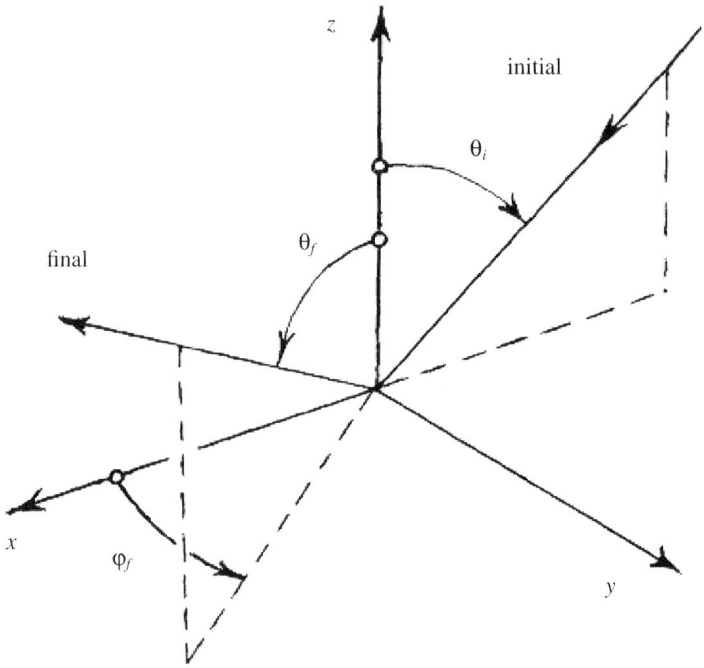

Fig. 7.1. Coordinate system for one- and two-dimensional scattering. The *X–Y* plane is the surface plane with air/vacuum above and material below. The planes of incidence and scattering are defined by the *Z* axis and the initial and the final directions of propagation.

where **f** is the spatial-frequency vector. The Jacobian relating the configuration, wave number, and frequency spaces is

$$\cos(\theta_f)d\boldsymbol{\omega}_f = d\hat{\mathbf{k}}_x d\hat{\mathbf{k}}_y = \lambda^2 d\xi_x d\xi_y. \tag{7.3}$$

The surface height is $Z(x, y)$. In the case of one-dimensionally rough or gratinglike surfaces, $Z(x, y) = Z(y)$ and the scattering is confined to the x–y plane, where $\varphi_f = 0$.

The electromagnetic field has inherent polarization properties. The most elemental description is in terms of two orthogonal degrees of linear polarization. In rationalized mks units, the relevant electric and magnetic vectors are

$$\mathbf{E} = E\hat{\mathbf{e}}, \quad \mathbf{H} = \frac{1}{\eta}\hat{\mathbf{k}} \times \mathbf{E}, \tag{7.4}$$

where η is the impedance of the medium,

$$\eta = \frac{E}{H} = \sqrt{\frac{\mu}{\varepsilon}}, \quad \eta_0 = \sqrt{\frac{\mu_0}{\varepsilon_0}} = 4\pi c \times 10^{-7} \approx 120\,\pi \approx 377\Omega, \tag{7.5}$$

and μ and ε are its magnetic permeability and electric permittivity, respectively. The subscripted zero values are the vacuum values: $\varepsilon_0 = 10^7/(4\pi c^2)$ F m^{-1}, $\mu_0 = 4\pi\ 10^{-7}$ H m^{-1}, and $c = 2.99792 \times 10^6$ m/s.

The states of linear polarization are described in terms of the direction of the electric vector with respect to the planes of incidence or scattering:

$$s\text{ (senkrecht)} = h\text{ (horizontal)} = \text{TM} = E = \| = -$$

→ Dirichlet (soft) boundary conditions for a perfect electrical conductor.

$$p\text{ (parallel)} = v\text{ (vertical)} = \text{TE} = H = \perp = +$$

→ Neumann (hard) boundary conditions for a perfect electrical conductor.

(7.6)

Although these symbols are common in the literature, they are not always used consistently.

In this chapter the states of linear polarization are in the notation *s, p*. Explicit expressions for the unit electrical-polarization vectors are then

$$\begin{pmatrix} \hat{\mathbf{e}}_{is} \\ \hat{\mathbf{e}}_{ip} \end{pmatrix} = \begin{pmatrix} \hat{\mathbf{y}} \\ \hat{\boldsymbol{\theta}}_i \end{pmatrix} = \begin{bmatrix} +\hat{\mathbf{y}} \\ -\hat{\mathbf{x}}\cos(\theta_i) - \hat{\mathbf{z}}\sin(\theta_i) \end{bmatrix},$$

$$\begin{pmatrix} \hat{\mathbf{e}}_{fs} \\ \hat{\mathbf{e}}_{fp} \end{pmatrix} = \begin{pmatrix} \hat{\boldsymbol{\varphi}}_f \\ \hat{\boldsymbol{\theta}}_f \end{pmatrix}$$

$$= \begin{bmatrix} -\hat{\mathbf{x}}\sin(\varphi_f) + \hat{\mathbf{y}}\cos(\varphi_f) \\ +\hat{\mathbf{x}}\cos(\theta_f)\cos(\varphi_f) + \hat{\mathbf{y}}\cos(\theta_f)\sin(\varphi_f) - \hat{\mathbf{z}}\sin(\theta_f) \end{bmatrix}. \tag{7.7}$$

Both triads, (p, s, k) and (E, H, k), form right-handed coordinate systems for the initial and the final states of propagation.

7.3. Scattering Amplitudes

7.3.1. *Definitions*

The elemental scattering process involves a plane wave striking the scattering object or surface:

$$\boldsymbol{\psi}_i = E_i \hat{\mathbf{e}}_i e^{+i(\mathbf{k}_i\mathbf{r} - \omega t)}. \tag{7.8}$$

The Fraunhofer or far-field region occurs for $r > 2L^2/\lambda$; r is the distance between the scatterer and the observer and L is a characteristic size of the illuminated area of the surface. In that region the scattered fields appear in the form of spherical or cylindrical waves. For a two-dimensional scatterer, the far-field form of the scattered field are the spherical waves,

$$\begin{bmatrix} E_s(\hat{\mathbf{k}}_f) \\ E_p(\hat{\mathbf{k}}_f) \end{bmatrix}_f = \frac{e^{+ikr}}{r} \begin{bmatrix} f_{ss}^{(2D)}(\hat{\mathbf{k}}_f, \hat{\mathbf{k}}_i) & f_{sp}^{(2D)}(\hat{\mathbf{k}}_f, \hat{\mathbf{k}}_i) \\ f_{ps}^{(2D)}(\hat{\mathbf{k}}_f, \hat{\mathbf{k}}_i) & f_{pp}^{(2D)}(\hat{\mathbf{k}}_f, \hat{\mathbf{k}}_i) \end{bmatrix} \begin{bmatrix} E_p(\hat{\mathbf{k}}_i) \\ E_s(\hat{\mathbf{k}}_i) \end{bmatrix}_i, \tag{7.9}$$

whereas for a one-dimensional scatterer they are the cylindrical waves,

$$\begin{bmatrix} E_s(\hat{\mathbf{k}}_f) \\ E_p(\hat{\mathbf{k}}_f) \end{bmatrix}_f = \sqrt{\frac{2}{i\pi k}} \frac{e^{+ikr}}{\sqrt{r}} \begin{bmatrix} f_{ss}^{(1D)}(\hat{\mathbf{k}}_f, \hat{\mathbf{k}}_i) & f_{sp}^{(1D)}(\hat{\mathbf{k}}_f, \hat{\mathbf{k}}_i) \\ f_{ps}^{(1D)}(\hat{\mathbf{k}}_f, \hat{\mathbf{k}}_i) & f_{pp}^{(1D)}(\hat{\mathbf{k}}_f, \hat{\mathbf{k}}_i) \end{bmatrix} \begin{bmatrix} E_s(\hat{\mathbf{k}}_i) \\ E_p(\hat{\mathbf{k}}_i) \end{bmatrix}_i. \tag{7.10}$$

These expressions involve 2×2 Mueller-like matrices that connect the incident and the final states of linear polarization. The matrix elements are the scattering amplitudes f, which contain all of the information necessary to describe scattering properties in the far field.

In the convention used here the spherical-wave scattering amplitudes have the dimensions of length and the cylindrical-wave amplitudes are dimensionless. Note also that the subscripts denoting the polarization are written in operator form; that is, from right to left. In other words, the subscript $\beta\alpha$ denotes $\alpha \rightarrow \beta$.

7.3.2. *General Properties*

The scattering amplitudes have a number of general properties that result from Maxwell's equations. The first is the *conservation of energy,* which, for a surface made of perfectly conducting material, takes the forms

$$\sum_{\beta} \int_{2\pi} |f_{\beta\alpha}^{(2D)}(\hat{\mathbf{k}}_f, \hat{\mathbf{k}}_i)|^2 = A \cos(\theta_i) d\omega_f,$$

$$\sum_{\beta} \int_{\pi} |f_{\beta\alpha}^{(1D)}(\hat{\mathbf{k}}_f, \hat{\mathbf{k}}_i)|^2 = \frac{\pi}{2} kL \cos(\theta_i) d\theta_f. \tag{7.11}$$

The expressions on the left of each equation represent the total scattered power, whereas those on the right represent the total power incident upon the surface area A or length L at the angle of incidence, θ_i.

A second general property is *reciprocity,* which requires that

$$f_{\beta\alpha}^{(nD)}(\hat{\mathbf{k}}_f, \hat{\mathbf{k}}_i) = \pm f_{\alpha\beta}^{(nD)}(-\hat{\mathbf{k}}_i, -\hat{\mathbf{k}}_f), \tag{7.12}$$

where the + sign applies for $\alpha = \beta$ and the – sign for $\alpha \neq \beta$. Reversing the direction of propagation converts $\hat{\mathbf{k}} = (\theta, \varphi)$ into $-\hat{\mathbf{k}} = (\pi - \theta, \pi + \varphi)$.

A third property, *duality,* requires that results are invariant under the replacements $\varepsilon \leftrightarrow \mu$, $\mathbf{E} \rightarrow +\mathbf{H}$, and $\mathbf{H} \rightarrow -\mathbf{E}$, where ε and μ are the permittivity and the permeability of the medium of propagation, respectively.

A fourth general property is the *optical theorem*—also known as the forward-scattering or extinction theorem. It relates the total scattering and absorption to the diminution of the incident field in accord with the conservation of energy. In the case of an object made of perfectly conducting material, it requires that

$$\sum_{\beta} \int_{2\pi} |f_{\beta\alpha}^{(2D)}(\hat{\mathbf{k}}_f, \hat{\mathbf{k}}_i)|^2 d\omega_f = +\frac{4\pi}{k} \operatorname{Im}[f_{\alpha\alpha}^{(2D)}(\hat{\mathbf{k}}_i, \hat{\mathbf{k}}_i)],$$

$$\sum_{\beta} \int_{\pi} |f_{\beta\alpha}^{(1D)}(\hat{\mathbf{k}}_f, \hat{\mathbf{k}}_i)|^2 d\theta_f = -2\pi \operatorname{Re}[f_{\alpha\alpha}^{(1D)}(\hat{\mathbf{k}}_i, \hat{\mathbf{k}}_i)]. \tag{7.13}$$

These simple forms, however, apply to only freestanding objects such as a sphere in space. Equivalent expressions can be written for surface scatterers, but the results depend on the specific geometries involved. In the case of objects made of imperfectly conducting material, Eqs. (7.11) and (7.13) are modified by terms that are included on the left-hand sides that represent absorption by the scatterer. The expressions for reciprocity and duality, however, are unaffected.

The use of each of these general properties can involve some subtleties, and the reader is referred to the Further Reading list for details.

7.3.3. *Surface Scattering*

The distribution of power scattered and reflected from a surface is written in a number of ways in the optics and radar literature:

$$\begin{aligned}
\frac{1}{\Phi_i}\left(\frac{d\Phi}{d\omega_f}\right)_{\alpha\rightarrow\beta}^{(2D)} &= \frac{1}{A\cos(\theta_i)} \lim_{r\rightarrow\infty}\left(r^2\left|\frac{E_f}{E_i}\right|^2\right) = \frac{|f_{\beta\alpha}^{(2D)}(\hat{\mathbf{k}}_f, \hat{\mathbf{k}}_i)|^2}{A\cos(\theta_i)} \\
&= \frac{\sigma_{\beta\alpha}^{(2D)}(\hat{\mathbf{k}}_f, \hat{\mathbf{k}}_i)}{A\cos(\theta_i)} = \frac{\mathrm{RCS}_{\beta\alpha}^{(2D)}(\hat{\mathbf{k}}_f, \hat{\mathbf{k}}_i)}{4\pi A\cos(\theta_i)} = \frac{\gamma_{\beta\alpha}^{(2D)}(\hat{\mathbf{k}}_f, \hat{\mathbf{k}}_i)}{4\pi\cos(\theta_i)} \\
&= \mathrm{DSC}_{\alpha\rightarrow\beta}^{(2D)}(\hat{\mathbf{k}}_f, \hat{\mathbf{k}}_i) = \cos(\theta_f)\mathrm{BRDF}_{\alpha\rightarrow\beta}^{(2D)}(\hat{\mathbf{k}}_f, \hat{\mathbf{k}}_i)
\end{aligned} \tag{7.14}$$

for a two-dimensional scatterer, and

$$\frac{1}{\Phi_i}\left(\frac{d\Phi}{d\theta_f}\right)^{(1D)}_{\alpha\to\beta} = \frac{1}{L\cos(\theta_i)}\lim_{r\to\infty}\left(r\left|\frac{E_f}{E_i}\right|^2\right) = \frac{2}{\pi k}\frac{|f^{(1D)}_{\beta\alpha}(\hat{\mathbf{k}}_f,\hat{\mathbf{k}}_i)|^2}{L\cos(\theta_i)}$$

$$= \frac{\sigma^{(1D)}_{\beta\alpha}(\hat{\mathbf{k}}_f,\hat{\mathbf{k}}_i)}{L\cos(\theta_i)} = \frac{\mathrm{RCS}^{(1D)}_{\beta\alpha}(\hat{\mathbf{k}}_f,\hat{\mathbf{k}}_i)}{2\pi L\cos(\theta_i)} = \frac{\gamma^{(1D)}_{\beta\alpha}(\hat{\mathbf{k}}_f,\hat{\mathbf{k}}_i)}{2\pi\cos(\theta_i)} \tag{7.15}$$

$$= \mathrm{DSC}^{(1D)}_{\alpha\to\beta}(\hat{\mathbf{k}}_f,\hat{\mathbf{k}}_i) = \cos(\theta_f)\mathrm{BRDF}^{(1D)}_{\alpha\to\beta}(\hat{\mathbf{k}}_f,\hat{\mathbf{k}}_i)$$

for a one-dimensional scatterer.

The quantity Φ_i in Eqs. (7.14) and (7.15) is $A\cos(\theta_i)|E_i|^2/(2\eta_0)$ in watts per square meter, where A is the total illuminated area of the surface for a two-dimensional scatterer and $A = 1 \times L$ for a one-dimensional or gratinglike scatterer.

In these expressions, σ is the *bistatic cross section,* RCS is the bistatic *radar cross section,* γ is the *bistatic scattering coefficient,* DSC is the *differential scattering cross section,* and BRDF is the *bidirectional reflectance distribution function.*

The bistatic cross section σ and the bistatic radar cross section have the dimensions of area and length or width. In practice, one frequently converts these quantities to dimensionless forms by normalizing them to the appropriate power of the radiation wavelength or a characteristic physical dimension of the scattering object.

On the other hand, γ and the bidirectional reflection function are dimensionless. The former is more appropriate for isolated objects, and the latter is more appropriate for distributed surface roughness. The γ's used here are the radar cross section normalized to the illuminated surface area or length, although alternative definitions appear in the literature.

7.3.4. *Total Integrated Scatter*

The total integrated scatter (TIS) or *albedo* are the dimensionless quantities

$$\mathrm{TIS}^{(2D)}_{\alpha}(\hat{\mathbf{k}}_i) = \frac{1}{\Phi_i}\int_{2\pi}\sum_{\beta}\left(\frac{d\Phi}{d\omega_f}\right)^{(2D)}_{\alpha\to\beta} d\omega_f = \frac{1}{A\cos(\theta_i)}\sum_{\beta}\int_{2\pi}|f^{(2D)}_{\beta\alpha}|^2 d\boldsymbol{\omega}_f,$$

$$\mathrm{TIS}^{(1D)}_{\alpha}(\hat{\mathbf{k}}_i) = \frac{1}{\Phi_i}\int_{\pi}\sum_{\beta}\left(\frac{d\Phi}{d\theta_f}\right)^{(1D)}_{\alpha\to\beta} d\theta_f$$

$$= \frac{1}{l\cos(\theta_i)}\frac{2}{\pi k}\sum_{\beta}\int_{\pi}|f^{(1D)}_{\beta\alpha}|^2 d\theta_f, \tag{7.16}$$

which are both unity for a perfect reflector, but less than unity for an imperfect one. One minus the total integrated scatter is the *absorbtivity.* Kirchhoff's law states that in thermal equilibrium the absorbtivity equals the *emissivity:*

$$\varepsilon_{\alpha}^{(\mathrm{nD})}(\hat{\mathbf{k}}_i) = 1 - \mathrm{TIS}_{\alpha}^{(\mathrm{nD})}(\hat{\mathbf{k}}_i). \tag{7.17}$$

The emissivity of a perfectly flat surface of arbitrary material is

$$\varepsilon_{\alpha} = 1 - |r_{\alpha}(\theta_i)|^2 = 1 - R_{\alpha}(\theta_i), \tag{7.18}$$

where r and R are its Fresnel amplitude- and intensity-reflection coefficients, respectively. In the case of a semi-infinite slab of homogeneous material these quantities are related to material parameters through the following well-known expressions:

$$\begin{aligned} r_s(\theta) &= \frac{\mu_r \sin(\theta) - \sqrt{\mu_r\varepsilon_r - \cos^2(\theta)}}{\mu_r \sin(\theta) + \sqrt{\mu_r\varepsilon_r - \cos^2(\theta)}} \longrightarrow -1, \\ r_p(\theta) &= \frac{\varepsilon_r \sin(\theta) - \sqrt{\mu_r\varepsilon_r - \cos^2(\theta)}}{\varepsilon_r \sin(\theta) + \sqrt{\mu_r\varepsilon_r - \cos^2(\theta)}} \longrightarrow +1. \end{aligned} \tag{7.19}$$

The subscript r on the permittivity and the permeability denote values relative to their vacuum values. The limiting values on the right are those of a perfect electric conductor.

7.3.5. *Stokes Matrix*

Polarization states are traditionally described by the *Stokes vector* $[I,Q,U,V]^T$ (the column vector is designated $[s_0, s_1, s_2, s_3]$ in Chap. 4, on Polarization), whereas applications involving radiation transport use the derived form:

$$\begin{bmatrix} I_p \\ I_s \\ U \\ V \end{bmatrix} = \begin{bmatrix} (I+Q)/2 \\ (I-Q)/2 \\ U \\ V \end{bmatrix} = \begin{bmatrix} |E_p|^2 \\ |E_s|^2 \\ 2\,\mathrm{Re}(E_pE_s^*) \\ 2\,\mathrm{Im}(E_pE_s^*) \end{bmatrix}. \tag{7.20}$$

The initial and the final states are connected by the 4×4 Stokes matrix:

$$\begin{bmatrix} I_p \\ I_s \\ U \\ V \end{bmatrix}_f =$$

$$\begin{bmatrix} |f_{pp}|^2 & |f_{ps}|^2 & +\mathrm{Re}(f_{pp}f_{ps}^*) & -\mathrm{Im}(f_{pp}f_{ps}^*) \\ |f_{sp}|^2 & |f_{ss}|^2 & +\mathrm{Re}(f_{ss}f_{sp}^*) & -\mathrm{Im}(f_{sp}f_{ss}^*) \\ +2\,\mathrm{Re}(f_{pp}f_{sp}^*) & +2\,\mathrm{Re}(f_{ps}f_{ss}^*) & +\mathrm{Re}(f_{pp}f_{ss}^* + f_{ps}f_{sp}^*) & -\mathrm{Im}(f_{pp}f_{ss}^*) - f_{ps}f_{sp}^*) \\ +2\,\mathrm{Im}(f_{pp}f_{sp}^*) & +2\,\mathrm{Im}(f_{ps}f_{ss}^*) & +\mathrm{Im}(f_{pp}f_{ss}^* + f_{ps}f_{sp}^*) & +\mathrm{Re}(f_{pp}f_{ss}^* - f_{ps}f_{sp}^*) \end{bmatrix}$$

$$\times \begin{bmatrix} I_p \\ I_s \\ U \\ V \end{bmatrix}_i . \quad (7.21)$$

The discussions in the preceding subsections are, in effect, concerned with only the components I_p and I_s. The full Stokes formalism is useful for exploring the polarimetric (ellipsometric) properties of rough surfaces.

7.3.6. *Diffuse and Specular Scattering*

Under some circumstances the scattered power distributions can be separated into *specular* and *diffuse* components, distinguished by their different angular distributions about the specular direction. In the statistical case subsequently discussed, the analogous separation is into *coherent* and *incoherent* components, distinguished by the coherence properties of their scattered fields.

That is,

$$\lim_{A\to\infty} \frac{1}{\Phi_i}\left(\frac{\mathrm{d}\Phi}{\mathrm{d}\boldsymbol{\omega}_f}\right)^{(2\mathrm{D})}_{\alpha\to\beta} = \left[\frac{1}{\Phi_i}\left(\frac{\mathrm{d}\Phi}{\mathrm{d}\boldsymbol{\omega}_f}\right)^{(2\mathrm{D})}_{\alpha\to\beta}\right]_{\mathrm{Diffuse}} + \rho_{\beta\alpha}^{(2\mathrm{D})}\delta_A(\theta_f - \theta_i)\delta(\varphi_f),$$

$$\lim_{L\to\infty} \frac{1}{\Phi_i}\left(\frac{\mathrm{d}\Phi}{\mathrm{d}\theta_f}\right)^{(1\mathrm{D})}_{\alpha\to\beta} = \left[\frac{1}{\Phi_i}\left(\frac{\mathrm{d}\Phi}{\mathrm{d}\theta_f}\right)^{(1\mathrm{D})}_{\alpha\to\beta}\right]_{\mathrm{Diffuse}} + \rho_{\beta\alpha}^{(1\mathrm{D})}\delta_L(\theta_f - \theta_i). \quad (7.22)$$

The diffuse terms generally have broad distributions about the specular direction, whereas the specular terms, as shown, are sharply peaked about that direction. The ρ's are the effective surface reflectivities, and the δ's are the point-spread functions of the scattering system, which become delta functions in the limit of infinitely broad illumination. In the literature, attention is frequently limited to the diffuse components of the total scattering.

7.4. Exact Results

7.4.1. *Introduction*

Rayleigh derived analytic expressions for the scattering of a semicircular ridge on a perfectly conducting plane by using the method of images, Twersky generalized those results to arbitrary elliptical ridges and hemispherical bumps, and Barrick investigated the scattering by elliptical grooves.

These exact results include all effects of multiple scattering and shadowing and are valid for all values of the parameter $x = ka$, where a is the radius of the object. Some of these results are given in the following subsections to illustrate general behaviors and for comparison with the approximate methods discussed in later sections.

7.4.2. *Scattering by a Semicircular Ridge*

The diffuse scattering amplitudes for a semicircular ridge of arbitrary radius a are

$$\begin{aligned} f_{ss}^{(1D)}(\hat{\mathbf{k}}_f, \hat{\mathbf{k}}_i) &= 2\sum_{n=0}^{\infty}(-1)^n \varepsilon_n A_n(ka)\sin(n\phi_1)\sin(n\phi_2), \\ f_{ps}^{(1D)}(\hat{\mathbf{k}}_f, \hat{\mathbf{k}}_i) &= f_{sp}^{(1D)}(\hat{\mathbf{k}}_f, \hat{\mathbf{k}}_i) = 0, \\ f_{pp}^{(1D)}(\hat{\mathbf{k}}_f, \hat{\mathbf{k}}_i) &= 2\sum_{n=0}^{\infty}(-1)^n \varepsilon_n B_n(ka)\cos(n\phi_1)\cos(n\phi_2), \end{aligned} \tag{7.23}$$

where

$$A_n(x) = -\frac{J_n(x)}{H_n^{(1)}(x)}, \quad B_n(x) = -\frac{J_n'(x)}{H_n'^{(1)}(x)}. \tag{7.24}$$

Here ε_n is the Neumann symbol (1 for $n = 0$ and 2 otherwise), J_n and H_n are standard Bessel and Hankel functions, and the primes denote differentiation with respect to the argument $x = ka$.

The cylinder lies along the Y axis, and the incident and the scattered directions are confined to the X–Z plane. The angles ϕ are the incident and the scattering angles measured relative to the X axis; that is, $\phi_1 = (\pi/2) - \theta_i$ and $\phi_2 = (\pi/2) + \theta_f$.

7.4.3. *Low-Frequency Limit,* $ka \ll 1$

This is also known as the *Rayleigh,* long-wavelength, or dipole limit. In this limit the diffuse scattering amplitudes for circular and ellipsoidal ridges and grooves have the following common form:

$$f_{ss}^{(1D)}(\hat{\mathbf{k}}_f, \hat{\mathbf{k}}_i) = \mp i \frac{\pi}{2} k^2 HW[K \cos(\theta_f)\cos(\theta_i)],$$
$$f_{ps}^{(1D)}(\hat{\mathbf{k}}_f, \hat{\mathbf{k}}_i) = f_{sp}^{(1D)}(\hat{\mathbf{k}}_f, \hat{\mathbf{k}}_i) = 0,$$
$$f_{pp}^{(1D)}(\hat{\mathbf{k}}_f, \hat{\mathbf{k}}_i) = \mp i \frac{\pi}{2} k^2 HW[1 - K \sin(\theta_f)\sin(\theta_i)]. \tag{7.25}$$

The minus–plus sign refers to ridges–grooves, H is the magnitude of the height or the depth of the surface defect, and W is its half-width in the surface plane. The constants K are given in Table 7.1.

Table 7.1. Values of the constant K in Eqs. (7.25)[a]

	Ridge	Groove
Narrow		
$W \ll H$	$1 + H/W \gg 1$	$0.3455 \cdots W/H$
Circular		
$W = H = a$	$1 + H/W = 2$	$0.3493 \ldots$
Shallow		
$W \gg H$	$1 + H/W = 1+$	$1-$

[a] The two decimal numbers are $28\,\zeta(3)/\pi^2$ and $(2\pi^2 - 8)/(3\,\pi^2 + 4)$.

The fact that $K > 1$ for ridges means that the pp scattering vanishes at the scattering angle where

$$\sin(\theta_f) = \frac{1}{K \sin(\theta_i)} \leq 1. \tag{7.26}$$

This null occurs because its direction lies, in effect, along the axis of the combined electric- and magnetic-dipole excitation of the ridge. In the limit of a narrow ridge, $K \gg 1$, the null lies at 0°, whereas in the limit of a shallow ridge, $K \to 1$, and the null moves to 90°. On the other hand, no nulls appear for grooves as they have $K < 1$.

This contrasting behavior offers a way of distinguishing ridges from grooves and determining their eccentricities. Although this is shown here for only elliptical shapes in perfectly conducting surfaces, minima in pp-scattering patterns have proven to be a practical method for distinguishing ridges from grooves under more realistic conditions.

7.4.4. Scattering by Hemispherical Bumps

In this case the low-frequency ($ka \ll 1$) forms of the diffuse scattering amplitudes are

$$f_{ss}^{(2D)}(\hat{\mathbf{k}}_f, \hat{\mathbf{k}}_i) = -a(ka)^2[\cos(\theta_f)\cos(\theta_i)\cos(\varphi_f)],$$
$$f_{ps}^{(2D)}(\hat{\mathbf{k}}_f, \hat{\mathbf{k}}_i) = -a(ka)^2[\cos(\theta_i)\sin(\varphi_f)],$$
$$f_{sp}^{(2D)}(\hat{\mathbf{k}}_f, \hat{\mathbf{k}}) = +a(ka)^2[\cos(\theta_f)\sin(\varphi_f)],$$
$$f_{pp}^{(2D)}(\hat{\mathbf{k}}_f, \hat{\mathbf{k}}_i) = +a(ka)^2[2\sin(\theta_f)\sin(\theta_i) - \cos(\varphi_f)], \tag{7.27}$$

where a is the radius. Note that, in the plane of incidence $\varphi_f = 0$, these results are essentially the same as those of a circular ridge; that is, Eqs. (7.25), with $H = W = a$ and $K = 2$.

Twersky gives expressions for arbitrary values of ka by using the standard Mie solution for an isolated sphere combined with its image in the surface plane. The results are rather complicated, however, and are not repeated here.

7.4.5. Illustrations

Figures 7.2 and 7.3 illustrate the different behaviors of the scattering cross sections in the Rayleigh or the long-wavelength limit.

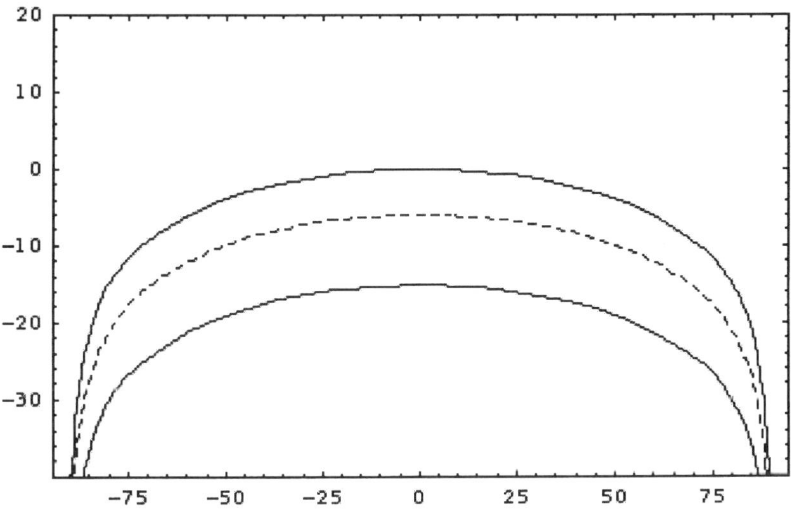

Fig. 7.2. Angular dependence of the s to s diffuse-scattering cross section of a small semicircular ridge (top) and a small semicircular groove (bottom). The dotted curve is the common ridge–groove pattern derived in the small-perturbation approximation discussed in Section 7.5. The quantity shown in 10 log 10[$K\cos(\theta_f)\cos(\theta_i)]^2$ versus the scattering angle θ_f, in degrees, for an angle of incidence θ_i of 60°.

In contrast, Fig. 7.4 shows the square magnitudes of the diffuse-scattering amplitudes calculated from Eqs. (7.23) for a large value rather than a small value of ka.

Two conspicuous features of the patterns in Fig. 7.4 are the spike in the

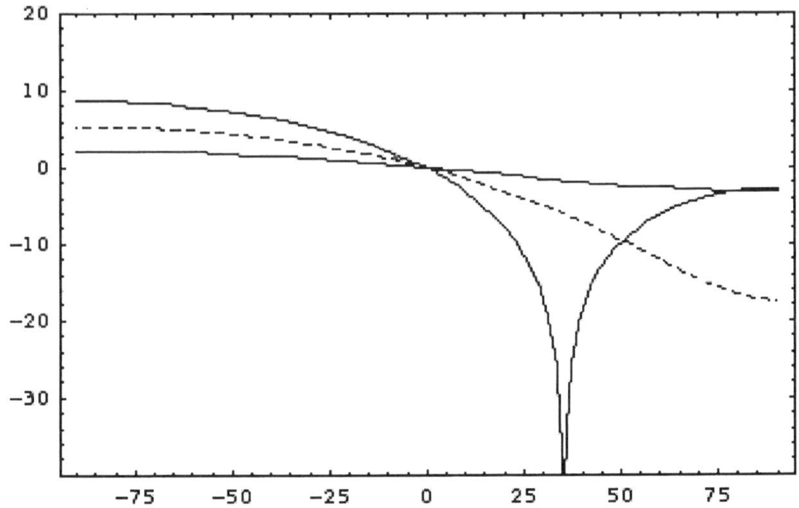

Fig. 7.3. Same as Fig. 7.2 but for p to p scattering. The ridge scattering shows a null at a scattering angle of ~35°. The quantity plotted is $10 \log_{10} [1 - K \sin(\theta_f) \sin(\theta_i)]^2$.

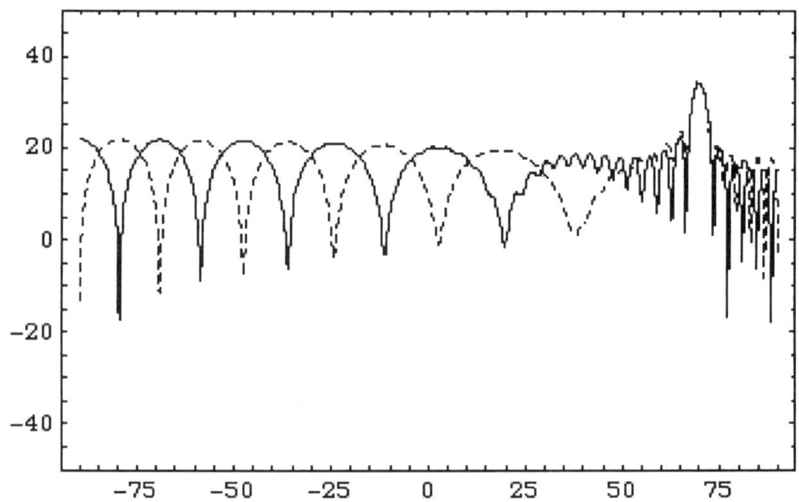

Fig. 7.4. 10 $\log_{10}$ of the square magnitude of the exact scattering amplitudes for ka = 50 and $\theta_i = 60°$. The solid curve is $p \rightarrow p$ scattering and the dashed curve is $s \rightarrow s$ scattering. The horizontal scale is the scattering angle θ_f, in degrees.

specularly reflected direction and the mutually "out-of-phase" interference patterns at lower scattering angles. The spike is, in effect, the diffraction pattern of the slit aperture occupied by the ridge, and the broad fringes come from the interference between single- and double-bounce scattering from the ridge as sketched in Fig. 7.5.

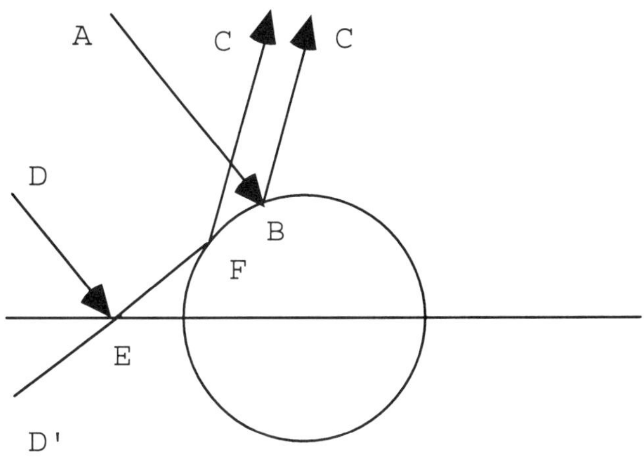

Fig. 7.5. Origin of single- and double-bounce interference effects. Ray *ABC* is single bounce and ray *DEFC* is double bounce. The double-bounce path involves an additional reflection coefficient of +1 for *p* scattering and –1 for *s* scattering, which lead to the out-of-phase patterns for the two polarizations. (Ray *D′EFC* is the double-bounce ray in the image representation.)

7.5. The Small-Perturbation Method

7.5.1. Introduction

The exact results in the preceding section offer insight into the surface-scattering process, but are severely limited representations of real surfaces. In this section the first of two approximate but more general methods is described—the small-perturbation method. This is valid providing the height fluctuations and the surface slopes and curvatures are small; that is, it is inherently a long-wavelength approximation.

The small-perturbation method is closely related to the Rayleigh method, which he introduced in his study of diffraction gratings, and was later generalized to arbitrary surface roughness by Rice. The Rayleigh and the small-perturbation methods are identical to lowest order but the former fails in second order. The common first-order analysis is known as the Rayleigh–Rice method.

7.5.2. Diffuse Scatter

The small-perturbation method forms of the diffuse-scattering amplitudes are

$$f^{(2D)}_{\beta\alpha}(\hat{\mathbf{k}}_f, \hat{\mathbf{k}}_i) = \frac{1}{\pi} k^2 \cos(\theta_s)\cos(\theta_i)^{\mathrm{SPM}} A^{(2D)}_{\beta\alpha}(\hat{\mathbf{k}}_f, \hat{\mathbf{k}}_i)$$

$$\times \iint_A \mathrm{d}x\mathrm{d}y \exp[i2\,\pi(f_x x + f_y y] Z(x, y),$$

$$f_{\beta\alpha}^{(1D)}(\hat{\mathbf{k}}_f, \hat{\mathbf{k}}_i) = k^2 \cos(\theta_s)\cos(\theta_i)\,{}^{\mathrm{SPM}}A_{\beta\alpha}^{(1D)}(\hat{\mathbf{k}}_f, \hat{\mathbf{k}}_i) \times \int_L \mathrm{d}x \exp(i2\pi f_x x) Z(x), \tag{7.28}$$

where the angular-polarization materials factors, ${}^{\mathrm{SPM}}A$, are

$$\begin{aligned}
&{}^{\mathrm{SPM}}A_{ss}^{(2D)}(\hat{\mathbf{k}}_f, \hat{\mathbf{k}}_i) = \\
&+\frac{(\mu_r-1)\left[\sqrt{\varepsilon_r\mu_r-\sin^2(\theta_f)}\sqrt{\varepsilon_r\mu_r-\sin^2(\theta_i)}\cos(\varphi_f)-\mu_r\sin(\theta_f)\sin(\theta_i)\right]-\mu_r^2(\varepsilon_r-1)\cos(\varphi_f)}{\left[\mu_r\cos(\theta_f)+\sqrt{\varepsilon_r\mu_r-\sin^2(\theta_f)}\right]\left[\mu_r\cos(\theta_i)+\sqrt{\varepsilon_r\mu_r-\sin^2(\theta_i)}\right]}, \\
&{}^{\mathrm{SPM}}A_{ps}^{(2D)}(\hat{\mathbf{k}}_f, \hat{\mathbf{k}}_i) = \\
&-\frac{\mu_r(\varepsilon_r-1)\sqrt{\varepsilon_r\mu_r-\sin^2(\theta_s)}-\varepsilon_r(\mu_r-1)\sqrt{\varepsilon_r\mu_r-\sin^2(\theta_i)}}{\left[\varepsilon_r\cos(\theta_f)+\sqrt{\varepsilon_r\mu_r-\sin^2(\theta_f)}\right]\left[\mu_r\cos(\theta_i)+\sqrt{\varepsilon_r\mu_r-\sin^2(\theta_i)}\right]}\sin(\varphi_s), \\
&{}^{\mathrm{SPM}}A_{sp}^{(2D)}(\hat{\mathbf{k}}_f, \hat{\mathbf{k}}_i) = \\
&-\frac{\varepsilon_r(\mu_r-1)\sqrt{\varepsilon_r\mu_r-\sin^2(\theta_f)}-\mu_r(\varepsilon_r-1)\sqrt{\varepsilon_r\mu_r-\sin^2(\theta_i)}}{\left[\varepsilon_r\cos(\theta_f)+\sqrt{\varepsilon_r\mu_r-\sin^2(\theta_f)}\right]\left[\mu_r\cos(\theta_i)+\sqrt{\varepsilon_r\mu_r-\sin^2(\theta_i)}\right]}\sin(\varphi_s), \\
&{}^{\mathrm{SPM}}A_{pp}^{(2D)}(\hat{\mathbf{k}}_f, \hat{\mathbf{k}}_i) = \\
&-\frac{(\varepsilon_r-1)\left[\sqrt{\varepsilon_r\mu_r-\sin^2(\theta_f)}\sqrt{\varepsilon_r\mu_r-\sin^2(\theta_i)}\cos(\varphi_f)-\varepsilon_r\sin(\theta_f)\sin(\theta_i)\right]-\varepsilon_r^2(\mu_r-1)\cos(\varphi_f)}{\left[\varepsilon_r\cos(\theta_f)+\sqrt{\varepsilon_r\mu_r-\sin^2(\theta_f)}\right]\left[\varepsilon_r\cos(\theta_i)+\sqrt{\varepsilon_r\mu_r-\sin^2(\theta_i)}\right]}.
\end{aligned} \tag{7.29}$$

Equations (7.28) and (7.29) satisfy both reciprocity and duality and, as expected, vanish for a vacuum, $\varepsilon_r = \mu_r = 1$. The corresponding one-dimensional forms are the two-dimensional ones evaluated in the plane of incidence $\varphi_f = 0$.

In the limit of a nonmagnetic perfect reflector Eqs. (7.29) reduce to simpler forms:

$$\begin{aligned}
&{}^{\mathrm{SPM}}A_{ss}^{(2D)}(\hat{\mathbf{k}}_f, \hat{\mathbf{k}}_i) = -\cos(\varphi_f), \quad {}^{\mathrm{SPM}}A_{ps}^{(2D)}(\hat{\mathbf{k}}_f, \hat{\mathbf{k}}_i) = -\frac{\sin(\varphi_f)}{\cos(\theta_f)}, \\
&{}^{\mathrm{SPM}}A_{pp}^{(2D)}(\hat{\mathbf{k}}_f, \hat{\mathbf{k}}_i) = -\frac{\cos(\varphi_f)-\sin(\theta_f)\sin(\theta_i)}{\cos(\theta_f)\cos(\theta_i)}, \quad {}^{\mathrm{SPM}}A_{sp}^{(2D)}(\hat{\mathbf{k}}_f, \hat{\mathbf{k}}_i) = +\frac{\sin(\varphi_f)}{\cos(\theta_i)}.
\end{aligned} \tag{7.30}$$

7.5.3. Small-Perturbation Method Value of the Bidirectional Reflection Distribution Function

The small-perturbation expressions for the diffuse component of the bidirectional reflection distribution function are

$$\mathrm{BRDF}^{(2D)}_{\alpha\to\beta}(\hat{\mathbf{k}}_f, \hat{\mathbf{k}}_i) = \frac{16\pi^2}{\lambda^4}\cos(\theta_f)\cos(\theta_i)Q^{(2D)}_{\beta\alpha}(\hat{\mathbf{k}}_f, \hat{\mathbf{k}}_i)W^{(2D)}(f_x, f_y),$$

$$\mathrm{BRDF}^{(1D)}_{\alpha\to\beta}(\hat{\mathbf{k}}_f, \hat{\mathbf{k}}_i) = \frac{16\pi^2}{\lambda^3}\cos(\theta_f)\cos(\theta_i)Q^{(1D)}_{\beta\alpha}(\hat{\mathbf{k}}_f, \hat{\mathbf{k}}_i)\cdot W^{(1D)}(f_x), \tag{7.31}$$

where

$$Q^{(nD)}_{\alpha\to\beta}(\hat{\mathbf{k}}_f, \hat{\mathbf{k}}_i) = |{}^{SPM}A^{nD}_{\beta\alpha}(\hat{\mathbf{k}}_f, \hat{\mathbf{k}}_i)|^2. \tag{7.32}$$

The W factors appearing here are the one- and the two-dimensional forms of the deterministic power spectral densities of the surface height distribution:

$$W^{(2D)}(f_x, f_y) = \frac{1}{A}\left|\iint_A \mathrm{d}x\mathrm{d}y \exp[i2\pi(f_x x + f_y y)]Z(x, y)\right|^2, \quad -\infty \le f_x, f_y \le +\infty,$$

$$W^{(1D)}(f_x) = \frac{1}{L}\left|\int_L \mathrm{d}x \exp[i2\pi f_x x]Z(x)\right|^2, \quad -\infty \le f_x \le +\infty, \tag{7.33}$$

where $Z(x, y)$ and $Z(x)$ denote the zero-mean surface height.

The one-dimensional power spectrum used here is *two sided;* that is, it includes negative as well as positive values of the surface frequency, even though the spectrum is independent of the sign of the frequency. In contrast, the metrology literature frequently uses a *one-sided* form, which is twice the result shown here, but for positive frequencies only.

Because of the wide application of Eqs. (7.31) and its statistical counterpart, which involves the ensemble average power spectrum, this simple first-order result has been called the *Golden Rule* of scattering.

7.5.4. Special Forms of the Q *Factors*

The Q values appearing in the bidirectional reflection distribution function are closely related to the Fresnel intensity-reflection coefficients and can be expressed in terms of them in two cases. First, in the forward-scattering direction,

$$Q^{(2D)}_{\alpha\to\beta}(\hat{\mathbf{k}}_i, \hat{\mathbf{k}}_j) = R_\alpha(\theta_i)\delta_{\alpha,\beta}, \tag{7.34}$$

and second, for $s \to s$ scattering at arbitrary angles of incidence and scattering,

$$Q^{(2D)}_{s\to s} = \sqrt{R_s(\theta_s)R_s(\theta_i)}\cos^2(\varphi_f). \tag{7.35}$$

These relations have some practical value as they do not require knowledge of the values of surface permittivity and permeability.

7.5.5. *Total Integrated Scatter of the Diffuse Component*

The total integrated scatter of the diffuse-scattering component can be written in the form

$$\mathrm{TIS}^{(\mathrm{nD})}_{\alpha\to\beta} = \left[4\pi\,\frac{\sigma^{(\mathrm{nD})}_{\mathrm{TIS}}}{\lambda}\cos(\theta_i)\right]^2 R(\theta_i),$$

where

$$[\sigma^{(2D)}_{\mathrm{TIS}}]^2_{\alpha\to\beta} = \iint_{2\pi} \mathrm{d}f_x \mathrm{d}f_y \left[\frac{\cos(\theta_f)}{\cos(\theta_i)}\,\frac{Q^{(2D)}_{\beta\alpha}(\hat{\mathbf{k}}_f, \hat{\mathbf{k}}_i)}{R(\theta_i)}\right] W^{(2D)}(f_x, f_y),$$

$$[\sigma^{(1D)}_{\mathrm{TIS}}]^2_{\alpha\to\beta} = \int_{\pi} \mathrm{d}f_x \left[\frac{\cos(\theta_f)}{\cos(\theta_i)}\,\frac{Q^{(1D)}_{\beta\alpha}(\hat{\mathbf{k}}_f, \hat{\mathbf{k}}_i)}{R(\theta_i)}\right] W^{(1D)}(f_x). \tag{7.36}$$

These small-perturbation method expressions for the mean-square surface roughness can be compared with the corresponding *intrinsic* values obtained if the square brackets in the integrands are replaced with unity and the ranges of integration are extended to cover the whole spatial-frequency plane:

$$[\sigma^{(2D)}_z]^2_{\mathrm{int}} = \int_{-\infty}^{+\infty}\int_{-\infty}^{+\infty} \mathrm{d}f_x \mathrm{d}f_y\, W^{(2D)}(f_x, f_y) = \lim_{A\to\infty}\frac{1}{A}\iint \mathrm{d}x\mathrm{d}y\, Z^2(x, y),$$

$$[\sigma^{(1D)}_z]^2_{\mathrm{int}} = \int_{-\infty}^{+\infty} \mathrm{d}f_x\, W^{(1D)}(f_x) = \lim_{L\to\infty}\frac{1}{L}\int \mathrm{d}x\, Z^2(x). \tag{7.37}$$

The intrinsic and the total integrated scatter values of the mean-square roughnesses differ because the square brackets in the integrands in Eqs. (7.36) act as a nonunit frequency filter factor and the hemispheric scattering geometry imposes finite limits in the range of integration. When the total integrated scatter values are determined from experiment, the spatial bandwidth is further reduced at scattering angles near glancing and about the specular direction. Nevertheless, the first line in Eqs. (7.36) is frequently used to estimate the magnitude of the roughnesses of surfaces from measurements of the diffuse-scattering intensity.

7.5.6. *Specular Scattering*

The preceding results refer to the diffuse component of small-perturbation method scattering. The corresponding expressions for the specularly scattered component are available in the literature for arbitrary surface materials, but are not given here. In the case of perfectly conducting material, the effective reflection coefficients can be obtained using the conservation of energy and appear in the form of 1 minus the expressions for the total integrated scatter given in Eqs. (7.36).

7.5.7. *Comparison of the Small-Perturbation Method and Exact Results*

The scattering amplitudes obtained by the small-perturbation method can be compared with the exact results for elliptical ridges, grooves, and hemispherical bumps given in Subsections 7.4.3 and 7.4.4.

The power spectra of such deterministic shapes are

$$\iint_A \exp[i2\pi(f_x x + f_y y)]Z(x, y)\,\mathrm{d}x\mathrm{d}y \rightarrow \iint_A Z(x, y)\,\mathrm{d}x\mathrm{d}y = \text{volume of defect},$$

$$\int_L \exp[i2\pi f_x x]Z(x)\,\mathrm{d}x \rightarrow \int_L Z(x)\,\mathrm{d}x = \text{Area of defect.} \tag{7.38}$$

where Z is the defect parameter. The area of an elliptical defect is $\mp\pi HW/2$, in which case small-perturbation method scattering amplitudes have the same form as the exact results, Eqs. (7.25), except that $K = 1$ appears rather than the exact value given in Table 7.1. (Irrelevant factors of modulo unity are ignored.) The small-perturbation method predictions have been included in Figs. 7.2 and 7.3 as dotted curves.

The small-perturbation method does not give the correct K values for elliptical ridges and grooves as it does not correctly include the effects of the infinite slopes and curvatures at their edges. However, it does give the correct values in the limit of very shallow shapes, where those edge contributions become relatively unimportant. Similar but larger deviations also appear for the hemispherical bump.

7.6. Physical-Optics Approximation

7.6.1. Introduction

The *physical-optics* approximation is also known as the tangent-plane or Huygens wave approximation. It is valid when the surface roughness has a large curvature, but places no direct restrictions on the its magnitude or slope. In this sense it is a high-frequency approximation.

Results of the physical-optics approximation are most simply expressed for perfect reflectors, although versions have been developed for surfaces with a short but nonvanishing skin depth (impedance boundary condition).

7.6.2. General Expressions

The physical-optics expressions of the total (diffuse plus specular) scattering amplitudes are

$$
\begin{aligned}
f^{(2D)}_{\beta\alpha}(\hat{\mathbf{k}}_f, \hat{\mathbf{k}}_i) &= i\,\frac{k}{2\pi}\int_S \mathrm{d}\mathbf{S}\; G^{(2D)}_{\beta\alpha}(\mathbf{r}_s)\exp[-i(\mathbf{k}_f-\mathbf{k}_i)\mathbf{r}_s] \\
&= i\,\frac{k}{2\pi}\iint_A \frac{\mathrm{d}x\mathrm{d}y}{n_z}\, G^{(2D)}_{\beta\alpha}(\mathbf{r}_s)\exp[-i2\pi(f_x x + f_y y)] \\
&\quad \times \exp[-i2\pi k_z Z(x, y)], \\
f^{(1D)}_{\beta\alpha}(\hat{\mathbf{k}}_f, \hat{\mathbf{k}}_i) &= \frac{k}{2}\int_C \mathrm{d}C G^{(1D)}_{\beta\alpha}(\mathbf{r}_s)\exp[-i(\mathbf{k}_f-\mathbf{k}_i)\mathbf{r}_s] \\
&= \frac{k}{2}\int_L \frac{\mathrm{d}x}{n_z}\, G^{(1D)}_{\beta\alpha}(\mathbf{r}_s)\exp(-i2\pi f_x x)\exp[-i2\pi k_z Z(x)]. \qquad (7.39)
\end{aligned}
$$

The integrals are taken over those portions of the shape (S) or contour (C) of the scattering object that are directly illuminated by the incident beam and are taken to be zero within the region of the geometric-optics shadow. The first forms of each preceding expression apply to isolated defects and the second to surfaces with roughness distributed over the surface area or line.

The angular-materials factor G is, for a perfect reflector,

$$G^{(2D)}_{\beta\alpha}(\mathbf{r}_s) = (\hat{\mathbf{n}}\hat{\mathbf{e}}_{i\alpha})(\hat{\mathbf{e}}_{f\beta}\hat{\mathbf{k}}_i) - (\hat{\mathbf{n}}\hat{\mathbf{k}}_i)(\hat{\mathbf{e}}_{f\beta}\hat{\mathbf{e}}_{i\alpha}), \qquad (7.40)$$

where $\hat{\mathbf{n}}$ is the upward-pointing unit surface normal:

$$\hat{\mathbf{n}} = \begin{pmatrix} n_x \\ n_y \\ n_z \end{pmatrix} = \frac{1}{\sqrt{1+(\partial Z/\partial x)^2+(\partial Z/\partial x)^2}}\begin{pmatrix} -\partial Z/\mathrm{d}x \\ -\partial Z/\mathrm{d}y \\ 1 \end{pmatrix} = n_z\begin{pmatrix} -\partial Z/\mathrm{d}x \\ -\partial Z/\mathrm{d}y \\ 1 \end{pmatrix}. \qquad (7.41)$$

If the roughness is distributed over the surface, the derivative terms can be evaluated by partial integration and neglecting the resulting "edge" terms. The scattering amplitudes then appear in the familiar form:

$$f_{\beta\alpha}^{(2D)}(\hat{\mathbf{k}}_f, \hat{\mathbf{k}}_i) = i\frac{k}{2\pi}\,{}^{PO}A_{\beta\alpha}^{(2D)}(\hat{\mathbf{k}}_f, \hat{\mathbf{k}}_i)\iint_A dxdy\,\exp[-i2\pi(f_x x + f_y y)] \times \exp[-i2\pi k_z Z(x, y)],$$

$$f_{\beta\alpha}^{(1D)}(\hat{\mathbf{k}}_f, \hat{\mathbf{k}}_i) = \frac{k}{2}\,{}^{PO}A_{\beta\alpha}^{(1D)}(\hat{\mathbf{k}}_f, \hat{\mathbf{k}}_i)\iint_L dx\,\exp[-i2\pi f_x x] \times \exp[-i2\pi k_z Z(x)], \tag{7.42}$$

where the physical-optics forms of the angular factors are

$${}^{PO}A_{ss}^{(2D)}(\hat{\mathbf{k}}_f, \hat{\mathbf{k}}_i) = -A_{pp}^{(2D)}(\hat{\mathbf{k}}_f, \hat{\mathbf{k}}_i) = \frac{[1+\cos(\theta_f)\cos(\theta_i)]\cos(\varphi_f) - \sin(\theta_f)\sin(\theta_i)}{\cos(\theta_f)+\cos(\theta_i)},$$

$${}^{PO}A_{ps}^{(2D)}(\hat{\mathbf{k}}_f, \hat{\mathbf{k}}_i) = +A_{sp}^{(2D)}(\hat{\mathbf{k}}_f, \hat{\mathbf{k}}_i) = \sin(\varphi_f). \tag{7.43}$$

These vector expressions are similar to the analogous scalar results discussed in the well-known book by Beckmann and Spizzichino (see Further Reading section). The fact that their magnitudes are independent of polarization mimics geometrical optics.

The physical-optics approximation violates both reciprocity and the conservation of energy, as it omits the effects of roughness-induced multiple scattering and shadowing. Despite these apparent deficiencies, the physical-optics approximation is immensely useful and surprisingly accurate under many circumstances.

7.6.3. Phase-Screen and Fourier Approximations

The dependencies of the scattering amplitudes on the surface roughness in Eqs. (7.42) are simpler if the k_z factor in the exponent is approximated. In particular,

$$\begin{aligned} k_z = 2\pi f_z &= k[\cos(\theta_f)+\cos(\theta_i)] \quad \text{(full physical optics)} \\ &\approx k[2\cos(\theta_i)] \quad \text{(phase-screen approximation)} \\ &\approx k(2) \quad \text{(Fourier approximation).} \end{aligned} \tag{7.44}$$

The phase-screen approximation is a paraxial approximation, which leads to simplified scaling properties of the scattering distribution. The Fourier approximation eliminates all geometrical dependencies from k_z, which permits the spatial integral involving the surface roughness to be written as a simple Fourier integral in which all geometrical and wavelength dependencies are contained in the magnitudes of the spatial frequencies f_x and f_y. This is similar to "shift invariance" in linear-system theory, although the connection is only formal. As shown in the following subsection, this same degree of invariance is achieved in the smooth-surface limit without such an approximation.

7.6.4. Comparison with Exact and Small-Perturbation Results

Equations (7.42) can be used to evaluate the diffuse scattering amplitudes for an isolated cylindrical ridge by use of image techniques, and results can be compared with the exact results in Eq. (7.26). The square magnitudes of the scattering amplitudes are found to be essentially identical for $ka \geq 10$, but the physical-optics results fail increasingly for smaller values of ka.

To compare the physical-optics results with the small-perturbation method results in Eqs. (7.31)–(7.33), the physical-optics expressions must be evaluated in the smooth-surface limit and their diffuse component isolated. This is done by replacing the second exponential in Eqs. (7.39) with the first three terms in its power-series expansion and by dropping the first and the third terms, which represent the specular-scattering components.

The second term, which involves the first power of the surface height, leads to an expression for the diffuse-scattering components of the scattering amplitudes that has the same form as that given by the small-perturbation method—an angular factor times the power spectral density of $Z(x, y)$. The angular factors of the physical-optics and the small-perturbation method calculations are also identical in the specular direction, although they differ, sometimes significantly, in other directions. This illustrates the general fact that the physical-optics approximation is essentially a paraxial approximation.

7.7. Statistical Methods

7.7.1. Introduction

Remarks to this point have been concerned with deterministic surfaces; that is, surface roughnesses described by an explicit function $Z(x, y)$. Statistical methods treat $Z(x, y)$ as a random variable and use deterministic calculations as a stepping stone to express the ensemble-average scattering properties of interest in terms of various statistics of the surface-height fluctuations $Z(x, y)$.

We obtain the ensemble average of a quantity Q, denoted by $\langle Q \rangle$, by weighting $Q(Z)$ by an appropriate distribution function of the random variable

Z and integrating over Z. The values of $\langle \text{BRDF} \rangle$ obtained in this way for different scattering theories and different distribution functions are discussed in the following subsections.

7.7.2. Distribution Functions

Because observable effects depend on products of scattering amplitudes, the most general distribution function required in the present context is the bivariate distribution, $p(Z_1, Z_2)$, where Z_1 and Z_2 are the surface heights at two arbitrary points on the surface. The corresponding univariate distribution $p(Z_1)$ is obtained from $p(Z_1, Z_2)$ by integration over Z_2.

The literature usually considers only Gaussian forms for p, but here we consider the more general νth-order gamma forms as well, which offer a means of exploring non-Gaussian effects in scattering. The Gaussian forms are the special case of the gamma forms obtained by taking the limit $\nu \rightarrow$ infinity.

The Gaussian bivariate and univariate distribution functions are

$$p_{\text{Gauss}}(Z_1, Z_2) = \frac{1}{2\pi\sqrt{\sigma^4 - C^2}} \exp\left[- \frac{Z_1^2 - 2CZ_1Z_2/\sigma^2 + Z_2^2}{2(\sigma^4 - C^2)} \right],$$

$$p(Z_1)_{\text{Gauss}} = \int dZ_2\, p_{\text{Gauss}}(Z_1, Z_2) = \frac{1}{\sqrt{2\pi}\sigma} \exp\left(- \frac{Z_1^2}{2\sigma^2} \right). \quad (7.45)$$

The corresponding νth-order gamma distribution functions are

$$p(Z_1, Z_2; \nu) = \frac{1}{\Gamma(\nu)} \frac{\sigma^2}{\sigma^2 - C} \left(\frac{Z_1Z_2}{C} \right)^{\frac{\nu-1}{2}} \exp\left(- \frac{Z_1 + Z_2}{\sigma^2 - C} \right)$$
$$\times I_{\nu-1}\left(\frac{2\sqrt{Z_1Z_2C}}{\sigma^2 - C} \right),$$

$$p(Z_1; \nu) = \int dZ_2\, p(Z_1, Z_2; \nu) = \frac{1}{\Gamma(\nu)} \left(\frac{Z_1}{\sigma} \right)^{\nu-1} \exp\left(- \frac{Z_1}{\sigma} \right), \quad (7.46)$$

where $0 < \nu < \infty$. The quantities C and σ appearing in Eqs. (7.45) and (7.46) are roughness statistics that are discussed later, and $I_{\nu-1}$ is a modified Bessel function.

7.7.3. Isolated Scatterers

The scattering amplitudes of a set of isolated scatterers randomly distributed on a plane is the sum of the individual patterns plus their interferences. If the distribution is sufficiently sparse, multiple-scattering effects can be neglected, and the average bidirectional reflection distribution function is the sum of the individual bidirectional reflection distribution functions. In the case of identical scatterers,

$$\begin{aligned}\langle \mathrm{BRDF}^{(2D)}_{\alpha\to\beta}(\hat{\mathbf{k}}_f, \hat{\mathbf{k}}_i)\rangle &= N_A A \mathrm{BRDF}^{(2D)}_{\alpha\to\beta}(\hat{\mathbf{k}}_f, \hat{\mathbf{k}}_i),\\ \langle \mathrm{BRDF}^{(1D)}_{\alpha\to\beta}(\hat{\mathbf{k}}_f, \hat{\mathbf{k}}_i)\rangle &= N_L L \mathrm{BRDF}^{(1D)}_{\alpha\to\beta}(\hat{\mathbf{k}}_f, \hat{\mathbf{k}}_i),\end{aligned} \tag{7.47}$$

where N_A is the average number of scatterers per unit area and N_L is the average number per unit length. This simple result is valid for any random distribution of scatterers providing they are sufficiently sparse, that is, that the fractional surface area that they occupy is much less than one.

7.7.4. *Small-Perturbation Method*

In this case the Golden Rule, Eqs. (7.31), shows that $\langle\mathrm{BRDF}\rangle$ is proportional to the ensemble-average power spectral density of the surface roughness W, which is denoted by the new symbol $S = \langle W\rangle$. This result is independent of the form of the distribution functions of the surface height.

7.7.5. *Physical-Optics Approximation*

The physical-optics expressions for $\langle\mathrm{BRDF}\rangle$ can be written as

$$\begin{aligned}\langle \mathrm{BRDF}^{(2D)}_{\alpha\to\beta}(\hat{\mathbf{k}}_f, \hat{\mathbf{k}}_i)\rangle &= \left(\frac{k}{2\pi}\right)^2 \frac{|{}^{PO}A^{(2D)}_{\beta\alpha}(\hat{\mathbf{k}}_f, \hat{\mathbf{k}}_i)|^2}{\cos(\theta_f)\cos(\theta_i)} \int_A d\boldsymbol{\tau}\, \exp[-i2\pi(f_x\tau_x + f_y\tau_y)]\\ &\quad\times \langle \exp[-i2\pi f_z Z(x, y)]\rangle_{\boldsymbol{\tau}}\\ &\to \left(\frac{k}{2\pi}\right)^2 \frac{|{}^{PO}A^{(2D)}_{\beta\alpha}(\hat{\mathbf{k}}_f, \hat{\mathbf{k}}_i)|^2}{\cos(\theta_f)\cos(\theta_i)}\, 2\pi \int_0^\infty \tau d\tau J_0(2\pi f\tau)\\ &\quad\times \langle \exp[-i2\pi f_z Z(x, y)]\rangle_{\boldsymbol{\tau}},\\ \langle \mathrm{BRDF}^{(1D)}_{\alpha\to\beta}(\hat{\mathbf{k}}_f, \hat{\mathbf{k}}_i)\rangle &= \frac{k}{2\pi} \frac{|{}^{PO}A^{(1D)}_{\beta\alpha}(\hat{\mathbf{k}}_f, \hat{\mathbf{k}}_i)|^2}{\cos(\theta_f)\cos(\theta_i)} \int_L d\tau_x \exp(-i2\pi f_x\tau_x)\\ &\quad\times \langle \exp[-i2\pi f_z Z(x)]\rangle_{\boldsymbol{\tau}_x}\\ &= \frac{k}{2\pi} \frac{|{}^{PO}A^{(1D)}_{\beta\alpha}(\hat{\mathbf{k}}_f, \hat{\mathbf{k}}_i)|^2}{\cos(\theta_f)\cos(\theta_i)}\, 2\int_0^\infty d\tau_x \cos[2\pi f_x\tau_x]\\ &\quad\times \langle \exp[-i2\pi f_z Z(x)]\rangle_{\boldsymbol{\tau}_x}.\end{aligned} \tag{7.48}$$

Here, τ is the vector distance between a pair of points in the surface plane and

$f = \sqrt{(f_x^2 + f_y^2)}$ is the magnitude of the spatial-frequency vector. The first line is the general two-dimensional result, and the second is the special form it assumes for surfaces that are statistically isotropic; that is, when the result is stastically independent of the direction of the lag vector.

Each of the expressions in Eqs. (7.48) involves the surface height in the final ensemble-average factor. The value of that average depends on the form of the bivariate distribution function involved. In the general two-dimensional case, the Gaussian and the gamma results are

$$\langle \exp[-i2\pi f_z Z(x, y)] \rangle_{\text{Gauss}} = \exp\left[-\frac{1}{2}(2\pi f_z)^2 D(\boldsymbol{\tau})\right],$$

$$\langle \exp[-i2\pi f_z Z(x, y)] \rangle_{\text{gamma}} = \left[1 + \frac{1}{2\nu}(2\pi f_z)^2 D(\boldsymbol{\tau})\right]^{-\nu}. \qquad (7.49)$$

The surface statistic that appears here is $D(\boldsymbol{\tau})$, the *structure function* of the surface-height fluctuations, $Z(x, y)$, which is defined and discussed in Section 7.8: The one-dimensional results are the same except that $\boldsymbol{\tau}$ is replaced with τ_x.

7.7.6. Coherent and Incoherent Scattering

The expressions for ⟨BRDF⟩ in Eq. (7.48) can be separated into coherent and incoherent components by use of the identity

$$\langle |f|^2 \rangle = |\langle f \rangle|^2_{\text{coherent}} + \langle |f - \langle f \rangle|^2 \rangle_{\text{incoherent}}, \qquad (7.50)$$

where f is the total scattering amplitude in Eqs. (7.42). This separation is analogous to the separation into specular and diffuse scattering in Eq. (7.21). As there, the coherent term is described in terms of an effective surface reflectivity ρ. The physical-optics values for a rough, perfectly conducting surface are

$$\rho_{ss} = \rho_{pp} = \left\langle \left| \exp\left[i4\pi \frac{Z}{\lambda} \cos(\theta_i) \right] \right|^2 \right\rangle, \quad \rho_{ps} = \rho_{sp} = 0. \qquad (7.51)$$

This factor, which represents the reduction of the coherent scattering due to the presence of surface roughness, is called the *Strehl factor* or the *Strehl ratio*. Its magnitude depends on the form of the univariate distribution of surface heights. In particular,

$$\text{Gaussian Strehl factor} = \exp\left\{-\left[4\pi \frac{\sigma}{\lambda} \cos(\theta_i)\right]^2\right\},$$

$$\text{gamma Strehl factor} = \left\{1 + \frac{1}{\nu}\left[4\pi\frac{\sigma}{\lambda}\cos(\theta_i)\right]\right\}^{-\nu}, \tag{7.52}$$

where σ is the intrinsic rms surface roughness.

In the smooth-surface limit the Strehl factors are independent of the form of the height distribution function because only the second moment of the height is involved. On the other hand, when the surface is rough enough to show a dependence on the form of the height distribution, the coherent component may be reduced to unobservability.

The literature frequently generalizes these results for a perfectly conducting material to surfaces with finite conductivities by multiplying the Gaussian results in Eqs. (7.49) and in (7.52) by $R(\theta_i)$.

7.8. Surface Statistics

7.8.1. Introduction

The average scattered power depends on different second-order statistical functions of the surface-height fluctuations—the power spectral density appears in the small-perturbation approximation and the structure function in the physical-optics calculations. Other discussions involve yet a third form, the autocovariance function. These functions are related by various mathematical transformations, as subsequently discussed.

7.8.2. Spectra, Structure, and Covariance Functions

The one- and two-dimensional average power spectral densities of the surface-height fluctuations are

$$S^{(2D)}(\mathbf{f}) = \lim_{A\to\infty} \langle W_A(\mathbf{f})\rangle,$$

$$S^{(1D)}(f_x) = \lim_{L\to\infty} \langle W_L(f_x)\rangle = \int_{-\infty}^{+\infty} S^{(2D)}(\mathbf{f})\mathrm{d}f_y, \tag{7.53}$$

where the W's are given by Eqs. (7.33). Note that the one- and two-dimensional spectra have different dimensions, m^3 and m^4, respectively, and that the one-dimensional spectrum is two sided; that is, $-\infty < f_x < +\infty$.

The one-dimensional spectrum has two interpretations: It is the power spectrum of a profile taken across the grain of a random gratinglike surface, $Z(x)$, and it is also the average spectrum of a one-dimensional profile taken in the x direction across the two-dimensional surface, $Z(x, y)$. On the other hand, it does not represent the plane-of-incidence slice of the two-dimensional scattering pattern that is given by $S^{(2D)}(f_x, 0)m^4$ and not $S^{(1D)}(f_x)m^3$.

The second surface statistic of importance in surface scattering is the *structure function*

$$D(\boldsymbol{\tau}) = \langle [Z(\mathbf{r} + \boldsymbol{\tau}) - Z(\mathbf{r})]^2 \rangle, \tag{7.54}$$

where **r** is the position vector in the x–y plane and $\boldsymbol{\tau}$ is the separation or lag between two surface points. For the structure function to exist, it is necessary only that the surface roughness has statistically stationary first differences, in the wide sense.

The structure function can be expressed in terms of the power spectrum and conversely according to

$$D(\boldsymbol{\tau}) = 2 \int_{-\infty}^{+\infty} [1 - \exp(i2\pi\mathbf{f} \cdot \boldsymbol{\tau})] S^{(2D)}(\mathbf{f}) \mathrm{d}\mathbf{f}.$$

$$S^{(2D)}(\mathbf{f}) = \frac{1}{2(2\pi f)^2} \int_{\infty} \exp(i2\pi\mathbf{f} \cdot \boldsymbol{\tau}) \nabla^2 D(\boldsymbol{\tau}) \mathrm{d}\boldsymbol{\tau}. \tag{7.55}$$

If the surface heights have statistically stationary magnitudes as well as stationary first differences, a third statistical function can be defined—the autocovariance function:

$$C(\boldsymbol{\tau}) = \langle Z(\mathbf{r} + \boldsymbol{\tau}) Z(\mathbf{r}) \rangle, \quad C(0) = \langle Z(\mathbf{r})^2 \rangle = \sigma^2, \tag{7.56}$$

where σ is the intrinsic rms surface roughness. It is related to the surface power spectra according to

$$\sigma^2 = \int_{\infty} S^{(2D)}(\mathbf{f}) \mathrm{d}\mathbf{f} = \int_{-\infty}^{+\infty} S^{(1D)}(f_x) \mathrm{d}f_x. \tag{7.57}$$

The autocovariance function is related to the power spectral density by the *Wiener–Khinchin theorem,*

$$C(\boldsymbol{\tau}) = \int_{\infty} \exp(-i2\pi\mathbf{f} \cdot \boldsymbol{\tau}) S^{(2D)}(\mathbf{f}) \mathrm{d}\mathbf{f},$$

$$S^{(2D)}(\mathbf{f}) = \int_{\infty} \exp(+i2\pi\, \mathbf{f} \cdot \boldsymbol{\tau}) C(\boldsymbol{\tau}) \mathrm{d}\boldsymbol{\tau}, \tag{7.58}$$

and to the corresponding structure function through

$$C(\boldsymbol{\tau}) = C(0) - \frac{1}{2} D(\boldsymbol{\tau}), \quad D(\boldsymbol{\tau}) = 2[C(0) - C(\boldsymbol{\tau})]. \tag{7.59}$$

The one-dimensional forms of Eqs. (7.58) and (7.59) follow by inspection. Again, the existence of the structure function and power spectra require only statistically stationary surface height differences, whereas the existence of the covariance function, and hence a finite rms surface roughness, also requires that the height fluctuations themselves be stationary.

If the two-dimensional surface is statistically isotropic, its spectrum depends on only the magnitude of the spatial frequency vector, $\mathbf{f} = \sqrt{(f_x^2 + f_y^2)}$, but not its direction. In that case the one- and two-dimensional spectra are related by the integral transforms:

$$S^{(1D)}(f_x) = 2\int_{f_x}^{\infty} \frac{f}{\sqrt{f^2 - f_x^2}} S^{(2D)}(f)\mathrm{d}f,$$

$$S^{(2D)}(f) = -\frac{1}{\pi}\int_{f}^{\infty} \frac{1}{\sqrt{f_x^2 - f^2}} \frac{\mathrm{d}}{\mathrm{d}f_x} S^{(1D)}(f_x)\mathrm{d}f_x. \tag{7.60}$$

The first is an Abel or half-integral transform and the second is an inverse-Abel or half-derivative transform, which can be written in a variety of algebraic forms. The second permits the ⟨BRDF⟩ of an isotropically rough surface to be predicted from optical or mechanical profile measurements of its surface topography. For simplicity, only isotropic surfaces are considered hereafter.

7.8.3. Analytic Models

Analytic models of statistical functions are useful for a number of reasons: When these functions are estimated from experimental measurements, models smooth the inherent data fluctuations and add physical information that permits the smoothed data to be extrapolated outside the measurement range. They also condense the description of the roughness into a few finish parameters and simplify the calculation of the average scattering patterns.

The power spectra of many smooth surfaces are fractallike—they tend to a finite value at low frequencies and fall off as an inverse power law at high frequencies. The simplest analytic model showing this behavior is the *ABC* model:

$$S^{(1D)}(f_x)_{ABC} = A[1 + (Bf_x)^2]^{-C/2}, \quad S^{(2D)}(f)_{ABC} = \bar{A}[1 + (Bf)^2]^{-(C+1)/2},$$

where

$$\bar{A} = \frac{\Gamma[(C+1)/2]}{\Gamma(1/2)\Gamma(C/2)} AB. \tag{7.61}$$

A, *B*, and $C > 1$ are the three adjustable parameters of this model. *A* and $\bar{A}$ are the finite low-frequency limits, B is a correlation length, and C determines the power-law falloff at high frequencies. The one-dimensional or profile spectrum falls off as $1/f_x^C$ and the isotropic two-dimensional spectrum falls off as $1/f^{C+1}$.

The autocovariance function corresponding to these spectra is

$$C(\tau) = 2\sqrt{2\pi}\,\frac{A}{B}\,\frac{2^{-C/2}}{\Gamma(C/2)}\left(2\pi\,\frac{\tau}{B}\right)^{\frac{C-1}{2}} K_{\frac{C-1}{2}}\left(2\pi\,\frac{\tau}{B}\right)$$

$$= C(0)\left\{1 - \frac{2^{2-C}}{C-1}\,\frac{\Gamma[(3-C)/2]}{\Gamma[(C-1)/2]}\left(2\pi\,\frac{\tau}{B}\right)^{C-1} + \cdots\right\},$$

$$C(0) = \sqrt{\pi}\,\frac{\Gamma[(C-1)/2]}{\Gamma(C/2)}\,\frac{A}{B} = \frac{2\pi}{C-1}\,\frac{\bar{A}}{B^2} = \sigma^2, \tag{7.62}$$

where K is a modified Bessel function. For this reason, the ABC model is sometimes called the K-correlation model. When $C = 2$, the imposing Bessel function form of $C(\tau)$ reduces to a simple decaying exponential.

A second model having fractallike spectra is the power-exponent model. This has the simple covariance function

$$C(\tau) = \sigma^2 \exp(-|\tau/\rho|^N) = \sigma^2(1 - |\tau/\rho|^N + \cdots), \tag{7.63}$$

where σ, ρ, and $0 < N < 2$ are the three adjustable parameters of this model: σ is the intrinsic rms surface roughness, ρ is the correlation length, and N determines the power-law falloff of the spectra at high frequencies. The profile spectrum falls off as $1/f_x^{N+1}$ and the isotropic two-dimensional form as $1/f^{N+2}$.

The power-exponential model has two disadvantages. First, it cannot model fractallike power spectra that fall off faster than $1/f_x^{3-}$ or $1/f^{4-}$. For example, when $N = 2$ both power-exponential spectra are Gaussians. The second disadvantage is that its power spectra cannot generally be expressed in closed form and must be evaluated numerically. One exception occurs when $N = 1$, in which case the power-exponential model is identical with the ABC model with $A = 2\sigma^2\rho$, $B = 2\pi\rho$, and $C = 2$.

In the low- and the high-frequency limits, however, the power-exponential spectra have simple algebraic values. In one dimension,

$$S^{(1D)}(f_x \ll 1/2\pi\rho)_{\mathrm{PEX}} = \sigma^2\rho\{2\Gamma[(1+N)/N]\},$$

$$S^{(1D)}(f_x \gg 1/2\pi\rho)_{\mathrm{PEX}} = \sigma^2\rho\left\{\sqrt{\pi}\,2^N N\,\frac{\Gamma[(1+N)/2]}{\Gamma[(2-N)/2]}\,\frac{1}{(2\pi f_x\rho)^{N+1}}\right\}, \tag{7.64}$$

and in two dimensions,

$$S^{(2D)}(f \ll 1/2\pi\rho)_{PEX} = \sigma^2\rho^2 \{\pi\Gamma[(2+N)/N]\},$$

$$S^{(2D)}(f \gg 1/2\pi\rho)_{PEX} = \sigma^2\rho^2 \left\{\pi 2^{N+1} N \frac{\Gamma[(2+N)/2]}{\Gamma[(2-N)/2]} \frac{1}{(2\pi f\rho)^{N+2}}\right\}. \quad (7.65)$$

The corresponding limits for the *ABC* model follow immediately from Eqs. (7.61).

7.8.4. Comparison of Models

Figures 7.6–7.8 compare the one-dimensional spectra of the *ABC* and the power-exponent models for $C = N + 1 = 1.5$, 2.0, and 2.5, in which the other model parameters have been adjusted so that the two models have the same low- and high-frequency limits.

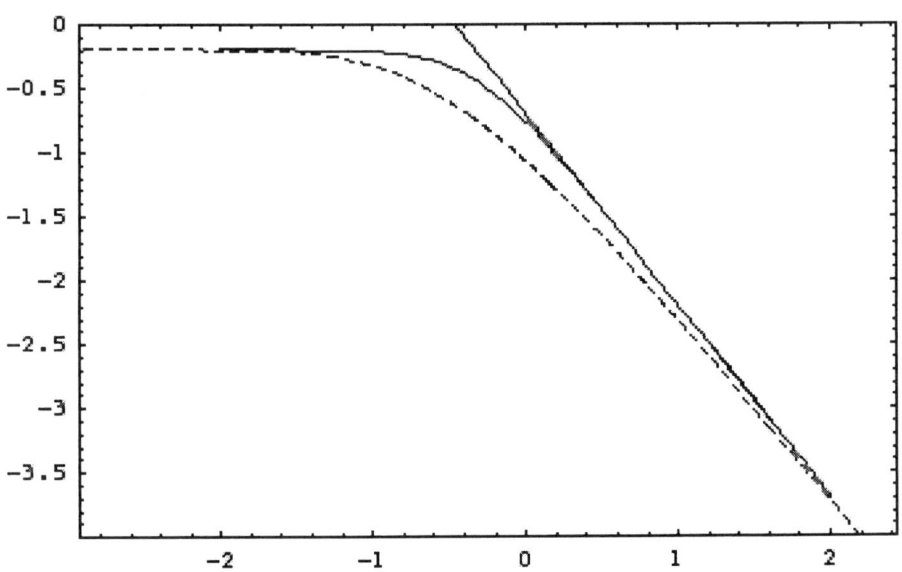

Fig. 7.6. A $\log_{10}$–$\log_{10}$ plot of one-dimensional power spectra for $C = N + 1 = 1.5$. The solid curve is the *ABC* model and the dashed curve is the power-exponent model. The straight line is their common asymptotic form.

Aside from the obvious differences in shape, the two models are essentially equivalent and the choice between them is usually determined by convenience. The *ABC* model is more convenient when one is working with spectra, where-

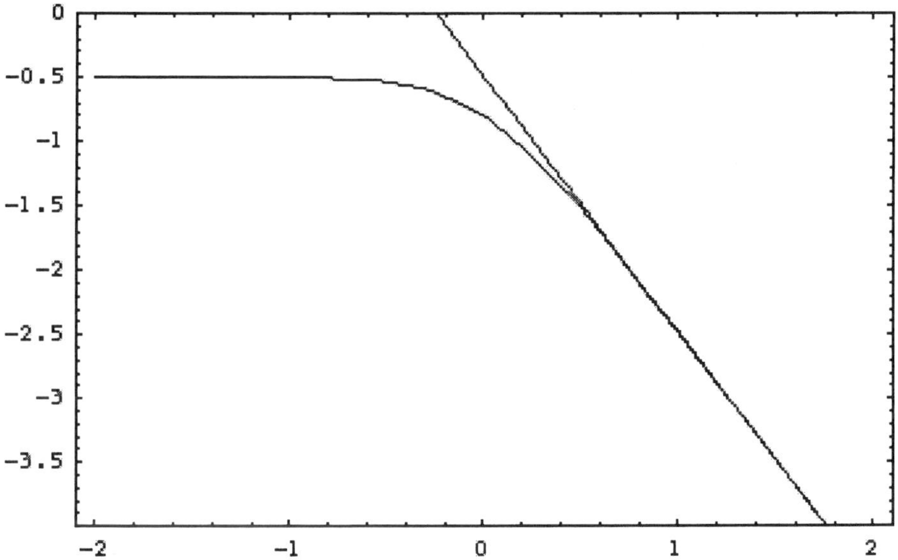

Fig. 7.7. Same as Fig. 7.6, but $C = N + 1 = 2$. In this case the *ABC* and the power-exponent models give the same results shown.

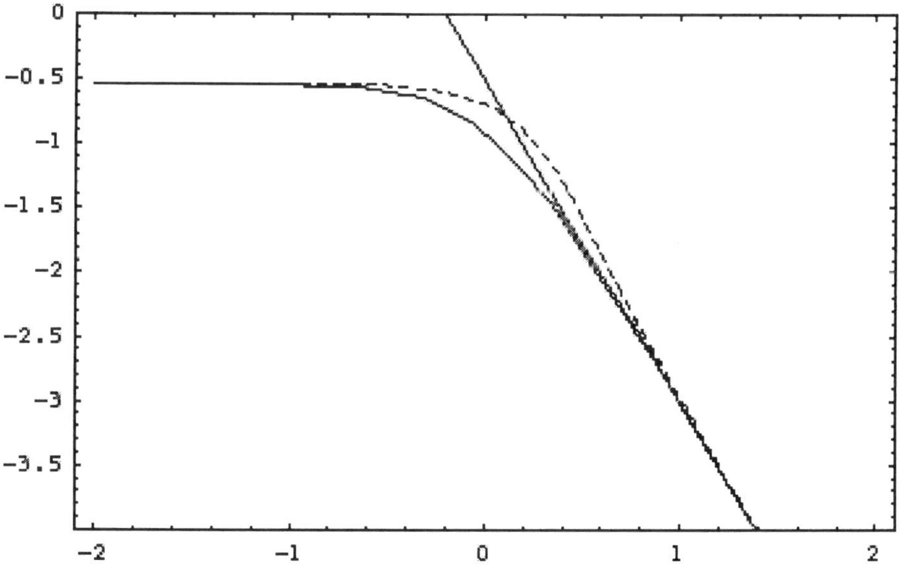

Fig. 7.8. Same as Fig. 7.6 but $C = N + 1 = 2.5$.

as the power-exponent model has a simpler covariance function. The *ABC* model also approaches its high-frequency asymptote more quickly than the power-exponent model does and from the same side and may therefore lead to more accurate estimations of the parameter $C = N + 1$.

7.8.5. Rough Surfaces

When the surface is very rough only the indicial or small τ form of the structure function is important in the expressions for the $\langle\text{BRDF}\rangle$ in Eqs. (7.49). In that limit the exponential factors appearing there become, for example,

$$\exp\left[-\frac{1}{2}(2\pi f_z)^2 D(\tau)\right] \rightarrow \exp\left[-(2\pi f_z T)^2 \left(\frac{\tau}{T}\right)^N\right], \tag{7.66}$$

where T is a new length parameter called the *topethesy* and $N = C - 1$ is the power-law parameter appearing in the *ABC* and the power-exponent models. The single parameter T is compounded out of the remaining two earlier model parameters.

It follows from the form of expression (7.66) that the resulting rough-surface spectra have the power-exponent form, but with model parameters that now depend on N, T, and λ. In other words, although both smooth and rough surfaces may appear fractallike in the physical-optics approximation, the parameters in the smooth-surface limit are intrinsic, whereas those appearing in the rough-surface limit depend on the radiation wavelength as well. It has been suggested that this situation might contribute to the apparent ubiquity of fractallike scattering patterns in nature.

In contrast, the structure functions of differentiable surfaces have structure functions with the indicial form, $D(\tau) \rightarrow (m\tau)^2$, where m is a random slope parameter. In that case, closed-form expressions for the $\langle\text{BRDF}\rangle$ can be obtained for each of the expressions in Eqs. (7.48). In particular, the Gaussian bivariate distribution leads to $\langle\text{BRDF}\rangle$s that have a Gaussian dependence on the statistics of m, whereas the gamma distribution leads to a dependence involving the modified Bessel function, $K_{\nu-1}$.

In addition, these closed forms of the $\langle\text{BRDF}\rangle$ are independent of the radiation wavelength and polarization. This reflects the fact that rough-surface scattering from rough differentiable surfaces is essentially a geometric-optics phenomenon and that the distribution of the scattered radiation is determined by the slope distribution function of the scattering surface.

The Gaussian and the Bessel slope distribution functions of differentiable surfaces can be derived directly from the corresponding bivariate height distribution functions by a limiting process.

7.9. Notation

Table 7.2. Constants

Notation	Definition	Equation of First Appearance	Unit
a	Radius	7.23	m
ABC	ABC spectral model	7.61	——
A, B, C	Model finish parameters	7.61	m^3, m, 0
$\bar{A}$	Derived finish parameter	7.61	m^4
A	Illuminated surface area	7.10	m^2
$A(\mathbf{k}_f, \mathbf{k}_i)_{\beta\alpha}$	Angular factor	7.29	0
α, β	Initial, final polarization states	7.11	——
c	Vacuum velocity of light	——	$m\ s^{-1}$
C	Contour	7.39	——
C	Covariance function (next)	7.45	m^2
$C(\boldsymbol{\tau})$	Covariance function	7.56	m^2
$D(\boldsymbol{\tau})$	Structure function	7.54	m^2
ε_0	Electrical permittivity of vacuum	7.5	$10^7/(4\pi c^2)$
ε_r	Relative electrical permittivity	7.19	0
ε_α	Surface emmissivity	7.17	0
η	Impedance of medium	7.4	Ω
η_0	Impedance of vacuum	7.5	Ω
$\mathbf{f}$	Spatial-frequency vector	7.2	m^{-1}
$f_{\beta\alpha}^{(1D)}, f_{\beta\alpha}^{(2D)}$	Scattering amplitudes	7.9, 7.10	0, m
$G_{\beta\alpha}^{(nD)}$	Angular factors	7.39	0
H_n, I_n, J_n, K_n, I_n	Bessel functions	7.24	0
H, W	Height and (half-) width parameters	7.25	m
$i = -j$	Imaginary unit	7.8	——
k	Wave number	7.1	m^{-1}
$\mathbf{k}_i, \mathbf{k}_f$	Initial and final wave vectors	7.1	m^{-1}
K	Shape parameter	7.25	0
L	Illuminated surface length	7.10	m
μ_0	Magnetic permeability of vacuum	7.19	$4\pi \times 10^{-7}$
μ_r	Relative magnetic permeability	7.19	0
$n = 1, 2$	Dimensional designations	7.9	0

(continued)

Table 7.2 *(continued)*

Notation	Definition	Equation of First Appearance	Unit
(nD)	*n*th dimensional surface	7.9	——
N	Model parameter	7.63	0
ν	Order parameter	7.46	0
PEX	Power-exponential spectral model	7.64	——
PO	Physical-optics approximation	7.42	——
r	Source–observer distance	7.8	m
$\mathbf{r}$	Distance vector in surface plane	7.28	m
$r_{\beta\alpha}(\theta_I)$	Amplitude-reflection coefficient	7.18	0
$R_{\alpha}(\theta_I)$	Intensity-reflection coefficient	7.18	0
ρ	Correlation length	7.63	m
$\rho_{\beta\alpha}$	Specular amplitude-refection coefficient	7.22	0
S	Surface	7.39	——
$S^{(\mathrm{nD})}$	Average power spectral density	7.53	m^3, m^4
SPM	Small-perturbation method	7.28	——
s, p	Polarization designations	7.6	——
σ	Intrinsic rms roughness	7.54	m
$\sigma^{(\mathrm{nD})}$	Bandwidth-limited rms roughnesses	7.36	m^2
T	Typothesy (finish parameter)	7.66	m
$\mathrm{TIS}_{\alpha}^{(\mathrm{nD})}$	Total integrated scatter including specular	7.16	0
$\mathrm{TIS}_{\beta\alpha}^{(\mathrm{nD})}$	Total integrated scatter excluding specular	7.36	0
$\boldsymbol{\tau}$	Lag or displacement vector	7.48	m
$W_{A,L}$	Deterministc power spectral densities	7.31	m^4, m^3
Z_j	Zero-mean surface height at point $j = 1, 2$	7.45	m
Z'_j	Nonzero mean surface height	7.46	m
$\langle - \rangle$	Ensemble average	7.47	——
$\int_{\infty} \mathrm{d}t$	Integration over the entire two-dimensional plane	7.37	m^2

Further Reading

V. Twersky, "On the nonspecular reflection of electromagnetic waves," J. Appl. Phys. **22,** 825–835, 1951.

V. Twersky, "Certain transmission and reflection theorems," J. Appl. Phys. **25,** 859–862, 1954.

V. Twersky, "On scattering and reflection of electromagnetic waves by rough surfaces," IEEE Trans. Antennas Propag. **5,** 81–90, 1957.

J. E. Burke and V. Twersky, "On scattering of waves by an elliptic cylinder and a semi-elliptic protuberance on a ground plane," J. Opt. Soc. Am. **54,** 732–744, 1964.

D. E. Barrick, "Low-frequency scatter from a semielliptic groove in a ground plane," J. Opt. Soc. Am. **60,** 625–634, 1970.

M. Born and E. Wolf, *Principles of Optics,* Cambridge Univ. Press, 1999.

A. Ishamaru, *Electromagnetic Wave Propagation, Radiation, and Scattering,* Prentice-Hall, 1991.

L. Tsang, J. A. Kong, and R. Shin, *Theory of Microwave Sensing,* Wiley, 1985.

L. Tsang, J. A. Kong, and K.-H. Ding, *Electromagnetic Scattering,* Wiley, 2000. (Two further volumes in 2001.)

F. T. Ulaby, R. K. Moore, and A. K. Fung, *Microwave Remote Sensing, Active and Passive,* 3 Vols., Artech, 1981.

R. Beckmann and A. Spizzichino, *The Scattering of Electromagnetic Waves from Rough Surfaces,* Pergamon, 1963.

A. G. Voronovich, *Wave Scattering from Rough Surfaces,* Springer, 1999.

The papers by Twersky and Burke and Twersky discuss the scattering by elliptical ridges and hemispherical bosses with considerable related material. Barrick considers the case of an elliptical groove. The book by Born and Wolf has much interesting collateral information. Ishamaru's paperback is an excellent engineering text. Both are well worth owning. The books by Tsang et al. are comprehensive and eminently readable physics-engineering texts. The books by Ulaby et al. are slightly more elementary engineering texts and include practical radar information. Beckmann and Spizzichino's book is classic but dated. Voronovich's book has good coverage of various theoretical issues but is terse and somewhat idiosyncratic.

8

Gaussian Beams

J. D. Gaskill

A Gaussian beam, also called a Gaussian spherical wave field, is a spherical wave field whose modulus in a plane transverse to the propagation direction varies in a Gaussian fashion. Propagation does not alter the fundamental nature of Gaussian beams; the effective width of a Gaussian beam may change as it propagates, as well as the curvature of its wave fronts, but it remains a Gaussian beam. The important features of Gaussian beam propagation are presented here; for those who desire more details, see the section on further reading at the end.

8.1. General Description of a Gaussian Beam

The Gaussian beams described here are those associated with lasers operating in the fundamental, or lowest-order, mode and propagating in homogeneous, isotropic media. Such a beam is depicted in Fig. 8.1, propagating in the z direction. The wave-field amplitude in the transverse plane $z = z_1$ may be described by a function of the form

$$u_1(r) = A_1 \operatorname{Gaus}\left(\frac{r}{b_1}\right) q\left(r; \frac{1}{\lambda R_1}\right), \tag{8.1}$$

where A_1 is the magnitude at the center of the beam, b_1 is the effective half-width of the beam, R_1 is the radius of curvature of the spherical wave front in the plane $z = z_1$, λ is the wavelength in the medium, and

$$\operatorname{Gaus}(r) = \exp(-\pi r^2), \tag{8.2}$$

$$q\left(r; \frac{1}{\lambda R_1}\right) = \exp\left(\frac{-j\pi r^2}{\lambda R_1}\right). \tag{8.3}$$

The beam is diverging if R_1 is positive and converging if R_1 is negative. Care

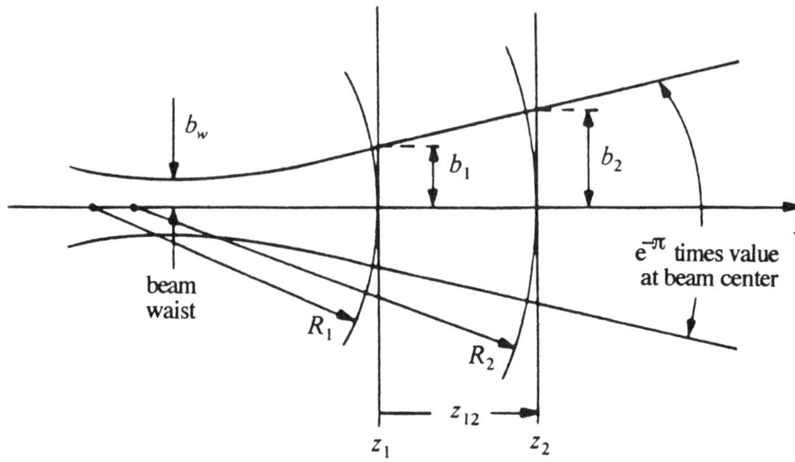

Fig. 8.1. Gaussian beam.

should be used when the results obtained here are compared with results of others, because the beam-width parameter b_1 is larger by a factor of π than the beam-width parameter found in much of the literature.

The Gaussian beam of Eq. (8.1) has a flux density given by

$$\begin{aligned} E_1(r) &= |u_1(r)|^2 \\ &= |A_1|^2 \operatorname{Gaus}\left(\frac{\sqrt{2}r}{b_1}\right), \end{aligned} \tag{8.4}$$

and its total power is the integral of $E_1(r)$ over the z_1 plane:

$$\begin{aligned} P_{\text{tot}} &= \int_0^{2\pi} \int_0^{\infty} E_1(r')r'\mathrm{d}r'\mathrm{d}\theta' \\ &= \frac{|A_1|^2 b_1^2}{2}. \end{aligned} \tag{8.5}$$

The power transmitted through a circular aperture of diameter d is given by

$$\begin{aligned} P\left(\frac{d}{b_1}\right) &= 2\pi|A_1|^2 \int_0^{d/2} \operatorname{Gaus}\left(\frac{\sqrt{2}r'}{b_1}\right) r'\mathrm{d}r' \\ &= P_{\text{tot}}\left[1 - \operatorname{Gaus}\left(\frac{d}{\sqrt{2}b_1}\right)\right]. \end{aligned} \tag{8.6}$$

An aperture of diameter $d = b_1$ transmits 79.3% of the total beam power, whereas an aperture of twice this diameter transmits 99.8% of the total power.

8.2. Effects of Diffraction on a Gaussian Beam

The Gaussian beam of Eq. (8.1) remains a Gaussian beam as it propagates from the plane $z = z_1$ to the plane $z = z_2$, where the wave field is described by

$$\begin{aligned} u_2(r) &= B_{12} q\left(r; \frac{1}{\lambda z_{12}}\right) H_0\left[u_1(r) q\left(r; \frac{1}{\lambda z_{12}}\right)\right]\Bigg|_{\rho = r/\lambda z_{12}} \\ &= B_{12} q\left(r; \frac{1}{\lambda z_{12}}\right) H_0\left[A_1 \operatorname{Gaus}\left(\frac{r}{b_1}\right) q\left(r; \frac{1}{\lambda R_1} + \frac{1}{\lambda z_{12}}\right)\right]\Bigg|_{\rho = r/\lambda z_{12}}. \end{aligned} \tag{8.7}$$

Here $H_0[\ ]$ denotes the zero-order Hankel transform and B_{12} is a constant. Equation (8.7) may be written as

$$u_2(r) = A_2 \operatorname{Gaus}\left(\frac{r}{b_2}\right) q\left(r; \frac{1}{\lambda R_2}\right), \tag{8.8}$$

where b_2, A_2, and R_2 are given by

$$b_2 = \left|\frac{\lambda z_{12}}{b_1}\right| \sqrt{1 + \left(\frac{b_1^2}{\lambda}\right)^2 \left(\frac{1}{z_{12}} + \frac{1}{R_1}\right)^2}, \tag{8.9}$$

$$A_2 = A_1 \left(\frac{b_1}{b_2}\right) e^{j(kz_{12}\phi_{12})}, \tag{8.10}$$

$$R_2 = z_{12} \left[\frac{1 + \left(\frac{b_1^2}{\lambda}\right)^2 \left(\frac{1}{z_{12}} + \frac{1}{R_1}\right)^2}{1 + \left(\frac{b_1^2}{\lambda}\right)^2 \left(\frac{1}{R_1}\right)\left(\frac{1}{Z_{12}} + \frac{1}{R_1}\right)}\right]. \tag{8.11}$$

The quantity ϕ_{12} denotes a phase shift given by

$$\phi_{12} = \tan^{-1}\left(\frac{-\frac{\lambda}{b_1^2}}{\frac{1}{z_{12}} + \frac{1}{R_1}}\right). \tag{8.12}$$

For the situation illustrated in Fig. 8.1, in which R_1 is depicted as positive, the beam is expanding at the plane $z = z_1$. Consequently, at any plane to the right of z_1, the beam width b_2 will be greater than the beam width at z_1. Not only that, but b_2 is a monotonically increasing function of z_{12}/R_1 in this region. However, the situation is different for observation planes lying to the left of z_1, i.e., for $z_{12} < 0$. As the observation plane is moved to the left of z_1, the beam width decreases at first. However, it reaches a minimum and then begins to increase again as z_2 is moved farther and farther to the left. The location of the minimum beam width is called the beam waist, and the plane containing the beam waist is denoted by z_w. The distance between z_1 and z_w, $z_{1w} = z_w - z_1$, is given by

$$z_{1w} = \frac{-R_1}{\left(\frac{\lambda R_1}{b_1^2}\right)^2 + 1}. \tag{8.13}$$

Thus, *the beam waist is not found exactly at the point where* $z_{12} = -R_1$, although it may be close to this point. If R_1 is negative, signifying a converging beam at $z = z_1$ as in Fig. 8.2, the beam waist will lie to the right of this plane. The beam width at the waist is given by

$$b_w = \left|\frac{\lambda R_1}{b_1}\right| \frac{1}{\sqrt{\left(\frac{\lambda R_1}{b_1^2}\right)^2 + 1}}. \tag{8.14}$$

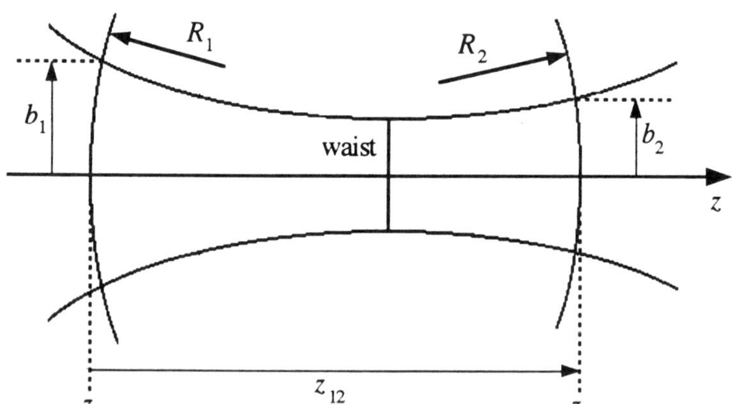

Fig. 8.2. Gaussian beam in the vicinity of the waist.

The behavior of R_2 with z_{12}/R_1 is more complicated than that of b_2. Under the assumption that R_1 is once again positive, as depicted in Fig. 8.1, R_2 will be positive at all observation planes to the right of z_1. In fact, it will be positive everywhere to the right of the waist and negative everywhere to the left of the waist. At the waist, R_2 has a value of $\pm\infty$, which indicates that the wave fronts are plane surfaces at that location. Thus, starting at the waist, $|R_2|$ decreases at first as the observation plane is moved away from the waist in either direction. At a certain distance from the waist, on either side, $|R_2|$ reaches a minimum value; finally, as the observation plane is moved still farther from the waist, it begins to increase once again (see Fig. 8.3).

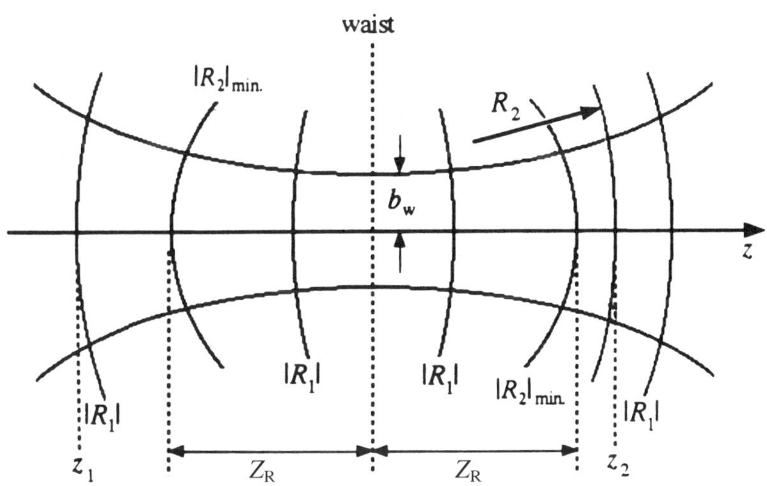

Fig. 8.3. Behavior of R_2 in the vicinity of the waist.

A curious property of Gaussian beams is as follows: Whenever the observation distance is equal to the negative of the initial radius of curvature, i.e., whenever $z_{12} = -R_1$, the beam width and radius of curvature are given by

$$b_2 = \left|\frac{\lambda z_{12}}{b_1}\right| = \left|\frac{\lambda R_1}{b_1}\right|, \tag{8.15}$$

$$R_2 = z_{12} = -R_1. \tag{8.16}$$

It is helpful to determine the behavior of b_2 and R_2 for the situation in which the initial plane $z = z_1$ is located at the beam waist. In this event we have $|R_1| = \infty$ *and* $b_1 = b_w$, which yields, with z_1 at the waist,

$$b_2 = b_w \sqrt{\left(\frac{\lambda z_{12}}{b_w^2}\right)^2 + 1}, \tag{8.17}$$

$$R_2 = z_{12}\left|\left(\frac{b_w^2}{\lambda z_{12}}\right)^2 + 1\right|. \tag{8.18}$$

Because the *Rayleigh range* is denoted by

$$Z_R = \frac{b_w^2}{\lambda}, \tag{8.19}$$

then

$$b_2 = b_w\sqrt{\left(\frac{z_{12}}{Z_R}\right)^2 + 1}, \tag{8.20}$$

$$R_2 = z_{12}\left|\left(\frac{Z_R}{z_{12}}\right)^2 + 1\right|, \tag{8.21}$$

when z_1 is at the waist.

At a distance of $z_{12} = \pm Z_R$ from the beam waist, the beam width has increased by a factor of $\sqrt{2}$ and the radius of curvature has a magnitude equal to just twice the value of Z_R; i.e., with z_1 at the waist and $z_{12} = \pm Z_R$,

$$b_2 = \sqrt{2}b_w, \tag{8.22}$$

$$R_2 = \pm 2Z_R. \tag{8.23}$$

The minimum value of $|R_2|$ occurs at a distance from the waist equal to the Rayleigh range. At large distances from the beam waist

$$b_2 \cong b_w\left|\frac{z_{12}}{Z_R}\right|, \tag{8.24}$$

$$R_2 \cong z_{12}. \tag{8.25}$$

8.3. Effects of Lenses on Gaussian Beam Propagation

The effects of lenses on the propagation of Gaussian beams are presented in this section (See Fig. 8.4). With the beam width b_1 and radius of curvature R_1 known at the plane $z = z_1$, a lens is placed at $z = z_2$. The lens has a focal length of f, which may be either positive or negative. The wave field incident upon the lens is given by

$$u_{\bar{2}}(r) = A_2 \,\mathrm{Gaus}\left(\frac{r}{b_2}\right) q\left(r; \frac{1}{\lambda R_{\bar{2}}}\right), \tag{8.26}$$

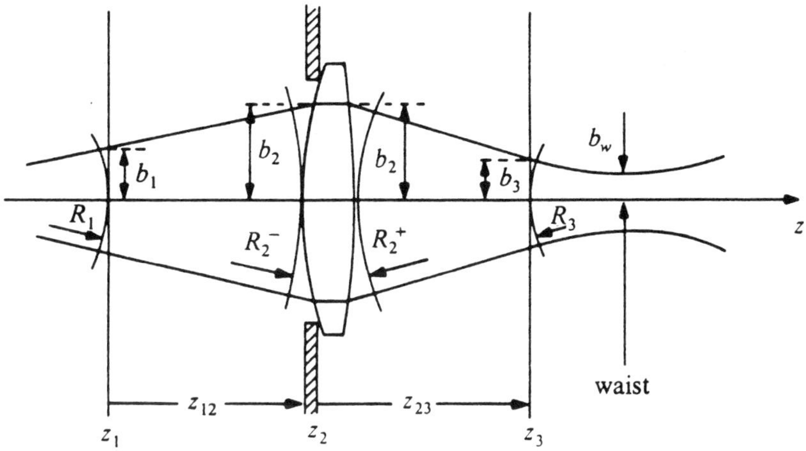

Fig. 8.4. Effect of a positive lens on a diverging Gaussian beam.

and that transmitted by the lens is

$$u_2^+(r) = A_2 B_l \,\mathrm{Gaus}\left(\frac{r}{b_2}\right) q\left(r; \frac{1}{\lambda R_2^-} - \frac{1}{\lambda f}\right) p_l(r), \tag{8.27}$$

where B_l is a constant and $p_l(r)$ describes the lens aperture.

If the lens aperture is large enough to transmit the entire beam (i.e., if d_l is the diameter of the aperture and $d_l > 2b_2$), then

$$u_3(r) = A_3 \,\mathrm{Gaus}\left(\frac{r}{b_3}\right) q\left(r; \frac{1}{\lambda R_3}\right), \tag{8.28}$$

where

$$b_3 = \left|\frac{\lambda z_{23}}{b_2}\right| \sqrt{1 + \left(\frac{b_2^2}{\lambda}\right)^2 \left(\frac{1}{z_{23}} + \frac{1}{R_2^-} - \frac{1}{f}\right)^2}, \tag{8.29}$$

$$A_3 = A_2 B_l \left(\frac{b_2}{b_3}\right) e^{j(kz_{23} + \phi_{23})}, \tag{8.30}$$

$$R_3 = z_{23} \left[\frac{1 + \left(\frac{b_2^2}{\lambda}\right)^2 \left(\frac{1}{z_{23}} + \frac{1}{R_2^-} - \frac{1}{f}\right)^2}{1 + \left(\frac{b_2^2}{\lambda}\right)^2 \left(\frac{1}{R_2^-} - \frac{1}{f}\right)\left(\frac{1}{z_{23}} + \frac{1}{R_2^-} - \frac{1}{f}\right)}\right], \tag{8.31}$$

$$\phi_{23} = \tan^{-1}\left(\frac{\dfrac{-\lambda}{b_2^2}}{\dfrac{1}{z_{23}} + \dfrac{1}{R_2^-} - \dfrac{1}{f}}\right). \tag{8.32}$$

The preceding expressions are simplified to some extent if we select the plane $z = z_1$ to coincide with the initial beam waist. To avoid ambiguities in notation we denote the width of the initial beam waist by b_{w1}, the Rayleigh range parameter associated with this waist by $Z_{R1} = b_{w1}^2/\lambda$, and the distance from this waist to the lens by z_{12}. With the distance to the observation plane again denoted by z_{23}, the observed beam width may be expressed by

$$b_3 = b_{w1}\left(\frac{z_{12}}{Z_{R1}}\right)\sqrt{\left(1 + \frac{z_{23}}{z_{12}} - \frac{z_{23}}{f}\right)^2 + \left(\frac{Z_{R1}}{z_{12}}\right)^2\left(1 - \frac{z_{23}}{f}\right)^2}. \tag{8.33}$$

This expression is an alternative form of that given by Dickson[1] [Eq. (33)]. Several special cases are of interest, and these are given by

$$b_3 = \begin{cases} b_{w1}\sqrt{\left(\dfrac{z_{23}}{Z_{R1}}\right)^2 + \left(1 - \dfrac{z_{23}}{f}\right)^2}, & z_{12} = 0 \qquad (8.34) \\[2ex] b_{w1}\sqrt{\left(\dfrac{z_{12}}{Z_{R1}}\right)^2 + 1}, & z_{23} = 0. \qquad (8.35) \\[2ex] \dfrac{\lambda f}{b_{w1}}, & z_{23} = f \qquad (8.36) \end{cases}$$

When the observation plane coincides with the back focal plane of the lens, the observed beam width is independent of the waist-to-lens distance z_{12}.

The waist-to-waist properties of a Gaussian beam that has passed through a lens are now presented. The initial beam waist lies in the plane $z = z_1$ and has a width of b_{w1}. In addition, the observation plane $z = z_3$ lies at the waist formed by the lens; this waist is located at a distance of z_{23} from the lens and has a width of b_{w3}. Finally, with waists at $z = z_1$ and $z = z_3$ then

$$z_{23} = f + \frac{f^2(z_{12} - f)}{Z_{R1}^2 + (z_{12} - f)^2}, \tag{8.37}$$

$$b_{w3} = b_{w1}\left[\frac{f}{\sqrt{Z_{R1}^2 + (z_{12} - f)^2}}\right], \tag{8.38}$$

[1] L. D. Dickson, "Characteristics of a propagating Gaussian beam," Appl. Opt. **9**, 1854–1861 (1970).

$$Z_{R3} = Z_{R1}\left[\frac{f^2}{Z_{R1}^2 + (z_{12} - f)^2}\right], \tag{8.39}$$

where Z_{R3} is the Rayleigh range parameter associated with the waist produced by the lens. Again, several special cases are of interest. For example, if the initial waist is placed in the front focal plane of the lens such that $z_{12} = f$,

$$z_{23} = f, \tag{8.40}$$

$$b_{w3} = \frac{\lambda f}{b_{w1}}, \tag{8.41}$$

$$Z_{R3} = \frac{f^2}{Z_{R1}}. \tag{8.42}$$

Thus a beam waist in the front focal plane *always* produces a beam waist in the back focal plane. Not only that, Dickson shows that the width of the waist produced by the lens is *always* a maximum when this waist is located in the back focal plane and its width decreases when it is moved to either side of the back focal plane. Examination of Eqs. (8.37)–(8.42) shows that beam waists are not *imaged* into beam waists in the usual sense. In fact, for certain values of the various parameters, the relationship between the locations of the waists is exactly opposite from the object-to-image relationship of normal imaging. For example, as the initial waist is moved toward the lens, the second waist may also move toward the lens rather than away from it, as might be expected from normal imaging considerations.

The location(s) of the initial beam waist that will produce a second waist of the *same size* is now determined (see Fig. 8.5). By setting $b_{w3} = b_{w1}$ in Eq.

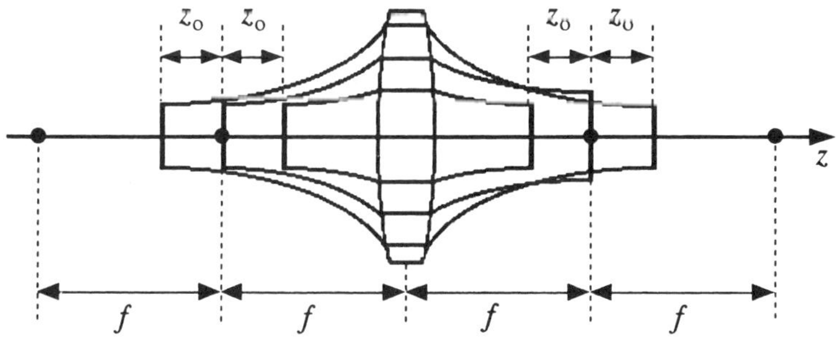

Fig. 8.5. Waist-to-waist mapping produced by a positive lens.

(8.38), we find the distances z_0 between the beam waists and the front and back focal planes to be

$$z_0 = \sqrt{f^2 - Z_{R1}^2}. \tag{8.43}$$

When the initial beam waist is located a distance of z_0 on either side of the front focal plane, the radius of curvature of the wave front incident upon the lens is $R_2^- = 2f$. The lens then converts this wave front into another one with radius $R_2^+ = -2f$, and the symmetrical nature of the situation becomes obvious. If the initial waist is placed in the front focal plane of a lens having a focal length $f = \pm Z_{R1}$, the lens will produce a second waist of the same size in its back focal plane.

8.4. Maximizing Beam Irradiance

To maximize the overall beam flux density on a target located at a fixed, predetermined distance, an optical relay system must produce a transmitted beam with radius of curvature R_2^+ that will, in turn, yield the minimum beam size on the target (see Fig. 8.6).

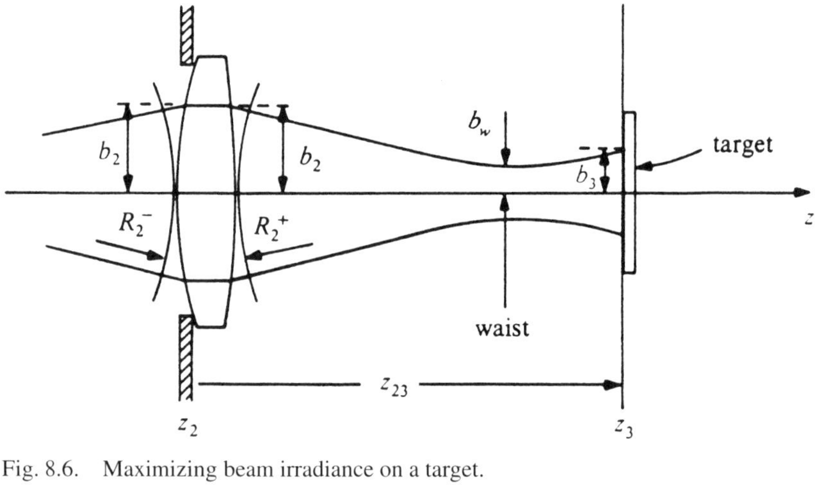

Fig. 8.6. Maximizing beam irradiance on a target.

It is assumed that the relay lens is located in the $z = z_2$ plane, the width of the beam incident upon the relay lens is fixed, the lens aperture is large enough to transmit the active beam, and the target is located in the $z = z_3$ plane. It is also assumed that the beam can be steered accurately, and the beam is therefore centered on the z axis for simplicity.

The optimum radius of curvature R_2^+ is that for which the beam width b_3 at the target is a minimum. At first glance it might appear that the optimum performance would be obtained if a value for R_2^+ is selected that produces a beam waist at the target; however, *that is not the case.* The beam width b_3 is minimized if

$$R_2^+ = -z_{23}, \tag{8.44}$$

and the resulting beam width for this value of $R^+{}_2$ is

$$(b_3)_{\min} = \left| \frac{\lambda z_{23}}{b_2} \right|. \tag{8.45}$$

Under these conditions, a beam waist will occur in front of the target; however, even though the width of this waist is smaller than $(b_3)_{\min}$, it will not remain smaller if it is relocated at the target. For the beam waist to occur at the target, the radius of curvature R_2^+ must satisfy the equation

$$\frac{z_{23}\lambda^2}{b_2^4} (R_2^+)^2 + R_2^+ + z_{23} = 0. \tag{8.46}$$

The solutions of this equation are

$$R_2^+ \; \frac{-1 \pm \sqrt{1 - 4\left(\dfrac{\lambda z_{23}}{b_2^2}\right)^2}}{\dfrac{2\lambda^2 z_{23}}{b_2^4}}. \tag{8.47}$$

The solutions for R_2^+ will be real valued only when

$$\left(\frac{\lambda z_{23}}{b_2^2}\right)^2 \leq 0.25, \tag{8.48}$$

and, if this condition is not met, *the beam waist cannot be located at the target.* However, even if this condition is met, *we do not obtain the minimum beam width at the target by locating the beam waist at the target.* Rather, a radius of curvature $R_2^+ = -z_{23}$ must be selected, which causes the beam waist to lie between the relay system and the target.

A slightly different problem is the maximization of the *peak* value of the irradiance at the target. The problem now is to select both the proper radius of curvature R_2^+ and the optimum value for the beam width b_2 relative to the fixed diameter d_l of the relay lens aperture. Given a Gaussian beam of fixed power, a relay lens aperture of fixed diameter, and a target located at a fixed distance, a beam expander that will maximize the *peak flux density* at the target must be designed. The desired effects can be obtained when the initial beam is passed through an appropriately selected lens that is located ahead of the final relay lens, as shown in Fig. 8.7. The beam incident upon the final relay lens will then be of the general form

$$u_2^-(r) = K_2\left(\frac{1}{b_2}\right)\mathrm{Gaus}\left(\frac{r}{b_2}\right)q\left(r;\frac{1}{\lambda R_2^-}\right), \tag{8.49}$$

where K_2 is a constant that specifies the phase of the incident beam and is re-

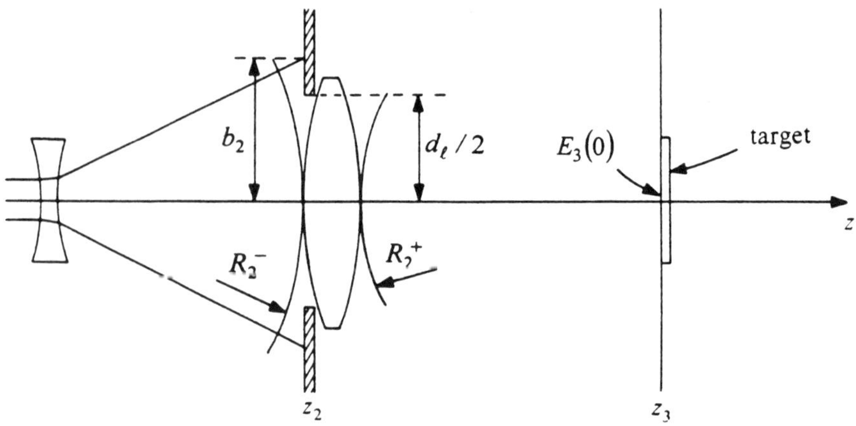

Fig. 8.7. Maximizing the peak irradiance on a target.

lated to the total beam power. The wave-field amplitude transmitted by the final relay lens is

$$u_2^+(r) = \frac{K_2 B_l}{b_2}\,\mathrm{Gaus}\left(\frac{r}{b_2}\right) q\left(r;\frac{1}{\lambda R_2^+}\right)\mathrm{cyl}\left(\frac{r}{d_l}\right), \tag{8.50}$$

where $\mathrm{cyl}(r/d_l)$ is an is an aperture function that is zero for $r > d_l/2$.

If the lens aperture now truncates the beam, the preceding results of this section are not applicable. For this particular problem it is reasonable to assume (although difficult to prove) that the peak flux density at the target will be obtained when $R_2^+ = -z_{23}$ and that it will occur on the z axis, i.e.,

$$\begin{aligned} E_3(r)_{\max} &= E_3(0) \\ &= |u_3(0)|^2. \end{aligned} \tag{8.51}$$

Thus it is necessary to maximize $|u_3(0)|$. Starting with

$$u_3(r) = B_{23}\, q\left(r;\frac{1}{\lambda z_{23}}\right) H_0\left[u_2^+(r')\, q\left(r';\frac{1}{\lambda z_{23}}\right)\right]\Bigg|_{\rho = r/\lambda z_{23}}, \tag{8.52}$$

choosing $R_2^+ = -z_{23}$, and setting $r = 0$, we find that the wave-field amplitude at the target is

$$u_3(0) = B_{23} H_0\left[\frac{K_2 B_l}{b_2}\,\mathrm{Gaus}\left(\frac{r'}{b_2}\right)\mathrm{cyl}\left(\frac{r'}{d_l}\right)\right]\Bigg|_{\rho=0}. \tag{8.53}$$

Then, from the central ordinate theorem of Fourier transforms,

$$\begin{aligned} u_3(0) &= K_2 B_l B_{23} 2\pi \int_0^{d_l/2} \left(\frac{1}{b_2}\right) \mathrm{Gaus}\left(\frac{r'}{b_2}\right) r' \mathrm{d}r' \\ &= K_2 B_l B_{23} b_2 \left[1 - \mathrm{Gaus}\left(\frac{d_l}{2b_2}\right)\right]. \end{aligned} \tag{8.54}$$

Finally,

$$|u_3(0)| = |K_2 B_l B_{23}| b_2 \left[1 - \mathrm{Gaus}\left(\frac{d_l}{2b_2}\right)\right], \tag{8.55}$$

which has a maximum when $b_2 = 0.791 d_l$. For this value of b_2 the power transmitted by the aperture will be somewhat less than the total incident-beam power; in fact, only 91.9% of the incident power passes through the aperture for this value of b_2.

The peak flux density at the target is given by

$$\begin{aligned} E_3(0) &= |K_2 B_l B_{23}|^2 b_2^2 \left[1 - \mathrm{Gaus}\left(\frac{d_l}{2b_2}\right)\right]^2 \\ &= 0.320 |K_2 b_l|^2 \left(\frac{d_l}{\lambda z_{23}}\right)^2. \end{aligned} \tag{8.56}$$

For comparison, the peak irradiance for a beam width of $b_2 = 0.5 d_l$, which allows 99.8% of the incident power to be transmitted, is found to be

$$\begin{aligned} E_3(0) &= |K_2 B_l|^2 \left(\frac{0.5 d_l}{\lambda z_{23}}\right)^2 [1 - \mathrm{Gaus}(1)]^2 \\ &= 0.229 |K_2 B_l|^2 \left(\frac{d_l}{\lambda z_{23}}\right)^2, \end{aligned} \tag{8.57}$$

which is only 71.5% of the peak flux density obtained with $b_2 = 0.791 d_l$. The net result is that, even though 8% of the transmitted beam power is lost when the relay lens aperture is overfilled, the peak flux density at the target is increased by nearly 40%. Results of related problems have been reported by Holms *et al.*,[2] who determined the on-axis flux density for Gaussian beams passed through annular apertures.

[2] D. A. Holmes, J. E. Korka, and P. V. Avizonis, "Parametric study of apertured focused Gaussian beam," Appl. Opt **11**, 565–574 (1992).

8.5. Elliptical Gaussian Beams

An elliptical Gaussian beam is one whose cross section is elliptical rather than circular, and such a beam can be produced by an optical element (e.g., a cylindrical lens) that causes the wave front of the beam to have different radii of curvature in the x and the y directions. Because the Gaussian beam is separable in x and y, the behavior of an elliptical Gaussian beam will be governed separately in each of the coordinate directions. The basic relationships are similar to those given in the preceding sections, but now the beam width and wavefront radii of curvature will generally be different in the x and the y directions at any given plane z = constant. It can be shown that a circular cross section will be found at only two points along the beam, and there is no point along the beam at which the wave fronts are planes. More information on this subject may be found in Yariv.[3]

Further Reading

J. D. Gaskill, *Linear Systems, Fourier Transforms and Optics,* Wiley Interscience, 1978.

A. E. Siegman, *An Introduction to Lasers and Masers,* McGraw-Hill, 1971.

[3] A. Yariv, *Introduction to Optical Electronics,* Holt, Rinehart & Winston, 1976.

9

Interferometers

J. C. Wyant and W. L. Wolfe

9.1. Types

Interferometers come in several different varieties. Some obtain the interference pattern with two beams, some with many. Some are common path; some use lateral shear; some radial shear and some use phase-shifting techniques. They are used for measuring length, figure, spectra, rotation, and more.

9.2. Two-Beam Interference

When two monochromatic waves interfere, the pattern is given by

$$I = I_1 + I_2 + 2\sqrt{I_1 I_2}\cos\delta,$$

where the I's represent the individual intensities (flux densities) and δ is the phase difference between the waves.

The maxima occur when $\delta = 2m\pi$, and m is an integer. Maxima occur when $\delta = (2m+1)\pi$. The visibility is given by

$$V = \frac{I_{\max} - I_{\min}}{I_{\max} + I_{\min}} = \frac{2\sqrt{I_1 I_2}}{I_1 + I_2}.$$

9.3. Fresnel Mirrors

Two mirrors set at a slight angle one to the other will produce an interference pattern. The fringes are equidistant straight lines separated by

$$\Delta x = \frac{\lambda}{2}\frac{D+d}{d}\frac{1}{\varepsilon},$$

where the distances are defined in Fig. 9.1 and the angle between the mirrors is ε.

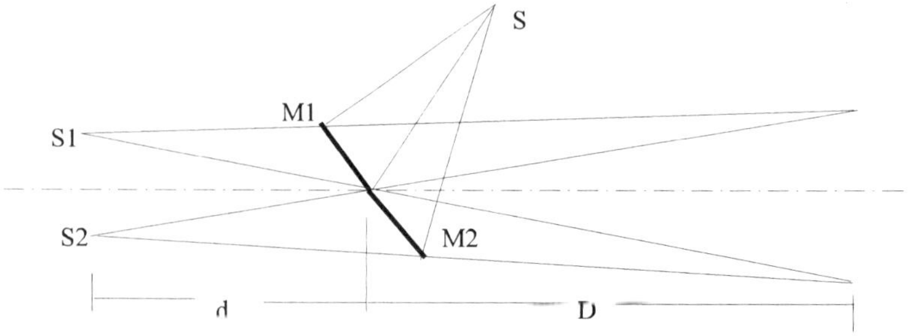

Figure 9.1. Fresnel mirrors.

9.4. Plane-Parallel Plate

When a monochromatic source illuminates a plane-parallel plate, as shown in Fig. 9.2, the interference fringes are circular. If the fringes are formed at infinity through use of a collimating lens, then the fringes are circles centered on the normal to the plate. The bright fringes occur when $2nd \cos \theta = (m \pm 1/2)\lambda$, whereas the dark fringes occur when $2nd \cos \theta = m\lambda$.

These are called Haidinger fringes, or fringes of equal inclination, as they depend on only the angle of incidence (for a plate of a given thickness). If the maximum or the minimum occurs at the center, the radii of the fringes are proportional to the square roots of the consecutive integers.

If the plate has a slight variation in thickness and is illuminated normally,

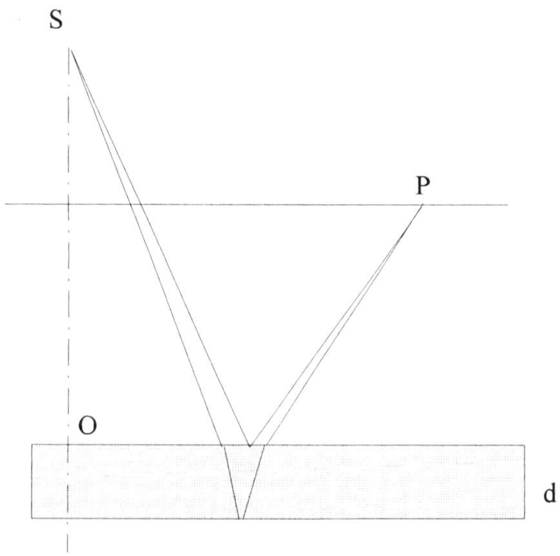

Figure 9.2. Plane-parallel plate geometry.

then the fringes are Fizeau fringes of equal thickness; that is, they show the contours of the plate.

9.5. Michelson Interferometer

The Michelson interferometer is shown in Fig. 9.3. Light comes from the extended source, shown at the left, and is split into two beams by the beam splitter (BS). The vertical beam goes through the beam splitter once, whereas the horizontal one passes through three times. Therefore a compensator plate is inserted in the vertical arm. (It could be the other way around.)

With this compensation, the path difference can be viewed as twice the difference between mirror M1 and the image of mirror M2, M2′, $2d$. With the mirrors parallel, the fringe pattern is given by $2nd \cos\theta = m\lambda$, and the fringes are localized at infinity. With the mirrors at a slight angle are straight lines parallel to the equivalent intersection of the mirrors and localized approximately at the intersection. When the mirrors are only a few wavelengths apart, white-light fringes appear, and can be used to determine their coincidence.

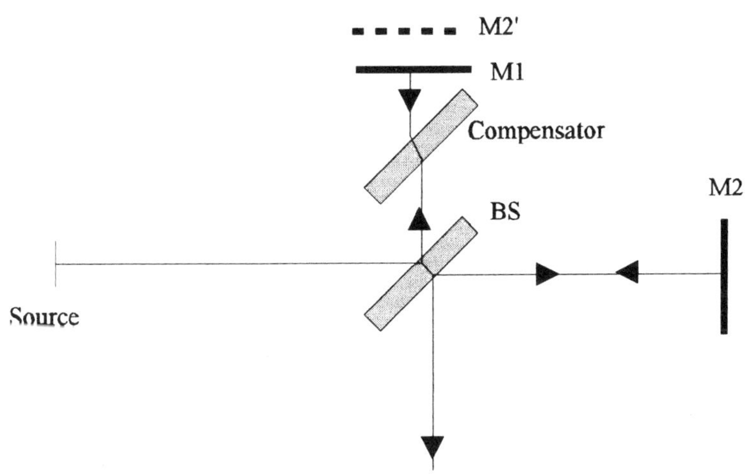

Figure 9.3. Michelson interferometer.

9.6. Twyman–Green Interferometer

This may be thought of as a Michelson interferometer with collimated light. Rather than an extended source, a "point" source is used at the focus of a lens or mirror. The output is a spot of varying intensity for a monochromatic source, the variation depending on the phase difference in the arms. The Twyman–Green interferometer can be used to test optical elements and as an interferometer spectrometer.

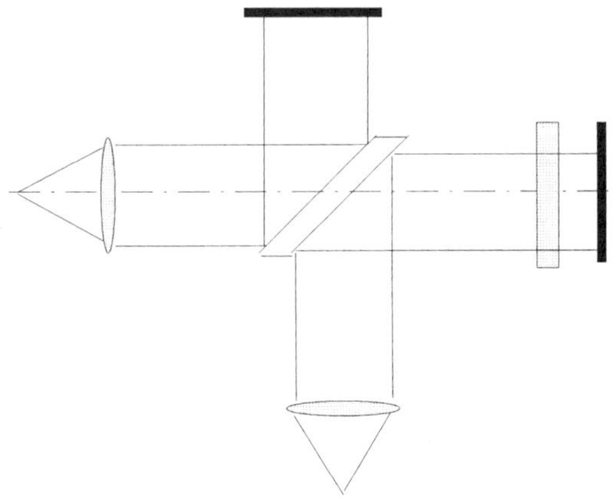

Figure 9.4. Twyman–Green interferometer.

9.7. Testing Optical Elements

We can test figures of optical elements with a Twyman–Green interferometer by inserting the element in one arm and altering the mirror shape in that arm. Figure 9.4 shows the Twyman–Green interferometer arranged to test a plane-parallel plate (not all refractions, nor the compensator plate, are shown for clarity). If a lens were to be tested, then the lens would be placed in the same position, but the mirror would be convex with its center of curvature at the focal point of the lens.

If a nonmonochromatic source is used with a Twyman–Green interferometer, it can be analyzed spectrometrically. If one mirror is moved and the output is recorded as a function of its position, an interferogram is obtained. The Fourier transform of the interferogram is the spectrum of the source:

$$S(\sigma) = \int_0^\infty P(\delta)\cos(2\pi\sigma\delta)d\delta,$$

where $S(\sigma)$ is the spectrum, $P(\delta)$ is the power as a function of path difference, and σ is the frequency in wave numbers. The resolution is given by

$$\Delta\sigma = \frac{1}{2\delta_{max}},$$

where δ_{max} is the maximum path difference. The resolving power is given by

$$Q = \frac{\sigma}{\Delta\sigma} = \frac{\lambda}{\Delta\lambda} = 2\delta\sigma.$$

9.8. Mach–Zehnder Interferometer

The Mach–Zehnder interferometer uses two mirrors and two beam splitters, as shown in Fig. 9.5. The two beam splitters need to be matched. If the light is monochromatic and collimated, then the two wave fronts will be different in phase by k times the optical path difference between the two arms, where k is the wave number. If the mirrors make a small angle with respect to each other, the fringes are unlocalized straight lines. Although the Mach–Zehnder interferometer is more versatile than the Twyman–Green interferometer, it is also more difficult to align, because it has two independent paths.

9.9. Sagnac Interferometer

This interferometer has beams traveling in opposite directions, as shown in Fig. 9.6. Because the two beams travel the same paths (in opposite directions),

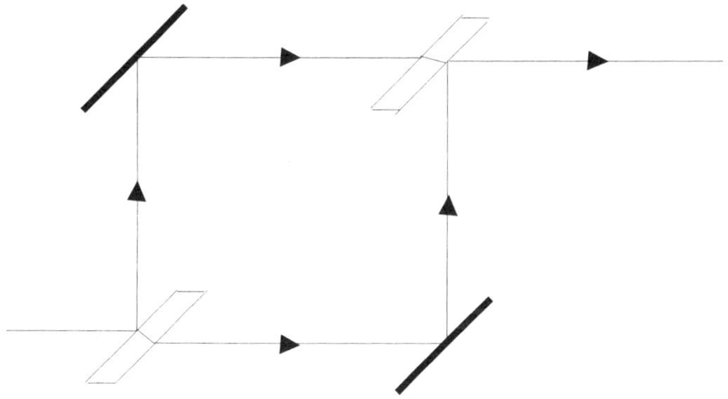

Figure 9.5. Mach–Zehnder interferometer.

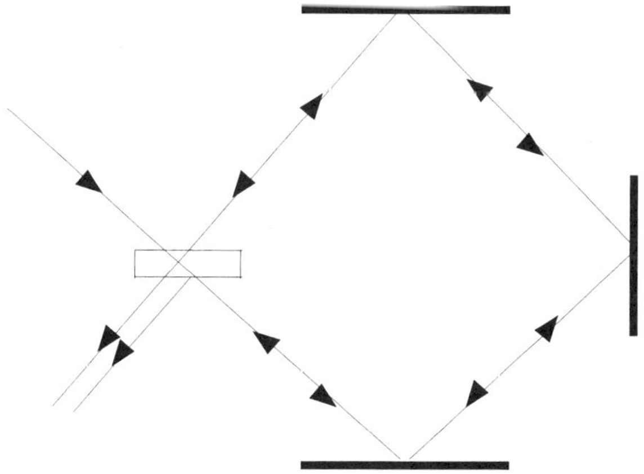

Figure 9.6. Sagnac interferometer.

this interferometer is highly stable and easy to align. In this form, the two beams formed from the separation of a single beam always exit parallel to each other, but may be displaced. The displacement adjustment is accomplished by displacement either of the mirrors or the beam splitter, or even by their rotation about an axis perpendicular to the plane of the paper. If the interferometer is rotated with an angular velocity ω, there will be a delay between the clockwise and the counterclockwise beam of $4A\omega/c^2$, where A is the area enclosed by the light path.

9.10. Jamin Interferometer

This interferometer, shown in Fig. 9.7, is used to measure the refractive index of gases. Because the two paths are identical with respect to the cells and the beam splitter and beam combiner, the optical path difference is the difference in the refractive index of the material in the two cells. Typically, one is evacuated. The gas is gradually admitted to the other cell and fringes are counted, or a compensation plate may be used and tipped gradually to change the optical path difference.

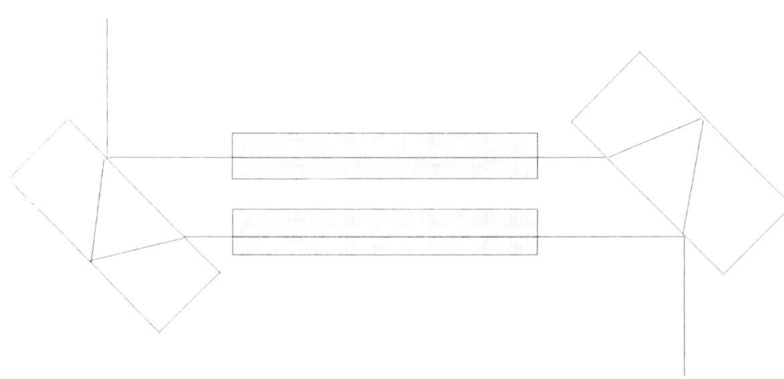

Figure 9.7. Jamin interferometer.

9.11. Rayleigh Interferometer

This interferometer (see Fig. 9.8) is similar to the Jamin, except that the two beams are sent into the cells by way of a lens and apertures.

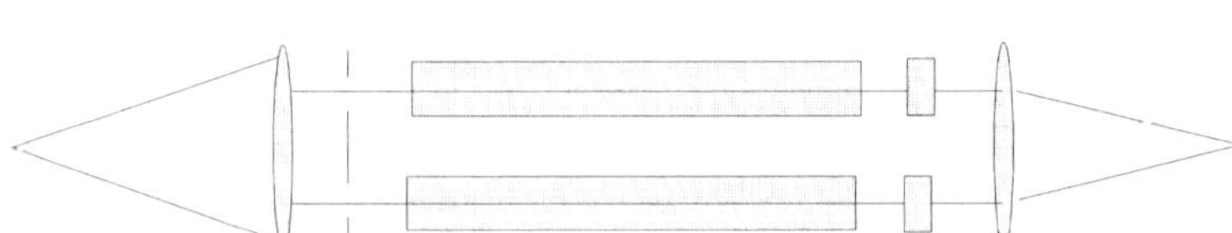

Figure 9.8. Rayleigh interferometer.

9.12. Fizeau Interferometer

This is a simple device (see Fig. 9.9) used for testing optical surfaces, especially flats. The pattern is fringes of equal thickness. It is sometimes useful to tilt the reference plate a little to get linear fringes.

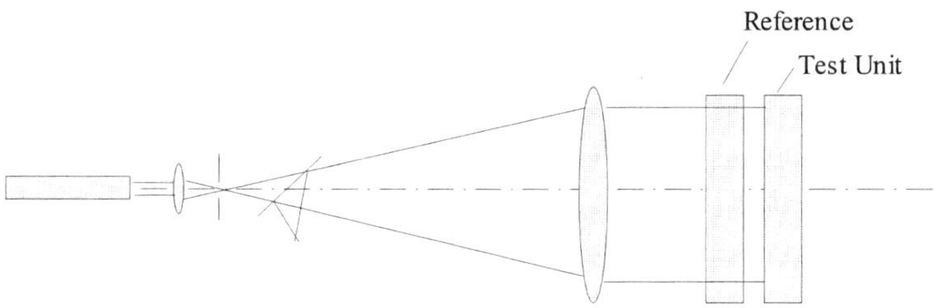

Figure 9.9. Fizeau interferometer.

9.13. Murty Shearing Interferometer

A laser can be used to illuminate a plane-parallel plate with a collimated beam (see Fig. 9.10). The sheared waves are those from the front and the back of the plate. This lateral shear is given by

$$\delta = \frac{d \sin 2\theta}{\sqrt{n^2 - \sin^2 \theta}}.$$

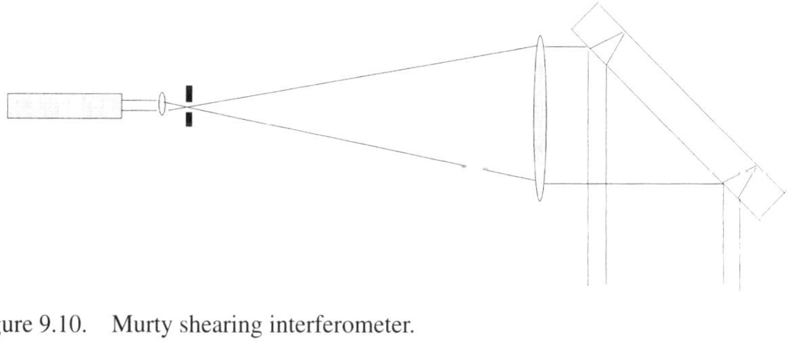

Figure 9.10. Murty shearing interferometer.

9.14. Radial-Shear Interferometer

Figure 9.11 shows how a given wave front can be interfered with a demagnified version of itself. The wave front to be tested is aimed at the center of the spherical surface. Part of it goes through the second surface unrefracted. The reflected part is aimed at the front, planar surface and reflects to a focus at the same center of curvature, from which it expands unrefracted out of the curved surface.

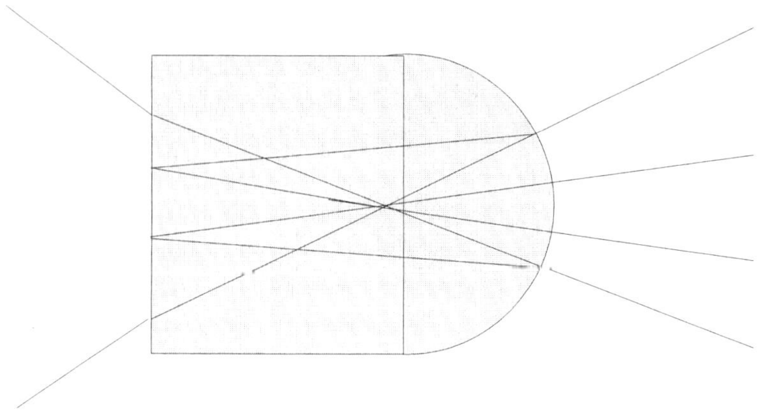

Figure 9.11. Radial-shearing interferometer.

9.15. Scatter-Plate Interferometer

This is often used to test optical surfaces. It is shown in this configuration in Fig. 9.12. The two scatter plates are weak diffusers so that both a diffuse beam and a specular beam are transmitted. The interference of interest takes place between the light not scattered by the first scatter plate and scattered by the second scatter plate and the light scattered by the first scatter plate and not scattered by the second scatter plate.

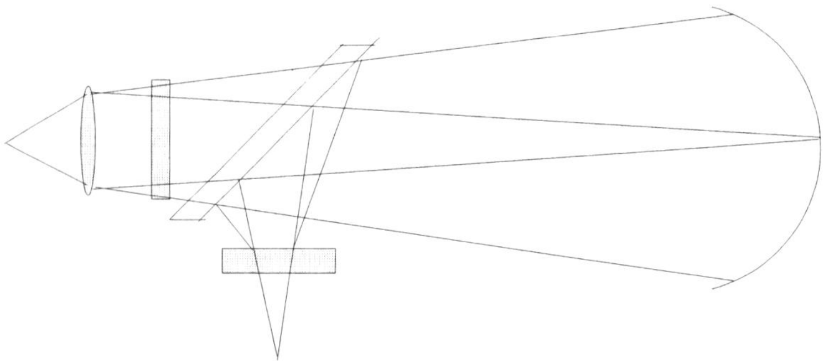

Figure 9.12. Scatter-plate interferometer.

9.16. Multiple-Beam Interferometers

In general, multiple-beam interferometers provide sharper lines than two-beam devices.

9.16.1. Plane-Parallel Plate

The plane-parallel plate has multiple reflections and refractions.

The amplitude pattern for the transmitted light is

$$A(\phi) = \frac{t_1 t_2 e^{-i\phi/2}}{1 - r_1 r_2 e^{-i\phi}},$$

where the t's are the amplitude transmittances, the r's are the amplitude reflectances, and $\phi = (4\pi/\lambda)\, nd \cos\theta$, where n is the refractive index of the plate and d is its thickness. The corresponding intensity is given by

$$I(\psi) = \frac{\tau_1 \tau_2}{1 + \rho_1\rho_2 - 2\sqrt{\rho_1\rho_2}\cos\psi},$$

where the τ's are the intensity transmittances, the ρ's are the intensity reflectances, and ψ is $\phi + \delta_1 + \delta_2$, where the δ's are the phase shifts on reflection.

The corresponding formulas for the reflected pattern are

$$A(\phi) = r_1 + \frac{t_1 t_2 r_2 e^{-i\phi/2}}{1 - r_1 r_2 e^{-i\phi}},$$

$$I(\phi) = 1 - \frac{\tau_1 \tau_2}{1 + \rho_1\rho_2 - 2\sqrt{\rho_1\rho_2}\cos\phi}.$$

If the two surfaces have the same transmittances and reflectances, then the intensity transmittance is given by

$$I(\psi) = \frac{\tau^2}{1 + \rho^2 - 2\rho\cos\psi} = \frac{\tau^2}{(1-\rho)^2 + 4\rho\sin^2(\psi/2)}.$$

The reflected intensity is given by

$$I(\phi) = \frac{2\rho(1 - \cos\phi)}{1 + \rho^2 - 2\rho\cos\phi} = \frac{4\rho\sin^2(\phi/2)}{(1-\rho)^2 + 4\rho\sin^2(\phi/2)}.$$

Fringes in transmission appear bright on a dark background and are the most useful. Fringes in reflection are the reverse and seldom used.

The maximum intensity on transmission is

$$I_{\max} = \frac{\tau^2}{(1-\rho)^2},$$

whereas the minimum is

$$I_{\min} = \frac{\tau^2}{(1+\rho)^2}.$$

The full width at half-maximum is given by $[2(1-\rho)]/\sqrt{\rho}$, and the finesse is $(\pi\sqrt{\rho})/(1-\rho)$.

9.16.2. Fabry–Perot Interferometer

The Fabry–Perot interferometer consists of two plane-parallel plates of material (often silica) that are separated by an air space. The interior surfaces have highly reflective coatings on them. The interference is thus generated by multiple-beam interference in a plane-parallel plate of air. The instrument function is

$$\tau(\nu) = \frac{\tau^2}{(1-\rho)^2 + 4\rho\sin^2(\psi/2)},$$

where

$$\psi = \frac{4\pi\nu nd}{c}\cos\theta + \delta,$$

τ is the (equal) reflectivity of the inside surfaces of the plates, ρ is their reflectivity, ν is the optical frequency, n is the refractive index, d is the thickness of the spacing, c is the speed of light, θ is the angle of the light relative to the normal to the plates inside the Fabry–Perot interferometer, and δ is the phase shift on reflection. The free spectral range is given by

$$\Delta\nu = \frac{1-\rho}{\pi\sqrt{\rho}}\frac{c}{2nd},$$

and the finesse is

$$\frac{\pi\sqrt{\rho}}{1-\rho}.$$

The maximum solid angle that the aperture can subtend at the source is $(2\pi)/Q$, where Q is the resolving power of the instrument. The etendue, or throughput, is given by this times the area of the aperture. The Fabry–Perot interferometer can scan the spectrum by either pressure scanning or mechanical scanning. The former provides a relative shift in wavelength of ~0.0003 wavelength per atmosphere for air. In mechanical scanning, care must be taken to preserve parallelism.

9.16.3. Spherical Fabry–Perot Interferometer

This version uses two, identical spherical mirrors separated by their radius. It is a confocal system. Any ray that has traversed the system four times is coincident with its input ray. The throughput is limited by the spherical aberration and is $2(d/D)^2$, where d is the thickness of the plane Fabry–Perot space and equal to twice the radius for the spherical and D is the useful aperture diameter in both cases. The spherical Fabry–Perot interferometer is easier to align and keep in alignment and has the same free spectral range, finesse, and resolving power as the plane Fabry–Perot interferometer.

9.17. Coherence Length and Time

The coherence time is given by

$$\Delta t = \frac{1}{\Delta\nu},$$

where $\Delta\nu$ is the frequency spread of the source.

The coherence length is given by

$$\Delta l = c\Delta t \approx \frac{c}{\Delta\nu} = \frac{\lambda_0^2}{\Delta\lambda_0},$$

where λ_0 is the mean wavelength *in vacuo*.

9.18. Coherence and Visibility

The mutual coherence function is defined as

$$\Gamma_{12}(\tau) = \langle E_1(t+\tau)E_2^*(t)\rangle$$

and can be normalized to give the complex degree of coherence:

$$\gamma_{12}(\tau) = \frac{\Gamma_{12}(\tau)}{\sqrt{I_1 I_2}} = |\gamma_{12}(\tau)|\, e^{[i\alpha_{12}(\tau)-2\pi\nu\tau]}.$$

The visibility is given by

$$V = \frac{2\sqrt{I_1 I_2}}{I_1 + I_2}|\gamma_{12}(\tau)| \rightarrow |\gamma_{12}(\tau)|,$$

where the arrow indicates the result when the two intensities are equal. In that case, the visibility is equal to the modulus of the complex degree of coherence.

9.19. Resolution and Resolving Power

The resolution of an interferometer used as a spectrometer is usually given as the full width at half-maximum of the spectral line.

The resolving power is defined as the ratio of the central wavelength to the wavelength resolution, or the equivalent ratio in frequency.

9.20. Free Spectral Range and Finesse

The free spectral range is the frequency or wavelength interval between different orders (of maxima). The finesse is the ratio of the free spectral range to the resolution.

9.21. Throughput

The throughput is a measure of the sensitivity of the instrument and is the product of the solid angle of collection and the aperture area. For instance, it is the area of a detector times the speed of the optics.

9.22. Phase Shifting

The equation for the intensity resulting from two-beam interference contains three unknowns, the two individual intensities and the phase difference between the two interfering beams. If three or more measurements are made of the intensity of the two-beam interference, all three unknowns can be determined. We can do this by varying the phase difference between the two beams in a known manner. This phase-shifting interferometry is a powerful tool for measuring phase distributions and related height distributions. Usually the phase difference is changed by 90° between consecutive measurements of the interference intensity. In this case, the three intensities can be written as

$$\begin{aligned} I_a &= I_1 + I_2 + 2\sqrt{I_1 I_2}\cos(\phi - \pi/2), \\ I_b &= I_1 + I_2 + 2\sqrt{I_1 I_2}\cos(\phi), \\ I_c &= I_1 + I_2 + 2\sqrt{I_1 I_2}\cos(\phi + \pi/2). \end{aligned}$$

Then the phase distribution is given by

$$\phi = \arctan\left(\frac{I_a - I_c}{-I_a + 2I_b - I_c}\right).$$

9.23. Van Cittert–Zernike Theorem

For small sources and large separation distances, the mutual coherence function and therefore the visibility are equal to the normalized Fourier transform of the source flux density distribution. This may be written as

$$\mu_{12} = \frac{e^{i\phi} \iint I(\xi,\eta) e^{-ik(p\xi+q\eta)} \mathrm{d}\xi \mathrm{d}\eta}{\iint e^{-ik(p\xi+q\eta)} \mathrm{d}\xi \mathrm{d}\eta},$$

where ϕ is a phase shift, ξ and η are spatial frequencies, and p and q are direction cosines.

9.24. Fringe Localization

The localization of fringes can be important if they are to be recorded, for instance, on film. Generally, interference patterns are not localized if they arise from point sources and are if they arise from extended sources.

Further Reading

P. Hariharan, *Optical Interferometry,* Academic, 1985.
P. Hariharan, *Basics of Interferometry,* Academic, 1992.
C. Candler, *Modern Interferometers,* 1951.
M. Born and E. Wolf, *Principles of Optics,* Pergamon, 1993.

10

Optical Coatings

Angus Macleod

10.1. Fundamentals

All phenomena are described with reference to a linearly polarized, plane, harmonic, electromagnetic wave. In the optical regime the relative permeability μ_r is unity so only the electric field takes part in interactions with the medium through which the wave propagates. Thus the wave amplitude and phase are those of the electric field. The magnetic-field amplitude may be derived from the electric field. The electric field, the magnetic field, and the direction of propagation form a right-handed set.

Complex waves are used. They are appropriate for all linear operations. The only nonlinear operation used is the multiplication of the electric and the magnetic fields in deriving the flux density of a wave, and then a special form of the operation will be used.

In an absorbing medium the complex index of refraction is used; $n \rightarrow n - i\kappa$ is the complex index of refraction, where $n = c/v$ and κ is the extinction coefficient.

The electric field strength of a linearly polarized plane, harmonic wave propagating in the z direction may be written as

$$\underline{E} = E\, \exp[i(\omega t - kz + \phi)] = E\, \exp(i\phi)\exp[i(\omega t - kz)],$$

where E is normal to the z axis and is fixed in the polarization direction, $E \exp(-i\phi)$ is the complex amplitude, $E \exp[i(\omega t - kz)]$ is the phase factor, $\omega = (2\pi/t)$ is the angular frequency, $k = \{[2\pi(n - i\kappa)]/\lambda\}$ is the (complex) wave number, λ is the free-space wavelength, and ϕ is the relative phase.

The magnetic field **H** has a direction that is normal to both **E** and z in the sense that E, H, and z form a right-handed set. The characteristic admittance of the medium is defined as

$$y = \frac{E}{H}. \tag{10.1}$$

At optical frequencies

$$y = (n - i\kappa)\, Y_0 = (n - i\kappa) \text{ free-space units,}$$

where Y_0 (= 1/377 S) is the characteristic admittance of free space. Flux density is the average power per unit area propagated by the wave. This is calculated from the product of the electric and the magnetic fields and is therefore a nonlinear quantity. For complex waves only, the magnitude of the average flux density of a linearly polarized, plane harmonic wave is given by

$$M = \frac{1}{2}\,\mathrm{Re}(EH^*) = \frac{1}{2}\,\mathrm{Re}(y)EE^*. \tag{10.2}$$

Admittances described by lowercase letters are inherent properties of materials; those described by uppercase letters are properties of structures that include interference effects. An exception is Y_0, which denotes the admittance of free space.

10.2. Simple Surface

Ideal surfaces, are simple, zero-thickness boundaries between two homogeneous and isotropic optical media. When a harmonic wave meets such a surface, some of the energy passes through the surface to form a transmitted wave and part of the energy is reflected to form a reflected wave. The boundary conditions determine how much light is reflected and how much is transmitted. The total components (including all reflections) of the electric and magnetic fields parallel to the surface, the tangential components, are continuous across the boundary.

10.2.1. Normal Incidence

It is conventional to arrange the coordinate system so that the propagation is along the z axis, which is normal to the surface. The phase factor of the reflected wave is of the form

$$\exp[i(\omega t + kz)], \tag{10.3}$$

and to avoid problems in applying the boundary conditions the surface is taken as $z = 0$. Then the phase factors of all the involved waves become $\exp(i\omega t)$ at the boundary and can be canceled out of the equations. The sign convention for positive directions of the vectors is shown in Fig. 10.1.

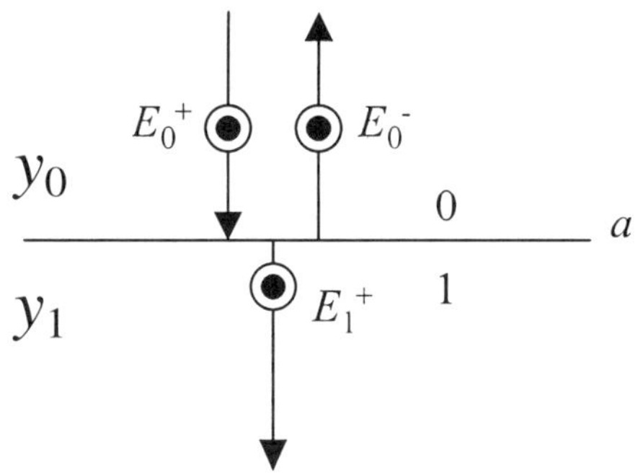

Figure 10.1. Sign convention for electric fields for normal incidence.

Surface quantities are denoted by the subscript *a:*

$$Y_a = \frac{H_a}{E_a}, \tag{10.4}$$

where

E_a, H_a = total tangential components of amplitude (including all beams), (10.5)

Y_a = surface admittance, (10.6)

$$r = \text{amplitude-reflection coefficient} = \frac{\mathscr{E}_{\text{reflected}}}{\mathscr{E}_{\text{incident}}} = \frac{(y_0 - Y_a)}{(y_0 + Y_a)}. \tag{10.7}$$

In the case of a simple surface,

$$Y_a = y_2, \tag{10.8}$$

so that

$$r = \frac{(y_0 - y_1)}{(y_0 + y_1)} = \frac{(n_0 - n_1) - i(\kappa_0 - \kappa_1)}{(n_0 + n_1) - i(\kappa_0 + \kappa_1)} \tag{10.9}$$

Then the amplitude-transmission coefficient is

$$\tau = \frac{\mathscr{E}_{\text{transmitted}}}{\mathscr{E}_{\text{incident}}}, \tag{10.10}$$

$$t = \frac{2y_0}{(y_0 + y_1)} = \frac{2(n_0 - i\kappa_0)}{(n_0 + n_1) - i(\kappa_0 + \kappa_1)}. \tag{10.11}$$

The phase shift on reflection is

$$\varphi = \arctan\left[\frac{2(n_0\kappa_1 - n_1\kappa_0)}{(n_0^2 - n_1^2 + \kappa_0^2 - \kappa_1^2)}\right], \tag{10.12}$$

and, on transmission, is

$$\zeta = \arctan\left[\frac{n_0\kappa_1 - n_1\kappa_0}{n_0(n_0 + n_1) + \kappa_0(\kappa_0 + \kappa_1)}\right]. \tag{10.13}$$

When y_0 is real, and only then, the flow of energy in the incident medium can be separated into an incident flux density and a reflected flux density. Then reflectance, defined as the ratio of the reflected and the incident flux densities, is given by

$$\begin{aligned}\rho &= rr^* \\ &= \frac{(y_0 - Y_a)(y_0 - Y_a)^*}{(y_0 + Y_a)(y_0 + Y_a)^*} \\ &= \frac{(y_0 - y_1)(y_0 - y_1)^*}{(y_0 + y_1)(y_0 + y_1)^*} \\ &= \frac{(n_0 - n_1 + i\kappa_1)(n_0 - n_1 + i\kappa_1)^*}{(n_0 + n_1 - i\kappa_1)(n_0 + n_1 - i\kappa_1)^*},\end{aligned} \tag{10.14}$$

and the transmittance, the ratio of the transmitted and the incident flux densities, is given by

$$\begin{aligned}\tau &= \frac{\mathrm{Re}(y_1)}{\mathrm{Re}(y_0)}\tau\tau^* \\ &= \frac{4\,\mathrm{Re}(y_0)\mathrm{Re}(y_1)}{(y_0 + y_1)(y_0 + y_1)^*},\end{aligned} \tag{10.15}$$

where y_0 must be real so that $\mathrm{Re}(y_0) = y_0$. In terms of n and κ, this is

$$\tau = \frac{4n_0n_1}{(n_0 + n_1 - i\kappa_1)(n_0 + n_1 - i\kappa_1)^*}. \tag{10.16}$$

Because only a single surface is involved, no energy is absorbed, so that

$$\rho + \tau = 1. \tag{10.17}$$

10.2.2. Oblique Incidence

At oblique incidence the effect of the surface is still to produce a reflected and a transmitted beam. Their directions are both in the plane of incidence. The reflected-beam direction is inclined to the surface normal at the same angle as that of the incident direction but on the opposite side, whereas the direction of propagation of the transmitted beam is related to the incident direction by Snell's Law:

$$n_0 \sin \vartheta_0 = n_1 \sin \vartheta_1 = (n = i\kappa)\sin \vartheta, \tag{10.18}$$

where the right-hand side refers to an absorbing emergent medium. For infinite plane waves we define reflectance and transmittance to involve the components of flux density normal to the surface so that

$$\rho + \tau = 1. \tag{10.19}$$

Because these flux densities involve the tangential components of the fields and because these fields are those taking part in the boundary conditions, we define the amplitude coefficients, r and t, in terms of the tangential components also. Next we must split the waves into two components, one with the electric vector parallel to the plane of incidence, known as p-polarized light, and one with the electric vector perpendicular to the plane of incidence, known as s-polarized light. Only in these two cases will the polarization state of the incident beam be preserved in the reflected and the transmitted beams. Any other polarization is expressed as a combination of s- and p-polarized components. We must also redefine our sign convention, as shown in Fig. 10.2.

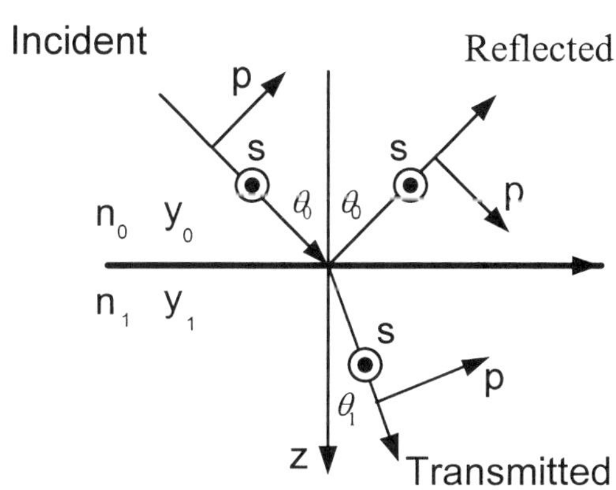

Figure 10.2. Sign convention for the positive directions of the electric field for oblique incidence; p and s indicate the two planes of polarization.

For s polarization, the optical admittance is

$$\eta_s = y \cos \vartheta = (n - i\kappa)\cos \vartheta. \tag{10.20}$$

For p polarization, the optical admittance is

$$\eta_p = y/\cos\vartheta = (n - i\kappa)/\cos\vartheta. \tag{10.21}$$

These admittances are measured in free-space units.

ϑ is given by Snell's law:

$$n_0 \sin\vartheta_0 = (n - i\kappa)\sin\vartheta, \tag{10.22}$$

where the absorbing alternative is used, and the phase thickness of the layers becomes, for each plane of polarization,

$$\delta = (2\pi/\lambda)(n - i\kappa)d\cos\vartheta. \tag{10.23}$$

In this general case, y and n are complex, implying that ϑ is also complex. It avoids difficulties over signs and quadrants if the following expressions are used:

$$(n - i\kappa)\cos\vartheta = (n^2 - \kappa^2 - n_0^2\sin^2\vartheta_0 - 2in\kappa)^{1/2}; \tag{10.24}$$

the correct solution is in the fourth quadrant. Then, from Eqs. (10.20) and (10.21),

$$\eta_s = (n^2 - \kappa^2 - n_0^2\sin^2\vartheta_0 - 2in\kappa)^{1/2}, \tag{10.25}$$

$$\eta_p = (n - i\kappa)^2/\eta_s. \tag{10.26}$$

It is important to note that the admittances of the incident and the exit media must also be converted according to Eqs. (10.20) and (10.21). The arrangement for the calculation parameters at oblique incidence then becomes

$$r = \frac{(\eta_0 - \eta_1)}{(\eta_0 + \eta_1)},$$

$$\rho = \frac{(\eta_0 - \eta_1)(\eta_0 - \eta_1)^*}{(\eta_0 + \eta_1)(\eta_0 + \eta_1)^*}, \tag{10.27}$$

where, as before, η_1 must be real, and

$$t = \frac{2\eta_0}{(\eta_0 + \eta_1)},$$

$$\tau = \frac{4\eta_0 \operatorname{Re}(\eta_1)}{(\eta_0 + \eta_1)(\eta_0 + \eta_1)^*}, \tag{10.28}$$

$$\rho + \tau = 1. \tag{10.29}$$

Note that the amplitude-transmittance and amplitude-reflection components are not strictly the Fresnel components. In fact, all except t_p do coincide with the Fresnel expressions.

10.3. Multilayer Coating

10.3.1. Normal Incidence

Note the numbering system. Layers are numbered in order from 1 at the incident-medium side.

In the case of a simple interface, Y, the admittance of the surface is just $n - i\kappa$, the optical admittance (in free-space units) of the second medium. Where the reflectance of a multilayer is concerned, we can use the same expressions, provided that Y is the optical admittance of the front surface of the multilayer. We calculate it from the following expressions:

$$Y = \frac{C}{B}, \tag{10.30}$$

where

$$\begin{bmatrix} B \\ C \end{bmatrix} = \left(\prod_{j=1}^{q} \begin{bmatrix} \cos \delta_j & \dfrac{i \sin \delta_j}{y_j} \\ i y_j \sin \delta_j & \cos \delta_j \end{bmatrix} \right) \begin{bmatrix} 1 \\ y_{\text{sub}} \end{bmatrix}, \tag{10.31}$$

$\delta_j = 2\pi(n_j - i\kappa_j)d_j/\lambda$,

δ = phase difference,

d_j = layer geometrical thickness

$y_j = (n_j - i\kappa_j)$ = film optical constants,

$y_{\text{sub}} = (n_{\text{sub}} - i\kappa_{\text{sub}})$ = substrate optical constants.

Provided the optical admittance of the incident medium y_0 is real, then

$$\rho = \frac{(y_0 B - C)(y_0 B - C)^*}{(y_0 B + C)(y_0 B + C)^*}, \tag{10.32}$$

$$\varphi = \arctan\left[\frac{iy_0(CB^* - BC^*)}{y_0^2 BB^* - CC^*}\right], \tag{10.33}$$

$$\tau = \frac{4y_0 \operatorname{Re}(y_{\text{sub}})}{(y_0 B + C)(y_0 B + C)^*}. \tag{10.34}$$

The quadrant of φ (see Table 10.1) is defined by the signs of the numerator and denominator in Eq. (10.33).

Table 10.1. Quadrant of Phase Shift

Numerator	+	+	–	–
Denominator	+	–	–	+
Quadrant	1st	2nd	3rd	4th

Equations (10.32) and (10.34) now represent quantities associated with a multilayer rather than with a simple boundary between two media. In this case $\rho + \tau$ is no longer necessarily unity and we write

$$\alpha = 1 - \rho - \tau, \tag{10.35}$$

where α is the absorptance of the multilayer.

Potential transmittance ψ is defined as the fraction of light entering a system that actually emerges. For an entire system we have

$$\psi = \frac{\tau}{1 - \rho}. \tag{10.36}$$

An alternative expression that applies either to a complete system or to a single layer that is part of the system is

$$\psi = \frac{\operatorname{Re}(y_{\text{exit}})}{\operatorname{Re}(BC^*)}, \tag{10.37}$$

where y_{exit} is the admittance at the rear of the assembly, subassembly, or layer and B and C refer to the front surface of the assembly, subassembly, or layer. It can be shown that for any assembly the potential transmittance of the entire assembly is the product of the individual potential transmittances.

10.3.2. Oblique Incidence

The expressions for the calculation of the properties of multilayers at oblique incidence are exactly the same as those for normal incidence but they use the appropriate *s* or *p* admittances. δ, the phase thickness of the layer, is given for both polarizations by

$$\begin{aligned}\delta &= 2\pi\frac{(n-i\kappa)d}{\lambda}\cos\vartheta,\\ &= \frac{2\pi d}{\lambda}(n^2-\kappa^2-n_0^2\sin^2\vartheta_0-2in\kappa)^{1/2}.\end{aligned} \tag{10.38}$$

Then

$$\begin{bmatrix}B\\C\end{bmatrix}=\left(\prod_{j=1}^{q}\begin{bmatrix}\cos\delta_j & \dfrac{i\sin\delta_j}{\eta_j}\\ i\eta_j\sin\delta_j & \cos\delta_j\end{bmatrix}\right)\begin{bmatrix}1\\ \eta_{\mathrm{sub}}\end{bmatrix}, \tag{10.39}$$

$$\rho=\frac{(\eta_0B-C)(\eta_0B-C)^*}{(\eta_0B+C)(\eta_0B+C)^*}, \tag{10.40}$$

$$\varphi=\arctan\left[\frac{i\eta_0(CB^*-BC^*)}{\eta_0^2BB^*-CC^*}\right], \tag{10.41}$$

$$\tau=\frac{4\eta_0\,\mathrm{Re}(\eta_{\mathrm{sub}})}{(\eta_0B+C)(\eta_0B+C)^*}, \tag{10.42}$$

$$\zeta=\arctan\left[\frac{i(\eta_0B+C-\eta_0B^*-C^*)}{\eta_0B+C+\eta_0B^*+C^*}\right], \tag{10.43}$$

$$\psi=\frac{\mathrm{Re}(\eta_{\mathrm{sub}})}{\mathrm{Re}(BC^*)}. \tag{10.44}$$

10.4. Metallic, Dielectric, and Semiconductor Films

The principal differences among the three classes of solid film materials are derived from the behavior of their electrons. In metals, the valence electrons are essentially free so they move through the solid. In dielectrics they are bound. In semiconductors the electrons are bound but the bonds are quite weak, and so there are always some free electrons at room temperature. The electric field of the light wave interacts with the electrons. In the case of the bound electrons, an oscillating dipole is produced that reradiates the energy obtained from the light. The radiated field combines with the primary field to produce a resultant wave that moves more slowly and therefore experiences an elevated refractive index. The greater the interaction, the greater the index.

Free electrons simply absorb the energy they extract from the field. Thus bound electrons increase n whereas free electrons increase κ. Dielectric materials therefore have vanishingly small κ and so their properties are determined by n, which varies only slightly with λ. Metals are characterized by a large κ, which increases roughly with λ. The optically useful metals have very small n. The semiconductors have some free electrons and so they have a κ usually slightly elevated compared with a dielectric, but the major effect is a large interaction with bound electrons that produces an exceedingly large refractive index. Because the binding is weak, short-wavelength photons can actually free the electrons and so they have high absorption at shorter wavelengths with a sharp transition between low- and high-loss regions. Thus they are useful mainly in the infrared region, and for thin-film design purposes they are treated as dielectrics.

The values of the optical constants for aluminum over the near-ultraviolet, visible, and near-infrared regions are shown in Fig. 10.3.

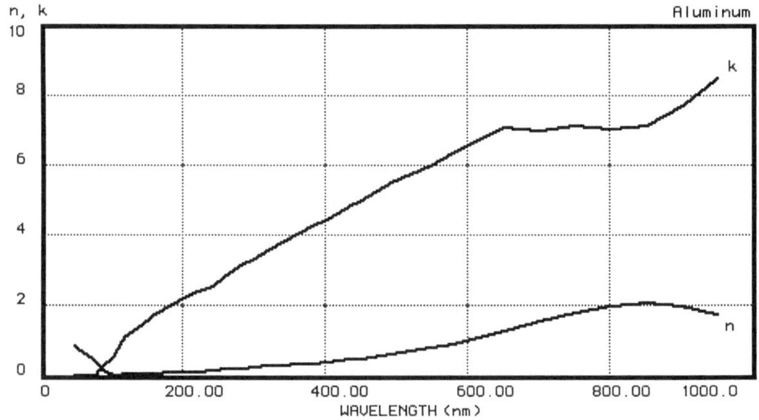

Figure 10.3. Optical constants n and κ of aluminum in the near-ultraviolet, visible, and near-infrared regions.

Note the rise in κ that is roughly proportional to λ. This means that the absorption coefficient, $\alpha = 4\pi\kappa/\lambda$, is largely independent of wavelength.

Table 10.2 lists the ideal properties of metals and dielectrics. Real materials depart more or less from these ideals.

Dielectric layers exhibit interference effects with properties that cycle as wavelength or thickness changes. The phase thickness $\delta = 2\,\pi nd/\lambda$ is the most important quantity. This is the change in phase suffered by a wave on one traversal of the film. As λ increases, δ becomes smaller, as n is changing only slightly, and so the properties of the film weaken. Metals ideally do not exhibit any cyclic properties. The reflectance simply increases or decreases in the same sense as β and/or κ. Because β is substantially constant whereas κ increases with λ, the properties of the metallic film become stronger with wave-

Table 10.2. Ideal Properties of Dielectrics and Metals

Dielectrics	Metals
$\kappa = 0$	$y = -i\kappa Y_0$
$y = nY_0$, n independent of λ	$\kappa \propto \lambda$
$\delta = \frac{2\pi nd}{\lambda} \propto \frac{1}{\lambda}$	$\beta = \frac{2\pi\kappa d}{\lambda} = \text{constant}$
$y = \text{constant}$	$y \propto \lambda$
Become weaker, i.e., τ increases as λ increases	Become stronger, i.e., ρ increases as λ increases
Progressive wave $Ee^{i(\omega t - \kappa z)}$	Evanescent wave $Ee^{(-\alpha z)/2}\, e^{i\omega t}$
Real dielectrics show lower losses than metals	Real metals show higher losses than dielectrics

length. Ideally, a dielectric layer should be most useful in a coating with high transmittance over an extended spectral region while a metal would be most appropriate in a coating with high reflectance over a wide region. If many layers are combined into a single coating then a dielectric assembly would lend itself to reflect a band of shorter wavelengths and transmit a wide range of longer wavelengths whereas a mixture of dielectric and metallic layers would be more suitable for a coating to transmit a shorter band of wavelengths and reflect all longer ones. These ideal properties are summarized in Fig. 10.4.

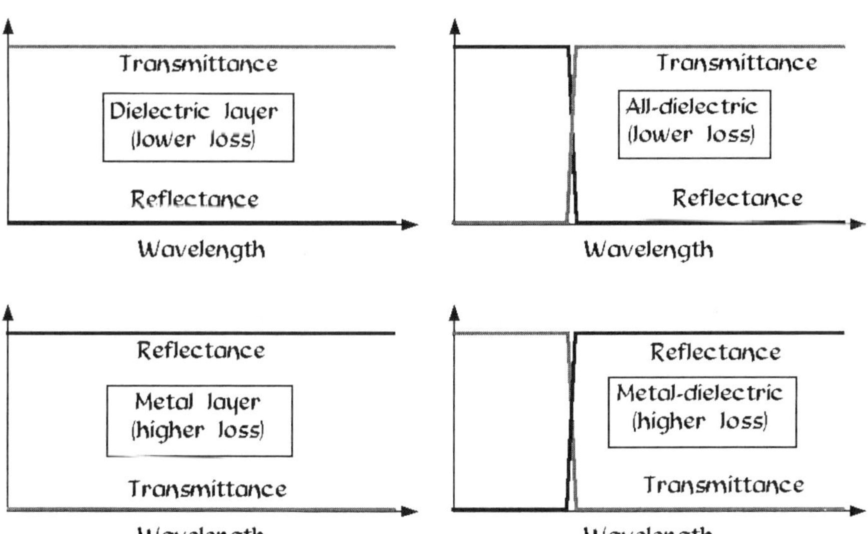

Figure 10.4. Broad characteristics of single-layer (left) and multilayer (right) coatings of ideal materials.

10.5. g and λ

A useful form for phase thickness involving dimensionless quantities is

$$\delta = 2\pi \frac{nd}{\lambda_0} \frac{\lambda_0}{\lambda}, \tag{10.45}$$

where λ_0 is the reference wavelength and $g = (\lambda_0/\lambda)$. Note that the relationship between δ and g is linear whereas that between δ and λ is nonlinear.

10.6. Quarter-Wave and Half-Wave Layers

The discussion in this section concerns lossless dielectric layers that represent the vast majority of thin films used in optical coatings.

Quarter-wave film: $\delta = \pi/2$. A quarter-wave layer of characteristic admittance y_f transforms the admittance of the substrate surface from y_{sub} to y_f^2/y_{sub}; this is the quarter-wave rule [which can be derived from Eqs. (10.30) and (10.31)]:

y_0

Quarter-wave thickness y_f $\equiv \dfrac{y_f^2}{y_{\text{sub}}}$

y_{sub}

The reflectance of a substrate coated with a single quarter-wave layer is

$$R = \left[\frac{(y_0 - y_f^2/y_{\text{sub}})}{(y_0 + y_f^2/y_{\text{sub}})} \right]^2. \tag{10.46}$$

Series of quarter waves:

y_0

Quarter-wave thickness y_1, y_2, y_3, y_4, y_5 $\equiv \dfrac{y_1^2\, y_3^2\, y_5^2}{y_2^2\, y_4^2\, y_{\text{sub}}}$

y_{sub}

If the number of layers is odd then the substrate will appear in the denominator, as shown. If the number is even then the substrate will appear in the numerator.

Reflectance of a series of quarter waves:

$$R = \left[\frac{\left(y_0 - \dfrac{y_1^2\, y_3^2\, y_5^2}{y_2^2\, y_4^2 y_{sub}}\right)}{\left(y_0 + \dfrac{y_1^2\, y_3^2\, y_5^2}{y_2^2\, y_4^2\, y_{sub}}\right)}\right]^2. \tag{10.47}$$

Half-wave layer: $\delta = \pi$. A half-wave layer has no effect on the admittance of the substrate (absentee layer). It can be considered as two identical quarter waves:

$$Y = \frac{y_f^2}{(y_f^2/y_{sub})} = y_{sub}. \tag{10.48}$$

Example: One quarter wave of MgF_2, of admittance 1.38, on a glass substrate, of admittance 1.52, in an incident medium of air, of admittance 1.00, gives

$$\rho = \left[\frac{1.00 - \dfrac{1.38^2}{1.52}}{1.00 + \dfrac{1.38^2}{1.52}}\right]^2 = 1.25\%. \tag{10.49}$$

Compare with 4.25% for uncoated glass. This is the simplest antireflection coating.

Notation: Because many coatings are based on quarter-wave layers, the shorthand notation used for filter designs is based on symbols for quarter waves. Capital letters are used to denote quarter waves at λ_0: H is usually of high index, L of low, but any letters can be used to designate any dielectric quarter wave. A multiplier can be used to change the thickness. 1.25L indicates a layer that is 5/16 wave in thickness, for example. HH or 2H denotes a half wave. A series of layers in vertical bars and raised to a power of x indicates that the structure is repeated x times. Often the incident and the emergent media will also be included in the expression.

Example:

$$\text{air}|\text{HLHLHLH}|\text{glass} = \text{air}|(\text{HL})^3\,\text{H}|\text{glass}.$$

10.7. Antireflection Coatings

Perfect single-layer quarter wave: The perfect antireflection coating (at a single wavelength) requires that $y_f = \sqrt{(y_0 y_{sub})}$ as a result of the quarter-wave rule. For glass (1.52) in air (1.00), the film must have a characteristic admittance of 1.23. It is impossible to find such an admittance in a sufficiently rugged film. Tougher MgF_2 layers, with poorer optical performances (see Fig. 10.5), are normally used. Residual reflectance at minimum is given by Eq. (10.46).

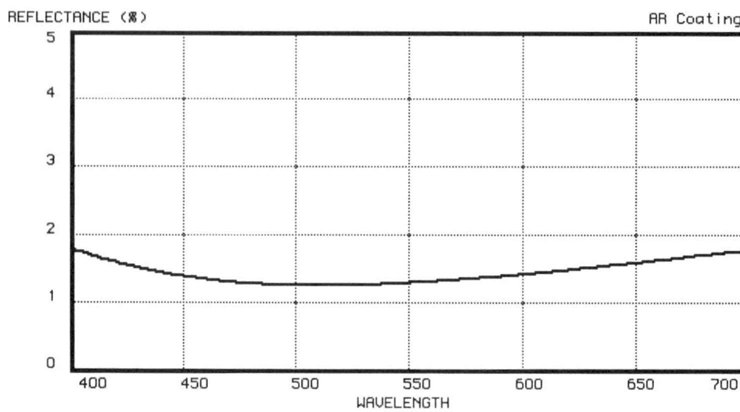

Figure 10.5. Performance of a single-layer antireflection (AR) coating consisting of MgF_2 on glass in air with a 510-nm λ_0.

Two-layer coatings: A perfect *quarter–quarter* coating (see Fig. 10.6) should satisfy

$$y_1^2/y_2^2 = y_0/y_{\text{sub}}. \tag{10.50}$$

A good rule of thumb is to pick a set of materials to give similar reflectances at each interface that give the broadest characteristic. If we keep 1.38 as the outermost layer then 1.70 should be the other.

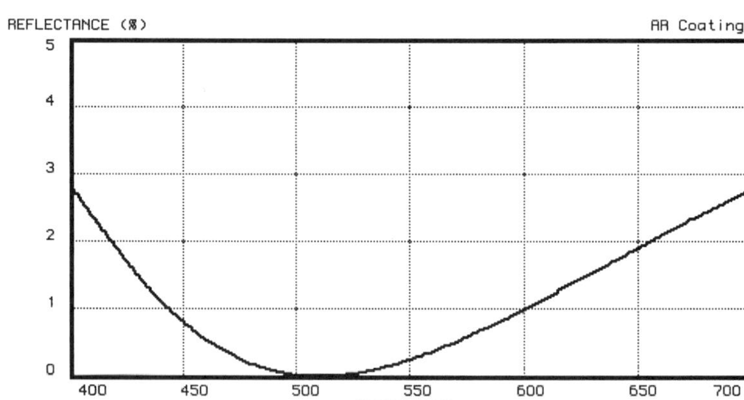

Figure 10.6. Antireflection coating of air | L(1.38) H(1.70) | glass plotted in terms of λ with a value of 510 nm for λ_0.

A *V coat* (see Fig. 10.7) uses prescribed materials and thicknesses that are not necessarily quarter waves. Equation (10.50) represents a limiting case. A more general expression calculates the necessary phase thicknesses for two layers that have prescribed characteristic admittances at the outset:

$$\tan^2 \delta_1 = \frac{(y_{sub} - y_0)(y_2^2 - y_0 y_{sub}) y_1^2}{(y_1^2 y_{sub} - y_0 y_2^2)(y_0 y_{sub} - y_1^2)}, \tag{10.51}$$

$$\tan^2 \delta_2 = \frac{(y_{sub} - y_0)(y_0 y_{sub} - y_1^2) y_2^2}{(y_1^2 y_{sub} - y_0 y_2^2)(y_2^2 - y_0 y_{sub})}. \tag{10.52}$$

Each equation has two solutions and they must be correctly paired, usually by trial and error.

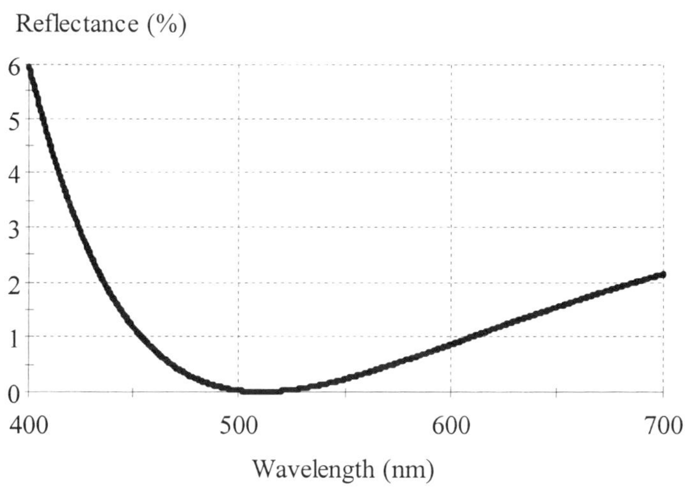

Figure 10.7. Typical V-coat characteristic with TiO_2 (2.349) and MgF_2 (1.385) as coating materials. Design: air | 1.301L 0.204H | glass with value of 510 nm for λ_0.

Quarter–half–quarter coating: A half-wave layer inserted between the two layers of the quarter–quarter coating gives a broader performance (see Fig. 10.8).

Four-layer coating: A high-index half-wave layer inserted inside the low-index layer of the V coat can also flatten the performance, and although the design now has four layers it uses only two materials (see Fig. 10.9).

10.8. Reflectance-Increasing Coatings

Quarter-wave stack: High reflectance should have alternate high- and low-index quarter waves in series:

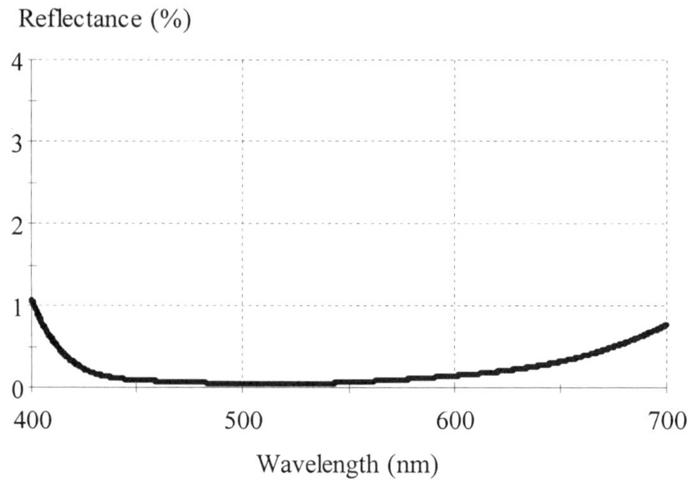

Figure 10.8. Quarter–half–quarter antireflection coating characteristic in which MgF_2(L), ZrO_2 (H), and Al_2O_3 (A) are used in the design air | LHHA | glass with a value of 510 nm for λ_0.

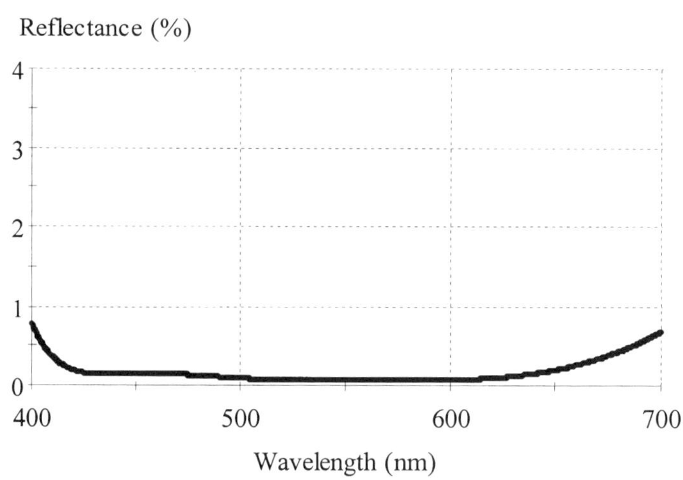

Figure 10.9. Four-layer antireflection coating in which MgF_2 and ZrO_2 are used. The starting design was air | LHH0.25L0.25H | glass with 510-nm λ_0. The design was then refined.

air | HLHLHLHL···H | glass = air | $(\mathrm{HL})^x$H | glass.

y_0

y_H

y_L

Quarter-wave thickness y_H $\equiv \dfrac{y_H^2\, y_H^2\, y_H^2}{y_L^2\, y_L^2\, y_{sub}}$

y_L

y_H

y_{sub}

$$\rho = \left[\frac{y_0 - \dfrac{y_H^{2x}}{y_L^{2x-2} y_{sub}}}{y_0 + \dfrac{y_H^{2x}}{y_L^{2x-2} y_{sub}}}\right]^2. \tag{10.53}$$

The characteristic of a typical quarter-wave stack is as shown in Fig. 10.10.

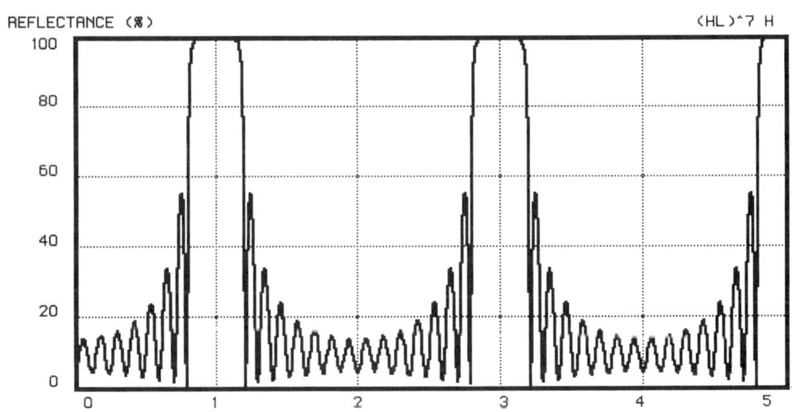

Figure 10.10. Reflectances of a quarter-wave stack of 15 layers of alternating admittances 2.4 and 1.38.

The width of the high-reflectance zone is given by $2\Delta g$, where

$$\Delta g = \frac{2}{\pi} \arcsin \frac{y_H - y_L}{y_H + y_L}. \tag{10.54}$$

Effect of absorption: When the extinction coefficients are not zero then the coating exhibits absorption. At λ_0 this is given by the following approximate expressions:

$$\alpha = \frac{2\pi n_0(\kappa_H + \kappa_L)}{(n_H^2 - n_L^2)} \text{ (final layer H)}, \qquad (10.55)$$

$$\alpha = \frac{2\pi(n_L^2\kappa_H + n_H^2\kappa_L)}{n_0(n_H^2 - n_L^2)} \text{ (final layer L)}. \qquad (10.56)$$

Extended-zone high-reflectance coatings: A typical design is

$$\text{air}|\text{HLHLHLHLHLHLH L}\mathit{HLHLHLHLHLHLH}|\text{glass}, \qquad (10.57)$$

where *H* and *L* indicate a different reference wavelength from that of H and L. The layer L is a quarter wave at an intermediate reference wavelength to suppress a reflectance dip that would otherwise occur. It is not critical. Performance is shown in Fig. 10.11. The oscillatory behavior is typical of extended-zone coatings.

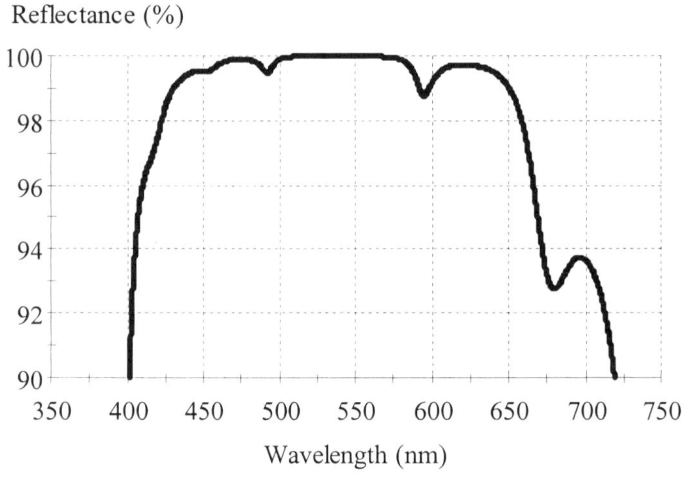

Figure 10.11. The characteristic curve of the extended-zone reflector, design (10.57). The design is air | $(HL)^5$ H1.15L1.3H$(1.3L1.3H)^5$ | glass with ZnS (H) and Na_3AlF_6 (L). λ_0 is 470 nm.

10.9. Filters Based on Quarter-Wave Stack

Transmittance of quarter-wave stack: Figure 10.12 shows various types of components that can be constructed from a quarter-wave stack. The main problem is the ripple, and this can be dealt with fairly satisfactorily.

The term *dichroic* refers to the fact that one wavelength range is split from another. Special types of dichroic filters reflect the infrared part and transmit the luminous part of the output from a light source and are known as *heat-reflecting filters* or, sometimes, *hot mirrors*. A component that reflects the luminous part and transmits the infrared is known as a *cold mirror*.

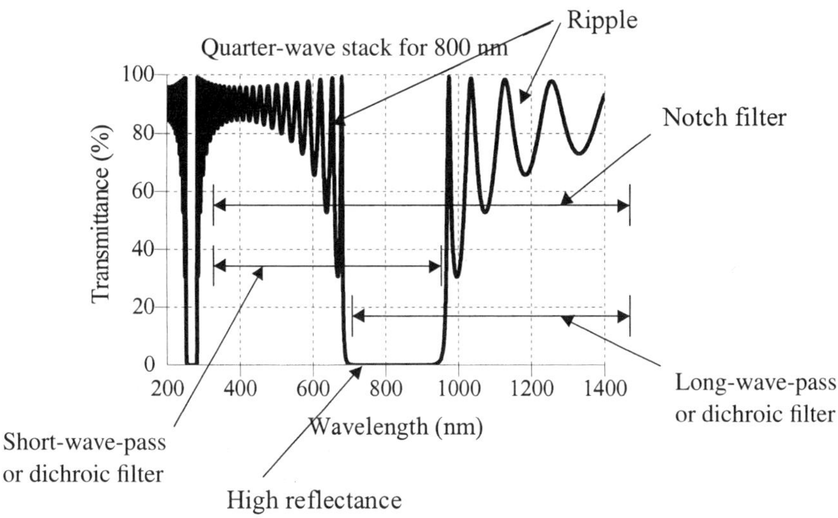

Figure 10.12. Transmittance characteristic of a typical quarter-wave stack, with 800-nm λ_0, showing some of the various basic components that can be constructed from it.

A simple technique for reducing ripple involves changing the thickness of the outermost layers so that they better match the inner core to the surrounding media. A particularly simple form of this uses H/2 (eighth-wave) layers to produce an improved long-wave-pass filter:

$$\text{air} \left| \frac{\text{H}}{2} \text{ LHLHLHLHLHLHLHL } \frac{\text{H}}{2} \right| \text{glass.} \tag{10.58}$$

The characteristic curve of this design is shown in Fig. 10.13.

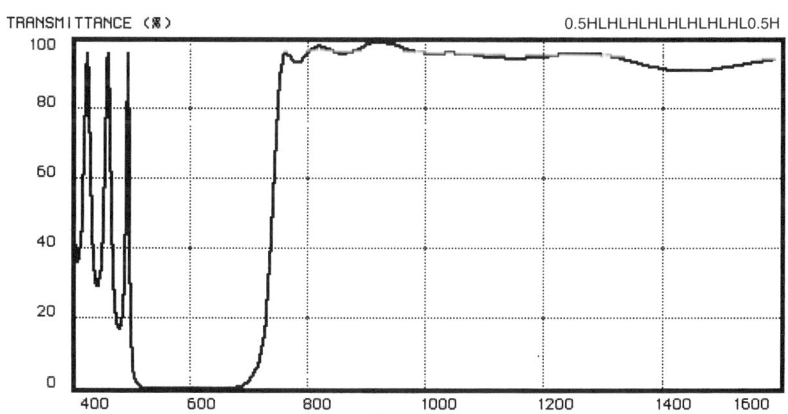

Figure 10.13. Characteristic curve of the long-wave-pass filter of design (10.58), where $y_H = 2.40$, $y_L = 1.38$, $y_{glass} = 1.52$, $y_{air} = 1.00$, and $\lambda_0 = 600$ nm.

For a performance that is even more improved, it is normal to retain the quarter-wave stack as the core of the filter and to add several matching layers on either side. These matching layers are designed by computer refinement to remove the ripple in the pass region.

10.10. Symmetrical Periods

Many of the most powerful analytical design techniques are based on the concept of symmetrical periods. It can be readily shown that any combination of thin films that is symmetrical, that is, one in which the sequence of layers is unchanged when they are listed in reverse order, can be represented by a single equivalent film having an optical admittance and phase thickness that can be calculated from the parameters of the individual layers. If we represent the product of the characteristic matrices of the layers of the symmetrical period by a single matrix $\mathcal{M}$, then we find that $M_{11} = M_{22}$ and that the determinant of the matrix is unity. The product matrix $\mathcal{M}$ can therefore be put in the following form:

$$\begin{bmatrix} M_{11} & M_{12} \\ M_{21} & M_{22} \end{bmatrix} = \begin{bmatrix} \cos\gamma & \dfrac{i\sin\gamma}{y_{\text{equiv}}} \\ iy_{\text{equiv}}\sin\gamma & \cos\gamma \end{bmatrix}, \tag{10.59}$$

where

$$y_{\text{equiv}} = (M_{21}/M_{12})^{1/2}, \tag{10.60}$$

$$\gamma = \arccos(M_{11}). \tag{10.61}$$

y_{equiv} is known as the equivalent admittance and γ as the equivalent phase thickness. The equivalence is valid for dielectric or metallic layers, but is most often applied to combinations of layers that are entirely dielectric.

When the equivalent parameters are calculated for an assembly of *dielectric* layers, it is found that, in some regions, the equivalent phase thickness and the equivalent admittance are imaginary and, in other regions, they are both real. There are no regions where they are not either both real or both imaginary. Regions where both are imaginary may be thought of as representing a pseudometallic behavior (only at optical frequencies) and are therefore associated with stop bands. Here the reflectance increases toward unity as the number of similar periods used in series in the multilayer increases. The regions in which the parameters are real represent dielectric behavior and are potential passbands.

The parameters γ and y_{equiv} are multivalued, and there is complete freedom of choice in choosing which particular value to use. It is most convenient to choose γ as that value nearest the total phase thickness of the symmetrical combination and also to choose the positive value for y_{equiv}. In the pass regions, the equivalent phase thickness of the period is found to be approximately equal to the total phase thickness of the layers of the period. If now q periods are placed in series, then, regardless of the actual number, the multilayer will act as a single layer with the equivalent admittance and with phase thickness q times that for a single period. To transform the regions of real admittance into passbands, all that is required is a coating at either end to match the equivalent layer into the substrate and into the surrounding medium.

Analytical expressions can be derived for y_{equiv} and γ but, even for a very small number of layers, they are somewhat complicated. For three layers of the form *ABA,* we have

$$M_{11} = \cos 2\delta_a \cos \delta_b - 0.5[(y_b/y_a) + (y_a/y_b)]\sin 2\delta_a \sin \delta_b, \tag{10.62}$$

$$\begin{aligned} M_{12} = (i/y_a)\{&\sin 2\delta_a \cos \delta_b + 0.5[(y_b/y_a) + (y_a/y_b)] \\ &\times \cos 2\delta_a \sin \delta_b + 0.5[(y_a/y_b) - (y_b/y_a)]\sin \delta_b\}, \end{aligned} \tag{10.63}$$

$$\begin{aligned} M_{21} = (iy_a)\{&\sin 2\delta_a \cos \delta_b + 0.5[(y_b/y_a) + (y_a/y_b)] \\ &\times \cos 2\delta_a \sin \delta_b - 0.5[(y_a/y_b) - (y_b/y_a)]\sin \delta_b\}. \end{aligned} \tag{10.64}$$

If straightforward evaluation is all that is required, then it is simpler to write a computer program in terms of the multiplication of the basic matrices.

Replacement of unobtainable admittance: The technique is also used in cases in which a layer with an unobtainable admittance is required in a design. It is sometimes possible to design a replacement consisting of a symmetrical period.

Analytical expressions for a quarter-wave replacement are simple:

$$\cos 2\delta_a = \frac{(y_b^2 + y_a^2)(y_{\text{equiv}}^2 - y_a^2)}{(y_b^2 - y_a^2)(y_{\text{equiv}}^2 + y_a^2)}, \tag{10.65}$$

$$\tan \delta_b = \frac{2y_a y_b}{(y_b^2 + y_a^2)} \frac{1}{\tan 2\delta_a}, \tag{10.66}$$

with optical thicknesses $n_a d_a/\lambda_0$ and $n_b d_b/\lambda_0$ given by $\delta_a/2\pi$ and $\delta_b/2\pi$, respectively. Note that the form makes it possible to achieve only equivalent admittances that lie within the range y_a–y_b.

10.11. Bandpass Filters

10.11.1. Single-Cavity Filter

Two quarter-wave stacks in series form a single-cavity filter. The central half-wave layer is the cavity layer:

$$\text{air}|\text{HLHLHLH HLHLHLH}|\text{glass}. \tag{10.67}$$

We can readily see that the entire structure is an absentee layer at λ_0 so the reflectance at λ_0 is that of the uncoated substrate. On either side of λ_0 the reflectance rises to what would be expected from the two quarter-wave stacks. The result is the characteristic shown in Figs. 10.14 and Figure 10.15, in which there are narrow bands of transmission surrounded by regions of high reflectance. This structure is a basic narrow-bandpass filter. The regions of high transmission outside the high-reflectance zones are once again termed sidebands and must be eliminated by additional components.

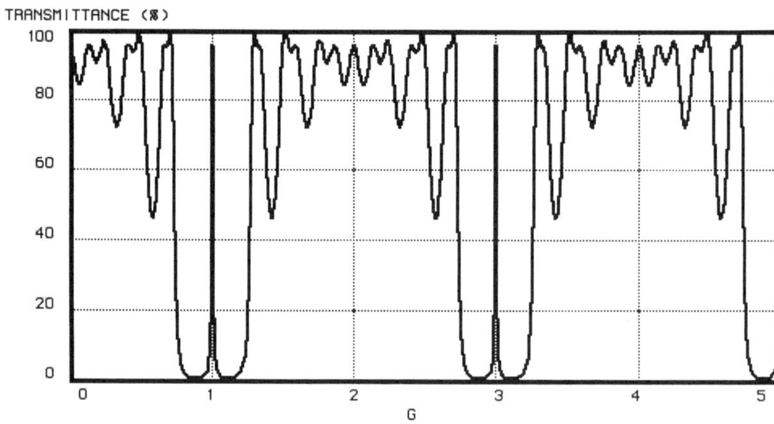

Figure 10.14. Basic narrow-band filter characteristic. The design is given by expression (10.67) with $y_H = 2.40$, $y_L = 1.38$, $y_{glass} = 1.52$, and $y_{air} = 1.00$.

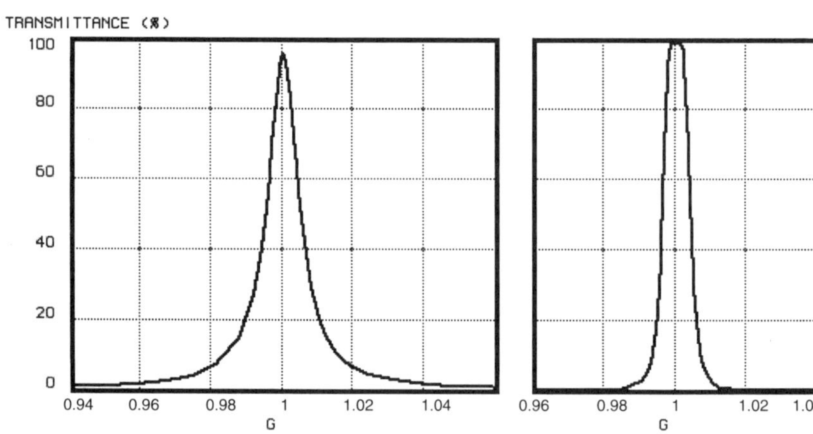

Figure 10.15. Comparison of the characteristic curves of design (10.67) and design (10.68). The width of the passbands is similar but the simpler design, (10.67) on the left, is more triangular than the squarer (10.68) on the right. The left-hand design is a single-cavity and the right-hand design a two-cavity design.

A basic design consists of two structures of the type of expression (10.67):

$$\text{air}|\text{HLHLHLHHLHLHLH L HLHLHLHHLHLHLH}|\text{glass}. \quad (10.68)$$

The central L layer is sometimes known as a coupling layer. If the absentee layers are eliminated then the structure reduces to

$$\text{air}|\text{L}|\text{glass}, \quad (10.69)$$

so that the reflectance at λ_0 is now that of a substrate with a single-layer antireflection coating. The performance of this structure is broadly similar to that of the original design but the passband is much more rectangular, as shown in Fig. 10.15. This design is known as a two-cavity filter.

Expression (10.70) continues the process. Yet another expression of the form of expression (10.67) is added. Canceling out the half waves shows that the entire structure disappears at $g = 1$, and an extra L layer at the end improves the transmittance, as in expression (10.71). This design is known as a three-cavity filter:

$$\begin{gathered}\text{air}|\text{HLHLHL HH LHLHLH L HLHLHL HH LHLHLH} \\ \text{L HLHLHL HH LHLHLH}|\text{glass},\end{gathered} \quad (10.70)$$

$$\begin{gathered}\text{air}|\text{L HLHLHLHHLHLHLH L HLHLHLHHLHLHLH} \\ \text{L HLHLHLHHLHLHLH}|\text{glass}.\end{gathered} \quad (10.71)$$

The two characteristics are shown in Fig. 10.16.

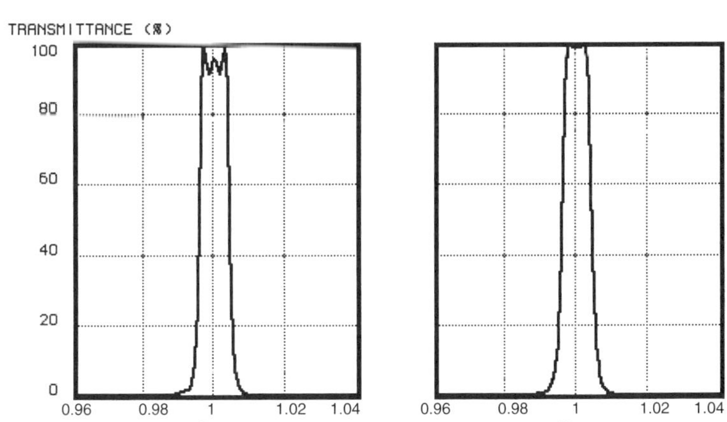

Figure 10.16. Three-cavity filters of design (10.70), left, and design (10.71) right. The extra matching layer improves the performance at the peak of the passband.

Such filters are used as beam splitters in dense-wavelength division multiplexing. A more fruitful way of analyzing their design is

$$\text{air}|\text{L HLHLHLH (HLHLHLH L HLHLHLH)}^2\text{ HLHLHLH}|\text{glass},$$

where the structure is of the form

$$\text{incident medium}|\text{matching (symmetrical structure)}^q\text{ matching}|\text{substrate}.$$

A more general symmetrical structure is of the form

$$m\text{H LHLHLHLH}\cdots\text{L }m\text{H or }m\text{L HLHLHLH}\cdots\text{H }m\text{L}, \tag{10.72}$$

where m is an integer and there are $2x + 1$ layers. The filter bandwidth for both structures in design (10.72) is given by

$$\left|\frac{\Delta\lambda_B}{\lambda_0}\right| = \frac{4}{m\pi}\left(\frac{n_\text{L}}{n_\text{H}}\right)^x \frac{(n_\text{H} - n_\text{L})}{(n_\text{H} - n_\text{L} + n_\text{L}/m)}. \tag{10.73}$$

The symmetrical structure has an equivalent admittance at λ_0 given by

$$\left(\frac{n_o}{n_i}\right)^x n_o \text{ for } m \text{ odd and } \left(\frac{n_i}{n_o}\right)^{x-1} n_i \text{ for } m \text{ even}. \tag{10.74}$$

This permits the design of the matching assemblies.

Absorption: For first-order filters and q cavities the absorption loss is given for a high-index cavity by

$$\alpha = q\pi\left(\frac{n_\text{H}}{n_\text{L}}\right)^x \frac{n_\text{L}(\kappa_\text{H} + \kappa_\text{L})}{(n_\text{H}^2 - n_\text{L}^2)} \tag{10.75}$$

and for a low index by

$$\alpha = q\pi\left(\frac{n_\text{H}}{n_\text{L}}\right)^x \frac{[n_\text{H}^2\kappa_\text{L} + n_\text{L}^2\kappa_\text{H}]}{n_\text{H}(n_\text{H}^2 - n_\text{L}^2)}. \tag{10.76}$$

10.11.2. Coatings with Metallic Layers

The reflectance of a metal surface (see Figs. 10.17 and 10.18) with admittance $n - i\kappa$ in an incident medium of admittance y_0 is given by

$$R = \frac{(y_0 - n)^2 + \kappa^2}{(y_0 + n)^2 + \kappa^2}. \tag{10.77}$$

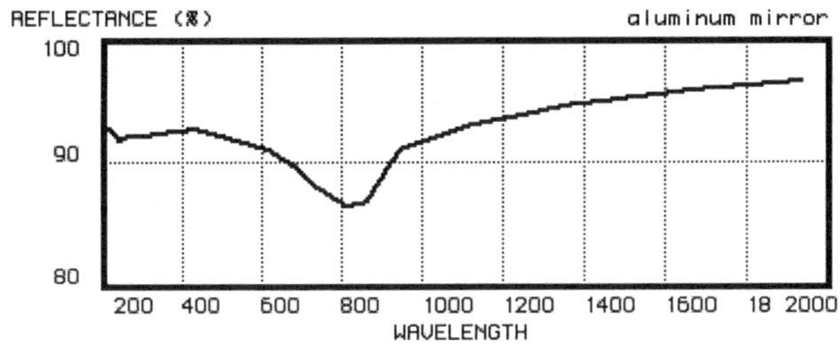

Figure 10.17. Reflectance of an aluminum coating calculated from the measured optical constants. See Fig. 10.26 for the reflectance of boosted aluminum.

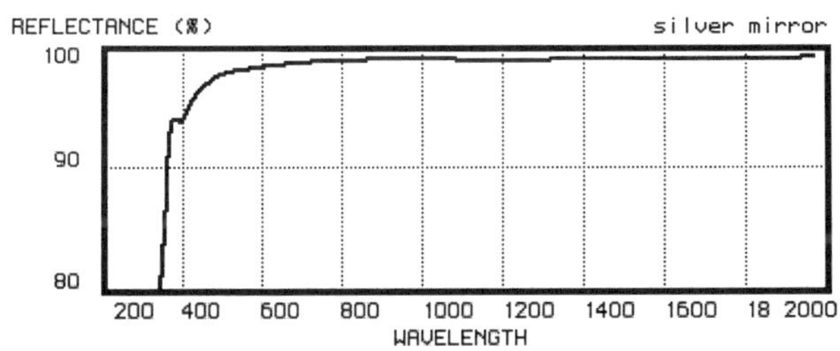

Figure 10.18. Reflectance of a silver coating. Note the steep dip in the near ultraviolet, which continues to virtually zero reflectance.

The effect of an added thin dielectric layer on the reflectance of an underlying metal is always to reduce the reflectance. The resultant reflectance is the result of multiple-beam interference in the dielectric layer, but a good approximation is given by the two-beam interference illustrated in Fig. 10.19. Beams *a* and *b* are almost in phase and, when the layer is of vanishing thickness, they add to give the reflectance of the uncoated metal. As the dielectric layer increases in

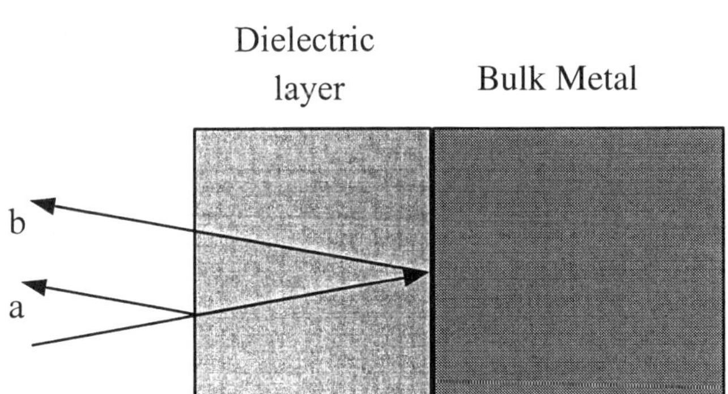

Figure 10.19. Reflectance of an overcoated metal is given approximately by the interference of the beams reflected from the outer dielectric surface and the inner dielectric–metal interface.

thickness, the phase difference between the beams increases and the interference becomes more and more destructive. Only when the layer is close to a half wave does the reflectance return to that of the uncoated metal.

In the case of metals that reflect poorly the two beams *a* and *b* can be of the same magnitude, and then the interference at the correct thickness of dielectric (close to a quarter wave) can result, theoretically in complete, destructive interference and therefore zero reflectance. The reflectance in practice can be <1%. Metals like titanium can be almost completely antireflected by an appropriate thickness of their own oxides. This leads to the appearance of brilliantly colored fringes that are often used artistically.

A high-performance metal can be antireflected in a similar way, but the outer surface of the dielectric layer has insufficient reflectance on its own. The addition of a thin metal layer to the outer dielectric surface gives a simple antireflection coating. Adjustment of both the outer-metal thickness and the dielectric-layer thickness permits perfect antireflection to be attained. Because it is used for matching the phases of the two beams, the dielectric layer is sometimes known as a phase-matching layer (see Fig. 10.20).

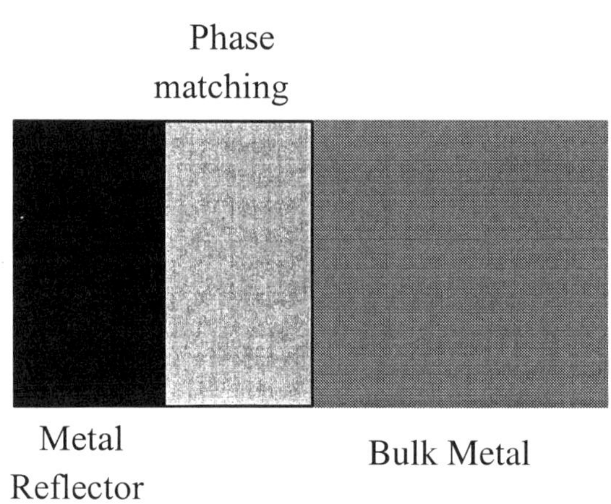

Figure 10.20. Antireflection coating for virtually any high-performance metal consists of a phase-matching dielectric layer and an outer metallic reflector. Variation of the thickness of the outer metal changes the magnitude of the reflected beam and alteration of the phase-matching layer adjusts the phase so that the interference with the beam from the bulk-metal surface can become completely destructive, giving perfect antireflection properties.

An antireflection coating for bulk aluminum, illustrated in Fig. 10.21, is constructed from a silica phase matcher followed by a thin chromium reflector. This is the principle behind many different types of coatings for optical data storage. For example, a thermoplastic phase-matching layer can become so altered by the application of heat from a laser beam that the antireflection properties are destroyed and the irradiated spot reflects strongly.

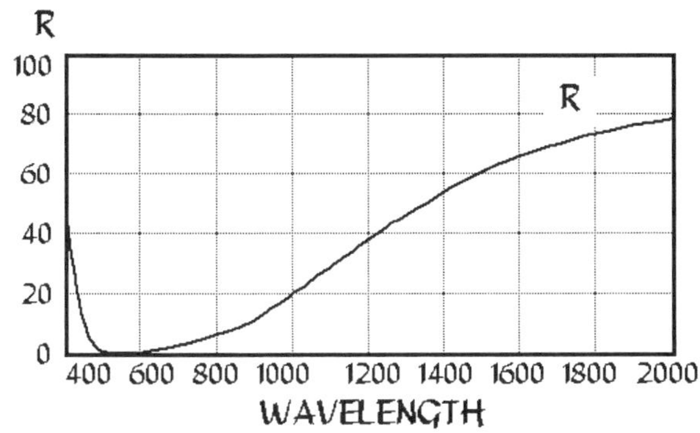

Figure 10.21. Performance of a two-layer dielectric–metal antireflection coating for aluminum.

The same techniques can be used to antireflect a thin-metal film on both sides, as in Fig. 10.22. The metal thickness must be such that the reflectances at the outer surface and at the metal–phase-matching layer interface are equal. The lower the outer reflectance, the thinner must be the metal layer. Because the transmittance of this component is important, the outer reflectors should not be absorbing. This implies the use of dielectric systems.

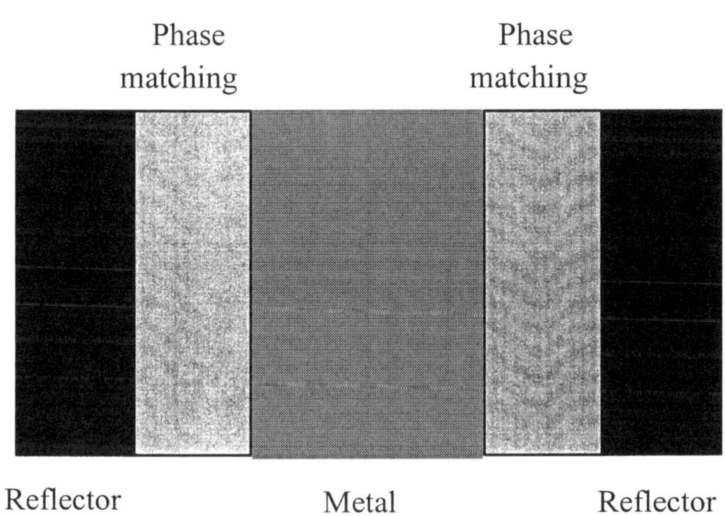

Figure 10.22. Thin-metal film antireflected on both sides by a phase-matching layer and reflector combination.

For a very thin metal the mismatch between high-admittance phase-matching layers and the surrounding media can give a sufficiently high matching reflectance that leads to a three-layer coating. The performance of such a coating is plotted in Fig. 10.23.

The performance shown in Fig. 10.23 is rather broad and can be narrowed

Incident medium	Phase matcher	Metal	Phase matcher	Emergent medium
Glass	TiO_2	Ag	TiO_2	Glass
Massive	28 nm	10 nm	28 nm	Massive

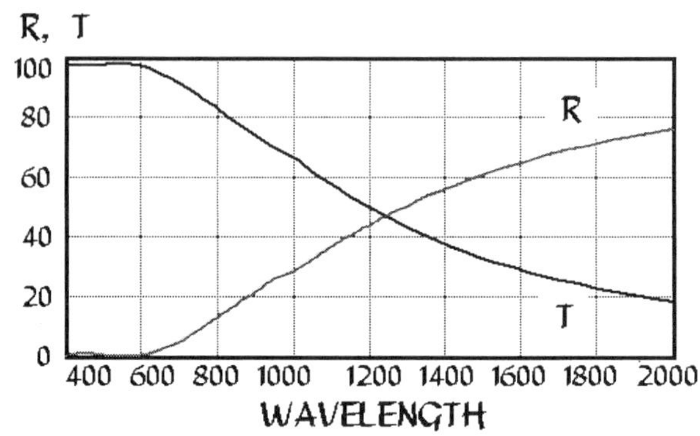

Figure 10.23. Three-layer coating consisting of a thin layer of silver with a phase-matching layer of TiO_2 on either side and sandwiched in glass. (Wavelength units are nanometers and reflectance and transmittance in are given as percentages.)

if more silver is used, which in turn demands a higher matching reflectance from the outer surface. An additional quarter wave of MgF_2 on either side, making a five-layer coating, increases the matching reflectance and permits the use of a thicker silver layer, as in Fig. 10.24.

Increasing the outer reflectance still further and therefore the silver thickness by the addition of an additional TiO_2 quarter wave narrows the transmittance zone and increases the reflectance. The coating now has seven layers, three dielectric layers on either side of the metal layer, as shown in Fig. 10.25.

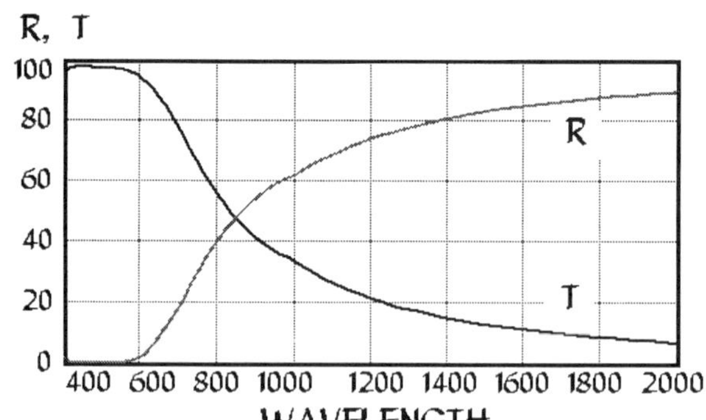

Figure 10.24. Performance of the five-layer coating.

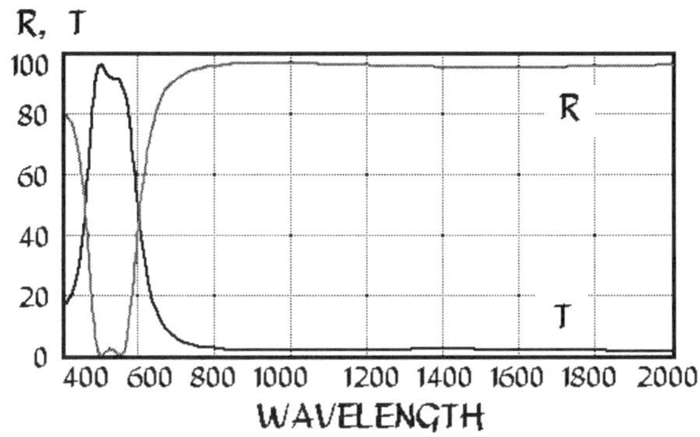

Figure 10.25. Performance of the seven-layer coating.

With a reflecting system consisting of four quarter waves of TiO_2 and MgF_2 on either side, together with the phase-matching layers, the coating is now a narrow-band filter, as shown in Fig. 10.26:

$$\text{glass}|\text{HLHLH}'\text{Ag H}'\text{LHLH}|\text{glass}. \tag{10.78}$$

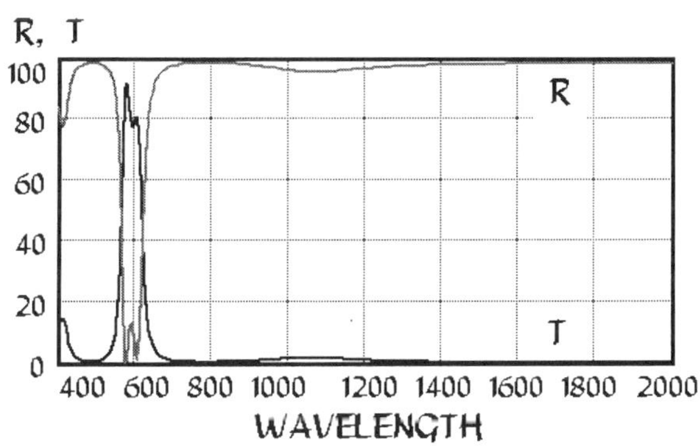

Figure 10.26. Performance of the metal–dielectric narrow-band filter of design (10.78).

It is possible to use a dielectric system to increase the reflectance of a metal rather than to reduce it. This can be regarded as a special case of the preceding coatings in which the phase-matching layer is adjusted to give constructive rather than destructive interference. When the phase-matching layer is low index and the outer reflector is a quarter-wave stack, the correct adjustment of the phase-matching layer is close to a quarter wave. In practice it is usually made to be exactly a quarter wave. Because this coating is now made up entirely of quarter waves we can use the quarter-wave rule to calculate its reflectance. The substrate can be the metal layer. The quarter-wave rule works just as well for absorbing substrates as it does for dielectric ones, although the layers must be dielectric. Thus the addition of four quarter-wave layers of alternating high and low admittance,

$$\text{air}|\text{HLHL}|\text{metal}, \tag{10.79}$$

leads to a reflectance (Fig. 10.27) of

$$\rho = \frac{[y_0 - (y_H^4/y_L^4)n]^2 + (y_H^4/y_L^4)\kappa^2}{[y_0 + (y_H^4/y_L^4)n]^2 + (y_H^4/y_L^4)\kappa^2}. \tag{10.80}$$

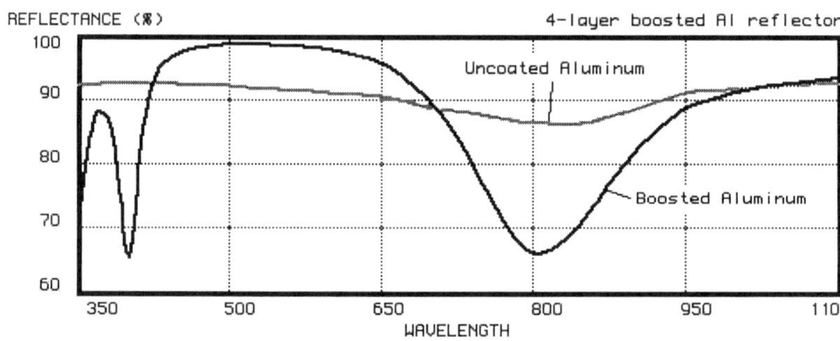

Figure 10.27. Comparison between the reflectance of aluminum boosted by four-layer SiO_2 TiO_2 centered on 510 nm and uncoated aluminum [design (10.79)]. Note the limited region over which enhanced performance is obtained.

Because n is small compared with κ, the effect of the great increase in the κ term is much greater than that of the term containing n and the reflectance increases. If the order of the layers were inverted, then the effect would be a reduction of reflectance. This dielectric enhancement of the reflectance of a metal layer is sometimes called boosting.

10.12. Potential Transmittance and Induced Transmission

Potential transmittance is the ratio of the flux density transmitted by the layer, or assembly of layers, to the flux density that actually enters:

$$\psi = \frac{\tau}{1 - \rho}. \tag{10.81}$$

To make the actual transmittance equal the potential transmittance, it is sufficient to make ρ zero. If we are dealing with dielectric materials, then $\tau = (1 - \rho)$ and the potential transmittance is always unity. It can be shown that, at any given wavelength, the potential transmittance of a layer depends on only the optical constants, the thickness of the layer, and the optical admittance of the assembly at the rear interface, the exit admittance. It does not depend on anything on the incident side of the layer. Further, the potential transmittance of an assembly of layers is the product of the potential transmittances of the individual layers, each calculated for the appropriate exit admittance. Because, once the parameters of a given layer have been defined, its potential transmittance is a function of only the exit admittance, one must be able to find a value of the exit admittance that maximizes the potential transmittance. A metal–dielectric filter that has been designed with this concept is known as an *induced-transmission filter* because it will have the maximum possible transmittance of which the metal layers are capable.

For one absorbing layer the expressions for ψ, ψ_{max}, and $(X_0 + iZ_0)$, the optimum exit admittance, are

$$\psi = \frac{\mathrm{Re}(Y_e)}{\mathrm{Re}(B_i C_i^*)}, \tag{10.82}$$

where

$$\begin{bmatrix} B_i \\ C_i \end{bmatrix} = [\mathcal{M}] \begin{bmatrix} 1 \\ Y_e \end{bmatrix}$$

and $[\mathcal{M}]$ is the characteristic matrix of the metal layer.

Let Y_e, the exit admittance, be $X + iZ$, and let the metal parameters be the usual n, κ, and d. Then

$$\delta = \frac{2\pi(n - i\kappa)d}{\lambda} = \frac{2\pi nd}{\lambda} - i\frac{2\pi\kappa d}{\lambda} = \alpha - i\beta,$$

$$\begin{aligned} \frac{1}{\psi} = {} & \frac{(n^2 - \kappa^2) - 2n\kappa Z/X}{(n^2 + \kappa^2)}(\sin^2\alpha\cosh^2\beta + \cos^2\alpha\sinh^2\beta) \\ & + (\cos^2\alpha\cosh^2\beta + \sin^2\alpha\sinh^2\beta) \\ & + (1/X)(n\sinh\beta\cosh\beta + \kappa\cos\alpha\sin\alpha) \\ & + \frac{X^2 + Z^2}{X(n^2 + \kappa^2)}(n\sinh\beta\cosh\beta - \kappa\cos\alpha\sin\alpha). \end{aligned} \tag{10.83}$$

This is a complicated expression, but it can be readily evaluated with a scientific calculator. The potential transmittance has been obtained, but the optimum exit admittance is also necessary. ψ is always positive and is a well-behaved function, so for an extremum in ψ there is also an extremum in $1/\psi$, of the opposite sense. Thus, the optimum exit admittance is given by

$$\left(\frac{\partial}{\partial X}\right)\left(\frac{1}{\psi}\right)=0, \quad \left(\frac{\partial}{\partial Z}\right)\left(\frac{1}{\psi}\right)=0,$$

$$X_{\text{opt}}=\left[\frac{(n^2+\kappa^2)(n\sinh\beta\cosh\beta+\kappa\sin\alpha\cos\alpha)}{2(n\sinh\beta\cosh\beta-\kappa\sin\alpha\cos\alpha)} - \frac{n^2\kappa^2(\sin^2\alpha\cosh^2\beta+\cos^2\alpha\sinh^2\beta)^2}{2(n\sinh\beta\cosh\beta-\kappa\sin\alpha\cos\alpha)^2}\right]^{1/2}, \tag{10.84}$$

$$Z_{\text{opt}}=\frac{n\kappa(\sin^2\alpha\cosh^2\beta+\cos^2\alpha\sinh^2\beta)}{(n\sinh\beta\cosh\beta-\kappa\sin\alpha\cos\alpha)}. \tag{10.85}$$

ψ_{max} is given by substitution of Eqs. (10.84) and (10.85) into Eqs. (10.83). Note that for β large, $X_{\text{opt}} \rightarrow n$ and $Z_{\text{opt}} \rightarrow \kappa$, that is, $Y_{\text{opt}} \rightarrow (n - i\kappa)^*$.

As a matching stack, an assembly of dielectric layers is needed to give an input admittance of $(X_{\text{opt}} + iZ_{\text{opt}})$ when deposited on the substrate y_{sub}. This is the same, except in opposite order, as adding a phase-matching dielectric layer to $(X_{\text{opt}} - iZ_{\text{opt}})$ to make the admittance real, followed by a series of quarter-wave layers to transform that real admittance into y_{sub}. A phase-matching layer of index n_f on $(X_{\text{opt}} - iZ_{\text{opt}})$ must have a thickness given by

$$\frac{n_f d_f}{\lambda}=\frac{1}{4\pi}\arctan\left[\frac{2Z_{\text{opt}}\,n_f}{(n_f^2-X_{\text{opt}}^2-Z_{\text{opt}}^2)}\right], \tag{10.86}$$

where the arctangent is to be taken in the first or the second quadrant. The real admittance given by the addition of this layer is

$$\mu=\frac{2X_{\text{opt}}\,n_f^2}{(X_{\text{opt}}^2+Z_{\text{opt}}^2+n_f^2)+[(X_{\text{opt}}^2+Z_{\text{opt}}^2+n_f^2)^2-4X_{\text{opt}}^2\,n_f^2]^{1/2}}. \tag{10.89}$$

μ will always be of lower admittance than the substrate because it is the first intersection of the locus of n_f with the real axis. A stack, usually quarter wave, of layers to match μ to y_{sub} is then added.

It can be shown that if the exit admittance is the optimum value, then the admittance presented by the front surface of the metal is the complex conjugate of the optimum. This is the same admittance as we have just matched. The entire matching structure is then added, in reverse order, to the front side of the metal layer.

10.13. Oblique Incidence

General: The phase thicknesses of all layers become smaller as tilt increases. This means that characteristics move to a shorter wavelength. For small tilts this is the principal effect. For larger tilts, the changes in the admittances become important and also the detuning of the layer thicknesses because high-index layers are affected less than low-index ones, and so there is a splitting of the performance for each plane of polarization and also a progressive degradation of the characteristic. The contrast between the admittances for *s*-polarized light becomes greater than at normal incidence whereas the contrast for *p* polarization becomes weaker. This implies that *s*-polarization characteristics become more pronounced, whereas *p*-polarization characteristics become weaker. A quarter-wave stack, for example, becomes narrower for *p* polarization and the reflectance falls, whereas exactly the opposite happens for *s* polarization. The performance of the long-wave-pass filter of Fig. 10.13 recalculated for a 45° angle of incidence is shown in Fig. 10.28, and the polarization splitting can be clearly seen.

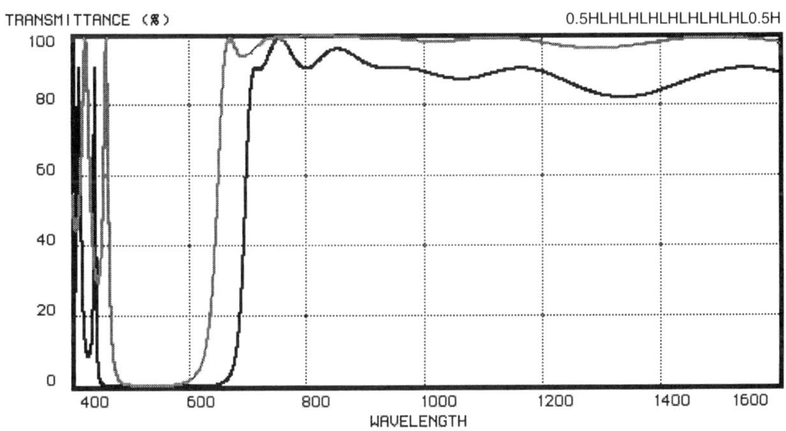

Figure 10.28. Performance of the long-wave-pass filter of Fig. 10.13 tilted to a 45° angle of incidence. The heavier curve represents *s* polarization, and the lighter curve represents *p* polarization.

As a rough guide, it is easier to achieve high transmission for *p* polarization and high reflectance for *s* polarization when the incidence is oblique.

Narrow-band filters—equivalent index: For small tilts a narrow-band filter characteristic moves to shorter wavelengths. The movement can be predicted by a model in which the entire change is due to a cavity layer's having an index called the equivalent index n^* that can be calculated from the properties of the filter design.

High-index cavity:

$$n^* = n_H \left[\frac{m - (m-1)(n_L/n_H)}{(m-1) - (m-1)(n_L/n_H) + (n_H/n_L)} \right]^{1/2}. \qquad (10.88)$$

Low-index cavity:

$$n^* = n_L \left[\frac{m - (m-1)(n_L/n_H)}{m - m(n_L/n_H) + (n_L/n_H)^2} \right]^{1/2}. \tag{10.89}$$

The shift is then given by

$$\frac{\Delta\lambda}{\lambda_0} = \frac{1}{2}\left(\frac{n_0\vartheta_0}{n^*}\right)^2, \tag{10.90}$$

or, with ϑ_0 in degrees,

$$\frac{\Delta\lambda}{\lambda_0} = 1.5 \times 10^{-4}\left(\frac{n_0\,\vartheta_0}{n^*}\right)^2. \tag{10.91}$$

Effect of an illumination cone: The expressions are less complicated in terms of wave number $\sigma = 1/\lambda$.

We assume first a cone axis at normal incidence:

$$\Theta = \text{cone semiangle,}$$

$$\sigma_0 = \text{normal-incidence peak,}$$

$$\Delta\sigma = \frac{\Theta^2}{2(n^*)^2}\sigma_0 = \sigma_\Theta - \sigma_0, \tag{10.92}$$

$$\sigma_{\text{peak}} = \sigma_0 + 0.5\Delta\sigma, \tag{10.93}$$

$$W = \text{half-peak bandwidth,}$$

$$W_\Theta^2 = W_0^2 + (\Delta\sigma)^2, \tag{10.94}$$

$$T_\Theta = \left(\frac{W_0}{\Delta\sigma}\right)\arctan\left(\frac{\Delta\sigma}{W_0}\right) \simeq 1 - \frac{1}{3}\left(\frac{\Delta\sigma}{W_0}\right)^2, \tag{10.95}$$

$$\Delta\sigma = \frac{\sigma_0}{8(n^*)^2 F^2}, \tag{10.96}$$

where F is the focal ratio.

For a cone of semiangle Θ tilted at angle χ, the result depends on whether or not the cone contains the normal. If the normal is within the cone then the results are the same as for a cone of semiangle $\chi + \Theta$. If the normal is outside the cone then we find the extreme values of σ, σ_1, and σ_2 associated with $\chi + \Theta$. Then

$$\sigma_{\text{peak}} = \frac{1}{2}(\sigma_1 + \sigma_2)$$

$$= \frac{\chi^2 + \Theta^2}{2(n^*)^2}\sigma_0 \quad (\chi \text{ and } \Theta \text{ in radians}) \tag{10.97}$$

$$= \frac{1.52 \times 10^{-4}(\chi^2 + \Theta^2)}{(n^*)^2}\sigma_0 \quad (\chi \text{ and } \Theta \text{ in degrees}).$$

The half-width is

$$W = [W^2{}_0 + (\sigma_2 - \sigma_1)^2]^{1/2}, \tag{10.98}$$

where

$$\frac{1}{2}(\sigma_2 - \sigma_1) = \frac{2\chi\Theta}{(n^*)^2}\sigma_0 \quad (\chi \text{ and } \Theta \text{ in radians}), \tag{10.99}$$

$$\frac{1}{2}(\sigma_2 - \sigma_1) = \frac{6.09 \times 10^{-4}\,\chi\Theta}{(n^*)^2}\sigma_0 \quad (\chi \text{ and } \Theta \text{ in degrees}). \tag{10.100}$$

Thin-film polarizers: Polarization splitting can be used to design polarizers. The long-wave-pass filter of Fig. 10.28 acts as a polarizer in the region of the edge where there is high p transmittance and high s reflectance. It is usual to enhance this effect in real components by the choice of a somewhat higher angle of incidence, corresponding to the Brewster angle between substrate and surrounding medium. [At the Brewster angle the p admittances for the two materials are equal and so there is no p reflectance. The Brewster angle referred to the incident medium is given by $\arctan(n_1/n_0)$]. The thicknesses of the layers are also adjusted so that the values of δ are in step, rather than detuned, avoiding any degradation of the shape of the characteristic that would otherwise occur.

Brewster angle polarizer: If the correct angle of incidence is chosen to correspond to the Brewster angle between the high- and the low-index layers of a coating, then any p reflectance disappears completely (except for the residual mismatch at substrate and incident media) and the s reflectance can be considerably enhanced. The coating must normally be immersed in a suitable prism.

The angle of incidence in the incident medium is given by

$$\sin^2\vartheta_0 = \frac{n_H^2\, n_L^2}{n_0^2\,(n_H^2 + n_L^2)}, \tag{10.101}$$

and, for a quarter-wave stack of x layers, where x is odd and the H layers are outermost,

$$\rho_p = \left[\left(\frac{n_0^2 \cos^2 \vartheta_H}{n_H^2 \cos^2 \vartheta_0} - 1 \right) \Big/ \left(\frac{n_0^2 \cos^2 \vartheta_H}{n_H^2 \cos^2 \vartheta_0} + 1 \right) \right]^2, \tag{10.102}$$

$$\rho_s = \left[\frac{n_0^2 \cos^2 \vartheta_0 - n_H^2 \cos^2 \vartheta_H (n_H \cos \vartheta_H / n_L \cos \vartheta_L)^{2(x-1)}}{n_0^2 \cos^2 \vartheta_0 + n_H^2 \cos^2 \vartheta_H (n_H \cos \vartheta_H / n_L \cos \vartheta_L)^{2(x-1)}} \right]^2, \tag{10.103}$$

and the degree of polarization in transmission and reflection is then

$$P_\tau = \frac{\tau_p - \tau_s}{\tau_p + \tau_s} = \frac{\rho_s - \rho_p}{2 - \rho_s - \rho_p},$$

$$P_\rho = \frac{\rho_s - \rho_p}{\rho_s + \rho_p}.$$

Such angles of incidence are impossible in air and can be achieved only if the incident medium is a material of elevated index, such as glass. This means that the construction of the device must be that of a cemented prism similar to that in Fig. 10.29. With correct choice of indices for prism and thin-film materials, a 45° angle of incidence becomes possible.

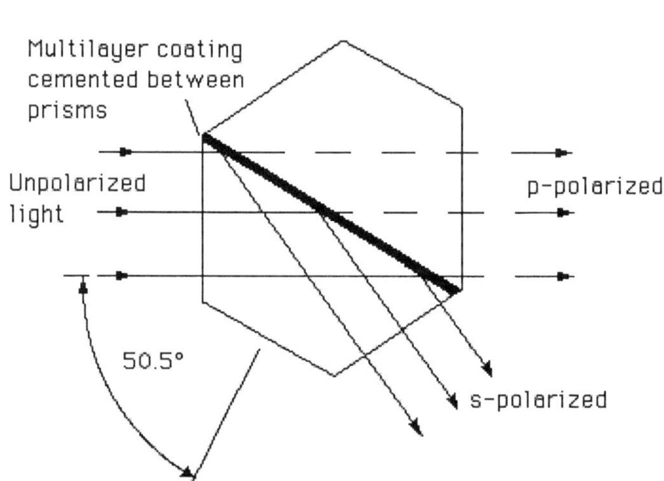

Figure 10.29. Construction of a polarizing beam splitter with a ZnS and cryolite multilayer in crown glass.

10.14. Short-Pulse Parameters

Pulse details: In this section, wherever the shape of the pulse is important, we assume a Gaussian pulse. A short pulse with a Gaussian envelope is shown in Fig. 10.30. At $z = 0$, we can write the temporal variation of the pulse envelope as

$$F(t) = \mathscr{A}e^{-[(t-t_0)^2/2\mu^2]}, \tag{10.104}$$

where 2μ is the width of the pulse measured at the $1/e$ power points. The Fourier transform gives the frequency distribution, and it is also a Gaussian function:

$$G(\omega) = \mathscr{B}e^{-[\mu^2(\omega-\omega_0)^2/2]}, \tag{10.105}$$

where $G(\omega)$ is the amplitude of the component wave of frequency ω and ω_0 is the carrier angular frequency.

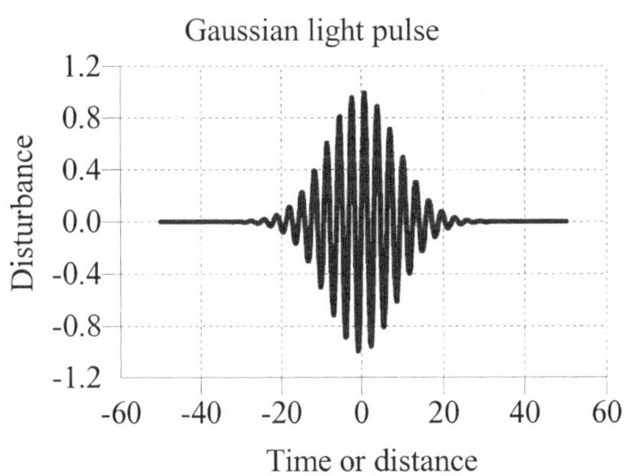

Figure 10.30. Short Gaussian-shaped pulse consisting of an envelope over a carrier of constant frequency. The carrier phase may move faster than the pulse, when it will appear to run through the envelope as it travels.

Phase velocity and group velocity: The carrier of the pulse moves with the velocity appropriate to a harmonic wave at the carrier frequency, and this is known as the phase velocity. The pulse envelope moves at a velocity that corresponds to the velocity of a phase coincidence among all the harmonic components of the pulse. This velocity is known as the group velocity:

$$\text{phase velocity} = v = \frac{\omega_0}{k_0}, \tag{10.106}$$

$$\text{group velocity} = \left(1 \bigg/ \frac{dk}{d\omega}\right)_{\omega=\omega_0}. \tag{10.107}$$

In a dispersive medium, the refractive index changes with frequency. The group velocity in terms of this change is

$$v_g = \frac{c}{n(\omega) + \omega \dfrac{dn(\omega)}{d\omega}}. \quad (10.108)$$

Group-velocity dispersion: There is no guarantee that the group velocity is independent of frequency. If the second derivative of k with respect to ω is nonzero, then there can be no exact phase coincidence of the harmonic components and the pulse will be perturbed. The second derivative is known as the group-velocity dispersion, even though it is strictly the reciprocal:

$$\text{GVD} = \frac{d^2k}{d\omega^2}. \quad (10.109)$$

For a dispersive material,

$$\frac{d^2k}{d\omega^2} = \frac{\lambda^3}{2\pi c^2}\frac{d^2n}{d\lambda^2}. \quad (10.110)$$

If the distance traveled by the pulse is z then

$$\tau_g^2 = \frac{d^2k}{d\omega^2}z = \text{GVD}z. \quad (10.111)$$

If Δt_0 is the initial width of the pulse, then the width after travel over a distance z is given by

$$\Delta t = \Delta t_0\left(1 + \frac{\Delta t_g^4}{\mu^4}\right)^{1/2}, \quad (10.112)$$

where

$$\Delta t_g = \frac{d^2k}{d\omega^2}z = \text{GVD}z.$$

Group delay, group-delay dispersion, and third-order dispersion: Let a pulse be reflected or transmitted by a coating and let the effect of the coating be limited to a change in phase, φ, of each harmonic component. Then we define

$$\text{group delay (GD)} = -\frac{\mathrm{d}\varphi}{\mathrm{d}\omega}, \tag{10.113}$$

$$\text{group-delay dispersion (GDD)} = -\frac{\mathrm{d}^2\varphi}{\mathrm{d}\omega^2}, \tag{10.114}$$

$$\text{third-order dispersion (TOD)} = -\frac{\mathrm{d}^3\varphi}{\mathrm{d}\omega^3}. \tag{10.115}$$

These quantities combine additively with the products of the various group-velocity terms with z as follows:

$$\text{total delay} = \Delta t = \frac{\mathrm{d}k}{\mathrm{d}\omega} z - \frac{\mathrm{d}\varphi}{\mathrm{d}\omega}, \tag{10.116}$$

$$\tau_g^2 = \frac{\mathrm{d}^2 k}{\mathrm{d}\omega^2} z - \frac{\mathrm{d}^2\varphi}{\mathrm{d}\omega^2}, \tag{10.117}$$

$$\tau_{\mathrm{TOD}}^3 = \frac{\mathrm{d}^3 k}{\mathrm{d}\omega^3} z - \frac{\mathrm{d}^3\varphi}{\mathrm{d}\omega^3}. \tag{10.118}$$

The group-velocity dispersion is

$$\frac{\mathrm{d}^2 k}{\mathrm{d}\omega^2} = \frac{\lambda^3}{2\pi c^2}\frac{\mathrm{d}^2 n}{\mathrm{d}\lambda^2}. \tag{10.119}$$

Chromatic dispersion: Chromatic dispersion in an optical medium refers to the change in group velocity with respect to wavelength rather than to angular frequency. As with group-velocity dispersion, the effect is to broaden a pulse:

$$\text{chromatic dispersion} = D = -\frac{\mathrm{d}}{\mathrm{d}\lambda}\left(\frac{1}{v_g}\right), \tag{10.120}$$

where the minus sign ensures that the effect has the same sign as the group-velocity dispersion does when multiplied by the corresponding $\Delta\lambda$. D has dimensions of time/(distance $\times$ wavelength) and for optical fibers usually has units of picoseconds times inverse kilometers times inverse nanometers.

The corresponding coating term is the chromatic dispersion coefficient, given by

$$\mathrm{CDC} = -\frac{\mathrm{d}^2\varphi}{\mathrm{d}\omega^2}\frac{2\pi c}{\lambda^2} = \frac{2\pi c}{\lambda^2}\mathrm{GDD}, \tag{10.121}$$

which can also be written as

$$\begin{aligned}\mathrm{CDC} &= -\frac{2\pi c}{\lambda^2}\frac{\mathrm{d}}{\mathrm{d}\lambda}\left(\frac{\mathrm{d}\varphi}{\mathrm{d}\lambda}\frac{\mathrm{d}\lambda}{\mathrm{d}\omega}\right)\frac{\mathrm{d}\lambda}{\mathrm{d}\omega} \\ &= -\frac{\lambda}{2\pi c}\left(2\frac{\mathrm{d}\varphi}{\mathrm{d}\lambda}+\lambda\frac{\mathrm{d}^2\varphi}{\mathrm{d}\lambda^2}\right).\end{aligned} \tag{10.122}$$

The spread in the group delay over the frequency range of the pulse is given by

$$\Delta t = \left(\frac{\mathrm{d}^2k}{\mathrm{d}\omega^2}z - \frac{\mathrm{d}^2\varphi}{\mathrm{d}\omega^2}\right)\Delta\omega, \tag{10.123}$$

which can be written in terms of the wavelength range $\Delta\lambda$ as

$$\begin{aligned}\Delta t &= -\left(Dz + \frac{\mathrm{d}^2\varphi}{\mathrm{d}\omega^2}\frac{\mathrm{d}\omega}{\mathrm{d}\lambda}\right)\Delta\lambda \\ &= -\left[Dz + \left(-\frac{\mathrm{d}^2\varphi}{\mathrm{d}\omega^2}\frac{2\pi c}{\lambda^2}\right)\right]\Delta\lambda \\ &= -\left[Dz + \left(\frac{2\pi c}{\lambda^2}\mathrm{GDD}\right)\right]\Delta\lambda.\end{aligned} \tag{10.124}$$

Note the minus sign when comparing Eq. (10.124) with Eq. (10.123). This is because greater frequencies correspond to smaller wavelengths and vice versa. So the comparison between the wavelengths separated by $\Delta\lambda$ in Eq. (10.124) is in the opposite sense to the comparison between frequencies in Eq. (10.123).

Further Reading

H. Anders, *Dünne Schichten für die Optik,* Wissenschaftliche Verlagsgesellschaft, 1965. (*Thin films in Optics,* Focal Press, 1967.)

H. Bach and D. Krause, eds., *Thin Films on Glass,* Springer-Verlag, 1997.

F. R. Flory, ed., *Thin Films for Optical Systems,* Vol. 1 of Optical Engineering Series, B. Thomson, Marcel Dekker, 1995.

H. Frey and G. Kienel, eds., *Dünnschicht Technologie,* VDI-Verlag, 1987.

S. A. Furman and A. V. Tikhonravov, *Basics of Optics of Multilayer Systems,* 1st ed., Editions Frontières, 1992.

H. L. Hartnagel, A. L. Dawar, A. K. Jain, and C. Jagadish, *Semiconducting Transparent Thin Films,* Institute of Physics, 1995.

O. S. Heavens, *Optical Properties of Thin Solid Films,* Butterworth, 1955.

L. Holland, *Vacuum Deposition of Thin Films,* Chapman & Hall, 1956.

R. E. Hummel and K. H. Guenther, eds., *Thin Films for Optical Coatings,* Vol. 1 of Handbook of Optical Properties Series, CRC, 1995.

M. R. Jacobson, ed., *Deposition of Optical Coatings,* Vol. MS06 of SPIE Milestone Series, SPIE, 1989.

M. R. Jacobson, ed., *Design of Optical Coatings,* Vol. MS26 of SPIE Milestone Series, SPIE, 1990.

M. R. Jacobson, ed., *Characterization of Optical Coatings,* Vol. MS63 of SPIE Milestone Series, SPIE, 1992.

Z. Knittl, *Optics of Thin Films,* Wiley, 1976.

H. M. Liddell, *Computer-Aided Techniques for the Design of Multilayer Filters,* Hilger, 1981.

P. H. Lissberger, "Optical applications of dielectric thin films," Rep. Prog. Phys. **33,** 197–268, 1970.

H. K. Pulker, *Coatings on Glass,* Elsevier, 1984.

J. D. Rancourt, *Optical Thin Films: Users' Handbook,* Macmillan, 1987.

A. Vasicek, *Optics of Thin Films,* North-Holland, 1960.

R. R. Willey, *Practical Design and Production of Optical Thin Films,* Marcel Dekker, 1996.

11

Holography

E. N. Leith

11.1. Introduction

Holograms are made by a two-step process. An interference pattern is recorded on a medium such as film. Then light is shone on the medium to produce a three-dimensional representation.

11.2. Recording

The basic expression for recording a hologram is

$$\vartheta(x,y) = g[I(x,y)] = g[(u_0 + u)(u_0 + u)^*], \tag{11.1}$$

where I is the intensity, which is equal to $|u_0 + u|^2$, ϑ is the transmittance of the hologram, u_0 is the reference beam, u is the object beam, and $g(\cdot)$ means function of and may indicate a nonlinear relation between intensity and transmittance of the recorded hologram. For analysis purposes, however, it is usual to assume a linear relation between intensity and the resulting transmittance:

$$\vartheta = k_0 - k_1 I \quad \text{or} \quad \vartheta = k_1 I, \tag{11.2}$$

depending on whether ϑ increases or decreases with increased exposure.

11.3. Recording Materials

Traditionally, high-resolution photographic film (e.g., Kodak 649F or Agfa-Gavaert) has been used. With the decreasing availability of photographic film and the increasing number of alternatives to film, holograms increasingly are being made on such materials as dichromated gelatin and photopolymers.

11.4. Reconstruction Process

In the reconstruction process, it is typical to illuminate the hologram with a readout beam that is a duplicate of the reference beam, both in wavelength and in geometry, giving

$$-k_1 u_0 u_0^* u = -k_1 a_0^2 a e^{i\phi}, \tag{11.3}$$
$$-k_1 u_0^2 u^* = -k_1 a_0^2 a e^{i(2ax-\phi)}, \tag{11.4}$$

where $u_0 = a_0 e^{i\alpha x}$ and $u = a e^{i\phi}$.

Equation (11.3) gives rise to what is called the true image, Eq. (11.4) to the conjugate image. If the true image is formed by such a reconstructing wave, i.e., one that duplicates the reference beam, the image will be of unity magnification and free from aberrations. If the recording and reconstruction geometry are different, the image position will change. The magnification will be other than unity, and aberrations will be present.

There are three levels of analysis for the holographic process. The exact analysis yields the conditions that produce perfect reconstruction, i.e., no aberrations. The first-order analysis tells where the image is formed, and its magnification, without regard to aberrations. The third-order analysis gives the aberrations. All of the following equations are in terms of the parameters defined in Fig. 11.1. In this analysis, the recording wavelength is λ_1, the reconstructing or

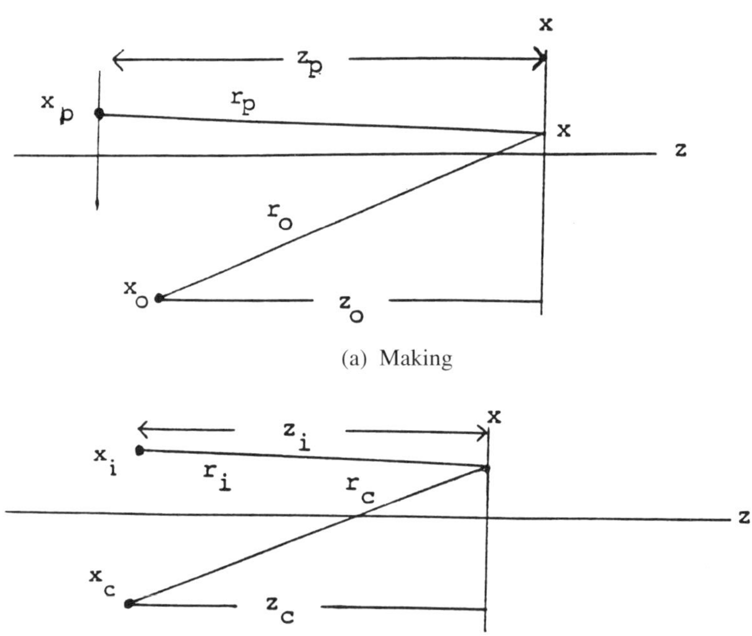

(a) Making

(b) Reconstruction

Figure 11.1. Diagrams for hologram analysis.

readout wavelength is λ_2, and the hologram, after being recorded, is scaled by a factor $1/p$.

11.5. Exact Analysis

$$\lambda_2 = \lambda_1/p, \tag{11.5}$$

$$z_c = z_0/p, \quad x_c = x_0/p,$$

$$z_i = z_p/p, \quad x_i = x_p/p,$$

$$M_x = 1/p = x_i/x_p, \tag{11.6}$$

where z is the distance along the optical axis and M is the lateral magnification.

These equations, taken together, indicate that all geometric parameters are scaled the same, by a factor of $1/p$, as well as the wavelengths.

11.6. First-Order Analysis

The first-order analysis yields the following equations:

$$1/z_i = 1/z_c \mp \lambda_2/\lambda_1 p^2\ 1/z_0 \pm \lambda_2/\lambda_1 p^2\ 1/z_p,$$

$$x_i/z_i = x_c/z_c \mp \lambda_2/\lambda_1 p\ x_0/z_0 \pm \lambda_2/\lambda_1 p x_p/z_p, \tag{11.7}$$

$$M_x = (\pm z_p/z_c\ \lambda_1/\lambda_2 - p z_p/z_0 + p)^{-1}. \tag{11.8}$$

The angular magnification is $\pm\lambda_2/\lambda_1 p$.

The longitudinal magnification is

$$M_z = \Delta z_i/\Delta z_p = \pm\lambda_1/\lambda_2\ M_x^2, \tag{11.9}$$

where Δz_p is the longitudinal separation of two object points and Δz_i is the longitudinal separation of the two image points.

The corresponding relation for conventional imagery, i.e., with lenses, is the same, except that the wavelength ratio is absent, as the wavelength cannot change in a conventional imaging process.

11.7. Third-Order Analysis

The third-order analysis is developed by the expansion of the various r's (r_p, r_0, etc.) to three terms instead of just two, as in the paraxial or first-order analysis. This leads to the third-order aberrations of the holographic process. These are of the same sort as for lens aberrations: spherical aberration, coma, astigmatism, distortion, and curvature of field. The mathematical expressions for these aberrations are complicated and are not given here. They can be found in Meier.[1]

[1] R. W. Meier, "Magnification and third-order aberrations in holography," J. Opt. Soc. Am. **55**, 987 (1965).

11.8. Volume Holograms

Holograms may be recorded in three-dimensional media as well as on surfaces. Photographic plates, with emulsion thicknesses ranging from 6 to 18 μm, would be regarded as a thin, or two-dimensional medium when the fringe patterns are coarse compared with this thickness. However, when the fringes become finer than the emulsion thickness, the medium must be considered as a volume structure, for which the conventional two-dimensional grating formulas do not apply and the theory of diffraction from three-dimensional structures applies. Volume holograms are of two basic types: (1) those in which the object and the reference beams impinge from the same side of the plate, producing fringes whose contours of constant phase cut across the emulsion; and (2) those in which the object and the reference beams impinge from opposite sides of the plate, producing fringes with contours running more nearly parallel to the emulsion surface.

Figure 11.2(a) shows a fringe pattern cutting across the emulsion. Contours of constant phase are shown as lines. The Bragg diffraction law is

$$\lambda = 2d \sin \theta. \tag{11.10}$$

Figure 11.2(b) shows fringes parallel to the emulsion surface. The Bragg law is the same as before. In Fig. 11.2(a) the angle is measured with respect to the normal to the emulsion surface; in Fig. 11.2(b) the angle is the complement of that angle. In each case, the angle is chosen to be the angle between the ray direction and the constant-phase surfaces.

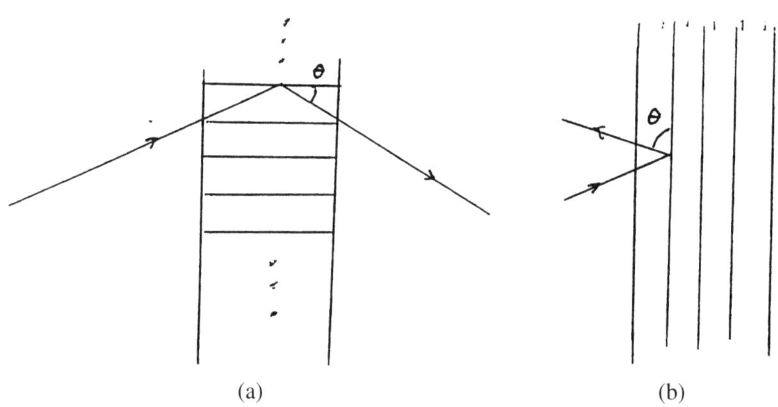

Figure 11.2. Diffraction from volume gratings.

It is the angle with respect to the fringes, not with respect to the emulsion surface, that is of interest and is important when the fringes are slanted, i.e., not parallel or perpendicular to the emulsion surface.

The same expression applies to the fringe-formation process (Fig. 11.3). Two beams interfere, as shown. The spacing of the fringes formed by interference of the two beams is readily shown from Fig. 11.3 to be

$$\lambda = 2d \sin \theta, \tag{11.11}$$

which is the same expression as that for diffraction from the recorded fringes.

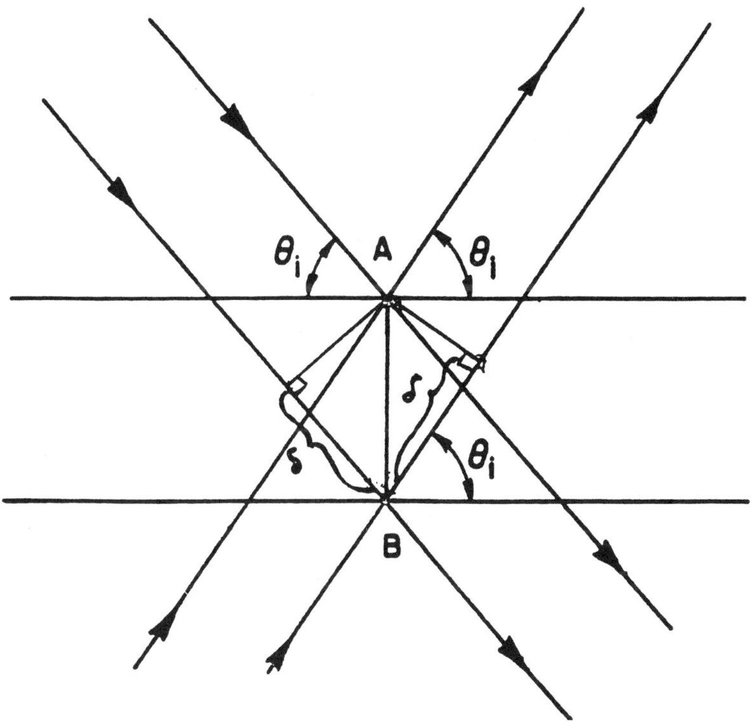

Figure 11.3. Fringe formation in a volume structure.

Note that the interference fringes bisect the angle between the two beams. The bandwidth of the Bragg diffracted (or Bragg reflected) light is

$$\Delta\lambda/\lambda_0 = 1/N, \tag{11.12}$$

where $\Delta\lambda$ is the free-space wavelength bandwidth, λ_0 is the beam midband wavelength, and N is the number fringes that the incident light beam intercepts as it passes through the grating.

The condition for a hologram to be considered thick (a volume hologram), as opposed to thin, is the Q factor,

$$Q = 2\pi\lambda_0 t/nd^2, \tag{11.13}$$

where λ_0 is the wavelength in air, n is the refractive index, t is the thickness, and d is the fringe spacing. The hologram can be considered thick (coupled wave theory applies) when $Q > 10$.

The maximum efficiencies of various holograms are given in Table 11.1. These efficiencies apply to simple, sinusoidal recorded interference fringes. If the fringes have modulations, these high efficiencies cannot be attained.

Table 11.1. Efficiencies of Various Hologram Types

Type	Max. Efficiency (%)
Thin amplitude transmission	6.25
Thin phase transmission	33.9
Thick amplitude transmission	3.7
Thick phase transmission	100
Thick amplitude reflection	7.2
Thick phase reflection	100

11.9. Wavelength and Angular Selectivity

The wavelength and the angular selectivities of thick holograms are in general not expressible in terms of simple formulas, the one exception being the wavelength selectivity when object and reference beams impinge from opposite sides of the recording plate [Eq. (11.12)]. Representative examples of these selectivities are given in the curves of Figs. 11.4 and 11.5, which are plotted for specific parameters. They show the general behavior of the selectivities.

11.10. Rainbow Holograms

The rainbow hologram-making process involves two steps. In the first step a conventional hologram is made; this is the so-called H_1 hologram. From this is made a second hologram, called the H_2 hologram. The formation of the H_2 hologram is shown in Fig. 11.6. A horizontal slit is place in front of the H_1 hologram and the conjugate image (a real image) is read out and is used as the object for the H_2 hologram. This two-step process involves various parameters that can be optimized:

1. The angle of the reference beam in the H_2, or the final, hologram (angle measured relative to the normal to the H_2 recording plate). If too small, there is inadequate dispersion of the slit image when the hologram is viewed, thus broadening the spectrum of the light entering the eye and degrading the resolution of the observed image. An angle too steep gives poor results, because most of the reference beam light reflects off the plate instead of exposing the emulsion and the effect of surface irregularities results in nonuniform exposure, giving the image an un-

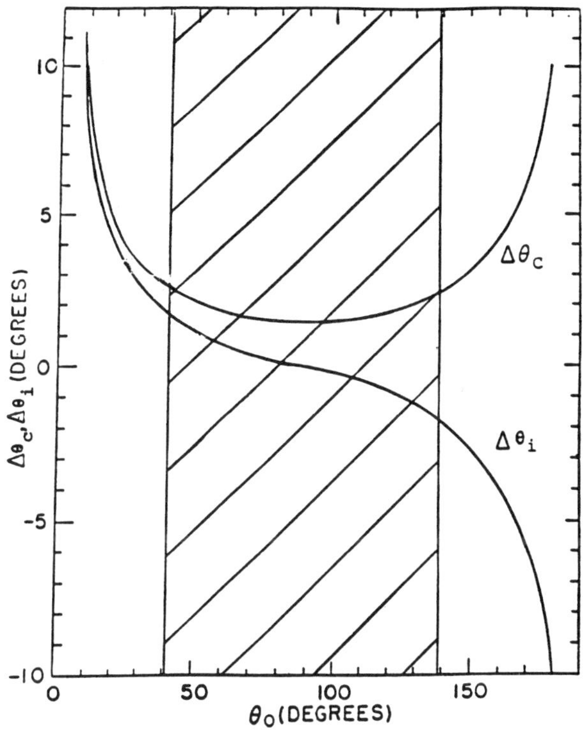

Figure 11.4. Plot of the change $\Delta\theta_c$ in readout needed to extinguish the diffracted wave as a function of θ_0. The two interfering plane waves used in creating the interference pattern are shown on the left. One beam impinges at normal incidence, and the other at an angle θ_0. The parameters are for Eastman Kodak 649F coated on glass plate; $\lambda = 6328$ Å and the emulsion thickness is 15.50 μm. Also shown is the change $\Delta\theta_i$ of the diffraction angle resulting from the change $\Delta\theta_c$. The angles are those within the emulsion (refractive index 1.52). The hatched area represents a region that is not readily accessible; that is, an angle of 41° within the emulsion corresponds to an angle of 90° for the beam in air.

even brightness. Angles of approximately 50° to 60° are generally satisfactory.

2. The H_1–H_2 separation. This is generally best set to the expected viewing distance, i.e., the viewer is best positioned at the image plane of the H_1 hologram, which is the position of the slit; this position gives uniform color illumination of the image. Otherwise, the image exhibits a rainbow of color, from top to bottom of the image.
3. The slit width at the H_1 hologram. This width determines the resolution of the observed image from the H_2 hologram but in a fairly complicated way. The formula

$$d' = dz_i/z_1 + \lambda z_1/d \qquad (11.14)$$

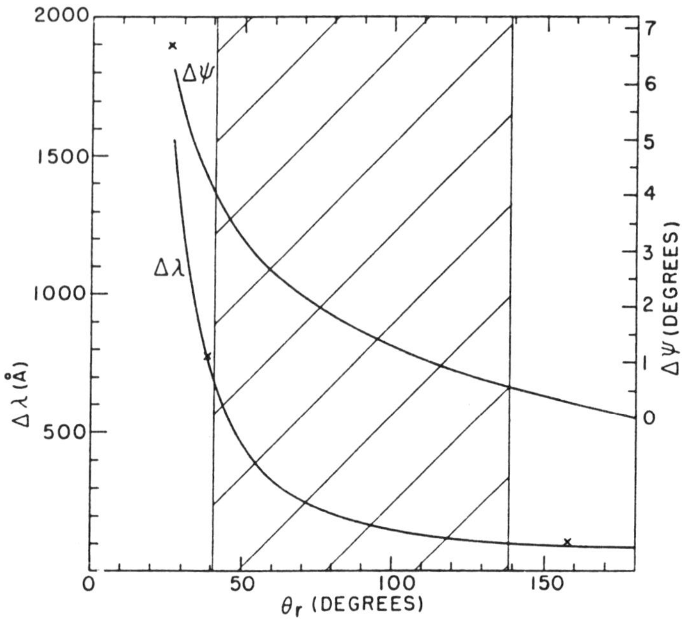

Figure 11.5. Plot of the wavelength change $\Delta\lambda$ needed to extinguish the diffracted wave while the readout angle θ_i is maintained constant. The parameters are for Eastman Kodak 649F emulsion coated on glass plate: $\lambda = 6328$ Å and $L = 15.5$ μm. Also shown is the variation of the diffraction angle associated with change $\Delta\lambda$ and some experimentally determined points (designated by ×). The calculated angles are those that exist within the emulsion (assumed to have a refractive index of 1.52). The hatched area is normally inaccessible, as explained in the caption of Fig. 11.4.

is useful in determining the optimum slit width. Here, d is the width of the slit, z_i is the distance of some image point of the holographic image from the H_2 hologram, z_1 is the distance from the H_1 hologram to the same object point when the H_2 hologram is made from the H_1 hologram, and d' is the resolution (full width at half-maximum) in the vertical direction for this image point.

The slit width can be optimized if the two terms on the right-hand side of Eq. (11.14) are set to be equal, a result that is readily verified by the differentiation of d' with respect to d. Which part of the holographic image to optimize is up to the discretion of the designer.

The optimum resolution in the horizontal direction is complicated and is a design issue that involves a number of factors.

11.11. Integral Holograms

These are holograms synthesized from a collection of ordinary photographs, each taken from a slightly different viewing position of the object, so that any

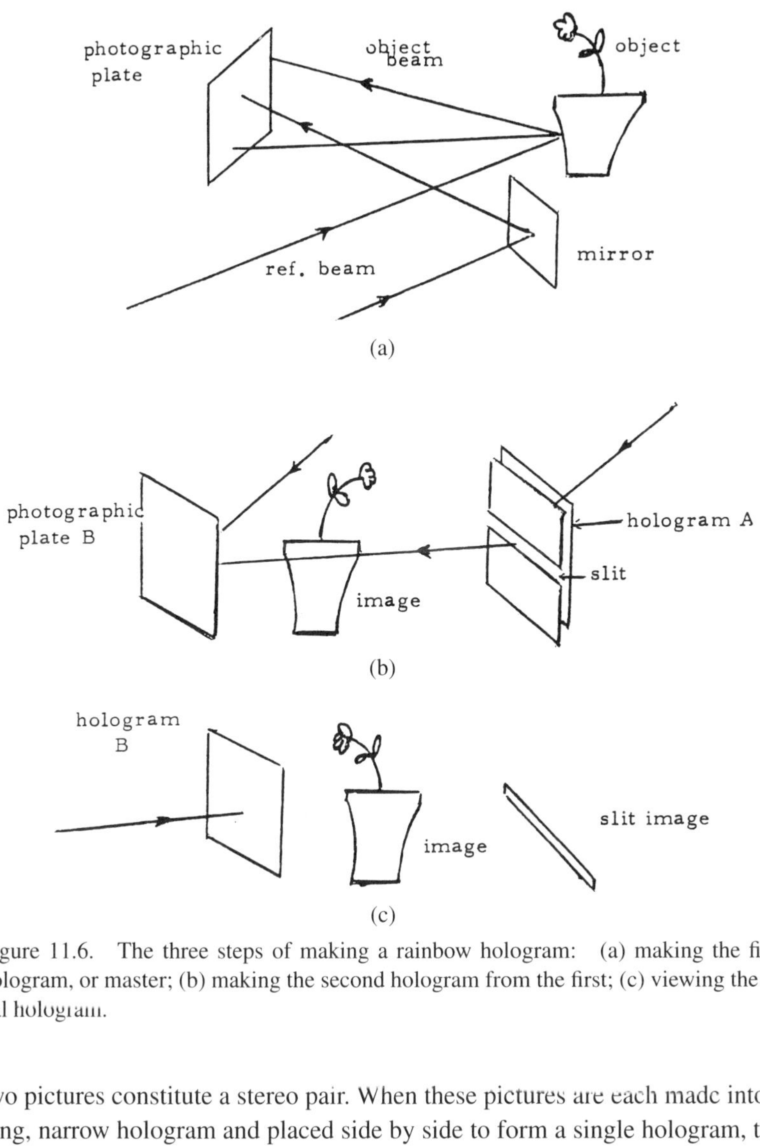

Figure 11.6. The three steps of making a rainbow hologram: (a) making the first hologram, or master; (b) making the second hologram from the first; (c) viewing the final hologram.

two pictures constitute a stereo pair. When these pictures are each made into a long, narrow hologram and placed side by side to form a single hologram, the result is an image that has complete three dimensionality and full parallax, but in the horizontal direction only. Typically they behave like rainbow holograms and are white-light viewable, provided the reference beam is introduced from above (or below), just as with rainbow holograms, except that there is no H_1 hologram and no horizontal slit at the H_1 hologram position. Instead, the recording of different photographs along the horizontal direction, but not in the vertical, becomes the equivalent of the slit. In their design, the same general conditions apply as with rainbow holograms and, in addition, there are considerations peculiar to integral holography. A major consideration is the width of the vertical sampling slit in the process of forming the hologram. If too broad, there are jumps in the parallax, i.e., the parallax is quantized. If too narrow, the image in the horizontal dimension will have low resolution because the width of the slit and thus the width of the individual holograms that form the integral hologram limit the resolution in the horizontal dimension. In general, the width of the slit is of the order of the aperture of the pupil of the eye.

The resolution in the horizontal direction is given by the simple expression for lens resolution: $\forall = f/D$, where f and D are the focal length and the horizontal aperture length of the individual holograms, respectively. The resolution in the vertical dimension is just that produced by conventional photography because, in the vertical dimension, the hologram is simply a conventional image.

11.12. Hologram Interferometry

Hologram interferometry has three basic forms: time average, double exposure, and real time. Other forms, a modification of these, include desensitized hologram interferometry and holographic contouring. In the time-average form it is generally assumed that the object (in the simplest case, a vibrating membrane) is vibrating sinusoidally and, as a consequence, the image from the hologram will have superimposed on it a fringe pattern of intensity given by

$$1 - J_0[(2\pi/\lambda D(\sin \phi_i + \phi_s)], \tag{11.15}$$

where J_0 is the zero-order Bessel function, D is the displacement of the membrane from its rest position, normal to the rest plane, and the ϕ's are the angles of incidence and reflection from the membrane, measured with respect to the membrane surface.

12

Geometrical Optics

Jasmin Côté

12.1. Optical Path Length

The optical path length is the equivalent distance in a vacuum that a light ray travels in a time t from point A to point B through an optical medium of refractive index $n(s)$:

$$\mathrm{OPL} = \int_A^B n(s)\mathrm{d}s = ct, \tag{12.1}$$

where $n(s)$ is the index of refraction encountered along the path from A to B, $\mathrm{d}s$ is an element on the path from A to B, and c is the speed of light in a vacuum.

12.2. Fermat's Principle

Fermat's principle states that the path taken by a light ray in going from point A to point B must traverse an optical path length that is stationary with respect to variations of that path.

For a function $f(x)$ that describes the optical path length of a light ray going from A to B, the stationary condition is satisfied if one sets $\mathrm{d}f(x)/\mathrm{d}x = 0$ and solves for x. Stationary implies that the optical path length is a maximum, minimum, or a point of inflection with a horizontal tangent with respect to the adjacent rays.

12.2.1. Snell's Law (Law of Refraction)

The law of refraction (see Fig. 12.1) is given by

$$n_1 \sin(\theta_i) = n_2 \sin(\theta_t), \tag{12.2}$$

where n_1 and n_2 are the refractive indices in respective media, θ_i is the ray angle of incidence, and θ_t is the angle the refracted ray makes with the surface normal.

12.2.2. *Law of Reflection*

The law of reflection (see Fig. 12.1) is given by

$$\theta_r = -\theta_i. \tag{12.3}$$

The angle of reflection equals the angle of incidence, but in the opposite direction and in the plane of incidence.

The signs in parentheses in Fig. 12.1 indicate the sign convention adopted throughout this chapter.

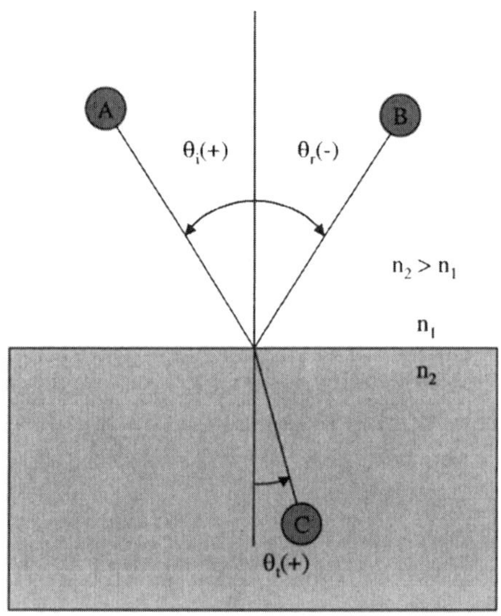

Figure 12.1. Laws of refraction and reflection.

12.3. Image Transformation

The image-transformation equations are a set of equations describing a collinear mapping between the object space and the image space. In a collinear mapping, there is a one-to-one correspondence among points, lines, and planes between the two spaces. Corresponding elements in the two spaces are conjugate elements.

12.3.1. *Cardinal points and planes*

The cardinal points and planes (Fig. 12.2) are conjugate elements in object space and image space. Their relation can be expressed by the magnification (see Table 12.1) between the two spaces.

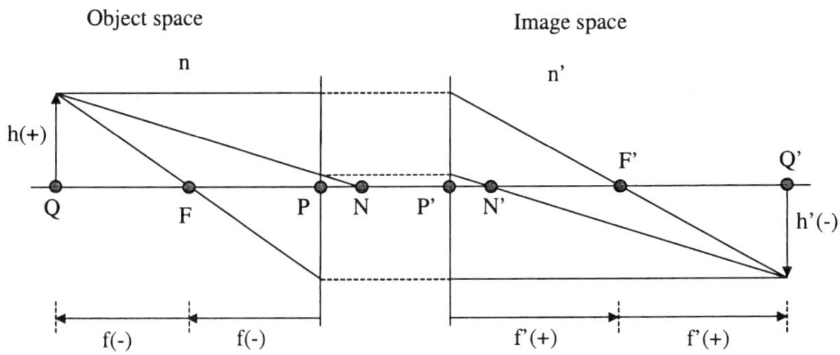

Figure 12.2. Cardinal points and planes.

Table 12.1. Magnification relations for cardinal points and planes[a]

Object Location	Image Location	Magnification ($m = h'/h$)
–Infinity	Rear focal plane (F')	0
Front focal plane (F)	Infinity	∞
Front principal plane (P)	Rear principal plane (P')	1
Front antiprincipal plane (Q)	Rear antiprincipal plane (Q')	–1
Front nodal plane (N)	Rear nodal plane (N')	n/n'

[a]The primed symbols represent image space, and the unprimed symbols represent object space.

12.3.2. *Nodal Points*

The nodal points are conjugate points of unit angular magnification:

$$PN = P'N' = f + f',$$

$$m_n = \frac{1}{k} = \frac{f}{f'} = \frac{n}{n'}, \tag{12.4}$$

where m_n is the front-nodal-plane to rear-nodal-plane magnification.

12.3.3. *Newtonian Equations*

The Newtonian equations (Fig. 12.3) describe the object and the image distances from the focal planes:

$$zz = ff',$$

$$\frac{z}{f} = \frac{-1}{m},$$

$$\frac{z'}{f'} = -m,$$

$$\frac{\Delta z'}{\Delta z} = -\frac{f'}{f} m m_\Delta = k m m_\Delta,$$

$$m_\Delta = -\frac{f}{z + \Delta z} = -\frac{z' + \Delta z'}{f'}, \tag{12.5}$$

where m is the object-to-image plane magnification and m_Δ is the object-to-image magnification if the object is displaced a distance Δz from its original position.

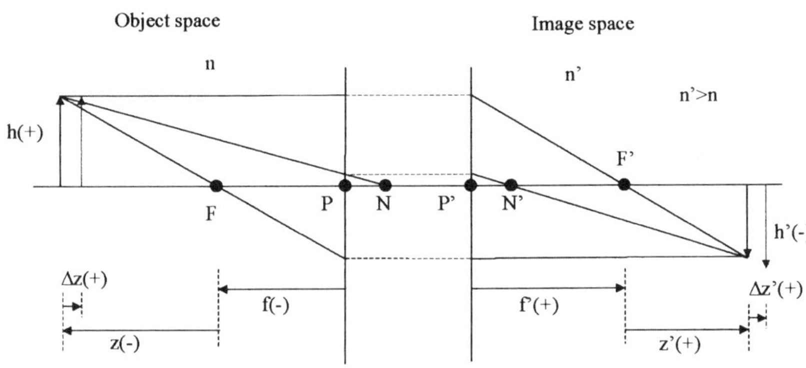

Figure 12.3. Newtonian equations.

12.3.4. Gaussian Equations

The Gaussian equations (Fig. 12.4.) describe the object and the image distance from the principal planes:

$$\frac{f}{t} + \frac{f'}{t'} = 1,$$

$$\frac{t}{f} = 1 - \frac{1}{m},$$

$$\frac{t'}{f'} = 1 - m,$$

$$\frac{t'}{t} = \left(\frac{-f'}{f}\right) m = km, \tag{12.6}$$

where m is the object-to-image magnification.

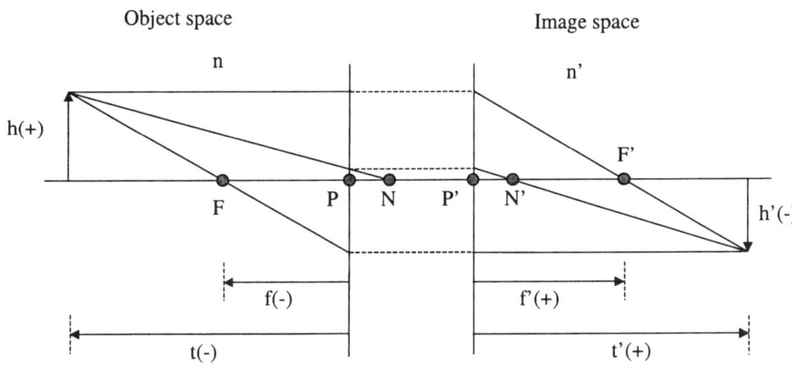

Figure 12.4. Gaussian equations.

12.3.5. Afocal Systems

Afocal systems (Fig. 12.5) are systems with constant transverse magnification and with longitudinal magnification that is proportional to the square of the transverse magnification:

$$m = \frac{f_2}{f'_1} = \text{constant},$$

$$\frac{\Delta z'}{\Delta z} = \left(\frac{-f'_1}{f_1}\right)\left(\frac{-f'_2}{f_2}\right) m^2, \tag{12.7}$$

where f_1 and f_2 are the front focal lengths of a two-element afocal system.

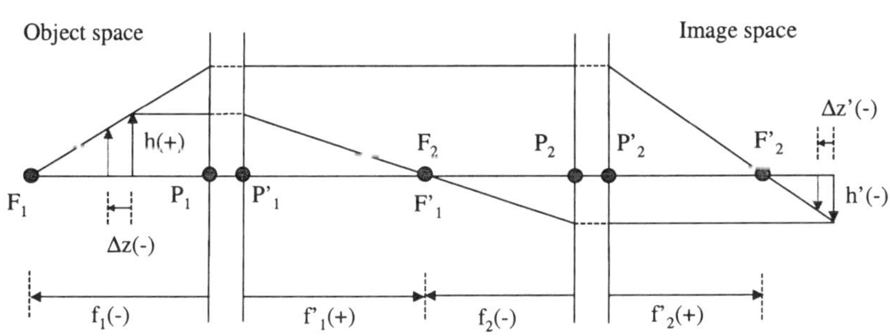

Figure 12.5. Afocal system.

12.4. Refraction and Reflection at a Spherical Interface

12.4.1. Sign Conventions

The sign conventions for a refractive surface are given in Fig. 12.6. Distances measured from left to right are positive. Angles measured counterclockwise are positive.

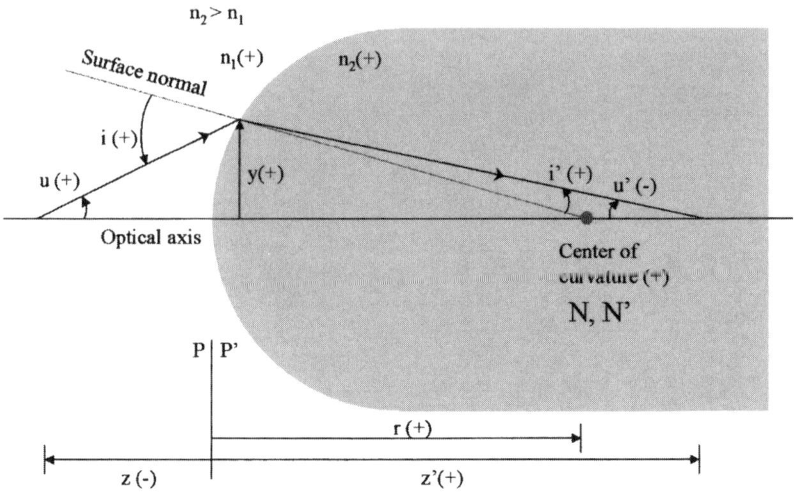

Figure 12.6. Sign conventions for a refractive surface.

For a reflective surface, the signs of all indices and spacings following a reflection are reversed.

12.4.2. Cardinal Points of a Refractive Surface

The principal points (P, P') are located at the refractive surface vertex, and the nodal points (N, N') are located at the center of curvature:

$$PN = P'N' = r,$$

$$c_s = \frac{1}{r},$$

$$\phi = (n' - n)c_s,$$

$$f_e = \frac{1}{\phi} = \frac{-f}{n} = \frac{f'}{n'},$$

$$f = \frac{-n}{\phi},$$

$$f' = \frac{n'}{\phi}, \tag{12.8}$$

where r is the surface radius of curvature, ϕ is the surface power, and f_e is the effective focal length.

12.4.2.1. Paraxial Refraction Equations

The paraxial refraction equations are

$$\begin{aligned} n'u' &= nu - y\phi, \\ \omega &= nu, \\ \omega' &= \omega - y\phi, \end{aligned} \tag{12.9}$$

where y is the ray height at the surface, and u and u' are the ray angles before and after the refractive surface, respectively.

12.4.2.2. Paraxial Transfer Equations

The paraxial transfer equations are

$$\begin{aligned} y' &= y + n'u'\left(\frac{t'}{n'}\right), \\ \omega' &= n'u', \\ \tau' &= \frac{t'}{n'}, \\ y' &= y + \omega'\tau', \end{aligned} \tag{12.10}$$

where t' is the distance to the next surface and τ' is the reduced distance.

12.4.3. Cardinal Points of a Reflective Surface

$$\begin{aligned} n' &= -n, \\ \phi &= -2nc_s, \\ f_e &= \frac{1}{\phi}, \\ f = f' &= \frac{-n}{\phi}. \end{aligned} \tag{12.11}$$

The paraxial reflection and the transfer equations for a reflective surface are given by

$$\begin{aligned} u' &= -u - 2yc_s, \\ y' &= y + t'u'. \end{aligned} \tag{12.12}$$

12.5. Thin Lens

A thin lens (Fig. 12.7) is a single surface or a system of surfaces for which the distances between the surfaces are considered negligible.

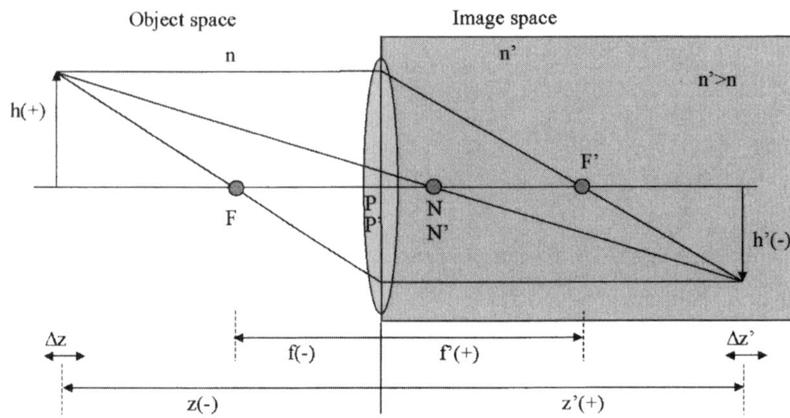

Figure 12.7. Thin lens.

12.5.1. Cardinal Points of a Thin Lens

$$PN = P'N' = f + f',$$

$$f = \frac{-n}{\phi} = -nf_e,$$

$$f' = \frac{n'}{\phi} = n'f_e, \qquad (12.13)$$

where PN and $P'N'$ are the distances between the principal plane and the nodal plane in the object and the image space, respectively.

12.5.2. Gaussian Equations for a Thin Lens

$$\frac{z}{n} = -\left(1 - \frac{1}{m}\right)f_e = \tau_z,$$

$$\frac{z'}{n'} = (1 - m)f_e = \tau_{z'},$$

$$m = \frac{z'n}{n'z} = \frac{\tau_{z'}}{\tau_z},$$

$$\frac{n'}{z'} = \frac{n}{z} + \phi,$$

$$\frac{\Delta z' n}{n' \Delta z} = m_1 m_2, \tag{12.14}$$

where m_1 and m_2 are the magnifications for the object position z and $(z + \Delta z)$, respectively.

12.6. System of Lenses or Optical Surfaces

Any system of lenses or optical surfaces can be reduced to an equivalent set of principal planes and an equivalent power. The two-component systems equations (Fig. 12.8) are given by

$$\phi = \phi_1 + \phi_2 - \phi_1 \phi_2 \tau,$$

$$\tau = \frac{t}{n_t},$$

$$f = \frac{-n}{\phi} = -n f_e,$$

$$f' = \frac{n'}{\phi} = n' f_e,$$

$$\frac{\delta'}{n'} = \frac{-\phi_1}{\phi} \tau,$$

$$\frac{\delta}{n} = \frac{\phi_2}{\phi} \tau,$$

$$\frac{\text{bfl}}{n'} = \frac{(1 - \phi_1 \tau)}{\phi} = \frac{(\delta' + f')}{n'},$$

$$\frac{\text{ffl}}{n} = \frac{(1 - \phi_2 \tau)}{-\phi} = \frac{(\delta + f)}{n},$$

$$PP' = t + \delta' - \delta,$$

$$m_n = \frac{-f}{f'} = \frac{n}{n'}, \tag{12.15}$$

where bfl and ffl are the back and the front focal lengths, respectively.

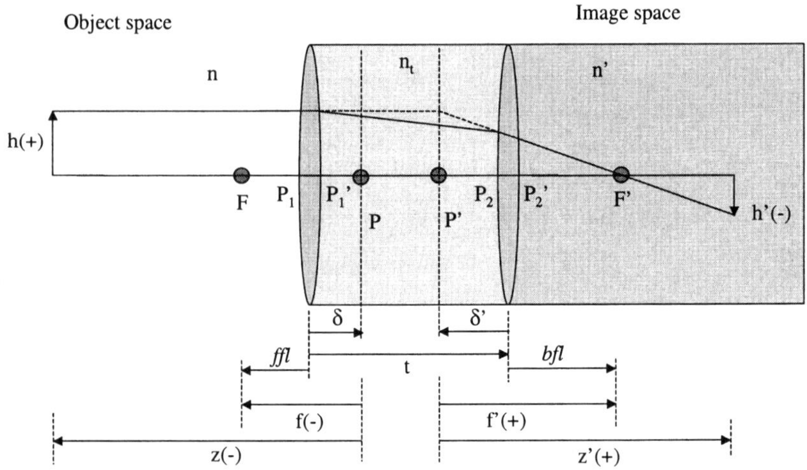

Figure 12.8. Two-component systems.

12.7. Thick Lens

A thick lens (Fig. 12.9) is reduced to a single set of principal planes and points by use of two-component system equations (12.15) and Eqs. (12.8) for a refractive surface:

$$\begin{aligned}\phi_1 &= (n_t - n)c_1,\\ \phi_2 &= (n' - n_t)c_2,\\ \phi &= \phi_1 + \phi_2 - \phi_1\phi_2\tau,\\ \tau &= \frac{t}{n_t}.\end{aligned} \tag{12.16}$$

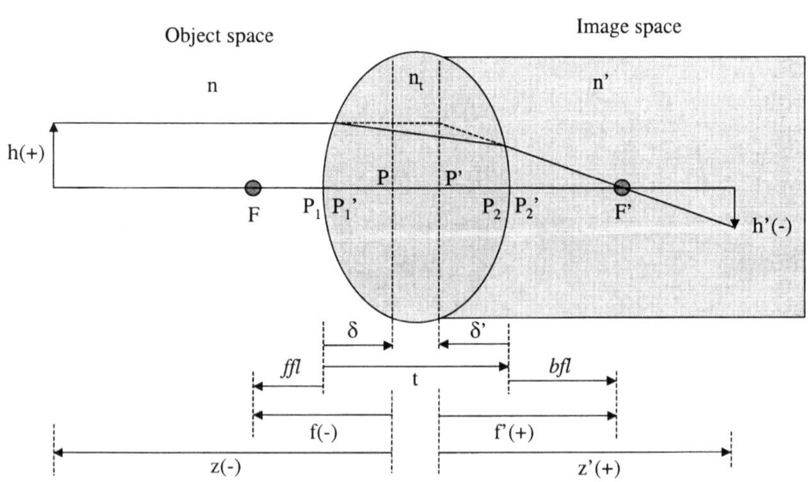

Figure 12.9. Thick lens.

12.7.1. Thick Lens in Air

A thick lens in air is a special case of the thick lens equations (12.16) in which $n = n' = 1$:

$$\phi = (n_t - 1)[c_1 - c_2 + (n_t - 1)c_1 c_2 \tau],$$

$$f = -f' = \frac{-1}{\phi},$$

$$\delta = \frac{-(n_t - 1)c_2}{\phi}\,\tau,$$

$$\delta' = \frac{-(n_t - 1)c_1}{\phi}\,\tau,$$

$$PP' = (n_t - 1)\tau\left[1 + (n_t - 1)\,\frac{c_1 c_2 \tau}{\phi}\right],$$

$$m_n = \frac{-f}{f'} = 1, \tag{12.17}$$

where PP' is the physical distance between the principal planes.

12.8. Stops and Apertures

12.8.1. Aperture and Field Stops

The aperture stop (see Fig. 12.10) is the aperture in the system that determines the diameter of the cone of light that the system will accept from an axial point

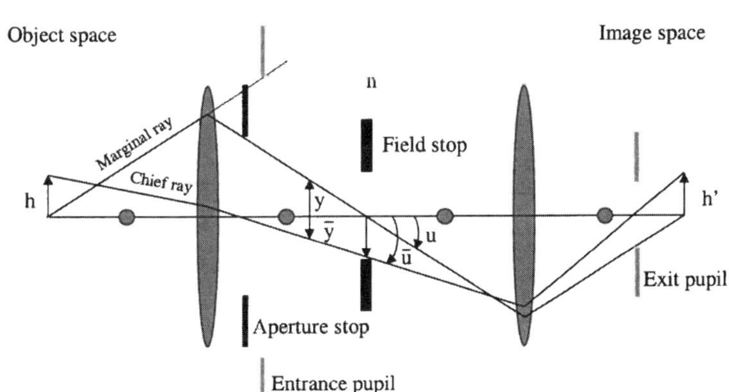

Figure 12.10. Stops and apertures.

on the object. The field stop is the aperture in the system that determines the field of view of the system.

12.8.2. Entrance and Exit Pupils

The entrance pupil is the image of the aperture stop in the object space. The exit pupil is the image of the aperture stop in the image space.

12.8.3. Marginal and Chief Rays

The marginal ray is a ray that originates from an object point on the optical axis and passes at the edge of the entrance pupil (see Fig. 12.10). The chief ray is the ray that originates from an object point at the edge of the field and passes through the center of the stop.

The Lagrange (optical) invariant is defined as

$$H = yn\bar{u} - \bar{y}nu, \tag{12.18}$$

where y and $\bar{y}$ are the marginal and the chief ray heights, respectively, and u and $\bar{u}$ are the marginal and the chief ray angles, respectively, at the plane where the Lagrange invariant is computed.

12.8.4. Optical Speed

The speed of an optical system is described by the numerical aperture NA and the F-number F, where

$$\begin{aligned} \text{NA} &= n' \sin(u'), \\ F &= \frac{1}{2\text{NA}} = \frac{f_e}{D}(1 - m), \end{aligned} \tag{12.19}$$

where f_e is the system effective focal length, D is the entrance pupil diameter, and m is the system magnification.

12.8.5. Various resolution criteria

The depth of focus (Fig. 12.11) is the amount the image may be defocused before the blur size (B') exceeds a specified size criterion:

$$|\delta'| = \frac{\beta L'^2}{D} = \frac{B'L'}{D} \approx B'(F). \tag{12.20}$$

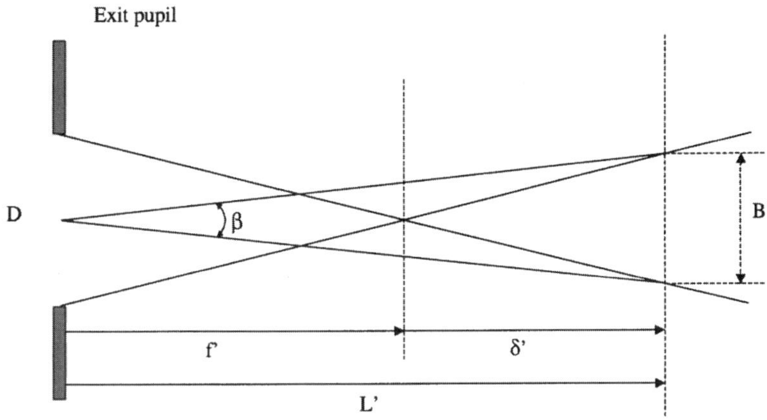

Figure 12.11. Depth of focus.

The depth of field (Fig. 12.12) is the amount the object may be shifted and remain within the blur criterion:

$$\delta^+ = \frac{\beta L^2}{D - \beta L},$$

$$\delta^- = \frac{-\beta L^2}{D + \beta L}. \tag{12.21}$$

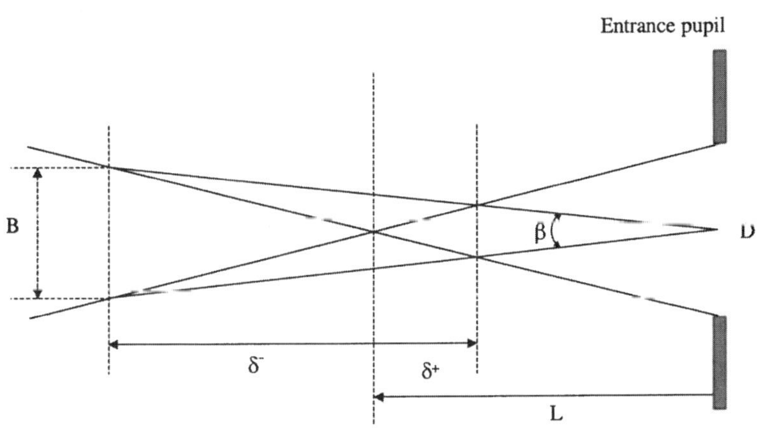

Figure 12.12. Depth of field.

The hyperfocal distance is the distance at which the system must be focused so that the depth of field extends to infinity:

$$L = \frac{-D}{\beta}. \tag{12.22}$$

The Rayleigh quarter-wave criterion for the depth of focus is defined as the amount of defocus that corresponds to an optical path difference of a quarter wave:

$$\delta = \frac{n'\lambda}{2\text{NA}^2} = 2n'\lambda(F)^2, \tag{12.23}$$

where n' is the index of refraction in the image space and λ is the wavelength.

The Rayleigh criterion (Fig. 12.13) for two equal-intensity point sources is such that when the maximum of one Airy disk falls on the first dark ring of the Airy disk of the other, the two sources are just resolved.

Airy disk radius:

$$r_a = \frac{0.61\lambda}{\text{NA}} = 1.22\lambda(\text{FN}). \tag{12.24}$$

Angular resolution:

$$\alpha = 1.22\,\frac{\lambda}{D}. \tag{12.25}$$

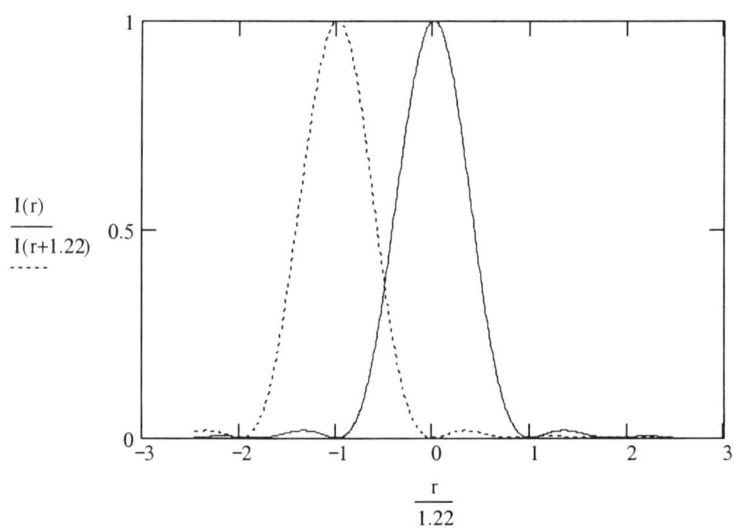

Figure 12.13. Rayleigh criterion for resolution.

Further Reading

F. A. Jenkins and H. E. White, *Fundamentals of Optics,* McGraw-Hill, 1957.

D. S. Goodman, Chap. 1, "General Principles of Geometric Optics," in *Handbook of Optics,* M. Bass, E. W. Van Stryland, D. R. Williams, and W. L. Wolfe, eds., McGraw-Hill, 1995.

W. J. Smith, *Modern Optical Engineering,* McGraw-Hill, 1990.
R. R. Shannon, *The Art and Science of Optical Design,* Cambridge, 1997.
R. E. Fischer and B. Tadic-Galeb, *Optical System Design,* McGraw-Hill, 2000.

The first reference is a basic book on optics that includes most of the material in this chapter. The second is a very nice treatment at a much higher level of the fundamentals of geometric optics. The next three cover the material in this chapter and much more about optical design.

13

Basic Lenses

R. Barry Johnson

13.1. Introduction

In this chapter a basic understanding is provided of using lenses for image formation and manipulation. The principles of image formation are reviewed first. The effects of lens shape, index of refraction, magnification, and F-number on the image quality of a singlet lens are discussed in some detail. Achromatic doublets are covered next. The chapter concludes with several techniques for rapid estimation of the performance of lenses.

13.2. Basic Image Formation

The equations for basic image formation are reviewed here. More information is in Chap. 12.

13.2.1. Magnification

Figure 13.1 illustrates an image being formed by a simple lens. The object height is h_o and the image height is h_i, with u_o and u_i being the corresponding slope angles. It follows from the Lagrange invariant that the *lateral magnification* is defined to be

$$m \equiv h_i/h_o$$

$$= \frac{(nu_o)}{(nu_i)}.$$

By convention, a height is positive if it is above the optical axis and a ray angle is positive if its slope angle is positive. Distances are positive if the ray propagates left to right.

If the object is moved a small distance s_o longitudinally, the corresponding displacement of the image s_i can be found by the differential form of the basic

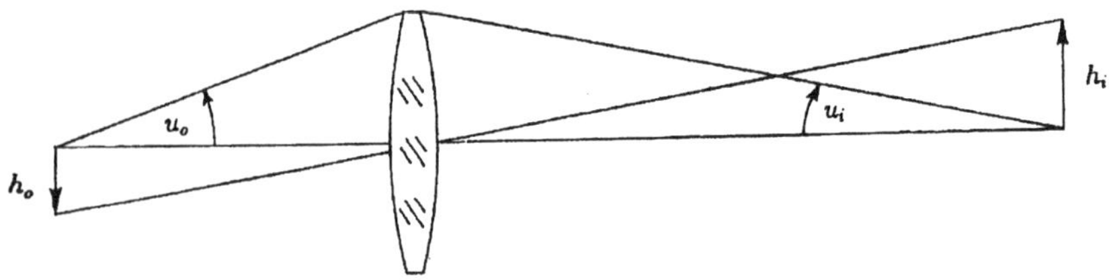

Fig. 13.1. Imaging by a simple lens.

imaging equation and leads to an equation analogous to the Lagrange invariant. The *longitudinal magnification* is then defined as

$$\bar{m} \equiv \frac{\partial s_i}{\partial s_o}$$

$$= \frac{(nu^2)_o}{(nu^2)_i}$$

$$= m^2 \left(\frac{n_i}{n_o} \right).$$

The following example illustrates one application of m and $\bar{m}$. A spherical object of radius r_o is to be imaged, as shown in Fig. 13.2. The equation of the ob-

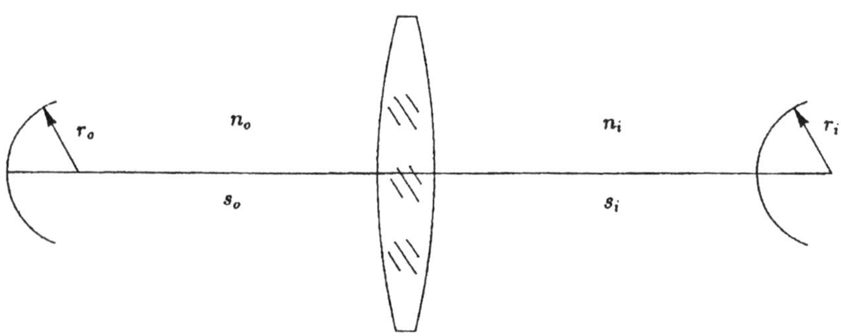

Fig. 13.2. Imaging of a spherical object by a lens.

ject is $r_o^2 = y_o^2 + z^2$, where z is measured along the optical axis and is zero at the object's center of curvature. Letting the surface sag as measured from the vertex plane of the object be denoted as ζ_o, the equation of the object becomes $r_o^2 = (r_o - \zeta_o)^2 + y_o^2$ because $z = r_o - \zeta_o$. In the region near the optical axis, $\zeta_o^2 \ll r_o^2$, which implies that $r_o = y^2/2\zeta_o$. The image of the object is expressed in the transverse or lateral direction by $y_i = my_o$ and in the longitudinal or axial direction by $\zeta_i = \bar{m}\zeta_o = \zeta_o m^2 (n_i/n_o)$. In a like manner, the image of the spherical object is expressed as $r_i \approx y_i^2/2\zeta_i$. By substitution, the sag of the image is expressed by

$$r_i \equiv \frac{n_o y_o^2}{2n_i \zeta_o}$$

$$= r_o \left(\frac{n_o}{n_i} \right).$$

Hence, in the paraxial region about the optical axis, the radius of the image of a spherical object is independent of the magnification and depends on only the ratio of the refractive indices of the object and the image spaces.

13.2.2. Scheimpflug Condition

When an optical system, as shown in Fig. 13.3, images a tilted object, the image will also be tilted. The intersection height of the object plane with the first

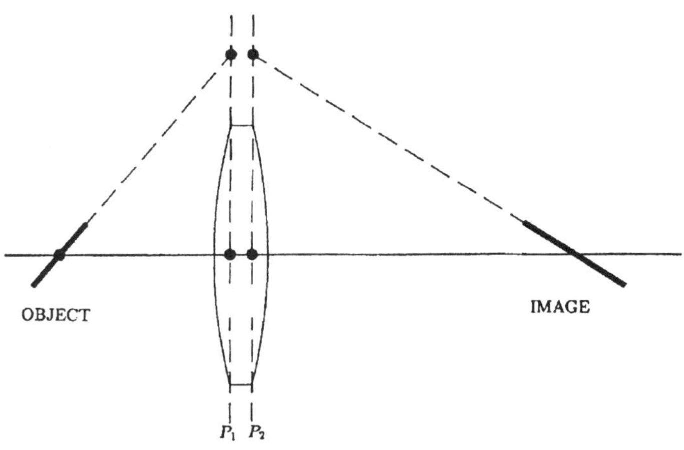

Fig. 13.3. Imaging of a tilted object, illustrating the Scheimpflug condition.

principal plane P_1 of the lens is the same as the intersection height of the image plane with the second principal plane P_2 of the lens. This principle is known as the *Scheimpflug condition.*

13.2.3. Image Position

The focal length of a lens is related to the power of the lens, ϕ, by

$$\phi = \frac{n_o}{f_o} = \frac{n_i}{f_i}.$$

The *lens law* can be expressed in several forms. If s_o and s_i are the distance from the object to the first principal point and the distance from the second principal point to the image, respectively, then the relationship between the object and the image is given by

$$\phi = \frac{n_i}{s_i} + \frac{n_o}{s_o}.$$

When the distances are measured with respect to the nodal points, the imaging equation becomes

$$\phi = \frac{n_o}{s_i} + \frac{n_i}{s_o}.$$

The power of a spherical refracting surface, with curvature c and n the refractive index following the surface, is given by

$$\phi = c(n - n_o).$$

The power of a single thick lens in air is

$$\phi_{\text{thick}} = \phi_1 + \phi_2 - \phi_1\phi_2 \frac{t}{n},$$

where t is the thickness of the lens. The distance from the first principal plane to the first surface is $-(t/n)\phi_2 f_1$, and the distance from the second principal point to the rear surface is $-(t/n)\phi_1 f_2$. The power of a thin lens ($t \rightarrow 0$) in air is given by

$$\phi_{\text{thin}} = 1/f_{\text{thin}} = (n - 1)(c_1 - c_2).$$

The power of a thin lens made of a given material is dependent on only the difference in the surface curvatures.

The object and the image points are conjugate points. If both points are located a finite distance from the lens system, then it is referred to as *finite-conjugate imaging*. If either of the points is essentially infinitely distant from the lens system, then it is typically called *infinite-conjugate imaging*. If both points are located infinitely distant from the lens system, then the lens system is referred to as *afocal*.

13.2.4. Graphical Ray Tracing

The principal planes and focal points can describe any rotationally symmetric lens element or system of lens elements. The image formed by a lens or system of lenses can be determined by the tracing of two rays from the object. The process for a system of lenses is to form an image with the first lens and then to designate this image as the object for the second lens. The image is again determined, and the process is repeated until the final image position is located and the image size is established. The process to graphically determine the size and the location of a single lens is illustrated in Fig. 13.4 and procedurally is as follows:

- Draw the optical system to scale, showing the location of the principal planes and the focal points for each lens (a thin lens has coincident principal planes).
- Draw a ray from the top of the object that passes through the front focal point (f_1) and continue it until it intersects the first principal plane.
- Draw a line parallel to the optical axis from this intersection point until the line intersects the second principal plane (principal planes are related by unit lateral magnification) and continue this line parallel to the optical axis.
- All rays passing through f_1 exit the lens parallel to the optical axis.
- Draw a second ray from the top of the object that is parallel to the optical axis and continue it until it intersects the first principal plane.
- Draw a line parallel to the optical axis from this intersection point until the line intersects the second principal plane.
- Draw a line from this intersection point that passes through the second focal point (f_2) until the line intersects the first line (all rays entering a lens parallel to the optical axis will pass through f_2).
- The intersection of these two rays is the top on the image, and the orthogonal projection to the optical axis establishes the axial location of the image.

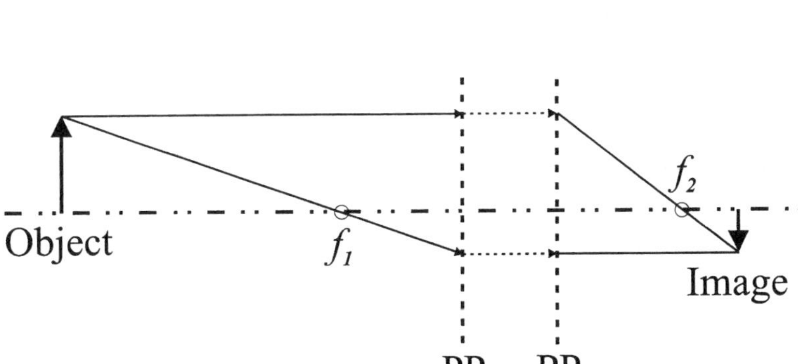

Fig. 13.4. Process of determining the size and the location of a single lens.

13.3. Stop and Pupils

Stops and their images, called pupils, control the size of the cone of light and the field of view.

13.3.1. Aperture Stop and Entrance–Exit Pupils

The *aperture stop* or *stop* of a lens is the limiting aperture associated with the lens that determines how large an axial beam may pass through the lens. Often an *iris* is the stop. The *marginal ray* is the extreme ray from the axial point of the object through the edge of the stop. The *entrance pupil* is the image of the stop formed by all lenses preceding it when viewed from object space. The *exit pupil* is the image of the stop formed by all lenses following it when viewed from image space. These pupils and the stop are all images of one another.

13.3.2. Principal and Chief Rays and Vignetting

The *principal ray* is defined as the ray emanating from an off-axis object point that passes through the center of the stop. In the absence of pupil aberrations, the principal ray also passes through the center of the entrance and the exit pupils. As the angle of the principal ray increases, the defining apertures of the components comprising the lens may limit the passage of some of the rays in the entering beam, thereby causing the stop not to be filled with rays. The failure of an off-axis beam to fill the aperture stop is called *vignetting.* The ray centered between the upper and the lower rays defining the oblique beam is called the *chief ray.* When the object moves to large off-axis locations, the entrance pupil often has a highly distorted shape and may be tilted and/or displaced longitudinally and transversely. Because of the vignetting and pupil aberrations, the chief and the principal rays may become significantly displaced from one another. In some cases, the principal ray is vignetted.

13.3.3. Field Stop and Entrance–Exit Windows

The *field stop* is an aperture that limits the passage of principal rays beyond a certain field angle. The image of the field stop when viewed from object space is called the *entrance window* and is called the *exit window* when viewed from image space. The field stop effectively controls the field of view of the lens system. Should the field stop be coincident with an image formed within or by the lens system, the entrance and the exit windows will be located at the object and/or image(s).

13.3.4. Telecentric Stop

A *telecentric stop* is an aperture located such that the entrance and/or exit pupils are located at infinity. This is accomplished when the aperture is placed in the appropriate focal plane. A stop placed at the front focal plane of a lens is imaged at infinity, and the principal ray exits the lens parallel to the optical axis. This feature is often used in metrology as the measurement error is less than that of conventional lens systems because the centroid of the blur remains at the same height from the optical axis even as the focus is varied.

13.4. *F*-Number and Numerical Aperture

The focal ratio or F-number F of a lens is defined as the effective focal length divided by the entrance pupil diameter D_{ep} (not the aperture-stop diameter). When the object is not located at infinity, the *effective* F-number is given by

$$F_{\mathrm{eff}} = F_{\infty}\,(1 - m),$$

where m is the magnification (be sure to use the correct sign). For example, a simple positive thin lens being used at unity magnification ($m = -1$) has $F_{\mathrm{eff}} = 2F_{\infty}$.

The *numerical aperture* of a lens is defined as

$$\mathrm{NA} = n_i \sin U_i,$$

where n_i is the refractive index in which the image lies and U_i is the slope angle of the marginal ray exiting the lens. If the lens is aplanatic (free of spherical aberration and linear coma), then

$$F_{\mathrm{eff}} = \frac{1}{2\mathrm{NA}}.$$

The term *speed* is frequently used to refer to a lens' numerical aperture in which a "fast" lens has a large final slope angle U_i. In air, the NA → unity as the marginal ray becomes orthogonal to the optical axis. As the speed of lens increases, so does the NA; however, F decreases. In photography, a fast lens is considered to have $F < 2$ whereas a slow lens has $F > 11$.

13.5. Magnifier or Eye Loupe

The typical magnifying glass, or *loupe,* comprises a singlet lens and is used to produce an erect but virtual magnified image of an object. The magnifying power of the loupe is stated to be the ratio of the angular size of the image when viewed through the magnifier to the angular size without the magnifier. Because a human eye is normally used at a distance of 25 mm, the magnifying power of the loupe is

$$\mathrm{MP} = \frac{25(\mathrm{cm})}{d_e + d_o - \phi d_e d_o},$$

where d_o is the distance from the object to the loupe, d_e is the separation of the loupe from the eye, and $\phi = 1/f$ is the power of the magnifier. When d_o is set to the focal length of the lens, the virtual image is placed at infinity and the magnifying power reduces to

$$\mathrm{MP} = \frac{25(\mathrm{cm})}{f}.$$

Should the virtual image be located at the near viewing distance of the eye (~25 cm), then

$$\mathrm{MP} = \frac{25(\mathrm{cm})}{f} + 1.$$

Typically, simple magnifiers are difficult to make with magnifying powers greater than ~10×.

13.6. Compound Microscopes

For magnifying power greater than that of a simple magnifier, a compound microscope, which comprises an objective lens and an eyepiece, may be used. The objective forms an areal image of the object at a distance s_{ot} from the rear focal point of the objective. The distance s_{ot} is called the *optical tube length* and is typically 160 mm. The objective magnification is

$$\mathrm{MP}_{\mathrm{obj}} = \frac{s_{\mathrm{ot}}}{f_{\mathrm{obj}}}.$$

The image formed is further magnified by the eyepiece, which has a $\mathrm{MP}_{\mathrm{ep}} = [250\ (\mathrm{mm})/f_{\mathrm{ep}}]$. The total magnifying power of the compound microscope is given by

$$\begin{aligned}\mathrm{MP} &= \mathrm{MP}_{\mathrm{obj}}\mathrm{MP}_{\mathrm{ep}}\\ &= \frac{160}{f_{\mathrm{obj}}}\frac{250}{f_{\mathrm{ep}}},\end{aligned}$$

where the focal lengths are specified in millimeters. Typically, f_{ep} = 25 mm, so its MP = 10. Should the objective have a focal length of 10 mm, the total magnifying power of the microscope is 16× multiplied by 10×, or 160×.

13.7. Field and Relay Lenses

These relocate a pupil or image.

13.7.1. Field Lenses

Field lenses are placed at (or near) an image location for the purpose of optically relocating the pupil or to increase the field of view of the optical system. For example, a field lens may be used at the image plane of an astronomical telescope such that the field lens images the objective lens onto the eyepiece. In general, the field lens does not contribute to the aberrations of the system except for distortion and field curvature. Because the field lens must be positive, it adds inward curving Petzval (curvature of the image). For systems having a small detector requiring an apparent increase in size, the field lens is a possible solution. The detector is located beyond the image plane such that it subtends the same angle as the objective lens when viewed from the image point. The field lens images the objective lens onto the detector.

13.7.2. Relay Lenses

Relay lenses are used to transfer an image from one location to another such as in a submarine periscope or borescope. They are also used as a means to erect an image in many types of telescopes and other such instruments. Often relay lenses are made with lens groups spaced about a stop, or an image of the system stop, in order to take advantage of the principle of symmetry, thereby minimizing the comatic aberrations and lateral color. The relayed image is frequently magnified. Field lenses are typically utilized at intermediate images to minimize the physical size of the relay lenses and to minimize vignetting. In a periscope, the first field lens images the stop (and the corresponding entrance and exit pupils) of the objective lens system onto the stop of the first relay lens. The second field lens images the stop of the first relay lens onto the stop of the second relay lens, and so on.

13.8. Aplanatic Surfaces and Immersion Lenses

Abbe called a lens an aplanat that has an equivalent refractive surface that is a portion of a sphere with a radius r centered about the focal point. Such a lens satisfies the Abbe sine condition. The lens is free of spherical aberration and coma near the optical axis. Consequently, the maximum possible numerical aperture of an aplanat is unity, or $F/0.5$. In practice, an F-number less than 0.6 is difficult to achieve. For an aplanat,

$$F = \frac{1}{2\text{NA}}.$$

It can be shown that three cases exist in which the spherical aberration is zero for a spherical surface. These are (1) the trivial case in which the object and the image are located at the surface, (2) the object and the image are located at the center of curvature of the surface, and (3) the object is located at the aplanatic point. The third case is of primary interest. If the refractive index preceding the surface is n_o and following the surface is n_i, then the object is located at a distance s_o from the surface, as expressed by

$$s_o = \frac{r(n_o + n_i)}{n_o},$$

and the image is located at

$$s_i = \frac{r(n_o + n_i)}{n_i}.$$

An immersion lens or contact lens can be formed from an aplanatic surface and a plano surface. Figure 13.5 illustrates a hemispherical magnifier that uses

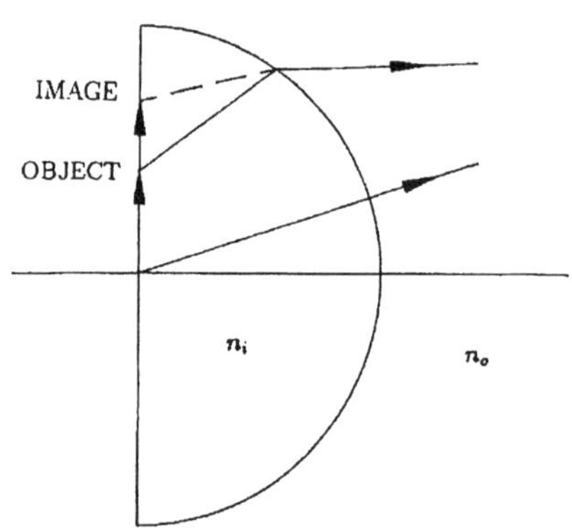

Fig. 13.5. Aplanatic hemispherical magnifier with the object and the image located at the center of curvature of the spherical surface. This type of magnifier has a magnification of n_i/n_o, which can be used as a contact magnifier or as an immersion lens.

the second aplanatic case. The resultant magnification is n_i if in air or n_i/n_o otherwise. A similar magnifier can be constructed with a hyperhemispherical surface and a plano surface, as shown in Fig. 13.6. The lateral magnification is

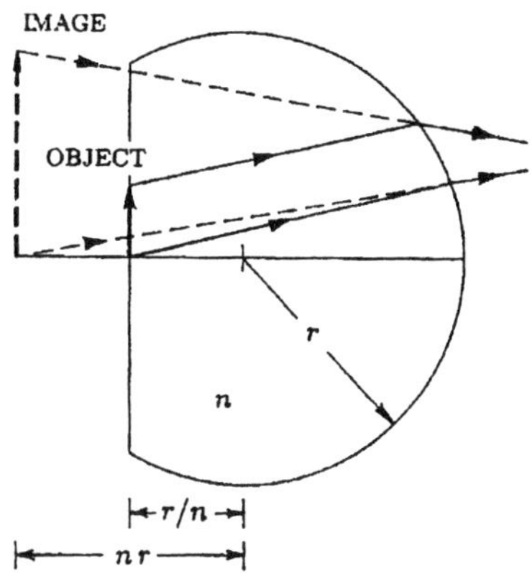

Fig. 13.6. Aplanatic hyperhemispherical magnifier or Amici lens has the object located at the aplanatic point. The lateral magnification is $(n_i/n_o)^2$.

n_i^2. This lens, called an *Amici lens,* is based on the third aplanatic case. The image is free of all orders of spherical aberration, third-order coma, and third-order astigmatism. Axial color is also absent from the hemispherical magnifier. These magnifiers are often used as a means to make a detector appear larger and as the first component in microscope objectives.

13.9. Single-Element Lenses

These have two surfaces and can be biconvex, biconcave, plano–convex, plano–concave, and meniscus.

13.9.1. Shape Factor

The spherical aberration of a lens is a function of its shape factor or bending. It is usually defined as

$$K = \frac{c_1}{c_1 - c_2},$$

where c_1 and c_2 are the curvatures of the lens with the first surface facing the object. When the lens bending is adjusted, the spherical aberration can be seen to have a minimum value.

The power of a thin lens, or the reciprocal of its focal length, is given by

$$\phi = \frac{(n-1)c_1}{K}.$$

13.9.2. Minimum Spherical Aberration

When the object is located at infinity, the shape factor for minimum spherical aberration can be represented by

$$K = \frac{n(2n+1)}{2(n+2)}.$$

The resultant third-order spherical aberration of the marginal ray in angular units is

$$\mathrm{SA3} = \frac{n(4n-1)}{64(n+2)(n-1)^2(F)^3},$$

where, for a thin lens, the F-number is the focal length f divided by the entrance-pupil diameter D_{ep}, which in this case is the same as the lens diameter.

When the object is located at a finite distance s_o, the equations for the shape factor and the residual spherical aberration are more complex. The relationship between the object distance and the magnification is

$$\frac{1}{s_o\phi} = \frac{m}{1-m},$$

where m is negative if the object distance and the lens power have opposite signs. The term $1/s_o\phi$ represents the reduced or ϕ-normalized reciprocal object distance υ, i.e., s_o is measured in units of focal length $1/\phi$. The shape factor for minimum spherical aberration is given by

$$K_m = \frac{n(2n+1)}{2(n+2)} + \frac{2(n^2-1)}{n+2}\left(\frac{m}{1-m}\right),$$

and the resultant third-order spherical aberration of the marginal ray in angular units is

$$\mathrm{SA3} = \frac{1}{16(n-1)^2(F)^3}\left[n^2 - (2n+1)K_m + \frac{n+2}{n}K_m^2\right.$$

$$(3n+1)(n-1)\left(\frac{m}{1-m}\right) - \frac{4(n^2-1)}{n}\left(\frac{m}{1-m}\right)K_m$$

$$\left.+\frac{(3n+2)(n-1)^2}{n}\left(\frac{m}{1-m}\right)^2\right],$$

where F is the effective focal length of the lens f divided by its entrance-pupil diameter. When the object is located at infinity, the magnification becomes zero and the preceding two equations reduce to those previously given.

13.9.3. Aberration Variations with m, n, and u

Figure 13.7 illustrates that lenses have a shape factor of 0.5 regardless of the refractive index when the magnification is –1 or $\upsilon = -0.5$. For this shape factor,

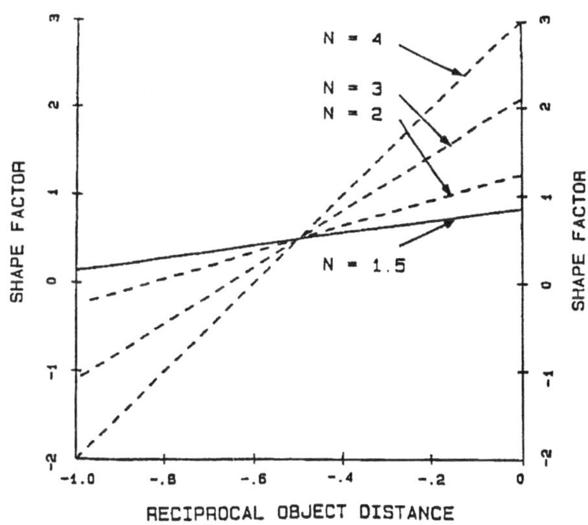

Fig. 13.7. Shape factor for a single lens is shown for several refractive indices as a function of reciprocal object distance v, where the distance is measured in units of focal length.

all lenses have biconvex surfaces with equal radii. When the object is at infinity and the refractive index is 4 (typical of germanium), lenses have a meniscus shape toward the image. For a lens with a refractive index of 1.5 (nominal value for glass), the shape is biconvex, with the second surface having a radius approximately six times greater than the first surface radius.

Because the minimum spherical lens shape is selected for a specific magnification, the spherical aberration will vary as the object and the image conjugates are adjusted, as shown in Fig. 13.8. For example, a lens having a refractive index of 1.5 and configured for $m = 0$ exhibits a substantial increase in spherical aberration when the lens is used at a magnification of –1. The ratio of the spherical aberration, when $m = -0.5$ and $m = 0$, increases as n increases.

Figure 13.9 shows the variation in angular spherical aberration when the lens bending is for minimum spherical aberration at a magnification of –1. In like manner, Fig. 13.10 presents the variation in angular spherical aberration for a convex–plano lens with the plano side facing the image. This figure can also be used when the lens is reversed by simply replacing the object distance

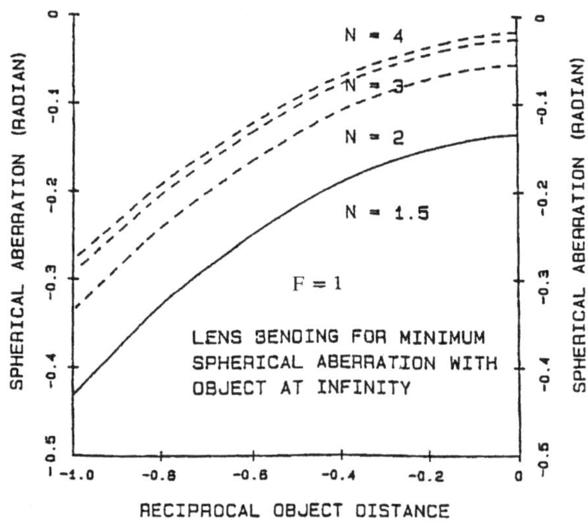

Fig. 13.8. Variation of angular spherical aberration as a function of reciprocal object distance v for various refractive indices when the lens is shaped for minimum spherical aberration with the object at infinity. Spherical aberration for a specific F-number is determined by dividing the the aberration value shown by $(F)^3$.

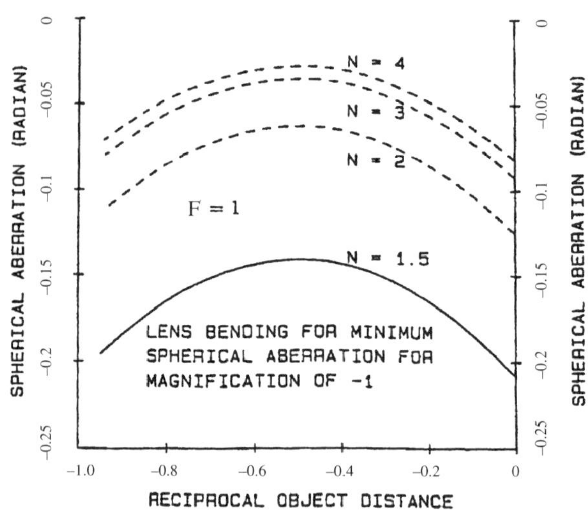

Fig. 13.9. Variation of angular spherical aberration as a function of reciprocal object distance v for various refractive indices when the lens is shaped for minimum spherical aberration for a magnification of –1. Spherical aberration for a specific F-number is determined by dividing the aberration value shown by $(F)^3$.

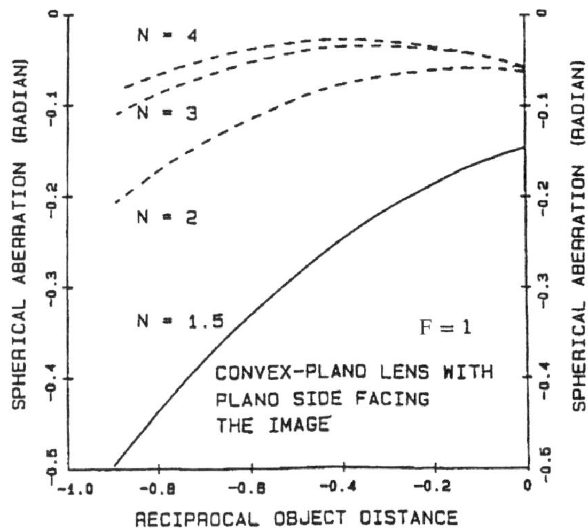

Fig. 13.10. Variation of angular spherical aberration as a function of reciprocal object distance v for various refractive indices when the lens has a convex–plano shape with the plano side facing the object. Spherical aberration for a specific F-number is determined by dividing the aberration value shown by $(F)^3$.

with the image distance. Figures 13.8–13.10 may provide useful guidance in setting up experiments when the three forms of the "off-the-shelf" lenses are available.

13.9.4. Off-Axis Aberrations

The off-axis performance of a lens shaped for minimum spherical aberration with the object at infinity can be estimated with the following equations. Assuming that the stop is in contact with the lens, the third-order angular sagittal coma is given by

$$\mathrm{CMA}_s = \frac{\theta}{16(n+2)(F)^2}$$

where the field angle θ is expressed in radians. The tangential coma is three times the sagittal coma, or $\mathrm{CMA}_t = 3\mathrm{CMA}_s$. The diameter of the angular astigmatic blur formed at best focus is expressed by

$$\mathrm{AST} = \frac{\theta^2}{F}.$$

The best focus location lies midway between the sagittal and the tangential foci. An estimate of the axial angular chromatic aberration is given by

$$\text{AChr} = \frac{1}{2V(F)},$$

where V is the Abbe number of the glass and $V = (n_d - 1)/(n_F - n_C)$ with n_d, n_F, and n_C at the helium d line, hydrogen F line, and hydrogen C line, respectively. The denominator of the Abbe number represents the relative *dispersion* of the glass, and the numerator indicates the relative power to bend to redirect the propagation direction of light rays.

Figure 13.11 shows the variation of the spherical aberration as a function of the shape factor for a thin lens with $n = 1.5$, stop in contact with the lens, and object at infinity.

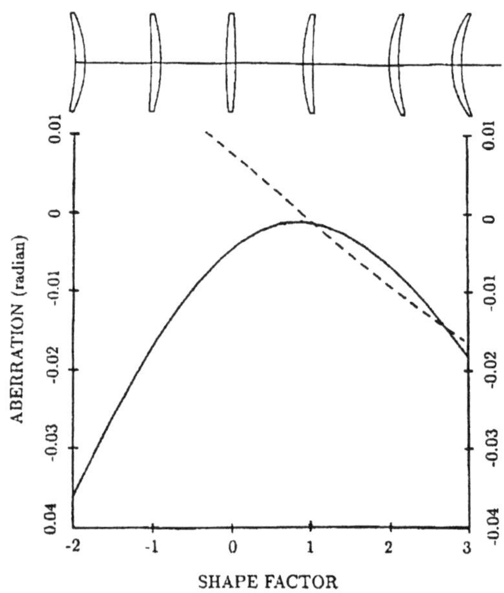

Fig. 13.11. Variation of spherical aberration (solid curve) and coma (dashed line) as a function of shape factor for a thin lens with a refractive index of 1.5, stop in contact with the lens, and the object at infinity. The shape of the lens as the shape factor changes is shown at the top of the figure.

13.9.5. Aberration Correction by Use of an Aspheric Surface

For certain cases, it is desirable to have a single lens with no spherical aberration. A useful form is the plano–convex with the plano side facing the object and the convex side being figured as a conic surface having a conic constant of $-n^2$. Caution should be exercised when using this lens form at other than infinite object distances; however, imaging at finite conjugates can be accomplished by use of two lenses with their plano surfaces facing one another and the magnification being determined by the ratio of the focal lengths. It should be noted that, for this lens form, the actual thickness of the lenses is not important and the inclusion of the conic surface does not alter the focal length.

If the object distance is at infinity and the image is located within the refractive medium n of the lens, then the convex surface of the lens (radius r) should face the object and have a conic constant of $-1/n^2$. The thickness t of the lens is the same as its focal length and specifies the location of the image, i.e., $rn/(n-1)$. A pair of these lenses, with the flat surfaces in contact with the object and the image, can also be used to form a relay lens for imaging at finite conjugates. Magnification is set by the ratio of the respective focal lengths. In principle, the distance between the lenses is not important when the field size is small.

If a singlet is made with a conic or fourth-order surface, then the spherical aberration can be corrected by the aspheric surface and the lens bending can be used to remove the coma. With the stop in contact with the lens, the residual astigmatism and chromatic errors remain as expressed by the preceding equations. The shape of the resulting lens is biconvex (the ratio of radii is ~10:1) for glass and meniscus for high-index material such as germanium.

13.9.6. *Influence of Aperture-Stop Position*

The first lens used for photography was designed in 1812 by the English scientist W. H. Wollaston approximately a quarter of a century before the invention of photography. This lens became known as the landscape lens and is illustrated in Fig. 13.12. If the stop is placed an appropriate distance in front of the lens and the F-number is modest, the image quality is better than that of the biconvex lens.

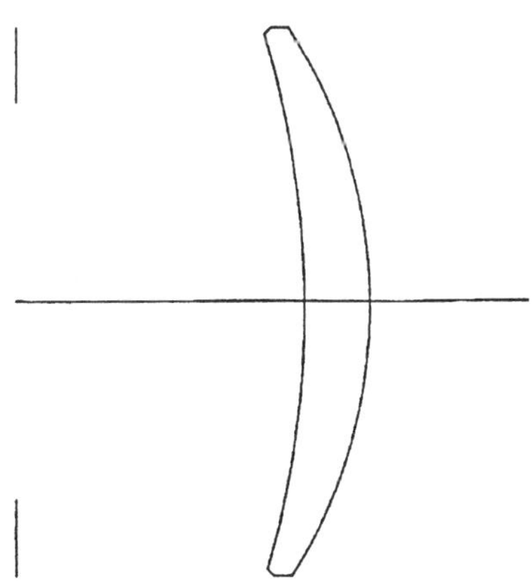

Fig. 13.12. Landscape lens with the aperture stop located to the left of the lens.

A graphical method was devised by Kingslake[1] in which he traced a number of rays in the meridional plane at a given obliquity angle, as illustrated in Fig. 13.13. A plot is made of the normalized intercept height of each real ray at

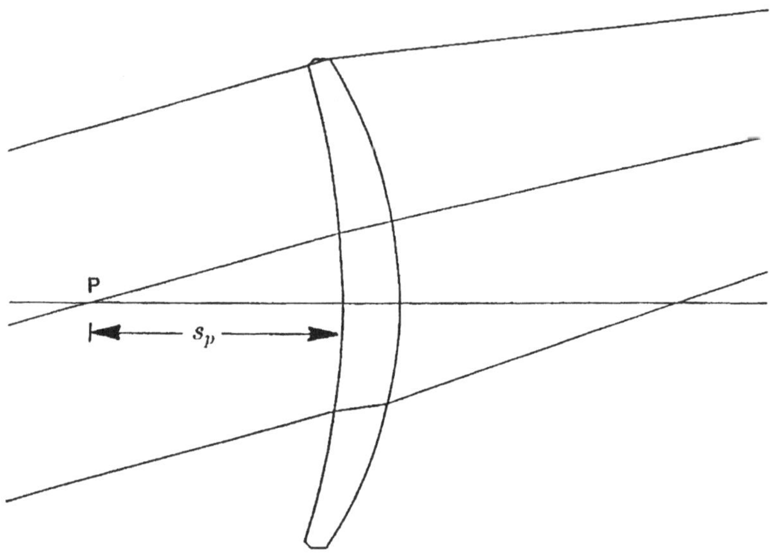

Fig. 13.13. Rays traced at a given obliquity where the intersection of a given ray with the optical axis is *P*, located a distance s_p from the front surface of the lens.

the image plane H_i a function of the distance s_p from the intersection of the ray with optical axis *P* to the front surface of the lens. Each ray can be viewed as the principal ray when the stop is located at the intersection point *P*. This H_i–s_p plot provides significant insight into the effect on image quality incurred by the placement of the stop. The shape of the curve provides information about the spherical aberration, coma, tangential field curvature, and distortion. Spherical aberration is indicated by an S-shaped curve, and the curvature at the principal-ray point is a gauge of the coma. The coma is zero at inflection points. When the curve is a straight line, both coma and spherical aberration are essentially absent. The slope of the curve at the principal-ray point is a measure of the tangential field curvature or the sag of the tangential field, i.e., astigmatism. The difference in height of the real and the Gaussian principal rays in the image plane is distortion. For situations in which the curve does not exhibit spherical aberration, it is impossible to correct the coma by shifting the stop. In addition, coma must be present if the astigmatism is to be affected by stop shift. Spherical aberration is field independent; i.e., it is present not only on the optical axis, but throughout the field of view in the same amount.

[1] R. Kingslake, *Lens Design Fundamentals,* Academic, 1978, pp. 209–215.

Because a simple meniscus lens has stop position and lens bending as degrees of freedom, only two aberrations can be corrected. Typically, coma and tangential field curvature are chosen to be corrected; adjustment of the F-number controls axial aberrations. The H_i–s_p plot for the lens is presented in Fig. 13.14. Examination of this graph indicates that the best stop location is at s_p = –0.1505 (left of the lens). For this selection, the coma and tangential astigmatism will be zero. The astigmatic field curves have a flat tangential image field

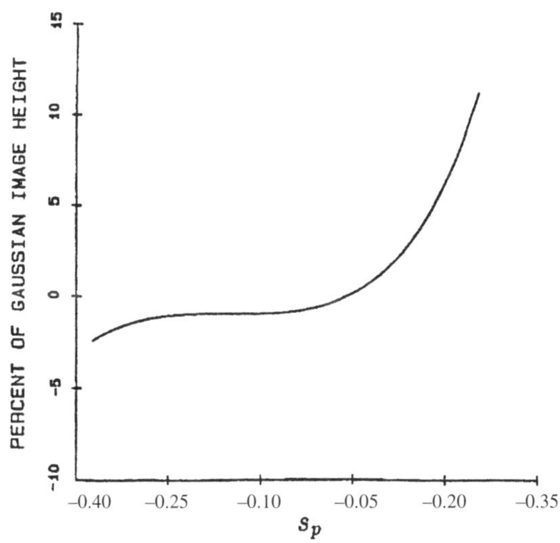

Fig. 13.14. The image height H_i of each ray traced in Fig. 13.13 is plotted against the intersection length s_p to form the H_i–s_p plot. H_i is expressed as a percentage of the Gaussian image height as a direct measure of distortion.

for all field angles, and the sagittal image field is inward curving. Other aberrations cannot be controlled and must consequently be tolerated. This condition is generally valid for only the evaluated field-angle obliquity. The Kingslake H_i–s_p method is applicable for any complexity lens. Frequently, the reference for measuring s_p is the first vertex of the first lens.

An alternative configuration can be used in which the lens is in front of the stop. Such a configuration is used to conserve space because the stop would be located between the lens and the image. The optical performance is typically poorer because of greater residual spherical aberration.

13.10. Two-Lens Systems

Figure 13.15 illustrates the general imaging problem in which an image is formed of an object by two lenses at a specified magnification and object-to-image distance. Most imaging problems can be solved by use of two equivalent lens elements. An equivalent lens can comprise one lens or multiple lens-

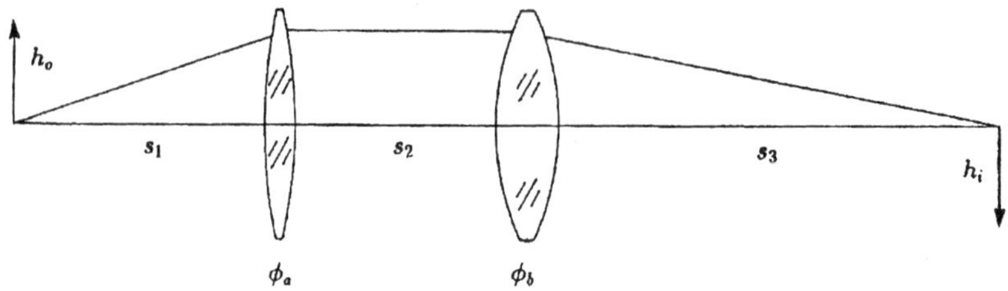

Fig. 13.15. General imaging problem in which the image is formed by two separated lenses.

es and may be represented by the principal planes and power of a single thick lens. All distances are measured from the principal points of each equivalent lens element. For simplicity, the lenses shown in Fig. 13.15 are considered thin lenses. If the magnification m, object–image distance $\mathscr{L}$, and lens powers ϕ_a and ϕ_b are known, then the equations for s_1, s_2, and s_3 are given by

$$s_1 = \frac{\phi_b(\mathscr{L} - s_2) - 1 + m}{m\phi_a + \phi_b},$$

$$s_2 = \frac{\mathscr{L}}{2}\left\{1 \pm \sqrt{1 - \frac{4[\mathscr{L}m(\phi_a + \phi_b) + (m-1)]}{\mathscr{L}^2 m\phi_a\phi_b}}\right\},$$

$$s_3 = \mathscr{L} - s_1 - s_2.$$

The equation for s_2 indicates that zero, one, or two solutions may exist.

If the magnification and the distances are known, then the lens powers can be determined by

$$\phi_a = \frac{\mathscr{L} + (s_1 + s_2)(m-1)}{ms_1s_2},$$

$$\phi_b = \frac{\mathscr{L} + s_1(m-1)}{s_2(\mathscr{L} - s_1 - s_2)}.$$

It can be shown that only certain pairs of lens powers can satisfy the magnification and separation requirements. Commonly, only the magnification and the object–image distance are specified, with the selection of the lens powers and locations to be determined. When the preceding equations are used, a plot of regions of all possible lens power pairs can be generated. Such a plot is shown as the shaded region in Fig. 13.16, where $\mathscr{L} = 1$ and $m = -0.2$. Examination of this plot can assist in the selection of lenses that may likely produce better performance if, for example, the minimum-power lenses are selected.

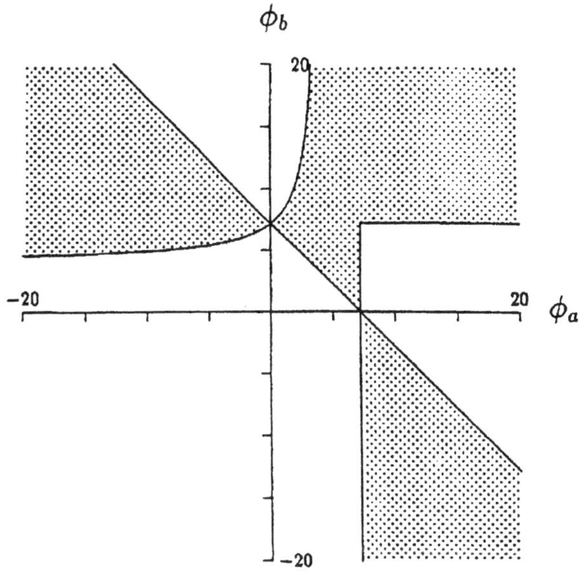

Fig. 13.16. Shaded regions indicate all possible power pairs for the two lenses used for imaging. The solution space may be limited by physical considerations such as maximum aperture.

The potential solution space may be limited if various physical constraints are placed on the lens system. For example, the allowable lens diameters can dictate the maximum powers that are reasonable. Lines of maximum power can then be plotted to show the solution space.

When s_1 becomes very large compared with the effective focal length efl of the lens combination, the optical power of the combination of these lenses is expressed by

$$\phi_{ab} = \phi_a + \phi_b - s_2\phi_a\phi_b.$$

The effective focal length is $1/\phi_{ab}$ or

$$f_{ab} = \frac{f_a f_b}{f_a + f_b - s_2},$$

and the back focal length bfl is given by

$$\text{bfl} = f_{ab}\left(\frac{f_a - s_2}{f_a}\right).$$

The separation s_2 between lenses is expressed by

$$s_2 = f_a + f_b - \frac{f_a + f_b}{f_{ab}}.$$

Figure 13.17 illustrates the two-lens configuration when thick lenses are used. The principal points for the lens combination are denoted by P_1 and P_2, P_{a1} and P_{a2} for lens *a*, and P_{b1} and P_{b2} for lens *b*. With the exception of the back focal length, all distances are measured from the principal points of each lens element or the combined lens system, as shown in the figure. For example, s_2 is

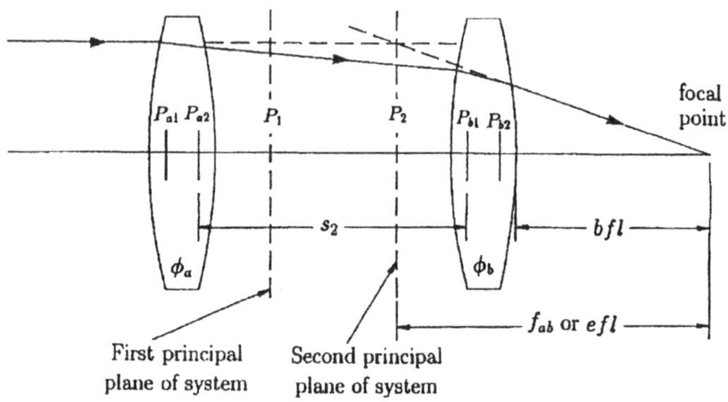

Fig. 13.17. Combination of two thick lenses illustrating the principal points of each lens and the system, the f_{ab} or efl, and the bfl. Distances are measured from the principal points, with the exception of the bfl.

the distance from P_{a2} to P_{b1}. The bfl is measured from the final surface vertex of the lens system to the focal point. When the equations in this section are used for equivalent lenses not having coincident principal planes (thick lenses or system of lenses), the distance $\mathcal{L}$ does not include the distance between the principal planes of an equivalent lens. Consequently, this fact should be considered whenever it is important that the physical (in contrast to optical) separation of the object and the image be a specific value.

13.11. Afocal Lenses

A special case of imaging occurs when both the object and the image are located at infinity. This class of optical system is called *afocal* to indicate that finite system focal points do not exist, i.e., parallel light into the system exits as parallel light although magnification of the incident beam may have occurred. Common examples of afocal systems are binoculars, Galilean telescopes, periscopes, beam expanders, etc. To create an afocal system, at least two lenses (each possibly comprising a group of lens elements and/or mirrors) are required with the focal point of one being coincident with the other. Consider two thin lenses (or groups in general) denoted as *a* and *b* having focal lengths of f_a and f_b, respectively. The afocal condition will be met when the second focal point of lens a is coincident with the first focal point of lens b. At least lens a or b must have positive optical power.

Case 1: $f_a > 0$ and $f_b > 0$. The lenses are separated by the distance $|f_a| + |f_b|$.
Case 2: $f_a > 0$ and $f_b < 0$. The lenses are separated by the distance $|f_a| - |f_b|$.

Assume that the diameter of the incident light beam is D_a and the diameter of the exiting beam is D_b. By use of principles previously present, it is evident that the ratio of the beam diameters is the same as the ratio of the focal lengths. These ratios are known as the (pupil) magnification, i.e.,

$$m = \frac{f_b}{f_a} = \text{sign}(f_a f_b)\,\frac{D_b}{D_a},$$

where the m is positive in Case 1 and negative in Case 2. This equation is useful in the specification of the lenses for such devices as a beam expander. Case 1 would afford the opportunity to incorporate a spatial filter to "clean" a laser beam. Case 2 does not have a real image formed so there cannot be a spatial filter. For the same magnification, this type of beam expander is shorter than that of the Case 1 by the factor

$$\frac{|f_a| - |f_b|}{|f_a| + |f_b|}.$$

If an off-axis beam of light is incident upon lens a with angle u_o, then light will exit the afocal system at angle u_i. The angular magnification of the system is the ratio of these angles, which can be determined with paraxial optics. However, because finite angles are involved in paraxial optics nu, this term must be interpreted as $n \tan u$. Objects viewed through a telescope will seem magnified by

$$\frac{\tan u_i}{\tan u_o},$$

where n is unity for a system in air and the image will appear inverted and reverted if this factor is negative. The relationship among the focal lengths, lens diameters, and field angles is given by

$$-\frac{\tan u_o}{\tan u_i} = \frac{f_b}{f_a} = \text{sign}(f_a f_b)\,\frac{D_b}{D_a}.$$

13.12. Achromatic Doublets

The singlet lens suffers from axial chromatic aberration, which is determined by the Abbe number V of the lens material and its F-number. A widely used lens form that corrects this aberration is the achromatic doublet, as illustrated in Fig. 13.18. An achromatic lens has equal focal lengths in c- and F-line

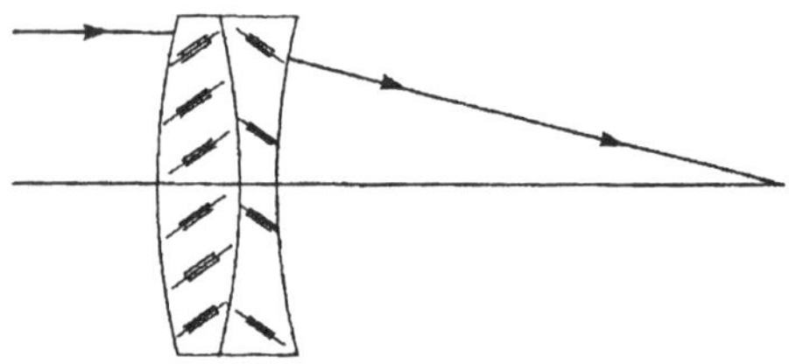

Fig. 13.18. Typical achromatic doublet lens.

light. This lens comprises two elements in which one element with a high V-number (crown glass) has the same power sign as the doublet and the other element has a low V-number (flint glass) with an opposite power sign. Three basic configurations are used. These are the cemented doublet, the broken-contact doublet, and the widely air-spaced doublet (dialyte). The degrees of freedom are two lens powers, glasses, and shape of each lens. There are two lenses, each with a power.

13.12.1. Primary Chromatic Aberration

The resultant power of two thin lenses in close proximity, $s_2 \to 0$, is $\phi_{ab} = \phi_a + \phi_b$, and the transverse primary chromatic aberration TPAC is

$$\mathrm{TPAC} = -y\,f_{ab}\left(\frac{\phi_a}{V_a} + \frac{\phi_b}{V_b}\right),$$

where y is the marginal ray height. Setting TPAC = 0 and solving for the powers of the lenses yields

$$\phi_a = \frac{V_a}{f_{ab}(V_a - V_b)},$$

$$\phi_b = \frac{-V_b\phi_a}{V_a}.$$

The bending or shape of a lens is expressed by $c = c_1 - c_2$ and affects the aberrations of the lens. The bending of each lens is related to its power by $c_a = \phi_a/(n_a - 1)$ and $c_b = \phi_b\,(n_b - 1)$. Because the two bendings can be used to correct the third-order spherical and coma, the equations for these aberrations can be combined to form a quadratic equation in terms of the curvature of the first surface c_1. Solving for c_1 will yield zero, one, or two solutions for the first lens. A linear equation relates c_1 to c_2 of the second lens.

While maintaining the achromatic correction of a doublet, the spherical aberration as a function of its shape (c_1) is described by a parabolic curve. Depending on the choices of glasses, the peak of the curve may be above, below, or at the zero spherical aberration value. When the peak lies in the positive spherical aberration region, two solutions with zero spherical aberration exist in which the solution with the smaller value of c_1 is called the left-hand solution (Fraunhofer or Steinheil forms) and the other is called the right-hand solution (Gaussian form). Two additional solutions are possible by reversal of the glasses. These two classes of designs are denoted as crown-in-front and flint-in-front designs. Depending on the particular design requirements, one should examine all four configurations to select the most appropriate. The spherical aberration curve can be raised or lowered by the selection of the V difference or the n difference. Specifically, the curve will be lowered as the V difference is increased or if the n difference is reduced. As for the thin singlet lens, the coma will be zero for the configuration corresponding to the peak of the spherical aberration curve.

13.12.2. Secondary Spectrum or Color

Although the primary chromatic aberration may be corrected, a residual chromatic error often remains and is called the secondary spectrum, which is the difference between the ray intercepts in d and c. Figure 13.19(a) illustrates an $F/5$ air-spaced doublet that exhibits well-corrected spherical light and primary chromatic aberrations and has notable secondary color. The angular secondary spectrum for an achromatic thin-lens doublet is given by

$$\text{SAC} = -\frac{P_a - P_b}{2F(V_a - V_b)},$$

where $P = (n_\lambda - n_c)/(n_f - n_c)$ is the partial dispersion of a lens material. In general, the ratio $(P_a - P_b)/(V_a - V_b)$ is nearly a constant, which means that little can be done to correct the SAC. A few glasses exist that allow $P_a - P_b \approx 0$, but $V_a - V_b$ is often small, which results in lens element powers of rather excessive strength in order to achieve achromatism. Figure 13.19(b) shows an $F/5$ air-spaced doublet that uses a relatively new pair of glasses that have a small $P_a - P_b$ and a more typical $V_a - V_b$. Both thc primary and the secondary chromatic aberrations are well corrected. Because of the relatively low refractive index of the crown glass, the higher power of the elements results in spherical aberration through the seventh order. Almost no spherochromatism (variation of

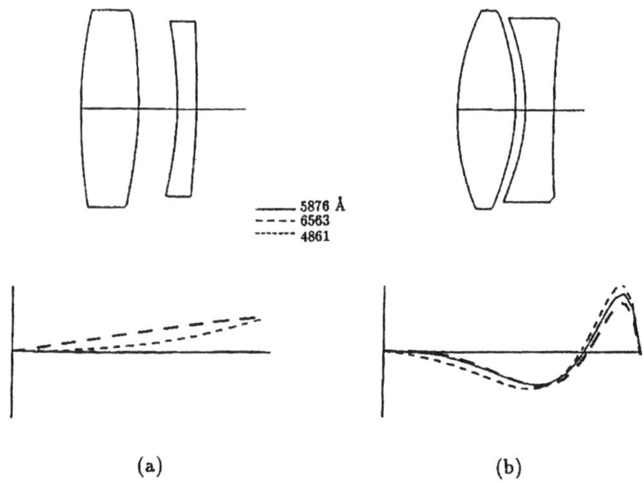

Fig. 13.19. An *F*/5 air-spaced doublet that uses conventional glasses is shown in (a) and exhibits residual secondary chromatic aberration. A similar lens is shown in (b) that uses a new glass to effectively eliminate the secondary color.

spherical aberration with wavelength) is observed. The 80% blur diameter is almost the same for both lenses.

13.12.3. Dialyte Lens

When the separation between the lens elements is nonzero, the resultant lens is known as a *dialyte,* as illustrated in Fig. 13.20. This type of lens is frequently

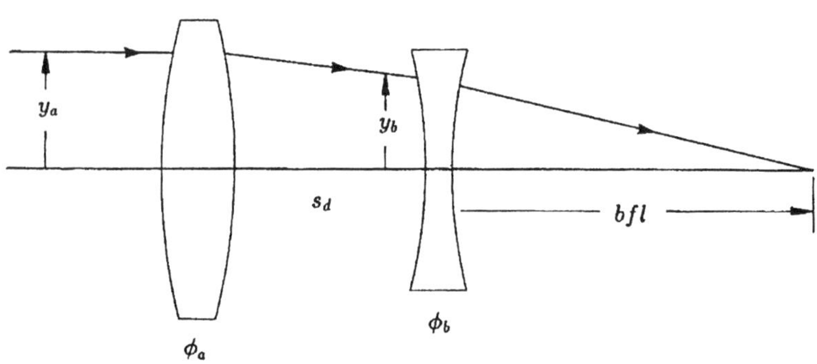

Fig. 13.20. Widely separated achromatic doublet known as the dialyte lens.

used as a simple telephoto lens or as the objective lens in an afocal telescope. As the lenses are separated by a distance s_d, the power of the flint or negative lens increases rapidly. The distance s_d may be expressed as a fraction of the crown–lens focal length by $p = s_d/f_a$. Requiring the chromatic aberration to be zero implies that

$$\frac{y_a^2}{f_a V_a} + \frac{y_b^2}{f_b V_b} = 0.$$

By inspection of the figure and the definition of p, it is evident that $y_b = y_a(1 - p)$, from which it follows that

$$f_b V_b = -f_a V_a (1 - p)^2.$$

The total power of the dialyte is

$$\phi_{ab} = \phi_a + \phi_b(1 - p).$$

Solving for the focal lengths of the lenses yields

$$f_a = f_{ab}\left[1 - \frac{V_b}{V_a(1 - p)}\right],$$

$$f_b = f_{ab}(1 - p)\left[1 - \frac{V_a(1 - p)}{V_b}\right].$$

The power of both lenses increases as p increases.

The typical dialyte lens suffers from residual secondary spectrum; however, it is possible to design an air-spaced achromatic doublet with only *one glass type* that has a significantly reduced secondary spectrum. Letting $V_a = V_b$ results in the former equations' becoming

$$f_a = \frac{p f_{ab}}{p - 1}, \quad f_b = -p f_{ab}(p - 1), \quad s_d = p f_a, \quad \text{bfl} = -f_{ab}(p - 1).$$

When $f_{ab} > 0$, then p must be greater than unity, which means that the lens is quite long. The focal point lies between the two lenses, which reduces its general usefulness. This type of lens is known as the *Schupmann* lens, based on his research in the late 1890s. Several significant telescopes, as well as eyepieces, have used this configuration. For $f_{ab} < 0$, the lens can be made rather compact and is sometimes used as the rear component of some telephoto lenses.

13.13. Rapid Estimation of Lens Performance

There are useful approximations for estimating simple lens performance.

13.13.1. Singlet

Figure 13.21 is a nomogram that allows quick estimation of the performance of a single refracting lens, with the stop at the lens, as a function of refractive

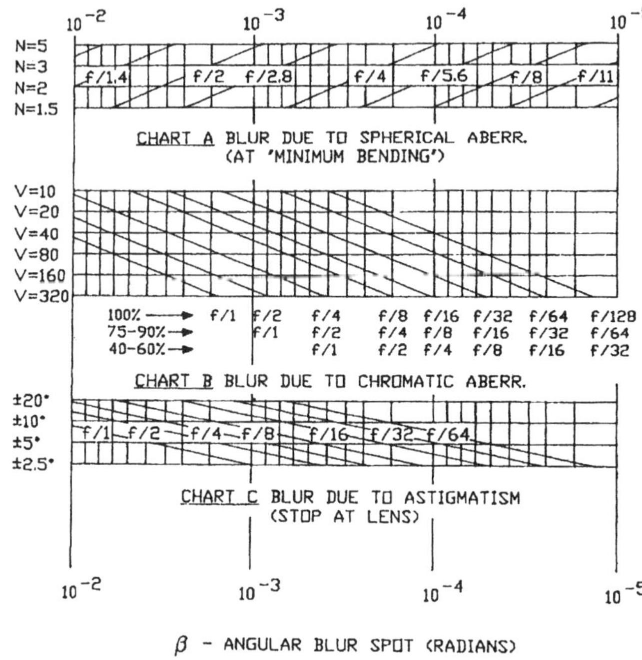

Fig. 13.21. Estimation of single-lens spot size as a function of refractive index, dispersion, F-number, and field of view. (Source: W. Smith, *Modern Optical Engineering,* McGraw-Hill, 1990, p. 458).

index N, dispersion V, F-number, and field of view θ. Chart A estimates the angular blur diameter β resulting from a singlet with bending for minimum spherical aberration. The angular chromatic blur diameter is given by Chart B. The three rows of F-number values below the chart represent the angular blur diameter that contains the indicated percentage of the total energy. Chart C shows the blur diameter that is due to astigmatism. Coma for a singlet bent for minimum spherical aberration with the stop at the lens is approximately

$$\frac{\theta}{16(N+2)(F)^2}.$$

13.13.2. Depth of Focus

The *depth of focus* of an optical system is expressed as the axial displacement that the image may experience before the resultant image blur becomes excessive. Figure 13.22 shows the geometric relationship of the angular blur tolerance $\Delta\theta$ to the depth of focus $\delta_{\pm}$. If the entrance-pupil diameter is D_{ep} and the image distance is s_i, then the depth of focus is

$$\delta_{\pm} = \frac{s_i^2\,\Delta\theta}{D_{\mathrm{ep}} \pm s_i\Delta\theta},$$

or, when $\delta \ll s_i$, the depth of focus becomes

$$\delta = \frac{s_i^2\Delta\theta}{D_{\mathrm{ep}}}.$$

When $s_i = f$, then

$$\delta = f\Delta\theta F.$$

The *depth of field* is the distance that the object may be moved without causing excessive image blur with a fixed image location. The distance at which a lens may be focused such that the depth of field extends to infinity is $s_o = D_{\mathrm{ep}}/\Delta\theta$ and is called the hyperfocal distance.

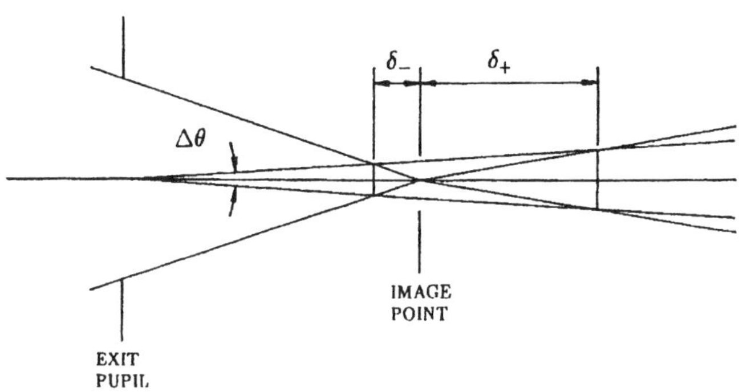

Fig. 13.22. Geometric relationships for determining the geometric depth of focus of a lens.

If the lens system is diffraction limited, then the depth of focus according to the Rayleigh criterion is given by

$$\delta = \pm\frac{\lambda}{2n_i\sin^2 u_i}.$$

13.13.3. *Diffraction-Limited Lenses*

The shape of the image irradiance of an incoherent, monochromatic point source formed by an aberration-free, circularly symmetric lens system is described by the Airy function

$$E(r) = C_0 \left[\frac{2J_1(kD_{ep}r/2)}{kD_{ep}r} \right]^2,$$

where J_1 is the first-order Bessel function of the first kind, D_{ep} is the diameter of the entrance pupil, k is $2\pi/\lambda$, r is the radial distance from the center of the image to the observation point, and C_0 is a scaling factor. The angular radius β_{DL} of the first dark ring of the image is $1.22\ \lambda/D_{ep}$. A common measure for the resolution is Rayleigh's criterion that asserts that two point sources are just resolvable when the maximum of one Airy pattern coincides with the first dark ring of the second Airy pattern, i.e., an angular separation of β_{DL}. Figure 13.23 presents a nomogram that can be used to make a rapid estimate of the diameter of angular or linear blur for a diffraction-limited system.

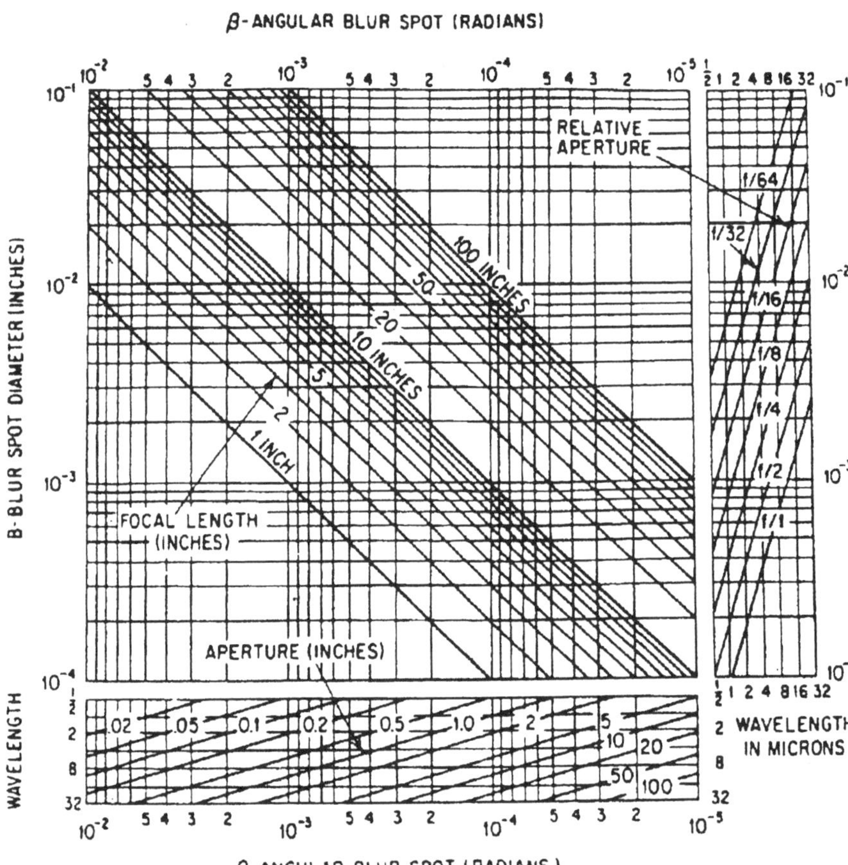

Fig. 13.23. Estimation of the spot diameter for a diffraction-limited lens system. The diameter is that of the first dark ring of the Airy disk. (Source: W. Smith, *Modern Optical Engineering*, McGraw-Hill, 1990, p. 458).

The modulation transfer function (MTF) at a specific wavelength λ for a circular entrance pupil can be computed by

$$\mathrm{MTF}_{\lambda}(\Omega) = \frac{2}{\pi}\left(\arccos \Omega - \Omega\sqrt{1-\Omega^2}\right) \quad \text{for} \quad 0 \le \Omega \le 1,$$

where Ω is the normalized spatial frequency (ξ/ξ_o) with the maximum or cutoff frequency ξ_o being given by $1/(\lambda_0 F)$. Should the source be polychromatic and the lens system be aberration free, then the perfect-image irradiance distribution of a point source can be written as

$$E(r) = C_1 \int_0^{\infty} \mathcal{R}(\lambda)\left[\frac{2J_1(kD_{\mathrm{ep}}r/2)}{kD_{\mathrm{ep}}r}\right]^2 \mathrm{d}\lambda,$$

where $\mathcal{R}$ is the peak normalized spectral weighting factor and C_1 is a scaling factor.

A quick estimation of this ideal irradiance distribution can be made by invoking the central-limit theorem to approximate this distribution by a Gaussian function, i.e.,

$$E(r) \approx C_2 e^{-(r^2/2\sigma^2)},$$

where C_2 is a scaling constant and σ^2 is the estimated variance of the irradiance distribution. When $\mathcal{R}(\lambda) = 1$ in the spectral interval λ_S to λ_L and zero otherwise with $\lambda_S < \lambda_L$, an estimate of σ can be written as

$$\sigma = \frac{\mathcal{M}\lambda_L}{\pi D_{\mathrm{ep}}},$$

where $\mathcal{M} = 1.335 - 0.625b + 0.25b^2 - 0.0465b^3$, with $b = (\lambda_L/\lambda_S) - 1$. Should $\mathcal{R}(\lambda) = \lambda/\lambda_L$ in the spectral interval λ_S to λ_L and zero otherwise, which approximates the behavior of a quantum detector, $\mathcal{M} = 1.335 - 0.65b + 0.385b^2 - 0.099b^3$. The Gaussian-estimated residual error is less than a few percent for $b = 0.5$ and remains useful even as $b \rightarrow 0$. Figure 13.24 contains plots of $\mathcal{M}$ for both cases of $\mathcal{R}(\lambda)$, where the abscissa is λ/λ_L.

A useful estimation of the modulation transfer function for this *polychromatic* lens system is given by

$$\mathrm{MTF}\,(\xi) \approx e^{-2(\pi\sigma\nu)^2},$$

where ξ is the spatial frequency. This approximation overestimates the modulation transfer function somewhat at lower spatial frequencies, while being rather a close fit at medium and higher spatial frequencies. The reason for this

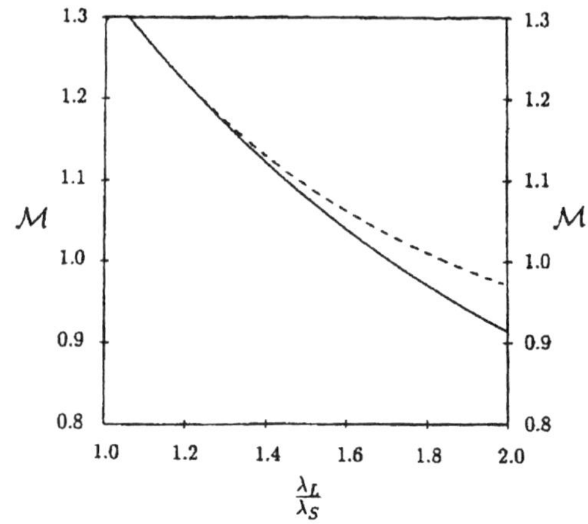

Fig. 13.24. Variation of $\mathcal{M}$ with λ_L/λ_S or $b + 1$ for $\tilde{\mathcal{R}}(\lambda) = 1$ as the solid curve and $\tilde{\mathcal{R}}(\lambda) = \lambda/\lambda_S$ as the dashed curve.

is that the central portion of the irradiance distribution is closely matched by the Gaussian approximation, whereas the irradiance estimation beyond several Airy radii begins to degrade, therefore having an impact on the lower spatial frequencies. Nevertheless this approximation can provide useful insight into expected performance limits for most well-corrected polychromatic systems.

Further Reading

Douglas S. Goodman, "Basic optical instruments," Chap. 4 in *Geometrical and Instrumental Optics,* Daniel Malacara, ed., *Methods of Experimental Physics,* 25, Academic Press, San Diego, 1988.

R. E. Hopkins, "Geometrical optics," Chap. 2 in *Geometrical and Instrumental Optics,* Daniel Malacara, ed., *Methods of Experimental Physics,* 25, Academic Press, San Diego, 1988.

R. E. Hopkins, "The components in the basic optical systems," Chap. 3 in *Geometrical and Instrumental Optics,* Daniel Malacara, ed., *Methods of Experimental Physics,* 25, Academic Press, San Diego, 1988.

R. Barry Johnson and C. Feng, "A history of IR lens designs," SPIE Critical Reviews CR37, 3–18, 1991.

Rudolf Kingslake, *A History of the Photographic Lens,* Academic Press, San Diego, 1989.

Rudolf Kingslake, *Lens Design Fundamentals,* Academic Press, New York, 1978.

Rudolf Kingslake, *Optical System Design,* Academic Press, New York, 1983.

Rudolf Kingslake, *Optics in Photography,* SPIE Press, Bellingham, 1992.

Rudolf Kingslake, "Basic geometrical optics," Chap. 6 in *Applied Optics and Optical Engineering,* 1, Academic Press, New York, 1965.

Milton Laikin, *Lens Design,* Marcel Dekker, New York, 1991.

MIL-HDBK-141, *Optical Design,* Defense Supply Agency, Washington, DC, 1962.

Warren J. Smith, *Modern Lens Design, A Resource Manual,* McGraw-Hill, New York, 1992.

Warren J. Smith, *Modern Optical Engineering,* 3rd ed., McGraw-Hill, New York, 2000.

14

Camera Lenses

Warren J. Smith

14.1. Introduction

This chapter gives the prescriptions and the performances of a number of popular camera and related lenses. A complete performance description is not possible for all of them in this short space, but enough is given for the readers to see if a particular lens is a good starting place for adaptation to their design requirements. Many of the designs are limited by curvature of field, in which case repositioning the image plane may produce a better overall result.

The lens prescriptions use the six-digit glass code, in which the first three digits are the first three digits of the refractivity and the next three are the Abbe number. For instance, a glass with a refractive index of 1.234 and an Abbe number of 77.6 has the six-digit code of 234776. The refractivity is the refractive index minus one, and the Abbe number is

$$V = \frac{n_d - 1}{n_F - n_C},$$

where the n's are those of the helium d line and the hydrogen F and C lines at 0.5876, 0.4861, and 0.6563 μm, respectively. The performances of the lenses were calculated for the three wavelengths of 0.455, 0.555, and 0.656, which are sufficiently close to these. The field angles in degrees are for the field in one direction. All lenses have focal lengths of 100 mm. The performance is described as the rms radius in microradians of the spot diameter of the spot diagram. Performance is usually given to field angles of 15°; some lenses are not designed for fields that large and become bad rapidly beyond their design values; some wide-angle lenses have good performances beyond that.

14.2. Achromatic Doublets

Doublets come in two basic forms—close spaced and separated. The close-spaced versions can be in contact or very close to it; the separated ones have a significant separation between the elements.

14.2.1. Close-Spaced Doublets

Close-spaced doublets can be in contact at the center (which is undesirable), the edge, or over the entire surface. They also can be slightly separated. Achromatic doublets are often used as telescope objectives, but also as long-focal-length (telephoto) camera lenses. In a meniscus configuration, they were also used as landscape camera lenses in the early days of photography. The two elements of a doublet are a flint glass (high dispersion) and a crown glass (low dispersion). The design can be accomplished by a choice of the powers of the *a* and *b* lenses as follows:

$$\varphi_a = \frac{V_a}{f(V_a - V_b)},$$

$$\varphi_b = \frac{V_b}{f(V_b - V_a)} = -\varphi_a \frac{V_b}{V_a},$$

where the φ's are the lens powers, V's are the Abbe numbers, and a and b designate the elements.

The Fraunhofer doublet can be in edge contact or cemented. The Fraunhofer lens has the crown glass as the first element. One example is a cemented doublet with a speed of *F*/7. The Steinheil lens has the opposite orientations of the flint and crown elements. One example is given in Table 14.1. The Gauss cannot be designed for speeds greater than *F*/5 in any reasonable way. One example at *F*/7 is given in Table 14.1. The lenses are shown in Figs. 14.1–14.3; the performances are given in Table 14.2.

14.2.2. Separated Doublets (Dialytes)

See Chap. 13, Section 13.10, for a description of these.

Table 14.1. Three Achromatic Doublets

Radius	Thickness	Material
F/7 Fraunhofer		
60.415	1.400	517642
–52.830	0.012	
–51.552	0.600	717295
–128.126	99.017	
F/7 Steinheil		
41.582	0.600	717295
27.865	0.019	
28.425	1.400	517642
–1936.302	98.514	
F/7 Gauss		
17.654	1.400	517642
53.898	0.100	
16.530	0.600	717295
13.075	94.931	

Table 14.2. Performances of Some Achromatic Doublets

	Angle (deg)							
Doublet	0	1	2	3	4	5	10	15
Fraunhofer	13	26	94	210	380	590	2410	5650
Steinheil	22	31	95	210	380	590	2420	5660
Gauss	73	78	106	171	273	408	1563	3533

Figure 14.1.

Figure 14.2.

Figure 14.3.

Frauhofer, Steinheil, and Gauss doublets, respectively.

14.3. Triplet Anastigmats

The eight design variables are just enough to correct for all the primary aberrations.

14.3.1. Cooke Triplet

This classic version is a negative element flanked by positive elements that are nearly plano–convex. These three designs have been designed for larger apertures and wider fields than are customary for Cooke triplets. Therefore they have a rather large residual aberration. Note how the vertex length and flint glass type vary with field and speed. The intended fields are ±16° for the *F*/2.5, ±23° for the *F*/4, and ±27° for the *F*/6.3. The lenses are shown in Figs. 14.4–14.6 and the prescriptions are given in Table 14.3 and 14.4; the performances are given in Table 14.5.

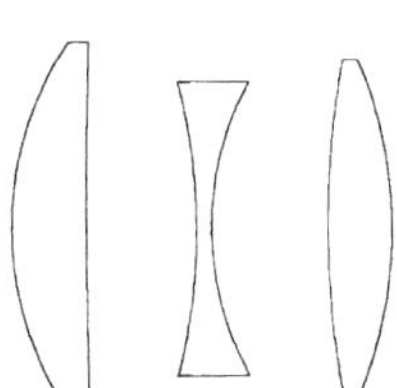

Figure 14.4. *F*/2.5 triplet.

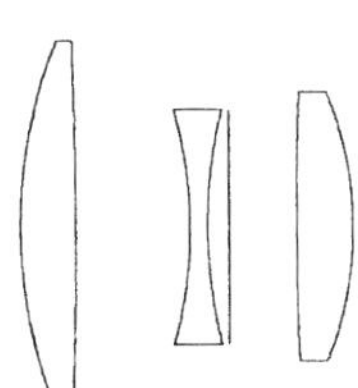

Figure 14.5. *F*/4.0 triplet.

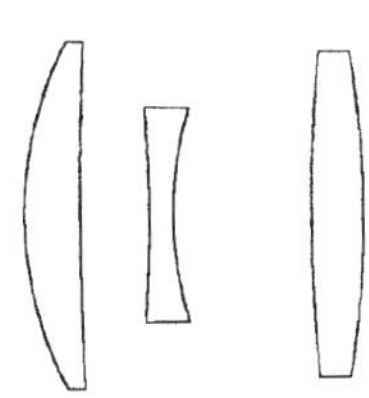

Figure 14.6. *F*/6.3 triplet.

Table 14.3. Triplets

Radius	Thickness	Material
	F/1.9 Hektor[a]	
48.8	18.0	624582
–69.3	5.7	622361
–208.4	7.9	
–53.7	8.2	624582
–27.6	4.9	567428
37.1	9.0	
81.7	11.4	624582
–47.8	4.1	541472
–63.2		
	F/2.5	
35.670	8.000	613586
675.100	11.900	
–61.660	1.600	648338
32.570	12.500	
98.520	7.000	613586
–45.500	78.183	
	F/4.0	
44.550	5.000	613586
–436.600	10.310	
–38.610	1.600	606378
42.620	1.980	
	6.060	
250.970	5.000	613586
–32.670	89.103	
	F/6.3	
24.110	3.700	613586
215.090	4.660	
–94.810	1.600	596392
23.760	2.370	
	6.760	
104.500	3.500	613586
–63.890	84.126	

[a] German patent 526308, 1930.

Table 14.4. Pestrecov and Schneider Triplets

Radius	Thickness	Material
	Pestrecov Triplet	
40.94	8.74	617550
plane	11.05	
–55.65	2.78	649338
39.75	7.63	
107.56	9.54	617550
–43.33	78.9261	
	Schneider Triplet	
26.160	4.916	678552
1201.700	3.988	
–83.460	1.038	648338
25.670	4.000	
	6.925	
302.610	2.567	651559
–54.790	81.433	

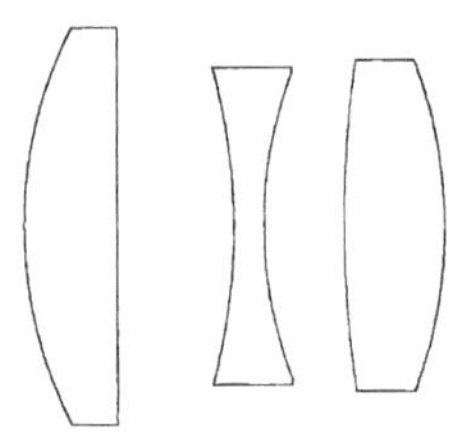

Figure 14.7. Pestrecov triplet.

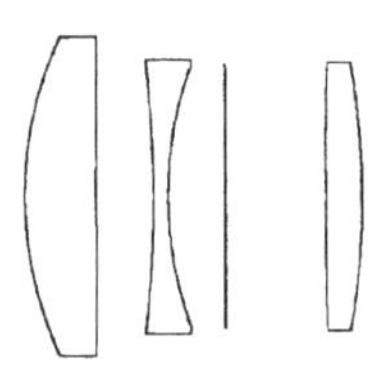

Figure 14.8. Schneider triplet.

Table 14.5. Performance of Some Triplets

Lens	Angle 0	1	2	3	4	5	10	15
F/2.5	1605	1613	1635	1676	1736	1821	2658	3699
F/4.0	236	240	253	274	299	330	521	772
F/6.3	52	61	82	109	138	168	296	324
Pestrecov	925	928	939	958	985	1023	1423	2212
Schneider	478	479	479	481	485	490	547	595

14.3.2. Pestrecov and Schneider Triplets

The Pestrecov triplet[1] has a speed of *F*/2.7 and is intended for a field of ±20°. The Schneider[2] is *F*/4.5 and ±25°. These lenses are shown in Figs. 14.7 and 14.8, the prescriptions are given in Table 14.4, and performances are given in Table 14.5.

14.4. Meniscus Lenses

These lenses derive their correction primarily from the use of one or more thick meniscus elements.

14.4.1. Hypergon

The Hypergon[3] consists of two identical meniscus elements with their concave surfaces facing each other and a stop between them. The inner and the outer radii of each element are almost the same curvature (0.5% different). It covers a field of ±67.5° with a speed of ~*F*/30, and is shown in Fig. 14.9.

14.4.2. Topogon

The Topogon[4] is a four-element Zeiss lens that can be designed for speeds of ~*F*/6 to *F*/11 and fields of 75° to 90°, and is shown in Fig. 14.10.

[1] K. Pestrecov, U.S. patent 2,453,260 (1948).
[2] Schneider, U.S. patent 1,987,878 (1935).
[3] U. S. patent 706,650 (1902).
[4] U. S. patent 2,031,792 (1936).

14.4.3. Protar

Lenses of the Zeiss protar[5] type cover fields of 60° to 90° at speeds of *F*/8 to *F*/18. The Protar is shown in Fig. 14.11; prescriptions for the Hypergon, Topogon, and Protar are given in Table 14.6, and performances are given in Table 14.8.

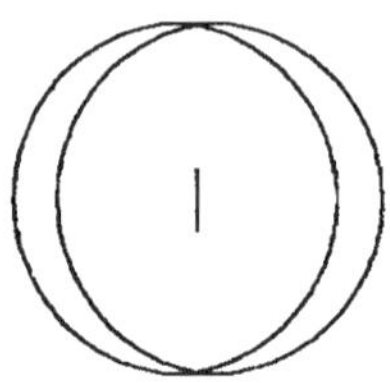

Figure 14.9. Hypergon.

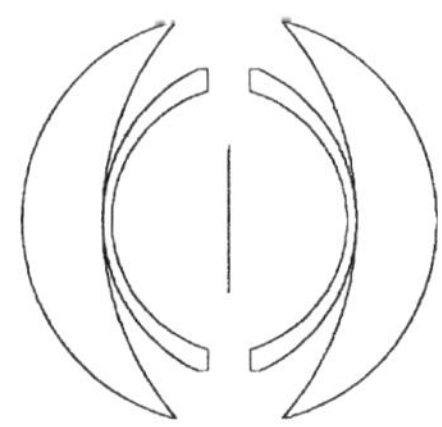

Figure 14.10. Topogon.

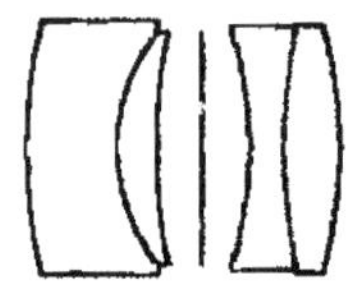

Figure 14.11. Protar.

Table 14.8. Performance of Some Meniscus Lenses

	Angle						
Lens	0	5	10	15	20	25	26.6
Hypergon	233	234	237	243	252	264	268
Topogon	636	565	541	934	1604	2357	2599
Protar	185	232	349	489	608	929	1209
Dagor[a]	210	200	270	420	560	610	600
Orthometar	54	69	148	282	520	1249	1742

[a] See Table 14.32 for an aspherized version.

14.4.4. Dagor and Orthometar

The Goerz Dagor (Fig. 14.12) uses six elements and covers fields of ~70° at speeds of ~*F*/5.6, but can be faster with smaller fields. The orthometar (Fig. 14.13) is designed for 34° at *F*/4.5. Both are described in Tables 14.7 and 14.8.

[5] U. S. patent 895,045 (1908).

Table 14.6. Hypergon, Topogon, and Protar

Radius	Thickness	Material
F/30 Hypergon		
8.57	2.2	510635
8.63	6.9	
stop	6.9	
–8.63	2.2	510635
–8.57	92.92	
F/6 Topogon		
16.875	6.66	620603
24.825	0.03	
13.641	0.75	717295
11.025	9.735	
(stop)	9.735	
–11.025	0.75	717295
–13.641	0.03	
–24.825	6.66	620603
–16.875	77.81	
Protar		
17.5	2.9	648338
5.8	1.3	603380
18.6	1.5	
(stop)	1.5	
–12.8	1.1	517522
18.6	1.8	615512
–14.3	95.01	

Table 14.7. Goerz Dagor and Orthometar

Radius	Thickness	Material
Goerz Dagor		
19.100	3.056	614564
–22.635	0.764	571529
8.272	1.910	519574
20.453	2.292	
	2.292	(stop)
–20.453	1.910	519574
–8.272	0.764	571529
22.635	3.056	614564
–19.10	96.267	
Orthometar		
25.9000	5.100	611559
–96.100	2.300	541472
18.400	0.800	
23.800	3.500	560612
35.500	2.850	
	2.850	(stop)
–33.100	4.000	560612
–22.700	1.600	
–18.100	1.900	541472
77.400	5.600	611559
–25.400	88.287	

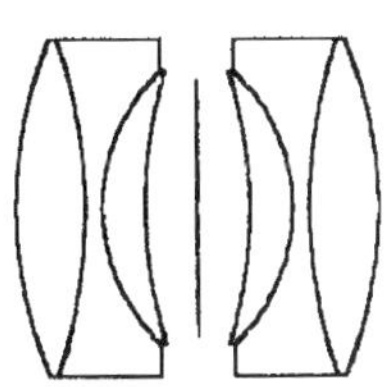

Figure 14.12. Dagor.

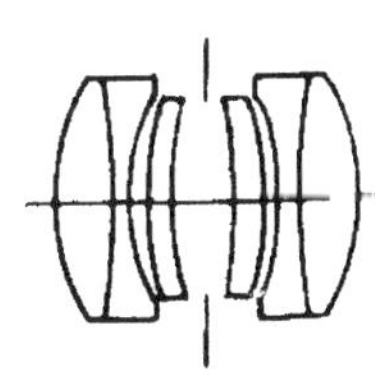

Figure 14.13. Orthometar.

14.4.5. Double Gauss (Biotar)

Three are shown, representing different speeds and fields, in Figs. 14.14–14.16; prescriptions are given in Table 14.9; and performances are given in Table 14.10.

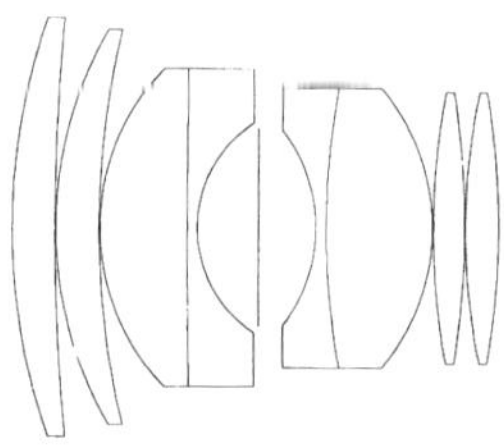

Figure 14.14. Angenieux.

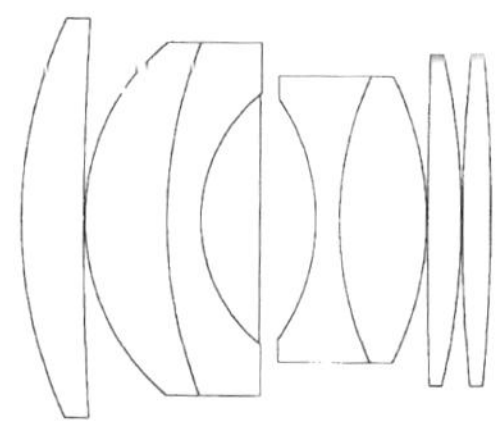

Figure 14.15. Leitz.

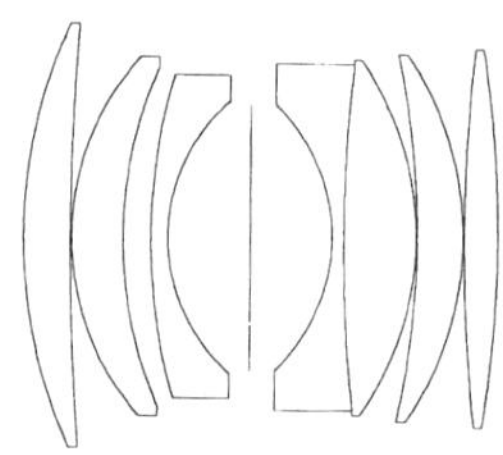

Figure 14.16. Fujioka.

14.4.6. Sonnars

Three Sonnars are shown in Figs. 14.17–14.19, prescriptions are given in Table 14.11, and performances are given in Table 14.12.

Table 14.9. Double Gauss

Radius	Thickness	Material
F/1.1 Angenieux		
164.120	10.990	673322
559.280	0.230	
100.120	11.450	670471
213.540	0.230	
58.040	22.950	691547
2551.000	2.580	673322
32.390	15.660	
	15.000	
–40.420	2.740	699301
192.980	27.920	620603
–55.5300	0.230	
192.98	7.980	691547
–225.280	0.230	
	8.480	691547
175.100	55.742	
–203.54		

Table 14.9. Double Gauss (cont.)

Radius	Thickness	Material	Radius	Thickness	Material
F/1.5 Leitz			*F*/1.4 Fujioka		
78.679	10.488	620603	88.811	8.73	773496
574.560	0.201		468.027	0.21	
38.323	13.718	620603	48.472	9.23	744449
80.693	5.643	673322	77.060	4.91	
26.733	9.861		104.392	3.03	689312
	9.500		32.532	14.77	
–34.181	4.028	548458	(stop)	15.00	
58.102	14.535	620603	–32.861	2.04	755275
–52.535	0.201		277.302	12.84	806407
1089.460	5.643	620603	–56.417	0.21	
–122.664	0.201		–152.163	8.24	804465
277.210	4.845	620603	–58.550	0.19	
293.740	56.104		290.780	6.21	720503
			–210.539	73.78	

Table 14.10. Performance of Some Double Gauss (Biotar) Lenses

	Angle							
Lens	0	1	2	3	4	5	10	15
Angenieux	240	360	560	760	950	1100	1390	1620
Leitz	630	727	955	1229	1499	1730	1873	10200
Fujioka	407	414	437	487	603	903	10500	12300

Table 14.11. Some Sonnars

Radius	Thickness	Material
F/1.2 ± 18°[a]		
121.480	8.810	613586
310.660	0.500	
73.720	8.380	613586
118.550	0.500	
50.420	33.700	607595
–105.160	3.390	689306
29.030	10.940	
	12.000	
59.390	13.720	613586
–190.620	31.406	
F/2 ± 19.8°[b]		
57.000	8.000	620603
146.300	0.400	
36.200	10.000	670471
110.000	6.000	465568
–300.000	6.800	689312
23.700	7.000	
	8.000	
200.000	2.000	569561
30.7000	12.000	670461
–152.640	48.771	
F/2.8 ± 23.8°[c]		
46.000	4.300	614552
110.000	0.210	
29.300	11.000	652449
–92.170	2.360	750349
21.680	5.650	
	1.300	
–90.000	7.000	501564
39.500	10.000	691547
–70	71.265	

[a] U. S. Patent 2,012,822.
[b] U. S. Patent 1,998,704.
[c] U. S. Patent 2,562,012.

Table 14.12. Performance of the Sonnars

	Angle							
Lens	0	1	2	3	4	5	10	15
F/1.2	238	246	271	314	378	467	1940	1110
F/2	480	430	390	370	350	330	480	1120
F/2.8	186	190	202	223	254	294	732	2390

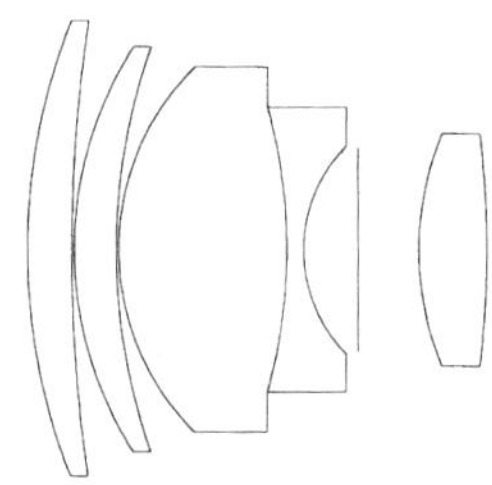

Figure 14.17. *F*/1.2 Sonnar.

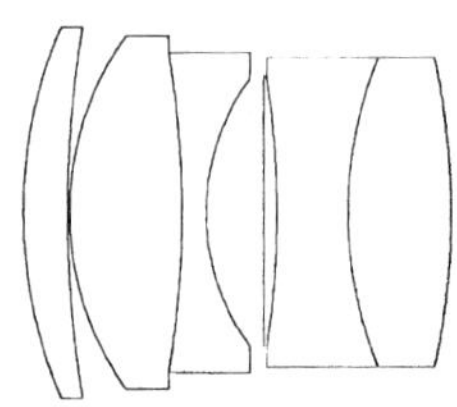

Figure 14.18. *F*/2.0 Sonnar.

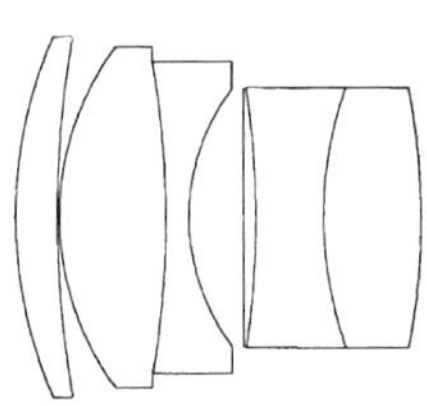

Figure 14.19. *F*/2.8 Sonnar.

14.5. Air-Spaced Anastigmats

Three air-spaced anastigmats are shown in Figs. 14.20–14.22, the prescriptions are given in Table 14.13, and the performances are given in Table 14.14.

14.5.1. *Tessars*

Table 14.13. Air-Spaced Anastigmats

Radius	Thickness	Material
F/1.8 ± 18.3° (Leitz[a])		
48.800	18.000	623581
–69.300	5.700	622360
–208.400	7.000	
	0.900	
–53.700	8.200	623581
–27.600	4.900	567428
37.100	9.000	
81.700	11.400	623581
–47.800	4.100	541472
–63.200	60.590	
F/2.8 ± 25.2° (Brendel[b])		
42.970	9.800	691547
–115.330	2.100	549454
306.840	4.160	
	4.000	
–59.060	1.870	640346
40.930	10.640	
183.920	7.050	691547
–48.910	79.831	
F/6.3 ± 30.1° (Rudolph, Zeiss)		
20.100	3.100	613586
	1.700	
–66.000	1.000	583465
18.800	2.400	
	1.900	
–81.100	1.000	523515
20.300	3.100	607567
–33.600	91.592	

[a] German patent 526308 (1930).
[b] U.S. patent 2,854,889.

Table 14.14. Performances of Some Tessars

	Angle							
Lens	0	1	2	3	4	5	10	15
Leitz	740	790	890	1060	1280	1540	2030	2080
Brendel	794	798	811	834	868	914	1413	2652
Zeiss	416	415	413	411	410	415	606	1119

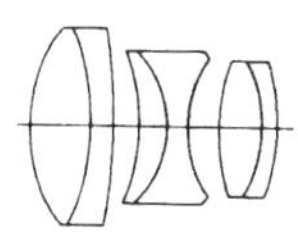

Figure 14.20. Leitz Hektor.

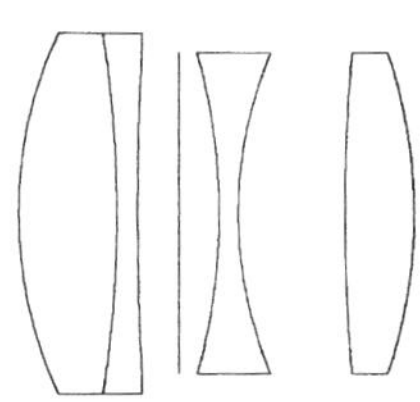

Figure 14.21. Brendel Tessar.

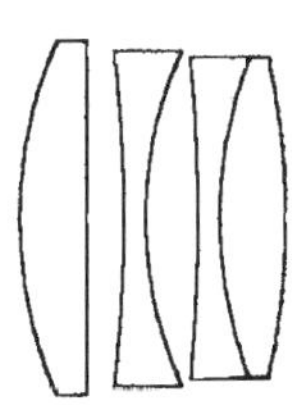

Figure 14.22. Zeiss Tessar.

14.5.2. Heliar

Two Heliars are shown in Figs. 14.23 and 14.24, prescriptions are given in Table 14.15, and performances are given in Table 14.16.

Table 14.15. Prescriptions of the Heliars

Radius	Thickness	Material
F/3.0 ± 23.3° (Booth-Dallmeyer[a])		
42.160	7.840	610567
–47.250	1.600	548458
–1505.600	4.800	
	3.520	
–43.840	1.600	548458
38.000	8.480	
838.400	1.600	548458
39.220	7.840	610567
–39.220	84.005	
F/3.5 ± 25° (Tronnier[b])		
30.81	7.70	652585
–89.35	1.85	603380
580.38	3.52	
–80.63	1.85	643480
28.34	4.18	
stop	3.00	
plano	1.85	581409
32.19	7.27	694533
–52.99	81.86	

[a] U.S. patent 1,421,156.
[b] U.S. patent 2,645,156.

Table 14.16. Performances of the Heliars

	Angle							
Lens	0	1	2	3	4	5	10	15
Booth	644	645	649	656	670	692	1036	1862
Angle	0	5	10	15	20	25		
Tronnier	220	210	240	290	350	610		

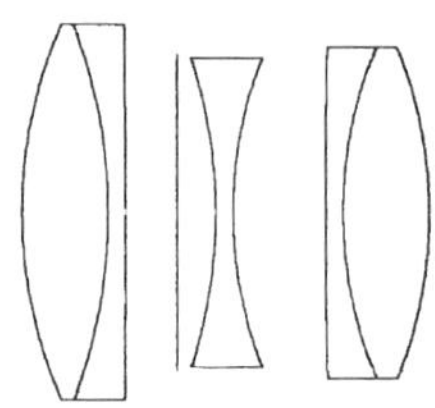

Figure 14.23. Booth–Dalmyer Heliar.

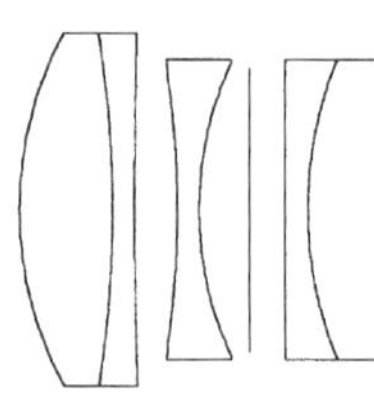

Figure 14.24. Tronnier Heliar.

14.6. Petzval Lenses

The original Petzval Portrait lens consisted of two spaced-apart achromatic doublets, and it operated at a speed of ~*F*/3.5 (fast for its day). The newer, faster Petzval projection lens is shown in Fig. 14.25, and the field-flattened version is shown in Fig. 14.26. The prescriptions are given in Table 14.17, and the performance data are given in Table 14.18.

Table 14.17. Petzval Lenses

Radius	Thickness	Material
F/1.6 Projection Lens		
81.504	16.4	523586
−76.719	4.0	617366
plano	82.0	
57.319	14.0	523586
−36.005	3.0	617366
−196.703	36.437	
F/1.6 Field-Flattened Projection Lens		
75.66	19.4	611588
−158.30	0.753	
−129.78	4.8	720293
600.68	72.7	
43.50	13.0	611588
−103.82	3.0	720293
plano	17.86	
−54.78	2.0	617366
243.12	13.201	

Table 14.18. Performances of the Petzvals

	Angle							
Lens	0	1	2	3	4	5	6	7
Projection	220	240	310	390	410	500	650	790
Flattened	200	200	210	230	260	280	320	380

Figure 14.25. Petzval projection lens.

Figure 14.26. Field-flattened Petzval.

14.7. Telephoto Lenses

If a positive front component and a negative rear component are used, it is possible to have a system with a long focal length in a compact package. The standard definition of a telephoto is a lens that has a front vertex-to-image distance that is less than its focal length. The ratio of these two is called the telephoto ratio; a small ratio is difficult to achieve. Two telephoto lenses are shown in Figs. 14.27 and 14.28, prescriptions are given in Table 14.19, and performances are given in Table 14.20. The first makes an excellent 200-mm *F*/5 (or a little faster) for a 35-mm camera; the second was designed for a focal length of 390 mm (the data given in Table 14.19 are for a focal length of 100 mm).

Table 14.19. Prescriptions of Telephoto Lenses

Radius	Thickness	Material
	F/5 ± 6°	
149.035	2.50	613586
–46.003	2.00	762265
–477.921	0.50	
26.522	2.50	613586
132.322	24.06	
–28.605	2.00	613586
22.989	1.05	
82.834	2.50	603380
–36.911	42.9	
	F/5.6 ± 6° (Kreitzer[a])	
33.072	2.386	518590
–53.387	0.077	
27.825	2.657	518590
–35.934	1.025	749350
40.900	22.084	
plano	1.794	785257
–16.755	.0641	835430
27.153	9.607	
–120.757	1.035	517522
–12.105	4.705	
–9.386	0.641	773496
–24.331	18.96	

[a] U.S. patent 4,359,272.

Table 14.20. Performance of Two Telephotos

	Angle						
Lens	0	1	2	3	4	5	6
F/5	8	10	15	22	24	36	51
Kreitzer	11	18	55	86			

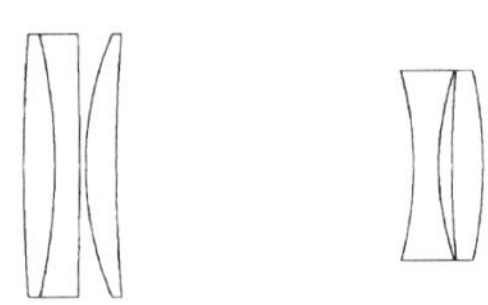

Figure 14.27. Telephoto lens.

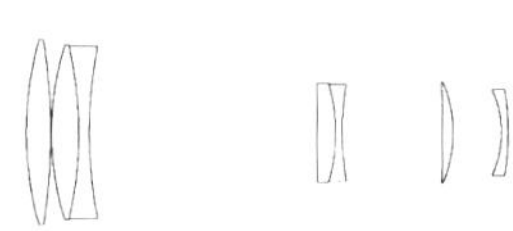

Figure 14.28. Kreitzer telephoto.

Table 14.21. Three Wide-Angle Lenses

Radius	Thickness	Material
$F/8.3 \pm 56.3°$ (Roosinov[a])		
50.610	3.080	639452
32.470	50.920	
68.060	13.160	613586
–35.120	4.390	548458
188.390	0.350	
	4.320	
–185.390	12.940	548458
34.490	50.000	613586
–66.830	3.020	
–31.870	52.779	626357
–49.440		
$F/3.4 \pm 40°$ (Biogon[b])		
138.198	3.400	504669
73.801	17.800	
162.285	3.400	487704
73.540	56.200	
76.383	40.100	720504
33.620	18.200	607494
–267.237	3.200	
	3.200	
–346.380	3.900	569561
41.930	34.300	623581
–31.480	19.600	722292
–76.359	69.600	
–55.279	12.500	642580
–93.624	23.093	
$F/5.0 \pm 50°$ (Bertele[c])		
100.060	3.900	516642
53.680	13.280	
107.260	2.880	516642
53.440	37.350	
56.550	24.880	720503
25.340	21.180	613586
–73.160	1.930	517522
1094.900	1.850	
	2.00	
–626.300	1.930	517522
57.760	27.340	615512
–25.850	15.560	720293
–56.300	36.200	
–50.250	2.880	500614
–98.960	13.280	
–51.890	3.030	500614
–94.240	39.570	

[a] U. S. patent 2,516,724.
[b] U. S. patent 2,721,499.
[c] U. S. patent 2,734,424.

14.8. Wide-Angle Lenses

Three wide-angle lenses are shown in Figs. 14.29–14.31, their prescriptions are given in Table 14.21, and their performances are given in Table 14.22.

Table 14.22. Performances of the Wide-Angle Lenses

Lens	Angle 0	10	20	30	40	50	60	70
Roosinov	2841	2939	3196	3488	3570	3010	1149	9771
Biogon	771	792	1278	2366	4069			
Bertele	160	120	160	250	280	620		

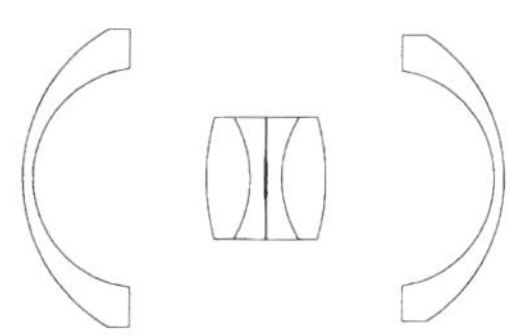

Figure 14.29. Roosinov.

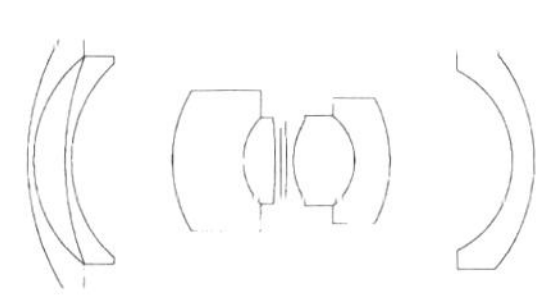

Figure 14.30. Biogon.

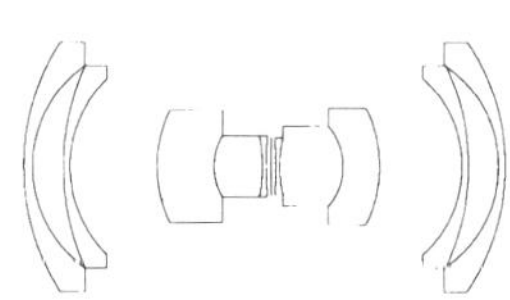

Figure 14.31. Bertele.

14.9. Reverse Telephotos (Retrofocus)

The retrofocus has a component arrangement that is reversed from that of the telephoto. The result is that the back focal length is longer than the effective focal length. This is useful when a large clearance or working distance is required (as with a 35-mm single-lens reflex so that the mirror can swing out of the way). Two are shown in Figs. 14.32 and 14.33. The prescriptions and performances are given in Tables 14.23 and 14.24, respectively.

Table 14.23. Reverse Telephotos

Radius	Thickness	Material
F/2.8 ± 37° (Mori[a])		
126.010	6.967	670516
52.254	57.332	
82.117	9.356	603425
−168.816	29.960	
−56.038	11.148	785261
226.239	2.588	
−208.842	7.266	804465
−64.382	0.398	
−1740.099	8.659	729547
−83.502	129.31	
F/2.0 ± 42° (Tsunashima[b])		
346.939	16.61	670573
1632.653	0.78	
193.678	15.84	773496
60.943	13.27	
120.408	5.14	697555
60.31	14.65	
239.559	27.71	668419
−2487.845	0.41	
152.959	4.37	697555
91.073	8.49	
121.633	29.39	749350
−906.759	5.00	
(stop)		
−1791.461	15.84	564608
−90.808	18.61	
−112.057	9.51	743493
−91.073	3.18	805255
220.816	5.55	
−676.045	13.47	743493
−87.347	0.41	
808.163	14.65	697555
−190.857	157.794	

[a] U. S. patent 2,203,653.
[b] U. S. patent 4,163,603.

Table 14.24. Performances of Two Reverse Telephotos

Lens	Angle 0	5	10	15	20	25	30	35
Mori	380	380	420	470	570	750	1000	1550
Tsunashima	550	690	930	1360	1870	2620	3480	4680

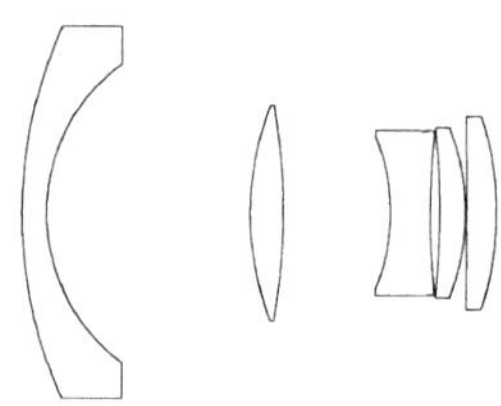

Figure 14.32. Mori.

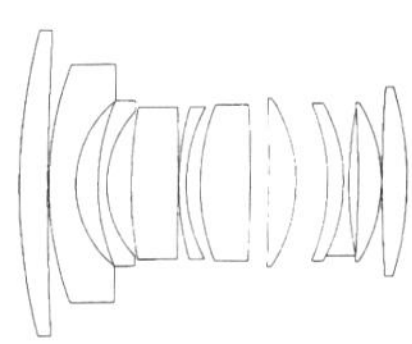

Figure 14.33. Tsunashima.

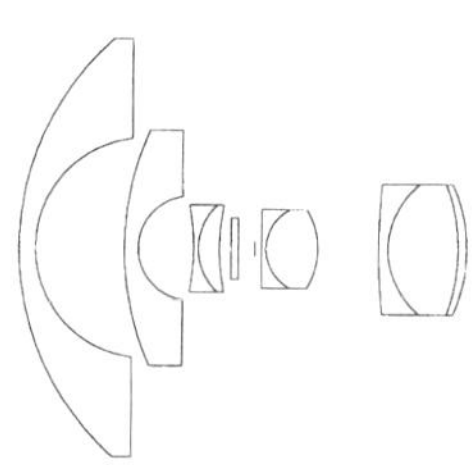

Figure 14.34. Miyamoto fish-eye lens.

14.10. Fish-Eye Lenses

A fish-eye lens may be thought of as a retrofocus lens carried to the extreme. By definition they cover a full field of at least 180°. The entrance pupil is different in position and orientation for large field angles (or no light would get into the lens). The prescription of the Miyamoto fish-eye lens is given in Table 14.25 and shown in Fig. 14.34. Note that columns 4, 5, and 6 of Table 14.25 continue the prescription.

Table 14.25. Miyamoto Fish-Eye Lens[a]

Radius	Thickness	Material	Radius	Thickness	Material
599.363	35.030	517642	stop	13.762	
235.825	190.161		38507.649	10.008	785261
605.513	30.025	487704	95.081	110.093	744447
111.094	120.102		–162.638	130.110	
–452.384	10.008	487704	1376.167	20.017	785261
127.733	45.038	785261	177.275	150.127	702410
462.892	25.021		–400.339	18.766	668419
plano	15.013	518590	–337.536	150.119	
plano	36.281				

[a] K. Miyamoto, "Fish eye lens," J. Opt. Soc. Am. **54,** 1060–1061, 1964.

14.11. Aspherics

In this section the improvements that can be obtained with aspheric surfaces are discussed. An achromatic doublet, a Cooke triplet, and a meniscus lens are used.

14.11.1. Achromatic Doublet

The prescription of the Fraunhofer doublet is transformed by aspherizing the surfaces. The reader can compare the results, which are given in Table 14.26.

This lens is not improved if we simply add conic surfaces to the existing radii of the finished design. The performance values are unchanged from Table 14.2. A faster lens (*F*/2.8), however, can be significantly improved by aspherizing just one surface. Table 14.27 gives the prescription for two *F*/2.8 doublets, one spherical surfaced and air spaced, the other a cemented form with the last surface aspherized. The performance improvement can be seen in Table 14.28.

Table 14.26. Aspherized Fraunhofer Doublet

Radius	Thickness	Material	Conic
60.415	1.400	517642	0.0359
–52.830	0.012		–0.0919
–51.552	0.600	717295	–0.4446
–128.126	99.017		–6.3918

Table 14.27. Two *F*/2.8 Fraunhofer Doublets

Radius	Thickness	Material	AD	AE	AF
82.773837	8.2	517642			
–47.064724	0.165419				
–45.099978	3.5	717295			
–81.323439	96.034332				
Cemented, Aspheric Doublet					
63.537038	8.2	517642			
–55.558872	3.5	717295			
–127.690991	94.648408		2.4063×10^{-7}	-1.4951×10^{-10}	4.8992×10^{-9}

Table 14.28. Performance of Spherical and Aspherized *F*/2.8 Fraunhofer Doublets

	Angle							
Lens	0	1	2	3	4	5	10	15
Spherical	152	176	244	354	505	698	2327	5177
Aspherized	68.1	80.6	124	202	316	464	1734	3969

14.11.2. Triplets

The prescription of the *F*/2.5 ± 6° triplet is transformed by aspherization of the surfaces. The reader can compare the results, which are given in Table 14.29. The radii and thicknesses are quite different from those of the spherical triplet; in fact, aspherizing causes a fairly radical redesign. The performance of the aspherized triplet is given in Table 14.30, as is the performance of the original spherical surfaced triplet. Note that three surfaces were aspherized so that all the aberrations could be affected.

14.11.3. Meniscus Lens

The prescription of the *F*/8 ± 26.6° Dagor meniscus lens is transformed by aspherizing of the surfaces. The reader can compare the results, which are given in Table 14.31.

Table 14.29. Aspherized Triplet

Radius	Thickness	Material	AD	AE	AF
40.4833	8	613586	-2.0918×10^{-6}	-3.313×10^{-10}	-1.2243×10^{-12}
–409.8544	9.877				
–62.5347	1.6	617366			
28.355	2.0		-8.2358×10^{-6}	-4.8992×10^{-9}	-7.139×10^{-12}
	6.713				
78.9174	7.0	613586			
–45.6699	82.4396		7.2728×10^{-7}	-6.3615×10^{-10}	3.01553×10^{-12}

Table 14.30. Performance of the Aspherized Triplet

	Angle				
Lens	0	4	8	12	16
Spherical surfaced triplet	111	91	87	97	79
Aspheric triplet	18	16	15	10	28

Table 14.31. Aspherized Dagor Lens

Radius	Thickness	Material	AD	AE	AF	AG
21.76	3.050	614564	-7.168×10^{-6}	2.5001×10^{-7}	-5.1421×10^{-9}	3.2018×10^{-11}
–18.498	2.352	571529				
9.179	1.900	519574				
23.236	1.493815					
	1.493815					
–23.236	1.900	519574				
–9.179	2.352	571529				
18.498	3.050	614564				
–21.760	92.659		7.168×10^{-6}	-2.5001×10^{-7}	5.1421×10^{-9}	-3.2018×10^{-11}

Table 14.32. Performance of the Aspherized Dagor Lens[a]

	Angle						
Lens	0	5	10	15	20	25	26.6
Dagor	120	160	230	290	360	380	360

[a] Compare this with the performance of the spherical Dagor as given in Table 14.8.

The performance given in Table 14.32 is for the prescription given in Table 14.31.

Further Reading

W. J. Smith, *Modern Optical Engineering,* 3rd ed., McGraw-Hill, 2000.
W. J. Smith, *Modern Lens Design,* McGraw-Hill, 1990.
A. Cox, *A System of Lens Design,* Focal Press, 1965 (out of print).
R. Kingslake, *A History of the Photographic Lens,* Academic, 1989.

The first text describes the design approach to many of these lenses and describes the lenses as well. The second is a compilation of ~275 designs and instruction on basic and advanced design techniques. The third has ~300 patent designs. In addition, most design programs have lens libraries.

15

Infrared Lenses

Robert E. Fischer and Michael E. Couture

15.1. Introduction

Infrared optical systems can have some significant differences from "comparable" visible ones. Detectors sometimes need to be cooled, many to temperatures of 100 K and below. Lens materials are expensive, brittle, and sometimes opaque. Thermal sensitivity of IR systems can be an order of magnitude or greater than that of similar visible systems. Narcissus and other effects can cause image anomalies. All of these effects make IR imaging systems substantially different from visible systems for the designer, manufacturer, assembler, tester, and end user.

15.2. Spectral Regions

For terrestrial applications the spectral regions are approximately 3–5 μm and 8–12 μm, depending on the application and the atmosphere. Therefore some representative designs are for these spectral bands.

15.3. Representative Infrared Materials

Materials used for IR lenses are made of single crystals, polycrystalline materials, special glasses, hot-pressed compacts, and those made by chemical vapor deposition. Table 15.1 provides some of the most important properties for some of the most important materials. Additional data may be found in *The Infrared Handbook.*[1]

Figure 15.1, called a glass map, is a plot of the refractive index as a function of the dispersion, as represented by the IR Abbe number or *V* number indicated at the bottom of the figure.

Several of the IR material characteristics offer distinct advantages over

[1] W. Wolfe and G. Zissis, eds., *The Infrared Handbook,* Society of Photo-Optical Instrumentation Engineers/Environmental Research Institute of Michigan, 1989.

Table 15.1. Principal Characteristics of Important IR Materials

Material	Wave Band (μm)	$n_{4.0}$, n_{10}[a]	$dn/dT^b(10^{-6}/C)$	Density (g/cc)	Other Characteristics
Ge	3–5, 8–12	4.025, 4.004	424 (4 μm) 404 (10 μm)	5.33	Brittle, semiconductor, can be diamond turned, visibly opaque, hard
Si	3–5	3.425, NA	159 (5 μm)	2.33	Brittle, semiconductor, diamond turned with difficulty, visibly opaque, hard
GaAs	3–5, 8–12	3.304, 3.274	150	5.32	Brittle, semiconductor, visibly opaque, hard
ZnS	3–5, 8–12	2.252, 2.200	43 (4 μm) 41 (10 μm)	4.09	Yellowish, moderate hardness and strength, can be diamond turned, scatters short wavelengths
ZnSe	3–5, 8–12	2.433, 2.406	63 (4 μm) 60 (10 μm)	5.26	Yellow–orange, relatively soft and weak, can be diamond turned, very low internal absorption and scatter
CaF_2	3–5	1.410, NA	–8.1 (3.39 μm)	3.18	Visibly clear, can be diamond turned, mildly hygroscopic
Sapphire	3–5	1.677 (n_o), 1.667 (n_e),	6 (*o*) 12 (*e*) NA	3.99	Very hard, difficult to polish because of crystal boundaries
AMTIR 1	3–5, 8–12	2.513, 2.497	72 (10 μm)	4.41	Amorphous IR glass, can be slumped to near-net shape
BK7 (Glass)	0.35–2.3	NA, NA	3.4	2.51	Typical optical glass

[a] W. Wolfe and G. Zissis, eds., *The Infrared Handbook,* Society of Photo-Optical Instrumentation Engineers/Environmental Research Institute of Michigan, 1989.

[b] P.J. Rodgers and M. Roberts, "Thermal compensation techniques," in M. Bass, E. Van Stryland, and W. Wolfe, *Handbook of Optics,* McGraw-Hill, 1995, Vol. I, Chap. 39.

"normal" optical glass, including high index of refraction and suitability for diamond machining for many of the materials. A high index of refraction generally means that fewer elements are required for achieving the ray bending required for achieving a quality image. In addition, diamond machining processes allow conic or general aspheric surface contours to be placed on elements, allowing superior image correction over an all-spherical system of the

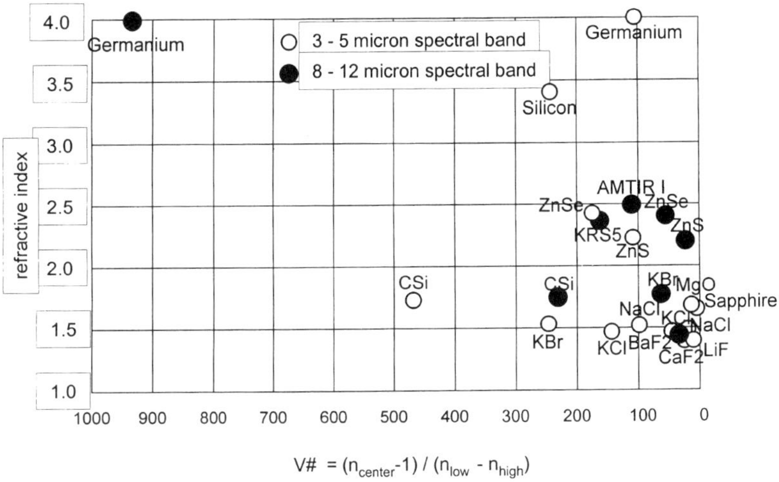

Fig. 15.1. Glass map of common IR materials.

same number of elements. Diamond machining also is used to produce kinoform diffractive surfaces for use in IR systems.

15.4. Thermal Sensitivity

$$\frac{1}{f}\frac{\mathrm{d}f}{\mathrm{d}T} = -\frac{\mathrm{d}n/\mathrm{d}T}{n-1} - \alpha = \gamma.$$

The temperature-induced focal shift[2] of a simple lens is approximated by the preceding equation, where df is the thermal focus shift, f is the ambient focal length, dn/dT is the thermo-optic coefficient for the lens material, dT is the temperature change, n is the index of refraction, α is the linear coefficient of expansion, and γ is the thermal glass constant.

Most infrared systems are more affected by temperature changes than are visible systems, as their thermo-optical coefficients are much higher. Some values are given in Table 15.1.

For a 75-mm-diameter $F/1.5$ germanium lens with a focal length of 112.5 mm, a temperature change of 40 °C induces a focus shift of 0.550 mm. The Rayleigh criteria for quarter wave of defocus is ± 0.046 mm, so the preceding defocus is approximately three waves of defocus or 12 Rayleigh criteria depths of focus.

A useful technique for athermalization is to make the mirrors of the all-reflective system of the same material as that of the optical bench. The change in

[2] P. J. Rodgers and M. Roberts, "Thermal compensation techniques," in M. Bass, E. Van Stryland, and W. Wolfe, eds., *Handbook of Optics*, McGraw-Hill, 1995, Vol. I, Chap. 39.

focus that is due to the change in mirror curvature is exactly offset by the change in length of the optical bench.

Generally, the best approach for thermal compensation in a refractive system is to move an element within the system by an appropriate amount. Trial and error—and some common sense—will indicate which element.

15.5. Transmission

The transparent elements will both reflect and absorb some of the incident light. The reflection can be estimated by the well-known expression for reflection at normal incidence:

$$\rho = \frac{(n-1)^2}{(n+1)^2}.$$

This is a good approximation to off-axis angles to ~25° (see Chap. 3, Electromagnetism Foundations, for the full equation).

The full expression for the transmission of a plane parallel plate, which applies quite well to lenses, is

$$\tau_e = \frac{(1-\rho)^2\,\tau_i}{1-\rho^2\,\tau_i^2},$$

where τ_e is the external transmittance, ρ is the single-surface reflectivity, and τ_i is the internal transmittance, given by

$$\tau_i = e^{-\alpha d},$$

where α is the absorption coefficient of the material and d is its thickness. If the absorption coefficient is zero the external transmittance becomes

$$\tau_e = \frac{2n}{n^2-1}.$$

It can be seen that, for germanium, this is ~0.5. Serious attention must be given to antireflection coatings! This approximate equation is a good approximation for most of the region over which a material is useful as a lens.

15.6. Absorption and Emission

By Kirchhoff's law, the emissivity of a material is equal to its emissivity under thermal equilibrium and for the same conditions (spectral band, direction, etc.). Both are important in IR lens design. If a material absorbs, it loses some transmission. This generally happens at the limits of the useful spectral band

and if there is some entrapped water, especially with calcium aluminate glasses and some synthetic quartzes. In addition, all semiconductors generate free electrons at elevated temperatures; the temperature at which this is significant depends on the size of the bandgap. For silicon and germanium this is ~400 K; for zinc selenide and zinc sulfide, it is somewhat higher.

The detector, especially for spaceborne systems, may be limited by the emission that is incident upon it. One source is the optics itself. In this case, both the emissivity and the temperature of the material are important in the design.

15.7. Objectives

These are the devices that form the images, along with any relay systems. In IR design, there is almost always a trade-off among lenses, mirrors, and combinations (catadioptrics). The latter two are covered in Chap. 16. Here several different designs are presented that can be adopted and adapted for different applications. Although the prescriptions are given to many significant figures, these are meant to be only examples or starting designs.

Three types of lenses are listed here, singlets, separated doublets, and triplets. They are for *F*/2, 50-mm focal-length lenses for the long-wave IR and mid-wave IR.

15.7.1. Long-Wave Infrared Lenses

The spectral band is 8–12 μm with equal weighting for 8, 10, and 12 μm and for field points at 0°, 1°, 2°, 3°, 4°, 5°, 10°, and 15°. The lenses will vary in design and performance for somewhat different bands, fields, and weightings.

15.7.1.1. Singlets

Here the choices are lens, radii, thickness, bending, and aspherics. Table 15.2 shows the prescriptions.

Table 15.2. Singlet Prescriptions

	Lens Material				
Parameter	Ge	AMTIR 1	ZnS	ZnSe	Si
Radius	100.3764	83.3944	76.6059	73.7221	90.6227
Thickness	2.2044	6.000	6.000	6.000	6.000
Radius	148.2928	206.0860	167.1212	119.3400	138.1659
Bfl	97.7631	85.778	94	94.0	94.5

Their performance in terms of the angular blur diameter are listed in Table 15.3. They have a speed of *F*/2.

It would appear that germanium is the best of these, but it is not *better* by

Table 15.3. rms Values of Blur-Circle Radii (mrad)

Lens Material	Half-Field Angle (deg)							
	0	1	2	3	4	5	10	15
Ge	0.515	0.553	0.702	1.009	1.484	2.124	7.740	17.7
AMTIR 1	1.502	1.527	1.623	1.838	2.222	2.799	8.561	19.2
ZnS	1.166	1.194	1.303	1.548	1.972	2.588	8.448	19.1
ZnSe	2.541	2.494	2.364	2.183	2.027	2.032	5.913	17.1
Si	0.588	0.605	0.704	0.966	1.421	2.058	7.785	18.0

much. All have rather poor performance at the larger field angles, mostly because of field curvature. If the speed of the germanium lens is changed to *F*/4, the performance is much better, but changed to *F*/1, it is much worse (see Table 15.4). Adding a conic aspheric also improves the performance. The first two have the same prescription; the prescription changes a little with the aspherics, with 80.9202 and 108.5070 as the radii, but the same thickness and back focal length.

Table 15.4. Performance of Other Germanium Singlets

Lens	Half-Field Angle (deg)							
	0	1	2	3	4	5	10	15
F/1 Ge		8.63	9.06	9.77	10.8	12.1	24.00	95.7
F/4 Ge		0.102	0.200	0.377	0.629	0.956	3.754	8.696
Aspheric *F*/2		0.223	0.298	0.603	1.094	1.746	7.382	17.4

15.7.1.2. Doublets

The design variables for singlets are doubled and there is an extra spacing for air-spaced doublets—as opposed to cemented doublets. Several prescriptions are given in Table 15.5 for the same 50-mm focal-length, *F*/2 system and spectral range, and fields. The performances are given in Table 15.6.

It can be seen that the different doublets have much the same resolution across the field. The germanium/zinc selenide doubler is a little better than the others, but cost considerations may dictate the choice.

15.7.1.3. Triplets

The prescriptions for triplets are shown in Table 15.7 and their performance in Table 15.8.

Fischer's lens has a more uniform field. Wolfe's is better near the center of the field. They are of very different shapes.

Table 15.5. Prescriptions for Doublets

	Lens				
Parameter	AMTIR 1/ZnS	Ge/AMTIR	Ge/ZnS	Ge/ZnSe	Ge/Ge
Radius	194.7722	167.8947	167.2510	189.9654	416.9942
Thickness	10.00	10	10	10	10
Radius	378.4872	218.1952	218.7854	255.2272	626.7167
Thickness	125.3315	10	10	10	10
Radius	64.1323	83.3987	83.3287	83.6571	95.4721
Thickness	10	10	10	10	10
Radius	140.394	111.7302	116.0185	118.3306	116.3475
Bfl	42.2835	83.0201	88.5743	83.2767	87.5623

Table 15.6. Performance of Some Doublets

Half-Field Angle (deg)	Lens Resolution (mrad)							
	AMTIR 1/ZnS	Ge/AMTIR 1		Ge/ZnS		Ge/ZnSe		Ge/Ge
0	0.883	0.47	0.34	0.22	0.33	0.72	0.47	0.08
1	0.895	0.51	0.45	0.34	0.44	0.76	0.56	0.22
2	0.928	0.67	0.69	0.63	0.72	0.91	0.79	0.49
3	0.980	1.00	1.08	1.05	1.13	1.23	1.18	0.89
4	1.050	1.51	1.61	1.61	1.68	1.72	1.71	1.43
5	1.134	2.19	2.29	2.32	2.37	2.38	2.38	2.11
10	1.689	8.16	8.05	8.34	8.13	8.32	8.12	7.95
15	2.996	18.8	18.03	19.0	18.2	19.0	18.2	18.3

Table 15.7. Two Triplets

Parameter	Wolfe	Fischer
Radius	18.3021	−18.2818
Thickness	0.9499	2
Radius	15.3831	−23.2325
Thickness	1.0000	2
Radius	19.9701	116.8642
Thickness	1.0072	2
Radius	18.5201	89.0374
Thickness	1.0000	2
Radius	17.6411	−107.8522
Thickness	1.1149	2
Radius	30.3345	−42.3473
Bfl	20.2265	2

Table 15.8. Triplet Performances

Half-Field Angle (deg)	Lens Resolution (mrad)	
	Wolfe	Fischer
0	0.000	0.256
1	0.056	0.256
2	0.213	0.258
3	0.476	0.262
4	0.847	0.268
5	1.326	0.278
10	5.446	0.452
15	12.9	1.06

15.7.2. Mid-Wave Infrared Lenses

The same field points, speed, and aperture are used as for the long-wave IR. Wavelengths are weighted equally for 3, 4.5, and 6 μm.

15.7.2.1. Singlets

The prescriptions are shown in Table 15.9 and their performances in Table 15.10.

Table 15.9. Singlets

Parameter	Si	Ge	AMTIR 1	ZnS	ZnSe
Radius	91.9021	94.2054	85.4989	82.9299	83.9548
Thickness	10	10	10	10	10
Radius	136.3206	125.5070	175.1231	197.1986	187.6366
Bfl	89.4965	89.3833	89.6107	89.6036	89.6036

Table 15.10. Singlet Performances

	Half-Field Angle (deg)							
Lens	0	1	2	3	4	5	10	15
Si	4.43	4.37	4.17	3.84	3.40	2.87	4.04	13.9
Ge	5.18	5.11	4.92	4.60	4.17	3.64	3.68	13.10
AMTIR 1	8.12	8.06	7.87	7.56	7.13	6.60	4.24	11.70
ZnS	3.95	3.88	3.68	3.35	2.92	2.40	4.66	15.00
ZnSe	4.75	4.69	4.49	4.15	3.70	3.16	4.11	14.10

15.7.2.2. Doublets

The prescriptions are in Table 15.11 and the performances in Table 15.12.

Table 15.11. Doublet Prescriptions

Parameter	*F*/2 Si/Ge	Ge/Si
Radius	59.7123	28.4894
Thickness	4	4
Radius	120.8308	31.4061
Thickness	3.77430	20.1033
Radius	−29.12117	−54.7862
Thickness	4	5
Radius	−31.93218	−56.3997
Thickness	46.49396	24

Table 15.12. Doublet Performances

	Half-Field Angle							
Lens	0	1	2	3	4	5	10	15
Si/Ge	1.295	1.244	1.120	0.926	0.748	0.829	4.565	12.322
Ge/Si	2.083	2.090	2.110	2.144	2.193	2.257	2.821	3.695

15.7.2.3. Triplets

Prescriptions are shown in Table 15.13 and performances in Table 15.14.

Table 15.13. Triplet Prescriptions

	Lens			
Parameter	*F*/2 Si/Ge/Si	*F*/2 ZnSe/ ZnS/Z *n*Se	ZnSe/ZnS/ Z*n*Se	ZnS/ZnSe/ Z*n*Se
Radius	22.44895	23.3036	56.44002	–27.92176
Thickness	4	4	4	5
Radius	33.09206	49.03778	–7481.77	–41.33857
Thickness	5	5	5	0.2499493
Radius	55.60298	2067.665	–205.7586	–161.891
Thickness	3	3	3	5
Radius	20.93529	21.42654	107.2641	–55.24695
Thickness	5	151.7503	42.55198	7
Radius	580.2887	4	29.44836	56.99487
Thickness	4	–57.3113	5	27.70128
Radius	–64.51205	34.15671	54.75175	
Thickness			13.5889	

Table 15.14. Triplet Performances

	Half-Field Angle							
Lens	0	1	2	3	4	5	10	15
Si/Ge/Si	0.420	0.440	0.449	0.450	0.474	0.516	0.860	1.475
ZnSe/ZnS/ZnS *e*-1	0.282	0.293	0.283	0.278	0.299	0.325	0.546	0.842
ZnSe/ZnS/ZnS *e*-2	0.255	0.248	0.226	0.211	0.225	0.290	0.970	1.977
ZnS/ZnSe/ZnZ *e*	0.220	0.393	0.391	0.407	0.414	0.450	0.762	1.266

15.8. Zoom Lenses

There are two types of zoom lenses, two-position and continuous. Only the second type is treated here. The first type is really just a combination of two designs. An excellent treatment by Mann includes both the techniques and many prescriptions and descriptions.[3]

15.9. Cold Stops

In IR systems that use cryogenically cooled detectors, the detectors or detector array are placed inside a Dewar, and the entire Dewar is kept at the temperature of the detector. An attempt is made such that the detector senses only scene energy or scene energy plus (the very low) radiation of the Dewar. There are two ways to do this. One is with a cold baffle; the other is a pupil manipulation technique. In the former, 100% efficiency cannot be obtained, but the system can often be simpler.

15.10. Baffle Technique

The baffle technique has been analyzed in considerable detail. The cold-shield efficiency for a baffle can be approximated in one dimension by the following formula:

$$\eta_{cs} = \frac{\arctan\left(\dfrac{D}{2f}\right)}{\arctan\left[\dfrac{fD - s(D-L)}{2f(f-s)}\right]},$$

where f is the focal length, D is the diameter of the cold lens, L is the length of the array, and s is the distance behind the lens to the baffle.

15.11. Pupil Technique

This is easy to state, but not so easy to do all the time. The lens that is the window of the Dewar must be the aperture stop of the system. This, of course, does not preclude other lenses from also being aperture stops and entrance pupils.

15.12. Design of Cold Optics

The flux density on a detector that is due to the radiation of the optics may be calculated by the following equation:

[3] A. Mann, *Infrared Optics and Zoom Lenses*, Vol. 42 of SPIE Tutorial Text Series, SPIE, 1999.

$$E_{\text{det}} = \varepsilon_1 L^{\text{BB}}(T_1)\Omega_1 + \tau_1\varepsilon_2 L^{\text{BB}}(T_2)\Omega_2 + \tau_1\tau_2\varepsilon_3 L^{\text{BB}}(T_3)\Omega_3 + \cdot + \cdot + \cdot,$$

where E is the radiant incidance on the detector, the ε's are emissivities, τ's are transmittances, L^{BB} are blackbody radiances, and T's are the temperatures of the elements, which are counted from the detector outward to the entrance pupil. The solid angles, in general, are the areas of the images of each of the elements in detector space divided by the square of the distances from the detector to those images. In some cases there will be a limiting solid angle that can be used for all of them.

If all the elements are at the same temperature, a frequently occurring case, then the equation simplifies to

$$E_{\text{det}} = (\varepsilon_1\Omega_1 + \tau_1\varepsilon_2\Omega_2 + \tau_1\tau_2\varepsilon_3\Omega_3)L^{\text{BB}}(T) + \cdot + \cdot + \cdot.$$

If there is a limiting solid angle, a further simplification is possible:

$$E_{\text{det}} = (\varepsilon_1 + \tau_1\varepsilon_2 + \tau_1\tau_2\varepsilon_3)L^{\text{BB}}(T)\Omega + \cdot + \cdot + \cdot.$$

If the materials are all the same, then

$$E_{\text{det}} = (\varepsilon + \tau\varepsilon + \tau\tau\varepsilon)L^{\text{BB}}(T)\Omega + \cdot + \cdot + \cdot = \varepsilon(1 + \tau + \tau^2 + \cdot + \cdot + \cdot)L^{\text{BB}}(T)\Omega.$$

15.13. Narcissus

Narcissus is an image anomaly unique to thermal imaging systems. The effect is named for the mythological character Narcissus, who fell in love with his own reflection while gazing in a pool. In IR imaging systems the effect refers to the detector's viewing itself from reflection off an optical surface in the lens train.

The following equation can be used as a rough estimate of the Narcissus contribution of a particular surface:

$$\Delta T_N = \frac{\rho\Delta T}{r^2},$$

where ΔT_N is the apparent temperature change in the image that is due to the Narcissus reflection, ΔT is the temperature difference between the scene and the detector, ρ is the surface reflectivity, and r is the ratio of the Narcissus reflected spot diameter to the normally imaged spot diameter.

If, for (a reasonable) example, $\Delta T = 223$ K (300 K ambient scene – 77 K detector temperature), $\rho = 0.05$, and $r = 3$, then

$$\Delta T_N = \frac{0.05 \times 223}{9} = 1.24 \text{ K}.$$

It is apparent that the effect is negligible for uncooled detectors and is reduced for larger spots.

There are two basic methods to minimize Narcissus. First, one can use very efficient, so-called *anti-Narcissus* coatings. These antireflection coatings typically have from 0.2% to 0.3% average reflectivity over the sensor spectral band. The second approach is to change the relative bending of the offending lens surfaces to minimize the solid angle of reflected radiation. This technique can usually maintain the desired level of optical performance, but may require the use of aspheric surfaces. Optical systems often have a flat protective window in front of the telescope. To avoid the Narcissus generated by the window, the window is tilted so that the reflected radiation falls out of the sensor's field of view.

15.14. Fabrication Considerations

Several of the crystalline materials are hygroscopic, which presents some challenges to the optical shop. The hygroscopic materials need to be appropriately coated to prevent damage from moisture and their housings often need to be dry nitrogen purged.

Some IR materials have an extremely high index of refraction; this requires antireflection coatings. Some are poisonous, requiring care in turning, grinding, and polishing. This is mostly a matter of eliminating any airborne powders and frequent washing.

Some IR materials can be single-point diamond turned. These include germanium, silicon, zinc sulfide, zinc selenide, AMTIR, and the fluorides. Sapphire cannot be diamond turned. Silicon can be diamond turned; however, the carbon in the silicon reacts with the carbon in the diamond, and this results in a shorter tool lifetime and thus higher cost.

15.15. Optical Testing

Most of the testing of IR lenses must be done in the IR! Even for the few good IR materials that transmit in the visible, the refractive indices are so different that visible tests are almost worthless.

Test equipment for IR lenses is similar in function to much of that used for visible optical systems. However, bar charts, for instance, are templates with bars of different temperature. Point sources are hot, and line sources are single bars at temperature. They may also be so-called blackbodies with cooler bars in front of them. IR interferometers have also been used.

Further Reading

W. E. Smith, *Modern Optical Engineering,* McGraw-Hill, 1990.
R. E. Fischer and B. Tadic-Galeb, *Optical System Design,* McGraw-Hill, 2000.
M. J. Riedl, *Optical Design Fundamentals for Infrared Systems,* Vol. 20 of SPIE Tutorial Text series, SPIE, 1995.
A. Mann, *Infrared Optics and Zoom Lenses,* SPIE, 2000.

The first two books deal with optical design in general. The third is specifically for IR optics. The last deals with IR optics and various zoom systems.

16

Reflecting and Catadioptric Objectives

Mary G. Turner

16.1. Reflecting Systems

Reflecting optical systems have several advantages over refracting systems. Mirrors can have significantly larger apertures than refractors. Mirrors have no chromatic aberration, neither lateral nor axial color nor chromatic variation of aberration (spherochromatism). For a given radius of curvature, the mirror surface has four times the surface power; as a consequence, the aberrations introduced by a reflective system are less than those of an equivalent refractor. Reflective systems can often be constructed in a more compact form than refractive systems. There are some drawbacks to reflective systems. These systems often require surfaces to be of a conic or a general aspheric shape, rather than spherical, and secondary mirrors can introduce obstruction into the optical system. Errors in the polishing of mirror surfaces have a greater impact than similar errors on a refractive surface.

Conic reflector surfaces include hyperboloids, paraboloids, ellipsoids, spheroids, and oblate ellipsoids. A conic surface is characterized by a local curvature that varies with position. The shape of the conic surface is described by the equation

$$z = \frac{c^2r^2}{1 + \sqrt{1 - (k + 1)c^2r^2}}, \tag{16.1}$$

where z is the sag of the surface, c is the base radius of curvature, r is the radial position across the aperture, and k is the conic constant.

A conic mirror surface will form a geometrically perfect image at the second focus of a point object located at the first focus. This stigmatic imaging condition holds for only one pair of conjugate points on axis. Images from other object points are aberrated.

Some surfaces may have a higher-order aspheric departure. These surfaces, referred to as general aspheres, have a surface sag that is described by the equation

$$z = \frac{c^2 r^2}{1 + \sqrt{[1 - (k+1)\, c^2 r^2]}} + \alpha_1 r^2 + \alpha_2 r^4 + \alpha_3 r^6 + \alpha_4 r^8 + \alpha_5 r^{10}, \quad (16.2)$$

where z, c, and r are the same as in the basic conic equation [Eq. (16.1)] and the remaining terms describe the rotationally symmetric deviation from a spherical or conical surface.

Many folded mirror systems have a variety of obscurations from the mirrors or mechanical components, such as spiders. These obscurations can result in a significant reduction in the contrast of the image and introduce scatter. Eccentric-pupil systems avoid these drawbacks.

There are many possible design forms for reflective telescopes. Most have one or two reflective surfaces, although there are some systems with more reflective surfaces.

16.1.1. Single-Mirror Telescopes

A single-mirror telescope is a telescope that has one powered mirror surface. The system may also include a flat mirror that acts as a fold mirror to direct the image away from the input optical axis. The powered mirror is usually a paraboloid (a parabola of revolution). A paraboloidal mirror has a shape defined by a conic constant of –1. It brings light from an infinitely distant on-axis object to a perfect image at the focal point of the mirror. This stigmatic imaging condition describes a system free of spherical aberration. As the object moves off axis, the image quality degrades because of coma, astigmatism, and field curvature. There are three distinct design forms for this telescope, the Newtonian, the Pfund, and the Herschelian.

The *Newtonian telescope* (Fig. 16.1) consists of a large parabolic mirror, which is usually the system stop. Additionally, a small, flat folding mirror is used to direct the final image to an off-axis location. The mirror, which is held in place by a mechanical spider, places the image in an accessible location but also obscures part of the incident light. This reduces the contrast in the image as well as introduces diffraction from the secondary and spider.

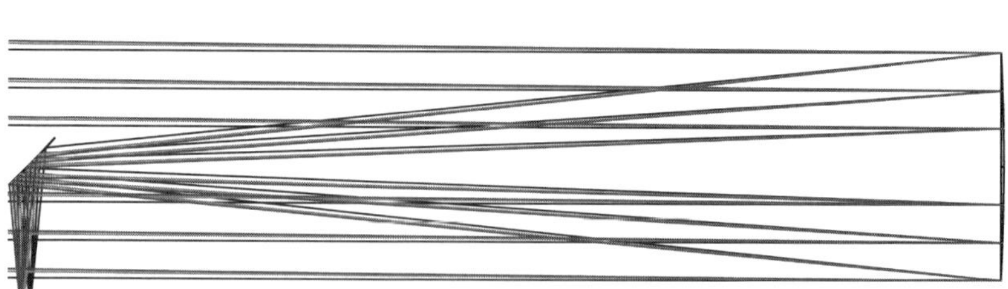

Fig. 16.1. Newtonian telescope.

In the *Pfund telescope* (Fig. 16.2) the primary mirror is a large folding flat located near the focus of the parabolic secondary. The image formed by the parabolic secondary is located along the optical axis, through a hole in the primary. The obscuration introduced in this system is less than that of the Newtonian. The support structure necessary to maintain the mirrors in position is not positioned within the optical path, eliminating the diffraction spikes that are seen in the image of the Newtonian telescope.

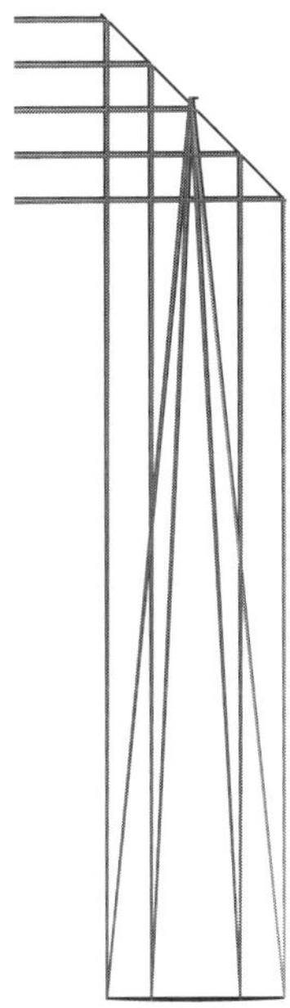

Fig. 16.2. Pfund telescope.

The *Herschelian telescope* (Fig. 16.3) does not require the use of a secondary mirror. As originally designed, the primary mirror was tilted relative to the optical axis, providing an unobscured, but aberrated, image for all object locations. The modern (preferred) form of this telescope uses an off-axis section of the paraboloidal mirror to form a perfect on-axis image.

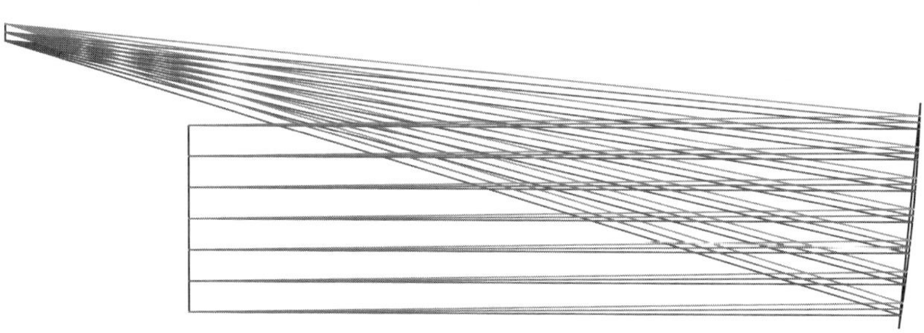

Fig. 16.3. Herschelian telescope.

16.1.2. Two-Mirror Telescopes

Most two-mirror telescopes can be classified into one of two distinct families, the Gregorian or the Cassegrain.

The *Gregorian telescope* (Fig. 16.4) is composed of two concave mirrors separated by a distance greater than the focal length of the primary. The primary mirror is a paraboloid and, as with the Newtonian, forms a perfect (on-axis) image, in this case at one focus of the ellipsoidal secondary. The secondary then forms the final image at the second ellipsoidal focus, located behind the primary mirror. This arrangement maintains the stigmatic condition for the on-axis point object, but suffers from coma, with some astigmatism and field curvature off axis. The off-axis performance is degraded slightly from that of the Newtonian, but splitting the required power between the two elements reduces the total system length.

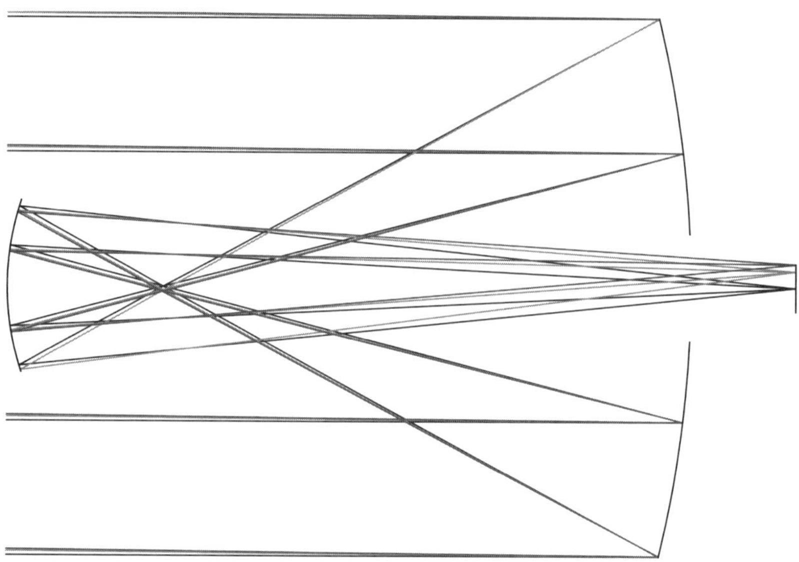

Fig. 16.4. Gregorian telescope.

The *Cassegrain telescope* (Fig. 16.5) consists of a concave paraboloidal primary and a convex hyperboloidal secondary. The primary forms a virtual image at one focus of the hyperboloid. The final image is formed at the second focus of the hyperboloid, located outside the primary mirror. As with the Gregorian, this arrangement maintains the stigmatic condition for an on-axis point object, but suffers from coma, with some astigmatism and field curvature off axis. The performance is similar to that of the Gregorian, with a smaller total length. Additionally, the ratio of the size of the secondary to the primary, which is a measure of the obscuration, slightly favors the Cassegrain form. This system also has a performance similar to that of the Newtonian. This sys-

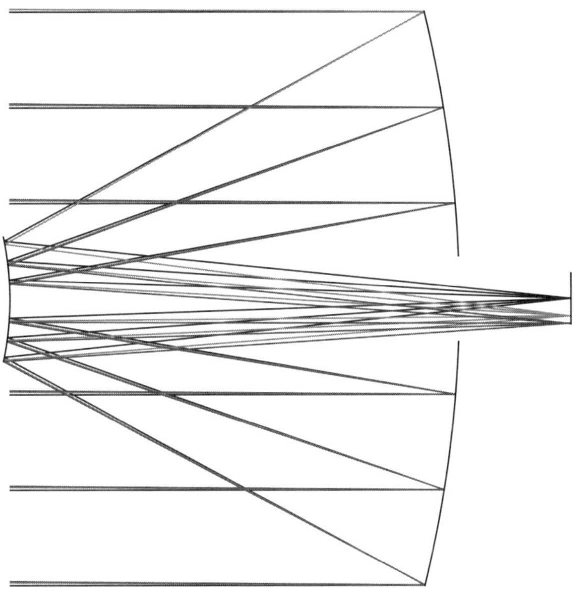

Fig. 16.5. Cassegrain telescope.

tem is more compact; but with two elements, it is more difficult to align than the Newtonian.

Two common afocal telescopes are the *Mersenne–Gregorian* and the *Mersenne–Cassegrain* (Fig. 16.6). The two paraboloidal mirrors in these tele-

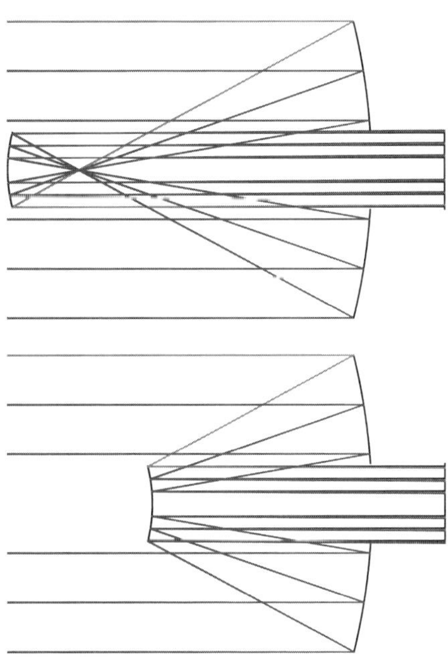

Fig. 16.6. Mersenne telescopes.

scopes serve as beam reducers, similar to a refracting telescope that has an objective and eyepiece. The Mersenne–Gregorian telescope is the reflective analog of the Keplerian-type refractor. There is an accessible prime focal point. The Mersenne–Cassegrain is analogous to the Galilean-type refractor. There is no accessible prime focal point. Both of these systems are aplanatic and anastigmatic. Both can be corrected for field curvature by repositioning of the stop. With an in-line configuration, both systems suffer from the contrast loss and diffraction effects common in systems with a centrally obscuring secondary. These systems can also be used with an eccentric primary, shown in Fig. 16.7, which results in the same image quality without any obscuration.

The maximum usable field of view in both the Gregorian and the Cassegrain telescope is limited by coma. Many other two-mirror telescopes are similar in form to either the Gregorian or the Cassegrain telescope, with the variations intended to produce a larger usable field of view or some other advantage.

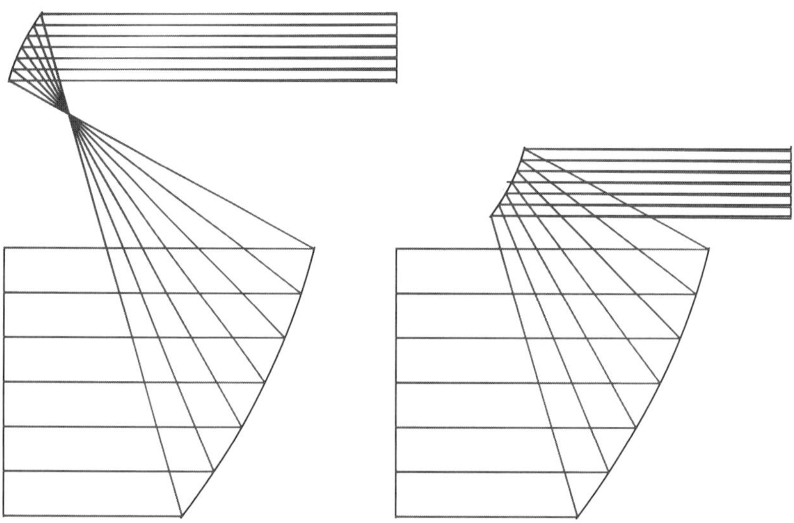

Fig. 16.7. Eccentric Mersenne telescopes.

The *Dall–Kirkham telescope*[1] is a Cassegrain-type telescope that consists of a concave paraboloidal primary and a convex spherical secondary. To eliminate third-order spherical aberration, a slight aspheric departure, use of the sixth-order aspheric term is required on the secondary mirror. This system suffers from considerable coma, which severely limits the usable field of view. The advantage of this system is the simplicity of the spherical primary mirror.

The *Gregorian aplanat*[2] is corrected for coma as well as for spherical aberration. The primary mirror is an ellipsoid, with a conic constant just less than

[1] A. R. Kirkham, Sci. Am. 1951.
[2] D. Korsch, *Reflective Optics,* Academic, 1991.

–1, representing only a small departure from a paraboloid. The secondary mirror is also a concave ellipsoid. The dominant remaining aberration is field curvature. The system suffers some astigmatism, although somewhat less than that found with the classical Gregorian. Distortion is very low. This telescope, as with all Gregorian derivatives, has an accessible focal point, which can be useful for stray-light suppression.

The aplanatic form of the Cassegrain telescope is the *Ritchey–Chretien.*[3] Both the concave primary and the convex secondary mirrors are hyperboloids. The conic constant of the primary mirror is only slightly less than –1, also just a slight departure from a paraboloid. Following from the work of Schwarzschild, the first telescope of this form, commonly referred to as the Ritchey–Chretien, was built by Ritchey. The Ritchey–Chretien suffers from less field curvature than the aplanatic Gregorian. The astigmatism in the Ritchey–Chretien is more than that in the Gregorian aplanat and slightly more than is found in the classical Cassegrain. Distortion is similar to that in the Gregorian aplanat.

A comparison between the these two aplanatic telescopes shows that, in terms of image quality as well as the size of the astigmatism-limited field, the Gregorian is the better optical system. However, other considerations favor the Ritchey–Chretien. In the Ritchey–Chretien, the secondary is slightly smaller, reducing the amount of obscuration. The distance from the primary to the secondary and from the secondary, to the image is significantly longer for the Gregorian than for the Cassegrain. The Gregorian has a significantly larger and more costly mechanical structure as a result. Further discussion on the basic two-mirror telescopes and their derivatives can be found in Korsch,[4] Schroeder,[5] and Smith.[6]

The *Schwarzschild objective*[7] (Fig. 16.8) is a system in which the focal length is shorter than the back focal distance. The system uses two spherical mirrors positioned so as to have a common center of curvature. Proper relative placement of the mirrors results in a cancellation of the spherical aberration from each surface. If the stop is placed at the center of curvature, the system will have no coma or astigmatism. The system is also free of distortion. Be cause of the residual higher-order aberrations, the curvatures and spacings may be adjusted from the ideal to provide better image quality[8] at the expense of introducing third-order aberrations. The major drawback to this design is

[3] H. Chretien, Rev. Opt. **1,** 1922.

[4] D. Korsch, *Reflective Optics,* Academic, 1991.

[5] D. Schroeder, *Astronomical Optics,* Academic, 2000.

[6] W. J. Smith, *Modern Lens Design,* McGraw-Hill, 1992.

[7] K. Schwarzschild, Untersuchungen zur Geometrischen Optik II. Theorie de Spiegelteleskope, Abhandlungen der Königichen Gesellschaft der Wissenshaften zu Göttingen, Neue Folge Band IV, Berlin, 1905.

[8] I. Abel, and M. Hatch, "The pursuit of symmetry" in *Wide-Angle Reflective Optical Designs,* Proc. SPIE **237,** 271, 1980.

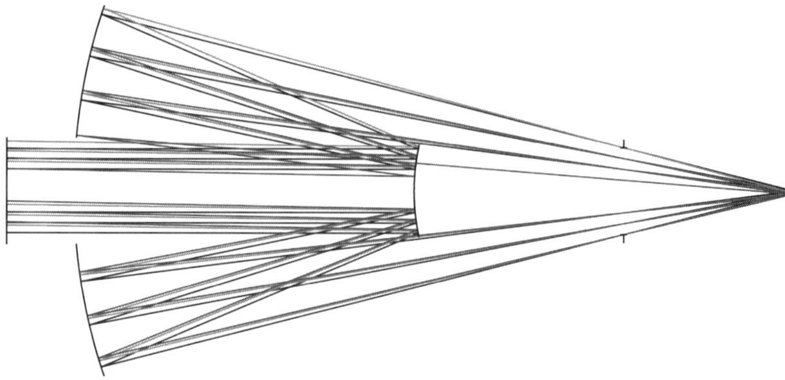

Fig. 16.8. Schwarzschild objective.

the size of the concave secondary as well as the system as a whole. The secondary mirror must be several times the size of the aperture and the total system length is more than four times the focal length. For these reasons, this system is generally used in applications such as microscope objectives, in which the size issues are minor.

The *Offner*[9] 1× relay (Fig. 16.9) is a finite-conjugate two-mirror, three-reflection objective. The design is aplanatic, anastigmatic, and free of distortion. This system, which images a ring field, is used in microlithographic applications in which high-resolution aberration-free imaging is required.

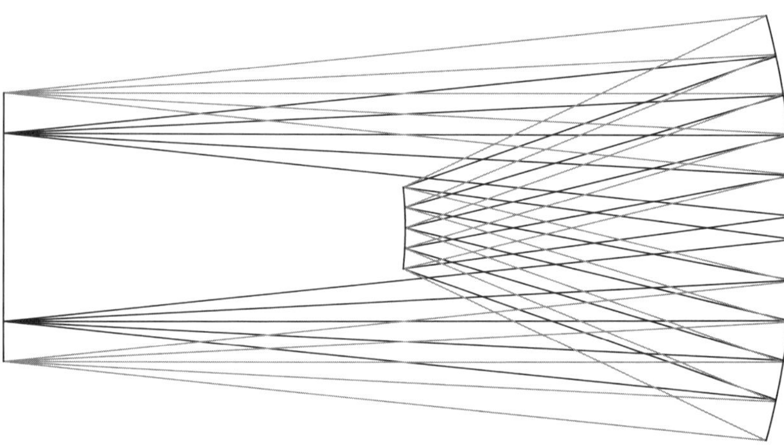

Fig. 16.9. Offner objective.

16.2. Catadioptric Systems

To correct for the aberrations remaining in the all-reflective systems, there are two alternatives. The first is to add more mirrors, providing more surface figures to be used for aberration correction. Although this is often an acceptable

[9] A. Offner, "Unit power imaging anastigmat," U.S. patent 3, 748, 015, 1973.

solution, the result is often increased obscuration or an increased footprint for the system, if eccentric mirror positions are used. The other alternative is to add refractive components to the system. Each additional refractive component has the advantage of introducing two new surfaces that can be used for aberration compensation at the cost of introducing chromatic aberration. Catadioptric systems are commonly used with many single- and two-mirror telescopes.

Three common types of refractive correctors are a meniscus spherical element referred to as a Maksutov corrector or a Bouwers corrector, the aspheric Schmidt corrector plate, and the Houghton corrector, which is an afocal system of two or three elements that correct the system for spherical aberration and coma.

16.2.1. Single-Mirror Catadioptric Systems

The *Newtonian–Maksutov*[10] (Fig. 16.10) consists of a meniscus corrector and a spherical primary mirror. The stop is located at the corrector lens. The corrector is adjacent to the fold mirror, allowing the fold mirror to be attached without the spider vanes, thereby reducing the amount of diffraction that occurs within the system. The field of view of this system is limited by coma; however, the coma is approximately one-fourth that of the Newtonian.

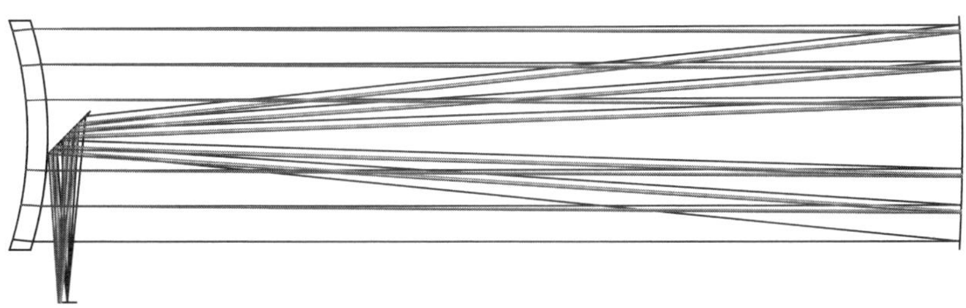

Fig. 16.10. Newtonian-Maksutov system.

A catadioptric form of the Hershelian telescope includes a *Houghton corrector* (Fig. 16.11) that has two corrector elements, each of which is tilted at a different angle, and a tilted spherical primary. The elements are shaped to compensate for the spherical aberration of the primary. By use of symmetric optics, the compensator lenses can also be self-correcting for polishing errors. Tilting the elements compensates for the coma as well as for astigmatism. Usually used for slower (≥*F*/8) systems, this design can provide near-diffraction-limited performance across the entire field of view.

The *Schmidt camera*[11] (Fig. 16.12) consists of an aspheric corrector plate

[10] H. Rutten and M. van Venrooij, *Telescope Optics*, Willmann-Bell, 1988.
[11] B. Schmidt, "Mitthandlungen," Sternwart **7**(36), 1932.

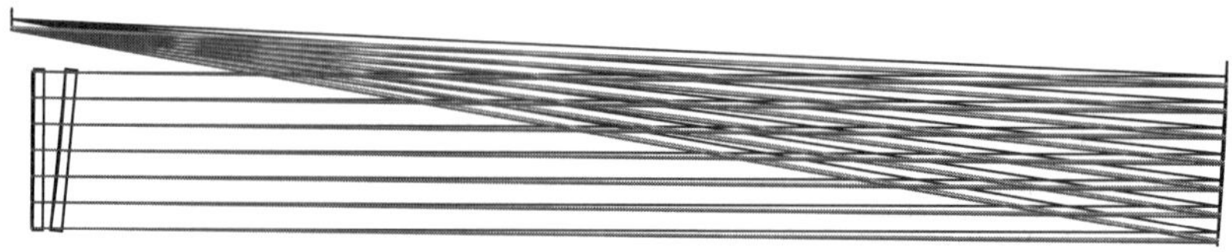

Fig. 16.11. Houghton corrector.

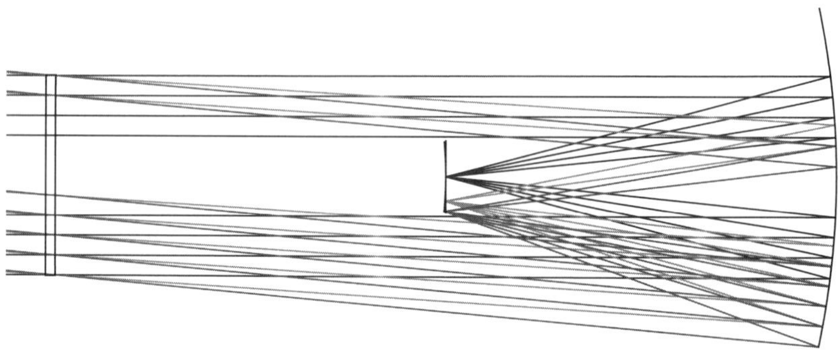

Fig. 16.12. Schmidt camera.

located at the center of curvature of a spherical primary mirror. The profile of the corrector is described by

$$z = \alpha_1 r^2 + \alpha_2 r^4 + \alpha_3 r^6 + \cdots . \tag{16.3}$$

The sag is similar to the aspheric surface, without the base curvature term. For most applications, only the first two terms are used. Higher-order terms are necessary for fast systems.

The best imaging occurs if the image surface is spherical with a radius equal to the focal length of the primary. If the stop is positioned at the center of curvature of the primary mirror, rays from each field position have their own axis of symmetry. The chief ray for each field point strikes the primary mirror at normal incidence. This symmetry results in a system that, to third order, is free of coma, astigmatism, and distortion. The aspheric shape of the corrector surface compensates for the spherical aberration introduced by the primary mirror. This also introduces higher-order asymmetry into the system. Off-axis ray bundles travel through asymmetric portions of the corrector, resulting in residual higher-order aberrations. Because of the position of the image, this telescope is not suited for visual work. As a camera, it can be used at speeds as fast as $F/1$ and at much larger field angles than other designs. Kingslake,[12] Schroeder,[13] and Laikin[14] discuss several systems based on this design.

[12] R. Kingslake, *Lens Design Fundamentals,* Academic, 1978.
[13] D. Schroeder, *Astronomical Optics,* Academic, 2000.
[14] M. Laikin, *Lens Design,* Marcel-Dekker, 2001.

The *Houghton camera* (1944) (Fig. 16.13) consists of a two-element corrector located near the focus of the spherical primary mirror. The stop is located at the corrector. The corrector elements compensate for the spherical aberration and coma introduced by the primary. If the corrector is positioned near the center of curvature of the mirror, it is also possible to correct for astigmatism. However, this would result in a system that is physically longer than for other stop positions. The performance is improved when the detector is curved to match the primary curvature. The corrector consists of a positive and a negative element. The elements are designed so that the curvatures of the negative element are of opposite sign to those of the positive element. This minimizes the effects of manufacturing errors, as long as the elements are tested relative to each other. Buchroeder[15] discusses variations on the design as well.

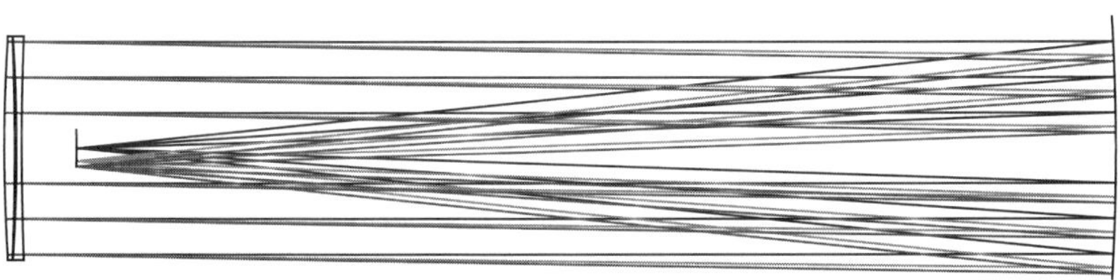

Fig. 16.13. Houghton camera.

The *Maksutov camera*[16] (Fig. 16.14) consists of a meniscus corrector element and a spherical primary mirror. The corrector can be located closer to the primary mirror than in the Schmidt or the Houghton camera without significant degradation in performance. The Maksutov corrector is designed so that

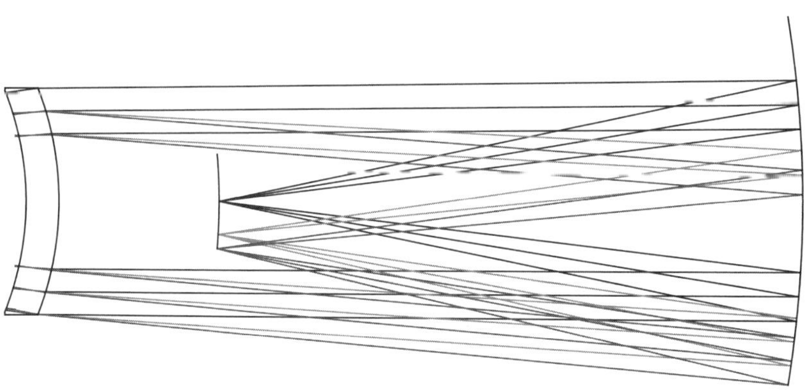

Fig. 16.14. Maksutov camera.

15 R. Buchroeder, "The Houghton telescope," Amateur Telescope Making, Summer, 4, 1980.
16 D. Maksutov, J. Opt. Soc. Am. **34,** 270, 1944.

the focal length does not change with index; that is, the system is self-correcting for axial chromatic aberration. This is done with the thickness set so that

$$t = \frac{n^2}{n-1}(R_1 - R_2), \tag{16.4}$$

where t is the element thickness, n is the index at the design wavelength, and R_1 and R_2 are the surface radii.

The three camera systems, Maksutov, Schmidt, and Houghton, all require a curved image surface for best performance because of the field curvature remaining in each of the designs. One solution for this is to position a two-element achromatic field flattener near the image surface to eliminate the Petzval curvature of the system, allowing for a flat image surface.

The *Mangin*[17] (Fig. 16.15) mirror is a meniscus element that has been aluminized on the back surface to produce a catadioptric system. The mirror contributes most of the power for the system. The purpose of the refractive element is to compensate for the spherical aberration introduced by the reflecting surface. The single-element Mangin design has considerable chromatic aberration, so an achromatic doublet is often used. The Mangin mirror system has approximately one-fourth the coma of the Newtonian system for the same size field. An air separation between the elements of the achromat can eliminate the coma. Mangin mirrors were initially designed for use in searchlights. These systems are generally used as part of a system, such as a catadioptric telephoto lens.

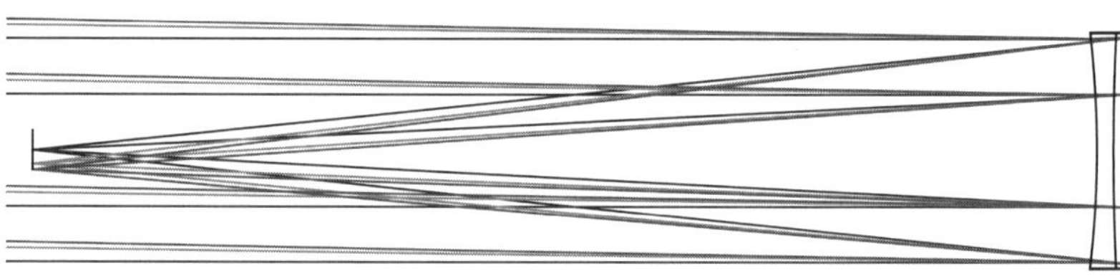

Fig. 16.15. Mangin mirror.

16.2.2. Two-Mirror Catadioptric Systems

As with single-mirror systems, catadioptric variations of the classical forms of two-mirror telescopes also exist. The additional elements can be used to correct for field aberrations as well as to reduce the number of conic or aspheric surfaces.

The *Maksutov–Cassegrain*[18] (Fig. 16.16) refers to any of several telescopes

[17] A. Mangin, "Memorial de l'officier du genie," **25,** 211–289, 1876.

[18] R. Sigler, "Compound catadioptric telescopes with all spherical surfaces," Appl. Opt. **17,** 1519–1526, 1978.

having a meniscus corrector and two spherical mirrors. In the Gregorian form, the secondary mirror is an aluminized section of the back surface of the meniscus element. The secondary and the meniscus surface have a common curvature, reducing the ability of the system to correct for field aberrations, especially coma.

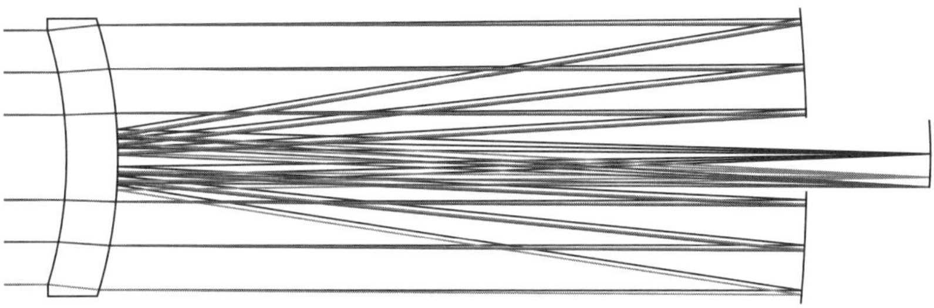

Fig. 16.16. Maksutov–Cassegrain telescope.

A variation of the Maksutov–Cassegrain (Fig. 16.17) has a spherical secondary mirror separate from the corrector. This additional design freedom allows for a system that is diffraction limited across the field of view. As with the Maksutov camera, it is not necessary to position the stop at the center of the mirror or to have the corrector surfaces concentric with the stop. The longer system has slightly less astigmatism.

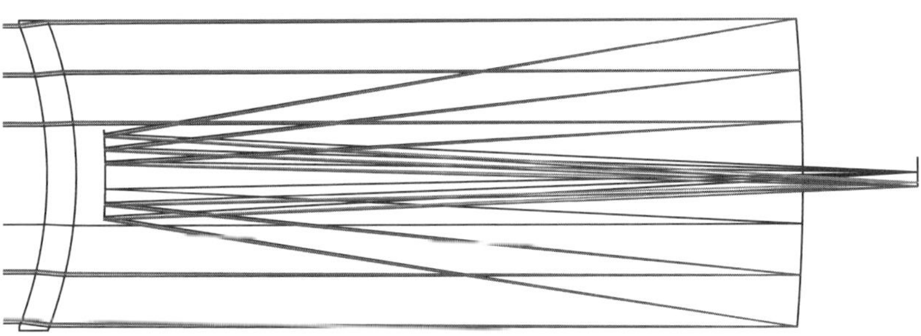

Fig. 16.17. Maksutov–Cassegrain variation.

The *Schmidt–Cassegrain telescope*[19] (Fig. 16.18) refers to any of several members of a family of telescopes having two mirror surfaces and a Schmidt corrector plate. The Schmidt–Cassegrain telescope can be designed with both the primary and the secondary mirror spherical. Systems faster than *F*/3 may use a conic secondary, often close to a paraboloid.

The Schmidt–Cassegrain telescope offers several advantages over the

[19] J. Baker, "A family of flat-field cameras equivalent in performance to the Schmidt camera," Proc. Am. Philos. Soc. **82,** 339, 1940.

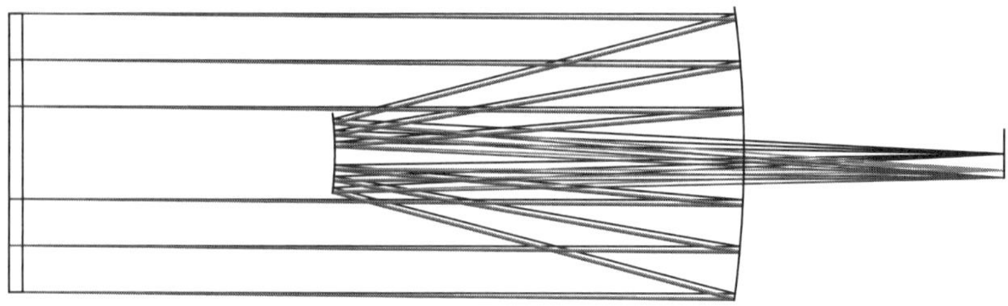

Fig. 16.18. Schmidt–Cassegrain telescope.

Schmidt camera. The image is accessible, allowing use as a visual instrument. It can be of shorter length as the corrector does not need to be at the center of curvature of the primary, and it is also possible to correct for the field curvature, although this is not necessary in many visual instruments. A Schmidt–Cassegrain telescope with spherical mirrors, of the same focal length and focal ratio, offers considerable improvement over any of the all-reflective two-mirror systems. The design is free of spherical aberration and coma. Best performance is obtained with a spherical image surface. Using an achromatic corrector plate can reduce the chromatic aberration. Moving the stop from the corrector location will result in greatly increased lateral chromatic aberration. An achromatic field flattener can also be added to reduce the Petzval curvature (Fig. 16.19). Two- and three-element corrector designs also exist. Schroeder[20] and Buchroeder[21] give an extensive analysis of these designs.

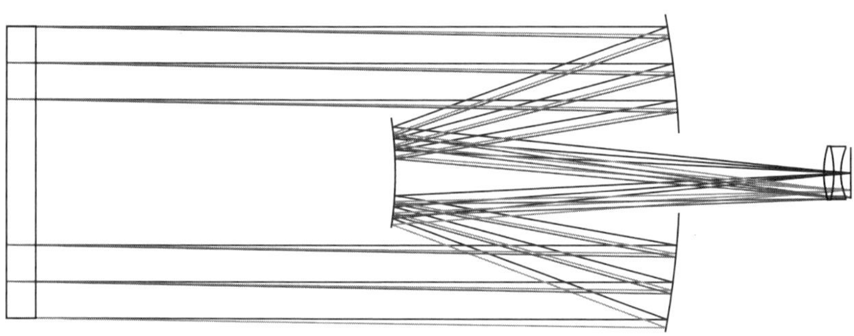

Fig. 16.19. Field-flattened Schmidt–Cassegrain system.

Catadioptric telephoto lenses often use Mangin-mirror-type components to reduce the overall length of the system. One example is shown in Fig. 16.20. These systems are designed to meet the needs of photography, including the reduction of chromatic aberrations. Several examples can be found in the literature or in patents. This design[22] uses two Mangin-type components. These

[20] D. Schroeder, *Astronomical Optics*, Academic, 2000.

[21] R. Buchroeder, "The Houghton Telescope," Amateur Telescope Making, Summer, 4, 1980.

[22] Y. Shimizu, "Catadioptric telephoto objective lens system," U.S. patent 3,632,190, 1972.

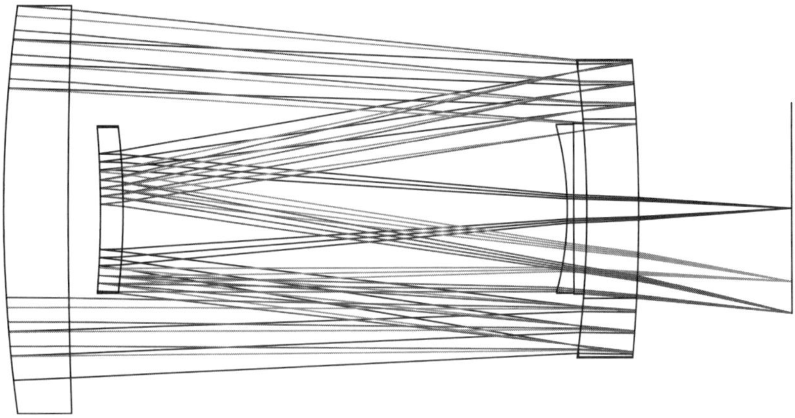

Fig. 16.20. Telephoto lens system.

surfaces introduce obscuration into the design. Other devices that use this form are solid and semisolid Schmidt and Schmidt–Cassegrain systems.[23]

16.3. Eccentric-Pupil Systems

When three or more mirrors are used in a single system, the mirrors are often centered off of the mechanical axis to limit or eliminate the effects of obscuration. However, three-mirror solutions can be found that have no spherical aberration, astigmatism, coma, or field curvature. These solutions are referred to as three-mirror anastigmatic systems. In the system described by Korsch,[24] shown in Fig. 16.21, the three surfaces are hyperboloids. The position of the image behind the secondary provides for suppression of stray light. There is considerable obscuration, but the image is diffraction limited.

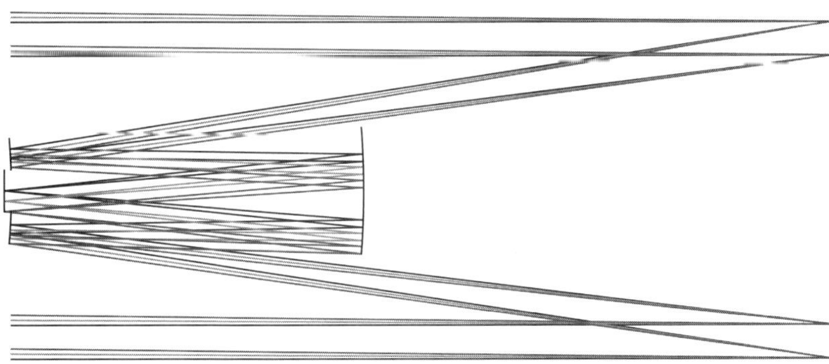

Fig. 16.21. Three-mirror anastigmatic system.

[23] J. Baker, "A family of flat-field cameras equivalent in performance to the Schmidt camera," Proc. Am. Philos. Soc. **82,** 339, 1940.

[24] D. Korsch, *Reflective Optics,* Academic, 1991.

A two-axis system adds a flat folding mirror to the three-mirror anastigmatic. The advantage of this design is that there is a reduced requirement for stray-light baffles to prevent stray light from reaching the detector. This variation of a Korsch two-axis design shows a 3.5-m *F*/10 three-mirror astigmatic covering a 1.5° full field. To minimize the stray-light effects, the fold mirror is placed near the exit pupil of the system, the small fold mirror, resulting in total obscuration of the on-axis field.

The *Paul three-mirror system* (also referred to as the Paul–Baker system, Fig. 16.22) is based on the Mersenne–Cassegrain afocal systems. The first two mirrors produce a reduced collimated output, which is reimaged by the tertiary

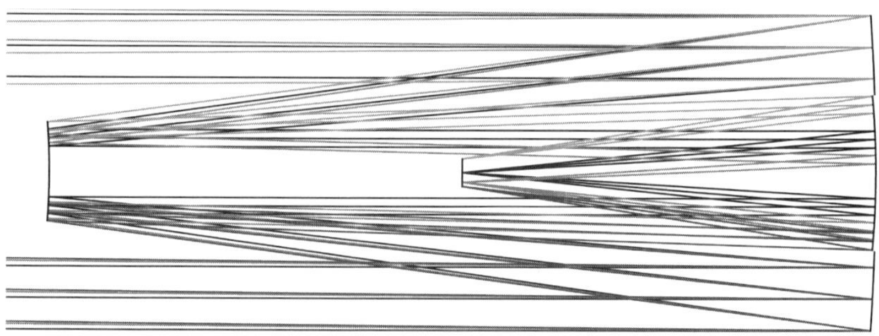

Fig. 16.22. Paul three-mirror system.

mirror onto the detector surface. If the secondary and the tertiary mirrors are both spherical and of common radius, then each will contribute equal amounts of spherical aberration, but opposite sign, making the system free of spherical aberration. Use of a curved focal surface can also eliminate coma and astigmatism. Allowing these two radii to have different values can eliminate the Peztval curvature. If the secondary is an ellipsoid, then all third-order aberrations can be eliminated.

Several variations of the Paul or Paul–Baker telescope have been investigated. These systems generally provide excellent image quality, covering a larger field than is usually available with the two-mirror designs. The major disadvantages to the in-line three-mirror system include the relatively high obscuration ratio that is due to the baffling necessary to prevent stray light from reaching the detector and the small available volume to place the detector. A flat-field, 3.5-m *F*/2.2 system provides excellent image quality, but with a obscuration of ~22%.

The eccentric-pupil off-axis *Paul–Gregorian* (Fig. 16.23) is a modification of the Mersenne–Gregorian form.[25] This system covers a 1° field that is centered 1° off axis. The secondary mirror is located at the center of curvature of

[25] D. Shafer, "New types of anastigmatic two mirror telescopes," J. Opt. Soc. Am. **66,** 1114, 1976.

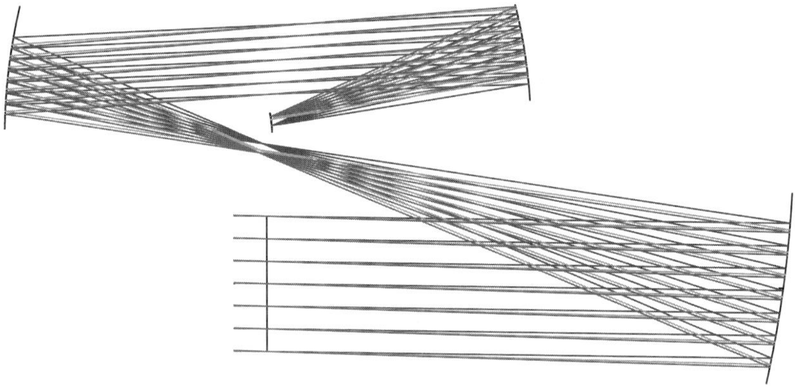

Fig. 16.23. Eccentric-pupil Paul–Gregorian system.

the spherical tertiary mirror. Adding higher-order aspheric terms to the paraboloid secondary can compensate for the spherical aberration introduced by the tertiary. Because of uncorrected field curvature, the image surface is curved. The image is diffraction limited at all field positions. Although the design is unobscured, the image location is difficult to obtain. Modifications to this design include decentering the tertiary to improve access to the image.

Another Korsch[26] three-mirror design is for a 1° field that is centered 5° off axis (Fig. 16.24). The primary mirror is an ellipsoid, the secondary a hyperboloid, and the tertiary a sphere. This unobscured system provides a readily accessible image. The internal focus, as well as the accessible pupil, provides a means for stray-light suppression.

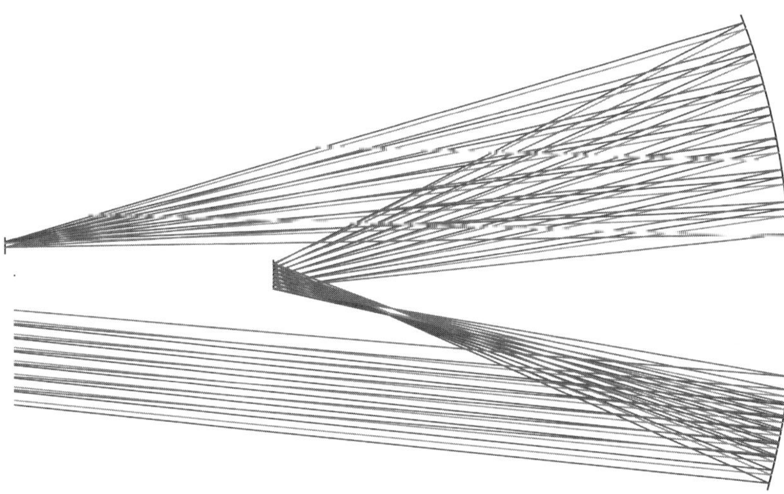

Fig. 16.24. Eccentric Korsch design.

[26] D. Korsch, *Reflective Optics,* Academic, 1991.

Several telescopes can be used over relatively wide fields (greater than 20°). The super-Schmidt camera uses two meniscus correctors positioned symmetrically around the stop. Additionally, a Schmidt-type aspheric corrector is positioned at the stop (Fig. 16.25). The single mirror is spherical. The second

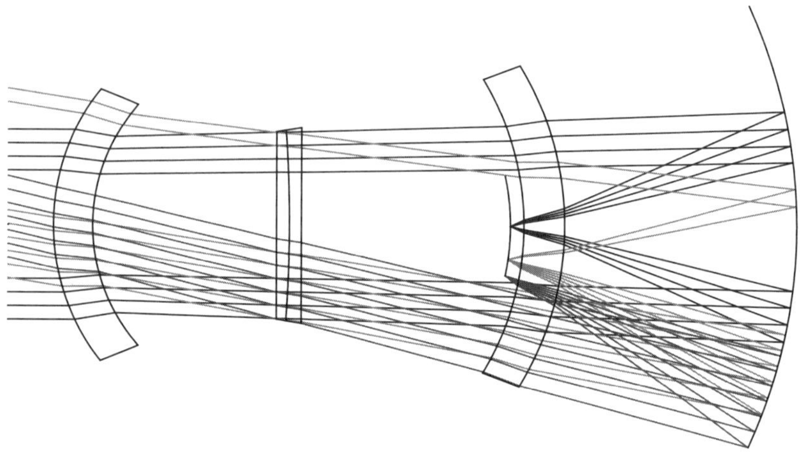

Fig. 16.25. Super-Schmidt camera.

meniscus corrector is encountered twice, before and after the primary mirror. The curved image surface is located between the stop and the second meniscus element. This location introduces some obscuration into the system. The corrector may be achromatized to limit chromatic aberrations. Because of the symmetry in this design, acceptable image quality can be achieved for fields of 40° and beyond. However, increasing the field size results in a larger image, introducing significantly more obscuration. For a 200-mm focal-length system operating at *F*/1.2, for a 45° field, the obscuration is ~80%; for a 40° field, it is ~70%; and for a 25° field, it is ~25%.

The *Baker–Nunn telescope*[27] (Fig. 16.26) is a variation on the Houghton design. A three-element corrector is used to eliminate spherical aberration and coma. This fast (*F*/1) system can also be used over a relatively large field of view; however, the size of the corrector optics also increases. The obscuration ratios for this 200-mm focal length design are 50% for a 40° field and 18% for a 25° field.

The *SEAL*[28] (which stands for spherical, elliptical, all-reflective, long) (Fig. 16.27) is a four-reflection, three-mirror system. The aspheric secondary mirror is also used as the quaternary surface. The stop is located at the plane tertiary mirror. The primary is spherical. This system has a wide, unobscured field of

[27] J. Baker, "A family of flat-field cameras equivalent in performance to the Schmidt camera," Proc. Am. Philos. Soc. **82,** 339, 1940.

[28] R. Owen, "Easily fabricated wide-angle telescope," Proc. SPIE **1354,** 430, 1990.

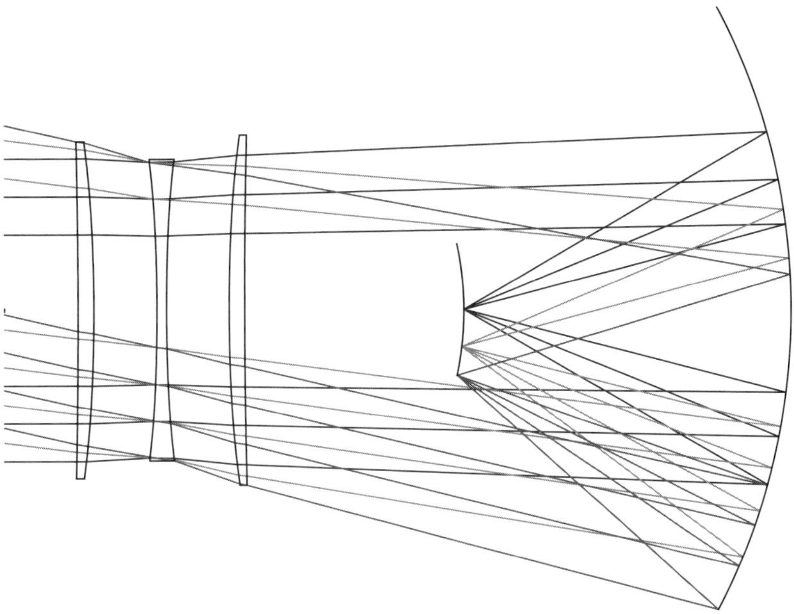

Fig. 16.26. Baker–Nunn telescope.

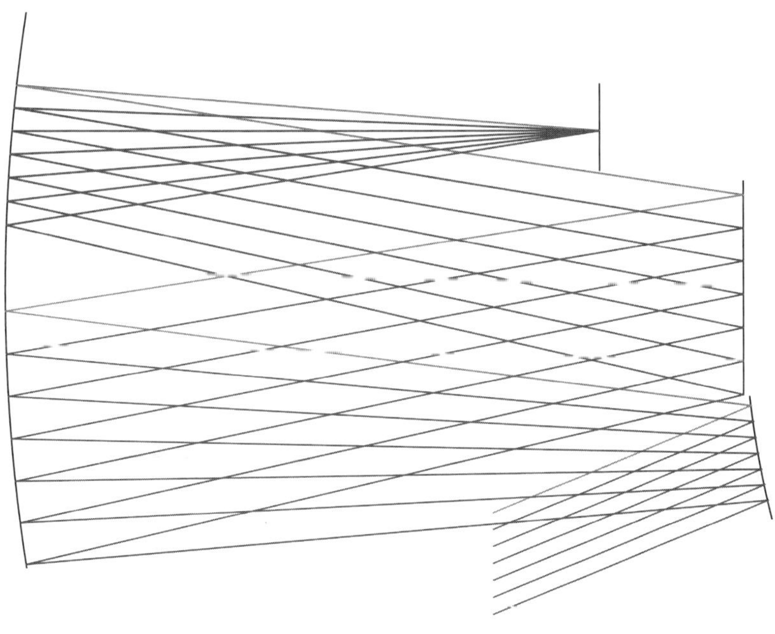

Fig. 16.27. SEAL system.

view, with an eccentric orientation. The total field is 10°, oriented at 22° to the axis. It is based on the Mersenne system, with the output from the secondary expanded, rather than compressed. The performance is limited by spherical aberration. If the flat mirror is aspherized, this can also be corrected. The second half of the design then acts as a reflective Schmidt.

The Schwarzschild objective (Fig. 16.28) can be used in conjunction with other optical systems to provide high-resolution imaging. This design[29] combines a Cassegrain system with a Schwarzschild objective. When the separation between the two modules is changed and the focus position is adjusted, this can act as a zoom-lens system.

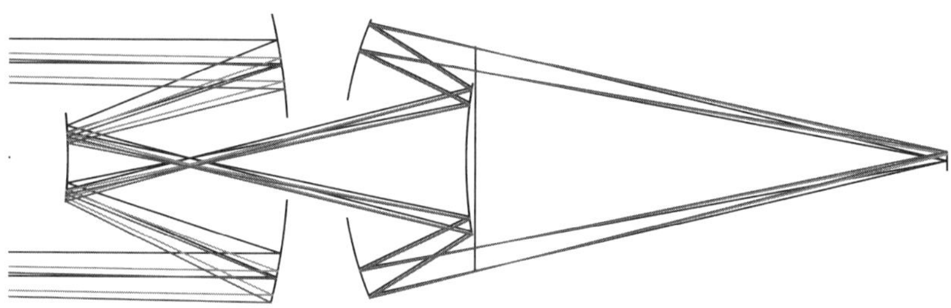

Fig. 16.28. Schwarzschild objective.

16.4. Comparisons

Comparison of the performances of these systems permits easy selection of a starting design for a particular application. Although there is an infinity of variations, a single set of variables provides the required information. The basic design parameters for this comparison are 1000-mm focal length, 200-mm aperture, used over the F-d-C wavelength range with a 0.5° semifield of view. Within these parameters, there are many possible system configurations. These different design forms depend on the power of the primary and the relative power of the secondary. With the curvatures specified, the separations and conic coefficients are defined. The effect of these different designs is seen in terms of aberration and obscuration. Design forms that are longer, with a greater separation between the primary and the secondary, have a tendency toward lower aberration across the field. However, the size of the secondary also increases, which increases the obscuration, reducing the overall contrast in the image. For these purposes, all Cassegrain and Gregorian telescopes and their derivatives were designed with the same basic form, so that the obscurations in the systems were held constant, and the impact of the design form on aberration can be demonstrated.

Almost all the designs were specified to exhibit perfect on-axis imagery.

[29] R. Johnson, J. Hadaway, and T. Burleson, "All-reflective four-element zoom telescope: design and analysis," SPIE **1354**, 669, 1990.

The increase in wave-front aberration is significantly slower in the Newtonian design than for either the Cassegrain or the Gregorian design. The secondary mirror in the Newtonian is also smaller than that found in either the Cassegrain or the Gregorian. Comparisons are shown in Fig. 16.29. A greater amount of wave-front aberration is seen in the Cassegrain than in the Gregorian, but the Gregorian form has the largest secondary mirror, indicating the largest reduction in image contrast.

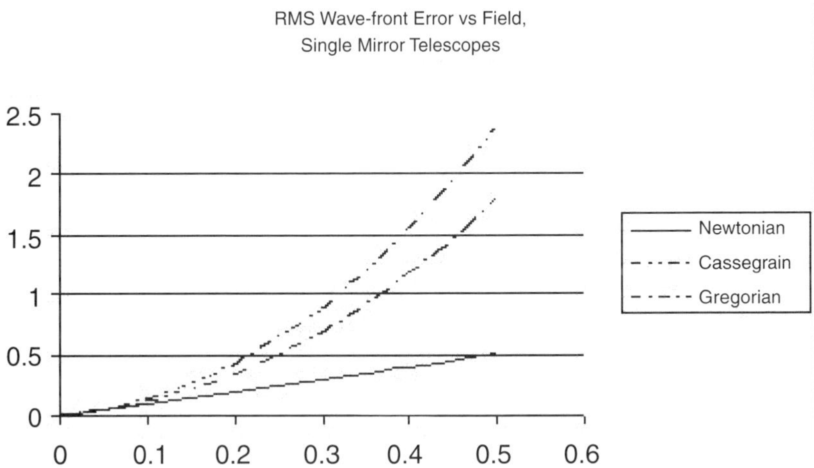

Fig. 16.29. Comparison of the performances of the Newtonian, Cassegrainian, and Gregorian.

Comparisons of the single-mirror systems, including the catadioptric variations, the aberrations of the Newtonian, the Herschel, and the Pfund, are very similar. The Herschel form has no obscuration, but has a larger mechanical footprint. The Newtonian form has the largest obscuration and the smallest mechanical structure. The Pfund form has an intermediate-sized obscuration as well as mechanical structure. These performances are shown in Fig. 16.30. The Newtonian–Maksutov design has an obscuration comparable with that of the classic Newtonian, but with significantly smaller aberration. The aberration correction available from the additional surfaces is readily seen in this design. The Herschel–Houghton design introduces no obscuration and, although this design has more residual aberration than does the Newtonian–Maksutov, it has the advantage of being almost uniform across the field of view. Although the catadioptric forms discussed show considerable improvement over the all-reflective designs, these forms are limited, as are other refractive systems, in terms of maximum available aperture.

Although all the variations of the classic designs can be made in either the Gregorian or the Cassegrain form, in general, because of the longer length and

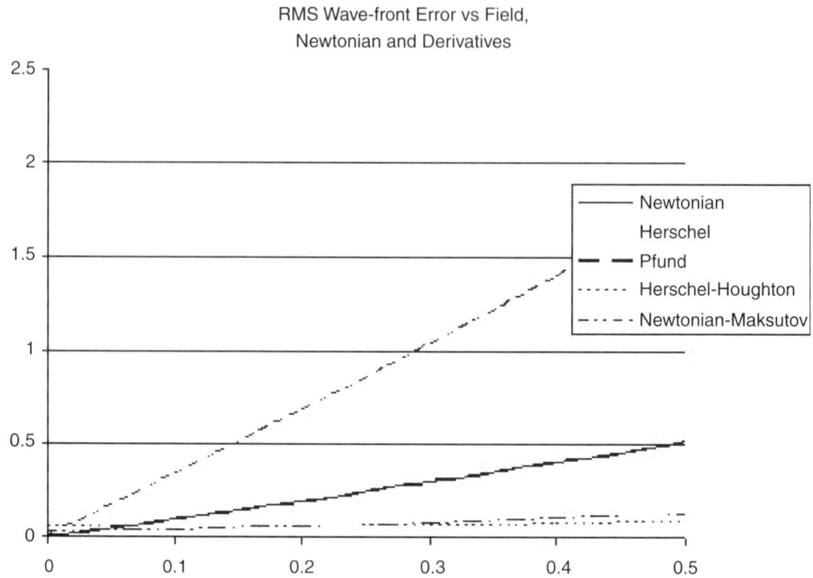

Fig. 16.30. Comparison of the performances of some single-mirror systems.

higher obscuration percentage, the Gregorian forms are often not used. The two common Gregorian telescopes are the classic Gregorian and the Gregorian aplanat. The performance of the two designs is similar, with the aplanat having marginally better performance. With the only difference in the designs being the conic constants, the obscuration effect is the same for the two designs.

Several forms of the Cassegrain telescope are found in common use, all reflective as well as catadioptric. The Dall–Kirkham form is useful for only very small fields because of the high coma. The other all-reflective systems all exhibit similar performance. Performances are shown in Fig. 16.31. For very small field angles, the Ritchey–Chretien form has the best performance, followed closely by that of the classic Cassegrain. As was the case with the single-mirror systems, adding one or more refractive components results in a significant improvement in the residual aberration. This is seen in the Schmidt–Cassegrain design.

It is also useful to compare the performances of several basic telescope designs at different apertures (Fig. 16.32). In comparing the rms wave-front errors for several 1000-mm focal-length telescopes with apertures from 100 to 400 mm (*F*/10 to *F*/2.5), it can be seen that the performances of the single-mirror designs degrade rapidly as the focal ratio becomes smaller. The Schmidt–Cassegrain shows the slowest performance degradation. One result of this is that properly designed Schmidt–Cassegrain telescopes can operate at speeds as fast as *F*/1.

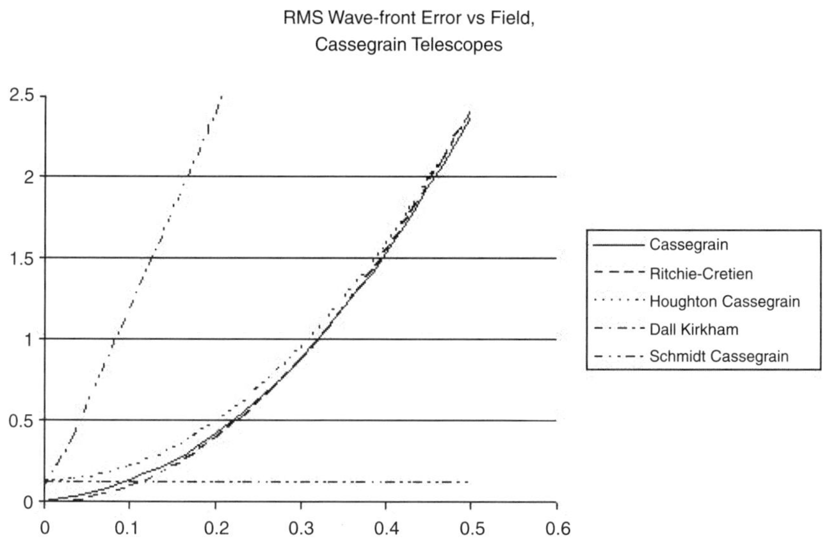

Fig. 16.31. Comparison of the performances of Cassegrain forms.

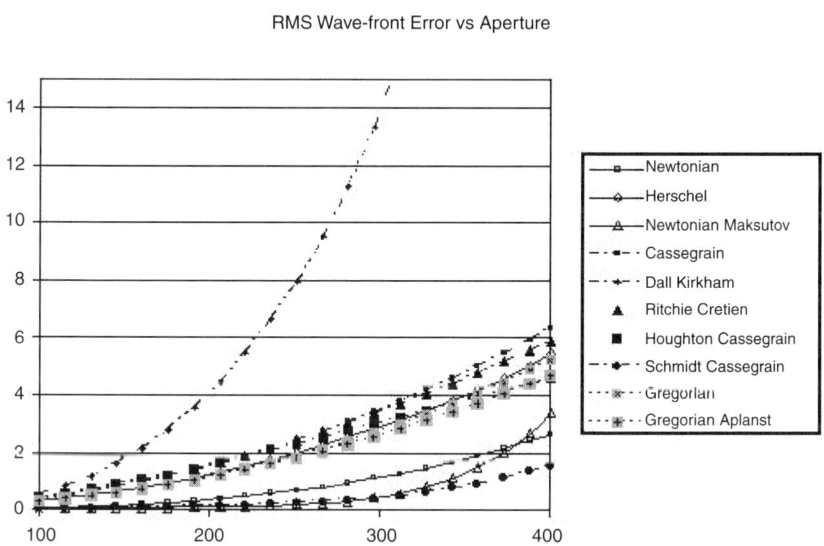

Fig. 16.32. Comparison of the performances of Cassegrain forms.

Further Reading

R. Kingslake, *Lens Design Fundamentals,* Academic, 1978.
W. Smith, *Modern Optical Engineering,* McGraw-Hill, 1990.
W. Smith, *Modern Lens Design,* McGraw-Hill, 1992.
D. Korsch, *Reflective Optics,* Academic, 1991.
D. Schroeder, *Astronomical Optics,* Academic, 2000.

A. Bouwers, *Achievements in Optics,* Elsevier, 1946.
W. Wolfe and G. Zissis, eds., *The Infrared Handbook,* SPIE, 1989.

The first two are classics in the field of optical design, and include mirror designs. The second book by Smith has many specific designs in it. Korsch has many mirror designs, some cited here, and some interesting design approaches to three-mirror systems. Schroeder's title speaks for itself. Bouwers treats several topics investigated during World War II in Holland, but notably has good treatments of the Schmidt and Bouwers cameras, including an early version of the super-Schmidt.

17

Diffractive Optical Elements

William C. Sweatt, Shanalyn A. Kemme, and Mial E. Warren

Diffractive optical elements are diffractive structures fabricated by lithographic methods. These typically are blazed gratings that are designed to direct almost all of the incident flux into one order. For a description of how diffractive optical elements are fabricated and how they work, see Chap. 8 of the OSA *Handbook of Optics* and other references.[1–5] Sometimes, diffractive optical elements are referred to as binary optical elements. The term *diffractive optical elements* may be preferable currently. In the past one could only approximate a ramp function in a blazed grating with steps created with several mask steps. Now there are several new lithographic "gray-scale" fabrication methods requiring only one mask.

17.1. Overview of Diffractive Optical Element Design

Section 17.2 begins with descriptions of ray tracing through diffractive optical elements. Ray tracing predicts where the diffracted orders go, but does not indicate how much light is in each order. The ray-tracing methods presented are accurate for any nonevanescent order. Section 17.2 also contains first-order and third-order properties of diffractive optical elements and resulting blur sizes, a compilation of the ray-tracing equations for transmissive and reflective elements, equations describing front- and back-surface reflective elements and a transmissive one, and a description of ray-based design methods, including

1 M. Bass, ed., *Handbook of Optics,* McGraw-Hill, 1995, Vol. 2, Chap. 8, pp. 8.1–8.19.

2 T. W. Stone and B. J. Thompson, eds., *Holographic and Diffractive Lenses and Mirrors,* Vol. MS34 of SPIE Milestone Series, SPIE, 1991.

3 I. Cindrich and S. H. Lee, eds., *Computer and Optically Generated Holographic Optics,* Proc. SPIE **1052,** 1989; **1211,** 1990; **1555,** 1991.

4 G. J. Swanson, "Binary optics technology. The theory and design of multi-layer diffractive optical elements," MIT Lincoln Laboratory Tech. Rep. 854, NTIS Pub. AD-A213-404.

5 M. W. Farn and W. B. Veldkamp, "Binary optics: Trends and limitations," in *Conference on Binary Optics,* NASA Conf. Pub. 3227, 1993, pp. 19–30.

Huygen's method, the use of a standard lens design code, and use of the Sweatt model.

When the rays are diffracted through small angles, the flux in the various orders can be calculated from scalar diffraction theory, as in Section 17.3. For diffraction angles of ~10° or less, the scalar calculation are fairly accurate. The efficiency of diffractive optical elements is described in Section 17.3 from a scalar point of view. For larger diffraction angles ($\theta > 10°$), scalar theory can give an indication of the flux in the orders, although for accurate results, one must resort to rigorous diffraction theory (Subsection 17.3.7). Subsection 17.3.7 contains a comparison of scalar theory and rigorous diffraction versus diffraction angle.

In Section 17.4 subwavelength diffractive structures and the effective medium theory are described. Some examples include a form birefringent wave plate, a polarizer, and effective-index high- and low-reflective structures.

The fabrication of diffractive optical elements is discussed in Section 17.5.

17.2. Ray Tracing Through Diffractive Optics

A general set of ray-tracing equations is presented in Eqs. (17.1a) and (17.1b). Equations for a radially symmetric diffractive optical element are presented in Eqs. (17.2a) and (17.2b).

17.2.1. Ray Tracing through a General Diffractive Optic

Ray tracing through a general diffractive optic can be done with the following formulas[6]:

$$L' = L + (p\lambda/2\pi)\xi(x,y)\cos\alpha, \tag{17.1a}$$

$$M' = M + (p\lambda/2\pi)\xi(x,y)\sin\alpha, \tag{17.1b}$$

where L, M, L', and M' are the ray's direction cosines in the x and the y directions before and after diffraction, measured in air (note that these are measured with respect to the vector normal to the diffracting surface), p is the order number and is usually $p = 1$, λ is the wavelength of the light in air, and $\xi(x,y)$ is the local spatial frequency of the grating, described in periods per millimeter or line pairs per millimeter. If Λ is the width of a period, then $\xi(x,y) = 1/\Lambda\,(x,y)$. α is the angle between the grating lines and the y axis measured in the plane of the diffracting surface.

[6] W. T. Welford, *Aberrations of Optical Systems,* Hilger, 1986, pp. 75–78.

17.2.2. Ray Tracing Through a Radially Symmetric Diffractive Optic

$$L' = L + (p\lambda/2\pi)\xi(r)x/r, \tag{17.2a}$$

$$M' = M + (p\lambda/2\pi)\xi(r)y/r, \tag{17.2b}$$

where L, M, L', M', p, and λ are as defined for Eqs. (17.1a) and (17.1b). $\xi(r) = r/(\lambda f_0) + C_3 r^3 + C_5 r^5 + \cdots$ is the grating's spatial frequency, where f_0 is the focal length of the element, and C_3, C_5, … , are higher-order terms available to optimize the image quality.

17.2.3. First-Order Properties of Radially Symmetric Diffractive Optical Elements

A diffractive optic used as a lens[7] will have a focal length f_0 defined for the first diffracted order and measured at wavelength λ_0. Note that both the object distance Z and the image distance Z' are defined as positive when they are on the same side of the diffractive optical element:

$$\frac{1}{Z'(\lambda)} - \frac{1}{Z} = \frac{p\lambda/\lambda_0}{f_0}. \tag{17.3}$$

The primary axial color of a diffractive lens is

$$\mathrm{PAC} = -f_0(\Delta\lambda/\lambda_0), \tag{17.4}$$

where $\Delta\lambda$ is the wavelength spread, $\Delta\lambda = \lambda_{\mathrm{max}} - \lambda_{\mathrm{min}}$.

The primary lateral color is the minimum diameter of the focused spot:

$$\mathrm{PLC} = 0.5D(\Delta\lambda/\lambda_0), \tag{17.5}$$

where D is the diameter of the diffractive optical element.

The dispersion of a diffractive optical element is negative (whereas that of glass is defined as positive). In the visible region, it is also an order of magnitude larger. The Abbe number of a diffractive optical element is

$$\nu = \lambda_D/(\lambda_F - \lambda_C) = -3.45, \tag{17.6}$$

where λ_F, λ_C, and λ_D are the blue, red, and yellow wavelengths, respectively, normally used to define the visible spectrum.

[7] E. B. Champagne, J. Opt. Soc. Am. **57,** 51, 1967.

17.2.4. Third-Order Aberrations of a Diffractive Lens

The third-order aberrations of a radially symmetric diffractive optic on a flat substrate described here are simplified from those in footnotes 1 and 7. Assume that the diffractive optical element, centered at (0,0,0), was designed to image perfectly an on-axis point at location $(0,0,Z_0)$ to another on-axis point located at $(0,0,Z_i)$, by use of a wavelength λ_0. This element is then used to image another object point located at $(0,\xi_0,L_0)$, at a new wavelength λ. The image of the new point is located at $(0,\xi_i,L_i)$, which can be found from

$$\frac{1}{L_i}-\frac{1}{L_0}=\frac{\lambda}{\lambda_0}\left[\frac{1}{Z_i}-\frac{1}{Z_0}\right], \tag{17.7}$$

$$\alpha=\frac{\xi_i}{L_i}=\frac{\xi_0}{L_0}, \tag{17.8}$$

where α is the field angle defined in Eq. (17.8).

The wave-front error that is due to third-order aberrations can be calculated for a point $(x,y,0)$ on the diffractive optical element from

$$\begin{aligned} W(x,y)=&\frac{1}{8}\left[\left(\frac{1}{L_i^3}-\frac{1}{L_0^3}\right)-\frac{\lambda}{\lambda_0}\left(\frac{1}{Z_i^3}-\frac{1}{Z_0^3}\right)\right](x^2+y^2)^2\\ &-\frac{\alpha}{2}\left(\frac{1}{L_i^2}-\frac{1}{L_0^2}\right)y(x^2+y^2)\\ &+\frac{\alpha^2}{4}\left(\frac{1}{L_i}-\frac{1}{L_0}\right)(3y^2+x^2). \end{aligned} \tag{17.9}$$

The term on the first line is spherical aberration, which is zero when the diffractive optical element is operated at the wavelength and the conjugates for which it was designed. The second and third lines contain the coma and astigmatism terms, respectively. Astigmatism and coma are not functions of either the wavelength (λ) or conjugate points (Z_i or Z_0). Coma will be present for all off-axis points when the magnification is not unity: $m=L_i/L_0=-1$. Astigmatism will be present in any off-axis image point. However, there is a quadratic dependence between astigmatism and the field angle α, so its effect is miniscule for small fields.

Note that the imagery at a given object point $(0,\xi_0,L_0)$ is commonly considered nearly diffraction limited if $W(x,y)\le\lambda/4$ for all x and y.

17.2.4.1. Peak Aberrations of a Diffractive Optical Element with Axial Symmetry

The peak wave-front error at best focus for each of the terms in Eq. (17.9) are

$$\text{Spherical}_{\max} = \pm\left[\left(\frac{1}{L_i^3} - \frac{1}{L_0^3}\right) - \frac{\lambda}{\lambda_0}\left(\frac{1}{Z_i^3} - \frac{1}{Z_0^3}\right)\right]\frac{D^4}{512},$$

$$\text{Coma}_{\max} = \pm(1 - m^2)\,\frac{D(\text{N.A.})^2\alpha}{4},$$

$$\text{Astigmatism}_{\max} = \pm(1 - m)\,\frac{D(\text{N.A.})\alpha^2}{8}. \tag{17.10}$$

The spherical aberration is zero if the diffractive optical element is used at the wavelength and conjugates for which it is designed. The constants in the coma and astigmatism terms are as follows: $m = L_i/L_0$ is the magnification, D is the diameter, N.A. is the numerical aperture of the output beam, and α is the field angle defined as $\alpha = \xi_i/L_i$.

17.2.4.2. Geometric Blur Sizes at the Best Focus Position

These blur sizes are given by

$$D_{\text{spherical}} = \left[\left(\frac{1}{L_i^3} - \frac{1}{L_0^3}\right) - \frac{\lambda}{\lambda_0}\left(\frac{1}{Z_i^3} - \frac{1}{Z_0^3}\right)\right]\frac{D^3 L_i}{256},$$

$$D_{\text{sagittal coma}} = (1 - m^2)\,\frac{D(\text{N.A.})\alpha}{2},$$

$$D_{\text{astigmatism}} = (1 - m)\,\frac{D\alpha^2}{2}. \tag{17.11}$$

The coefficients are defined as in Eqs. (17.10).

17.2.5. Sketches of Blazed Reflective and Transmissive Diffractive Optical Elements

Figures 17.1, 17.2, and 17.3 show the geometry of blazed diffractive optics for front-surface reflection, back-surface reflection, and transmission, respectively. The equations for the period Λ are exact for all cases. The spatial frequency ξ (lines per unit length) is the inverse of the period width, i.e., $\xi = 1/\Lambda$. The expressions given for the average slope of the facets θ_{ave} and the height of the blaze h are approximate, although fairly accurate, when $\Lambda \gg \lambda$.

17.2.6. Designing Systems Containing Diffractive Optical Elements

There are three methods of designing lenses. For point-to-point imaging, one may use the Huygen's method described in Subsection 17.2.6.1. It allows the calculation of the period positions for a diffractive optical element operating as

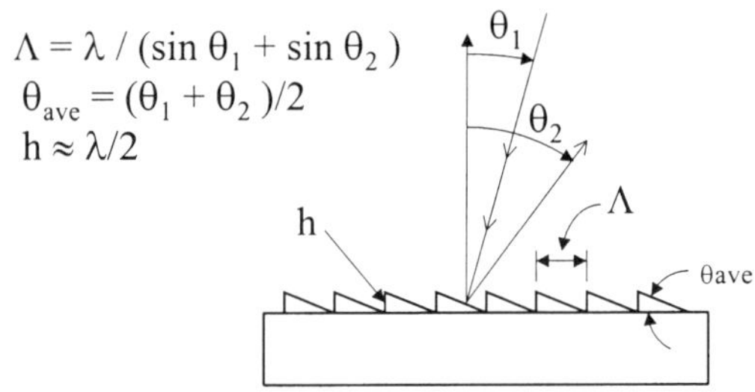

Figure 17.1. Front-surface reflection diffractive optical element

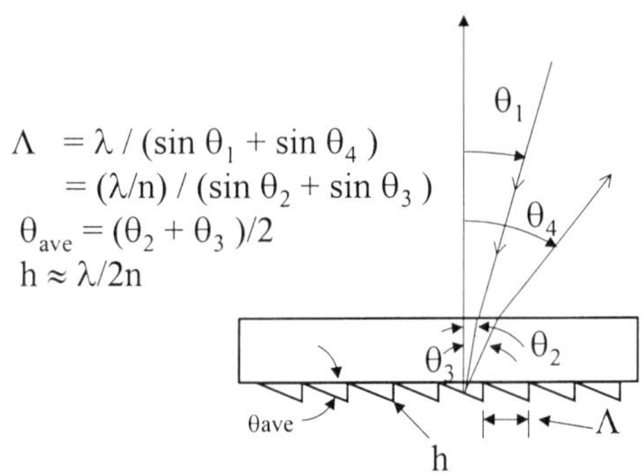

Figure 17.2. Back-surface reflection diffractive optical element.

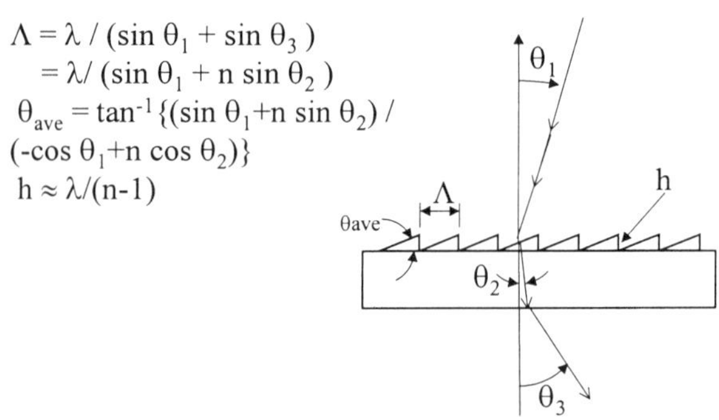

Figure 17.3. Transmissive diffractive optical element.

a simple lens. If the system contains two or more elements with focal power (diffractive optical elements, lenses, or curved mirrors), one wants to use a lens design code, such as ZEMAX™,[8] CODE V™,[9] or SuperOslo™,[10] as described in Subsection 17.2.6.2. The Sweatt model (Subsection 17.2.6.3) was

[8] ZEMAX, Version 9.0, Focus Software, San Diego, CA.
[9] CODE V, Optical Research Associates, Pasadena, CA.
[10] SuperOslo, product of Sinclair Optics, Fairport, NY.

developed for describing holographic optical elements as extremely high-index lenses, although it also works for diffractive optical elements. It is useful for designing diffractive optical elements on oddly shaped surfaces and for developing intuition about how diffractive optical elements work.

17.2.6.1. Huygen's Method

The Huygen's method does not require a lens design code for the design of a simple diffractive optical element for point-to-point imaging. However, away from the design point, the image will suffer coma, much as a simple parabolic mirror does. And, as with a simple parabola, the coma can be severe if the element is very fast (see Subsection 17.2.4). The Huygen's method,[11] which is based on the Huygens–Fresnel derivation of zone plates, traces rays from the object to the image. As one moves outward from the most direct ray, the optical path length, that is, the distance from the object point to the diffractive optical element plus the distance from the diffractive optical element to the image, increases, as shown in Fig. 17.4.

17.2.6.2. Using a Lens Design Code to Design a Diffractive Optical Element

This method is powerful and quite straightforward. A modern lens design code allows the diffractive structure to be optimized by itself or in conjunction with other diffractive optical elements, lenses, mirrors, etc. Furthermore, the image quality can be optimized over several points within the field simultaneously and over a range of wavelengths, if appropriate. Also, these design programs allow other constraints to be held during the optimization. This is decidedly the best way to design an optical system in almost all situations. The following paragraphs give two examples.

The first example is a 25-mm-diameter, 1° field-of-view, *F*/4 cemented doublet lens, optimized for the visible spectrum. After it was designed, a radially symmetric diffractive element was added to it (cemented on the back), and the system was reoptimized. With the diffractive optical element added, the rms wave-front error averaged across the field dropped from $\lambda/4$ to $\lambda/30$. The design code allows us to optimize the rms wave-front error (the sum of all the aberrations) at several points in the field and at three wavelengths at each point. This optimization could not possibly have been done any other way—even the types of glass in the doublet were optimally chosen. The design code can optimize the rms. The wavelengths, aperture diameter, fields of view, and a representative pair of glasses (a crown and a flint) were entered into the code. The back focal distance was set to 100 mm, and the maximum and the minimum glass thicknesses (5 and 2 mm) were entered. It was assumed that the dif-

[11] E. Hecht and A. Zajac, *Optics,* Addison-Wesley, 1979, pp. 375–376.

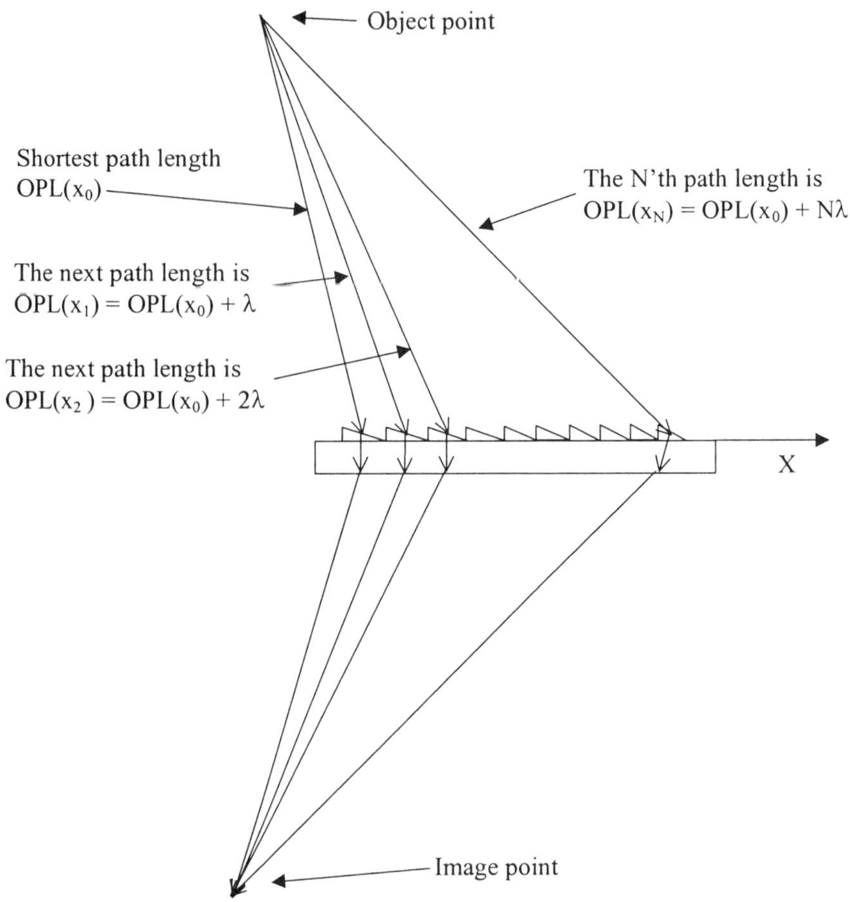

Figure 17.4. Graphic description of Huygen's method for calculating the period locations (x_1, x_2, … ,) on the diffractive optical element. OPL, optical path length.

fractive optical element would be fabricated on a silica plate and glued on the back of the cemented doublet. The diffractive optical element was described by a radially symmetric diffractive model (known in ZEMAX™ as a "binary 2" surface). During the optimization, the diffractive optical element's coefficients for the r^2, the r^4, and the r^6 terms, the lens thicknesses, and the radii of curvature were varied. Also, the computer was allowed to substitute other glass types in the lens. The standard merit function (rms wave-front option) was used. The result is shown in Fig. 17.5.

The second example is shown in Fig. 17.6. It is a microscanner, also designed with ZEMAX™. When a 1.55-μm laser source is moved around in the 1-mm-diameter object plane, the output beam is scanned over a 20° field of view. The diffractive optical elements are written on both the front and the

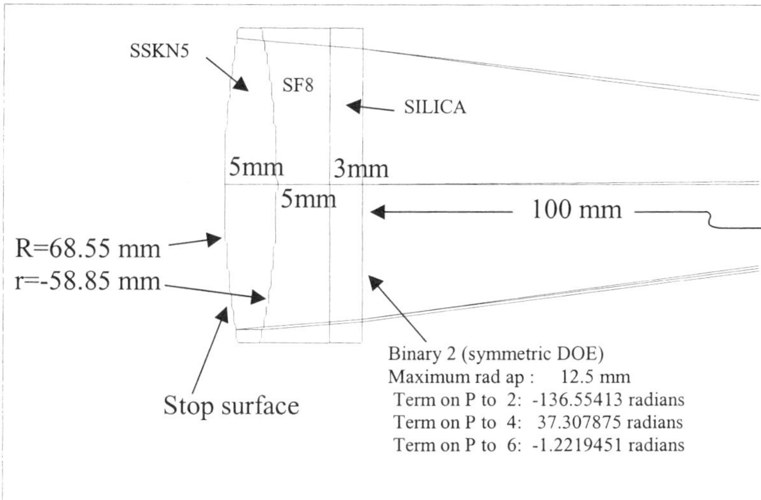

Figure 17.5. Diffractive optical element design example 1: doublet with diffractive optical element.

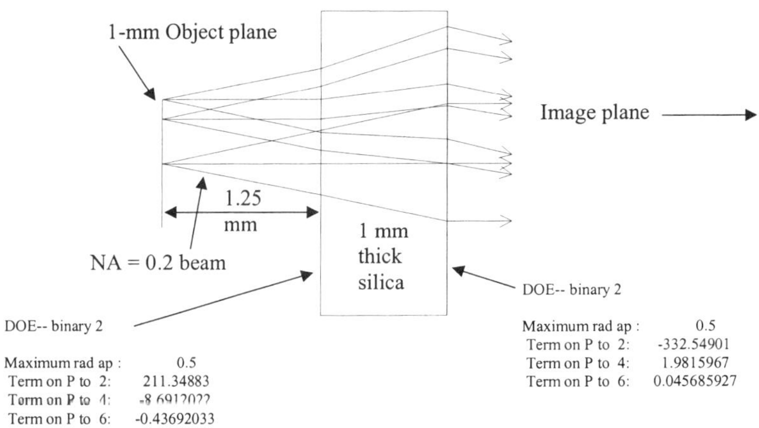

Figure 17.6. Diffractive optical element design example 2: microscanner with diffractive optical element.

back of the fused-silica plate. The rms wave-front error in this design is approximately $\lambda/20$ over the whole field.

In these rather simple designs, only radially symmetric diffractive elements on flat surfaces have been needed to satisfy the system requirements. All of the major lens design codes have other diffractive models that allow fringes to be described with a general polynomial. These polynomials are typically of the form

$$\xi(x,y) = \sum \sum C_{i,j} x^i y^j,$$

where x and y are surface coordinates and the indices $i = 0,1,2,3, \ldots$, and $j = 0,1,2,3, \ldots$. The lens design codes typically allow the surfaces upon which the diffractive optical elements are fabricated to be radially symmetric aspheres.

17.2.6.3. Sweatt Model

The Sweatt model is an ultrahigh-index refractive lens that has the same macroscopic optical properties, as does a focusing diffractive optical element.[12–14] It is useful to system designers in two ways. First, it allows optical designers to apply their intuition about refractive components to diffractive optical elements. For example, the aberrations introduced into a converging beam by a linear grating are identical to those introduced by a wedged plate. Similarly, spherochromatism introduced by a lens and a focusing diffractive optical element are one and the same when viewed this way. Second, the Sweatt model can be used to design or analyze diffractive optical elements of any type fabricated on oddly shaped surfaces. This scheme is useful for designing diffractive optical elements on cylinders, cones, toroids, and other strange shapes. The analogy between an ultra-high-index lens and a diffractive element reduces to

$$n - 1 = (\lambda/\lambda_0)(n_0 - 1), \tag{17.12}$$

$$t(x,y) = \lambda_0\varphi_0(x,y)/[(n_0 - 1)2\pi] + t_0. \tag{17.13}$$

The lens refractive indices n_0 and n are associated with wavelengths λ_0 and λ. The thickness and the phase of the diffractive optical element at point x,y on the lens' surface are $t(x,y)$ and $\varphi_0(x,y)$, respectively. The thickness t_0 is added largely for cosmetic value, so that the high-index lens always has positive thickness. Figure 17.7 shows the parameters for a simple lens.

A diffractive element on a substrate with a nonstandard surface profile can be modeled in several steps. First, the profile must be described mathematically. Then the high-index lens can be added. The second surface of the high-index lens should initially have the same shape as the substrate. Thus, before optimization, the element has no power. Then the shape of the second surface of the high-index lens can be varied, thereby simulating power in the equivalent diffractive optical element. Finally, the thickness profile of the high-index lens must be determined and, from it, the properties of the diffractive optical element can be calculated as previously described.

[12] W. C. Sweatt, "Mathematical equivalence between a holographic optical element and an ultra-high-index lens," J. Opt. Soc. Am. **69,** 486–487, 1979.

[13] M. W. Farn, "Quantitative comparison of the general Sweatt model and the grating equation," Appl. Opt. **31,** 5312–5316, 1992.

[14] W. C. Sweatt, "Describing holographic optical elements as lenses," J. Opt. Soc. Am. **67,** 803–808, 1977.

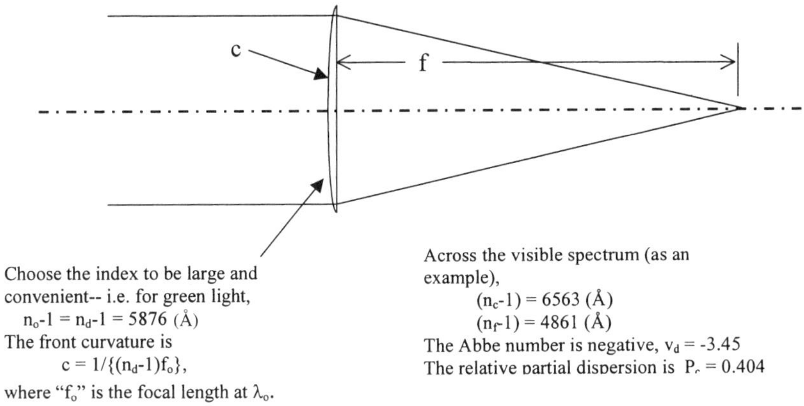

Figure 17.7. Diffractive optical element design example 3: simple lens diffractive optical element.

17.3. Scalar Theory for Phase Diffractive Optical Elements

Many diffractive optical element designs can be obtained by scalar theory. A brief outline of the equations used in scalar theory are presented here; more detailed discussions may be found elsewhere.[1,4,15,16] In general, the following conditions make the scalar theory application valid:

1. The diffractive optical element modifies the phase front of an incident beam by imparting a local phase delay, $\varphi(x,y)$, in one plane.
2. The local period or pitch, $\Lambda(x,y)$, must be large compared with the wavelength, λ:

$$\Lambda(x,y) \geq 10\lambda. \tag{17.14}$$

3. Vector field effects, such as polarization information, are not included in the model.

The phase diffractive optical element is described by the complex transmittance and Fig. 17.8:

$$t(x,y) = e^{i\varphi(x,y)}. \tag{17.15}$$

Typically a blazed diffractive optical element is approximated as an N-level diffractive optical element, where $N = 2^M$ and M is the number of mask process steps, as shown in Fig. 17.9.

[15] J. W. Goodman, *Introduction to Fourier Optics,* McGraw-Hill, 1968 (2nd ed., 1995).
[16] G. J. Swanson, "Binary optics technology: Theoretical limits on the diffraction efficiency of multilevel diffractive optical elements," MIT Lincoln Laboratory Tech. Rep. 914, 1991.

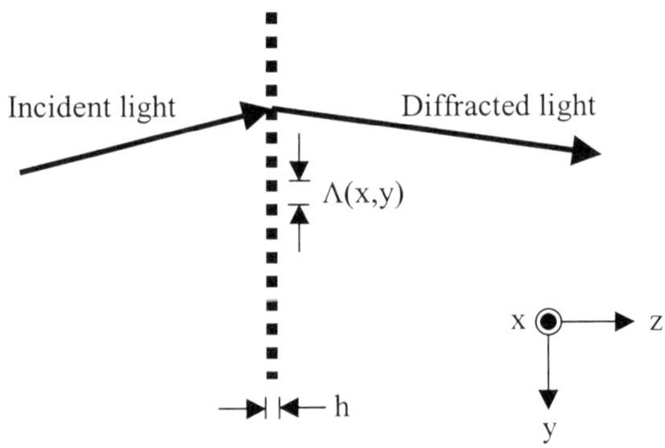

Figure 17.8. Grating notation and coordinate system.

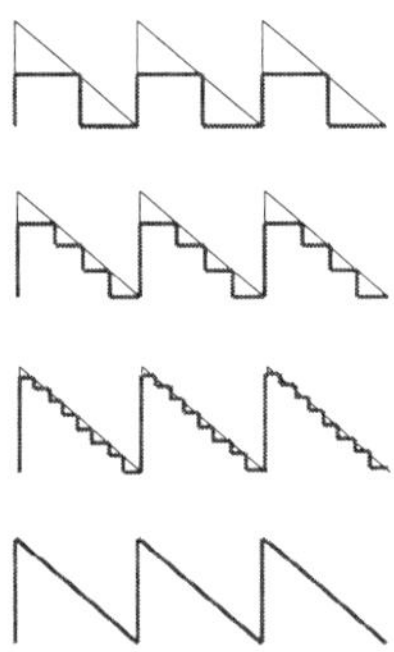

In general, for Pitch>Wavelength

# Levels	# Masks	Efficiency
2	1	41%
4	2	82%
8	3	95%
16	4	99%

Figure 17.9. Blazed gratings and stepped approximations.

Scalar theory approximates the efficiency of the $p = \pm 1$ diffraction order for an N-level diffractive optical element as

$$\eta_N = \left| \frac{\sin\left(\frac{\pi}{N}\right)}{\left(\frac{\pi}{N}\right)} \right|^2 . \tag{17.16}$$

With this approximation, a two-level grating (binary grating) order will have an efficiency of 40.5%. In the same way, the efficiencies for 4-, 8-, and 16-level gratings are 81.6%, 95%, and 99%, respectively.

If the preceding conditions are not met, then the efficiencies predicted by scalar theory are generally larger than measured efficiencies. For example, a surface-relief grating with a low-refractive-index contrast ratio necessitates a rather thick grating region h to effect a π phase difference that may detrimentally affect the measured efficiency compared with that predicted by scalar theory. High-aspect-ratio features (large h/Λ) can produce significant multiple reflections among the diffractive optical element features, again resulting in lower measured efficiencies than those predicted by scalar theory.

Although scalar theory may exaggerate the efficiency of a diffractive optical element, it is a useful method to determine component depths and transition points. Moreover, it often produces a conceptual layout as well as a good initial input to a more rigorous, iterative process.

17.3.1. *Fresnel Phase-Plate Transition Radii*

Transition radii for a radially symmetric Fresnel phase plate with a normal-incidence plane wave may be calculated by use of the right triangle in Fig. 17.10. The term in λ^2 can be dropped, and a good approximation retained:

$$f^2 + r_q^2 = (f + q\lambda)^2,$$
$$r_q^2 = 2fq\lambda + q^2\lambda^2. \qquad (17.17)$$

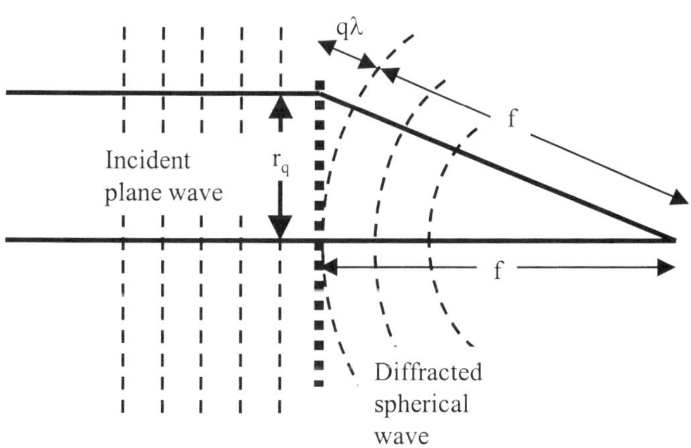

Figure 17.10. Notation for a plane wave incident upon a Fresnel phase plate.

17.3.2. *Fresnel Phase-Plate Minimum Feature Size*

The most important fabrication condition that limits the diffractive optical element design is the minimum feature size. In this example, the minimum feature size corresponds to min(Δr), which is obtained from the differential of the second equation of Eqs. (17.17):

$$2r_q\Delta r = 2f\Delta q\lambda + 2q\Delta q\lambda^2,$$

$$\min(\Delta r) = \frac{f\lambda + q_{\max}\lambda^2}{R}. \tag{17.18}$$

$\Delta q = 1$ when Δr = feature size. Also, $q_{\max}$ is the index when $r_q = R$ = the radius of the lens.

17.3.3. N-Level Fresnel Phase-Plate Transition Radii

Typically a blazed diffractive optical element is approximated as an N-level diffractive optical element, where $N = 2^M$ and M is the number of mask process steps. In this case, the transition radii for an N-level Fresnel phase plate are given by

$$r_{qN} = \sqrt{\frac{2fq\lambda}{N} + \frac{q^2\lambda^2}{N^2}}. \tag{17.19}$$

17.3.4. N-Level Fresnel Phase-Plate Minimum Feature Size

The minimum feature size for an N-level Fresnel phase plate is obtained with the same method as that for Eqs. (17.18):

$$\min(\Delta r_{qN}) = \frac{f\lambda}{NR} + \frac{q_{\max}\lambda^2}{N^2R}, \tag{17.20}$$

where $q_{\max}$ is the index when $r_q = R$ = the radius of the lens.

17.3.5. N-Level Fresnel Phase-Plate Minimum F Number

It is useful to know the smallest F number (fastest lens) that may be specified for a fabrication-limited minimum feature size, $\min(\Delta r)$. This is given by

$$F = \frac{N\min(\Delta r)}{2\lambda}. \tag{17.21}$$

17.3.6. N-Level Diffractive Optical Element Etch Depth

The physical height of an individual feature in an N-level diffractive optical element in air is obtained by the slicing of the desired phase delay into layers:

$$\varphi = \frac{2\pi}{\lambda}(n-1)h,$$

$$\frac{2\pi}{N} = \frac{2\pi}{\lambda}(n-1)h_N,$$

$$h_N = \frac{\lambda}{N(n-1)},$$

$$h_M = \frac{\lambda}{2^M(n-1)}, \tag{17.22}$$

where $N = 2^M$ and M is the number of mask process steps.

17.3.7. Rigorous Coupled-Wave Analysis

When the constraints of scalar theory are no longer valid for a component design, as is the case for gratings with $\Lambda < \sim 5\lambda$, rigorous coupled-wave analysis may be used to numerically predict grating efficiencies and polarization properties.[17–21] This algorithm solves Maxwell's equations, with the appropriate boundaries above and below the grating region, with a linear superposition of plane waves. The permitivity in the grating region is written as a constant plus an expansion term describing the modulation of the permitivity. The algorithm is exact for an infinitely periodic grating and an infinite number of orders, but more practically will converge to a grating efficiency prediction by use of a finite number of orders. The boundary conditions are applied to the Helmholtz equation for each of the six components of the **E** and the **H** fields. These components are coupled when Maxwell's equations are utilized. Standard eigenmatrix methods result in a plane-wave solution set.

As an example, Fig. 17.11 shows rigorous coupled-wave analysis grating efficiency predictions for a normally incident TE and TM plane wave on a binary (two-level) fused-silica grating of depth $h = 1.0$ µm as a function of normalized pitch.

[17] M. G. Moharam and T. K. Gaylord, "Diffraction analysis of dielectric surface relief grating," J. Opt. Soc. Am. **72,** 1383–1392, 1982.

[18] T. K. Gaylord and M. G. Moharam, "Analysis and applications of optical diffraction by gratings," Proc. IEEE **73,** 894–937, 1985.

[19] L. Li, "A modal analysis of lamellar diffraction gratings in conical mountings," J. Mod. Opt. **40,** 553–573, 1993.

[20] R. Petit, ed., *Electromagnetic Theory of Gratings,* Springer, 1980.

[21] D. Maystre, "Rigorous vector theories of diffraction gratings," in *Progress in Optics,* E. Wolf, ed., North-Holland, 1984, Vol. XXI, pp. 1–67.

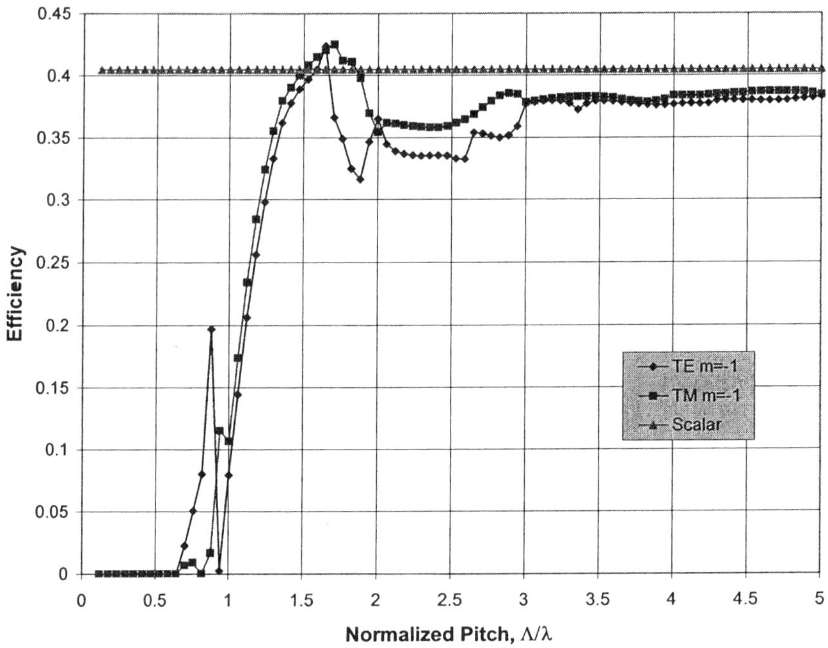

Figure 17.11. Scalar and rigorous coupled-wave analysis efficiency predictions for a two-level fused-silica grating in air with depth $h = 1.0$ μm.

17.4. Subwavelength Structures

A subwavelength structure propagates only the zero transmissive and/or reflective orders. This occurs when the structure, such as a grating, has a pitch Λ that is smaller than the wavelength λ of the incident light.

17.4.1. Effective Medium Theory for Subwavelength Structures

Effective medium theory is an approximating theory that assigns a propagation constant β to the propagating harmonic field. The ratio of this propagation constant to the amplitude of the **k**-vector in vacuum leads to the effective refractive index of the grating region. A brief outline of the necessary equations is given here; more detailed discussions may be found elsewhere.[22–26]

[22] P. Yeh, A. Yariv, and C. S. Hong, "Electromagnetic propagation in periodic stratified media. I. General theory," J. Opt. Soc. Am. **67,** 423–438, 1977.

[23] A. Yariv and P. Yeh, "Electromagnetic propagation in periodic stratified media. II. Birefringence, phase matching, and x-ray lasers," J. Opt. Soc. Am. **67,** 438–448, 1977.

[24] M. Born and E. Wolf, *Principles of Optics,* 6th ed., Pergamon, 1980, Section 14.4.2.

[25] J. Turunen and F. Wyrowski, eds., *Diffractive Optics for Industrial and Commercial Applications,* Akademie Verlag, 1997, pp. 303–323.

[26] H. Kikuta, Y. Ohira, and K. Iwata, "Achromatic quarter-wave plates using the dispersion of form birefringence," Appl. Opt. **36,** 1566–1572, 1997.

17.4.1.1. Effective Medium Theory TE Incident Light

In Fig. 17.12, the **E**-vector for the TE incident plane wave is parallel to $\hat{\mathbf{x}}$ and incident at an angle θ. In the TE configuration, solve Eq. (7.23) for β_{TE}:

$$\cos(k_{1Z}a)\cos(k_{2Z}b) - \frac{1}{2}\left(\frac{k_{2Z}}{k_{1Z}} + \frac{k_{1Z}}{k_{2Z}}\right)\sin(k_{1Z}a)\sin(k_{2Z}b) = \cos(k_V n_1 \Lambda \sin\theta), \tag{17.23}$$

where $k_{1Z} = \sqrt{n_1^2 k_V^2 - \beta_{TE}^2}$ and $k_{2Z} = \sqrt{n_2^2 k_V^2 - \beta_{TE}^2}$. Here the amplitude of the **k**-vector in vacuum is $k_V = 2\pi/\lambda_0$. For a normally incident plane wave, the term on the right-hand side of Eq. (7.23) is unity. The effective index of refraction for the TE configuration is

$$n_{\text{eff TE}} = \frac{\beta_{TE}}{k_V}. \tag{17.24}$$

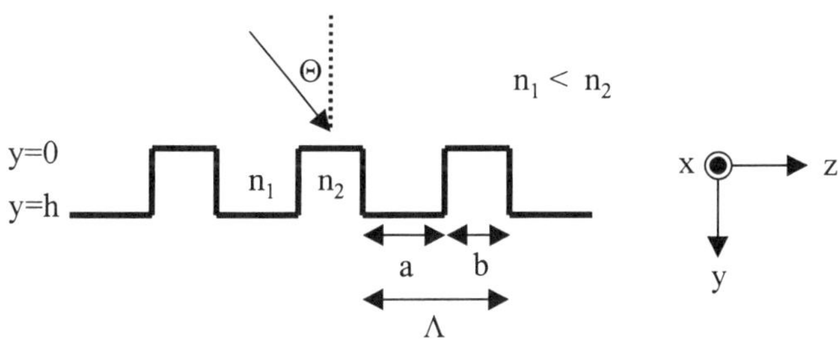

Figure 17.12. Grating configuration and notation for effective medium theory.

17.4.1.2. Effective Medium Theory TM Incident Light

In Fig. 17.12 the **E**-vector for the TM incident plane wave is in the *yz* plane and incident at an angle θ.

In the TM configuration, solve Eq. (17.25) for β_{TM}:

$$\cos(k_{1Z}a)\cos(k_{2Z}b) - \frac{1}{2}\left(\frac{n_1^2 k_{2Z}}{n_2^2 k_{1Z}} + \frac{n_2^2 k_{1Z}}{n_1^2 k_{2Z}}\right)\sin(k_{1Z}a)\sin(k_{2Z}b) = \cos(k_V n_1 \Lambda \sin\theta), \tag{17.25}$$

where $k_{1Z} = \sqrt{n_1^2 k_V^2 - \beta_{TM}^2}$ and $k_{2Z} = \sqrt{n_2^2 k_V^2 - \beta_{TM}^2}$. Again, the amplitude of the **k**-vector in vacuum is $k_V = 2\pi/\lambda_0$. For a normally incident plane wave, the term on the right-hand side of Eq. (17.25) is unity. Finally, the effective index of refraction for the TM configuration is

$$n_{\text{eff TM}} = \frac{\beta_{TM}}{k_V}. \tag{17.26}$$

17.4.2. The Quasi-Static Limit in Effective Medium Theory

In the limit where $\Lambda/\lambda \rightarrow 0$ or, more practically, when $\Lambda \leq \lambda/10$, then the effective refractive indices as a function of fill factor $\sigma = b/\Lambda$ become[27]

$$n_{\text{eff TE}} = \sqrt{\sigma n_2^2 + (1-\sigma)n_1^2},$$

$$n_{\text{eff TM}} = \frac{1}{\sqrt{\dfrac{\sigma}{n_2^2} + \dfrac{(1-\sigma)}{n_1^2}}}. \tag{17.27}$$

17.4.3. Effective-Index High- and Low-Reflective Coatings

Clearly manipulation of subwavelength grating parameters can lead to tailored effective refractive indices in the grating region. This can be utilized to fabricate effective antireflective coatings, graded antireflective coatings, and two-dimensional antireflective coatings.

17.4.4. Form-Birefringent Wave Plate

Because the effective refractive index of a grating is different for TE and TM incident plane waves, the grating behaves as a uniaxial birefringent medium. This is termed *form birefringence*.[27] When a plane wave is incident upon a grating where the **E**-vector is 45° with respect to the grating vector then a phase delay results between the TE and the TM components as

$$\Delta\varphi = \varphi_{\text{TE}} - \varphi_{\text{TM}} = \frac{2\pi(n_{\text{eff TE}} - n_{\text{eff TM}})h}{\lambda}. \tag{17.28}$$

Effective medium theory and even the quasi-static approximations are useful in obtaining a rough estimate of grating depth h required for a desired wave-plate retardance. However, for visible and near-IR wavelength regimes where grating depths may be large, grating aspect ratios high, and semimetallic materials utilized, it is advisable to arrive at a final answer with the use of rigorous coupled-wave analysis.

In the regime where the grating pitch is just less than the incident wavelength, the induced form birefringence is a strong function of wavelength.[26] A set of parameters resulting in a linear response of Δn_{eff} with respect to wavelength produces an achromatic phase delay. An achromatic wave plate typically requires tight fabrication control on grating depth h and grating duty cycle σ.

[27] M. Born and E. Wolf, *Principles of Optics,* 6th ed., Pergamon, 1980, Section 14.5.2.

17.4.5. *Polarizers*

The polarization dependence of the effective refractive indices in a subwavelength grating may be enhanced by fabricating a metal amplitude grating. This is the visible and near-IR component equivalent to a standard IR wire-grid polarizer. Once again a precise set of grating parameters, pitch and duty cycle, should be obtained with a rigorous treatment. However, some insight is gained with the quasi-static approximation in which the metal grating refractive index is complex:

$$\tilde{n}_2 = n_2 + i\kappa_2. \tag{17.29}$$

If we assume the grating metal is a good conductor such that $\tilde{n}_2 \approx i\kappa_2$ and $\kappa_2 \gg n_1$, then Eqs. (17.27) become

$$n_{\text{eff TE}} \approx i\kappa_2\sqrt{\sigma},$$

$$n_{\text{eff TM}} \approx \frac{n_1}{\sqrt{1-\sigma}}. \tag{17.30}$$

In the limit where κ_2 is large, the TE component of the incident plane wave is reflected whereas the TM component is transmitted through a dielectric substrate.

17.5. Fabrication of Diffractive Optical Elements

17.5.1. *Overview*

Diffractive optical elements are generally fabricated by lithographic processes. In mass production, a master may be made by lithography and replicated by various means. An exception is the use of diamond turning to produce diffractive optical elements on the surfaces of refractive or reflective elements, including spherical surfaces. Diamond turning is not discussed in this chapter. The fabrication of a diffractive optical element by lithographic processes actually is performed in three, sometimes four, steps.[28] First is the generation of a photomask or data file representing the physical diffractive optical element dimensions for the lithography process to be used. Second is the generation of the pattern by that lithographic process. Third is the transfer of that pattern into a second, usually more robust, optical material. In some cases step three is not necessary if the photoresist or other lithographic patterning material is suitable as an optical medium. Steps two and three are often repeated in a multiphase

[28] B. Kress and P. Meyruesis, *Digital Diffractive Optics*, Wiley, 2000.

level design, as discussed in Section 17.3. Fourth is transfer of the finished relief pattern into another substrate by a replication process. This is an optional step, as direct lithographic production of diffractive optical elements is often preferable for high-performance or prototype systems.

17.5.2. Photomask Design

It is assumed that the result of the diffractive optical element optical design process is a phase function describing the desired phase shift of the diffractive optical element as a function of a two-dimensional coordinate system. The phase function is typically a polynomial that may be quite complex. One of the advantages of diffractive optics is the ability to create an optical surface that is unrealizable by any conventional fabrication approach. The polynomial describing the diffractive optical element surface will typically be a smooth, idealized function. To realize a physical device, the phase function must first be transformed into a geometrical layout and then converted into a data file in a format suitable for the particular mask-generation process to be used. For a conventional photomask, this is a file to be used by a photomask shop to generate a mask by an electron-beam pattern generator. In other processes, the data format may be different, but the layout process is similar.

The first step in the layout process is to convert the phase function to a discrete, modulo 2π function that is reduced in optical thickness, making use of the fact that a monochromatic wave is insensitive to phase shifts of integral multiples of 2π. For broadband applications, larger phase intervals are sometimes used.[29] The phase function is then of the form

$$\varphi_0(x_k, y_k) = p\lambda, \tag{17.31}$$

where p has integral values. If plotted, the function is a set of contour plots on phase intervals of 2π. This will define the mask geometry for fabrication of a simple binary grating (as in two discrete optical thickness values). The conventional approach to a diffractive optical element (and the reason for the phrase binary optics) is the use of a binary coding scheme that allows M overlapping mask levels to define 2^M discrete phase levels in the optic. Therefore if the diffractive optical element is to be a multiphase element, composed of multiple surface-relief heights, as illustrated in Fig. 17.9, there is an additional binning of the phase function required:

$$\varphi_0(x_k, y_k) = \frac{p\lambda}{M}. \tag{17.32}$$

[29] M. Rossi, R. E. Kunz, and H. P. Herzig, "Refractive and diffractive properties of planar micro-optical elements," Appl. Opt. **34,** 5996–6007, 1995.

The lithographic fabrication approaches require that the real surface be a stepwise approximation of the ideal surface. The gray-scale processes, which are briefly discussed in Subsection 17.5.3, still require this binning of the phase function to realize discrete exposure levels in the photomask, with a much larger number of levels. The implication of this is that the photomask generation will require a solution of the phase function polynomial to define the exact location of the features in the mask. This may done with software written to perform the numerical solutions and output a data file in a format that may be imported by the computer-aided design software used for photomask design. Some computer-aided design tools have powerful programming capabilities that will allow the calculations to be performed within the application. Two common data file formats used by the photomask industry are GDSII and CIF.[28] Unfortunately, these formats do not allow for continuous curves, based on historical limitations in photomask-generation tools, and allow only polygonal boundaries with limited numbers of vertices (typically 200). This can be a severe limitation for diffractive optical elements, and various programming tricks, including breaking the shape into multiple polygons and defining nonstandard features with the cooperation of the mask vendor, may be necessary. The pattern-generating tool that physically writes the pattern on the photomask will usually require an additional data transformation from the data file to a proprietary data format that defines the individual machine exposures performed by the tool. This process is commonly known as fracturing. In the case of electron-beam pattern generators this process consists of fracturing the polygonal features of the layout into adjacent trapezoidal elements that will be written. The vertices of the trapezoids fall on a rectangular grid that defines the precision limit of the writing process. Similarly, if a direct-write process is being used, the pattern is written directly on the substrate without the use of an intermediate photomask. It is important to work closely with the mask vendor or lithography engineer to get the desired results.

17.5.3. Microlithography

The mask layout process has resulted in a geometric pattern representing discrete phase steps in the diffractive optical element. A lithographic process is then used to create a physical structure on an optical surface. The primary process is photolithography in which a UV-sensitive photoresist on the surface of the part is exposed by use of a photomask (opaque metal patterns on a glass or fused-silica plate). There are different versions of photolithography, with different capabilities. The process most commonly used in the integrated circuit industry is projection lithography, in which an optical system projects a reduced image of the photomask onto the photoresist-coated surface. Because the exposed pattern is a reduced version of the mask, the pattern is repeatedly exposed across the surface, as a stage is precisely stepped in intervals of the exposure field. These systems are capable of high throughput and extreme precision. Minimum linewidths or critical dimensions are typically submicrometer, with the state of the art progressing rapidly. The alignment precision from layer to layer is in turn a fraction of the critical dimension. These systems are expensive to own and operate, are sensitive to surface topology of the substrate because of their shallow depth of field, and can expose only a limited field. A second process that is less expensive and more often seen in research facilities, and still used in the production of discrete electronic and optoelectronic devices, is contact photolithography, in which the photomask is directly exposed in physical contact or close proximity to the surface of the photoresist. The critical dimensions allowed by this process are usually greater than 1μm, and alignment precision is limited to similar dimensions. Large fields, up to the size of the photomask, can be exposed. The third process is direct-write processing with an optical pattern generator. This is less commonly used for direct exposure of a sample and more often used for the generation of lower-cost photomasks. The advantage of this approach is the flexibility of writing one-off patterns without the costs of manufacturing photomasks. Electron-beam lithography is the method of choice for the manufacture of high-quality photomasks. The electron-beam lithography process can also be used as a direct-write process on optical substrates. This gives the ultimate in dimensional resolution, with critical dimensions in the tens of nanometers and, in sequential processes, alignment of one level to the next can be within tens of nanometers. This is a superb method for diffractive optical elements that require very fine linewidths because of short wavelengths or large diffraction angles. The disadvantage of the process is the cost, slow throughput, and limitations of the resists that are used. The thin resists used for high resolution are not suitable for many deep-etch processes. A variation of photolithography uses a gray-scale mask that transmits varying amounts of UV light in photolithography by a photomask that can be written with varying optical density. The most popular version of this uses a proprietary photomask glass that is exposed with varying electron-beam doses in an electron-beam lithography tool. The glass darkens in a well-characterized dependence on the electron flux. The advantage of this process is that a single exposure of photoresist will result in a practically continuous relief profile in the photoresist with appropriate process control. [30]

[30] W. Daschner, P. Long, R. Stein, C. Wu, and S. H. Lee, "Cost-effective mass fabrication of multilevel diffractive optical elements by use of a single optical exposure with a gray-scale mask on high-energy beam-sensitive glass," Appl Opt. **36,** 4675–4680, 1997.

17.5.4. Pattern Transfer

To realize a working diffractive optical element, the geometric mask pattern must be realized as a physical height or thickness profile. For a transmissive optic the phase function $\varphi_0(x,y)$ is transformed into a thickness function $h(x,y)$ by

$$h(x,y) = \frac{\varphi_0(x, y)\lambda}{2\pi(n-1)}. \tag{17.33}$$

For a reflective optic the transformation is

$$h(x,y) = \frac{\varphi_0(x, y)\lambda}{4\pi}. \tag{17.34}$$

If the application of the diffractive optical element is suitable for a polymeric material, the photoresist or electron-beam resist pattern itself may be used as the optical structure if the thickness of the resist material is a correct optical thickness after processing.[31] In most cases, it is desired to transfer the resist pattern into a better optical medium. This may be done in a subtractive process by using an etch process to transfer the resist geometry into the underlying substrate material. In the case of multiple-phase-level diffractive optical elements, this process may be applied with a sequence of masks and varying etch depths to realize an approximation to the ideal phase function profile.[32] In the case of gray-scale lithography, the pattern transfer must etch away the resist material in a controlled and repeatable manner that will transfer the surface-relief profile of the resist into the substrate with fidelity. Another approach to pattern transfer is to use an additive process in which a thin film of optical material is deposited on top of the resist pattern.[33] The thin film of material becomes a coating of the underlying substrate where there are clear areas in the resist. The areas where resist is present have an overcoating of the optical material that will be removed when the underlying resist is removed. This process is known as liftoff. Sequences of masking, deposition, and liftoff can produce a thickness profile with multiple phase levels as in multiple etch processes.

[31] E.-B. Kley, "Continuous profile writing by electron and optical lithography," Microelectron. Eng. **34,** 261–298, 1997.

[32] M. B. Stern, "Pattern transfer for diffractive and refractive microoptics," Microelectron. Eng. **34,** 299–319, 1997.

[33] J. Jahns and S. J. Walker, "Two-dimensional array of diffractive microlenses fabricated by thin film deposition," Appl. Opt. **29,** 931–936, 1990.

17.5.5. Replication

Low-cost production of diffractive optical elements can be accomplished by using the processes described in the preceding subsections to produce a master template that can be replicated by various processes into formable optical materials such as common optical-grade plastics. One of the processes is casting, in which the profile is used as mold for a liquid medium that can be hardened by chemical means and then released from the mold surface. This process has historically been used for replication of extremely precise diffraction gratings. Another approach is injection molding, in which a liquid thermoplastic is forced into a mold under pressure and then cooled. This process is capable of efficient mass production and the precision of the molds is rapidly progressing. The third approach is embossing, in which the master template is applied to a plastic layer under pressure and heat, and the layer conforms to the master surface.[34]

Further Reading

M. Bass, ed., *Handbook of Optics,* McGraw-Hill, 1995, Vol. 2, Chap. 8.

M. W. Farn and W. B. Veldkamp, "Binary optics: Trends and limitations," in *Conference on Binary Optics,* NASA Conf. Pub. 3227, 1993, pp. 19–30.

W. T. Welford, *Aberrations of Optical Systems,* Hilger, 1986, pp. 75–78.

ZEMAX, Version 9.0, Focus Software, San Diego, CA.

CODE V, Optical Research Associates, Pasadena, CA.

SuperOslo, product of Sinclair Optics, Fairport, NY 14450.

W. C. Sweatt, "Mathematical equivalence between a holographic optical element and an ultra-high-index lens," J. Opt. Soc. Am. **69,** 486–487, 1979.

M. W. Farn, "Quantitative comparison of the general Sweatt model and the grating equation," Appl. Opt. **31,** 5312–5316, 1992.

The first two references are general treatments of the subjects. The next four are lens design programs that incorporate the use of diffractive optics in their software. The last two are discussions of the Sweatt high-index model for diffractive lenses.

[34] H. P. Herzig, ed., *Micro-optics,* Taylor & Francis, 1997, pp. 153–177.

18

Aberrations

Virendra N. Mahajan

18.1. Introduction

In this chapter the aberrations of rotationally symmetric imaging systems are described in terms of a power series and Zernike polynomials. Analytical expressions are given for the primary aberrations of simple systems such as a thin lens, a plane-parallel plate, a conic mirror, Schmidt and Bouwers–Maksutov cameras, and a Mangin mirror. Aberrations of two-mirror astronomical telescopes are also given, including those introduced by misalignments of one mirror with respect to the other. Detailed derivations of the expressions given here can be found in the author's *Optical Imaging and Aberrations.*[1] The chapter ends with aberration tolerances based on the Strehl ratio for systems with circular and annular pupils.[2] The chromatic aberrations representing the variation of image distance and height with wavelength are not discussed in this chapter.

18.2. Gaussian Imaging

A Cartesian sign convention is used and light is assumed to be incident upon a system from left to right. When light travels from right to left, as when it is reflected by an odd number of mirrors, then the refractive index and the spacing between two adjacent surfaces are given a negative sign. Any quantities that are numerically negative in the figures are indicated by a parenthetical negative sign (–).

Figure 18.1 illustrates Gaussian imaging of an object. The distances S and S' of the object and its image from the principal points H and H' of a system with a refracting power K and equivalent focal length f_e are related to each other according to

[1] V. N. Mahajan, *Optical Imaging and Aberrations, Part I: Ray Geometrical Optics*, SPIE, 1998.
[2] V. N. Mahajan, *Optical Imaging and Aberrations, Part II: Wave Diffraction Optics*, SPIE, 2001.

$$\frac{n'}{S'} - \frac{n}{S} = \frac{n'}{f'} = -\frac{n}{f} = K = \frac{1}{f_e}, \tag{18.1}$$

where n and n' are the object- and image-space refractive indices and f and f' are the corresponding focal lengths. The transverse, angular, and linear magnifications of the image are given by

$$M_t = h'/h = nS'/n'S = n\beta_0/n'\beta_0', \tag{18.2}$$

$$M_\beta = \beta_0'/\beta_0 = S/S', \tag{18.3}$$

$$M_l = \Delta S'/\Delta S = (n/n')(S'/S)^2 = (n'/n)M_t^2 = M_t/M_\beta, \tag{18.4}$$

respectively. The Lagrange invariant of the system is $nh\beta_0$, i.e.,

$$n'h'\beta_0' = nh\beta_0. \tag{18.5}$$

When an object lies at infinity at a certain angle β from the optical axis, the object-space Lagrange invariant $nh\beta_0$ approaches $-nx_0\beta$, where x_0 is the height at which the axial ray is incident upon the system.

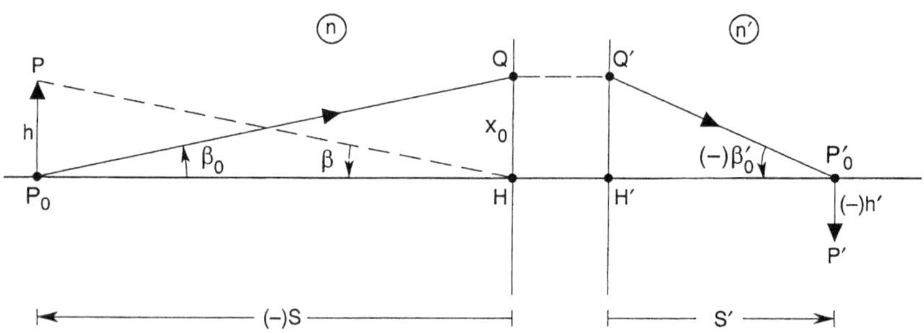

Figure 18.1. Gaussian imaging of an object P_0P of height h lying at a distance S from the object-space principal point H of an imaging system. P'_0P' is the Gaussian image of height h' lying at a distance S' from the image-space principal point H'. The Lagrange invariant of the image is $nh\beta_0$. An axial ray is shown incident at an angle β_0. The corresponding image ray makes an angle β'_0. Numerically negative quantities are indicated with a parenthetical negative sign (–).

18.3. Wave and Ray Aberrations

In Gaussian optics, the rays diverging from a point object and transmitted by an imaging system converge to its Gaussian image point. In a real system, they generally do not converge to a point and their displacements from the Gaussian image point represent the ray aberrations. A spherical wave incident upon the system exits from it only as an approximately spherical wave, and its deviations from being spherical represent the wave aberrations. The wave aberrations of a multielement system are additive but the ray aberrations are not; i.e., the wave aberration of a ray for the entire system is equal to the sum of its wave aberrations for each of the elements, but its displacement in the final image plane cannot be obtained by adding its values in the image planes for the elements.

18.3.1. Relationship between Wave and Ray Aberrations

Figure 18.2 illustrates a Gaussian reference sphere *S* of radius of curvature *R* and an aberrated wave front *W* for a system imaging an off-axis point object. The reference sphere is centered at the Gaussian image point *P′* with coordinates $(x_g, 0)$, and the wave front passes through the center *O* of the exit pupil ExP. A ray such as GR (general ray) from the object passing through the system is shown intersecting the Gaussian image plane at *P″*. The ray aberration is the displacement *P′P″*. The ray intersects the wave front and the reference sphere at points $\bar{Q}$ and *Q*, respectively, and $n_i\bar{Q}Q$ is its wave aberration, where n_i is the refractive index of the image space. It is numerically positive because it travels a longer optical path length compared with the chief ray (passing through *O*) to reach the reference sphere.

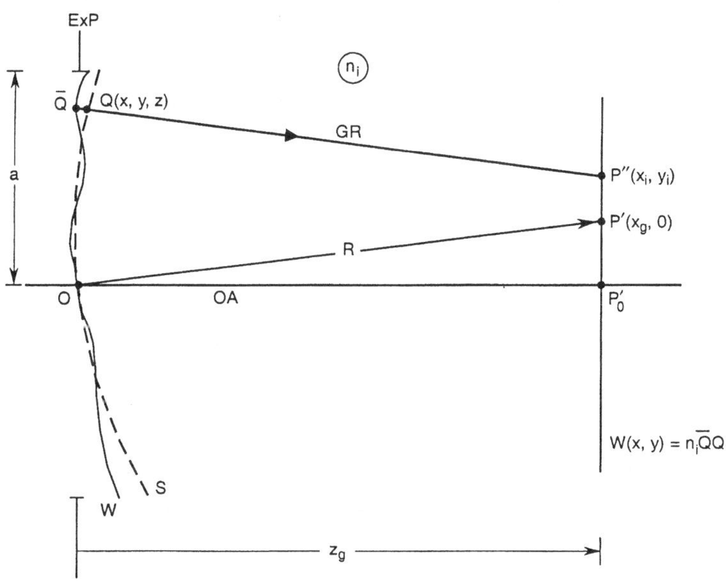

Figure 18.2. Aberrated wave front *W* for an off-axis point object. The reference sphere *S* of radius of curvature *R* is centered at the Gaussian image point *P′*. Both *W* and *S* pass through the center *O* of the exit pupil ExP. GR is a general ray intersecting the Gaussian image plane at the point *P″*. P'_0 is the Gaussian image of an axial point object. The wave aberration $n_i\bar{Q}Q$, as shown, is numerically positive. The ray aberration is *P′P″*. The two aberrations are related to each other according to Eq. (18.6).

The wave aberration $W(x, y)$ at a point Q on the reference sphere with coordinates (x, y, z) and the ray aberration $(x_i - x_g, y_i)$ in the Gaussian image plane are related to each other according to

$$(x_i - x_g, y_i) = \frac{R}{n_i}\left(\frac{\partial W}{\partial x}, \frac{\partial W}{\partial y}\right), \tag{18.6}$$

where (x_i, y_i) represent the coordinates of P''. For systems with narrow fields of view, P' lies close to P'_0 and R may be replaced with the the distance z_g between the pupil and image planes. Of course, for an axial point object, $R = z_g$. For a radially symmetric aberration $W(r)$, a ray of zone r in the exit-pupil plane intersects the image plane at a distance r_i from P', given by

$$r_i = \frac{R}{n_i}\frac{\partial W}{\partial r}. \tag{18.7}$$

18.3.2. Defocus Aberration

When one element (or more) of a system is displaced slightly along its optical axis, the image is displaced longitudinally from a Gaussian image point P_1 to a point P_2 [see Fig. 18.3(a)]. The wave front W is spherical with its center of curvature at P_2 instead of P_1. The defocus wave aberration $n_i Q_2 Q_1$ of a ray of zone r for a system with a circular exit pupil of radius a is given by

$$W(\rho) = B_d\,\rho^2, \tag{18.8}$$

where $\rho = r/a$ is the normalized distance of Q_1 from the axis,

$$B_d = \frac{n_i}{2}\left(\frac{1}{z} - \frac{1}{R}\right) a^2 \simeq -n_i\frac{\Delta R}{8F^2} \tag{18.9}$$

is the peak value of the defocus aberration, and $z \simeq R$. The quantity $\Delta R = z - R$ is the longitudinal defocus, and $F = R/2a$ is the focal ratio of the image-forming light cone. The radius of the geometrical image centered at P_1 is equal to $4B_dF$.

A defocus aberration is also introduced if the system is assembled properly, but the image is observed in a plane other than the Gaussian image plane. If a system forms an aberration-free image at a Gaussian image point P_2 but the image is observed in a defocused plane passing through a point P_1, then the aberration of the wave front at a point Q_1 on the reference sphere is given by Eq. (18.8).

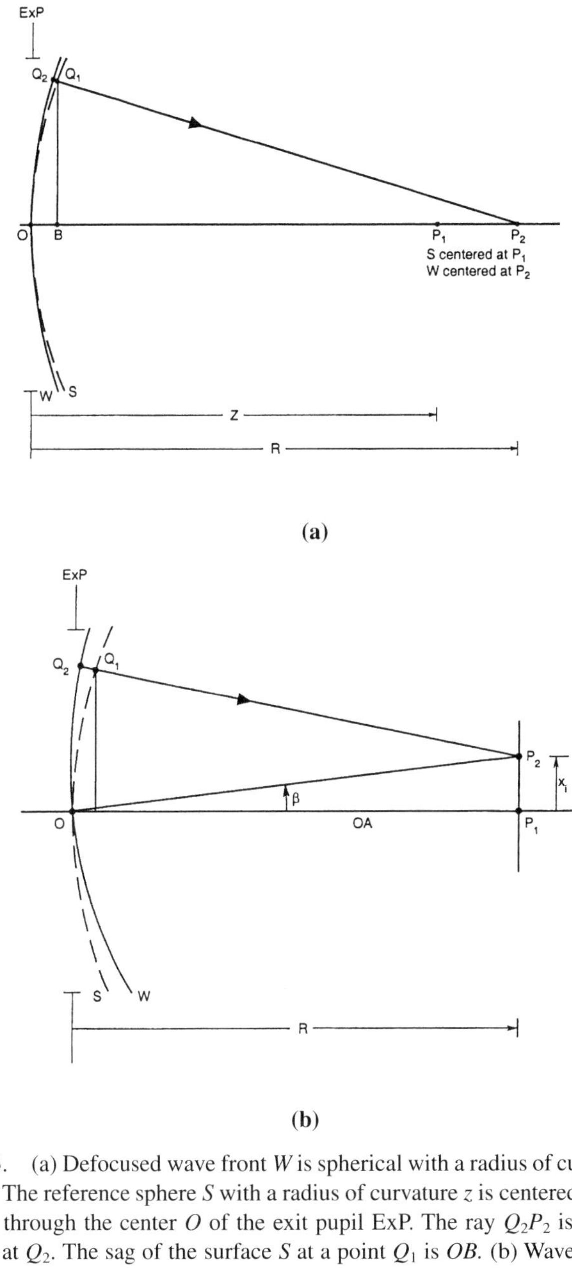

Figure 18.3. (a) Defocused wave front W is spherical with a radius of curvature R centered at P_2. The reference sphere S with a radius of curvature z is centered at P_1. Both W and S pass through the center O of the exit pupil ExP. The ray Q_2P_2 is normal to the wave front at Q_2. The sag of the surface S at a point Q_1 is OB. (b) Wave-front tilt. The spherical wave front W is centered at P_2 and the reference sphere S is centered at P_1. For small values of P_1P_2, the two spherical surfaces are tilted with respect to each other by an angle $\beta = P_1P_2/R$, where R is their radius of curvature.

18.3.3. Wave-Front Tilt Aberration

When one element (or more) of a system is inadvertently tilted and/or decentered slightly, the image P_1 of a point object is displaced tranversally to P_2, as illustrated in Figure 18.3(b). The ray aberration is given by $x_i = R\beta$, where β is a small angle by which the wave front is tilted with respect to the reference sphere. For a system with a circular pupil, the corresponding wave aberration $n_i Q_2 Q_1$ is given by

$$W(\rho, \theta) = B_t \rho \cos\theta, \tag{18.10}$$

where (ρ, θ) are the polar coordinates of the point Q_1 projected onto the plane of the exit pupil and $B_t = n_i a\beta$ is the peak value of the tilt aberration. Thus, if an aberration-free wave front is centered at P_2, then an observation with respect to P_1 as the origin implies that a tilt aberration of $B_t\rho\cos\theta$ has been introduced.

18.4. Aberrations of a Rotationally Symmetric System

18.4.1. Power-Series Expansion

The aberration function of a rotationally symmetric system depends on three rotational invariants, h^2, r^2, and $hr\cos(\theta - \theta_o)$, where (h, θ_o) and (r, θ) are the polar coordinates of the object and the pupil points, respectively. A power-series expansion of the aberration function for a point object along the x_0 axis, so that $\theta_0 = 0$, may be written as

$$W(h'; r, \theta) = \sum_{l=0}^{\infty}\sum_{n=1}^{\infty}\sum_{m=0}^{n} {}_{2l+m}a_{nm} h'^{2l+m} r^n \cos^m\theta, \tag{18.11}$$

where n is a positive integer *not* including zero, h' is the height of the Gaussian image point, ${}_{2l+m}a_{nm}$ are the expansion coefficients, and $n - m \geq 0$ and even. The order i of an aberration term, which is equal to its degree in the object and the pupil coordinates, is given by $i = 2l + m + n$. The order i is always even, and the number of terms N_i of this order is given by

$$N_i = (i + 2)(i + 4)/8. \tag{18.12}$$

This number includes a term with $n = 0 = m$, called piston aberration, although such a term does not constitute an aberration (as it corresponds to the chief ray that, by definition, has a zero aberration associated with it). The aberration terms are referred to as the classical aberrations.

The primary or Seidel aberration function ($i = 4$) consisting of five terms is given by

$$W_P(h'; r, \theta) = {}_0a_{40}r^4 + {}_1a_{31}h'r^3 \cos\theta + {}_2a_{22}h'^2r^2 \cos^2\theta + {}_2a_{20}h'^2r^2 + {}_3a_{11}h'^3 r \cos\theta, \quad (18.13)$$

where ${}_0a_{40}$, ${}_1a_{31}$, ${}_2a_{22}$, ${}_2a_{20}$, and ${}_3a_{11}$ represent the coefficients of spherical aberration, coma, astigmatism, field curvature, and distortion, respectively. The chief ray intersects the Gaussian image plane at a distance ${}_3a_{11}h'^3R$ from the Gaussian image point. The secondary or Schwarzschild aberration function ($i = 6$), consisting of nine terms, is given by

$$\begin{aligned} W_S(h'; r, \theta) = {}& {}_0a_{60}r^6 + {}_1a_{51}h'r^5 \cos\theta + {}_2a_{42}h'^2r^4 \cos^2\theta \\ & + {}_3a_{33}h'^3r^3 \cos^3\theta + {}_2a_{40}h'^2r^4 + {}_3a_{31}h'^3r^3 \cos\theta \quad (18.14) \\ & + {}_4a_{22}h'^4r^2 \cos^2\theta + {}_4a_{20}h'^4r^2 + {}_5a_{11}h'^5 r \cos\theta, \end{aligned}$$

where the first four terms with $l = 0$ are the secondary aberrations [spherical, coma, astigmatism (also called wings), and arrows], and the last five are the lateral aberrations (spherical, coma, astigmatism, field curvature, and distortion). The lateral aberrations are similar to the primary aberrations except for their dependence on the image height h'. The lateral spherical aberration ${}_2a_{40}h'^2r^4$ is also called the oblique spherical aberration.

18.4.2. No Explicit Dependence on Object Coordinates

If the aberration terms that have different dependencies on the object coordinates but the same dependence on pupil coordinates are combined so that there is only one term for each pair of (n, m) values, Eq. (18.11) for the power-series expansion of the aberration function may be written as

$$W(\rho, \theta) = \sum_{n=1}^{\infty} \sum_{m=0}^{n} a_{nm}\rho^n \cos^m\theta, \quad (18.15)$$

where an expansion coefficient

$$a_{nm} = a^n \sum_{l=0}^{\infty} {}_{2l+m}a_{nm}h'^{2l+m} \quad (18.16)$$

represents the peak value of an aberration term. The number of terms through a certain order i in the reduced power-series expansion of the aberration function is also given by Eq. (18.12). This number includes a nonaberration piston term corresponding to $n = 0 = m$. The terms of Eq. (18.11) through a certain order i correspond to those terms of Eq. (18.15) for which $n + m \le i$.

The primary aberration function ($n + m \le 4$) in this simplified form is given by

$$W_P(\rho, \theta) = a_{11}\rho \cos\theta + a_{20}\rho^2 + a_{22}\rho^2 \cos^2\theta + a_{31}\rho^3 \cos\theta + a_{40}\rho^4. \quad (18.17)$$

Distortion and wave-front tilt aberration have the same dependence on the pupil coordinates, but their coefficients depend on the image height differently. The distortion coefficient a_{11} varies with h' as h'^3, but the tilt coefficient B_t in Eq. (18.10) is independent of h'. Similarly, field curvature and defocus wave aberration have the same dependence on pupil coordinates. However, whereas the field curvature coefficient a_{20} varies with h' as h'^2, the defocus coefficient B_d in Eq. (18.8) is independent of h'.

The aberration function through the sixth order ($n + m \le 6$) may be written as

$$\begin{aligned} W(\rho, \theta) = {} & a_{11}\rho \cos\theta + a_{20}\rho^2 + a_{22}\rho^2 \cos^2\theta + a_{31}\rho^3 \cos\theta \\ & + a_{33}\rho^3 \cos^3\theta + a_{40}\rho^4 + a_{42}\rho^4 \cos^2\theta \\ & + a_{51}\rho^5 \cos\theta + a_{60}\rho^6. \end{aligned} \quad (18.18)$$

Because the dependence of a term on the image height h' is contained in the coefficient a_{nm}, the coefficients in Eqs. (18.17) and (18.18) that seem to be the same are actually different from each other. For example, a_{11} is equal to ${}_3a_{11}h'^3a$ in Eq. (18.17) but equal to $({}_3a_{11}h'^3 + {}_5a_{11}h'^5)a$ in Eq. (18.18).

18.4.3. Zernike Polynomial Expansion

The aberration function for a system with a circular exit pupil can also be expanded in terms of a complete set of Zernike circle polynomials $R_n^m(\rho) \cos m\theta$ that are orthogonal over a unit circle in the form[3]

$$W(\rho, \theta) = \sum_{n=0}^{\infty} \sum_{m=0}^{n} c_{nm}[2(n+1)/(1+\delta_{m0})]^{1/2} R_n^m(\rho) \cos m\theta, \quad (18.19)$$

where c_{nm} are the expansion coefficients that depend on the image height h', n and m are positive integers including zero, $n - m \ge 0$ and even, δ_{ij} is a Kronecker delta, and

$$R_n^m(\rho) = \sum_{s=0}^{(n-m)/2} \frac{(-1)^s (n-s)!}{s!\left(\dfrac{n+m}{2} - s\right)!\left(\dfrac{n-m}{2} - s\right)!} \rho^{n-2s} \quad (18.20)$$

[3] V. N. Mahajan, *Optical Imaging and Aberrations, Part II: Wave Diffraction Optics,* SPIE, 2001.

is a polynomial of degree n in ρ containing terms in ρ^n, ρ^{n-2}, ... , and ρ^m. The radial polynomials $R_n^m(\rho)$ are even or odd in ρ, depending on whether n (or m) is even or odd. Also $R_n^n(1) = 1$ and $R_n^n(\rho) = \rho^n$. The orthonormal Zernike polynomials and the names associated with some of them when identified with aberrations are listed in Table 18.1 for $n \leq 8$. The Zernike expansion coefficients are given by

$$c_{nm} = (1/\pi)[2(n+1)(1+\delta_{m0})]^{1/2} \int_0^1 \int_0^{2\pi} W(\rho, \theta) R_n^m(\rho) \cos m\theta \rho d\rho d\theta, \quad (18.21)$$

Table 18.1. Orthonormal Zernike Circle Polynomials and Their Names When Identified with Aberrations

n	m	Polynomial $\left[\frac{2(n+1)}{1+\delta_{m0}}\right]^{1/2} R_n^m(\rho)\cos m\theta$	Aberration Name[a]
0	0	1	Piston
1	1	$2\rho\cos\theta$	Distortion (tilt)
2	0	$\sqrt{3}(2\rho^2-1)$	Field curvature (defocus)
2	2	$\sqrt{6}\rho^2\cos 2\theta$	Primary astigmatism
3	1	$\sqrt{8}(3\rho^3-2\rho)\cos\theta$	Primary coma
3	3	$\sqrt{8}\rho^3\cos 3\theta$	
4	0	$\sqrt{5}(6\rho^4-6\rho^2+1)$	Primary spherical
4	2	$\sqrt{10}(4\rho^4-3\rho^2)\cos 2\theta$	Secondary astigmatism
4	4	$\sqrt{10}\rho^4\cos 4\theta$	
5	1	$\sqrt{12}(10\rho^5-12\rho^3+3\rho)\cos\theta$	Secondary coma
5	3	$\sqrt{12}(5\rho^5-4\rho^3)\cos 3\theta$	
5	5	$\sqrt{12}\rho^5\cos 5\theta$	
6	0	$\sqrt{7}(20\rho^6-30\rho^4+12\rho^2-1)$	Secondary spherical
6	2	$\sqrt{14}(15\rho^6-20\rho^4+6\rho^2)\cos 2\theta$	Tertiary astigmatism
6	4	$\sqrt{14}(6\rho^6-5\rho^4)\cos 4\theta$	
6	6	$\sqrt{14}\rho^6\cos 6\theta$	
7	1	$4(35\rho^7-60\rho^5+30\rho^3-4\rho)\cos\theta$	Tertiary coma
7	3	$4(21\rho^7-30\rho^5+10\rho^3)\cos 3\theta$	
7	5	$4(7\rho^7-6\rho^5)\cos 5\theta$	
7	7	$4\rho^7\cos 7\theta$	
8	0	$3(70\rho^8-140\rho^6+90\rho^4-20\rho^2+1)$	Tertiary spherical

[a] The words "orthonormal Zernike" are to be associated with these names, e.g., *orthonormal Zernike primary astigmatism*.

as may be seen by substitution of W from Eq. (18.19).

The number of Zernike (or orthogonal) aberration terms in the expansion of an aberration function through a certain order n is given by

$$N_n = \left(\frac{n}{2} + 1\right)^2 \quad \text{for even } n, \tag{18.22a}$$

$$= (n+1)(n+3)/4 \quad \text{for odd } n. \tag{18.22b}$$

The mean value of a Zernike aberration is zero. An aberration coefficient c_{nm} (except c_{00}) represents the standard deviation of the corresponding aberration term across the exit pupil. The variance of the aberration function is given by

$$\sigma_w^2 = \langle W^2(\rho, \theta)\rangle - \langle W(\rho, \theta)\rangle^2 = \sum_{n=1}^{\infty} \sum_{m=0}^{n} c_{nm}^2, \tag{18.23}$$

where σ_W is its standard deviation. Unless the mean value of the aberration $\langle W\rangle = 0$, $\sigma_w \neq W_{\text{rms}}$, where $W_{\text{rms}} = \langle W^2\rangle^{1/2}$ is the root-mean-square (rms) value of the aberration.

The polynomials suitable for systems with annular pupils are the Zernike annular polynomials.[4,5] The orthonormal Zernike annular polynomials for $n \leq 6$ are listed in Table 18.2, where ε is the linear obscuration ratio of the pupil. The radial circle polynomials $R_n^m(\rho)$ are replaced with the corresponding annular polynomials $R_n^m(\rho; \varepsilon)$, which obey the orthogonality relation

$$\int_{\varepsilon}^{1} R_n^m(\rho; \varepsilon)\, R_{n'}^m(\rho; \varepsilon)\rho d\rho = \frac{1-\varepsilon^2}{2(n+1)}\, \delta_{nn'}. \tag{18.24}$$

For $m = 0$, the radial polynomials are equal to the Legendre polynomials $P_n(\cdot)$ according to

$$R_{2n}^0(\rho; \varepsilon) = P_n\left[\frac{2(\rho^2-\varepsilon^2)}{1-\varepsilon^2} - 1\right]. \tag{18.25a}$$

They can be obtained from the circle polynomials $R_{2n}^0(\rho)$ by replacing ρ with $[(\rho^2 - \varepsilon^2)/(1-\varepsilon^2)]^{1/2}$. Moreover,

4 V. N. Mahajan, *Optical Imaging and Aberrations, Part II: Wave Diffraction Optics*, SPIE, 2001.

5 V. N. Mahajan, "Zernike annular polynomials for imaging systems with annular pupils," J. Opt. Soc. Am. **71,** 75–78, 1408, 1981; J. Opt. Soc. Am. A **1,** 685, 1984; also "Zernike annular polynomials and optical aberrations of systems with annular pupils," Appl. Opt. **33,** 8125–8127, 1994.

$$R_n^{n-2}(\rho;\varepsilon)=\frac{n\rho^n-(n-1)[(1-\varepsilon^{2n})/(1-\varepsilon^{2(n-1)})]\rho^{n-2}}{\{(1-\varepsilon^2)^{-1}[n^2(1-\varepsilon^{2(n+1)})-(n^2-1)(1-\varepsilon^{2n})^2/(1-\varepsilon^{2(n-1)})]\}^{1/2}},\tag{18.25b}$$

$$R_n^n(\rho;\varepsilon)=\rho^n\bigg/\left(\sum_{i=0}^{n}\varepsilon^{2i}\right)^{1/2},\tag{18.25c}$$

Table 18.2. Orthonormal Zernike Annular Polynomials

n	m	$[2(n+1)/(1+\delta_{m0})]^{1/2}\,R_n^m(\rho;\varepsilon)\cos m\theta$
0	0	1
1	1	$[2/(1+\varepsilon^2)^{1/2}]\rho\cos\theta$
1	1	$\left[\sqrt{3}/(1-\varepsilon^2)\right](2\rho^2-1-\varepsilon^2)$
2	2	$\left[\sqrt{6}/(1+\varepsilon^2+\varepsilon^4)\right]^{1/2}\rho^2\cos 2\theta$
3	1	$\dfrac{\sqrt{8}[3(1+\varepsilon^2)\rho^3-2(1+\varepsilon^2+\varepsilon^4)\rho]\cos\theta}{(1-\varepsilon^2)[(1+\varepsilon^2)(1+4\varepsilon^2+\varepsilon^4)]^{1/2}}$
3	3	$\left[\sqrt{8}/(1+\varepsilon^2+\varepsilon^4+\varepsilon^6)^{1/2}\right]\rho^3\cos 3\theta$
4	0	$\left[\sqrt{5}/(1-\varepsilon^2)^2\right][6\rho^4-6(1+\varepsilon^2)\rho^2+1+4\varepsilon^2+\varepsilon^4]$
4	2	$\dfrac{\sqrt{10}\{4\rho^4-3[(1-\varepsilon^8)/(1-\varepsilon^6)]\rho^2\}\cos 2\theta}{\{(1-\varepsilon^2)^{-1}[16(1-\varepsilon^{10})-15(1-\varepsilon^8)^2/(1-\varepsilon^6)]^{1/2}\}}$
4	4	$\left[\sqrt{10}/(1+\varepsilon^2+\varepsilon^4+\varepsilon^6+\varepsilon^8)^{1/2}\right]\rho^4\cos 4\theta$
5	1	$\dfrac{\sqrt{12}[10(1+4\varepsilon^2+\varepsilon^4)\rho^5-12(1+4\varepsilon^2+4\varepsilon^4+\varepsilon^6)\rho^3+3(1+4\varepsilon^2+10\varepsilon^4+4\varepsilon^6+\varepsilon^8)\rho]\cos\theta}{(1-\varepsilon^2)^2[(1+4\varepsilon^2+\varepsilon^4)(1+9\varepsilon^2+9\varepsilon^4+\varepsilon^6)]^{1/2}}$
5	3	$\dfrac{\sqrt{12}\{5\rho^5-4[(1-\varepsilon^{10})/(1-\varepsilon^8)]\rho^3\}\cos 3\theta}{\{(1-\varepsilon^2)^{-1}[25(1-\varepsilon^{12})-24(1-\varepsilon^{10})^2/(1-\varepsilon^8)]\}^{1/2}}$
5	5	$\left[\sqrt{12}/(1+\varepsilon^2+\varepsilon^4+\varepsilon^6+\varepsilon^8+\varepsilon^{10})\right]^{1/2}\rho^5\cos 5\theta$
6	0	$\sqrt{7}[20\rho^6-30(1+\varepsilon^2)\rho^4+12(1+3\varepsilon^2+\varepsilon^4)\rho^2-(1+9\varepsilon^2+9\varepsilon^4+\varepsilon^6)]/(1-\varepsilon^2)^3$
6	2	$\dfrac{\sqrt{14}[15(1+4\varepsilon^2+10\varepsilon^4+4\varepsilon^6+\varepsilon^8)\rho^6-20(1+4\varepsilon^2+10\varepsilon^4+10\varepsilon^6+4\varepsilon^8+\varepsilon^{10})\rho^4+6(1+4\varepsilon^2+10\varepsilon^4+20\varepsilon^6+10\varepsilon^8+4\varepsilon^{10}+\varepsilon^{12})\rho^2]\cos 2\theta}{(1+\varepsilon^2)^2[(1+4\varepsilon^2+10\varepsilon^4+4\varepsilon^6+\varepsilon^8)(1+9\varepsilon^2+45\varepsilon^4+65\varepsilon^6+45\varepsilon^8+9\varepsilon^{10}+\varepsilon^{12})]^{1/2}}$
6	4	$\dfrac{\sqrt{14}\{6\rho^6-5[(1-\varepsilon^{12})/(1-\varepsilon^{10})]\rho^4\}\cos 4\theta}{\{(1-\varepsilon^2)^{-1}[36(1-\varepsilon^{14})-35(1-\varepsilon^{12})^2/(1-\varepsilon^{10})]\}^{1/2}}$
6	6	$\left[\sqrt{14}/(1+\varepsilon^2+\varepsilon^4+\varepsilon^6+\varepsilon^8+\varepsilon^{10}+\varepsilon^{12})\right]^{1/2}\rho^6\cos 6\theta$

$$R_n^m(1;\varepsilon) = 1, \quad m = 0, \tag{18.26a}$$

$$\neq 1, \quad m \neq 0. \tag{18.26b}$$

An annular aberration function can be expanded in terms of Zernike annular polynomials in a manner similar to that of Eq. (18.19). The expansion coefficients are likewise given by an equation similar to Eq. (18.21), except that the lower limit of the radial integration is ε instead of zero.

18.4.4. Seidel Sums and Coefficients

The primary aberration function of Eq. (18.13) may be written as

$$\begin{aligned} W(r, \theta; h') = {} & a_s r^4 + a_c h' r^3 \cos\theta + a_a h'^2 r^2 \cos^2\theta + a_d h'^2 r^2 \\ & + a_t h'^3 r \cos\theta, \end{aligned} \tag{18.27}$$

where a_s, a_c, a_a, a_d, and a_t represent the coefficients of spherical aberration, coma, astigmatism, field curvature, and distortion, respectively. Similarly, Eq. (18.17) in terms of the peak aberration coefficients may be written as

$$W(\rho, \theta) = A_s\rho^4 + A_c\rho^3 \cos\theta + A_a\rho^2 \cos^2\theta + A_d\rho^2 + A_t\rho \cos\theta. \tag{18.28}$$

The various aberration coefficients are related to each other according to

$$\begin{aligned} & A_s \equiv a_{40} = a_s a^4, \quad A_c \equiv a_{31} = a_c h' a^3, \quad A_a \equiv a_{22} \equiv a_a h'^2 a^2, \\ & A_d \equiv a_{20} = a_d h'^2 a^2, \quad A_t \equiv a_{11} = a_t h'^3 a. \end{aligned} \tag{18.29}$$

Seidel sums and Seidel coefficients have been used in the literature to describe the primary aberration function of an optical system. (See Mouroulis and Macdonald,[6] Born and Wolf,[7] Welford,[8] and Shannon.[9] Note, however that in footnotes 6 and 8, the sign convention for wave aberration is opposite to that given in this chapter.) The aberration at point (r, θ) in the plane of the exit pupil for a point object with a Gaussian image height h' is written in the form

$$\begin{aligned} W(\rho, \theta; h') = {} & (1/8)S_{\mathrm{I}}\rho^4 + (1/2)S_{\mathrm{II}}(h'/h'_{\max})\rho^3 \cos\theta \\ & + (1/2)S_{\mathrm{III}}(h'/h'_{\max})^2\rho^2 \cos^2\theta \\ & + (1/4)\,(S_{\mathrm{III}} + S_{\mathrm{IV}})(h'/h'_{\max})^2\rho^2 \\ & + (1/2)S_{\mathrm{V}}(h'/h'_{\max})^3\rho \cos\theta, \end{aligned} \tag{18.30}$$

[6] P. Mouroulis and J. Macdonald, *Geometrical Optics and Optical Design,* Oxford Univ. Press, 1997.

[7] M. Born and E. Wolf, *Principles of Optics,* 7th ed., Oxford Univ. Press, 1999.

[8] W. T. Welford, *Aberrations of the Symmetrical Optical System,* Academic, 1974.

[9] R. R. Shannon, *The Art and Science of Optical Design,* Cambridge Univ. Press, 1997.

where $h'_{\max}$ is the maximum image height and S_{I}, S_{II}, S_{III}, S_{IV}, and S_{V} are the Seidel sums of the optical system representing contributions of its surfaces to spherical aberration, coma, astigmatism, field curvature, and distortion, respectively. The quantity S_{IV} is the Petzval sum. When a Seidel sum is multiplied by the focal ratio of the image-forming light cone, it gives the spot radius in the Gaussian image plane in the case of S_{I}, sagittal coma in the case of S_{II}, radius of the circle of least confusion in the case of S_{III}, spot radius in the case of $S_{\mathrm{III}} + S_{\mathrm{IV}}$, and image displacement in the case of S_{V}.

If the chief ray from a point object passing through the center of the entrance pupil makes an angle β with the optical axis, then the aberration at a point (r_{en}, θ_{en}) in the plane of this pupil is written in the form

$$\begin{aligned} W(r_{en}, \theta_{en}; \beta) = &-(1/4)Br_{en}^4 + F\beta r_{en}^3 \cos\theta_{en} - C\beta^2 r_{en}^2 \cos^2\theta_{en} \\ &- (1/2)D\beta^2 r_{en}^2 + E\beta^3 r_{en}\cos\theta_{en}, \end{aligned} \tag{18.31}$$

where *B, F, C, D,* and *E* are the Seidel coefficients and correspond to spherical aberration, coma, astigmatism, field curvature, and distortion, respectively. The coordinates (r_{en}, θ_{en}) and (r, θ) of a ray in the planes of the entrance and exit pupils, respectively, are related to each other according to

$$r(\cos\theta, \sin\theta) = mr_{en}(\cos\theta_{en}, \sin\theta_{en}), \tag{18.32}$$

where m is the pupil magnification. If a_{en} represents the radius of the entrance pupil, then

$$\rho = r/a = r_{en}/a_{en}. \tag{18.33}$$

Accordingly, the aberration function in terms of the coordinates of the ray in the plane of the exit pupil may be written as

$$\begin{aligned} W(\rho, \theta; \beta) = &-(1/4)B\,(a/m)^4\,\rho^4 + F\,(a/m)^3\beta\rho^3\cos\theta \\ &- C\,(a/m)^2\beta^2\rho^2\cos^2\theta - (1/2)D(a/m)^2\beta^2\rho^2 \\ &+ E(a/m)\beta^3\rho\cos\theta. \end{aligned} \tag{18.34}$$

The various coefficients are related to each other according to

$$A_s = (1/8)S_{\mathrm{I}} = -(1/4)(a/m)^4\,B = a^4 a_s, \tag{18.35a}$$

$$A_c = (1/2)(h'/h'_{\max})S_{\mathrm{II}} = \beta(a/m)^3 F = h' a^3 a_c, \tag{18.35b}$$

$$A_a = (1/2)\,(h'/h'_{\max})^2\,S_{\mathrm{III}} = -\beta^2(a/m)^2 C = h'^2 a^2 a_a, \tag{18.35c}$$

$$\begin{aligned} A_d = &(1/4)(h'/h'_{\max})^2\,(S_{\mathrm{III}} + S_{\mathrm{IV}}) = -(1/2)\beta^2\,(a/m)^2\,D \\ &= h'^2 a^2 a_d, \end{aligned} \tag{18.35d}$$

$$A_t = (1/2)(h'/h'_{\max})^3 S_{\mathrm{V}} = \beta^3(a/m)E = h'^3 a a_t. \tag{18.35e}$$

18.5. Spot Sizes

The spot sizes for primary aberrations and balanced aberrations for the smallest rms spot radius are given in this section.

18.5.1. Spherical Aberration

Figure 18.4 illustrates a wave front W aberrated by spherical aberration $A_s\rho^4$ with respect to a reference sphere S centered at the Gaussian image point P_0' of an axial point object P_0. The aberration function for determining the ray distribution in a slightly defocused image plane lying at a distance z from the plane of the exit pupil may be written as

$$W(\rho) = A_s\rho^4 + B_d\rho^2, \tag{18.36}$$

where B_d is given by Eq. (18.9). If $n_i = 1$, as is often the case in practice, the rays of zone ρ lie in the defocused image plane on a circle of radius

$$r_i = 8FA_s|\rho^3 + (B_d/2A_s)\rho|. \tag{18.37}$$

For the marginal rays, i.e., for $\rho = 1$, $r_i \rightarrow 0$ if $B_d = -2A_s$. They intersect the axis

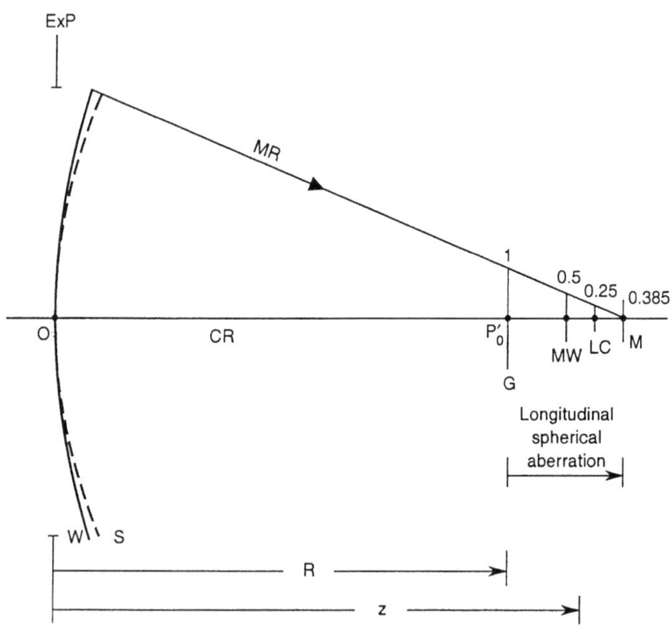

Figure 18.4. Ray spot radii in various image planes for a wave front W aberrated by spherical aberration. G, Gaussian or paraxial, M, marginal, MW, midway, LC, least confusion. The reference sphere S is centered at a Gaussian image point P_0'.

at a distance $\Delta R = 16F^2A_s$ from P_0', called the longitudinal spherical aberration. The image plane MW lying midway between the Gaussian and the marginal planes corresponds to $B_d = -A_s$. The spot radius in this plane is half that in the Gaussian image plane G and corresponds to marginal rays. The image plane that has the smallest spot radius corresponds to that value of B_d that minimizes the maximum value of r_i as ρ varies from 0 to 1 in Eq. (18.37). The spot radius is minimum in a plane LC (least confusion) corresponding to $B_d = -3A_s/2$, i.e., a plane that is 3/4 of the way from the Gaussian image plane to the marginal image plane. The spot radius in this case is 1/4 of the Gaussian spot radius and corresponds to the rays of zone $\rho = 1/2$ and 1. This spot is the circle of least confusion. The spot radii in the various image planes considered here are listed in Table 18.3. The optimum amount of defocus based on diffraction corresponds to the midway plane (see Section 18.12) because, in that case, the balanced aberration giving minimum variance is $A_s(\rho^4 - \rho^2)$, similar to the Zernike polynomial $R_4^0(\rho)$.

Table 18.3. Spot Radius in Units of 8 FA_s for a System with Peak Spherical Aberration A_s

Image Plane	Balancing Defocus B_d/A_s	Spot Radius $r_{i\max}$	rms Radius $r_{i\text{rms}}$
Gaussian	0	1	0.5
Marginal	–2	0.385	0.289
Midway	–1	0.5	0.204
Minimum rms radius	–4/3	1/3	0.167
Least confusion	–3/2	0.25	0.177

18.5.2. Coma

For a system with comatic aberration $A_c\rho^3 \cos\theta$, the tangential marginal rays intersect the Gaussian image plane farthest from the Gaussian image point at a distance $6FA_c$, called tangential coma. The sagittal marginal rays intersect at a distance $2FA_c$, called sagittal coma. The centroid of the image lies at $(2FA_c, 0)$. Its rms radius is minimum and equal to $2\sqrt{2/3}FA_c$ when measured with respect to the centroid; the corresponding balanced aberration is $A_c(\rho^3 - \rho)\cos\theta$ (see Subsection 18.5.4). The balanced aberration with minimum variance is $A_c[\rho^3 - (2/3)\rho]\cos\theta$ (see Subsection 18.12.2), similar to the Zernike polynomial $R_3^1(\rho)\cos\theta$.

18.5.3. Astigmatism and Field Curvature

The aberration function for the image of a point object aberrated by astigmatism and field curvature observed in a defocused plane may be written as

$$W(\rho,\theta) = A_a\rho^2 \cos^2 \theta + A_d\rho^2 + B_d\rho^2, \tag{18.38}$$

where A_a and A_d are both proportional to h'^2 and the balancing defocus coefficient B_d is related to the longitudinal defocus ΔR according to Eq. (18.9). The spot shape and size, including its rms radius, in an image plane defined by the balancing defocus are given in Table 18.4. The two line images are the astigmatic focal lines, and the distance $8F^2A_a$ between them is the longitudinal astigmatism. Halfway between the two line images lies the circle of least (astigmatic) confusion. Substituting $B_d = -(A_a + 2A_d)/2$ into Eq. (18.38) yields the balanced aberration $(A_a/2)\rho^2 \cos 2\theta$, similar to the Zernike polynomial $R_2^2(\rho)\cos 2\theta$. Astigmatism balanced in this manner not only gives the smallest spot but also yields minimum aberration variance (see Section 18.12). This image is referred to as the best image.

Table 18.4. Spot Shape, Size, and rms Radius for Astigmatism A_a and Field Curvature A_d in Various Image Planes Defined by Defocus B_d

Image Plane	Balancing Defocus B_d	Spot Shape and Size[a]	rms Radius $r_{irms}/2FA_a$
General	B_d	Elliptical, $8F(A_a + A_d + B_d) \times 8F(A_d + B_d)$	$\left[1 + 2\dfrac{A_d + B_d}{A_a} + 2\left(\dfrac{A_d + B_d}{A_a}\right)^2\right]^{1/2}$
Gaussian	0	Elliptical, $8F(A_a + A_d) \times 8FA_d$	$\left[1 + 2\dfrac{A_d}{A_a} + 2\left(\dfrac{A_d}{A_a}\right)^2\right]^{1/2}$
Sagittal	$-A_d$	Line along x_i axis, $8FA_a$	1
Tangential	$-(A_d + A_a)$	Line along y_i axis, $8FA_a$	1
Best	$-(A_d + A_a/2)$	Circular, $4FA_a$	$1/\sqrt{2}$

[a] Spot sizes are full major and minor axes of an elliptical image, full length of a line image, and diameter of the best image.

The sagittal, tangential, and best images of a plane object are approximately spherical with radii of curvature given by

$$R_s = h'^2/16F^2A_d = 1/4R^2a_d, \tag{18.39a}$$

$$R_t = h'^2/16F^2(A_a + A_d) = 1/4R^2(a_a + a_d), \tag{18.39b}$$

$$R_b = h'^2/8F^2(A_a + 2A_d) = 1/2R^2(a_a + 2a_d), \tag{18.39c}$$

respectively. Accordingly,

$$3R_s^{-1} - R_t^{-1} = 4R^2(2a_d - a_a) = 2R_p^{-1}, \tag{18.40}$$

where $R_p \equiv R_{ik}$ is the radius of curvature of the Petzval image surface (see Section 18.6). Thus the Petzval surface is three times as far from the tangential surface as it is from the sagittal surface. Moreover, the sagittal surface lies between the tangential and the Petzval surfaces. When astigmatism is zero, the sagittal and the tangential surfaces reduce to the Petzval surface. Moreover,

$$R_b^{-1} = (R_s^{-1} + R_t^{-1})/2, \tag{18.41}$$

i.e., the curvature of the best-image surface is equal to the mean value of the curvatures of the sagittal and the tangential surfaces.

18.5.4. Balanced Aberrations for Minimum rms Spot Radius

A balanced aberration, giving the smallest rms radius, in terms of Zernike circle polynomial $R_n^m(\rho)\cos m\theta$ is given by $B_n^m(\rho)\cos m\theta$, where[10]

$$B_n^m(\rho) = R_n^m(\rho) - R_{n-2}^m(\rho). \tag{18.42}$$

These polynomials are listed in Table 18.5 and may be obtained from the Zernike polynomials given in Table 18.1. If the aberration function is written in terms of these polynomials, e.g.,

$$W(\rho, \theta) = \sum_{n=0}^{\infty} \sum_{m=0}^{n} b_n^m \, B_n^m(\rho)\cos m\theta, \tag{18.43}$$

then the rms radius of the image spot is given by

$$r_{irms} = 2F \left\{ \sum_{n/2=1}^{\infty} 4n(b_n^0)^2 + \sum_{m=1}^{\infty} \left[m(b_m^m)^2 + \sum_{i-1}^{\infty} 2(2i+m)(b_{2i+m}^m)^2 \right] \right\}^{1/2}. \tag{18.44}$$

The polynomials $B_4^0(\rho)$, $B_3^1(\rho)\cos\theta$, and $B_2^2(\rho)\cos 2\theta$ represent balanced spherical aberration, balanced coma, and balanced astigmatism, respectively, giving a minimum rms spot radius.

18.6. Petzval Image Surface

The Petzval image of an off-axis point object is obtained by the sequential application of the Gaussian imaging equation on the auxiliary axis connecting

[10] J. Braat, "Polynomial expansion of severely aberrated wavefronts," J. Opt. Soc. Am. A **4**, 643–650, 1987.

Table 18.5. Balanced Aberration Polynomials for Minimum rms Spot Radius

n	m	$B_n^m(\rho)\cos m\theta$	Balanced Aberration
0	0	1	Piston
1	1	$\rho \cos \theta$	Tilt
2	0	$2(\rho^2 - 1)$	Defocus
2	2	$\rho^2 \cos 2\theta$	Primary astigmatism
3	1	$3(\rho^3 - \rho)\cos \theta$	Primary coma
3	3	$\rho^3 \cos 3\theta$	
4	0	$2(3\rho^4 - 4\rho^2 + 1)$	Primary spherical
4	2	$4(\rho^4 - \rho^2)\cos 2\theta$	Secondary astigmatism
4	4	$\rho^4 \cos 4\theta$	
5	1	$5(2\rho^5 - 3\rho^3 + \rho)\cos \theta$	Secondary coma
5	3	$5(\rho^5 - \rho^3) \cos 3\theta$	
5	5	$\rho^5 \cos 5\theta$	
6	0	$2(10\rho^6 - 18\rho^4 + 9\rho^2 - 1)$	Secondary spherical
6	2	$3(5\rho^6 - 8\rho^4 + 3\rho^2)\cos 2\theta$	Tertiary astigmatism
6	4	$6(\rho^6 - \rho^4) \cos 4\theta$	
6	6	$\rho^6 \cos 6\theta$	
7	1	$7(5\rho^7 - 10\rho^5 + 6\rho^3 - \rho)\cos \theta$	Tertiary coma
7	3	$7(3\rho^7 - 5\rho^5 + 2\rho^3)\cos 3\theta$	
7	5	$7(\rho^7 - \rho^5) \cos 5\theta$	
7	7	$\rho^7 \cos 7\theta$	
8	0	$2(35\rho^8 - 80\rho^6 + 60\rho^4 - 16\rho^2 + 1)$	Tertiary spherical

the object point and the vertex center of curvature of an imaging surface. The radius of curvature R_{ik} of the Petzval image surface for a system of k refracting surfaces with vertex radii of curvature $R_1, R_2, \ldots, R_k$ is given by

$$\frac{1}{R_{ik}} = n_k \sum_{j=1}^{k} \frac{1}{R_j} \left(\frac{1}{n_j} - \frac{1}{n_{j-1}} \right), \tag{18.45}$$

where n_{j-1} and n_j are the refractive indices of the media to the left and the right of the jth surface, respectively. If the jth surface of a system is a reflecting one, then $n_{j-1} = 1$ and $n_j = -1$ when rays are incident upon it from left to right. However, if they are incident from right to left as, for example, on the secondary mirror of a two-mirror system, then $n_{j-1} = -1$ and $n_j = 1$. Similarly, the radius of curvature of the Petzval surface formed by a system consisting of k mirrors with radii of curvature R_j, $j = 1, 2, \ldots, k$, is given by

$$R_{ik}^{-1} = 2(-1)^k \sum_{j=1}^{k} (-1)^j R_j^{-1}. \tag{18.46}$$

18.7. Stop-Shift Equations

Figure 18.5 illustrates an optical imaging system forming the image of an off-axis point object P at P' at a height h' from its optical axis. The peak aberration coefficients of the aberration functions at two exit pupils ExP_1 and ExP_2 (such that they subtend equal angles at the axial image point P'_0, yielding $a_2/a_1 = L_2/L_1$, thereby conserving the amount of light in the image) with respect to their centers O_1 and O_2 are related to each other according to the stop-shift equations:

$$A_{s2} = A_{s1}, \tag{18.47a}$$

$$A_{c2} = A_{c1} - 4\zeta A_{s1}, \tag{18.47b}$$

$$A_{a2} = A_{a1} - 2\zeta A_{c1} + 4\zeta^2 A_{s1}, \tag{18.47c}$$

$$A_{d2} = A_{d1} - \zeta A_{c1} + 2\zeta^2 A_{s1}, \tag{18.47d}$$

$$A_{t2} = A_{t1} - 2\zeta(A_{a1} + A_{d1}) + 3\zeta^2 A_{c1} - 4\zeta^3 A_{s1}, \tag{18.47e}$$

where $\zeta = (L_1 - L_2)h'/a_1L_2$. Equation (18.47a) shows that the peak value of spherical aberration of a system is independent of the position of its aperture

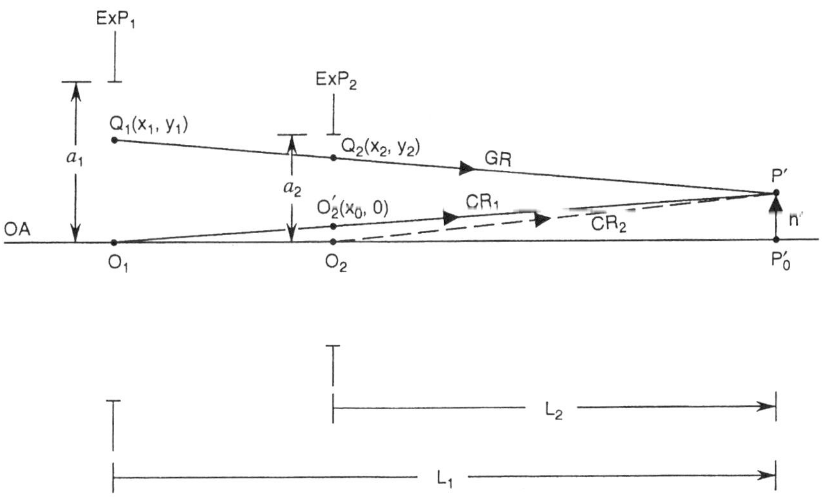

Figure 18.5. Exit pupils ExP_1 and ExP_2 corresponding to two positions of the aperture stop of an optical system forming a Gaussian image P' of an off-axis point object P. The chief rays CR_1 and CR_2 are for the pupils ExP_1 and ExP_2, respectively. The coordinates of the corresponding points Q_1 and Q_2 where a ray intersects the two pupils are related to each other according to $(x_1, y_1) \simeq (L_1/L_2)(x_2 - x_0, y_2)$, where $x_0 = h'(L_1 - L_2)/L_1$ and $(x_0, 0)$ are the coordinates of O'_2 where CR_1 intersects ExP_2.

stop. Equation (18.47b) shows that, if a system is free of spherical aberration, then the peak value of its coma is independent of the position of its aperture stop. It also shows that when the spherical aberration is not zero, its coma can be made zero if an aperture stop position is chosen that corresponds to

$$\zeta = A_{c1}/4A_{s1} \quad \text{or} \quad L_1/L_2 = 1 + a_{c1}/4a_{s1}. \tag{18.48}$$

The corresponding value of the astigmatism coefficient given by $A_{a2} = A_{a1} - A_{c1}^2/4A_{s1}$ is minimum or maximum, depending on whether A_{s1} is positive or negative, as may be seen from Eq. (18.47c).

If a system is aplanatic (i.e., free of spherical aberration and coma) for a certain position of the aperture stop, it is aplanatic for any position of the aperture stop. Moreover, for such a system, Eqs. (18.47c) and (18.47d) show that the peak values of its astigmatism and field curvature are also independent of the position of its aperture stop. However, the peak value of its distortion changes with a change in the position of its aperture stop unless the sum of the peak values of its astigmatism and field curvature is zero, as may be seen from Eq. (18.47e). This equation shows that the peak value of distortion depends on the position of the aperture stop unless spherical aberration, coma, and the sum of the peak values of astigmatism and field curvature are each equal to zero, or the sum of the three terms depending on ζ on its right-hand side is zero. An example of the latter is a spherical mirror with a collocated aperture stop compared with a remote stop located at its center of curvature. As expected, the coefficient of Petzval field curvature, which depends on $2A_d - A_a$, does not change when the position of the aperture stop is changed.

For an anastigmatic system (i.e., free of spherical aberration, coma, and astigmatism), the field curvature aberration is due to only the Petzval curvature, which, in turn, is independent of the position of its aperture stop. Its distortion depends on the position of its aperture stop unless its Petzval curvature is also zero. Of course, if all of the primary aberrations of a system are zero for a certain position of its aperture stop, then they are zero for any other position as well.

18.8. Aberrations of Simple Refracting Systems

The quality of an image formed by a system depends on its wave aberrations for the point object under consideration. In this section, expressions for the primary aberrations of a thin lens and a plane-parallel plate are given.

18.8.1. Thin Lens

Figure 18.6 illustrates a thin lens of refractive index n consisting of two spherical surfaces of radii of curvature R_1 and R_2 with a collocated aperture stop so

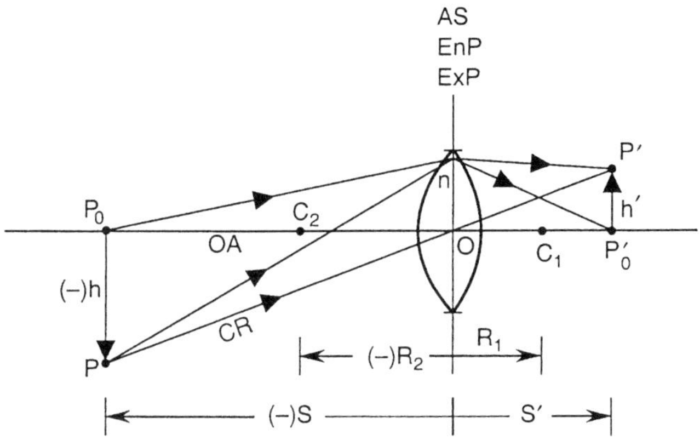

Figure 18.6. Imaging by a thin lens of refractive index n and two spherical surfaces of radii of curvature R_1 and R_2 with their centers of curvature at C_1 and C_2. $P_0'P'$ is the Gaussian image of the object P_0P. The object and image heights are h and h', respectively. The aperture stop AS, entrance pupil EnP, and exit pupil ExP are all located at the lens.

that the entrance and the exit pupils are also located there. The object and the image distances S and S' for a lens of focal length f' are related to each other by the Gaussian imaging equation

$$\frac{1}{S'} - \frac{1}{S} = \frac{1}{f'} = (n-1)\left(\frac{1}{R_1} - \frac{1}{R_2}\right). \tag{18.49}$$

The object and image heights h and h' are related to each other by the magnification equation

$$M = h'/h = S'/S. \tag{18.50}$$

The aberration coefficients of the lens are given by

$$a_s = -\frac{1}{32n(n-1)f'^3}\left[\frac{n^3}{n-1} + (3n+2)(n-1)p^2 + \frac{n+2}{n-1}q^2 + 4(n+1)pq\right], \tag{18.51a}$$

$$a_c = -\frac{1}{4nf'^2S'}\left[(2n+1)p + \frac{n+1}{n-1}q\right], \tag{18.51b}$$

$$a_a = -1/2f'S'^2, \tag{18.51c}$$

$$a_d = -(n+1)/4nf'S'^2, \tag{18.51d}$$

$$a_t = 0,$$

where

$$p = -1 - 2f'/S = 1 - 2f'/S', \tag{18.52}$$

$$q = \frac{R_2 + R_1}{R_2 - R_1} \tag{18.53}$$

are the position and the shape factors of the lens, respectively. Distortion is generally nonzero when the aperture stop does not lie at the lens. The radius of curvature R_p of the Petzval image surface is given by $-nf'$. The spherical aberration of a thin lens (with spherical surfaces) cannot be zero when both the object and its image are real. A doublet designed to correct for spherical aberration can at the same time be corrected for coma.[11,12]

18.8.1.1. Aplanatic Lens

The aplanatic conjugate points of a thin lens are given by

$$(S, S') = [(n-1)/nf', (n-1)f']. \tag{18.54}$$

The magnification of the image is $1/n$. An aplanatic lens can only make a diverging beam less divergent or a converging beam more convergent. It cannot, for example, convert a diverging beam into a converging beam.

18.8.1.2. Concentric Lens

A concentric lens behaves as a negative thin lens placed at the common center of curvature of its two surfaces with a focal length that is n times the focal length of a thin lens of the same refractive index and surfaces with the same radii of curvature. It introduces only astigmatism and distortion when placed in a converging beam so that it is concentric with the beam focus. Neither surface of the lens introduces any spherical aberration, coma, or axial color. It can be used for controlling the tangential field curvature of an imaging system.[12]

18.8.1.3. Field-Flattening Lens

A field-flattening lens, typically planoconvex or planoconcave, is placed in the image plane of an optical system to flatten the curvature of the image surface. An example of such a lens is a planoconvex lens placed at the image plane of a Schmidt camera to flatten the Petzval curvature of its spherical mirror. Because the field-flattening lens is placed at the image plane of an optical system, the distances of the object and the image for it are (practically) zero. Consequently the expressions for the aberrations of a thin lens with a collocated aperture

[11] V. N. Mahajan, *Optical Imaging and Aberrations, Part I: Ray Geometrical Optics,* SPIE, 1998.
[12] V. N. Mahajan, *Aberration Theory Made Simple,* Vol. TT06 of SPIE Tutorial Text Series, SPIE, 1991.

stop cannot be used. A stop placed in the image plane cannot act as an aperture stop because it cannot control the focused imaging beams. Because the object distance for the two lens surfaces is zero, each surface is anastigmatic; i.e., neither surface introduces any spherical aberration, coma, or astigmatism. Hence the field-flattening lens is also anastigmatic.

The Petzval curvature of the lens, which is independent of the object location, is used to cancel or flatten the curvature of an image surface produced by an imaging system. Thus an image surface of radius of curvature R_i is flattened if a lens of refractive index n and focal length $f' = R_i/n$ is placed at the image plane. However, a field flattener also introduces distortion. If s_1 is the distance of an image of height h' from the exit pupil of a system, a lens with surfaces of radii of curvature R_1 and R_2 displaces it by an amount

$$\Delta h' = \frac{1}{2f'}\left(\frac{1}{nR_1} + \frac{1}{R_2} - \frac{1}{ns_1}\right)h'^3. \tag{18.55}$$

18.8.1.4. Thin Lenses in Contact

The radius of curvature of the Petzval surface of a system consisting of a series of m thin lenses of refractive indices n_j and focal lengths f_j' in contact with each other, where $j = 1, 2, \ldots, m$, is given by

$$1/R_p = \sum_{j=1}^{m} 1/n_j f_j'. \tag{18.56}$$

The peak value of its astigmatism for an object height h at a distance S is given by

$$A_a = -(ah/S)^2/2f', \tag{18.57}$$

where a is the radius of the aperture stop located at the lens system and f' is the focal length of the system given by

$$1/f' = \sum_{j=1}^{m} 1/f_j'. \tag{18.58}$$

18.8.1.5. Lens with Conic Surfaces

Spherical aberration of a thin lens with conic surfaces of eccentricities e_1 and e_2 is given by

$$a_{sc} = a_s + \frac{n-1}{8}\left(\frac{e_1^2}{R_1^3} - \frac{e_2^2}{R_2^3}\right). \tag{18.59}$$

It can be made zero by an appropriate choice of the eccentricities.

18.8.2. Plane-Parallel Plate

Because the thickness of a thin lens is neglected by its definition, the imaging and aberration equations for a plane-parallel plate cannot be obtained from those for such a lens by letting the radii of curvature of its two surfaces approach infinity. Figure 18.7(a) illustrates a plate placed in the path of a converging beam displacing its focus from P_1 to P_2 besides introducing aberrations. Figure 18.7(b) shows a right-angle reflecting prism as an example of a plane-parallel plate. It is used in optical systems to deviate the path of a beam by 90°. Its diagonal face acts like a mirror because the rays incident upon it undergo a total internal reflection. The "unfolded" path of the rays, called a tunnel diagram, illustrates that the prism *ABC* is equivalent to a plane-parallel plate *ABCD* in terms of their optical path lengths.

Figure 18.7(c) illustrates a circular plane-parallel plate of radius *a*, thickness *t*, and refractive index *n* forming the image of a point object *P* lying at a distance *S* from its front surface and at a height *h* from its axis. The plate displaces the image by an amount

$$PP'' = t(1 - 1/n), \tag{18.60}$$

which is independent of the object distance *S*. For a plate with an aperture stop AS located at its front surface, its entrance pupil EnP is also located there. Its exit pupil ExP, which is the image of EnP by the plate, lies to the right of the front surface at a distance given by Eq. (18.60). The aberration function at the exit pupil is given by

$$W(r, \theta; h) = \frac{(n^2 - 1)t}{8n^3 S^4}(r^4 - 4hr^3 \cos\theta + 4h^2r^2 \cos^2\theta + 2h^2r^2 - 4h^3 r \cos\theta). \tag{18.61}$$

The aberration increases linearly with the plate thickness *t*, but reduces to zero for a collimated incident beam as may be seen by letting $S \rightarrow -\infty$.

18.9. Aberrations of Simple Reflecting Systems

18.9.1. Imaging Relations

Figure 18.8 illustrates a spherical reflecting surface (mirror) of radius of curvature *R* imaging a point object. The line joining the vertex V_0 and the center of curvature *C* of the surface defines its optical axis. The distances *S* of a point object P_0 and S' of its Gaussian image P_0' are related to each other according to

$$\frac{1}{S'} + \frac{1}{S} = \frac{2}{R} = \frac{1}{f'}, \tag{18.62}$$

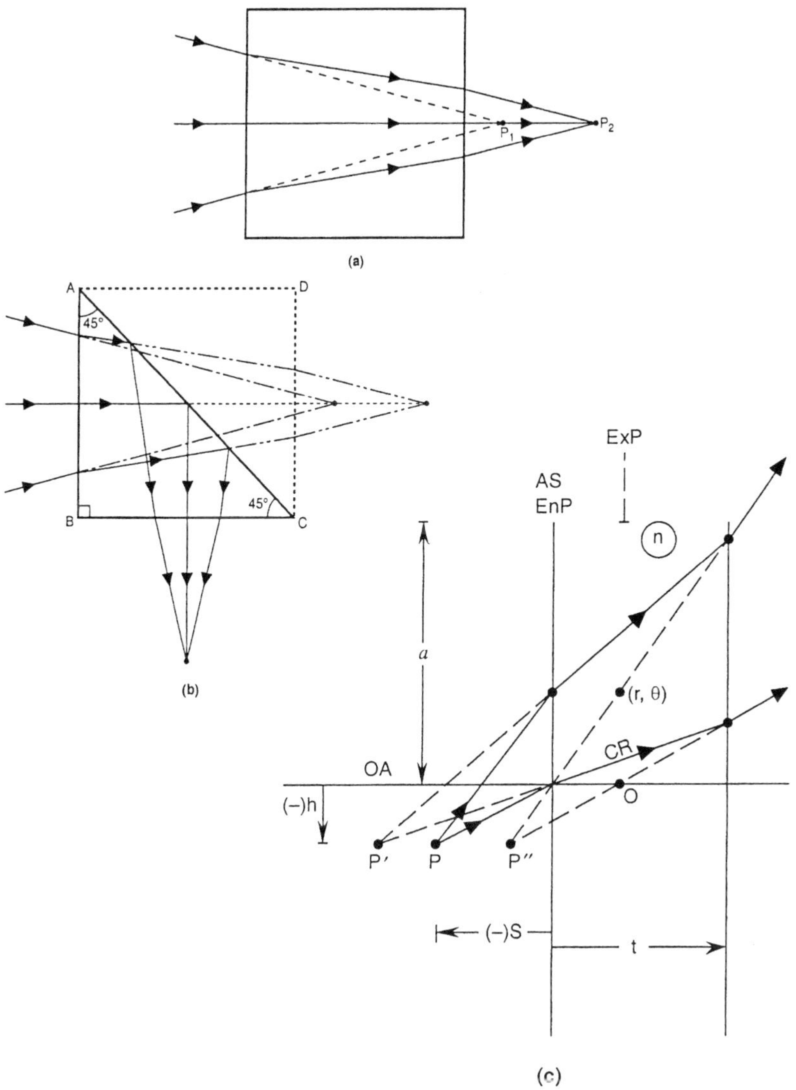

Figure 18.7. (a) Plane-parallel plate placed in the path of a converging beam of light. Rays incident upon the plate converging toward P_1 converge toward P_2 after refraction by it. (b) A right-angle reflecting prism placed in the path of a converging beam. The optical path lengths of the rays for the prism are equivalent to those for a plane-parallel plate, where the virtual portion *ADC* of the equivalent plate is obtained by a reflection of its real portion *ABC* by the reflecting surface *AC*. (c) Imaging of a point object *P* by a plane-parallel plate of refractive index *n* and thickness *t*. *P*′ is the image of *P* formed by the first surface, and *P*″ is the image of *P*′ formed by the second surface of the plate. The aperture stop AS and therefore the entrance pupil EnP of the plate are located at the first surface. The exit pupil ExP, which is the image of the EnP by the plate, lies at a distance $t(1 - 1/n)$ from it. The chief ray CR and a general ray appearing to intersect the exit pupil at a point (r, θ) are shown.

where $f' = R/2$ is the focal length of the mirror. The quantities R, S, S', and f' are numerically negative in the figure. If an off-axis point object P is located at a height h, its Gaussian image P' lies at a height h' given by

$$M = h'/h = -S'/S, \tag{18.63}$$

where M is the magnification of the image.

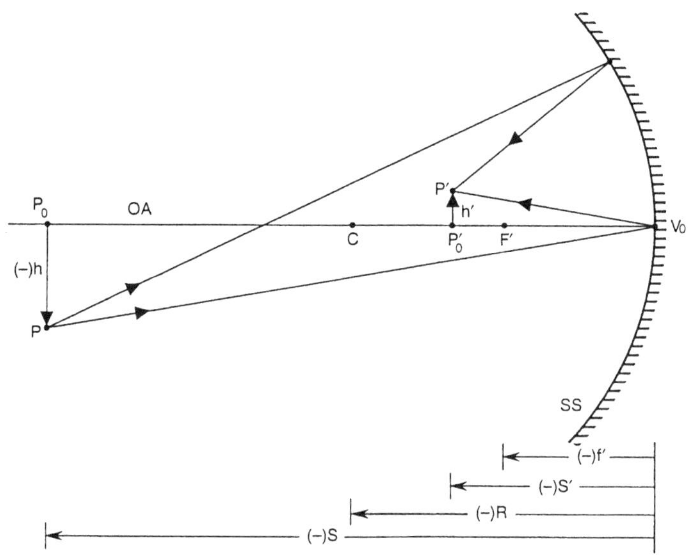

Figure 18.8. Imaging by a spherical reflecting surface SS of radius of curvature R, center of curvature C, vertex V_0, and focal point F'. The axial point object P_0 and its Gaussian image P_0' lie at distances S and S' from the vertex, respectively. The point object P and its Gaussian image P' lie at heights h and h', respectively, from the optical axis OA. The quantities f', R, S, S', and h are all numerically negative. GR is a general ray, and CR is the chief ray.

18.9.2. Conic Mirror

The primary aberration function of a conic mirror of eccentricity e and vertex radius of curvature R with its aperture stop located at its geometrical or conic focus near its vertex, i.e., at a distance $R/(1+e)$, imaging an object lying at infinity at an angle β from the axis of the mirror, is given by

$$W_c(r_{\text{en}}, \theta_{\text{en}}; \beta) = \frac{1-e^2}{4R^3} r_{\text{en}}^4 - \frac{e\beta}{R^2} r_{\text{en}}^3 \cos\theta_{\text{en}} - \frac{\beta^2}{2R} r_{\text{en}}^2 + \frac{e\beta^3}{(1+e)^2} r_{\text{en}} \cos\theta_{\text{en}}, \tag{18.64}$$

where (r_{en}, θ_{en}) are the coordinates of a point in the plane of the entrance pupil (which is the same as the aperture stop). Astigmatism is zero for any conic mirror, and coma and distortion are zero for a spherical mirror ($e = 0$).

If the aperture stop is collocated with the mirror so that the entrance and exit pupils are also collocated, then its primary aberration function in the plane of the exit pupil is given by

$$W_c(r, \theta; \beta) = \frac{1-e^2}{4R^3} r^4 + \frac{1}{R^2} \beta r^3 \cos\theta + \frac{1}{R} \beta^2 r^2 \cos^2\theta. \quad (18.65)$$

When the aperture stop is located at the mirror, its primary aberrations, other than spherical aberration, are identical with those for a spherical mirror. For a paraboloidal ($e = 1$) mirror, spherical aberration is zero regardless of the location of its aperture stop. Therefore the peak value of its coma is also independent of the position of its aperture stop. Its astigmatism vanishes if it is placed in the focal plane of the mirror. Of course, it forms the image of an axial point object ($\beta = 0$) aberration free.

18.9.3. Aspheric Plate and a Conic Mirror

The spherical aberration and coma of a conic mirror with a collocated aperture stop can be compensated for if a plate of refractive index n and thickness

$$t(r) = t_0 - \frac{1-e^2}{4(n-1)R^3} r^4 \quad (18.66)$$

is placed at a distance $d_o = R/(1 - e^2)$ from the mirror [see Figure 18.9(a)], where t_0 is a constant. Because d_o is numerically negative, e and d_o must satisfy the inequalities $e < 1$ and $|d_o| > |R|$. The aberration function of the combined system is given by

$$W_s(r, \theta; \beta) = -\frac{e^2}{R(1-e^2)} \beta^2 r^2 \cos^2\theta - \frac{1}{2R(1-e^2)} \beta^2 r^2 - \frac{1}{(1-e^2)^2} \beta^3 r \cos\theta. \quad (18.67)$$

For a spherical mirror ($e = 0$), the plate lies at its center of curvature and compensates for its astigmatism also.

Spherical aberration and coma of a mirror can also be compensated for if an aspheric plate of thickness

$$t(r) = t_0 - \frac{(1+e^2)^4 r^4}{4R^3(n-1)(1-e^2)^3} \quad (18.68)$$

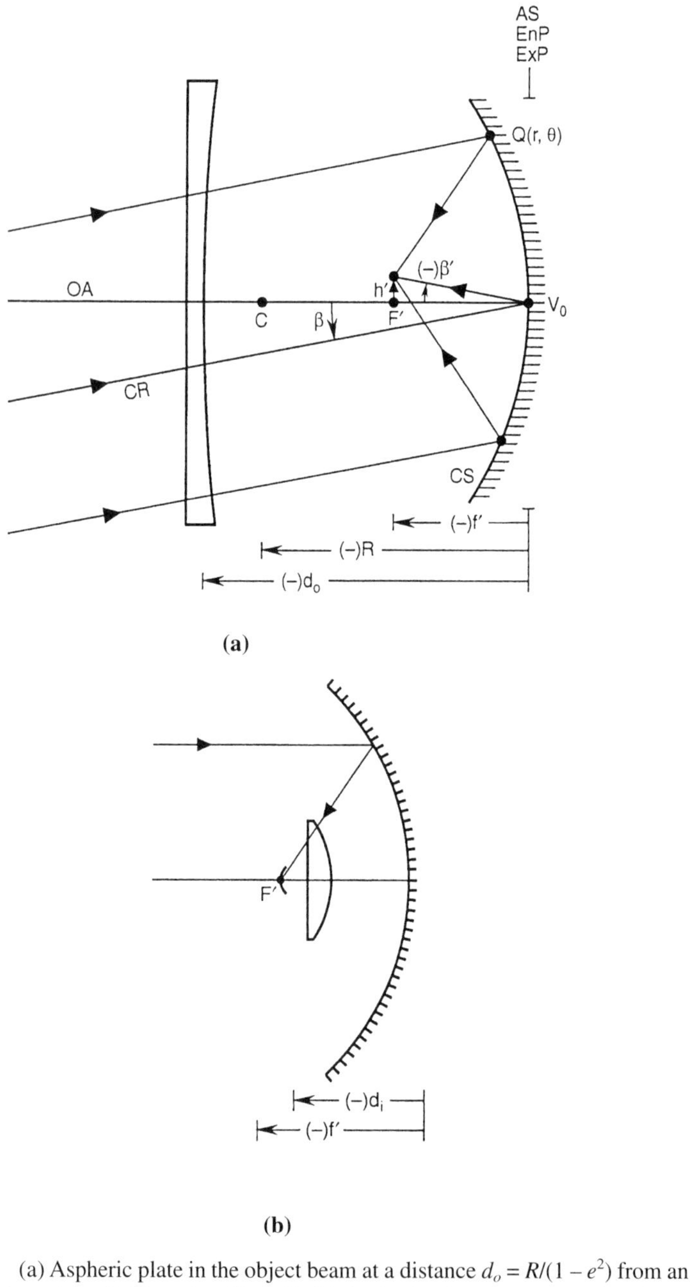

Figure 18.9. (a) Aspheric plate in the object beam at a distance $d_o = R/(1 - e^2)$ from an ellipsoidal mirror ($e < 1$). Note that $\beta' = -\beta$. (b) Aspheric plate in a converging beam formed by a hyperboloidal ($e > 1$) mirror at a distance $d_i = R/(1 + e^2)$ from it.

is placed at a distance $d_i = R/(1 + e^2)$ from a mirror in a converging image beam, as illustrated in Figure 18.9(b). The aberration function of the combined system is again given by Eq. (18.67). Because $|d_i|$ must be less than $|f'|$ for a plate lying in a converging beam, both spherical aberration and coma of an ellipsoidal ($e < 1$) mirror cannot be corrected with a single plate. For $e > 1$, the sign of r^4 term in Eq. (18.68) is opposite to that in Eq. (18.66), which is valid for $e < 1$. Moreover, as e increases (beyond a value of unity), $|d_i|$ decreases, and the aberration coefficients decrease numerically. Hence the size of the plate and/or the field of view of the system for a given image quality increase. The effect of the field curvature term as an aberration is reduced to zero if the image is observed on a spherical surface curved as shown in Figure 18.9(b) with a radius of curvature $R(1 - e^2)/2$.

18.9.4. Schmidt Camera

The aberration function for an object lying at infinity at an angle β from the axis of a spherical mirror with an aperture stop located at its center of curvature C, as illustrated in Fig. 18.10(a), is given by

$$W_{\text{sm}}(r;\beta) = \frac{r^4}{4R^3} - \frac{\beta^2 r^2}{2R}. \tag{18.69}$$

In a Schmidt camera, a transparent plate is placed at C. For a plate of refractive index n, its thickness for eliminating spherical aberration while minimizing the spherochromatism introduced by it is given by

$$t(r) = t_0 - \frac{1}{4(n-1)R^3}\left(r^4 - \frac{3}{2}a^2r^2\right). \tag{18.70}$$

A ray incident normal to the plate at a height of $r = \sqrt{3}a/2$, called the neutral zone, passes through it undeviated, as indicated in Fig. 18.10(b). The spherical Petzval image surface can be flattened by placing a thin lens of positive focal length near it, i.e., by canceling the Petzval curvature of the mirror with that of the lens. For a planoconvex field-flattening lens of refractive index n, the radius of curvature of its curved surface is given by $-(n - 1)R/2n$. Its planar surface is placed against the image or film, as illustrated in Fig. 18.10(b).

18.9.5. Bouwers–Maksutov Camera

In a Bouwers–Maksutov camera,[13,14] a thin spherical meniscus corrector lens is used to compensate for the spherical aberration of a spherical mirror, as il-

[13] A. Bouwers, *Achievements in Optics,* Elsevier, 1949, Chap. 1.
[14] D. D. Maksutov, "New catadioptric meniscus systems," J. Opt. Soc. Am. **34,** 270–284, 1944.

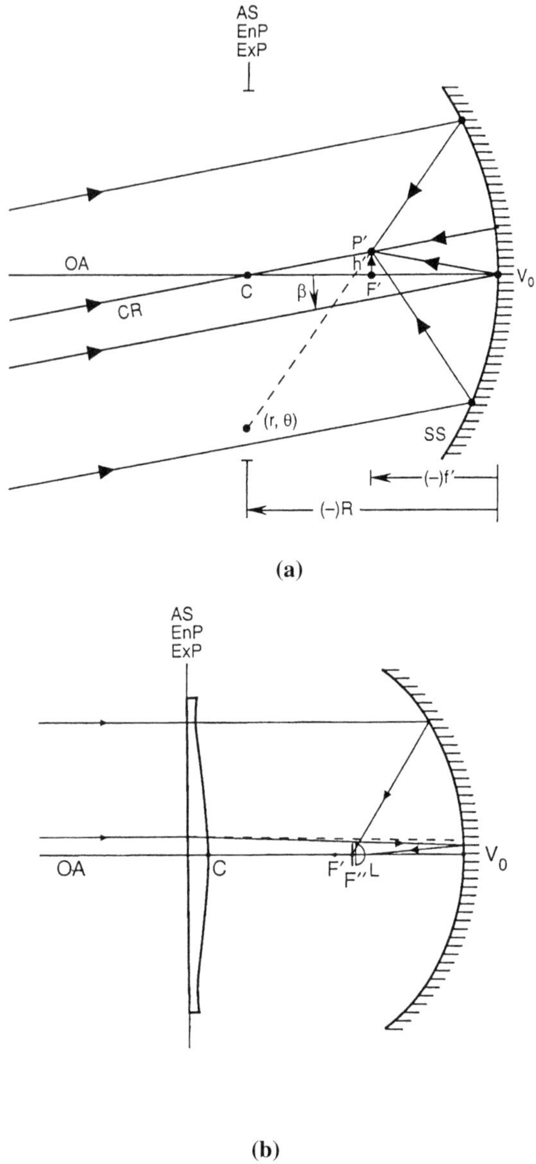

Figure 18.10. (a) Spherical mirror with its aperture stop and therefore its entrance pupil lying at its center of curvature *C*. Its exit pupil also lies there with a magnification of −1. (b) Schmidt camera consisting of a spherical mirror and a transparent plate placed at its center of curvature *C*. The dashed lines indicate the path of a ray in the absence of the Schmidt plate. All rays passing through the plate and reflected by the concave spherical mirror are focused at *F″*, where the ray passing through the neutral zone of the plate is focused by the mirror. A planoconvex lens *L* flattens the spherical Petzval image surface.

lustrated in Fig. 18.11. There is no unique position for the placement of the lens as long as it is concentric with the mirror. Its focal length is given by $f_l' = -[nR_1R_2/t(n-1)]$, where n is its refractive index, R_1 and R_2 are the radii of curvature of its surfaces, and $t = R_1 - R_2$ is its thickness. The spherical aberration of the lens–mirror system for an object lying at infinity at angle β from its axis is given by

$$a_{ss} = \frac{(n-1)(n+2)}{8n^2} \frac{t}{R_1^3 R_2} + \frac{1}{4R^3}. \tag{18.71}$$

For a given value of the radius of curvature R of the mirror, the spherical aberration given by Eq. (18.71) is made zero by a suitable choice of R_1, t, and n. The image of an extended object formed on a surface concentric with the mirror is free of primary aberrations. As in the case of a Schmidt camera, here too the spherical aberration is zero only at a wavelength corresponding to the refractive index n.

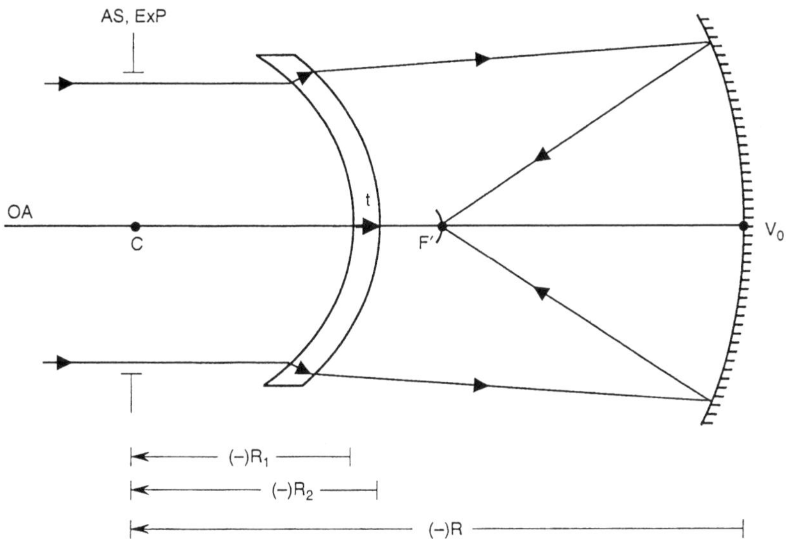

Figure 18.11. Bouwers–Maksutov camera showing a concave spherical mirror and a concentric meniscus corrector lens. F' is the Gaussian focus of the mirror.

18.9.6. Mangin Mirror

A Mangin mirror consists of a thin negative meniscus lens with a silvered back surface, as illustrated in Fig. 18.12. If R_1 and R_2 are the radii of curvature of the lens surfaces and n is its refractive index, its focal length is given by

$$1/f' = (2n/R_2) - 2(n-1)/R_1. \tag{18.72}$$

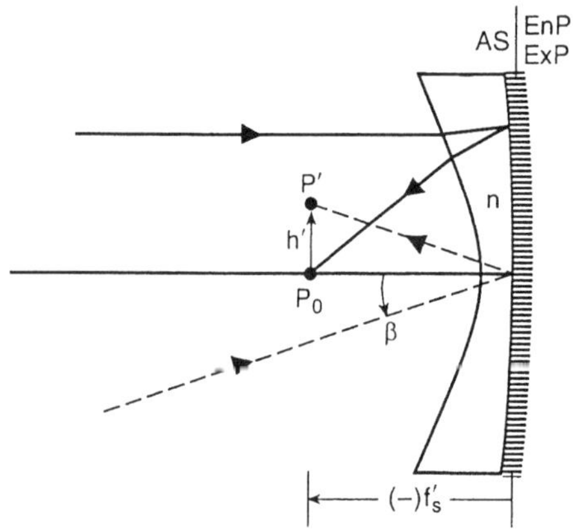

Figure 18.12. Mangin mirror consisting of a thin negative meniscus lens with a silvered back surface.

Its longitudinal chromatic aberration (or axial color) is given by

$$\delta S' = [S'^2(2f'_s - R_1)/nR_1f'_s]\delta n, \tag{18.73}$$

where S' is the image distance. For an aperture stop located at the mirror, its lateral color is zero. Its primary aberration coefficients for imaging an object lying at infinity are given by

$$a_s = -\frac{n-1}{4n^2}\left[\frac{1}{R_1^3} - \frac{n+3}{2R_1^2f'_s} + \frac{4n+5}{4R_1f'^2_s} - \frac{4n^2-3}{8(n-1)f'^3_s}\right], \tag{18.74a}$$

$$a_c = [2(n^2-1)f'_s - (2n^2-1)R_1]/4n^2R_1f'^3_s, \tag{18.74b}$$

$$a_a = 1/2f'^3_s, \tag{18.74c}$$

$$a_d = (n^2-1)(R_1 - 2f'_s)/4n^2R_1f'^3_s, \tag{18.74d}$$

and $a_t = 0$. Its spherical aberration is made zero by solving the cubic equation in R_1.[15] For example, it is zero for $R_1 = f'_s$ and $n = 1.5$. The radius of curvature of its Petzval surface is given by

$$R_p^{-1} = [R_1 + 2(n^2-1)f'_s]/n^2R_1f'_s. \tag{18.75}$$

For primary aberrations of the Mangin mirror for finite conjugates, the reader may refer to a paper by Gelles.[16]

[15] M. J. Reidl, "The Mangin mirror and its primary aberrations," Appl. Opt. **13,** 1690–1694, 1974.

[16] R. Gelles, "Aberrations of the Mangin mirror," Opt. Eng. **24,** 322–325, 1985.

18.10. Two-Mirror Astronomical Telescopes

Expressions for the aberrations of two-mirror astronomical telescopes are given in this section.

18.10.1. Gaussian Imaging and System Aberrations

Figure 18.13 illustrates a two-mirror astronomical telescope[17–21] forming the image of an object lying at infinity. The axial image formation is illustrated by the solid-line rays and the off-axis by the dashed-line rays. The focal length of the telescope is given by

$$f' = f_1' f_2' / (f_1' - f_2' - t), \tag{18.76}$$

where f_1' and f_2' are the focal lengths of the mirrors and t is the distance of the secondary mirror M_2 from the primary mirror M_1. For an object lying at an angle β from the optical axis of the system, the magnification M_2 of the image formed by the secondary mirror, called the secondary magnification, is given by

$$M_2 = h_2' / h_1' = -f' / f_1', \tag{18.77}$$

where $h_1' = -\beta f_1'$ and $h_2' = \beta f'$ are the heights of the images formed by the primary and secondary mirrors, respectively. The final image lies at a distance $S_2' = f'(f_1' - t)/f_1'$ from the secondary mirror.

The aperture stop of the telescope lies at its primary mirror M_1. Consequently its entrance pupil EnP also lies at M_1. Its exit pupil ExP is the image of M_1 by the mirror M_2. The pupil magnification is given by

$$m_2 = f'(f_1' - t)/f_1'(f' + t). \tag{18.78}$$

If D_1 is the diameter of the primary mirror, then the diameter of the exit pupil is $D_{ex} = |m_2| D_1$. The focal ratio (or the F-number) of the image-forming light cone is given by

$$F = |f'| / D_1. \tag{18.79}$$

[17] V. N. Mahajan, *Optical Imaging and Aberrations, Part I: Ray Geometrical Optics,* SPIE, 1998.

[18] S. C. B. Gascoigne, "Recent advances in astronomical optics," Appl. Opt. **12,** 1419–1429, 1973; also, "Some recent advances in the optics of large telescopes," Q. J. R. Astronom. Soc. **9,** 98–115, 1968.

[19] W. B. Wetherell and M. P. Rimmer, "General analysis of aplanatic Cassegrain, Gregorian, and Schwarzchild telescopes," Appl. Opt. **11,** 2817–2832, 1972.

[20] D. J. Schroeder, *Astronomical Optics,* 2nd ed., Academic, 2000.

[21] R. N. Wilson, *Reflecting Telescope Optics* I, Springer, 1996, Sections 3.7 and 3.8.

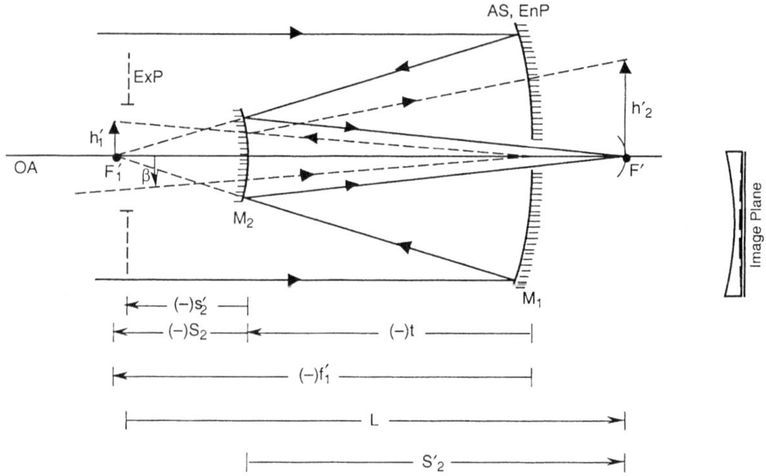

(a)

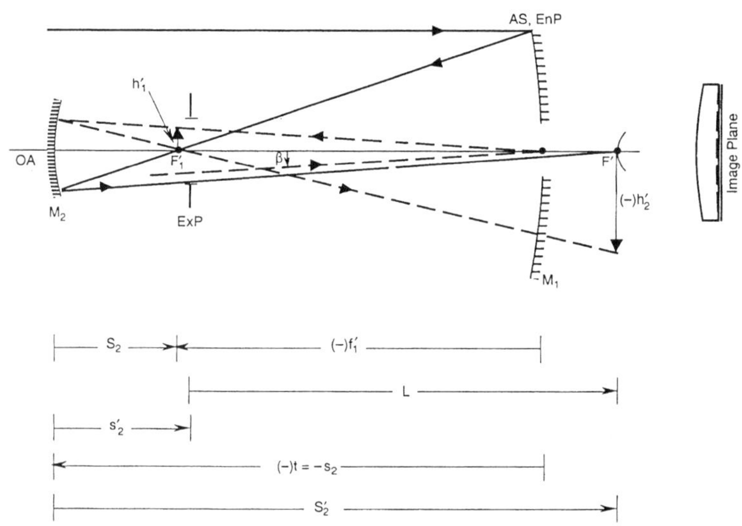

(b)

Figure 18.13. Astronomical telescope consisting of two conic mirrors M_1 and M_2. (a) Cassegrain and (b) Gregorian forms are shown in this figure. The aperture stop of the telescope lies at its primary mirror M_1. The spherical image surface passing through the focal point F' is an illustration of the Petzval image surface. The axial image formation is illustrated by the solid-line rays and off-axis by the dashed-line rays. A field-flattening lens is shown on the right-hand side of a telescope, where the Petzval image surface is indicated by a dashed line.

The radius of curvature of the Petzval image surface of the telescope is given by

$$R_p = f_1' f_2' / (f_1' - f_2'). \tag{18.80}$$

This surface shown passing through F' in Fig. 18.13 is curved toward the primary mirror in the case of a Cassegrain telescope and away from it in a Gregorian telescope. The Petzval surface is flattened by use of a planoconcave field lens in a Cassegrain telescope and a planoconvex lens in a Gregorian, with the planar surface facing the image surface. The focal length and the radius of curvature of the curved surface of the field-flattening lens are given by $f_l' = R_p/n$ and $R = R_p\,(n-1)/n$, respectively. The signs of various quantities for the two types of telescopes are summarized in Table 18.6.

The primary aberration function at the exit pupil of the telescope for mirrors with eccentricities e_1 and e_2 is given by

$$\begin{aligned} W_{cs}(r, \theta; h_2') = {} & a_{scs} r^4 + a_{ccs} h_2' r^3 \cos\theta + a_{acs} h_2'^2 r^2 \cos^2\theta \\ & + a_{dcs} h_2'^2 r^2 + a_{tcs} h_2'^3 r \cos\theta, \end{aligned} \tag{18.81}$$

where

$$\begin{aligned} a_{scs} = {} & \frac{1}{32 m_2^4 f_1'^3} \left\{ 1 - e_1^2 + \frac{(f_1' - t)(f' + f_1')^3}{f'^3 f_1'} \right. \\ & \left. \times \left[e_2^2 - \left(\frac{f' - f_1'}{f' + f_1'} \right)^2 \right] \right\}, \end{aligned} \tag{18.82a}$$

$$a_{ccs} = \frac{1}{4 m_2^3 f'^3} \left\{ 1 - \frac{t(f' + f_1')^3}{2 f' f_1'^3} \left[e_2^2 - \left(\frac{f' - f_1'}{f' + f_1'} \right)^2 \right] \right\}, \tag{18.82b}$$

$$\begin{aligned} a_{acs} = {} & \frac{1}{2 m_2^2 f'^4 (f_1' - t)} \left\{ -f_1'(f' + t) + \frac{t^2 (f' + f_1')^3}{4 f' f_1'^2} \right. \\ & \left. \times \left[e_2^2 - \left(\frac{f' - f_1'}{f' + f_1'} \right)^2 \right] \right\}, \end{aligned} \tag{18.82c}$$

$$\begin{aligned} a_{dcs} = a_{dc2} & = \frac{1}{2} \left(a_{ac2} + \frac{1}{2 m_2^2 f_2' f'^2} \right) \\ & = \frac{1}{2} a_{acs} + \frac{1}{4 m_2^2 f'^2} \left(\frac{1}{f_2'} - \frac{1}{f_1'} \right). \end{aligned} \tag{18.82d}$$

The primary mirror does not introduce any distortion because the aperture stop

Table 18.6. Signs of Focal Lengths, etc., for Cassegrain and Gregorian Telescopes

Quantity	Cassegrain	Gregorian
f'_1	–	–
f'_2	–	+
f'	+	–
M_2	+	–
m_2	+	–
S'_2	+	+
s'_2	–	+
t	–	–
R_p	–	+

is collocated with it. Moreover, its eccentricity does not affect its coma and astigmatism.

Some specific telescopes are discussed in the following five subsections.

18.10.2. *Classical Cassegrain and Gregorian Telescopes*

The classical Cassegrain and Gregorian telescopes consist of a concave paraboloidal primary mirror ($e_1 = 1$) and a secondary mirror of eccentricity given by

$$e_2 = (f' - f'_1)/(f' + f'_1) = (M_2 + 1)/(M_2 - 1) \tag{18.83}$$

so that the spherical aberration of the teslescope is zero. The secondary mirror is convex and hyperboloidal ($e_2 > 1$) in a Cassegrain telescope and concave and ellipsoidal ($e_2 < 1$) in a Gregorian. The coma and astigmatism of these telescopes is given by

$$W_{c/a}(r, \theta; h'_2) = (1/4m_2^3 f'^3)h'_2 r^3 \cos\theta - (1/2\, m_2^3 f'^3)h_2'^2 r^2 \cos^2\theta. \tag{18.84}$$

Coma of these telescopes is the same as that of a paraboloidal mirror with the same diameter as the primary mirror and a focal length equal to that of the telescope. However, their astigmatism is larger by a factor of $1/m_2$ (as $|m_2| < 1$ for these telescopes). The field of view of these telescopes is limited by their coma. If the paraboloidal mirror and the telescopes have the same focal ratio, then the angular sizes of the coma ray spots are the same for both, but the angular size of the astigmatism spot is larger by a factor of $1/|m_2|$ for the telescopes compared with that for the paraboloidal mirror.

18.10.3. Aplanatic Cassegrain and Gregorian Telescopes

In the aplanatic Cassegrain and Gregorian telescopes, the eccentricities of the two mirrors that yield zero spherical aberration and coma are given by

$$e_1^2 = 1 + 2f_1'^2(f_1' - t)/tf'^2, \tag{18.85}$$

$$e_2^2 = \left(\frac{f' - f_1'}{f' + f_1'}\right)^2 + \frac{2f'f_1'^3}{t(f' + f_1')^3}. \tag{18.86}$$

The corresponding astigmatism is given by

$$a_{acs} = -f_1'(2f' + t)/4m_2^2 f'^4(f_1' - t), \tag{18.87}$$

which sets the limit on an acceptable field of view. Because $e_1 \neq 0$, the primary mirror does not form an aberration-free image of an on-axis point object. In the aplanatic Cassegrain telescope, called the Ritchey–Chrétien (e.g., the Hubble Space Telescope), $f_1' - t < 0$ and $f' > |f_1'|$, and both e_1 and e_2 are greater than one, i.e., the mirrors are hyperboloidal. In the aplanatic Gregorian telescope, $f_1' - t > 0$, $f' < 0$, and both e_1 and e_2 are less than one, i.e., the mirrors are ellipsoidal.

18.10.4. Afocal Telescope

Equations (18.82) show that an afocal (or a Mersenne) telescope consisting of a pair of confocal paraboloidal mirrors, i.e., one for which $f' = \infty$, or $t = f_1' - f_2'$, and $e_1 = 1$ and $e_2 = 1$, is an anastigmat. Because the two mirrors have a common focus, a collimated beam incident upon it is focused by the first mirror and recollimated by the second; hence it is also a beam expander (or reducer when used in reverse). Figure 18.14 illustrates a beam expander with its aperture stop and exit pupil located at the secondary mirror M_2. A collimated beam of diameter D_1 incident at an angle β from the optical axis emerges as a collimated beam of diameter D_2 at an angle β'. The beam-expansion ratio is given by $M = \beta'/\beta = f_2'/f_1' = D_2/D_1$. The primary aberration function is given by

$$\begin{aligned} W(r, \theta; \beta') = (1/4)[&\beta'^2(f_2'^{-1} - f_1'^{-1})r^2 \\ &- \beta'^3(M^2 + 2M - 3)r\cos\theta]. \end{aligned} \tag{18.88}$$

The consequence of nonzero field curvature is that the output beam is actually not quite collimated, but is instead focused at a distance of $2/\beta'^2(f_2'^{-1} - f_1'^{-1})$ from the exit pupil of the system. Similarly, the consequence of nonzero distortion is that the output beam makes an angle of $\beta' - \beta'^3\,(M^2 + 2M - 3)/4$ with the optical axis.

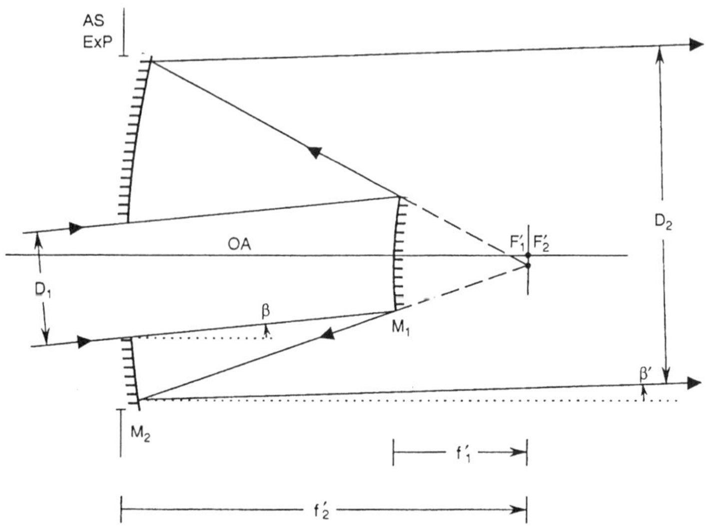

Figure 18.14. Schematic of a beam-expander system consisting of two confocal paraboloidal mirrors M_1 and M_2 with their focal points F'_1 and F'_2. Its aperture stop and therefore its exit pupil are located at M_2. The dotted lines shown are parallel to the optical axis OA.

18.10.5. Couder Anastigmatic Telescopes

The telescope is anastigmatic if $t = -2f'$ in addition to the mirror eccentricities given by Eqs. (18.85) and (18.86). A special case is the Schwarzschild telescope consisting of two concentric spherical mirrors, i.e., one for which $e_1 = 0 = e_2$ and $t = 2(f'_1 - f'_2) = -2f'$. The ratio of the radii of curvature of the two surfaces is given by

$$R_2/R_1 = f'_2/f'_1 = (3 \mp \sqrt{5})/2 = 0.382 \text{ or } 2.618. \tag{18.89}$$

The corresponding secondary magnification M_2 is equal to 0.618 or –1.618. The image observed on a spherical surface that is concentric with the mirrors is free of primary aberrations, except distortion. For a concave primary mirror, a value of $M_2 = 0.618$ yields a virtual final image, as illustrated in Fig. 18.15(a), which is not a practical solution. For a convex primary mirror, $M_2 = -1.618$ yields a real image, as illustrated in Fig. 18.15(b). In either case, the diameter of the larger mirror is 4.236 times that of the smaller mirror and 20% of the central area of the axial image-forming light cone is obscured. The exit pupil lies between the common center of curvature and the focal plane in Fig. 18.15(a) and to the right of the focal plane in Fig. 18.15(b). Its diameter is equal to that of the secondary mirror. A lens of refractive index n flattens the the spherical Petzval surface if its focal length is given by $f'_1 = -f'/n = R_p/n$. Its

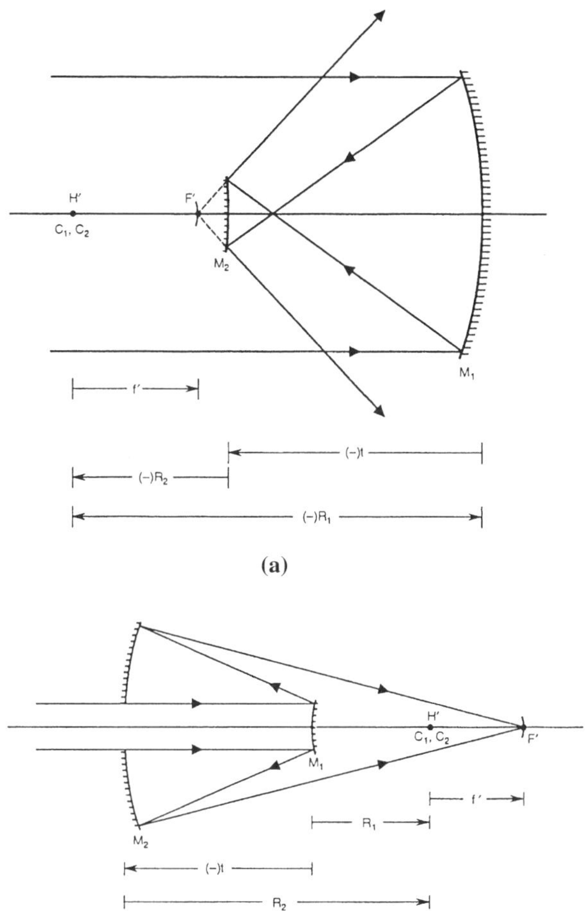

(b)

Figure 18.15. Schwarzschild telescope with a concave primary mirror. C_1 and C_2 are the centers of curvature of the mirrors M_1 and M_2, respectively. The principal point H' of the system coincides with C_1 and C_2. The final image is virtual in (a) and real in (b).

distortion vanishes if the radii of curvature of its surfaces are given by $R_{1l} = (n^2 - 1)f/[1 - n(n + 1)]$ and $R_{2l} = (n^2 - 1)f$.

18.10.6. Dall–Kirkham Telescope

In a Dall–Kirkham telescope, the secondary mirror is spherical ($e_2 = 0$) and its spherical aberration is made zero by letting

$$e_1^2 = 1 - (f_1' - t)(f' + f_1')(f' - f_1')^2/f_1' f'^3 < 1. \tag{18.90}$$

Its coma coefficient is given by

$$a_{ccs} \simeq (1 + M_2^2)/8m_2^3 f'^3, \tag{18.91}$$

which is $(1 + M_2^2)/2$ times the coma of a classical Cassegrain telescope. However, it is less sensitive to misalignments and relatively easy to fabricate and test.

18.11. Misaligned Telescopes

When a telescope is misaligned, the image of an object formed by it is displaced and aberrations are introduced.[22–24] Expressions for the aberrations introduced by misalignments of the secondary mirror of a two-mirror telescope with respect to its primary mirror are given in this section.

18.11.1. Decentered Secondary Mirror

Figure 18.16(a) shows a properly aligned two-mirror telescope. When the secondary mirror is decentered by a small amount Δ along the x axis, as in Fig. 18.16(b), the image is displaced by an amount $(1 - M_2)\Delta$. The axial coma introduced by it is given by

$$\delta W_{cd}(r, \theta; h_2') = -a_{cd} r^3 \cos\theta, \tag{18.92}$$

where

$$a_{cd} = \frac{(M_2 - 1)^3}{8m_2^3 f'^3}\left(e_2^2 + \frac{M_2 + 1}{M_2 - 1}\right)\Delta. \tag{18.93}$$

The results obtained on substituting the value of e_2 for the various telescopes into Eq. (18.93) are given in Table 18.7. It can be seen that an aplanatic telescope is somewhat more sensitive to decenter than a classical telescope.

18.11.2. Tilted Secondary Mirror

When the secondary mirror is tilted with respect to the primary mirror by an angle β, as in Fig. 18.16(c), the image is displaced by an amount $2\beta S_2'$. The axial coma introduced by it is given by

$$\delta W_{ct}(r, \theta; \beta) = -(f_2'/4m_2^3 f'^3)(M_2 - 1)^2(M_2 + 1)\beta r^3 \cos\theta, \tag{18.94}$$

[22] V. N. Mahajan, *Optical Imaging and Aberrations, Part I: Ray Geometrical Optics,* SPIE, 1998.
[23] D. J. Schroeder, *Astronomical Optics,* 2nd ed., Academic, 2000.
[24] R. N. Wilson, *Reflecting Telescope Optics I,* Springer, 1996, Sections 3.7 and 3.8.

(a)

(b)

(c)

(d)

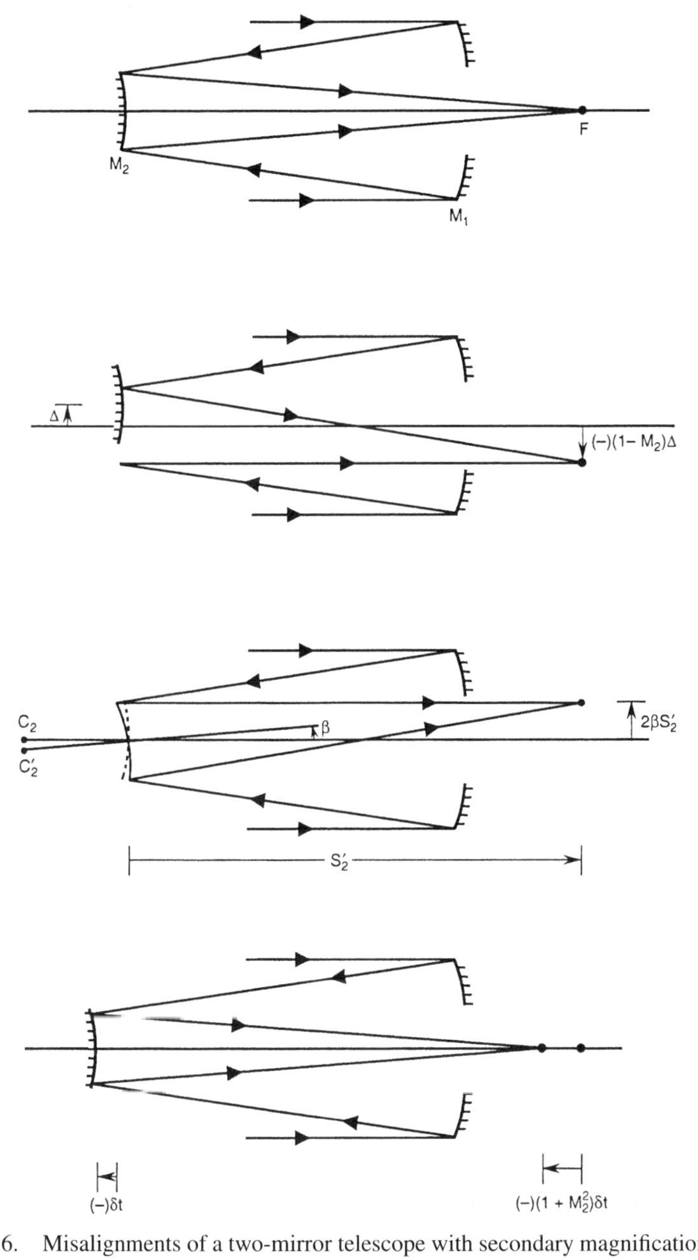

Figure 18.16. Misalignments of a two-mirror telescope with secondary magnification M_2. (a) Aligned telescope with its axial focus at F. (b) Secondary mirror decentered along x axis by Δ. (c) Secondary mirror tilted by an angle β so that its center of curvature is displaced from C_2 to C_2'. (d) Secondary mirror despaced by δt, which is numerically negative in the figure.

Table 18.7. Coma Due to a Decenter Δ or a Despace δt of the Secondary Mirror

Telescope	a_{cd}	$\delta(h_2' a_{ccs})$
Classical Cassegrain and Gregorian	$\dfrac{M_2(M_2^2-1)\Delta}{4m_2^3 f'^3}$	$\dfrac{\beta_o[(M_2-1)f_1' + t(M_2^2-1)]\delta t}{2m_2^3 M_2^2 f_1'^3 (f_1' - t)}$
Aplanatic Cassegrain and Gregorian	$\dfrac{[M_2(M_2^2-1)-(f'/t)]\Delta}{4m_2^3 f'^3}$	$\dfrac{\beta_o\{f_1'[2(M_2+1)-(f_1'/t)] + 2t(M_2^2-1)\}\delta t}{4m_2^3 M_2^2 f_1'^3 (f_1' - t)}$
Dall–Kirkham ($e_2 = 0$)	$\dfrac{(M_2-1)(M_2^2-1)\Delta}{8m_2^3 f'^3}$	Equation (18.105) with $e_2 = 0$
Afocal telescope ($M_2 \to \infty$)	$-\dfrac{(1+e_2^2)\Delta}{8m_2^3 f_1'^3}$	$\dfrac{\beta_o[f_1' - e_2^2(f_1' - 4t)]\delta t}{8m_2^3 f_1'^3 (f_1' - t)}$
Beam expander (Mersenne telescope)	$-\dfrac{\Delta}{4m_2^3 f_1'^3}$	$-\dfrac{\delta t}{8m_2^4 f_1'^4}$

which is independent of its eccentricity. For the afocal telescope, $M_2 \to \infty$ and Eq. (18.94) reduces to

$$\delta W_{ct}(r, \theta; \beta) = (f_2'/8m_2^3 f_1'^3)\beta r^3 \cos\theta. \tag{18.95}$$

18.11.3. Decentered and Tilted Secondary Mirror

If the decenter Δ and tilt β of the secondary mirror are such that the axes of the two mirrors are coplanar, the resulting axial coma is zero if $\beta = \Delta/d$, where

$$d = 2f_2'\left(1 + e_2^2\,\frac{M_2-1}{M_2+1}\right)^{-1}. \tag{18.96}$$

It is equivalent to rotating the axis of the secondary mirror about a fixed point, called the neutral point,[25] on the axis of the primary mirror lying at a distance d from the vertex of the secondary mirror. No axial coma is introduced when the axes of the two mirrors intersect at this point. If the decenter at a distance d resulting from tilt is not equal to $-\Delta$, there will be axial coma. The axial coma is proportional to the separation of the axes of the mirrors in a transverse plane passing through the neutral point (see Fig. 18.17). Thus, if Δ_d and $\Delta_t = \beta d$ are the decenters in the plane of the neutral point because of surface decenter and tilt, respectively, the net decenter in this plane is given by

$$\Delta_{np} = \Delta_d + \Delta_t = \Delta_d + \beta d \tag{18.97}$$

[25] W. B. Wetherell and M. P. Rimmer, "General analysis of aplanatic Cassegrain, Gregorian, and Schwarzchild telescopes," Appl. Opt. **11,** 2817–2832, 1972.

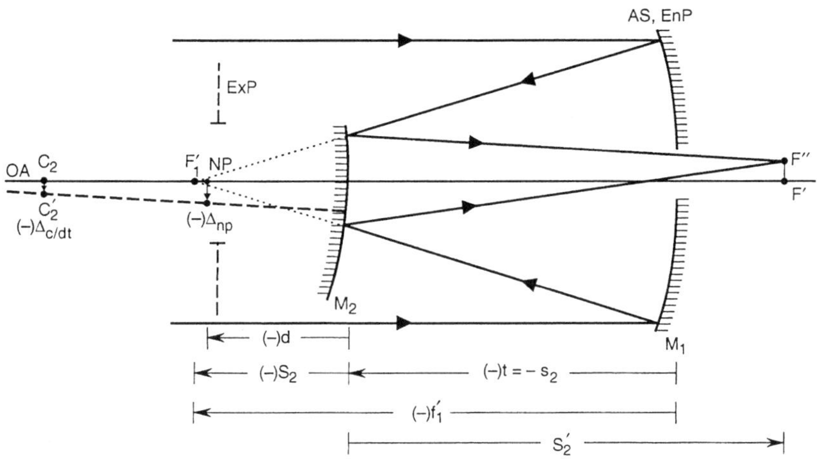

Figure 18.17. Cassegrain telescope with a decentered and tilted secondary mirror. The image displacement $F'F''$ is proportional to the transverse displacement C_2C_2' of the center of the curvature of the secondary mirror. The axial coma is proportional to the separation of the axes of the primary and secondary mirrors in the plane of the neutral point NP. The size of the secondary mirror is exaggerated in the figure for convenience.

and the coefficient of axial coma is given by

$$a_{c/dt} = \frac{(M_2 - 1)^3}{8M_2^3 f'^3}\left(e_2^2 + \frac{M_2 + 1}{M_2 - 1}\right)\Delta_{np}. \tag{18.98}$$

Adding the displacements of the center of curvature of the secondary mirror resulting from the surface decenter and tilt, we may write its total displacement as

$$\Delta_{c/dt} = \Delta_{c/d} + \Delta_{c/t} = \Delta_d + \beta R_2, \tag{18.99}$$

where R_2 is the vertex radius of curvature of the mirror. The total image displacement is given by

$$F'F'' = (1 - M_2)\Delta_{c/dt}. \tag{18.100}$$

The image displacement is zero if the misalignment is only a rotation of the mirror about its center of curvature. This is understandable because the image location depends on the vertex radius of curvature of a mirror, which is not changed by a rotation about its center of curvature. The image displacement is proportional to the separation of the axes of the two mirrors in a transverse plane passing through the center of curvature of the secondary mirror. For zero axial coma, the surface decenter and tilt are related to each other according to $\Delta_d = -\beta d$. The corresponding image displacement is given by

$$F'F'' = (1 - M_2)(R_2 - d)\beta. \tag{18.101}$$

Thus no axial coma is introduced if a moving object is tracked by rotating the secondary mirror about the neutral point.

In the case of a Dall–Kirkham telescope, the secondary mirror is spherical; i.e., $e_2 = 0$, and Eq. (18.96) yields $d = 2f'_2$. Thus the neutral point lies at the center of curvature of the secondary mirror. If this mirror is decentered and tilted so that its center of curvature is not displaced from the optical axis of the primary mirror, then no axial coma is introduced and, of course, there is no image displacement. In the case of a classical Cassegrain telescope, substituting the value of e_2 given by Eq. (18.83) and $M_2 = -f'/f'_1$ into Eq. (18.97), we obtain

$$d = f'_2(f' + f'_1)/f' = f'_1 - t. \tag{18.102}$$

Thus the neutral point lies at the focus of the primary mirror. The value of e_2 for an aplanatic Cassegrain (or a Ritchey–Chrétien) telescope is given by Eq. (18.86). For the Hubble telescope, the neutral point lies between the center of the exit pupil and the focus of the primary mirror, as shown in Fig. 18.17. In this figure, for an oblate spheroid secondary mirror ($e_2^2 < 0$), the neutral point lies to the left of its center of curvature until it approaches $-\infty$ for $e_2^2 = -(M_2 + 1)/(M_2 - 1)$. According to Eq. (18.93), the axial coma resulting from a surface decenter is zero for this value of e_2. As e_2^2 decreases further, the neutral point moves to the right of the secondary mirror.

18.11.4. Despaced Secondary Mirror

If the secondary mirror moves away from the primary mirror by a (numerically negative) amount δt, as indicated in Figure 18.16(d), the image formed by it moves by an amount $(1 + M_2^2)\delta t$. Hence, for a fixed observation plane, there is a longitudinal defocus of $(1 + M_2^2)\delta t$. Besides defocus, because the aberrations of the unperturbed system depend on t, additional aberrations are also introduced. The change in spherical aberration a_{scs} of a telescope with a small change δt in the spacing t is given by

$$\delta a_{scs} = -\frac{1}{8m_2^4 M_2^3 f_1'^4}\left\{(M_2 - 1)^3\left[e_2^2 - \left(\frac{M_2 + 1}{M_2 - 1}\right)^2\right] + M_2(M_2^2 - 1)\right\}\delta t. \tag{18.103}$$

Substituting the value of e_2, we obtain the aberration for various telescopes given in Table 18.8. In order that a telescope despacing yield zero spherical aberration, the value of e_2, according to Eq. (18.103), must be given by

$$e_2 = (M_2 + 1)^{1/2}/(M_2 - 1). \tag{18.104}$$

Table 18.8. Spherical Aberration Due to a Despace δt of the Secondary Mirror

Telescope	δa_{scs}
Classical Cassegrain and Gregorian	$-\dfrac{(M_2^2 - 1)}{8m_2^4 M_2^2 f_1'^4}\,\delta t$
Aplanatic Cassegrain and Gregorian	$\dfrac{M_2(M_2^2 - 1) - \dfrac{2f'}{t}}{8m_2^4 M_2^3 f_1'^4}\,\delta t$
Dall–Kirkham telescope ($e_2 = 0$)	$-\dfrac{M_2^2 - 1}{8m_2^4 M_2^3 f_1'^4}\,\delta t$
Afocal telescope ($M_2 \to \infty$)	$-\dfrac{e_2^2}{8m_2^4 f_1'^4}\,\delta t$
Beam expander (Mersenne telescope) ($e_2 = 1$)	$-\dfrac{\delta t}{8m_2^4 f_1'^4}$

For other primary aberrations for an object with a field angle β_o, the variation of the final image height $h_2' = \beta_o f' = -M_2 \beta_o f_1'$ with respect to t must also be taken into account. For example, the additional coma introduced by a despace error can be obtained by taking the derivative of $h_2' a_{ccs}$, where a_{ccs} is given by Eq. (18.82b). Thus

$$\frac{\partial(h_2' a_{ccs})}{\partial t} = \frac{\beta_o(M_2 - 1)}{8m_2^3 M_2^3 f_1'^3 (f_1' - t)} \Bigg\{ 4M_2 f_1' + 4tM_2(M_2 + 1) - (f_1' - 4t)(M_2 - 1)^2 \left[e_2^2 - \left(\frac{M_2 + 1}{M_2 - 1} \right)^2 \right] \Bigg\}. \tag{18.105}$$

Its value for various telescopes is given in Table 18.7. The corresponding wave aberration is given by $\delta(h_2' a_{ccs}) r^3 \cos\theta$. For the classical telescopes, coma resulting from despace is in addition to the coma for the properly spaced telescope [see Eq. (18.84)]. Similarly, an improperly spaced aplanatic telescope is not aplanatic; its coma is given in Table 18.7.

18.12. Strehl Ratio and Aberration Tolerance

Tolerances for primary aberrations in systems with circular, annular, and Gaussian pupils are given in this section for a Strehl ratio of 0.8.[2,18,19] For large aberrations, the tolerance may be obtained from the geometrical spot sizes discussed in Section 18.5. For example, one wave of defocus, or equivalently $8\lambda F^2$ longitudinal defocus, gives a spot of radius $4\lambda F$ (compared with $1.22\lambda F$ for the radius of the Airy disk). Aberration tolerance may also be obtained for a certain value of Hopkins or Struve ratio.[26–28]

18.12.1. Strehl Ratio

The Strehl ratio is the ratio of irradiances at the center of the image of a point object with and without aberrations. Rayleigh showed that a quarter wave of primary spherical aberration reduces the irradiance by 20%; i.e., the Strehl ratio for this aberration is 0.8. This result has brought forth Rayleigh's $\lambda/4$ rule, namely, that a Strehl ratio of approximately 0.8 is obtained if the maximum absolute value of the aberration at any point in the pupil is equal to $\lambda/4$. A variant of this definition is that an aberrated wave front that lies between two concentric spheres that are spaced a quarter wave apart will give a Strehl ratio of approximately 0.8. Thus, instead of $|W_p| = \lambda/4$, we require $W_{p-v} = \lambda/4$, where $|W_p|$ is the peak absolute value and W_{p-v} is the peak-to-valley value of the aberration. For small aberrations, the Strehl ratio is approximately given by

$$S \simeq (1 - \sigma_\Phi^2/2)^2 \simeq 1 - \sigma_\Phi^2 \simeq \exp(-\sigma_\Phi^2), \tag{18.106}$$

where σ_Φ^2 is the variance of the phase aberration. The aberration tolerance for a Strehl ratio of 0.8 is given by $\sigma_\Phi = \sqrt{0.2}$ or $\sigma_w = (\lambda/2\pi)\sqrt{0.2} \simeq \lambda/14$, where σ_w is the standard deviation of the wave aberration. Thus, for example, the random figure errors of a mirror with a standard deviation of $\lambda/28$ will reduce the Strehl ratio of an image by 20% under normal incidence.

18.12.2. Circular Pupils

Table 18.9 gives the form as well as the standard deviation σ_Φ of a primary aberration, where an aberration coefficient A_i represents its peak value. It also

[26] V. N. Mahajan, *Optical Imaging and Aberrations, Part II: Wave Diffraction Optics,* SPIE, 2001.

[27] V. N. Mahajan, "Strehl ratio for primary aberrations: some analytical results for circular and annular pupils," J. Opt. Soc. Am. **72,** 1258–1266, 1982; Errata, **10,** 2092, 1993; also "Strehl ratio for primary aberrations in terms of their aberration variance," J. Opt. Soc. Am. **73,** 860–861, 1983.

[28] V. N. Mahajan, "Uniform versus Gaussian beams: A comparison of the effects of diffraction, obscuration, and aberrations," J. Opt. Soc. Am. A **3,** 470–485, 1986; also "Zernike–Gauss polynomials and optical aberrations of systems with Gaussian pupils," Appl. Opt. **34,** 8057 1995.

Table 18.9. Primary Aberration Standard Deviation and Tolerance for a Strehl Ratio of 0.8

Aberration	$\Phi(\rho, \theta)$	σ_Φ	Diffraction Focus	W_p	W_{p-v}	$S = 0.8$		
						A_i	W_p	W_{p-v}
Spherical	$A_s\rho^4$	$\frac{2A_s}{3\sqrt{5}}$		A_s	A_s	0.25	0.25	0.25
Balanced spherical	$A_s(\rho^4 - \rho^2)$	$\frac{A_s}{6\sqrt{5}}$	$(0, 0, 8F^2A_s)$	$A_s/4$	$A_s/4$	1	0.25	0.25
Coma	$A_c\rho^3 \cos\theta$	$\frac{A_c}{2\sqrt{2}}$		A_c	$2A_c$	0.21	0.21	0.42
Balanced coma	$A_c(\rho^3 - 2\rho/3)\cos\theta$	$\frac{A_c}{6\sqrt{2}}$	$(4FA_c/3, 0, 0)$	$A_c/3$	$2A_c/3$	0.63	0.21	0.42
Astigmatism	$A_a\rho^2 \cos^2\theta$	$\frac{A_a}{4}$		A_a	A_a	0.30	0.30	0.30
Balanced astigmatism	$A_a\rho^2(\cos^2\theta - 1/2) = (A_a/2)\rho^2 \cos 2\theta$	$\frac{A_c}{2\sqrt{6}}$	$(0, 0, 4F^2A_a)$	$A_a/2$	A_a	0.37	0.18	0.37

lists the tolerance, i.e., the value of the aberration coefficient A_i, for a Strehl ratio of 0.8. The aberration tolerance is for the wave (as opposed to phase) aberration coefficient, as is customary in optics. A Strehl ratio of 0.8 is obtained for $|W_p| = \lambda/4 = W_{p-v}$ for spherical aberration only. For other primary aberrations, distinctly different values of $|W_p|$ and W_{p-v} give a Strehl ratio of 0.8. The values of $|W_p|$ and W_{p-v} are also given in terms of the aberration coefficient A_i. Thus it is advantageous to use σ_w for estimating the Strehl ratio. A Strehl ratio of ≥ 0.8 is obtained for $\sigma_w \lesssim \lambda/14$.

Table 18.9 also lists the form of a balanced primary aberration, its standard deviation, and its tolerance for a Strehl ratio of 0.8. Also listed in the table is the location of the diffraction focus, i.e., the point with respect to which the aberration variance is minimum so that the Strehl ratio at it is maximum. For example, spherical aberration is balanced with an equal and opposite amount of defocus. The standard deviation is reduced, and aberration tolerance is increased, by a factor of 4. The corresponding defocused plane is different from the one that contains the circle of least confusion. In the case of coma, the balancing aberration is a wave-front tilt with a coefficient that is minus two-thirds of the coma coefficient. Thus the maximum Strehl ratio is obtained at a point that is displaced from the Gaussian image point but lies in the Gaussian image plane. The standard deviation for balanced coma is 1/3 of its corresponding value when not balanced. In the case of astigmatism, the best Strehl ratio is obtained in the plane of the circle of least confusion. The standard deviation of balanced astigmatism is only 1/1.225 of its corresponding value for classical astigmatism (i.e., without any balancing). The balanced aberrations can be easily identified with Zernike circle polynomials, e.g., balanced spherical with $R_4^0(\rho)$, balanced coma with $R_3^1(\rho)\cos\theta$, and balanced astigmatism with $R_2^2(\rho)$ $\cos 2\theta$.

For large aberrations, the Strehl ratio is maximum for a nonoptimally balanced aberration. For example, when $A_s = 3\lambda$, the optimum amount of defocus is $B_d = -3\lambda$, but the Strehl ratio is a minimum and equal to 0.12. The Strehl ratio is maximum and equal to 0.26 for $B_d \simeq -4\lambda$ or -2λ. For $A_s \lesssim 2.3\lambda$, the axial irradiance is maximum at a point with respect to which the aberration variance is minimum. In the case of coma, the point with respect to which the aberration variance is minimum lies in the image plane at a distance of $4A_cF/3$ from the origin, where A_c is in length units. The maximum irradiance occurs at this point only if $A_c \lesssim 0.7\lambda$, which in turn corresponds to $S \gtrsim 0.76$. For larger values of A_c, the distance of the point of maximum irradiance does not increase linearly with its value and even fluctuates in some regions. Thus, only for large Strehl ratios, the irradiance is maximum at the point associated with minimum aberration variance. Moreover, it is found that for $A_c > 2.3\lambda$, the Seidel coma gives a larger Strehl ratio than the balanced coma; i.e., the irradiance at the origin is larger than at the point with respect to which the aberration variance is minimum.

18.12.3. *Annular Pupils*

The standard deviation of primary aberrations in systems with annular pupils having a ratio of ε between their inner and outer diameters is given in Table 18.10. The corresponding balanced aberrations, their standard deviation, and the location of diffraction focus are given in Table 18.11. The standard deviation of spherical and balanced spherical aberrations and defocus decreases as ε increases. Correspondingly, the tolerance in terms of their aberration coefficient A_s and A_d, for a given Strehl ratio, increases. The standard deviation of coma, astigmatism, balanced astigmatism, and tilt increases as ε increases. The standard deviation of balanced coma first slightly increases, achieves its maximum value at $\varepsilon = 0.29$, and then decreases rapidly as ε increases. The factor by which the standard deviation of an aberration is reduced by balancing it with another aberration is reduced in the case of spherical aberration, but increases in the case of astigmatism, as ε increases.

Table 18.10. Primary Aberrations and Their Standard Deviation for a System with a Uniformly Illuminated Annular Pupil

Aberration	$\Phi(\rho, \theta)$	σ_Φ
Spherical	$A_s\rho^4$	$(4 - \varepsilon^2 - 6\varepsilon^4 - \varepsilon^6 + 4\varepsilon^8)^{1/2} A_s/3\sqrt{5}$
Coma	$A_c\rho^3 \cos\theta$	$(1 + \varepsilon^2 + \varepsilon^4 + \varepsilon^6)^{1/2} A_c/2\sqrt{2}$
Astigmatism	$A_a\rho^2 \cos^2\theta$	$(1 + \varepsilon^4)^{1/2} A_a/4$
Field curvature (defocus)	$A_d\rho^2$	$(1 - \varepsilon^2) A_d/2\sqrt{3}$
Distortion (tilt)	$A_t\rho \cos\theta$	$(1 + \varepsilon^2)^{1/2} A_t/2$

Just as we identified balanced aberrations in systems with circular pupils with Zernike circle polynomials, we can identify them in systems with annular pupils with Zernike annular polynomials,[29] e.g., balanced spherical aberration with $R_4^0(\rho;\, \varepsilon)$, balanced coma with $R_3^1(\rho;\, \varepsilon)\cos\theta$, and balanced astigmatism with $R_2^2(\rho;\, \varepsilon)\cos 2\theta$.

18.12.4. *Gaussian Pupils*

In optical systems with Gaussian pupils, i.e., those with a Gaussian amplitude across the wave front at their exit pupils, the standard deviation of an aberration is smaller compared with those with a uniform pupil because the wave amplitude is smaller away from the center where the aberration is larger. Accordingly, the Strehl ratio for a Gaussian pupil for a given amount of a primary aberration is higher than that for a uniform pupil, or the aberration tolerance

[29] V. N. Mahajan, *Optical Imaging and Aberrations, Part II: Wave Diffraction Optics,* SPIE, 2001.

Table 18.11. Balanced Primary Aberrations, Their Standard Deviation, and Diffraction Focus for a System with a Uniformly Illuminated Annular Pupil

Balanced Aberration	$\Phi(\rho, \theta; \varepsilon)$	σ_Φ	Diffraction Focus
Spherical	$A_s[\rho^4 - (1+\varepsilon^2)\rho^2]$	$\frac{1}{6\sqrt{5}}(1-\varepsilon^2)^2 A_s$	$[0, 0, 8(1+\varepsilon^2)F^2 A_s]$
Coma	$A_c\left(\rho^3 - \frac{2}{3}\frac{1+\varepsilon^2+\varepsilon^4}{1+\varepsilon^2}\rho\right)\cos\theta$	$\frac{(1-\varepsilon^2)(1+4\varepsilon^2+\varepsilon^4)^{1/2}}{6\sqrt{2}(1+\varepsilon^2)^{1/2}}A_c$	$\left[\frac{4(1+\varepsilon^2+\varepsilon^4)}{3(1+\varepsilon^2)}FA_c, 0, 0\right]$
Astigmatism	$A_a\rho^2(\cos^2\theta - 1/2)$	$\frac{1}{2\sqrt{6}}(1+\varepsilon^2+\varepsilon^4)^{1/2}A_a$	$(0, 0, 4F^2 A_a)$

for a given Strehl ratio is higher. The balanced aberrations in this case are the Zernike–Gauss polynomials that can be obtained from the Zernike polynomials by the Gram–Schmdit orthogonolization process.[30,31] Table 18.12 gives the standard deviation of primary aberrations for a Gaussian pupil with an amplitude variation as $\exp(-\rho^2)$. The balanced aberrations are also included in this table. In each case, the standard deviation is smaller than that for a uniformly illuminated pupil, as may be seen by comparing with the corresponding results given in Table 18.9. The diffraction focus in the case of astigmatism is independent of whether the pupil is Gaussian or uniform.

Table 18.12. Primary Aberrations and Their Standard Deviation for Systems with Gaussian Circular Pupils with an Amplitude Variation as $\exp(-\rho^2)$

Aberration	$\Phi(\rho, \theta)$	σ_Φ	Diffraction Focus
Spherical	$A_s\rho^4$	$A_s/3.67$	
Balanced spherical	$A_s(\rho^4 - 0.933\rho^2)$	$A_s/13.71$	$(0, 0, 7.46F^2A_s)$
Coma	$A_c\rho^3 \cos\theta$	$A_c/3.33$	
Balanced coma	$A_c(\rho^3 - 0.608\rho)\cos\theta$	$A_c/8.80$	$(1.22FA_c, 0, 0)$
Astigmatism	$A_a\rho^2\cos^2\theta$	$A_a/4.40$	
Balanced astigmatism	$A_a\rho^2(\cos^2\theta - 1/2)$	$A_a/5.61$	$(0, 0, 4F^2A_a)$
Defocus	$B_d\rho^2$	$B_d/3.55$	
Tilt	$B_t\rho\cos\theta$	$B_t/2.19$	

18.12.5. Fabrication or Deformation Errors

When an optical element is fabricated, its shape will differ slightly from its prescription. If δS is the surface or figure error at a point along the surface normal, the corresponding wave-front error is given by

$$\delta W = \delta S(n\cos\theta - n'\cos\theta'), \tag{18.107}$$

where θ and θ' are the angles of incidence and refraction of the ray at the point. For a reflecting surface, Eq. (18.107) reduces to $\delta W = 2\delta S\cos\theta$. Because of the random nature of the fabrication errors, the total error variance is equal to the sum of the variances of the wave-front errors contributed by the surfaces of a system, and its tolerance is obtained from the desired value of the Strehl ratio. For example, the tolerance for the standard deviation of the figure errors of

[30] V. N. Mahajan, *Optical Imaging and Aberrations, Part II: Wave Diffraction Optics,* SPIE, 2001.

[31] V. N. Mahajan, "Uniform versus Gaussian beams: a comparison of the effects of diffraction, obscuration, and aberrations," J. Opt. Soc. Am. A **3,** 470–485, 1986; also "Zernike–Gauss polynomials and optical aberrations of systems with Gaussian pupils," Appl. Opt. **34,** 8057, 1995.

mirrors in a two-mirror astronomical telescope for a Strehl ratio of 0.8 is approximately $\lambda/40$.

The wave aberration of a ray for a deformed multielement system can be obtained from the ray-trace data of the undeformed system, i.e., without ray tracing the deformed system.[32] If δS_j is the deformation of jth surface of the system for a certain ray i, then the wave aberration δW_i of the ray for the system is given by the sum of the wave aberrations δW_{ij} contributed by all the surfaces, i.e.,

$$\delta W_i = \sum_j \delta W_{ij} = \sum_j \delta S_{ij}(n_j \cos\theta_{ij} - n'_j \cos\theta'_{ij}), \tag{18.108}$$

where θ_{ij} is the angle of incidence of the ray in a medium of refractive index n_j and θ'_{ij} is the angle of refraction in a medium of refractive index n'_j in the undeformed state of the system.

Further Reading

V. Mahajan, *Aberration Theory Made Simple,* Vol. TT06 of SPIE Tutorial Text Series, SPIE, 1991.

W. T. Welford, *Aberrations of Optical Systems,* Hilger, 1986.

R. R. Shannon, *The Art and Science of Optical Design,* Cambridge Univ. Press, 1997.

R. E. Fischer and B. Tadic-Galeb, *Optical System Design,* McGraw-Hill, 2000.

W. J. Smith, *Modern Optical Engineering,* 2nd ed., McGraw-Hill, 1990.

F. A. Jenkins and H. E. White, *Fundamentals of Optics,* 3rd ed., McGraw-Hill, 1957.

D. C. O'Shea and M. E. Harrigan, *Aberration Curves in Lens Design,* Chap. 33 in M. Bass, E. Van Stryland, D. Parker, and W. L. Wolfe, *Handbook of Optics,* McGraw-Hill, 1995.

The first reference, by the author, is a more comprehensive treatment of the material presented here. The second reference is a similar, but older treatise. The next three deal more with the correction of aberrations than with their description. The sixth reference contains a simpler, less detailed discussion of aberrations. The last reference relates aberrations to various aberration curves generated by design programs.

[32] V. N. Mahajan and J. Govignon, "Computer simulation of an adaptive optical system," Proc. SPIE **172,** 439–451, 1980.

19

Image Quality

Robert R. Shannon

19.1. Introduction

The fundamental purpose of an imaging system is to collect light from each illuminated point on an object and convey that light to a corresponding image point. Image quality is a measure of the extent to which the irradiance distribution in the image is a satisfactory representation of the original object radiance distribution.

19.2. Defocus

The geometrical image is defined by the distribution of the intersections of rays on the chosen image surface. If the image surface is located at a conjugate image point, then, for a perfect optical system, each point on the object is represented by an identical point in the image. If the image location is defocused from this conjugate image location, the spreading of the convergent cone of rays causes the image to grow to a size that is characterized by the diameter d of the so-called circle of confusion, where d is given by

$$d = \frac{\delta}{F},$$

where δ is the focal error and F is the F number of the image-forming beam. In an approximate sense, this dimension represents the minimum resolvable dimension of the image. When scaled by the magnification from the image to the object, this represents the minimum resolvable distance on the object.

The depth of focus Δl_{def} is the distance on the optical axis over which the diameter of the blur circle lies within a specified tolerable value:

$$\Delta l_{\mathrm{def}} = \pm d_a F.$$

The dimensions of the depth of focus are those of the allowable blur diameter

d_a and are usually symmetric about the position of best focus. The corresponding depth of field ΔL is the allowable distance the object may range such that the image location remains within the prescribed depth of focus:

$$\Delta L = M^2 \Delta l_{\text{def}}.$$

The magnification M is that from the image to the object. For the object at infinity, the hyperfocal distance defined by the circle of confusion is given by

$$L_{\text{hyp}} = \frac{f(f + \Delta l_{\text{def}})}{\Delta l_{\text{def}}} \approx \frac{f^2}{\Delta l_{\text{def}}}.$$

The hyperfocal distance is the closest distance at which the object can be located from the front focal point such that it is acceptably in focus if the detector is at the nominal back focal point.

19.3. Point-Spread Function

The units used to describe the light distribution in the object or the image are either radiance or irradiance. In this chapter they are used interchangeably, as the relative distribution is the important consideration in image quality.

A detailed description of the imaging properties of the system may be obtained from a calculation or measurement of the point-spread function psf(x, y), which is the light distribution in the image of a point object as a function of the local image coordinates x and y. When this is obtained, the distribution $i(x, y)$ in the image of any arbitrary object $o(x, y)$ can be found by the application of the convolution process:

$$i(x, y) = \iint o(x - x', y - y') f(x', y') \mathrm{d}x' \mathrm{d}y' = o(x, y) \otimes f(x, y),$$

where $\otimes$ represents convolution, and x, x', y, y' are the coordinates in the image space. As an example, a point object of unit radiance located at (α, β) on the object surface would be represented by

$$o(x, y) = \delta(x - \alpha, y - \beta),$$

where δ is the Dirac delta function. A point that is not unit radiance (or irradiance) would be represented by the same function multiplied by the radiance (or irradiance) value. Application of the convolution would result in placing a single spread function at the coordinates (α, β). Ignored here is that a complete interpretation of the imaging function would include the magnification between the image and the object. The user of any of these equations must examine the location in the image or the object coordinates and appropriately scale the coordiinates.

19.4. Modulation Transfer Function

Any arbitrary object radiance that can be represented by a function can be used in these imaging calculations. Some are of more general interest. The sinusoidal object has particular relevance because Fourier techniques can be applied to such object structures. A continuous sinusoidal object of spatial frequency with peak radiance a and contrast m can be represented in one dimension by

$$o(x, y) = \frac{a}{2}[1 + m\cos(2\pi\xi x)].$$

The important features are the spatial frequency ξ and the modulation (contrast) m. The modulation is

$$m = \frac{L_{max} - L_{min}}{L_{max} + L_{min}},$$

where L is the radiance. There are other definitions of contrast. One is

$$C_R = \frac{L_{max}}{L_{min}} = \frac{1 + m}{1 - m}.$$

This contrast ratio can become infinite for m equal to unity.

A full two-dimensional sinusoidal object having modulation $m(\xi_x, \xi_y)$ at the stated spatial frequency is written as

$$o(x, y) = \frac{a}{2}\{1 + m(\xi_x, \xi_y)\cos[2\pi(x\xi_x + y\xi_y)]\}.$$

The actual spatial frequency is $\xi = \sqrt{\xi_x^2 + \xi_y^2}$ with a pattern having a period $p = 1/\xi$ directed at an angle $\Psi = \arctan(\xi_y/\xi_x)$ with respect to the x axis. The peak radiance is located at the origin of the spatial coordinates, and by this selection has a zero spatial phase shift in the base coordinate set.

If the sinusoidal object or image has a nonzero spatial phase shift, the describing equations contain a spatial phase shift $\phi(\xi_x, \xi_y)$, which is usually stated in radians, but sometimes in degrees. The lateral distance shift from the origin is to coordinates given by

$$(\Delta_{v_x}, \Delta_{v_y}) = \left[\frac{\phi(\xi_x, 0)}{2\pi\xi_x}, \frac{\phi(0, \xi_y)}{2\pi\xi_y}\right].$$

In any physically realizable optical imaging system, the point-spread function is determined by a combination of diffraction and aberrations. The diffraction-limited point-spread function for a system with a circular aperture and no aberrations is given by the Airy function:

$$i(r) = \left[\frac{2J_1\left(\frac{\pi r}{\lambda F}\right)}{\frac{\pi r}{\lambda F}}\right]^2,$$

where J_1 is the Bessel function of the first kind and r is the radial coordinate, the distance from the center of the spread function, F is the F number, and λ is the wavelength of the light. The radial distance to the first zero occurs when the Bessel function argument is 3.833 and $r = 1.22\lambda F$.

19.5. Rayleigh and Sparrow Resolution Criteria

The Rayleigh resolution criterion states that, for the images of two equal radiance point sources to be distinguishable, the maximum of one must fall on the first zero of the other. The actual resolution distance will usually be somewhat smaller when the two-dimensional nature of the image is considered and somewhat larger if the sources are of different radiance. Other criteria depend on the characteristics of the sources, noise, and the specific detection process. One of these is the Sparrow criterion that states that the equal sources are just resolved when there is no dip distinguishable. This occurs at ~83% of the Rayleigh separation.

The Rayleigh criterion may also be expressed in angular units as

$$\alpha = \frac{1.22\lambda}{D}.$$

The main lobe of the Airy pattern has a diameter twice this value. A rectangular aperture has a diffraction pattern given by

$$i(x, y) = \frac{\sin\frac{\pi x}{\lambda F}}{\frac{\pi x}{\lambda F}}\frac{\sin\frac{\pi y}{\lambda F}}{\frac{\pi y}{\lambda F}} = \operatorname{sinc}\frac{\pi x}{\lambda F}\operatorname{sinc}\frac{\pi y}{\lambda F}.$$

The first zeros in each of the directions are given by

$$\alpha = \frac{\lambda f}{A}, \quad \beta = \frac{\lambda f}{B},$$

where A and B are the lengths of the two sides of the aperture and f is the focal length. Subsequent zeros occur at 2α, 2β, 3α, 3β, and 4α, 4β.

19.6. Encircled Energy

Another approach to describing the imaging characteristics of an optical system is the encircled energy, a plot of the integrated fraction of light received from a point object that lies within a specified radius r on the image surface. For diffraction-limited imagery, the formula is

$$E(r) = 1 - J_0^2\left(\frac{\pi r}{\lambda F}\right) - J_1^2\left(\frac{\pi r}{\lambda F}\right).$$

The preceding equations describe the situation for which the image is at the conjugate image position and there is no aberration. If the image position is moved from the perfect location, then a defocused point-spread function occurs. An equivalent statement is that the aberration of defocus is introduced. The description of the point-spread function in the presence of defocus does not readily submit to simple analytic formulas. Obvious extensions of the concept of encircled energy include ensquared energy or even enrectangled energy, both of which can be used to describe the flux on a single pixel or detector element.

There are several ways of describing the aberrations in an optical system. The most useful concept for calculating image quality is to use the wave-front aberrations in the exit pupil. The optical path is calculated along rays from an object point to the intersection of the ray with the exit-pupil reference sphere. This reference sphere is centered about the desired image reference point, usually determined by the paraxial optics of the lens system. (See Chap. 12, Geometric Optics).

The aberrations are described by a function $W(x_p, y_p)$ in the exit pupil:

$$W(x_p, y_p) = \frac{\text{OPD}(x_p, y_p)}{\lambda}.$$

The defocus aberration in wavelength units is

$$W(x_p, y_p) = w_d \rho^2,$$

where ρ is the normalized radial distance in the exit pupil that has a diameter D,

$$\rho = \frac{2}{D}\sqrt{x_p^2 + y_p^2}.$$

The linear measure of defocus δ is related to the defocus wave-front aberration by

$$\delta = 8\lambda w_d F^2.$$

One quarter wavelength of defocus error corresponds to the Rayleigh depth of focus distance:

$$\delta_{\text{Rayleigh}} = 2\lambda F^2.$$

This is generally taken as the diffraction-limited depth of focus error over which the change in image quality is considered negligible.

The pupil function for the system is then written as

$$A(x_p, y_p)e^{2\pi i W(x_p,y_p)}.$$

The point-spread function can be calculated by use of the diffraction integral:

$$\text{psf}(x, y) = \left[\iint_{\text{pupil}} A(x_p, y_p)e^{2\pi i W(x_p,y_p)} e^{\frac{-2\pi i}{\lambda R}(xx_p + yy_p)}\,\mathrm{d}x_p\mathrm{d}y_p\right]^2,$$

where R is the distance from the pupil to the image reference point. This integral is written in the form of a Fourier transform of the pupil function with respect to the image and the pupil coordinates, with the region of integration over the pupil. The square of the integral converts the amplitude to power in some geometric form.

19.7. Strehl Ratio

The Strehl ratio is the ratio of the flux density in the center of the diffraction image to the flux density that would occur there if there were no aberrations. It is unity for a perfect aberration-free system and decreases with increasing aberrations. It is frequently used as an index of image quality of systems with small amounts of aberration. A Strehl ratio of greater than 0.7 is indicative of a “substantially perfect system.”

19.8. Optical Transfer Function

The optical transfer function provides an approach to the description of the complex general imaging process that permits easy inclusion of aberrations, diffraction, and defocus. It is defined as

$$\mathrm{OTF}(\xi_x, \xi_y) = \mathrm{MTF}(\xi_x, \xi_y)\mathrm{PTF}(\xi_x, \xi_y),$$

where MTF is the modulation transfer function and PTF is the phase transfer function. For a linear, stationary system the optical transfer function is the normalized Fourier transform of the point-spread function:

$$\mathrm{OTF}(\xi_x, \xi_y) = \frac{\iint \mathrm{psf}(x, y)e^{-2\pi i(\xi_x x+\xi_y y)}\mathrm{d}x\mathrm{d}y}{\iint \mathrm{psf}(x,y)\mathrm{d}x\mathrm{d}y}.$$

The region of integration is formally over all space, but for practical purposes it need be only over a region that is large compared with the area covered by the point-spread function. The value of the optical transfer function at zero spatial frequency is unity.

The modulation transfer function represents the contrast (modulation) transfer at each spatial frequency, and the phase transfer represents the coordinate shift for each spatial frequency:

$$\mathrm{PTF}(\xi_x, \xi_y) = e^{i\phi(\xi_x,\xi_y)}.$$

For a rotationally symmetric point-spread function, the spatial phase shift can be either 0 or a multiple of π. A linear shift with frequency indicates a simple displacement of the origins at a particular field position that is due to scaling of distortion errors. A nonlinear spatial phase shift with frequency accompanies a point-spread function that has a lateral asymmetry.

The optical transfer function is the Fourier transform of the point-spread function, and the point-spread function is the complex square of the Fourier transform of the pupil function. The optical transfer function can be written as a convolution integral of the pupil function over the region of the pupil:

$$\mathrm{OTF}(\xi_x, \xi_y) = \iint_{\mathrm{pupil}} A\left(s - \frac{\xi_x}{2}, t - \frac{\xi_y}{2}\right) A\left(s + \frac{\xi_x}{2}, t + \frac{\xi_y}{2}\right) \times e^{ikW\left(s-\frac{\xi_x}{2},t-\frac{\xi_y}{2}\right)} e^{-ikW\left(s-\frac{\xi_x}{2},t-\frac{\xi_y}{2}\right)} \mathrm{d}s\mathrm{d}t.$$

In this rather imposing equation, s and t are the pupil coordinates with the shifted pupil area leading to an overlap area that gives a finite contribution to the integral. The special case of zero aberration with a circular pupil leads to a much simpler expression:

$$\mathrm{OTF}_{\mathrm{diff}} = \frac{2}{\pi}\left[\arccos\chi - \chi\sqrt{1-\chi^2}\right],$$

where χ is a spatial frequency normalized to the cutoff spatial frequency:

$$\chi = \frac{\sqrt{\xi_x^2 + \xi_y^2}}{\xi_{\mathrm{max}}} = \lambda F\sqrt{\xi_x^2 + \xi_y^2}.$$

The cutoff spatial frequency is $1/\lambda F$. The optical transfer function is zero outside the circular region defined by this cutoff frequency. The spatial period is λF and the Rayleigh resolution distance is 1.22 times this.

The calculation of optical transfer function curves for specific aberrations must be done numerically. Modern design programs do this conveniently. Some sources of plots of optical transfer functions are given in the Further Reading section.

One useful modulation transfer function plot is for defocusing. The out-of-focus geometrical image appears as a uniformly illuminated disk; the modulation transfer function is given by

$$\mathrm{MTF}_{\mathrm{geom}}(\xi) = \frac{2J_1(\pi\xi d)}{\pi\xi d},$$

where d is the diameter of the out-of-focus blur. This can also be written as

$$\mathrm{MTF}_{\mathrm{geom}}(\xi) = 2j\mathrm{inc}\left(\frac{\pi\xi\delta}{F}\right).$$

19.9. Root-Mean-Square Wave-Front Errors

Aberrations can be described in terms of root-mean-square (rms) values. The nth average of the wave-front aberration over the pupil is

$$\langle W^n\rangle = \frac{\iint_{\mathrm{pupil}} [W(x_p, y_p)]^n \mathrm{d}x_p\mathrm{d}y_p}{\iint_{\mathrm{pupil}} \mathrm{d}x_p\mathrm{d}y_p}.$$

The rms wave-front error is

$$W_{rms} = \sqrt{\langle W^2 \rangle - \langle W \rangle^2}.$$

This quantity is dimensionless because it is defined with respect to a wavelength of interest. It can also be expressed in micrometers or nanometers if this is more convenient. There is a fixed relationship between the peak and rms values for well-known aberrations, which are given in Table 19.1.

Table 19.1. Peak and rms Aberration Ratios

Aberration	Peak/rms	rms/Peak
Defocus	3.5	0.286
Third-order spherical at best focus	13.4	0.075
Fifth-order balanced spherical	57.1	0.017
Third-order coma	8.6	0.116
Third-order astigmatism, best focus	5.0	0.200
Smooth random errors	~5	~0.2

19.10. Other Useful Relations

Useful equations result from the use of rms wave-front aberrations. One of the earliest and most frequently used is the Marechal relation:

$$i_{\text{Strehl}} \equiv 1 - (2\pi W_{rms})^2.$$

The Marechal relation applies very well for W_{rms} <0.07 waves for which the Strehl ratio is approximately 0.8. This is the "generally accepted" definition of a diffraction-limited optical system. The relation becomes less accurate for larger amounts of aberration. The formula is useful as a quality index for Strehl ratios as low as ~0.25 or rms wave-front errors as great as 0.135 waves for most applications. For a broader range, the actual central irradiance in the point image is more dependent on the specific form of aberration.

For defocus, the Rayleigh defocus distance corresponds to 0.25 waves of peak focus error or to 0.07 waves of rms error.

Another useful empirical relation is

$$i_{\text{Strehl}} \equiv e^{-46W_{rms}^2}.$$

This works fairly well for rms wave-front errors as large as ~0.25 waves, depending on the type of aberration.

These wave-front errors are useful in conjunction with interferograms used to test optical systems. A useful link between the rms wave-front error and the optical transfer function is the approximate relation

$$\mathrm{MTF}_{\mathrm{app}}(\chi) = \mathrm{OTF}_{\mathrm{diff}}(\chi)\left\{1 - \left(\frac{W_{\mathrm{rms}}}{0.18}\right)^2 [1 - 4(\chi - 0.5)^2]\right\},$$

where χ is the normalized spatial frequency, $\mathrm{MTF}_{\mathrm{app}}$ is the approximate modulation transfer function, and $\mathrm{MTF}_{\mathrm{diff}}$ is the diffraction-limited modulation transfer function. This provides a reasonably good value for the modulation transfer function for average types of aberrations up to the relative spatial frequency where the modulation transfer function first reaches zero. This relationship is useful in tolerancing, comparing optical designs, and in developing specifications. It is, of course, suggested that exact calculations be carried out occasionally when possible to confirm that the approximations are valid over the range of aberrations that exist in the optical system.

Further Reading

M. Born and E. Wolf, *Principles of Optics,* Pergamon, 1959.
J. Goodman, *Introduction to Fourier Optics,* McGraw-Hill, 1968.
C. Williams and O. Becklund, *Introduction to the Optical Transfer Function,* Wiley, 1989.
R. Shannon, *The Art and Science of Optical Design,* Cambridge, 1997.
R. Fischer and B. Tadec-Galeb, *Optical System Design,* McGraw-Hill, 2000.
W. J. Smith, *Modern Optical Engineering,* McGraw-Hill, 1990.

The first of these is comprehensive on electromagnetic theory and other aspects of optical foundations. Chapter 9 is most pertinent here. The second is a basic treatment of optical performance in the spatial frequency domain, especially Chaps. 2 and 5. The third is specific on the optical transfer function, especially Chaps. 2, 5, 9, and the appendix. Chapters 4 and 6 of the next text are most pertinent. The fifth, especially Chap. 10, has information on lens-design program outputs involving the modulation transfer function, the point-spread function, etc. The final book includes Chap. 11, Image Evaluation, that covers much of the material here and shows modulation transfer functions for various aberrations.

20

Design and Engineering of Fiber-Optic Systems

Casimer DeCusatis

20.1. Introduction

Optical fibers and waveguides were first envisioned as practical devices in the 1960s; the suggestions that they could be useful as an alternative to copper wires for communication systems was first made by Kao and Hockham in 1966.[1] Early optical fibers, however, exhibited prohibitively high loss, well in excess of 10 dB/km; low-loss silica fibers were first introduced by Corning in 1970,[2] and further advances in optical material led to attenuation well below 1 dB/km in subsequent years. The development of the laser in the early 1960s and advances in semiconductor device processing have led to a revolution in the use of optical fibers for communication, sensing, image processing, metrology, medicine, and other applications. Because of the wide range of applications and the interdisciplinary nature of this field, a comprehensive treatment would easily fill many volumes; indeed, there are numerous reference works, textbooks, monographs, and technical papers that have been published in this area. The scope of this chapter is to provide an introduction to the topic and an overview of some important engineering design principles that are common to many different applications. First, optical waveguides used for integrated optics and similar applications, including the characteristic equations governing mode propagation, are discussed. This is then extended to a description of mode propagation in both single-mode and multimode optical fibers, including fundamental fiber properties such as numerical aperture, attenuation, dispersion, step-index and graded-index fibers, spectral dispersion, and fiber-bending effects. With this background, other concepts can be discussed, such as optical-fiber tapers, fiber coupling, and bundles, as well as a more detailed treatment of fibers for communication systems, including wavelength multi-

[1] K. C. Kao and G. A. Hockham, "Dielectric fiber surface waveguides for optical frequencies," Proc. IEE **113,** 1151–1158, 1966.

[2] F. P. Kapron, D. B. Keck, and R. D. Maurer, "Radiation losses in glass optical wavegudies," Appl. Phys. Lett. **17,** 423, 1970.

plexing. Measurement procedures for critical fiber parameters, such as mode field diameter, cutoff wavelength, and chromatic dispersion will also be discussed. Finally, some nonlinear optical properties of fibers are reviewed, including four-wave mixing, self-phase modulation, stimulated Brillouin and Raman scattering, and how they are used in applications such as optical amplifiers and fiber Bragg gratings.

20.2. Surface Waveguides

Optical signals can be guided in either a cylindrical fiber-optic cable or in a planar substrate such as a printed circuit board. The latter case is also called a slab waveguide or surface waveguide and is used for integrated optics applications. A typical surface waveguide is shown in Fig. 20.1; to guide light, the refractive index of the middle layer, n_1, must be higher than the refractive index of both the surrounding materials, n_2 and n_3. For simplicity, first consider the case of a symmetric light guide. A light ray incident upon the face of the guide at an angle θ will be refracted by Snell's law into an angle θ_1; the ray is subsequently incident upon the boundary between n_1 and n_2. For the ray to be guided or totally internally reflected back into the waveguide from this interface, we require that the angle $(90 - \theta_1)$ be less than the critical angle. It can be shown[3] that a cone of incident light rays will be guided along the waveguide provided that they satisfy the condition

$$\sin\theta \approx \theta = \sqrt{n_1^2 - n_2^2} = \text{NA},$$

where θ is known as the numerical aperture (NA) of the waveguide.

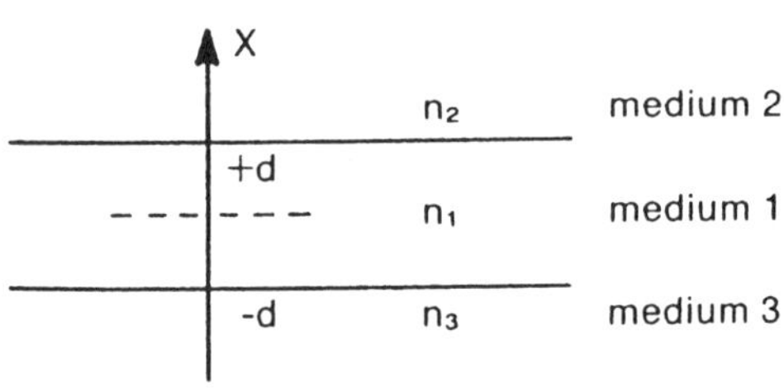

Fig. 20.1. Geometry of a typical optical surface waveguide (adapted from P. Das, *Optical Signal Processing,* Springer-Verlag, 1991, Fig. 2.45, p. 79).

When Maxwell's equations are solved with the appropriate boundary conditions, it is possible to derive expressions for the electric field of the light[4]; only the important results are discussed here. Light propagates in different optical paths, or modes, within the waveguide; these are known as transverse

[3] P. Das, *Optical Signal Processing,* Springer-Verlag, 1997.
[4] H.A. Haus, *Waves and Fields in Optoelectronics,* Prentice-Hall, 1984.

electric (TE) and transverse magnetic (TM) modes. The TE mode means that there is no electric-field component in the direction of propagation; similarly, for the TM mode there is no magnetic field in the direction of propagation. For the case of a symmetric waveguide, the modes can be further divided into even and odd modes. It can be shown (see footnotes 3 and 4) that the electric field for the even TE modes is given by

$$E_y = E_{y0} \cos \alpha_1 x e^{-jk_1 z}, \quad |x| \leq d,$$

$$E_y = \frac{\alpha_1}{\alpha_2} E_{y0}\, e^{-\alpha_2(x-d)} e^{-jk_1 z}, \quad |x| \geq d,$$

where

$$\alpha_1 = \frac{\omega^2}{c^2} n_1^2 - k_1^2,$$

$$\alpha_2 = k_1^2 - \frac{\omega^2}{c^2} n_2^2,$$

$$k_1 = \frac{2\pi n_1}{\lambda} = \text{propagation constant},$$

$$\omega = 2\pi \nu,$$

ν is the frequency, c is the speed of light, E_{y0} is a constant representing the electric-field strength, and d is the width of the waveguide. Similar results are obtained for the odd TE modes, which are given by

$$E_y = E_{y0} \sin \alpha_1 x e^{-jk_1 z}, \quad |x| \leq d.$$

Figure 20.2 plots the first four TE modes as a function of x in the waveguide. Note that, although there is total internal reflection, there is still a finite electric field in the medium n_2 that decays exponentially as we move away from the interface. This is known as an evanescent mode. The parameters α_1 and α_2 are determined from the so-called characteristic equation given by

$$\tan(\alpha_1 d) = \frac{\alpha_2}{\alpha_1}$$

for the even modes and

$$\tan(\alpha_1 d) = -\frac{\alpha_1}{\alpha_2}$$

for the odd modes. It is customary to define the quantities

$$\alpha_1 d = X, \quad \alpha_2 d = Y,$$

$$R = \frac{\omega^2 d^2}{c^2}(n_1^2 - n_2^2)$$

so that the characteristic equations become

$$Y = X \tan X \quad \text{for even modes,}$$

$$Y = -X \cot X \quad \text{for odd modes,}$$

$$X^2 + Y^2 = R^2.$$

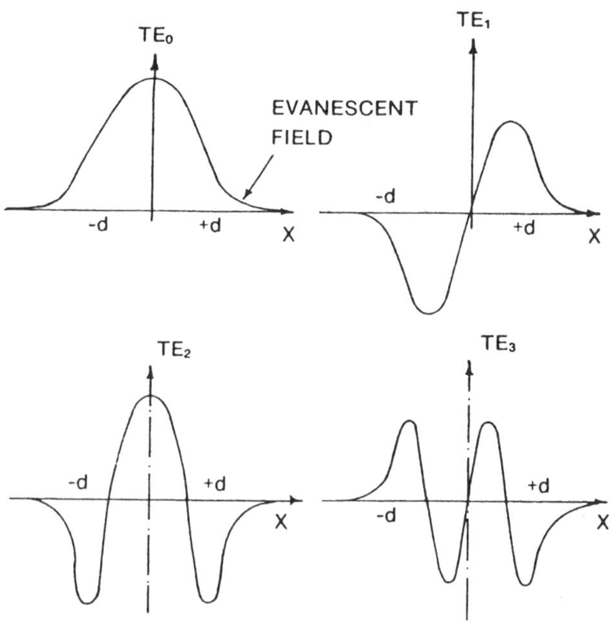

Fig. 20.2. TE modes in a dielectric slab waveguide as a function of x (adapted from P. Das, *Lasers and Optical Engineering,* Springer-Verlag, 1991, Fig. 4.3.4, p. 321).

For the parallel planar waveguide case, analytical solutions can be found; otherwise we must solve the preceding transcendental equation. These expressions are plotted in Fig. 20.3; for a particular value of R, we obtain the values for X and Y, where the circle of radius R intersects the set of curves shown in the figure. Note that the number of propagating modes increases as R increases; for $R < \pi/2$, only two modes can propagate, one even and one odd. For a multimode planar waveguide, $R \gg 2\pi$ or

$$d \gg \frac{\lambda_0}{\sqrt{n_1^2 - n_2^2}} = \frac{\lambda_0}{\text{NA}}.$$

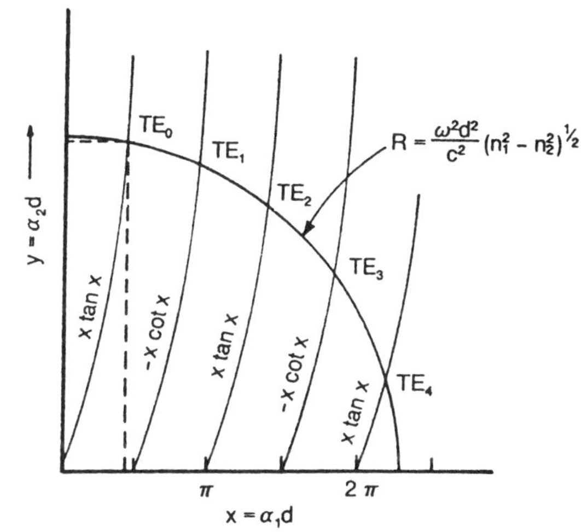

Fig. 20.3. Characteristic equation plot of TE modes in a dielectric slab waveguide (adapted from P. Das, *Optical Signal Processing,* Springer-Verlag, 1991, Fig. 2.47, p. 82).

For the case of nonsymmetric waveguides, n_2 is not equal to n_3, and the characteristic equation becomes

$$\tan(2\alpha_1 d) = \frac{\alpha_2 + \alpha_3}{1 - \alpha_2\alpha_3/2\alpha_1},$$

where

$$\alpha_3 = k_1^2 - \frac{\omega^2}{c^2} n_3^2.$$

The results for TM mode propagation are quite similar to those for TE mode propagation; the following set of characteristic equations are obtained:

$$\tan(\alpha_1 d) = \frac{n_1^2}{n_2^2}\frac{\alpha_2}{\alpha_1} \qquad \text{for even modes,}$$

$$\tan(\alpha_1 d) = -\frac{n_2^2}{n_1^2}\frac{\alpha_1}{\alpha_2} \qquad \text{for odd modes.}$$

Waveguides with a square cross section will be insensitive to the polarization of the incident light, whereas rectangular cross-sectional waveguides will in general be polarization sensitive. Finally, note that, because the propagation constant K is a function of ω, surface waveguides are in general dispersive. It can be shown (see footnotes 3 and 4) that for a length L of symmetric multimode waveguide the difference in propagation delays between the low- and the high-frequency light waves is given by

$$\Delta\tau_g = \frac{L}{c}(n_1 - n_2),$$

where we have neglected the material dispersion.

20.3. Propagation in Single-Mode Fibers

In the design of fiber-optic systems it is important to know how many modes can propagate in the fiber, the phase constants of the different modes, and their spatial profiles. To do this we need to solve the wave equation for a particular fiber geometry. The solution depends on the specific refractive-index profile of the fiber. For the case of step-index fiber profiles, a complete set of analytical solutions exists[5]; these can be grouped into three different types of modes, depending on the direction of the electric-field vector relative to the direction of propagation. They are called TE, TM, and hybrid modes. The hybrid modes can be further separated into two classes, depending on whether the electric field E or magnetic field H is largest in the transverse direction; these are called EH and HE modes, respectively. In practice, the refractive-index difference between the core and the cladding of an optical fiber is so small (~0.002 to 0.009) that most of these modes are degenerate, and it is sufficient to use a single notation for all modes, called the linearly polarized notation. A linearly polarized mode is denoted by two subscripts, which refer to the radial and the azimuthal zeros of the particular mode; for example, the fundamental mode is the LP_{01} mode. This is the only mode that will propagate in a single-mode fiber.

Optical fibers with a uniform refractive index in the core and a different index in the cladding are called step-index fibers. To determine quickly if a particular linearly polarized mode will propagate in a step-index fiber, it is useful to define two dimensionless parameters, V and b, as follows:

$$V = ka\sqrt{n_1^2 - n_2^2} = \frac{2\pi a}{\lambda}\,\mathrm{NA} \approx \frac{2\pi}{\lambda}\,an_1\sqrt{2\Delta},$$

where a is the radius of the fiber core, λ is the wavelength of light, the core and the cladding refractive indices are given by n_1 and n_2, respectively, and

[5] T. Okoshi, *Optical Fibers*, Academic, 1982.

$$\Delta \approx \frac{n_1 - n_2}{n_1}.$$

The V number is also called the normalized frequency. The normalized propagation constant b is defined as

$$b = \frac{(\beta^2/k^2) - n_2^2}{n_1^2 - n_2^2},$$

where β is the phase constant of a particular linearly polarized mode and k is the propagation constant in vacuum. This expression is cumbersome to use, as b has to be calculated directly from the wave equation; however, Marcuse et al.[6] have shown that, to a good approximation, the following formulas can be used to calculate b for different linearly polarized modes:

$$\text{LP}_{01}{:}b_{01} = 1 - \left[\frac{1 + \sqrt{2}}{1 + (4 + V^4)^{1/4}}\right]^2,$$

$$\text{LP}_{lm}{:}b_{lm} = 1 - \frac{U_c^2}{V^2} \exp\left[\frac{2}{S} \arcsin\left(\frac{S}{U_c}\right) - \arcsin\left(\frac{S}{V}\right)\right],$$

where

$$S = \sqrt{U_c^2 - l^2 - 1},$$

$$U_c = T - \frac{W - 1}{8\text{T}} - \frac{4(W - 1)(7W - 31)}{3(8T)^3},$$

$$T = \pi[m + 1/2(l - 1) - 1/4],$$

$$W = 4(l - 1)^2.$$

Using these equations, we can generate a series of graphs similar to those shown in Fig. 20.4. The wavelength for which b is zero is known as the cutoff wavelength; we can solve the preceding equation for the cutoff wavelength to obtain

$$\lambda_{\text{co}} = \frac{2\pi}{V_{\text{co}}} a n_1 \sqrt{2\Delta},$$

where the V number at cutoff is given by V_{co}. For wavelengths less than the cutoff, modes cannot propagate in the optical fiber. Note that b can vary only

6 D. Marcuse, D. Gloge, and E. A. Marcatini, "Guiding properties of fibers," in *Optical Fiber Telecommunications,* Academic, 1979, Chap. 3, pp. 38–100.

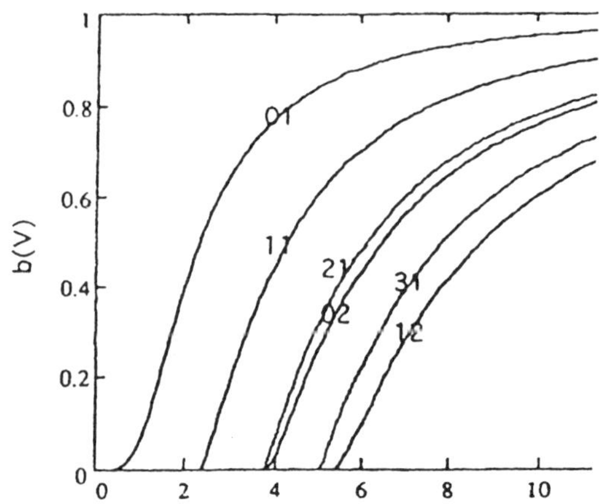

Fig. 20.4. Cutoff frequencies for the lowest-order linearly polarized modes (adapted from U. Österberg, "Optical fiber, cable, and connectors," in *Handbook of Fiber Optic Data Communication,* Academic, 1998, Chap. 1, Fig. 1.4, p. 8).

between 0 and 1. Cutoff values for the V number for a few linearly polarized modes are given in Table 20.1. For good single-mode performance, the V number should lie between 1.8 and 2.4; for values greater than 2.4, the fiber is multimode. In the range of single-mode propagation, a good analytic approximation for $b(V)$ is given by

$$b(V) = \left(1 - 1.1428 - \frac{0.996}{V}\right)^2.$$

Table 20.1. Cutoff Frequencies of Various LP_{lm} Modes in a Step-Index Fiber[a]

$l = 0$ Modes	$J_1(V_{co}) = 0$	$l = 1$ Modes	$J_0(V_{co}) = 0$
Mode	V_{co}	Mode	V_{co}
LP_{01}	0	LP_{11}	2.4048
LP_{02}	3.8317	LP_{12}	5.5201
LP_{03}	7.0156	LP_{13}	8.6537
LP_{04}	10.1735	LP_{14}	11.7915
$l = 2$ Modes	$J_1(V_{co}) = 0;\ V_{co} \neq 0$	$l = 3$ Modes	$J_2(V_{co}) = 0_i;\ V_{co} \neq 0$
LP_{21}	3.8317	LP_{31}	5.1356
LP_{22}	7.0156	LP_{32}	8.4172
LP_{23}	10.1735	LP_{33}	11.6198
LP_{24}	13.3237	LP_{34}	14.7960

[a] Adapted from U. Österberg, "Optical fiber, cables, and connectors," in *Handbook of Fiber Optic Data Communication,* Academic, 1998, Chap. 1, Table 1.1, p. 9.

Near the single-mode cutoff ($V_{co} = 2.405$) the core diameter and wavelength of a step-index fiber can be related as follows:

$$\lambda_{co} = \frac{\pi}{2.405} 2an_1\sqrt{2\Delta}.$$

20.4. Propagation in Multimode Fiber

Optical fibers in which the core is so small that only one mode can propagate are known as single-mode fibers. The core diameter in a typical single-mode fiber is ~9–10 μm. Fibers with a larger core are known as multimode; common types of multimode fiber have a core diameter of either 50 or 62.5 μm, although other sizes are also available. So far, we have assumed that the fiber core has a uniform refractive index n_1 and the cladding has a uniform refractive index n_2. This is known as a step-index fiber. Because different modes in a step-index fiber have different propagation times, this type of fiber is not commonly used for multimode applications. Instead, a fiber with a graded refractive-index core has become more common. The various graded-index profiles can be generated with the expressions

$$n^2(r) = n_1^2 \left[1 - 2\Delta\left(\frac{r}{a}\right)^q\right] \quad \text{for } 0 < r < a,$$

$$n^2(r) = n_1^2(1 - 2\Delta) = n_2^2 \quad \text{for } r > a,$$

where q is called the profile exponent and r is the radial distance away from the center of the fiber core. Many different index profiles can be generated in the fiber core.[7–10] The optimum profile occurs for q slightly less than 2, which results in the minimal dispersion. The total number of modes that can propagate in a multimode fiber N is given by the following expression, valid for large V numbers only:

$$N = 1/2 \frac{q}{q+2} V^2.$$

Note that, for the special case of a step-index fiber, q tends to infinity and $N = V^2/2$.

7 J. P. Powers, *An Introduction to Fiber Optic Systems,* Asken, 1993.
8 P. Das, *Lasers and Optical Engineering,* Springer-Verlag, 1991.
9 E. Hect and A. Zajac, *Optics,* Addison-Wesley, 1979.
10 U. Österberg, "Optical fiber, cable, and connectors," in *Handbook of Fiber Optic Data Communication,* C. DeCusatis, E. Maass, D. Clement, and R. Lasky, eds., Academic, 1st ed., 1988, 2nd ed., 2001.

20.5. Fiber Coupling

Geometric optics may be used to a good approximation when we are calculating the amount of light that can be coupled into an optical fiber. Light rays will be confined to the fiber core only if they undergo total internal reflection at the core–cladding interface. As illustrated in Fig. 20.5, this will occur only for light rays entering the fiber within an acceptance cone defined by the angle θ. It is conventional to define the numerical aperture of the fiber as

$$\mathrm{NA} = n_0 \sin\theta,$$

where n_0 is the refractive index of the medium where the light originates (usually the air, for which $n_0 = 1$). The numerical aperture can also be expressed in terms of the dimensionless parameter Δ as

$$\mathrm{NA} = n_1\sqrt{2\Delta}.$$

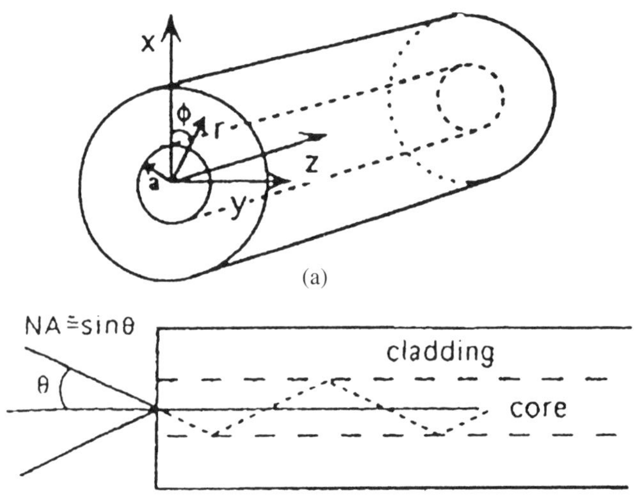

(a)

(b)

Fig. 20.5. (a) Typical geometry of an optical fiber (adapted from U. Österberg, "Optical fiber, cable, and connectors," in *Handbook of Fiber Optic Data Communication,* Academic, 1998, Chap. 1, Fig. 1.1, p. 5); (b) acceptance angle of an optical fiber (U. Österberg, "Optical fiber, cables, and connectors," in *Handbook of Fiber Optic Data Communication,* Academic, 1998, Chap. 1, Fig. 1.8, p. 13).

In general, larger numerical apertures allow more light to couple into the fiber. We may define a coupling efficiency as a ratio of the amount of optical power launched into the fiber, P_{input}, to the amount of optical power accepted into the LP_{lm} mode, P_{lm}, as

$$\eta = \frac{P_{lm}}{P_{\text{input}}}.$$

It is usually more convenient to convert coupling efficiency into units of decibels for link budget analysis; the attenuation is then given by

$$\text{attenuation} = 10 \log \eta.$$

There are two common light sources for optical fibers, incoherent (light-emitting diodes or LEDs) and coherent (lasers). Incoherent light sources are most often used with multimode fibers. It can be shown (see footnotes 8–10) that for an incoherent light source the total optical power accepted by the fiber, P_{acc}, is given by

$$P_{\text{acc}} = \pi L_s A_f (\text{NA})^2,$$

where L_s is the radiance of the light source in watts per area and steradian and A_f is the area of the fiber core. Coherent laser light can often be approximated by a Gaussian beam; for the case of single-mode fiber, the LP_{01} mode can also be approximated by a Gaussian field to better than 96% accuracy with the function

$$E(r) = E_o \exp\left[-\left(\frac{r}{w_g}\right)^2\right],$$

where E_o is the amplitude and $2w_g$ is known as the spot size or mode field diameter. As shown in Fig. 20.6, the mode field diameter is not the same as the

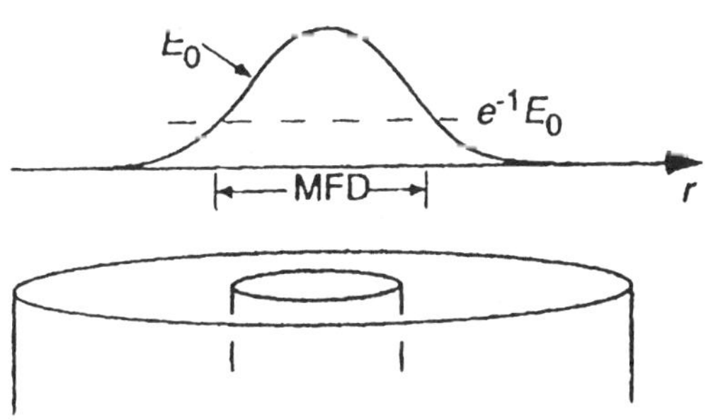

Fig. 20.6. Gaussian electric field and mode field diameter for an optical fiber. The mode field diameter is determined by the points where the optical power decreases by e^{-2} or where the amplitude decreases by e^{-1}; it is not necessarily the same diameter as the fiber core (adapted from U. Österberg, "Optical fiber, cable, and connectors," in *Handbook of Fiber Optic Data Communication,* Academic, 1998, Chap. 1, Fig. 1.5, p. 9).

diameter of the fiber core; for a single-mode fiber the mode field diameter is larger than the actual geometric diameter of the fiber whereas for a multimode fiber the mode field diameter is typically smaller than the fiber diameter. For wavelengths between 0.8 and 2.0 times the cutoff, it is possible to calculate the optimal mode field diameter[11] as

$$\frac{w_g}{a} = 0.65 + 1.619V^{-3/2} + 2.87V^{-6}.$$

In this case, it is possible to solve analytically for the coupling efficiency of a coherent source into a single-mode fiber[11,12]:

$$\eta = \left(\frac{4D}{B}\right) e^{-AC/B},$$

where

$$A = \frac{(ks_1)^2}{2},$$

$$B = G^2 + (D+1)^2,$$

$$C = (D+1)F^2 + 2DFG\sin\theta + D(G^2 + D + 1)\sin^2\theta,$$

$$D = \left(\frac{s_2}{s_1}\right)^2,$$

$$F = \frac{2\Delta}{ks_1^2},$$

$$G = \frac{2\Delta Z}{ks_1^2},$$

$$k_0 = \frac{2\pi n_0}{\lambda},$$

ΔZ is the separation distance between the source and the fiber, Δx is the lateral displacement, θ is the angular displacement, s_1 is the mode field diameter of the source, s_2 is the mode field diameter of the fiber, and n_0 is the refractive index of the medium between the source and fiber.

This expression takes into account four different coupling situations, as il-

[11] D. Marcuse, "Loss analysis of single-mode fiber splices," Bell Syst. Tech. J. **56,** 703–718, 1977.

[12] S. Nematomo and T. Makimoto, "Analysis of splice loss in single-mode fibers using a Gaussian field approximation," Opt. Quantum Electron. **11,** 447–457, 1979.

lustrated in Fig. 20.7. If only one of these is present at a time, the expressions simplify as follows:

- Spot-size mismatch ($s_1 \neq s_2$):

$$\eta = \left(\frac{2s_1 s_2}{s_1^2 + s_2^2} \right)^2 .$$

- Transverse offset Δx:

$$\eta = e\left[-\left(\frac{\Delta x}{s_2}\right)^2\right].$$

- Longitudinal offset ΔZ:

$$\eta = \frac{1}{1 + \left(\frac{\Delta Z}{2Z_0} \right)^2},$$

  where Z_0 is the Rayleigh range (see footnote 10).
- Angular misalignment θ from an original angle θ_0:

$$\eta = e\left[-\left(\frac{\theta}{\theta_0}\right)^2\right].$$

Coupling efficiency can be improved by use of a lens between the source and the fiber, as shown in Fig. 20.8.[13] The lens matches the output angle of the source to the acceptance angle of the fiber. In this case, the previous formulas must be multiplied by the lens magnification factor,

$$M = \frac{d_{\text{receive}}}{d_{\text{emit}}},$$

which is the ratio of the received and the emitted spot diameters. The formulas also need to be corrected for reflection losses; if the refractive index of the material between the source and the fiber is n_0, the coupled power into the fiber must be reduced by a factor ρ, given by

$$\rho = \left(\frac{n_1 - n_0}{n_1 + n_0} \right)^2 .$$

To make a worst-case estimate of the coupling loss in a fiber joint, the fiber-coupling-loss nomograph of Fig. 20.9 may be used. The only specifications re-

[13] J. R. Webb and U. Österberg, "Fiber, cable, and coupling," in *Optoelectronics for Data Communication,* R. Lasky, U. Österberg, and D. Stigliani, eds., Academic, 1997, pp. 7–76.

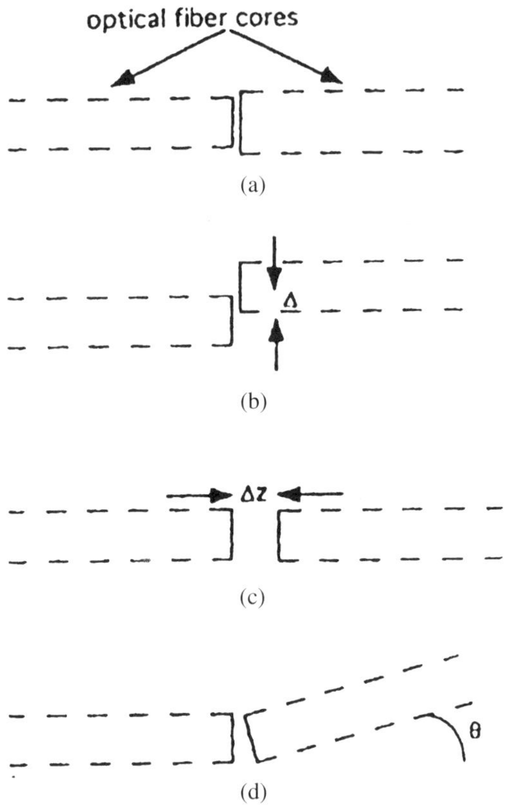

Fig. 20.7. Examples of different types of coupling misalignment between optical fiber cores: (a) spot-size mismatch (b) transverse misalignment (c) longitudinal misalignment (d) angular misalignment (adapted from U. Österberg, "Optical fiber, cable, and connectors," in *Handbook of Fiber Optic Data Communication,* Academic, 1998, Chap. 1, Fig. 1.9, p. 15).

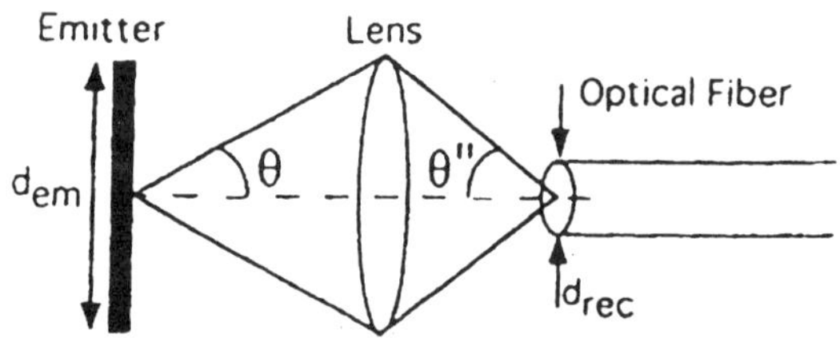

Fig. 20.8. Coupling from an extended optical source into a fiber by use of a lens (adapted from U. Österberg, "Optical fiber, cable, and connectors," in *Handbook of Fiber Optic Data Communication,* Academic, 1998, Chap. 1, Fig. 1.10, p. 16).

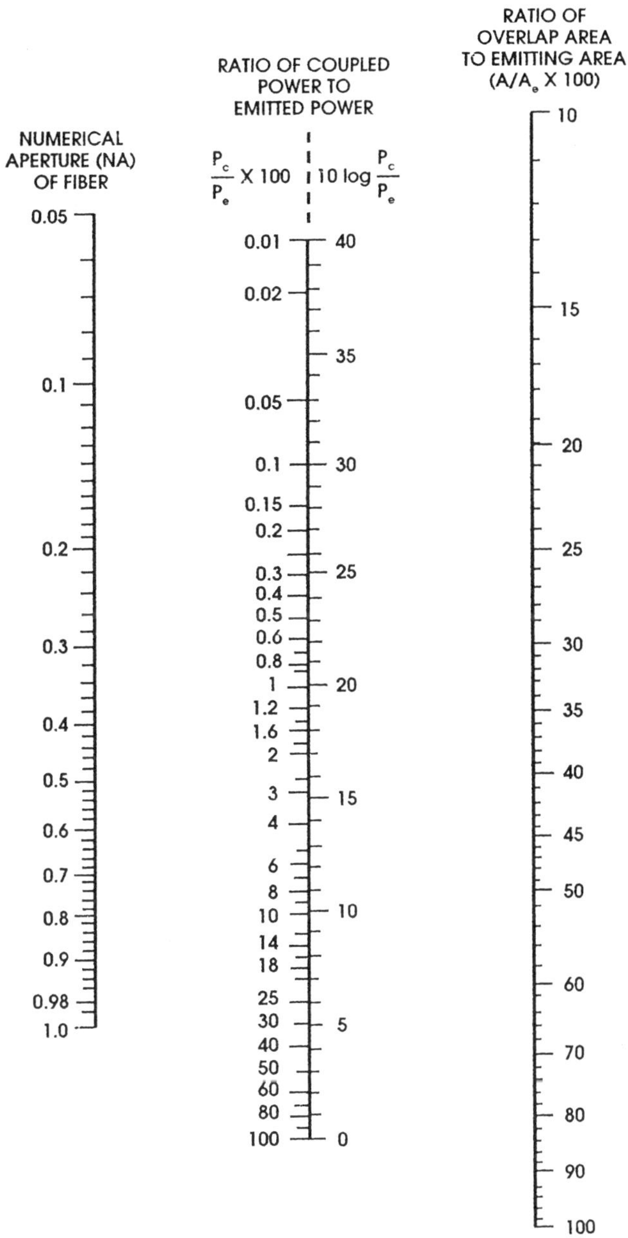

Fig. 20.9. Fiber-coupling-loss nomograph; given specifications from optical-fiber and light-source data sheets plus the overlap area between the emitter and the fiber core, we can estimate the worst-case coupling loss from a butt joint by putting a straightedge across the nomograph. We assume that all specified power is emitted from the front of the light source and that the source is perfectly butted to the fiber with only an infinitesimal air gap; a step-index fiber is assumed; for a graded-index fiber the loss increases by ~1 dB (adapted from R. Heyler, "Photonics packaging, solving the manufacturing bottleneck," in *The Photonics Design and Applications Handbook,* Lauren, 2000, p. H-173).

quired are commonly available on engineering data sheets, plus the area of overlap between the emitting area and the fiber core; coupling loss can then be estimated by placing a straightedge across this figure. In the case of an optical transmitter (LED or laser diode) butt coupled to a fiber, the overlap region must be calculated with planar geometry; this is illustrated in Fig. 20.10 for the

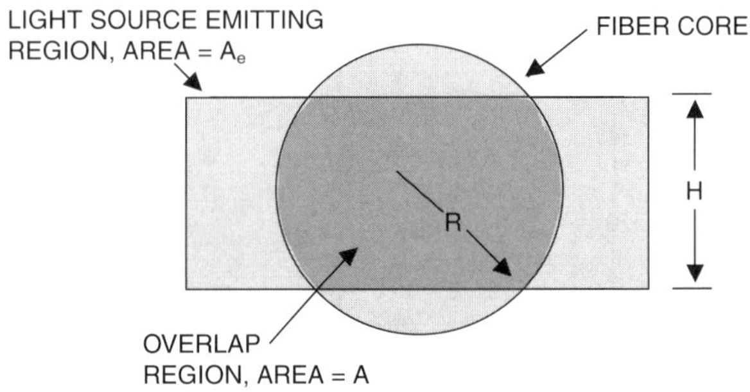

Fig. 20.10. Overlap area between a light emitter and a fiber core, used in calculating parameters for the nomograph of Fig. 20.9 (adapted from R. Heyler, "Photonics packaging, solving the manufacturing bottleneck," in *The Photonics Design and Applications Handbook,* Lauren, 2000, p. H-173).

common case of a rectangular transmitter coupled into a circular fiber. This figure allows the designer to compute trade-offs in a fiber system, such as numerical aperture vs. fiber core diameter, in order to optimize the system's coupling loss. In Fig. 20.10 we assume that all specified power is emitted from the front of the light source where the butt is made and that the source is tightly butted against the fiber, leaving only an infinitesimal air gap. Step-index fibers are assumed; for dimensionally similar graded-index fibers, the loss will be ~1 dB greater than the figure indicates.

20.6. Optical Transmission

Most optical fibers are based on silica glass; as such, there are various absorption and scattering properties of optical signals in these fibers. For example, impurities in the fiber core contribute to absorption of the light, structural imperfections in the waveguides contribute to losses at the core–cladding interface, and fundamental mechanisms such as Rayleigh scattering and infrared absorption bands in the glass set limitations on the amount of light that can be transmitted through an optical link. Attenuation is perhaps the most important factor in the design of effective fiber-optic communication links; Table 20.2 shows some of the most common factors affecting attenuation, and Fig. 20.11 illustrates the transmission loss in silica optical fibers. These loss factors are either intrinsic properties of the glass or extrinsic properties resulting from the

Table 20.2. Factors Affecting Attenuation in Optical Fibers[a]

Intrinsic loss mechanisms
Tail of infrared absorption by Si—O coupling
Tail of ultraviolet absorption that is due to electron transitions in defects
Rayleigh scattering that is due to spatial fluctuations of the refractive index
Absorption by impurities
Absorption by molecular vibration of OH
Absorption by transition metals
Structural imperfections
Geometrical nonuniformity at core–cladding boundary
Imperfection at connection or splicing between fibers

[a] Adapted from U. Österberg, "Optical fiber, cable, and connectors," in *Handbook of Fiber Optic Data Communication,* Academic, 1998, Chap. 1, Table 1.4, p. 18.

manufacturing processes (which have improved steadily over the years). We can see from Fig. 20.11 that there are three transmission windows in the optical fiber, in the wavelength ranges near 800, 1300, and 1550 nm. The minimum attenuation occurs near 1550 nm, ~0.2 dB/km; the loss near 1300 nm is ~0.5 dB/km, and near 800 nm is ~2–3 dB/km. Although it is tempting to always use the minimum attenuation wavelengths, other factors influence the selection of an operating wavelength, including dispersion, compliance with industry standards, and cost of the source and fiber.

A useful property of optical fibers is that different wavelengths of light do not interfere with each other when transmitted through the same fiber; this principle, which is similar to that of an FM radio, is known as wavelength division multiplexing (WDM). Many different kinds of optical multiplexing are possible, and wavelength-multiplexing systems may be classified as follows.[14,15]

- Coarse wavelength-division-multiplexing systems typically used only 2–3 wavelengths widely spaced, for example, 1300 and 1550 nm. Applications of this technology are somewhat limited, although recently coarse wavelength-division-multiplexing systems with four to eight channels have been used in small networks, and two wavelength-multiplexing systems are also used to remotely pump some types of optical amplifiers.
- Wide-spectrum wavelength-division-multiplexing systems can support

[14] "Market Report on Future Fiber Optic Technologies," available from Frost and Sullivan, New York, May, 1999.

[15] C. DeCusatis and P. Das, "Subrate multiplexing using time and code division multiple access in dense wavelength division multiplexing networks," in *Optical Networking and Communications,* I. Chlamtac, ed., Proc. SPIE **4233,** 3–11, 2000.

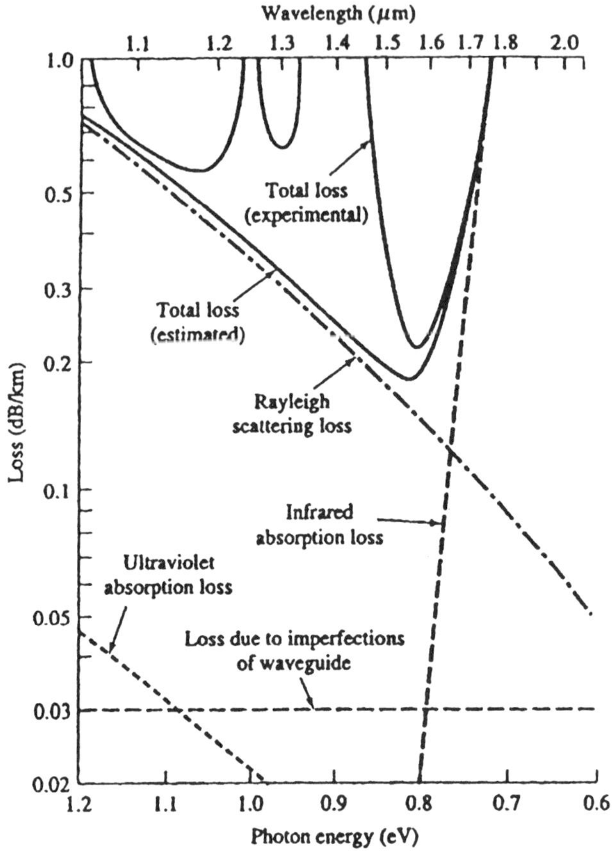

(a)

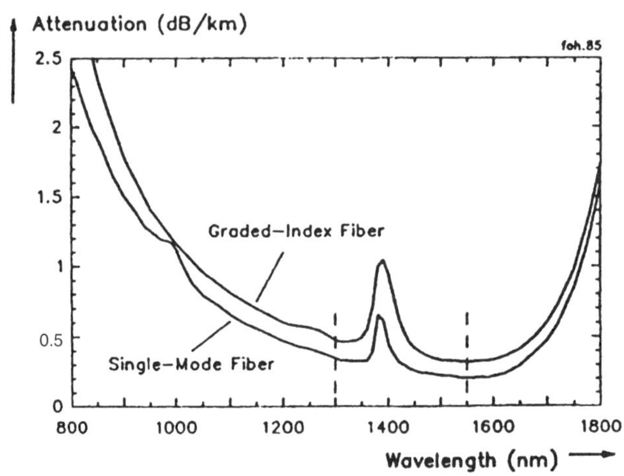

(b)

Fig. 20.11. (a) Transmission loss in silica-based fibers; (b) wavelength dependence of fiber attenuation (adapted from U. Österberg, "Optical fiber, cable, and connectors," in *Handbook of Fiber Optic Data Communication,* Academic, 1998, Chap. 1, Fig. 1.12, p. 19, and Fig. 1.13, p. 20).

up to 16 channels, using wavelengths that are spaced relatively far apart; there is no standardized wavelength spacing currently defined for such systems, although spacings of 1–30 nm have been used. They are meant to serve as a low-cost alternative to more dense wavelength-division-multiplexing systems for applications that do not require large numbers of channels on a single fiber path. For example, a form of wide-spectrum wavelength division multiplexing is being considered as an option for the emerging 10-Gbit/s Ethernet data communication standards. These are also known as sparse wavelength-division-multiplexing systems.

- Dense wavelength-division-multiplexing systems use wavelengths spaced much closer together, typically following multiples of the International Telecommunication Union industry standard wavelength grid,[16] with wavelengths near 1550 nm and a minimum wavelength spacing of 0.8 nm (100 GHz). A list of common International Telecommunication Union grid wavelengths is given in Table 20.3. First-generation dense wavelength-division-multiplexing systems typically use up to eight full duplex channels multiplexed into a single duplex channel.
- Second-generation dense wavelength-division-multiplexing systems use up to 16 channels.
- Third-generation dense wavelength-division-multiplexing systems use up to 32 channels; this is the largest system currently in commercial production for data communication applications.
- Fourth-generation or ultradense wavelength-division-multiplexing systems are expected to use 40 channels or more and may deviate from the current International Telecommunication Union grid wavelengths; channel spacing as small as 0.4 nm (50 GHz) has been proposed.[17] These systems are not yet commercially available and are the subject of much ongoing research.

If the optical fiber is improperly cabled or installed, additional loss can be experienced because of bending of the fibers. This falls into two categories, microbending (which is due to nanometer-scale variations in the fiber) and macrobending (which is due to much larger, visible bends in the fiber). Figure 20.12 shows qualitatively how both types of bending loss may contribute to the attenuation of a single-mode or multimode fiber; it is obviously desirable to minimize bending whenever possible. Most qualified optical-fiber cables spec-

[16] "Optical interface for multichannel systems with optical amplifiers," draft standard G.MCS, annex A4 of standard COM15-R-67-E, available from the International Telecommunication Union, 1999.

[17] M. Ferries, "Recent developments in passive components and modules for future optical communication systems," in Digest of OSA Annual Meeting, Optical Society of America, 1999, paper MI1, p. 61.

Table 20.3. International Telecommunication Union Grid Standard Wavelengths for Dense Wavelength-Division-Multiplexing Systems[a]

Wavelength (nm)	Frequency (THz)	Wavelength (nm)	Frequency (THz)
1528.77	196.1	1567.13	191.3
1529.55	196.0	1567.94	191.2
1530.33	195.9	1568.77	191.1
1531.12	195.8	1569.59	191.0
1531.90	195.7	1570.42	190.9
1532.68	195.6	1571.24	190.8
1533.47	195.5	1572.06	190.7
1534.25	195.4	1572.05	190.6
1535.04	195.3	1573.71	190.5
1535.82	195.2	1574.54	190.4
1536.61	195.1	1575.37	190.3
1537.40	195.0	1576.19	190.2
1538.19	194.9	1577.03	190.1
1538.98	194.8	1577.85	190.0
1539.77	194.7	1578.69	189.9
1540.56	194.6	1579.51	189.8
1541.35	194.5	1580.35	189.7
1542.14	194.4	1581.18	189.6
1542.94	194.3	1582.02	189.5
1543.73	194.2	1582.85	189.4
1544.53	194.1	1583.69	189.3
1545.32	194.0	1584.52	189.2
1546.12	193.9	1585.36	189.1
1546.92	193.8	1586.19	189.0
1547.72	193.7	1587.04	188.9
1548.51	193.6	1587.88	188.8
1549.32	193.5	1588.73	188.7
1550.12	193.4	1589.56	188.6
1550.92	193.3	1590.41	188.5
1551.72	193.2	1591.25	188.4
1552.52	193.1	1592.10	188.3
1553.33	193.0	1592.94	188.2
1554.13	192.9	1593.80	188.1
1554.94	192.8	1594.64	188.0
1555.75	192.7	1595.49	187.9
1556.55	192.6	1596.34	187.8
1557.36	192.5	1597.19	187.7
1558.17	192.4	1598.04	187.6
1558.98	192.3	1598.89	187.5
1559.79	192.2	1599.74	187.4
1560.61	192.1	1600.60	187.3
1561.42	192.0	1601.45	187.2
1562.23	191.9	1602.31	187.1
1563.05	191.8	1603.16	187.0
1563.86	191.7	1604.02	186.9
1564.67	191.6	1604.88	186.8
1565.48	191.5	1605.73	186.7
1566.31	191.4		

[a] The C band (or blue band) extends from ~1528.77 nm (196.1 THz) to 1566.31 nm (191.4 THz), and the L band (or red band) extends from 1565.48 nm (191.5 THz) to 1605.73 nm (186.7 THz). The minimum channel spacing is 0.8 nm (100 Ghz) anchored to a 193.1-THz reference (after International Telecommunication Union standard G.MCS Annex A of COM15-R 67-E).

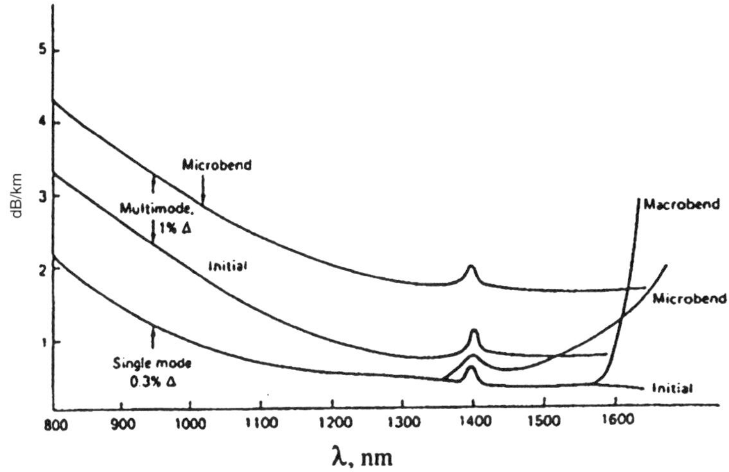

Fig. 20.12. Bend-induced losses of optical fibers (adapted from U. Österberg, "Optical fiber, cable, and connectors," in *Handbook of Fiber Optic Data Communication,* Academic, 1998, Chap. 1, Fig. 1.9, p. 15).

ify a maximum bend radius to limit macrobending effects, typically around 10–15 mm.[10]

Another important factor in fiber-optic link design is dispersion, which refers to the fact that different wavelengths of light will experience different propagation constants and therefore travel with different velocities along the fiber. This does not alter the wavelength (frequency) content of the light, but rather makes a narrow optical pulse broaden in time as it propagates; dispersion therefore affects the bit rate in communication systems. When an optical pulse propagates in a dispersive media, it travels with the group velocity

$$v_g = \frac{\mathrm{d}\omega}{\mathrm{d}k_1},$$

which determines the dispersion properties of the fiber. The phase travels with the phase velocity given by

$$v_p = \frac{\omega}{k_1} = \frac{c}{n_1}.$$

There are three major components to dispersion. First, material dispersion occurs because the refractive index is a function of wavelength, which can be calculated from the so-called Sellmeier equation[18]:

[18] N. Shibata and T. Edahiro, in Trans. Inst. Electron. Commun. Eng. Jpn. **65,** 166–172, 1982.

$$n_1^2(\lambda) = 1 + \sum_{i=1}^{3} \frac{a_i \lambda^2}{\lambda^2 - b_i},$$

where the constants a_i and b_i vary with the particular fiber composition.[10,18] Material dispersion is given by

$$D_{\text{mat}} = -\frac{\lambda}{c}\frac{\mathrm{d}^2 n_1}{\mathrm{d}\lambda^2}.$$

Second, waveguide dispersion is caused by the different propagation characteristics of light in the core and cladding. This can be approximated by

$$D_{wg} \approx -\frac{\lambda}{c} n_1 \Delta \frac{\mathrm{d}^2 b}{\mathrm{d}\lambda^2}.$$

Together, material and waveguide dispersion are called chromatic dispersion; the total dispersion is plotted in Fig. 20.13. Finally, intermodal dispersion arises from different modes having different travel times through the fiber; this dispersion is defined[19] as the difference between the propagation times for the modes traveling the longest and the shortest distances in the fiber. We can approximate the modal dispersion in a step-index fiber as

$$D_{\text{mod}} = -\frac{N_1 \Delta}{c},$$

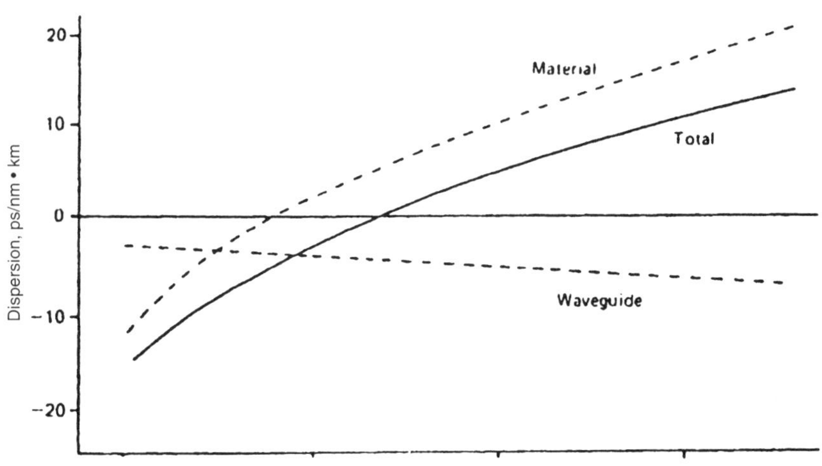

Fig. 20.13. Chromatic dispersion vs. wavelength for a single-mode fiber (adapted from U. Österberg, "Optical fiber, cable, and connectors," in *Handbook of Fiber Optic Data Communication,* Academic, 1998, Chap. 1, Fig. 1.15, p. 22).

[19] M. M. K. Liu, "Fiber optic cable and coupling," in *Principles and Applications of Optical Communications,* R. Lasky, U. Österberg, and D. Stigliani, eds., Irwin, 1996.

where N_1 is the group refractive index, and in a graded-index fiber as

$$D_{\text{mod}} = -\frac{N_1 \Delta^2}{8c}.$$

The total dispersion in a multimode fiber is then given by

$$D^2_{\text{total}} = D^2_{\text{mat}} + D^2_{\text{wg}} + D^2_{\text{mod}}.$$

This can be translated into bandwidth of the fiber, which is specified in megahertz times kilometers according to

$$\text{BW} = \frac{350}{D_{\text{total}}}.$$

Although most optical fibers are made of doped silica glass, plastic optical fibers are also available; these are commonly used in applications that do not require long transmission distances, such as medical instrumentation, automobile and aircraft control systems, and consumer electronics. Both step-index and graded-index plastic fiber are available, although only the step-index fiber is considered a commercial product at this time. Although there are many potential plastics that could be considered, the most commonly used is poly(methyl methacrylate). Doping the poly(methyl methacrylate) fibers can dramatically change their transmission properties; for example, Fig. 20.14 shows typical plastic-fiber attenuation vs. wavelength for both a standard poly(methyl methacrylate) fiber and a more recently introduced fiber in which deuterium replaces hydrogen in some parts of the polymer molecules. Attenuation is very high compared with that of glass fiber at all wavelengths; trans-

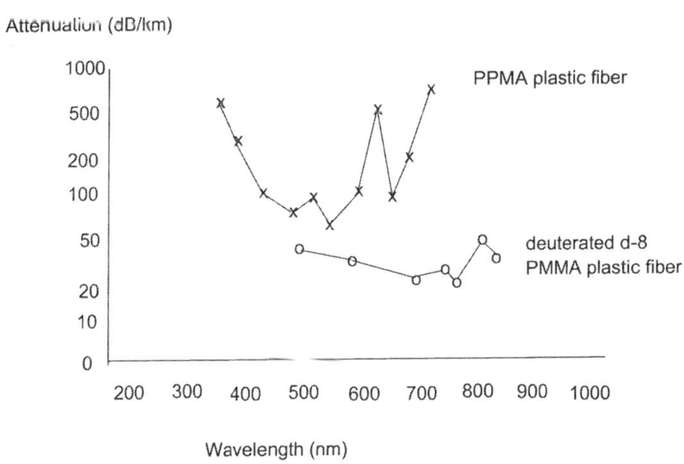

Fig. 20.14. Attenuation vs. wavelength for plastic optical fibers and the effects of regular vs. deuterated materials.

mission windows that use visible light near 570 and 650 nm are feasible for short distances (up to 100 m). At this distance, a step-index fiber has a bandwidth of ~125 MHz, whereas for graded-index fibers bandwidths of better than 500 MHz have been reported.[20] Tables 20.4 and 20.5 give some typical

Table 20.4. Typical Specifications of Plastic Optical Fiber

Parameter	Specification
Core diameter	980 μm
Cladding diameter	1000 μm (1 mm)
Jacket diameter	2.2 mm
Attenuation (at 850 nm)	< 18 dB/100 m (180 dB/km)
Numerical aperture	0.30
Bandwidth (at 100 m)	Step index: 125 MHz Graded index: 500 MHz

Table 20.5. Typical Specifications of Single-Mode Glass Optical Fiber [per Consultative Committee for International Telephone and Telegraph (CCITT) Recommendation G.652]

Parameter	Specification
Core diameter (mode field diameter)	9–10 μm
Cladding diameter	125 μm
Cutoff wavelength	1100–1280 nm
Bend loss at 1550 nm	< 1 dB for 100 turns around a 7.5-cm-diameter mandrel
Dispersion	< 3.5 ps/nm km between 1285 and 1330 nm < 6 ps/nm km between 1270 and 1340 nm < 20 ps/nm km at 1550 nm
Dispersion slope	< 0.095 ps/nm^2 km

properties of a plastic optical fiber and a single-mode glass optical fiber, respectively; note the large core diameter of the plastic fiber, which is 100 times bigger than that of a single-mode fiber. The combination of simplified alignment with optical sources and detectors as well as the low cost of visible optical sources makes a plastic optical fiber cost competitive for some applications. A major drawback to plastic fiber is the difficulty in creating fusion splices with acceptable attenuation; typical splice losses are ~5 dB. Other combinations of fiber materials are the subject of ongoing research, in particular hard-polymer-clad (glass) fiber.

[20] H. Dutton, *Optical Communications*, Academic, 1999.

20.7. Optical Cable, Couplers, and Splitters

There are many different types of fiber-optic cable; as shown in Fig. 20.15, it is possible to package multiple fibers into a single unibody cable or into a ribbon or zipcord structure. Bundles of fibers whose ends are bound together, ground, and polished can form flexible light pipes. Of course, it is possible to bundle the fibers in such a way that there is no fixed relationship between the location of an input fiber and an output fiber; the principal purpose of such structures is to conduct light from one location to another, for illumination as an example; these are sometimes referred to as incoherent bundles, although they have little to do with optical coherence theory. A more interesting case is that in which the fibers are carefully arranged so that the fibers occupy the same relative positions at both ends of the bundle; such bundles are said to be coherent. A coherent bundle of single-mode fiber is capable of conducting a high-quality image even when the bundle is made highly flexible; such fiber arrays have many applications in remote vision systems and are used in fiber-optic endoscopes for medical applications. Not all fiber arrays are made flexible; fused, rigid bundles or mosaics can be used to replace low-resolution sheet glass in cathode ray tubes. Mosaics consisting of hundreds to millions of individual fibers with their claddings fused together have mechanical properties very much like those of homogeneous glass. Another common application of mosaics is as a field flattener. If the image formed by a lens system falls on a curved surface, it is often desirable to reshape it into a plane, for example, to match a photographic film plate. A mosaic can be ground and polished on one end surface to correspond with the contours of the image and on the other surface to match the configuration of the detector. Similarly, a sheet of fused tapered fibers can be used either to magnify an image or to miniaturize an image, depending on whether the light enters the smaller or the larger end of the fibers.

Many simple devices such as fiber-optic splitters, couplers, and combiners have been manufactured; the most common techniques include fiber tapering and fusion splicing,[21–24] etching,[25] and polishing.[26–28] Other fabrication tech-

[21] D. Nolan, "Tapered fiber couplers, mux and demux," in *Handbook of Optics*, Optical Society of America, 2000, Vol. IV, Chap. 8.

[22] T. Ozeki and B. S. Kawaski, "New star coupler compatible with singe multimode fiber links," Electron. Lett. **12,** 151–152, 1976.

[23] B. S. Kawaski and K. O. Hill, "Low loss access coupler for multimode optical fiber distribution networks," Appl. Opt. **16,** 1794–1795, 1977.

[24] G. E. Rawson and M. D. Bailey, "Bitaper star couplers with up to 100 fiber channels," Electron. Lett. **15,** 432–433, 1975.

[25] S. K. Sheem and T. G. Giallorenzi, "Singlemode fiber optical power divided; encapsulated teching technique," Opt. Lett. **4,** 31, 1979.

[26] Y. Tsujimoto, H. Serizawa, K. Hatori, and M. Fukai, "Fabrication of low loss 3 dB couplers with multimode optical fibers," Electron. Lett. **14,** 157–158, 1978.

[27] R. A. Bergh, G. Kotler, and H. J. Shaw, "Singlemode fiber optic directional coupler," Electron. Lett. **16,** 260–261, 1980.

[28] O. Parriaux, S. Gidon, and A. Kuznetsov, "Distributed coupler on polished singlemode fiber," Appl. Opt. **20,** 2420–2423, 1981.

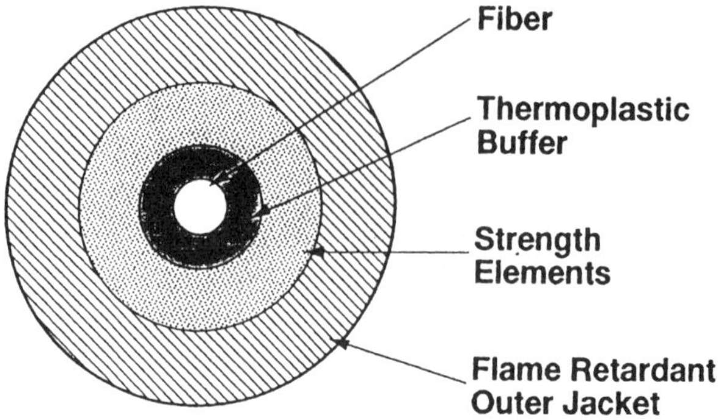

(a)

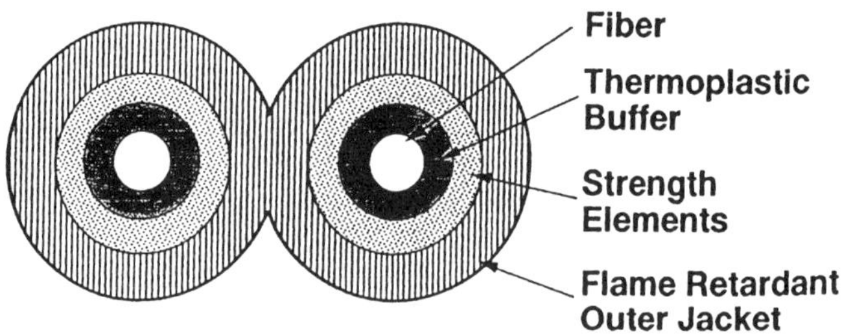

(b)

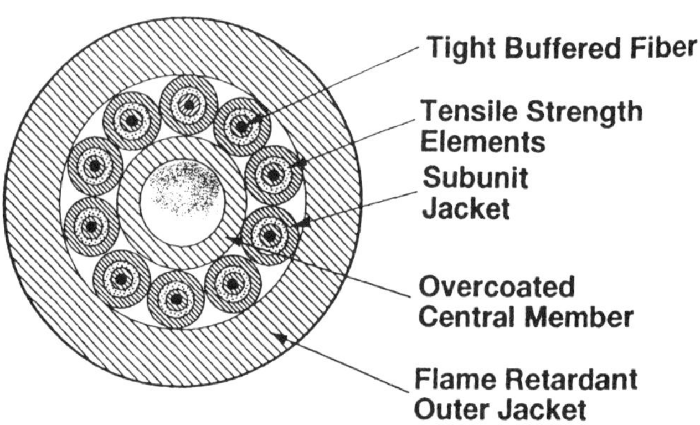

(c)

Fig. 20.15. Different types of optical-fiber cables: (a) single-fiber unibody cable (can also be used for dual-fiber cables), (b) two-fiber zipcord cable, (c) tight buffered fan-out cable with strength elements and multiple fibers (adapted from R.T. Hudson, D.R. King, T.H. Rhyne, and T.A. Torchia, "Planning and building the optical link," in *Handbook of Fiber Optic Data Communication,* Academic, 1998, Chap. 8, Figs. 8.9, 8.10, and 8.11, p. 272).

niques can also be used, including micro-optics and integrated optical components; however, optical-fiber devices are particularly useful because they can be inserted into existing networks as just another piece of cable. One of the most common devices is a tapered fiber-optic power splitter, often implemented in a single-mode fiber.[29] In this process, two glass fibers with their protective jackets removed are brought close together and parallel to each other, then fused and stretched by a torch or a similar heat source. Light that is initially launched into only one fiber will be partially coupled into the adjacent fiber as it propagates through the tapered region. Light propagating in the single-mode fiber is not confined to the core but extends into the surrounding cladding. In the case of a fiber taper it has been shown[30] that light propagating through the input fiber core is initially transferred to the cladding interface as it enters the tapered region, then to the core–cladding mode of the adjacent fiber. The light transfers back to the core modes as it exits the tapered region. This is known as a cladding-mode coupling device. Light that is transferred to a higher-order mode of the core–cladding structure is readily stripped away by the higher refractive index of the fiber coating, resulting in excess attenuation. The simplest case of light coupling from the cladding of one fiber into another through a fused taper can be described to a good approximation by the scalar wave equation and first-order perturbation theory[31]; if light is propagating along the z axis, then the exchange of optical power P is given by

$$P = \sin^2(\kappa z),$$

where

$$\kappa = \frac{2\pi}{\lambda k_1^2}\sqrt{1-(n_1/n_2)^2}\,\frac{n_1^2-n_2^2}{(n_1^2-n_2^2)^{1.5}}\,\frac{K_0[2k_1+(2\pi z_d/\lambda)]\sqrt{n_1^2-n_2^2}}{(k_1 n_1^2-n_2^2)}.$$

Here, K_0 is a constant[21] and z_d is the overlap distance between the two fibers. Although this is only an approximation and neglects higher-order terms, it does reflect the sinusoidal dependence of coupled power on wavelength and the dependence of power transfer on cladding diameter and other effects. Tapered couplers can be used to separate wavelengths by use of this dependence; by proper choice of the device length and taper ratio, two wavelengths can be made to emerge from two different output ports. Some applications include filters for wavelength-division-multiplexing systems, or multiplexing signal and pump beams in an erbium-doped fiber amplifier. In some cases, such as optical

[29] B. S. Kawaski, K. O. Hill, and R. G. Lamont, "Biconical-tapered singlemode fiber coupler," Opt. Lett. **6,** 327, 1981.

[30] R. G. Lamont, D. C. Johnson, and K. O. Hill, Appl. Opt. **24,** 327–332, 1984.

[31] A. Snyder and J. D. Love, *Optical Waveguide Theory,* Chapman & Hall, 1983.

power splitters, it is more desirable to remove the dependence of coupled power on wavelength; achromatic couplers can be fabricated from two fibers with different propagation constants. These are known as dissimilar fibers; in most cases, fibers are made dissimilar by a change in their cladding diameters or cladding indices. In this case, the preceding equations for coupled power P_1 can be modified to

$$P_1(z) = P_1(0) + \{F_0^2[P_2(0) - P_1(0) + (k_1\, k_2)/\kappa][P_1(0)P_2(0)]^5\}\sin^2(\kappa z/F_0),$$

where

$$F_0 = \frac{1}{\sqrt{1 + (k_1^2 - k_2^2)/4\kappa^2}}.$$

Other approaches are also possible, such as tapering the device such that the modes expand well beyond the cladding boundaries[32] or encapsulating the fibers in a third material with a different refractive index.[33,34] Often it is desirable to taper multiple fibers together so that an input signal is split between many output fibers. Typically, a single input is split into 2^M outputs (i.e., 2, 4, 8, or 16), where the configuration of fibers in the tapered region affects the output power distribution; care must be taken to achieve a uniform optical power distribution among the output fibers.[35] The optical power coupled from one fiber into another can also be changed if the tapered device is bent at its midpoint; this frustrates coupled power transfer. For example, displacing one end of a 1-cm-long taper by only 1 mm can change the coupled power by over 30 dB.[21] Applications for this effect include variable optical attenuators and optical switches.

20.8. Performance Measurements

In this section are briefly introduced the techniques used for measurement of attenuation, mode field diameter, cutoff wavelength, chromatic dispersion, fiber loss over distance, and bit error rates of a fiber-optic communication link. More details descriptions can be found in the industry standardized fiber-measurement procedures; these include the fiber-optic test procedures and Elec-

32 D. L. Weidman, "Achromat overclad coupler," U.S. patent 5,268,979, 14 December 1993.

33 C. M. Truesdale and D. A. Nolan, "Core-clad mode coupling in a new three-index structure," presented at the European Conference on Optical Communication, 3–7 May 1986, Barcelona, Spain, 1986.

34 D. B. Keck, A. J. Morrow, D. A. Nolan, and D. A. Thompson, J. Lightwave Technol. **7,** 1623–1633, 1989.

35 W. J. Miller, D. A. Nolan, and G. E. Williams, "Method of making a 1 × *N* coupler," U.S. patent 5,017,206, 22 April 1991.

tronics Industry Association/Telecommunications Industry Association standards, as well as standards defined by the American National Standards Institute and the International Standards Organization.[36]

Attenuation is a fundamental parameter of any fiber-optic system; one common measurement technique is the cutback method. Fiber transmission as a function of wavelength, $P_A(\lambda)$, is measured for the full length of the fiber. The fiber is then cut ~1–2 mm away from the input end, and the transmission is measured once again. If this second measurement is called $P_B(\lambda)$, then the fiber attenuation constant α_f may be calculated from

$$\alpha_f = 10 \log \left[\frac{P_A(\lambda)}{P_B(\lambda)} \right].$$

Note that this is a destructive method, although the fiber length is not reduced by a significant amount. The cutback method is typically used during fiber manufacturing. To measure the total attenuation of a connectorized fiber link nondestructively in the field or laboratory, other methods may be used. For example, an optical power meter and source (either a laser or an LED, depending on the fiber type and intended application) can be used.[37] First, a reference reading is taken with the source directly connected to the power meter by a reference fiber cable. The reference cable should be fabricated and characterized to strict manufacturing tolerances to remove potential error sources such as eccentricity in the fiber connectors. Once this reading is established, the source can be used to measure a desired section of connectorized fiber cable; the difference between the two power readings gives the total cable loss. There are two types of measurements, known as single-ended and double-ended loss. We measure the single-ended loss by keeping the reference cable attached to the light source and mating the fiber under test to the reference cable on one end and the power meter on the other. This approach measures the loss of the connector mated to the launch cable as well as the loss of the fiber cable under test and any splices or connectors in the cable under test. Note that the cable under test can be reversed to measure the loss of each connector end individually. In a double-ended test, two reference cables are used, one attached to the light source and one to the power meter; the cable under test is inserted between them. In this way, the loss of two connectors is included in the measurement of the fiber under test. There are various sources of uncertainty in this measure-

[36] FOTP, ANSI and EIA/TIA standards may be ordered from Global Engineering Documents, 3130 South Harbor Road Blvd. #330, Santa Anna, Calif. (FOTP-XXX is equivalent to EIA/TIA-455-XXX).

[37] J. Hayes and G. LeCheminant, "Testing fiber optic local area networks," in *Handbook of Fiber Optic Data Communication,* C. DeCusatis, E. Maass, D. Clement, and R. Lasky, eds., Academic, 1st ed., 1988, 2nd ed., 2001, Chap. 9.

ment, including tolerances on the power meter and source and statistical variation of the optical connector loss; still, it should be possible to achieve tolerances within 0.5 dB under good conditions.

The source–meter measurement technique has the disadvantage of requiring access to both ends of the fiber link; sometimes this is not practical, if the link is very long or has intermediate connection points that are not readily accessible. Another measurement technique estimates the total fiber loss by launching a narrow optical pulse into the fiber and measuring the strength of light reflected back from any discontinuities such as connectors, splices, or the far end of the link. It is then possible to generate a graph of the link attenuation as a function of distance and to locate high-loss connectors or splices with access to only one end of the fiber link. Commercial measurement equipment is available for this procedure, known as an optical time-domain reflectometer. Note that the optical time-domain reflectometer does not measure fiber loss, but implies it by looking at the backscattered light from the fiber; this measurement cannot always be correlated with the optical power budget of a communication link. An optical time-domain reflectometer is similar in concept to a radar system, in that it analyzes the small fraction of light scattered back up the fiber link, assuming that the backscatter coefficient is constant (only approximately one-millionth of the launched signal is back reflected to the optical time-domain reflectometer). Because the group velocity of light in the fiber can be determined, the round-trip time for a light pulse can be expressed in terms of distance. Recently, optical time-domain reflectometers have become available with very narrow source pulses that are capable of resolving discontinuities in the fiber spaced as little as a few millimeters apart. However, an optical time-domain reflectometer also has a "dead zone" of several meters or more near the source that cannot be accurately measured because of timing considerations and high reflections from connectors near the source that can saturate the optical time-domain reflectometer's receiver; most optical time-domain reflectometers are designed to measure cable plants that are several kilometers or more in length. This measurement is often misunderstood and can suffer from several sources of error. The backscatter coefficient is actually a function of the material properties of the fiber core and the core diameter; variations in these introduce uncertainty in the measurement and can actually show a relative optical gain at a connection point. When used with a multimode fiber, optical time-domain reflectometers tend to underfill the higher-order modes, which are the greatest loss; this makes it difficult to correlate optical time-domain reflectometer measurements with source–meter measurements of attenuation. The optical time-domain reflectometer can underestimate the cable plant loss by as much as 3 dB. Although the optical time-domain reflectometer remains a useful tool for troubleshooting problems in a large cable plant, there is a corresponding uncertainty in locating the position of a connection loss, especially because manufactured fiber cable often has as much as 1% more fiber than the length of the cable to prevent stress on the fiber when the cables are pulled.

Many other characteristics of an optical fiber are measured during manufacturing of the fiber cable, rather than in the field, because of the complexity of the measurement equipment involved. For example, the mode field diameter is commonly measured with the near-field method. The fiber end face is imaged onto a small-area photodetector by lenses to expand, collimate, and focus the light. The measured spot size can be fitted to the equations previously given, from which the mode field diameter may be calculated. As another example, the cutoff wavelength is measured with the single-bend attenuation method. To do this, optical power transmission as a function of wavelength is measured under two different fiber-bending conditions. First, the fiber is measured with a bend diameter of 28 cm; then the same fiber is remeasured with an additional fiber loop with a diameter of 4–6 cm, which is intended to attenuate the higher-order modes as much as possible. The cutoff wavelength is then defined as that wavelength for which the normalized power becomes linear to within 0.1 dB, as illustrated in Fig. 20.16. The apparatus for measuring chro-

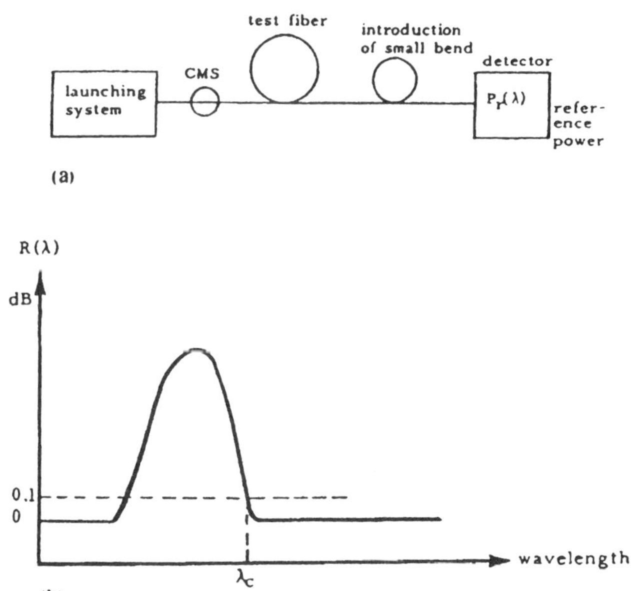

Fig. 20.16. Typical apparatus for cutoff wavelength measurement: (a) test equipment, (b) typical measurement of cutoff wavelength, single-mode reference (bending method) (adapted from U. Österberg, "Optical fiber, cable, and connectors," in *Handbook of Fiber Optic Data Communication,* Academic, 1998, Chap. 1, Fig. 1.20, p. 25).

matic dispersion is shown in Fig. 20.17; it requires an interferometer isolated from mechanical vibrations by some means such as an optical table. The measurement source is a high-power (100-W) incoherent high-pressure arc lamp in conjunction with a monochrometer to select different optical wavelengths. Light from this source is split between two paths; one passes through the fiber

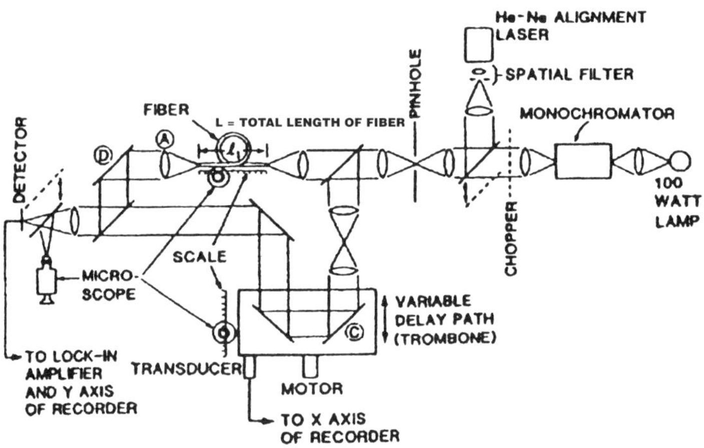

Fig. 20.17. Apparatus for dispersion measurement by the interferometric method (adapted from U. Österberg, "Optical fiber, cable, and connectors," in *Handbook of Fiber Optic Data Communication,* Academic, 1998, Chap. 1, top of Fig. 1.21, p. 26).

under test, and the second passes through a controlled variable optical delay line. The two beams are recombined and focused onto a detector, which is used to measure the relative delay through the fiber at a given wavelength. A typical plot of the delay curve resulting from this approach is shown in Fig. 20.18.

In a digital optical communication link, a fundamental measurement is the probability that a transmitted data bit will be correctly received. This is described by the bit-error-rate test, which measures the ratio of the number of bits received incorrectly to the total number of bits transmitted over a specified time interval. For example, if we measure a data link operating at 1 Gbit/s and observe 1000 errors in a period of 1 h, then the corresponding bit error rate would be 2.77×10^{-10}. Typical communication links specify target bit error rates between 10×10^{-9} and 10×10^{-12} or lower. As we might expect, the bit error rate is related to the signal-to-noise ratio Q on the link according to

$$\mathrm{BER} = \frac{1}{2\pi} \int_{Q}^{\infty} e^{-Q^2/2}\, \mathrm{d}Q \cong \frac{1}{Q\sqrt{2\pi}}\, e^{-Q^2/2}.$$

Note that we must conduct a test for a sufficient time in order to receive enough errors that the result is statistically significant; ideally, several hundred

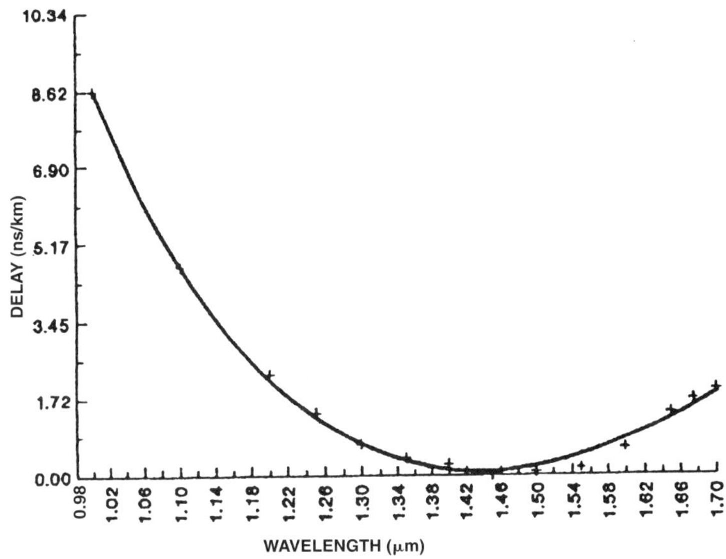

Fig. 20.18. Typical delay curve resulting from the measurement shown in Fig. 20.17 (adapted from U. Österberg, "Optical fiber, cable, and connectors," in *Handbook of Fiber Optic Data Communication,* Academic, 1998, Chap. 1, bottom of Fig. 1.21, p. 26).

to a thousand errors should be allowed to accumulate during a bit-error-rate test. This leads to a fundamental problem, as it can take hours to accumulate enough errors to collect a single bit-error-rate measurement; in the case of low data rate links operating a low bit error rate, the measurement becomes impractical because it could take months! For this reason, the bit error rate is often measured with variable attenuation in the optical link to reduce the measurement time and then is extrapolated to lower levels. The basic apparatus for a bit-error-rate measurement consists of a data pattern generator that can produce a known data sequence, an optical transmitter, a fiber link with variable attenuation, and an optical receiver. The received data are compared with the transmitted data, and the number of errors is counted over time. Note that the error detector must be synchronized with the data source to perform this measurement. Various types of data patterns may be used to simulate actual link conditions or to stress the performance of the link by aggravating timing jitter or making clock recovery difficult. Because the bit error rate increases near the edges of the data eye window, it can be used to indirectly measure transmit and receive eye openings.

20.9. Nonlinear Fiber Optics

The interaction of light with the optical-fiber material is typically very small, particularly at low optical-power levels. However, as the level of optical power in the fiber is increased, nonlinear effects can become significant; this is especially important for long-distance fiber links, which provide the opportunity for smaller effects to build up over a distance. In an optical communication system, nonlinear effects can induce transmission errors that place fundamental limits on system performance, in much the same fashion as attenuation or dispersion effects. At the same time, some important components, such as optical amplifiers, rely on nonlinear effects for their operation.

One of the most common nonlinear interactions is known as four-wave mixing, which occurs when two or more optical signals propagate in the same direction along a common single-mode fiber. As illustrated in Fig. 20.19, the

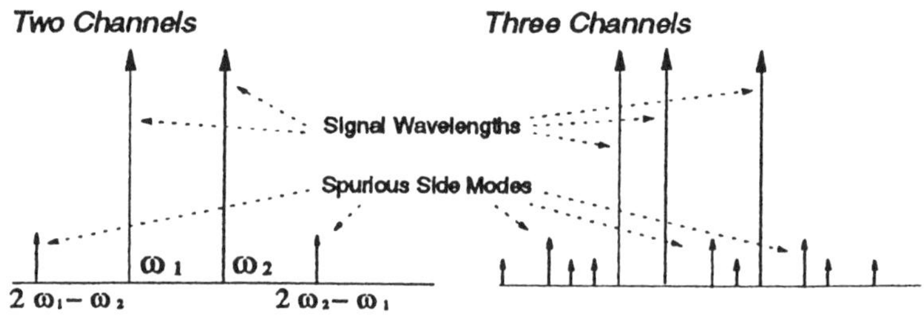

Fig. 20.19. Four-wave-mixing effects.

signals can mix to produce new signals at wavelengths that are spaced at the same intervals as those of the original signals. The effect can also occur among three or more signals, making the overall effect quite complex. Four-wave mixing increases exponentially with signal power and becomes greater as the channel spacing is reduced; in particular, it is a concern with dense wavelength-division-multiplexing systems. If wavelength-division-multiplexing channels are evenly spaced, then the spurious four-wave-mixing signals will appear in adjacent wavelength channels and act as noise. One method of dealing with this problem is to space the channels unevenly to reduce the effect of added noise on adjacent channels; however, four-wave mixing still removes some optical power from the desired signal levels. Because four-wave mixing is caused by signals that remain in phase with each other over a significant propagation distance, the effect is stronger for lasers with a long coherence length. Also, four-wave mixing is strongly influenced by chromatic dispersion; because dispersion ensures that different signals do not stay in phase with each other for very long, it acts to reduce the effect of four-wave mixing.

Another nonlinear optical phenomenon is known as frequency chirping, or inducing a linear frequency sweep in an optical pulse. Until now, we have assumed that the refractive index of the fiber core is a constant, independent of the optical power. Actually, sufficiently high optical-power levels can affect the material properties of the glass and induce small changes in the refractive index. The frequency chirp is generated by self-phase modulation, which arises from the interaction of the propagating light and the intensity-dependent portion of the fiber's refractive index.[38] Because these effects are caused by the propagating signal itself, they are known as carrier-induced phase modulation. The fiber's refractive index can be expressed as follows:

$$n_1 = n_{10} + n_{12}\, I(t),$$

where n_{10} is the refractive index under low optical-power conditions (for this case $n_1 = n_{10}$), $I(t)$ is the intensity profile of the propagating light, and n_{12} is a positive material constant.[38] The propagation constant is then given by

$$k_1 = \frac{\omega_0 n_{10}}{c} + \Gamma I(t),$$

where the constant Γ represents a collection of terms.[38] The phase of the optical pulse now becomes

$$\psi = \omega_0 t - \frac{\omega_0 n_{10} z}{c} - \Gamma I(t),$$

where z is the propagation distance. The instantaneous frequency of the light is thus proportional to the negative time derivative of the intensity profile,

$$\omega_i = \mathrm{d}\psi/\mathrm{d}t = \omega_0 - \Gamma\, \mathrm{d}I(t)/\mathrm{d}t,$$

and the properties of the resulting chirp depend on the time-varying light intensity; for many pulse shapes, including Gaussian, this results in a nonuniform frequency chirp. Another property of interest is the group-velocity dispersion that can be calculated by an expansion of the propagation constant $k(\omega)$ about its center frequency and retention of the second-order derivative.[39] Optical fibers with a positive group-velocity dispersion cause the frequency components to spread out as the light propagates along the fiber; by contrast, in a fiber with a negative group-velocity dispersion the frequency components move closer together as the light propagates. The specific material properties of the fiber and the wavelength of light determine whether the group-velocity dispersion is positive or negative. The combined effects of self-phase modulation and group-velocity dispersion are called dispersive self-phase modulation; this effect has applications in optical-pulse compression systems.[39]

[38] C. DeCusatis and P. Das, "Spread spectrum techniques in optical communication using transform domain processing," IEEE J. Sel. Areas Commun. **8,** 1608–1616, 1990.

[39] T. Brown, "Optical fibers and fiber optic communications," in *Handbook of Optics,* Optical Society of America, 2000, Chap. 1.

When there are multiple signals at different wavelengths in the same fiber, nonlinear phase modulation in one signal can induce phase modulation of the other signals. This is known as cross-phase modulation; in contrast with other nonlinear effects, it does not involve power transfer between signals. Cross-phase modulation can introduce asymmetric spectral broadening and distortion of the pulse shapes.

At higher optical-power levels, nonlinear scattering may limit the behavior of a fiber-optic link. The dominant effects are stimulated Raman and Brillouin scattering. When incident optical power exceeds a threshold value, significant amounts of light may be scattered from small imperfections in the fiber core or by mechanical (acoustic) vibrations in the transmission media. These vibrations can be caused by the high-intensity electromagnetic fields of light concentrated in the core of a single-mode fiber. Because the scattering process also involves the generation of photons, the scattered light can be frequency shifted.[20] Put another way, we can think of the high-intensity light as generating a regular pattern of very slight differences in the fiber refractive index; this creates a moving diffraction grating in the fiber core, and the scattered light from this grating is Doppler shifted in frequency by ~11 GHz. This effect is known as stimulated Brillouin scattering; under these conditions, the output light intensity becomes nonlinear as well. Stimulated Brillouin scattering will not occur below the optical power threshold defined by

$$P_{\mathrm{Bcrit}} = \frac{21A_c}{G_b L_c}\ \mathrm{W},$$

where L_c is the effective interaction length, A_c is the cross-sectional area of the guided mode, and G_b is the Brillouin gain coefficient.[40,41] Brillouin scattering has been observed in single-mode fibers at wavelengths greater than cutoff with optical power as low as 5 mW; it can be a serious problem in long-distance communication systems when the span between amplifiers is low and the bit rate is less than ~2 Gbit/s, in wavelength-division-multiplexing systems up to ~10 Gbit/s when the spectral width of the signal is very narrow (less than 0.2 nm), or in remote pumping of some types of optical amplifiers.[20] In general, stimulated Brillouin scattering is worse for narrow laser linewidths (and is generally not a problem for a channel bandwidth greater than 100 MHz), wavelength (stimulated Brillouin scattering is worse near 1550 nm than near 1300 nm), and signal power per unit area in the fiber core. All of these factors are

[40] C. DeCusatis, "Fiber optic link requirements," in *Handbook of Fiber Optic Data Communication*, C. DeCusatis, D. Clement, E. Maass, and R. Lasky, eds., Academic, 1st ed., 1988, 2nd ed., 2001, Chap. 7.

[41] D. Cotter, "Observation of stimulated Brillouin scattering in low loss silica fibers at 1.3 microns," Electron. Lett. **18,** 105–106, 1982.

summarized in the preceding expression; Fig. 20.20 shows the thresholds for stimulated Brillouin scattering as a function of wavelength. Stimulated Brillouin scattering can be a concern in long-distance communication systems when the span between amplifiers is large and the bit rate is below ~2.5 Gbit/s, in wavelength-division-multiplexing systems in which the spectral width of the

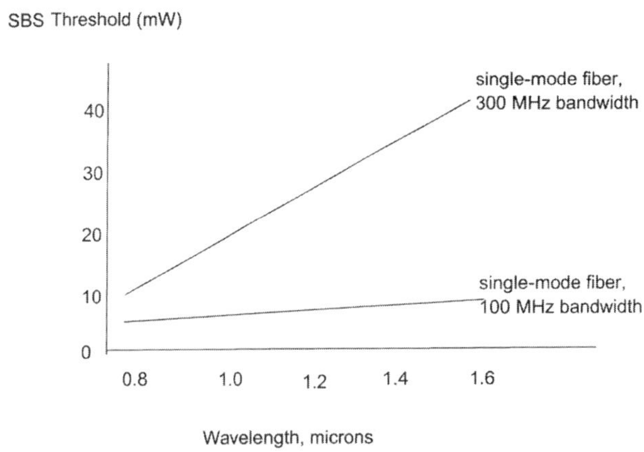

Fig. 20.20. Stimulated Brillouin scattering threshold vs. wavelength.

signal is very narrow (less than 0.2 nm), and in remote pumping of optical amplifiers with narrow linewidth sources. In cases in which stimulated Brillouin scattering could be a problem, the source linewidth can be intentionally broadened by an external modulator or additional rf modulation on the laser injection current. However, this is a trade-off against long-distance transmission, as broadening the linewidth also increases the effects of chromatic dispersion.

When the scattered light experiences frequency shifts outside the acoustic phonon range that are due instead to modulation by impurities or molecular vibrations in the fiber core, the effect is known as stimulated Raman scattering. The mechanism is similar to that of stimulated Brillouin scattering, and scattered light can occur in both the forward and the backward directions along the fiber. The threshold below which Raman scattering will not occur is given by

$$P_{\mathrm{Rcrit}} = \frac{16A_c}{G_r L_c}\ \mathrm{W},$$

where G_r is the Raman gain coefficient.[42] Note that stimulated Raman scattering is also influenced by fiber dispersion, and the standard fiber reduces the effect of stimulated Raman scattering by half (3 dB) compared with that ob-

[42] R. Stolen, "Nonlinear properties of optical fiber," in *Optical Fiber Communications,* S. E. Miller and A. G. Chynoweth, eds., Academic, 1979, Chap. 5.

tained with a dispersion-shifted fiber. As a rule of thumb, the optical-power threshold for Raman scattering is approximately three times larger than that for Brillouin scattering. Another good rule of thumb is that stimulated Raman scattering can be kept to acceptable levels if the product of total power and total optical bandwidth is less than 500 GHz W. This is quite a lot; for example, consider a 10-channel dense wavelength-division-multiplexing system with standard wavelength spacing of 1.6 nm (200 GHz). The bandwidth becomes $200 \times 10 = 2000$ GHz, so the total power in all 10 channels would be limited to 250 mW in this case (in most dense wavelength-division-multiplexing systems, each channel will be well below 10 mW for other reasons, such as laser safety considerations). In a single-mode fiber, typical thresholds for Brillouin scattering are ~10 mW and for Raman scattering ~35 mW; these effects rarely occur in a multimode fiber, in which the thresholds are ~150 and 450 mW, respectively. In general, the effect of stimulated Raman scattering becomes greater as the signals are moved farther apart in wavelength (within some limits); this introduces a trade-off with four-wave mixing, which is reduced as the signal spacing increases.

We can construct optical amplifiers by using the principle of stimulated Raman scattering; if a pump signal with relatively high power (half a watt or more) and a frequency 13.2 THz higher than the signal frequency is coupled into a sufficiently long length of fiber (~1 km), then amplification of the signal will occur. Unfortunately, more efficient amplifiers require that the signal and pump wavelengths be spaced by almost exactly the Raman shift of 13.2 THz; otherwise the amplification effect is greatly reduced. It is not possible to build high-power lasers at arbitrary signal wavelengths; one possible solution is to build a pump laser at a convenient wavelength, then wavelength shift the signal by the desired amount.[20] However, another good alternative to stimulated Raman scattering amplifiers are the widely used erbium-doped fiber amplifiers. These allow the amplification of optical signals along their direction of travel in a fiber without the need to convert back and forth from the electrical domain. Although there are other types of optical amplifiers based on other rare-earth elements such as praseodymium or neodymium, and even some optical amplifiers based on semiconductor devices, the erbium-doped amplifiers are the most widely used because of their maturity and good performance at wavelengths of interest near 1550 nm.

An erbium-doped fiber amplifier operates on the same principle as that of an optically pumped laser; it consists of a relatively short (~10-m) section of fiber doped with a controlled amount of erbium ions. When this fiber is pumped at high power (10–300 mW) with light at the proper wavelength (either 980 or 1480 nm), the erbium ions absorb the light and are excited to a higher energy state. Another incident photon around the 1550-nm wavelength will cause stimulated emission of light at the same wavelength, phase, and direction of travel as those of the incident signal. Erbium-doped fiber amplifiers are often characterized by their gain coefficient, defined as the small signal gain divided by the pump power. As the input power is increased, the total gain of the erbium-doped fiber amplifier will slowly decrease; at some point, the erbium-doped fiber amplifier enters gain saturation, and further increases to the input power cease to result in any increase in output power. Because erbium-doped fiber amplifiers do not distort the signal, unlike electronic amplifiers, they are often used in gain saturation. The gain curve of a typical erbium-doped fiber amplifier is shown in Fig. 20.21; note that the gain at 1560 nm is

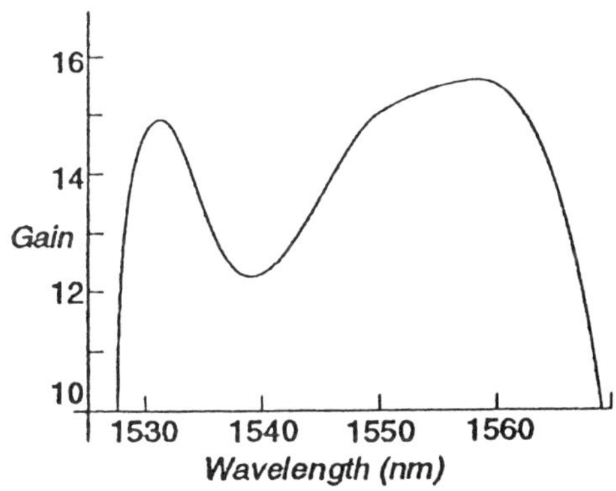

Fig. 20.21. Typical gain curve of an erbium-doped fiber amplifier showing nonuniform gain vs. wavelength (adapted from H. Dutton, *Optical Communications,* Academic, 1999, Fig. 47, p. 68).

approximately twice as large as the gain at 1540 nm. This can be a problem when one is operating wavelength-division-multiplexing systems; some channels will be strongly amplified and dominate over other channels that are lost in the noise. Furthermore, a significant complication with erbium-doped fiber amplifiers is that their gain profile changes with input signal power levels; so, for example, in a wavelength-division-multiplexing system, the amplifier response may become nonuniform (different channels have different effective gains) when channels are added or dropped from the fiber. This requires some form of equalization to achieve a flat gain across all channels. There has been a great deal of research in this area; some proposals include adding an extra wavelength-division-multiplexing channel locally at the erbium-doped fiber amplifier to absorb excess power (gain clamping) and manipulating either the fiber doping or the core structure. Another concern with erbium-doped fiber amplifiers is that some of the excited erbium undergoes spontaneous emission, which can create light propagating in the same direction as the desired signal. This random light is amplified and acts as background noise on the fiber link; the effect is known as amplified spontaneous emission. Because amplified spontaneous emission can be at the same wavelength as that of the desired signal, it may be difficult to filter out; furthermore, amplified spontaneous emission accumulates in systems with multiple amplifier stages and is proportional to the amplifier gain.

Other nonlinear effects can be used to produce useful devices by a change in the properties of the fiber itself; one of the most common examples is fiber Bragg gratings. When an optical fiber is exposed to ultraviolet light, the fiber's refractive index is changed; if the fiber is then heated or annealed for a few hours, the index changes can become permanent. This phenomenon is called photosensitivity.[43,44] The magnitude of the index change depends on many factors, including the irradiation wavelength, intensity, and total dose, the composition and doping of the fiber core, and any materials processing done either before or after irradiation. In germanium-doped single-mode fibers, index differences between 10^{-3} and 10^{-5} have been obtained. If this effect is used, periodic diffraction gratings can be written in the core of an optical fiber. This was first achieved by interference between light propagating along the fiber and its own reflection from the fiber end face[45]; this is known as the internal writing technique, and the resulting gratings are known as Hill gratings. Another approach is the transverse holographic technique, in which the fiber is irradiated from the side by two beams that intersect at an angle within the fiber core. Gratings can also be written in the fiber core by irradiation of the fiber through a phase mask with a periodic structure. These techniques can be used to write fiber Bragg gratings in the fiber core; such gratings reflect light in a narrow bandwidth centered around the Bragg wavelength λ_B, which is given by

$$\lambda_B = 2\, N_{\text{eff}} \Lambda,$$

where Λ is the spatial period, or pitch, of the periodic index variations and N_{eff} is the effective refractive index for light propagating in the fiber core. There are many applications for fiber Bragg gratings in optical communications and optical sensors, such as tapped optical-delay lines, filters, multiplexers, optical strain gauges, and others; an extensive review is provided in footnotes 46 and 47. Fiber Bragg gratings function in reflection, whereas many applications re-

[43] K. Hill, "Fiber Bragg gratings," in *Handbook of Optics,* Optical Society of America, 2000, Vol. IV, Chap. 9.

[44] B. Poumellec, P. Niay, M. Douay, "The UV induced refractive index grating in Ge:SiO_2 preforms: Additional CW experiments and the macroscopic origins of the change in index," J. Phys. D **29,** 1842–1856, 1996.

[45] K. O. Hill, B. Malo, F. Bilodeau, et al., "Photosensitivity in optical fibers," Annu. Rev. Mater. Sci. **23,** 125–157, 1993.

[46] B. S. Kawaski, K. O. Hill, D. C. Johnson, et al., "Narrowband Bragg reflectors in optical fibers," Opt. Lett. **3,** 66–68, 1978.

[47] K. Hill and G. Meltz, "Fiber Bragg grating technology: Fundamentals and overview," J. Lightwave Technol. **15,** 1263–1276, 1997.

quire a transmission effect; this conversion is accomplished an optical circulator, a Michelson or a Mach–Zehnder interferometer, or a Sagnac loop.[43]

Further Reading

T. Okoshi, *Optical Fibers,* Academic, 1982.

J. P. Powers, *An introduction to fiber optic systems,* Asken, Homewood, IL, 1993.

H. A. Haus, *Waves and Fields in Optoelectronics,* Prentice-Hall, 1984.

D. Marcuse, D. Gloge, and E. A. Marcatini, "Guiding properties of fibers," in *Optical Fiber Telecommunications,* Academic, 1979.

C. DeCusatis, E. Maass, D. Clement, and R. Lasky, eds., *Handbook of Fiber Optic Data Communication,* Academic, 2nd ed., 2001.

The first two references deal in considerable detail with the properties of optical fibers. The next two treat the modes and field in the fibers and some of the properties useful for communications. The final reference is a compendium of articles dealing with the title subject.

21

Radiometry and Photometry

W. Wolfe

21.1. Introduction

This chapter provides information about radiative transfer and radiometric measurements and the equivalents for photometry.

21.2. Solid Angle

The solid angle is the area of an object projected onto a sphere of radius R divided by R squared. It is most often measured in steradians, but square degrees are sometimes used.

21.3. Frequency and Wavelength Units

The frequency of radiation can be measured in frequency ν (cycles per second or hertz), radian frequency ω (radians per second), and wave number σ (cycles per centimeter). The symbol $\tilde{\nu}$ is also used for wave number. The wavelength is indicated by λ and is usually given in either nanometers or micrometers.

21.4. Spectral, Total, and Band-Limited Quantities

The radiometric quantities listed in Table 21.1 can be spectral, total, or band limited. As an example, the spectral radiance may be written as

$$L_\lambda = \frac{\partial L}{\partial \lambda}, \quad L_\sigma = \frac{\partial L}{\partial \sigma}. \tag{21.1}$$

The total quantities are usually written without subscripts. The band-limited quantities are written as

$$L_\Delta \lambda = \int_{\Delta\lambda} L_\lambda \, \mathrm{d}\lambda = \int_{\Delta\sigma} L_\sigma \mathrm{d}\sigma. \tag{21.2}$$

21.5. Symbols, Units, and Nomenclature

There are recommended symbols and units for radiative terms. Table 21.1 gives some of them.

Table 21.1. Definitions, Units, and Symbols

Term	Symbol	Definition	Radiometric	Photometric
Energy	U		joule (J)	
Flux	Φ	$\frac{\partial U}{\partial t}$	watt (W)	
Flux density				
Incidance	E	$\frac{\partial \Phi}{\partial A}$	W m^{-2}	Illuminance
Exitance	M	$\frac{\partial \Phi}{\partial A}$	W m^{-2}	Luminous exitance
Intensity	I	$\frac{\partial \Phi}{\partial \Omega}$	W sr^{-1}	Luminous intensity
Sterance	L	$\frac{\partial^2 \Phi}{\cos\theta\, \partial A \partial \Omega}$	W m^{-2}sr^{-1}	Luminance

21.6. Radiative Transfer

The fundamental equation of radiative transfer between two differential elements of area a distance R apart is given by

$$d\Phi = L\,\frac{dA_1 \cos\theta_1 dA_2 \cos\theta_2}{R^2}, \tag{21.3}$$

where the angles are those between the line of sight and the local surface normals.

This can also be written as

$$\begin{aligned} d\Phi &= L dA_1 \cos\theta_1 d\Omega_{12} = L dA_2 \cos\theta_2 d\Omega_{21} \\ &= L dA_1 d\Omega'_{12} = L dA_2 d\Omega'_{21}, \end{aligned} \tag{21.4}$$

where $d\Omega$ is the solid angle and $d\Omega'$ is the projected solid angle. The most compact form is

$$d\Phi = L dG, \tag{21.5}$$

where dG is the differential throughput (also called etendué and grasp).

A useful quantity for transfer with Lambertian surfaces is the geometrical configuration factor, also known as view factor, interaction factor, angle factor, etc. It is the etendué divided by π.

21.6.1. Transfer with an Isotropic Radiator

An isotropic radiator is defined as one that has a radiance that is independent of angle. The relationship between the radiance and the radiant exitance for such a radiator is

$$L = \frac{M}{\pi}, \tag{21.6}$$

where the radiance and the radiant exitance can have any spectral modification described in Section 21.2.

21.6.2. Transfer with a Circular Disk

The incidance upon a differential element from a circular disk (see Fig. 21.1) of radius r at a central distance R is given by

$$E = \frac{LA}{R^2}\frac{r^2}{R^2 + r^2} = \frac{I}{R^2}\left[1 - \left(\frac{r}{R}\right)^2 + \left(\frac{\mathrm{r}}{\mathrm{R}}\right)^4 - + \cdots\right]. \tag{21.7}$$

This shows that the inverse square law applies only to sources that are small with respect to the distance.

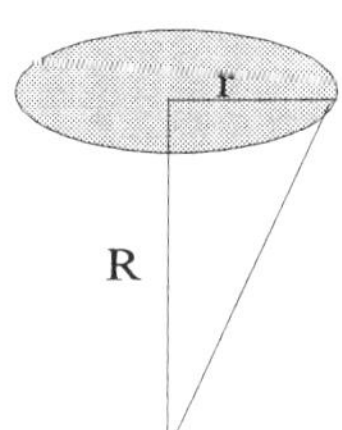

Fig. 21.1. Circular disk.

21.6.3. Transfer between Rectangular Plates

In the following examples it is assumed that the radiance is Lambertian. If not, the radiance can be placed inside the integral with the proper angular variation.

The transfer between a differential element and a parallel, rectangular plate (see Fig. 21.2) a distance Z away is given by

$$E = L \int_0^X \int_0^Y \frac{Z^2}{\rho^4} \, \mathrm{d}x\mathrm{d}y, \tag{21.8}$$

where the geometry and terms are defined in Fig. 21.3.

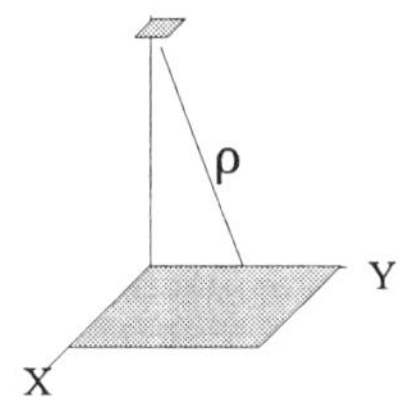

Fig. 21.2. Differential element and parallel rectangle.

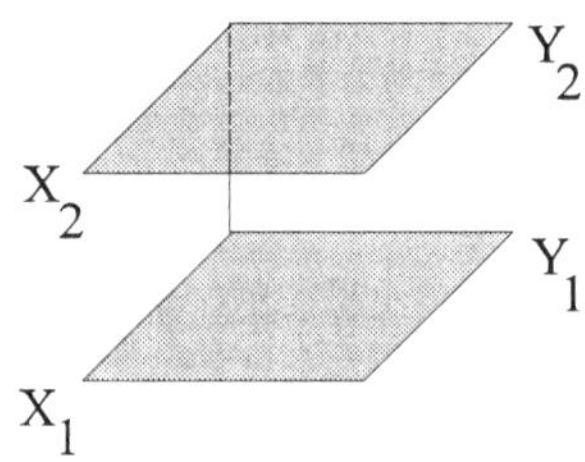

Fig. 21.3. Parallel plane rectangles.

The transfer between two parallel rectangles is given by

$$E = LZ^2 \int_0^{X_1} \int_0^{Y_1} \int_0^{X_2} \int_0^{Y_2} \frac{\mathrm{d}x_1 \mathrm{d}x_2 \mathrm{d}y_1 \mathrm{d}y_2}{[(x_2 - x_1)^2 + (y_2 - y_1)^2 + Z^2]^2}. \tag{21.9}$$

The transfer between two perpendicular rectangles (see Fig. 21.4) is given by

$$\Phi = L \int_0^X \int_0^Y \int_0^X \int_0^Z \frac{z\sqrt{(x_2 - x_1)^2 + y_2^2}}{[(x_2 - x_1)^2 + y^2 + z^2]^2} \, \mathrm{d}x\mathrm{d}y\mathrm{d}x\mathrm{d}z. \tag{21.10}$$

The transfer between two plane-parallel circular disks (see Fig. 21.5) is

$$E = 4\pi^2 LZ^2 \int_0^{R_1} \int_0^{R_2} \frac{r_1 \mathrm{d}r_1 r_2 \mathrm{d}r_2}{Z^2 + (r_2 \sin\phi_2 - r_1 \cos\phi_1)^2 + (r_2 \cos\phi_2 - r_1 \sin\phi_1)}, \tag{21.11}$$

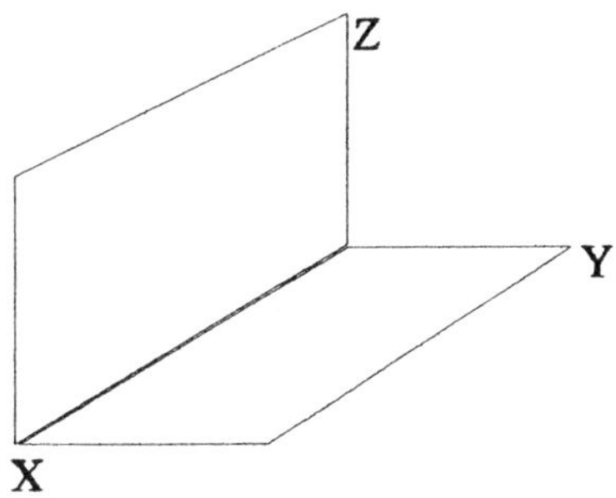

Fig. 21.4. Perpendicular rectangles.

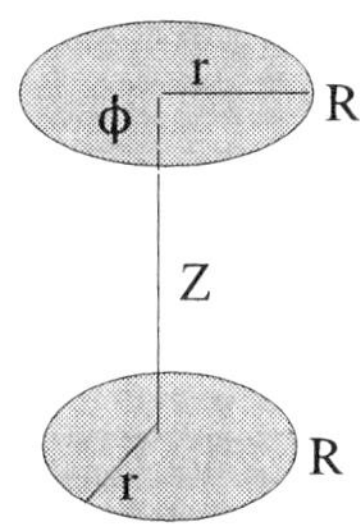

Fig. 21.5. Parallel circular disks.

where R is the maximum radius, r is the radial variable, and ϕ is the polar angle. One disk has the subscript 1, the other 2.

The transfer between two perpendicular plane-circular disks (see Fig. 21.6) is

$$E = L \iiiint \frac{r_1 \sin \phi_1 \sqrt{(r_1 \cos \phi_1 - r_2 \cos \phi_2)^2 + r_2^2 \sin^2 \phi_2}}{[(x_2 - x_1)^2 + y^2 + z^2]^2} \times r_1 \mathrm{d}r_1 r_2 \mathrm{d}r_2 \mathrm{d}\phi_1 \mathrm{d}\phi_2, \tag{21.12}$$

where the integrations extend from 0 to the full radii and to 2π.

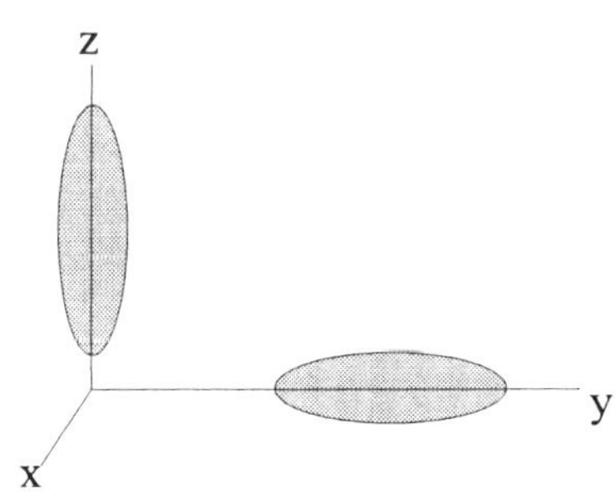

Fig. 21.6. Perpendicular circular disks.

21.7. Invariants

In any medium of refractive index n, the optical throughput $n^2 A\Omega$ is invariant. The optical radiance L/n^2 is also invariant.

21.8. Material Properties

The optical properties of materials useful for radiometry include those related to emission, reflection, transmission, and absorption.

21.8.1. Nomenclature

Some workers use the *ion* ending (transmission, etc.) to indicate the process, the *ance* ending to indicate the property of a sample, and the *ivity* ending to indicate a generic property. Only generic properties are considered here.

21.8.2. Emissivities

Directional, spectral emissivity is defined as the ratio of the spectral radiance from a real material to that of a blackbody:

$$\epsilon(\lambda, d) = \frac{L_\lambda(T, \lambda)}{L_\lambda^{BB}(T, \lambda)}. \tag{21.13}$$

The integrated, directional emissivity is given by

$$\epsilon(\lambda, d) = \frac{\int L_\lambda(T, \lambda) d\lambda}{\int L^{BB}{}_\lambda (T, \lambda) d\lambda} = \frac{\int \epsilon(\lambda, T) L^{BB}{}_\lambda(T, \lambda) d\lambda}{\int L^{BB}{}_\lambda(T, \lambda) d\lambda}. \tag{21.14}$$

If the integration is over the total spectrum, the emissivity is called total.

Hemispherical emissivities are defined as the ratio of exitances rather than radiances.

21.8.3. Absorptivities

All definitions that apply to emissivity apply to absorptivity, with absorption rather than emission as the process.

21.8.4. Reflectivities

The reflectivity of a material is defined as the ratio of the light that is reflected by the material to that which is incident upon it.

Specular reflectivity is the ratio of flux or flux density at the output to that at the input for a collimated beam. The measurements are made in the collimated beam. The specular reflectivity of a plane-parallel plate is given by

$$\rho_\infty = \rho + \frac{(1-\rho)^2\tau^2}{1-\rho^2\tau^2}, \tag{21.15}$$

where ρ_∞ represents the reflectivity for an infinite number of reflections in the plate, ρ is the Fresnel reflectivity of a single surface, and τ is the internal transmittance, given by

$$\tau = e^{-\alpha d}, \tag{21.16}$$

where α is the absorption coefficient and d is the thickness. The reflectivity of a single surface at normal incidence is given by

$$\rho = \frac{(n-1)^2}{(n+1)^2}. \tag{21.17}$$

If the absorption is negligible, then the reflectivity of the plate is given by

$$\rho_\infty = \frac{n^4 + 6n^2 + 2n + 1}{n^4 + 2n^3 + 2n^2 + 2n + 1} \tag{21.18}$$

(which really doesn't help much).

These are good approximations for the transmission of curved windows and lenses as well (if the concentration of energy by the lens is taken into account).

Directional–hemispherical reflectivity is the ratio of the light collected over the hemisphere to that of a beam that is incident upon it. The incident beam is usually collimated. If not, its cone of concentration needs to be specified.

Hemispherical–directional reflectivity incorporates the opposite geometry and is equal to directional–hemispherical reflectivity.

Bidirectional reflectivity is defined as the ratio of the reflected radiance to that of the incident incidance:

$$\rho_{\mathrm{bd}}(\theta_i, \phi_i; \theta_r, \phi_r) = \frac{L(\theta_r, \phi_r)}{E(\theta_i, \phi_i)}. \tag{21.19}$$

This can be spectral or over any spectral band.

21.8.5. Transmissivities

All the definitions concerning reflectivity apply to transmissivity with transmission rather than reflection. The specular transmission of a plane-parallel plate is given by

$$\tau_\infty = \frac{(1-\rho)^2\tau}{1-\rho^2\tau^2}, \tag{21.20}$$

where the symbols are the same as those in Subsection 21.8.4. If the absorption is negligible, the expression reduces to

$$\tau_\infty = \frac{2n}{n^2+1}. \tag{21.21}$$

The simpler approximation,

$$\tau_\infty = (1-\rho)^2\tau, \tag{21.22}$$

can be used if $\rho^2\tau^2$, the error, is small enough. This means that, for a relatively transparent sample, the reflectivity must be low. For Fresnel reflectivity at normal incidence the reflectivity is $[(n-1)/(n+1)]^4$ and is 0.16% for $n = 1.5$, 1.2% for $n = 2$, 6.25% for $n = 3$, and 12.96% for $n = 4$.

21.8.6. Relationships among Absorptivity, Emissivity, Transmissivity, and Reflectivity

Light incident upon a plane-parallel plate is reflected, absorbed, and transmitted such that

$$\alpha + \rho + \tau = 1.$$

This applies to the total values, spectral values, and each component of polarization, but only for directional–hemispherical quantities. One important example relates to the determination of an emissivity from a measurement of a reflectivity. For an opaque material, $\epsilon = \alpha = 1 - \rho$, but the reflectivity must be directional–hemispherical and not bidirectional—in general.

For a plane-parallel plate, the absorptivity is

$$\alpha_\infty = \frac{(1-\rho)(1-\tau)}{1-\rho\tau}, \tag{21.23}$$

and this is equal to the emissivity.

21.9. Kirchhoff's Law

Under conditions of thermal equilibrium, the absorptivity of a given material is equal to its emissivity. This applies to the spectral absorptivity and emissivity, the integrated value over the same spectral band, and each component of polarization (if the polarization is conserved and the temperature is the same).

21.10. Ideal Radiators

An ideal radiator, often called a blackbody, is one that radiates according to Planck's law. The spectral distributions can be expressed in terms of wavelength and various frequencies. The values can be expressed in terms of the flux of energy or photons. There are also useful expressions for the total values and the cumulative increases in flux with wavelength or wave number.

21.10.1. Spectral Radiant Exitance

The spectral radiant exitance can be expressed in wavelength units as

$$M_\lambda = \frac{2\pi c^2 h}{\lambda^5 (e^{hc/\lambda kT} - 1)} = \frac{c_1}{\lambda^5 (e^{c_2/\lambda T} - 1)} = \frac{1}{\lambda^5 (e^{1.4438/\lambda T} - 1)}, \tag{21.24}$$

where c is the speed of light, h is Planck's constant, k is the Boltzmann constant, c_1 and c_2 are the first and the second radiation constants, respectively, and T is the temperature in degrees kelvin.

This can also be expressed as a function of wave number,

$$M_\sigma = \frac{c_1 \sigma^3}{(e^x - 1)}, \tag{21.25}$$

frequency,

$$M_\nu = \frac{c_1 \nu^3}{e^x - 1}, \tag{21.26}$$

and normalized frequency x, which is defined as $x = hc/\lambda kT$.

The photon expressions are

$$M_{q,\lambda} = \frac{2\pi c}{\lambda^4 (e^x - 1)},$$

$$M_{q,\sigma} = \frac{2\pi c \sigma^2}{e^x - 1}. \tag{21.27}$$

21.10.2. Maxima

The different Planck curves have different values for their maxima. These are

$$\lambda_{\max} T = 2897.756\ \mu\text{m K} \approx 3000\ \mu\text{m K},$$

$$\frac{T}{\sigma_{\max}} = 5099.4149\ \mu\text{m K}. \tag{21.28}$$

The equivalent expressions for photon distributions are

$$\lambda_{\max} T = 3669.72613\ \mu\text{m K},$$

$$\frac{T}{\sigma_{\max}} = 11076.30384\ \mu\text{m K}. \tag{21.29}$$

21.10.3. Total Quantities

The two total quantities are the integrals from zero to infinity over either wavelength or frequency of the power or the photon distributions:

$$M = \sigma T^4 = 5.66962 \times 10^{-8} T^4\ \text{W m}^{-2}, \tag{21.30}$$

$$M_q = 1.5202 \times 10^7 T^3\ \text{s}^{-1}\ \text{m}^{-2}. \tag{21.31}$$

21.10.4. Differential Quantities

The change in blackbody radiation with respect to temperature is useful in some applications. A simple way to handle this is a general way. If P is any Planck function (for wavelength or frequency and power or photons) then

$$\frac{\partial P}{\partial T} = P \frac{xe^x}{e^x - 1} \frac{1}{T}. \tag{21.32}$$

This factor can be used to evaluate the spectral and integrated values previously given.

The maxima for the change of exitance are

$$\lambda_{\max} T = 2410.2369\ \mu\text{m K}, \quad \frac{T}{\sigma_{\max}} = 0.375656\ \text{cm K}. \tag{21.33}$$

The maxima for the change of photon exitance are

$$\lambda_{\max} T = 2919.5093\ \mu\text{m K}, \quad \frac{T}{\sigma_{\max}} = 0.5585979\ \text{cm K}. \tag{21.34}$$

21.10.5. *Cumulative Functions*

Cumulative functions give the flux density from zero wavelength to some specified value or from zero frequency to some specified value. They are most useful in calculating the flux density in the band by subtraction. However, math programs such as Mathcad can do this most efficiently.

21.10.6. *Other Properties of Blackbody Radiation*

The radiation inside a blackbody cavity has homogeneous energy density. As it radiates out of the aperture, it is both isotropic and unpolarized and has a coherence length that is approximately equal to the peak of the spectral distribution of the radiation, that is, ~3000 divided by the temperature in degrees kelvin.

The blackbody radiance in a medium of refractive index n is given by

$$L^{\mathrm{BB}} = n^2 L_0^{\mathrm{BB}}.$$

21.11. Evaluation of Blackbody Simulator Shapes

Much has been written on this subject. The simplest and reasonably accurate one is that of Gouffé:

$$\epsilon_c = \left[\frac{\epsilon}{\epsilon\left(1 - \dfrac{a}{A}\right) + \dfrac{a}{A}}\right]\left[1 + (1-\epsilon)\left(\frac{a}{A} - \frac{a}{A_0}\right)\right], \tag{21.35}$$

where ϵ_c is the cavity emissivity, ϵ is the emissivity of the material covering the inside of the cavity, a is the area of the aperture, A is the area of the interior surface, and A_0 is the area of the sphere that has a diameter equal to the length of the interior of the cavity.

21.12. Photometry

Photometry deals with radiometric measurements that are related to the response of the human eye. That response is called the visibility curve or the luminous efficacy. Table 21.2 gives values for it; tables with more spectral resolution are also available.

The response can also be nicely approximated with a Gaussian curve:

$$V(\lambda) = e^{-\left(\frac{\lambda - 555}{60}\right)^2}.$$

For the 37 points used here, the mean rms error is less than 7%. If the denominator is taken as 55 (because it is easier to remember), the rms error is 7.7%.

21.13. Photometric Units and Nomenclature

The basic units and their relationship to radiometric units are given in Table 21.2. Additional units and conversions are in Chap. 34.

Table 21.2. Photometric Units and Quantities

Radiometric Quantity	Photometric Quantity	Units
Energy	Exposure	
Flux	Luminosity	lumen
Exitance	Luminous exitance	lux
Incidance	Luminous incidance	lux
Intensity	Luminous intensity	candela
Sterance	Luminance	nit

Further Reading

C. DeCusatis, *Handbook of Applied Photometry,* American Institute of Physics, 1997.

F. Grum and R. J. Becherer, *Optical Radiation Measurements,* Academic, 1979.

R. McCluney, *Introduction to Radiometry and Photometry,* Artech House, 1994.

J. H. Taylor, *Radiation Exchange, An Introduction,* Academic, 1990.

W. L. Wolfe, *Introduction to Radiometry,* Society of Photo-Optical Instrumentation Engineers, 1998.

C. L. Wyatt, *Radiometric Calibration and Methods,* Academic, 1978.

J. M. Palmer, "Radiometry and Photometry: Units and conversions," in M. Bass,

J. M. Enoch, E. Van Stryland, and W. L. Wolfe, *Handbook of Optics,* Optical Society of America, 2001, Vol. III.

J. W. T. Walsh, *Photometry,* Dover, 1958.

nvl.nist.gov/pub/nistpubs/jres/jres.htm.

22

Stray-Light Suppression

Robert P. Breault

22.1. Introduction

The goal of stray-light suppression in optical systems is to reduce the amount of undesirable radiation in specific locations to a desired level. It is not necessary to reduce it to zero. Stray light should not be an afterthought. It should be considered when the earliest optical designs are considered.

The process takes into account the following considerations:

- the purpose or mission of the sensor
- the environment in which it will function
- the optical design and performance requirements of the system
- the mechanical design, size, and shape of the structures of the system
- the field of view of the detector
- the thermal-emission characteristics of some systems
- the scattering and reflection characteristics of all surfaces.

For the optical characteristics, the spectral, spatial, angular, and polarization distributions must be considered and known.

22.2. Process of Stray-Light Control

There are three steps in controlling stray light in optical systems: preliminary design, detailed design, and test. The preliminary design can often be done without complicated software and will greatly reduce the costs of such detailed analyses later. The testing should be done to confirm the results of the first two steps; it should not be done without the first two steps.

22.2.1. Preliminary Design

This step is done after the optical design and some of the mechanical designs are well on the way. It involves both visualization and simple calculations.

The first step is to determine the *direct* critical surfaces. These are the parts of the system that can be seen from the detector (image) surface. The next step is to determine the *imaged* critical surfaces, that is, those surfaces that can be seen by the detector through an optical element. The critical surfaces can be determined to a large extent by use of layout drawings, but even better are the three-dimensional computer-aided design visualization techniques. At the preliminary stage the diagram may be enough. Critical surfaces may be mirror or lens surfaces, their mounts, the optical bench, the outer barrel, structural supports, and baffle vanes and surfaces. In some cases the detector also sees itself.

The second step is to determine which surfaces are imaged through the optical elements on the detector. This can be accomplished with first-order optics to image each of the surfaces into image space, the optical space of the detector.

The third step is to determine the surfaces that can scatter light to any of these critical surfaces. If the critical objects are directly illuminated by the scene, the preliminary geometry is complete. Otherwise one more level of calculation is necessary.

The critical surfaces should be reduced to a minimum before the actual calculations of transfer are made.

The fourth step is to approximate the scattered radiation. The transfer calculation is performed with the fundamental equation of radiative transfer:

$$d\Phi = L\,\frac{dA_1\cos\theta_1 dA_2\cos\theta_2}{R^2},$$

where $d\Phi$ is the differential flux, L is the radiance, the dA's are the differential areas, R is the line-of-sight distance between surfaces i and $i+1$, and the θ's are the angles between the surface normals and the line of sight. In the first instance, the i+ 1 surface is the detector and the ith surfaces are the critical areas. Integration over the surfaces should be performed. In most cases the integral will not be able to be integrated in closed form, but it may be approximated at this stage. This equation may be rewritten in a more useful form for these calculations:

$$d\Phi = \rho E\,\frac{dA_2\cos\theta_2 dA_2\cos\theta_2}{R^2},$$

where ρ is the bidirectional reflectance distribution function (see Chap. 21), a function of two angles of incidence and two angles of reflection. The irradiance E on surfaces $i+1$, the critical surfaces, is determined the same way, but with $i+1$ substituted for i and $i+2$ for $i+1$. An alternative representation is in terms of the differential geometrical configuration factor:

$$d\Phi = \pi \mathrm{GCF} L,$$

where the (integrated) geometrical configuration factor is defined as

$$\mathrm{GCF} = \frac{A_2 \cos \theta_2 A_2 \cos \theta_2}{\pi R^2}.$$

The fifth step, if needed, is to attempt to reduce the stray light on the detector. The form of the geometrical configuration factor shows one way to do this: increase the angles of emission and incidence, decrease the full area, and increase the distance. A second way is to reduce the bidirectional reflectance distribution function. Because the bidirectional reflectance distribution function can never go to zero, but the limits of integration can, it is more effective to reduce the geometrical configuration factor.

Sometimes the process of determining the surfaces that illuminate the critical surfaces can be easier if we determine the surfaces that are initially illuminated by looking into the optical system. These steps usually have to be iterated.

22.2.2. Detailed Designs

There are several computer analysis programs available to the designer. Their operation is complicated and takes experience. The designer needs to decide whether to buy or lease and learn the program or have the analysis done. Some programs are APART and ASAP, available from Breault Research, Light Tools, available from CODE V, and TRACEPRO, available from Lambda Research.

The chief shortcomings of these detailed computer analyses are inadequate input data and user error. The bidirectional reflectance distribution function values of samples may not be the same as those of the final instrument. Not all scatter paths may be included. The system may not be completely characterized.

22.2.3. Testing

Although the computer programs are now quite reliable, there is no substitute for testing. As with all such testing, considerable analysis of the validity and totality is necessary. A particularly vexing problem is the degree to which the environment influences the test. This can be stray light from the walls of a lab or the difference in configuration that is due to operation in a cold vacuum. No specifics can be given about all the tests, but consider testing over a limited range of angles, wavelengths, temperatures, and so on.

22.3. Performance Measures

The critical performance measures describe how much light gets to the focal plane compared with that on the front end of the system, usually the front of the baffle tube. This is usually described in terms of the normalized detector irradiance, formerly called the point-source transmittance, the ratio of the radiant incidance at the image to that of the beam at the entrance to the baffle tube. There are similar measures that can be confused; care must be taken to determine exactly which measure is used. The four possibilities are

- the ratio of the flux density on the image to that of the incident beam at the plane of the baffle
- the ratio of the flux density on the image to that on the entrance plane of the baffle
- the ratio of the flux on the image to that on the entrance plane of the baffle
- the ratio of the flux on the image to that of the incident beam at the plane of the baffle.

The difference between the first two is a cosine (small); the differences between the first two and the last two is the ratio of the areas of the aperture and the image (large). The first one is preferred.

22.4. System Design

Obtaining a low flux density on the focal plane from undesirable sources is a function of the entire system design. The recommended system is reflective, as mirrors typically have lower scatter than lenses. It should be unobscured because obscurations scatter and diffract and are difficult to eliminate or largely reduce with Lyot stops. The system should have a prime focus for a field stop; that is, a Gregorian form is better than a Cassegrain form. The system should have a Lyot stop, that is, an aperture stop as near to the image surface as possible.

22.3. Baffles

Although the literature is not unanimous on this, baffles are defined as conical or cylindrical tubes that shade the primary from unwanted sources and often contain vanes on their inner surface (see Fig. 22.1).

The first step in baffle design is the calculation of its length. Usually a baffle is designed to reject light from a source (like the Sun or the Moon) a given angle θ from the optical axis of the optical system. The baffle length with respect to the primary diameter, the *L*/*D* ratio, is given by cot θ. A simple geo-

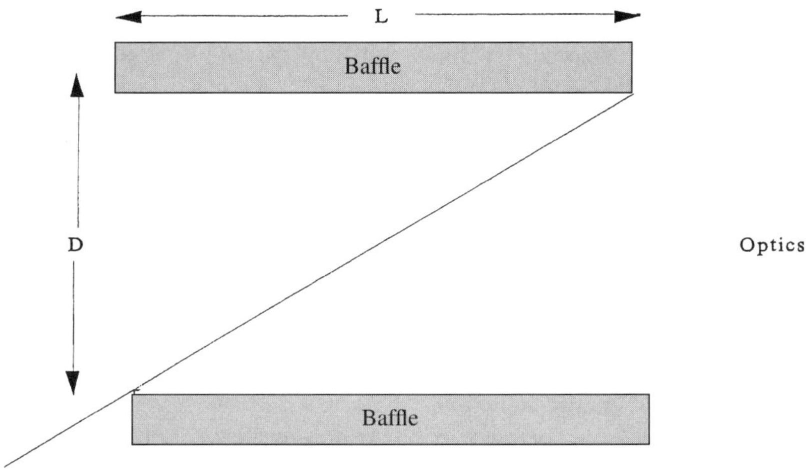

Fig. 22.1. Baffle schematic.

metric construction can be used to determine the placement of the baffles. In Fig. 22.1 the vanes are assumed to be perpendicular to the optical axis and the tube is a right circular cylinder. Other geometries are adaptations of this. The requirement is that any ray must reflect off the front of a vane or the side of the tube and then to the back of another vane before it can get to the primary. The key is to get two low-scatter transfers before going to a critical object. That is the criterion for the construction subsequently described and shown in Fig. 22.2.

The vane depth *d* is chosen, and two lines are drawn. The first line, 1, is drawn to the edge of the primary at the allowed angle. The second line, 2, is parallel to it, but to the intersection of the extension of the primary to the baffle tube base. Then a line, 3, is drawn parallel to the optical axis at the edge of

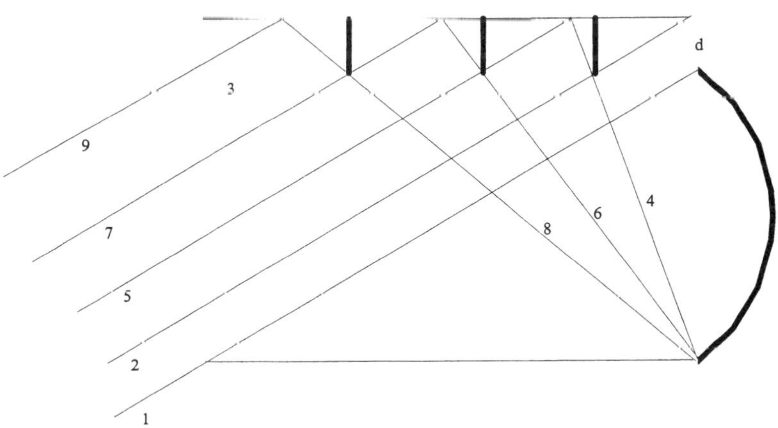

Fig. 22.2. Baffle construction.

the primary; this is the vane tip line. The next line is drawn from the bottom of the primary to the intersection of line 2 with line 3. A vane is placed there. Line 4 is drawn from the bottom of the primary to the tip of the vane and onto the base of the tube. Line 5 is drawn parallel to line 1 and to the point where line 4 meets the base of the vane. The next line goes from the bottom of the primary to the intersection of lines 3 and 5. A vane is placed there and the rest of the lines continue in that order.

Occasionally angled and nonplanar vanes are considered. Some of the trade-offs can be seen from Fig. 22.2. If the baffle has a larger diameter D, then d is larger. The intersection of line 4 with the baffle base will be farther to the left, and the baffles will be spaced farther apart. However, that means a bigger, not a longer, baffle. The baffle must retain its L/D ratio. It is rare that such vanes have superior performance, but there are some cases, and these are described in some of the further reading.

The tips of the vanes should be sharp to avoid reflection off the ends. The beveled side should be away from the mirror.

The vanes may be tilted, but it can be seen that the tips must still be at the intersections of lines 4, 6, and 8 with 3. They will be longer if tilted. Forward tilting, that is, with the tips farther to the left than the bases, means that the reflection off the vane will be more oblique and consequently more likely to be specular and larger. The reverse is true for backward tilt.

One nice technique for a sun baffle is a rotating tube with only half a shield. The obvious prices to pay are the mechanics and reliability.

22.4. Aperture Stops

The aperture stop may be determined by the optical design for the reduction of aberrations. For stray-light purposes it eliminates radiation from all elements preceding it from reaching the image plane, except imaging elements, central obscurations, or vignetting objects. If the stop is at the image plane, there is very little scattered light! (This is one technique for cold shielding in infrared systems.)

Figure 22.3 shows a two-mirror design; it has the stop at the first element, the primary. The second element is larger than the beam on the secondary to accommodate the field of view. The on-axis rays are solid. The rays from a field point are dashed (and exaggerated). Part of the main baffle following the first element is a critical surface. Judicious design of the inner conical and secondary baffles can keep this to a minimum—perhaps at the expense of some obscuration.

If the stop is moved to the second element, then that portion of the baffle will not be seen. The first element then must be larger, but this is part of the trade-off. Direct radiation from optical elements cannot be blocked by an aperture stop, but central obscurations in front of the stop can be blocked by a cen-

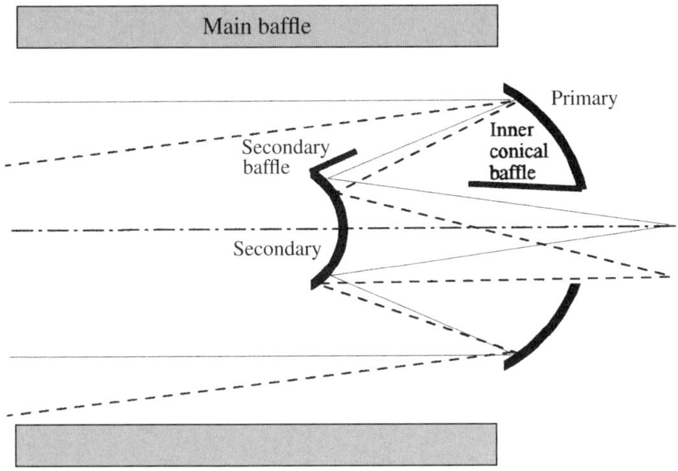

Fig. 22.3. Two-mirror design.

tral obscuration at the aperture stop. Parallax may require a larger obscuration at the stop than the real obscuration.

Figure 22.4 shows a refractive equivalent. This is the part of the baffle that is a critical surface because rays from it can pass through the field stop, get to the secondary, and be focused onto the image surface.

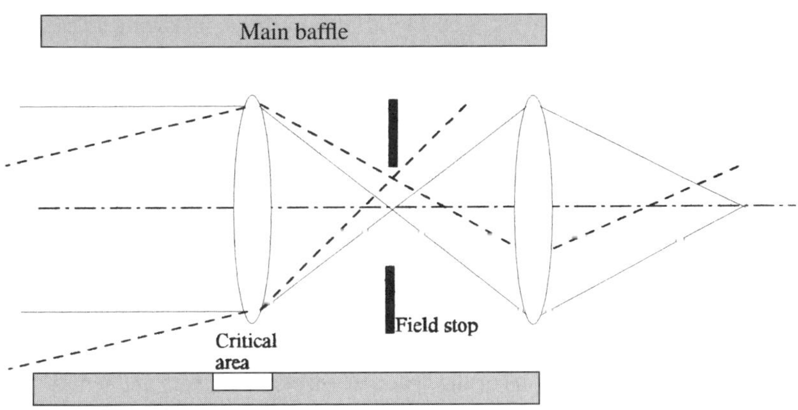

Fig. 22.4. Refractive equivalent.

Figure 22.5 shows the point-source transmittance curves for the mirror system that has the stop on the primary and on the secondary.

22.5. Lyot Stops

A limiting aperture placed at the image of an aperture is sometimes called a glare stop or a Lyot stop and has the same properties as the aperture stop itself.

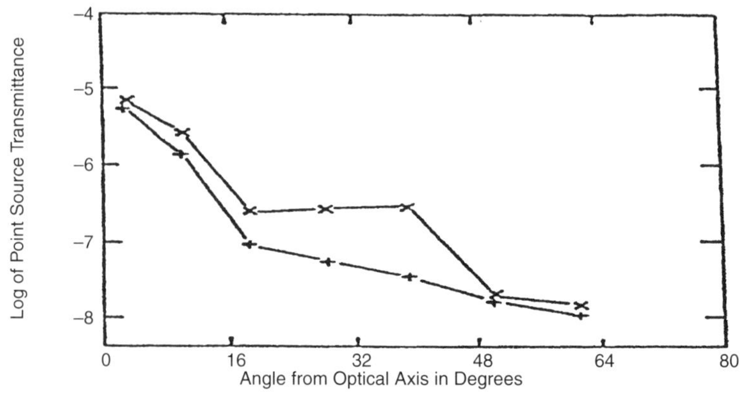

Fig. 22.5. Point source transmittance.

Because Lyot stops are close to the image plane, they tend to further reduce the number or extent of critical surfaces. Reimaging the largest optical element as the stop takes full advantage of the light-gathering power and stray-radiation features.

22.6. Field Stops

Often the detector is the field stop. Other apertures can be used that are images of the detector and are field stops as well. Baffle surfaces following a field stop cannot be seen outside the field of view in object space. Optical elements may allow out-of-field critical surfaces to be seen through the field stop from within the field in the image plane. Field stops do not remove critical surfaces but limit the propagation of power to them. Aperture stops restrict the flow of power from surfaces to the detector. Field stops limit the flow of power from sources to the surfaces.

22.7. Surface Properties

The scattering characteristics of the specular surfaces (mirrors and lenses) and of absorbing surfaces (blacks) need to be input into the programs.

22.7.1. Blacks

Although it is sometimes assumed that these absorbers have isotropic reflectivity, they never do for all angles. Typically, such surfaces become rather specular at larger angles of incidence.

The directional emissivity has angular characteristics similar to those of the directional reflectivity. An extensive treatment of blacks is in the *Handbook of Optics.*[1]

22.7.2. *Mirrors*

The bidirectional reflectance distribution function is a function of the angle of incidence and both the polar and the azimuthal angles of reflection. A full representation is therefore cumbersome. It is usually permissible to use a single profile in the plane of incidence for satisfactory results, although the theory of microroughness scattering denies this.

22.7.3. *Lenses*

The proper function for characterizing lenses is the bidirectional transmittance distribution function, the transmissive form of the bidirectional reflectance distribution function. Typically the bidirectional transmittance distribution function of a lens is larger than the bidirectional reflectance distribution function of a mirror.

Further Reading

S. M. Pompea and R. P. Breault, "Black surfaces for optical systems," in M. Bass, E. Van Stryland, D. R. Williams, and W. L. Wolfe, eds., *Handbook of Optics,* McGraw-Hill, 1995, Chap. 37.

S. H. C. McCall, SORIC, 78 Normark Drive, Thornhill, Ontario L3T 3R1, Canada.

The following SPIE Proceedings are all useful: Vol. 107, April 1977; Vol. 181, February 1979; Vol. 216, February 1980; Vol. 257, September 1980; Vol. 384, 1983; Vol. 511, August 1984; Vol. 675, August 1986; Vol. 967, August 1988; Vol. 1331, July 1990; Vol. 1753, July 1992.

[1] S. M. Pompea and R. P. Breault, "Black surfaces for optical systems," in M. Bass, E. Van Stryland, D. R. Williams, and W. L. Wolfe, eds., *Handbook of Optics,* McGraw-Hill, 1995, Chap. 37.

23
Thermal Detectors

William L. Wolfe

23.1. Introduction

Thermal detectors receive radiation power, absorb it, change temperature, and change a property that can then be detected electrically. Thermal detectors consist principally of bolometers, thermocouples, and pyroelectrics. There are several types of bolometers and variations of pyroelectrics. Thermal detectors are used in arrays for uncooled imaging devices, spectrometers, some astronomy applications, and in solid-state physics measurements.

23.2. Responsivity

The responsivity in terms of the spectrum of the temperature change ΔT per input flux Φ is given by

$$\mathcal{R}_T = \frac{\Delta T}{\Phi} = \frac{\eta}{G\sqrt{1 + (\omega\tau)^2}} = \frac{\eta}{G\sqrt{1 + \omega^2\tau^2}},$$

where η is the fraction of light absorbed, G is the thermal conductance, τ is the time constant (C/G), and ω is the radian frequency, where C is the thermal capacitance. The ideal thermal detector has a responsivity that is independent of optical wavelength. The responsivity is also given in terms of voltage and current; these depend on how the change in temperature is converted to a change in electrical signal.

23.3. Bolometers

The voltage responsivity of a resistive bolometer is given by

$$\mathcal{R}_V = \frac{I_b \alpha R_e \eta}{G(1 + \omega^2\tau^2)},$$

where I_b is the bias current, α is the temperature coefficient of resistance,

$\frac{1}{R}\frac{dR}{dT}$, R is the electrical resistance, η is the fraction of light absorbed, G is the thermal conductance, ω is the radian frequency, and τ is the time constant, C/G, where C is the thermal capacitance. The fraction of light absorbed, η, is equal to the absorptivity, which is equal to the emissivity.

Some, but not many, bolometers are made of metal. The relative temperature change of resistance for a semiconductor is

$$\alpha = \frac{1}{R}\frac{dR}{dT} = -\frac{\beta}{T_d^2},$$

where β is a positive constant of ~3400 K. For thermistor material (and many other semiconductors) α is approximately –0.040. GaAs is –0.09, Te and Bi from –0.003 to –0.01, and VO_2 is –0.02. Thus the critical temperature for a thermistor bolometer operated at 300 K is 325 K and is near that for these others. Some of these semiconductors have a semiconductor–metal phase transition, usually above room temperature. The values cited here are for temperatures below this transition.

23.4. Thermocouples and Thermopiles[1]

The voltage responsivity $\mathcal{R}_V(\omega)$ of thermocouples in the frequency domain is

$$\mathcal{R}_V(\omega) = \frac{V}{\Phi} = \frac{n\Delta S}{G\sqrt{1 + \omega^2\tau^2)}},$$

where V is the output voltage, Φ is the input flux, n is the number of junctions in the thermopile, ΔS is the difference in the Seebeck coefficients of the two metals, ω is the radian frequency, and τ is the time constant. The current responsivity in the frequency domain is

$$\mathcal{R}_I(\omega) = \frac{I}{\Phi} = \frac{n\Delta S}{G(R + R_l + R_D)}\frac{1}{\sqrt{1 + \omega^2\tau^2}},$$

where R is the thermocouple electrical resistance, R_l is the load resistance, and R_D is the dynamic resistance given by $R_D = S^2RT$ and arises from the Peltier effect.

[1] M. R. Holter, S. Nudelman, G. H. Suits, W. L. Wolfe, and G. J. Zissis, *Fundamentals of Infrared Technology,* Macmillan, 1962.

23.5. Pyroelectrics

There are three types of pyroelectrics[2] in current use: unbiased, dc biased, and pulsed biased. They are also called the inherent pyroelectric mode, the induced pyroelectric mode (also called dielectric bolometer mode), and the pulsed pyroelectric mode, respectively. The voltage responsivity in the frequency domain for the unbiased pyroelectric is

$$\mathcal{R}_V(\omega) = \frac{\eta\omega pAR}{G\sqrt{(1+\omega^2\tau_e^2)(1+\omega^2\tau_T^2)}},$$

where η is the absorption efficiency, ω is the radian frequency, p is the pyroelectric coefficient, A is the sensitive area, R is the electrical resistance, G is the thermal conductance, and the τ's are the electrical and the thermal time constants. The expression for the biased pyroelectric is very complicated.

The inherent pyroelectric mode has no bias field so that its conductivity is not critical, but it must operate below the transition point and therefore is material limited. It has a large D and small p and concomitant small current responsivity but large voltage responsivity.

The induced pyroelectric mode has a dc bias field, a high D and high voltage responsivity, is stable, and can operate near the critical point so that there is a wider range of materials and opportunity for optimization. It must be temperature controlled, and leakage current can be critical.

The pulsed mode does not need a chopper (but it is subject to the referencing required in all starers), and it has twice the responsivity of the induced mode (because it switches polarity).

23.6. Thermal Detector Noises

Thermal detectors are affected (afflicted) with radiation or photon noise, Johnson noise, and current or excess noise.

23.6.1. *Photon (Thermal) Noise*[3]

The mean-square fluctuation in power between a pixel and its background is

$$\bar{\Phi}^2 = 4kT^2GB = 16\eta A\sigma kT^5B,$$

2 E. H. Putley, "The pyroelectric detector," in R. K. Willardson and A. C. Beer, eds., *Semiconductors and Semimetals,* Academic, 1970, Vol. 5, Chap. 6.

3 P. W. Kruse, R. D. McGlauchlin, and R. B. McQuistan, *Elements of Infrared Technology,* Wiley, 1962.

where A is the sensitive area, η is the absorption efficiency, σ is the Stefan–Boltzmann constant, k is the Boltzmann constant, T is absolute temperature, B is the noise bandwidth, and the radiative conductivity $G = 4\eta\sigma T^3$. The specific detectivity is

$$D^* = \frac{\eta}{\sqrt{8\eta\sigma k(T_d^5 + T_b^5)}} = \frac{4.0 \times 10^{16}\sqrt{\eta}}{\sqrt{T_d^5 + T_b^5}},$$

where D^* is the specific detectivity in centimeters times root hertz per watt and the subscripts represent the detector and the background. Table 23.1 provides some representative values for thermal detectors (see also footnote 1).

23.7. Arrays

Thermal detectors also exist in arrays. Arrays have been constructed of bolometers, pyroelectrics, ferroelectrics, and thermoelectrics.

The sensitivity of these arrays are usually characterized by a noise-equivalent thermal detector:

$$\text{NETD} = \frac{4F^2 V_n}{A_d \tau_a \tau_o \mathcal{R}(T)\mathrm{d}M/\mathrm{d}T} = \frac{4F^2\sqrt{B}}{D^* \tau_a \tau_o \pi D \alpha \mathcal{R}(T)\mathrm{d}L/\mathrm{d}T},$$

where F is the focal ratio for the proper image distance, that is, the image distance divided by the entrance pupil diameter, V_n is the noise voltage in the frequency band B, A_d is the detector area, τ_a is the atmospheric transmission, τ_o is the optics transmission, $\mathcal{R}$ is the detector responsivity, M is the radiant exitance of the source in the spectral band, T is the absolute temperature of the source, D^* is the specific detectivity, D is the diameter of the (circular) aperture, α is the angular subtense of the detector, and L is the radiance of the source in the spectral band.

Most bolometer arrays[4] use a silicon microstructure as the base for a detector of vanadium oxide that rests on a silicon nitride base. The present arrays have 240 × 336 pixels, each 50 × 50 μm. The thermal conductance is 0.1 μW K^{-1}. The frame rate is 30 Hz, and the noise-equivalent thermal detector for $F/1$ optics at 300 K is 0.04 K.

[4] R. A. Wood, C. J. Han, and P. W. Kruse, "Integrated uncooled infrared detector imaging array," in *Proceedings of the 1992 IEEE Solid State Sensor and Actuator Workshop,* IEEE, 1992; R. A. Wood, "Uncooled thermal imaging with monolithic silicon focal plane arrays," in *Infrared Technology XIX,* B. F. Andresen, ed., Proc. SPIE **2020,** 1993.

Table 23.1. Representative Detector Properties

Type	Operating Temperature (K)	D* (10^8 cm $Hz^{1/2}$/W)	Noise-Equivalent Power (10^{-10} W)	Time Constant (ms)	Size (mm^2)
Silicon bolometer	1.6		3×10^{-5}	8	0.25–0.70
Metal bolometer[a]	2–4	1		10	
Germanium bolometer[b]	2–4		0.005	0.4	1.5
Carbon bolometer[c]	2–4		0.03	10	20
Superconductor bolometer[d]	15		0.2	0.5	5×0.25
Thermistor bolometer[e]	300	1–6		1–8	0.01–10
Thermocouple[f]	300		2–10	10–40	0.1×1–0.3×3
Thermopile[g]	300			3–10	1–100
Pyroelectric	300	2–5		10–100[h]	4

[a] W. G. Langton, "A fast sensitive metal bolometer," J. Opt. Soc. Am. **36,** 355, 1946; B. H. Billings, W. L. Hyde, and E. E. Barr, "An investigation of the properties of evaporated metal bolometers," J. Opt. Soc. Am. **37,** 123, 1947.

[b] F. J. Low, "Low temperature infrared bolometer," J. Opt Soc. Am. **51,** 1300, 1961; S. Zwerdling, R. A. Smith, and J. P. Thierault, "A fast high responsivity bolometer for the very far infrared," Infrared Phys. **8,** 271, 1968; N. Coron, "Infrared helium cooled bolometers in the presence of background radiation: Optimal parameters and ultimate performance," Infrared Phys. **16,** 411, 1976; N. Coron, G. Dambier, J. Le Blanc, and J. P. Moliac, "High performance, far infrared bolometer working directly in a helium bath," Rev. Sci Instrum. **46,** 492, 1975.

[c] W. S. Boyle and K F. Rodgers, "Performance characteristics of a new low temperature bolometer," J. Opt Soc. Am. **49,** 66, 1959.

[d] W. H. Andrews, R M. Milton, and W. De Sorbo, "A fast superconducting bolometer," J. Opt. Soc. Am. **36,** 518, 1946; R. Milton, "A superconducting bolometer for infrared measurements," Chem. Rev. **39,** 419, 1946; N. J. Fuson, "The infrared sensitivity of superconducting bolometers," J. Opt. Soc. Am. **38,** 845, 1948.

[e] W. H. Brattain and J. A. Becker, "Thermistor bolometer detecting system for infrared spectrometers," J. Opt. Soc. Am. **36,** 354, 1946; R. DeWaard and E. M. Wormser, "Description and properties of various thermal detectors," Proc. IRE **47,** 1508, 1959.

[f] P. B. Felgett, "Dynamic impedance and sensitivity of radiation thermocouples," Proc. Phys. Soc. B **62,** 351, 1949.

[g] D. F. Hornig and B. J. O'Keefe, "Design of fast thermopiles and ultimate sensitivity of thermal detectors," Rev. Sci. Instrum. **18,** 7, 1947.

[h] Smaller values may be obtained with higher noise-equivalent powers.

Superconducting arrays that operate between 70 and 90 K have been constructed of YBCO.[5] A 240 × 336 array of 50-μm pixels and *F*/1 optics has a calculated noise-equivalent thermal detector of 2 mK, assuming no excess noise.

The pyroelectric arrays[6] are a hybrid construction in which the detector material is lithium tantalate or lead zirconate that is thinned and bump-bonded to a silicon substrate with the readouts. An array with 100 × 100 pixels and a noise-equivalent thermal detector, under the same conditions as those of the preceding paragraph, is 0.35 K.

The ferroelectric array,[7] an induced pyroelectric array, is made of barium strontium titanate with a Curie point at 22 C, and the array stabilized to that temperature. The array has 80,000 pixels, 50 × 50 μm, and obtains a noise-equivalent thermal detector of 0.05 K (under the same conditions as those already given).

Many thermoelectric arrays consist of a thermopile on a silicon substructure.[8] Both 64- and 96-element linear arrays have been made that attained a noise-equivalent thermal detector of 0.10 K under the same conditions, except for an *F*/0.73 lens.

Further Reading

W. L. Wolfe and P. W. Kruse, "Thermal detectors," in M. Bass, E. Van Stryland, D. Williams, and W. L. Wolfe, eds., *Handbook of Optics,* McGraw-Hill, 1995, Chap. 19.

W. L. Wolfe and G. J. Zissis, eds., *The Infrared Handbook,* Society of Photo-Optical Instrumentation Engineers, 1978.

W. D. Rogatto, ed., *Electro-Optical Components,* Vol. 3 in J. S. Accetta and D. L. Shumaker, eds., *The Infrared and Electro-Optical Systems Handbook,* Society of Photo-Optical Instrumentation Engineers.

P. W. Kruse, L. D. McGlauchlin, and R. B. McQuistan, *Elements of Infrared Technology,* Wiley, 1962.

M. R. Holter, S. Nudelman, G. H. Suits, W. L. Wolfe, and G. J. Zissis, *Fundamentals of Infrared Technology,* Macmillan, 1962.

R. A. Smith, F. E. Jones, and R. P. Chasmar, *The Detection and Measurement of Infrared Radiation,* Oxford Univ. Press, 1968.

[5] B. R. Johnson, T. Ohnstein, C. J. Han, R. Higashi, P. W. Kruse, R. A. Wood, H. Marsh, and S. B. Dunham, "High-Tc superconducting microbolometer arrays fabricated by silicon micromachining," IEEE Trans. Applied Supercond. **3,** 2856, 1993; B. R. Johnson and P. W. Kruse, "Silicon microstructure superconducting microbolometer infrared arrays," in *Infrared Technology XIX,* B. F. Andresen, ed., Proc. SPIE **2020,** 1993.

[6] N. Butler and S. Iwasa, "Solid state pyroelectric imager," in *Infrared Technology XIX,* B. F. Andresen, ed., Proc. SPIE **2020,** 1993.

[7] C. Hanson, H. Beratan, R. Owen, M. Corbin, and S. McKenney, "Uncooled thermal imaging at Texas Instruments," in *Infrared Detectors: State of the Art,* W. H. Makky, ed., Proc. SPIE **1735,** 1992; C. M. Hanson, "Uncooled ferroelectric thermal imaging," in *Infrared Technology XIX,* B. F. Andresen, ed., Proc. SPIE **2020,** 1993.

[8] I. H. Choi and K. D. Wise, "A silicon-thermopile-based infrared sensing array for use in automated manufacturing," IEEE Trans. Electron Dev. **ED-33,** 72, 1986.

R. C. Jones, "The ultimate sensitivity of infrared detectors," J. Opt Soc. Am. **37,** 879, 1974.
S. Nudelman, "The detectivity of infrared detectors," Appl. Opt. **1,** 627, 1962.
R. D. Hudson and J. W. Hudson, *Infrared Detectors,* Halsted, 1975.
F. R. Arams, *Infrared to Millimeter Wave Detectors,* Artech House, 1973.
R. B. Johnson and W. L. Wolfe, eds., *Selected Papers on Infrared Design,* Vol. 513 of SPIE Milestone Series, Society of Photo-Optical Instrumentation Engineers, 1985.
E. L. Dereniak and G. W. Boreman, *Infrared Detectors and Systems,* Wiley, 1996.
P. W. Kruse and D. D. Skatrud, eds., *Uncooled Infrared Imaging Arrays and Systems.*
R. K. Willardson and A. C. Beer, eds., *Semiconductors and Semimetals,* Academic, 1997.
P. W. Kruse, *Uncooled Thermal Imaging Arrays, Systems and Applications,* Society of Photo-Optical Instrumentation Engineers, 2001.

The first three are relatively current handbooks that include information on thermal detectors and more. The next three are relatively old textbooks that may be out of print, but have good sections on thermal detector theory. The next two are articles on aspects of the theory, and the next three are collections of reprints. The last is a modern book on uncooled arrays and systems. Pertinent journals to consult are *Infrared Physics and Technology, Optical Engineering, Applied Optics,* and *Infrared and Millimeter Waves.* SPIE volumes 2552, 2746, 3379, 3436, and 3698 provide useful, unrefereed material.

I would like to acknowledge the many helpful criticisms and contributions of Paul Kruse. Most of what is good in this chapter was due to him. Any misteaks are mine.

24

Photon Detectors

Carolyn J. Sher DeCusatis

24.1. Introduction

Photon detectors convert incident photons directly to electrical signals by a variety of technologies. The types of photon detectors that are currently available are subsequently reviewed; the descriptions include their detection scheme, relevant equations, list of applications, and charts of their figures of merit.

24.2. Photomultiplier Tube

The photomultiplier tube is a photoemissive device. The device consists of a vacuum tube with metal elements called the photocathode and the dynodes, which are under a high voltage (see Fig. 24.1). The first element, the photocathode, is negatively biased and will eject photons when light is incident upon it because of the photoelectric effect. The energy of the electrons is determined by

$$^{1}/_{2}\, mv^2 = h\nu - W, \tag{24.1}$$

where m is the mass of the electron, v is its velocity, ν is its frequency, W is the work function of the photocathode, and h is Planck's constant.

The photocurrent I is

$$I = \frac{M\eta e\Phi}{h\nu} = \frac{M\eta e\Phi\lambda}{hc}, \tag{24.2}$$

where M is the multiplication factor, η is the quantum efficiency of the cathode, Φ is the incident power, and λ is the wavelength.

The multiplication factor M can be calculated from the number of dynodes multiplying electrons in the tube:

$$M = \delta^N, \tag{24.3}$$

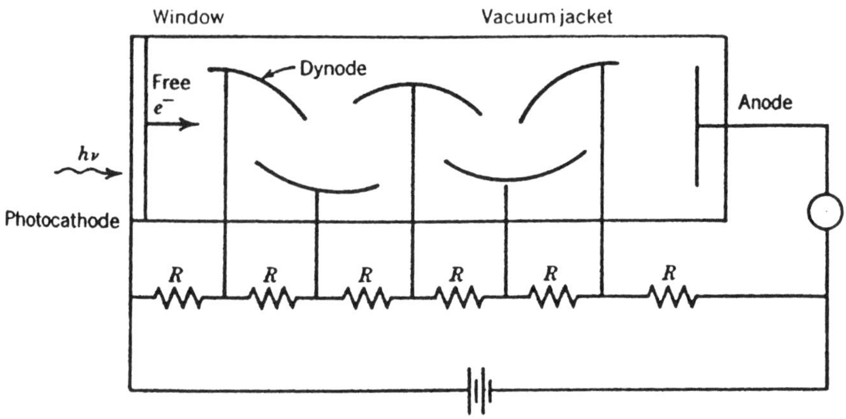

Fig. 24.1. Photomultiplier tube schematic. (Adapted from E.L. Dereniak and D.G. Crowe, *Optical Radiation Detectors,* Wiley, 1984. Reprinted with permission of Wiley.)

where δ is the gain at each dynode and N is the number of dynodes.

The responsivity $\mathcal{R}$, which is the ratio of the photocurrent to the power of the incident light, is

$$\mathcal{R} = \frac{M\eta e}{h\nu} = \frac{M\eta e\lambda}{hc}. \tag{24.4}$$

The gain of a photomultiplier tube is determined by the applied voltage; because the device can take thousands of volts before the breakdown voltage between the insulating material and spacers is reached, amplifications of 10^6 can be achieved.[1] The response speed is of the order of 10^{-10} seconds.[2] The response is proportional to light intensity over 6 to 8 orders of magnitude.[3] The responsivities of typical photomultiplier tubes are shown in Fig. 24.2; their supply voltages are given in Table 24.1.

24.3. Photoconductive Detector

Photoconductive detectors change their resistance when incident photons raise electrons into the conduction band, thereby reducing the resistance. The detector is usually connected in a voltage divider network and supply voltage. Typical photoconductive detectors are made from cadmium or lead crystals (CdS, CdSe, PbS, and PbSe) or mercury crystals (HgCdTe). A schematic of a photo-

[1] *Photomultiplier Tubes,* Hamamatsu Corp., 1990, p. 8.
[2] W. R. McCluney, *Introduction to Radiometry and Photometry,* Artech House, 1994, p. 196.
[3] *Oriel Catalog,* Oriel Corp., 1994, Vol. II, Chap. 3, pp. 3–9.

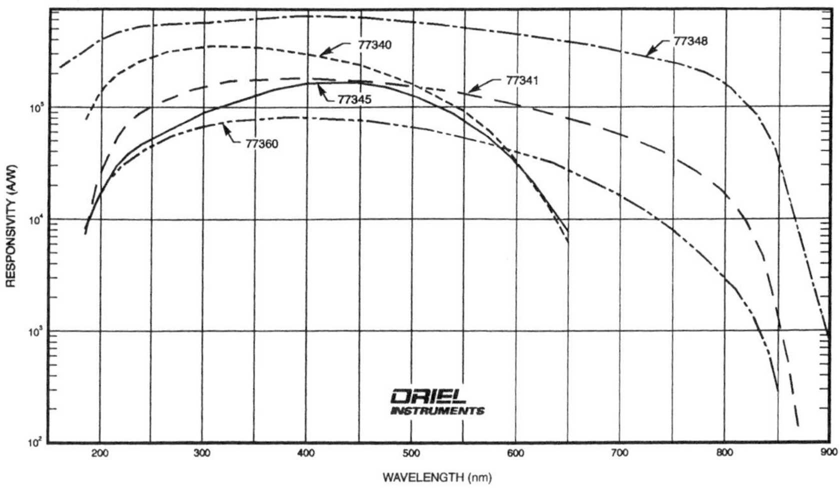

Fig. 24.2. Typical responsivity of photomultiplier tubes operating at the supply voltages specified in Table 24.1 (Models 77361 and 77343 come with borosilicate glass windows that limit their usefulness below 400 nm.) (Reprinted with permission from *The Book of Photon Tools,* Oriel Instruments, 2000.)

Table 24.1. Supply Voltages for the Photomultipliers of Fig. 24.2[a]

Current Amplification	Anode Responsivity (A/W)	Anode Dark Current (nA)	Rise Time (ns)	Electron Transit Time (ns)	Noise-Equivalent Power (W $Hz^{-1/2}$)	Model Number
7.5×10^6	3.6×10^5	1	2.2	22	1.4×10^{-16}	77340
5.0×10^6	2.0×10^5	2	2.2	22	2.8×10^{-16}	77341
2.0×10^5	380	30	2.0	20	1.2×10^{-13}	77343
1.8×10^5	1.1×10^4	2	2.0	20	2.2×10^{-16}	77344
1.0×10^7	6.8×10^5	2	2.2	22	1.2×10^{-16}	77348
1.3×10^6	9.1×10^4	5	1.2	18	5.0×10^{-16}	77360
2.1×10^6	1.8×10^5	2	12	60	2.0×10^{-16}	77345
5.3×10^5	3.4×10^4	3	15	60	6.6×10^{-16}	77346
5.0×10^5	950	1000	10	50	4.2×10^{-13}	77361

[a] Reprinted with permission from *The Book of Photon Tools*, Oriel Instruments, 2000.

conductive detector is shown in Fig. 24.3. They are often fabricated with a lens to increase their noise-equivalent power and generally make the physically small detector look optically larger. Figure 24.4 shows the schematic of an optically immersed detector.

The photocurrent I and responsivity $\mathcal{R}$ for photoconductors are given by

$$I = \frac{\eta e \phi}{h\nu} \frac{\tau_0}{\tau_d} = \frac{\eta e \phi \lambda}{hc} \frac{\tau_0}{\tau_d}, \tag{24.5}$$

$$\mathcal{R} = \frac{\eta e}{h\nu} \frac{\tau_0}{\tau_d} = \frac{\eta e \lambda}{hc} \frac{\tau_0}{\tau_d}, \tag{24.6}$$

where τ_0 is the recombination time and τ_d is the drift time for a carrier across the length *d*. Another way of looking at it is that τ_0/τ_d is the fraction of the crystal length drifted by the average excited carrier before recombining.

To lower the drift time, practical photoconductive detectors are often structured with interdigitized metal contacts, as shown in Fig. 24.5. If the interdigi-

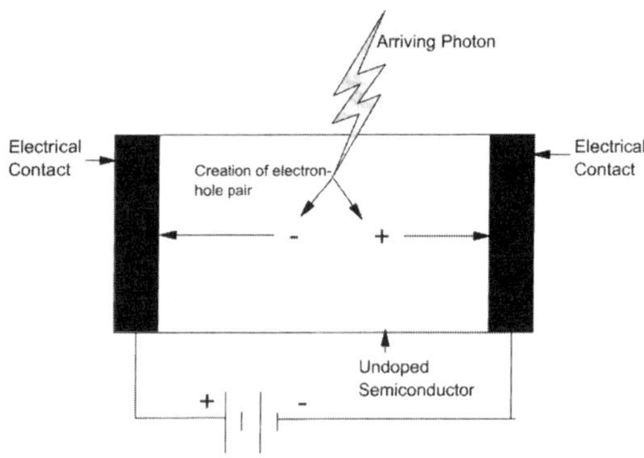

Fig. 24.3. Photoconductive detector schematic.

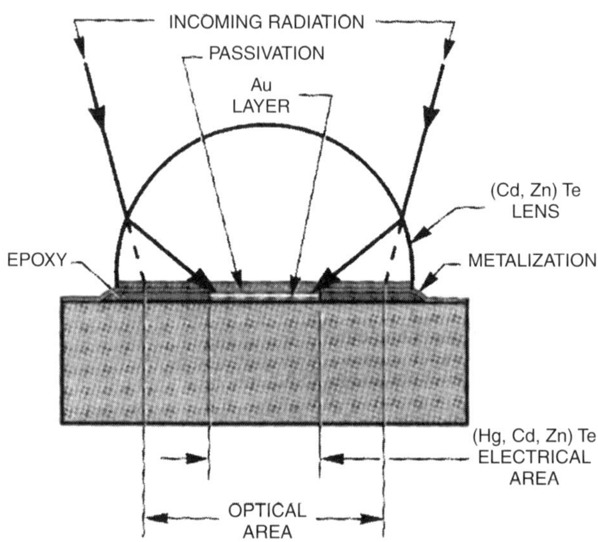

Fig. 24.4. Optically immersed HgCdZnTe detector schematic. (Reprinted with permission from *The Book of Photon Tools,* Oriel Instruments, 2000.)

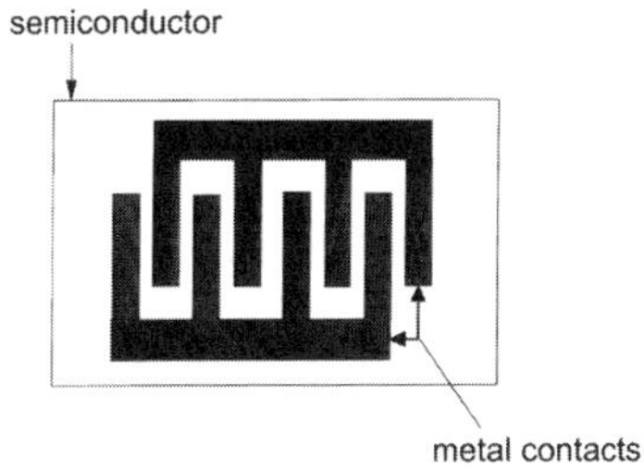

Fig. 24.5. Practical photoconductive detector schematic.

tized metal contacts are Schottky-barrier diodes, the detector is called a metal–semiconductor–metal detector; this is treated in Section 24.9.

The noise-equivalent power values of typical PbS and PbSe detectors are shown in Fig. 24.6.

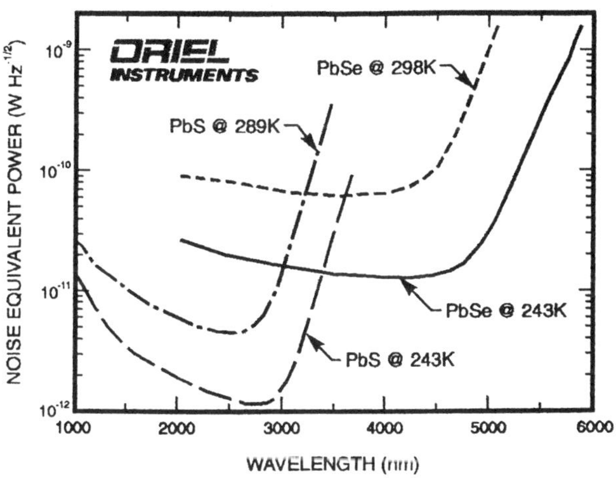

Fig. 24.6. Noise-equivalent powers of PbS and PbSe detectors. (Reprinted with permission from *The Book of Photon Tools*, Oriel Instruments, 2000.)

24.4. The *p-i-n* Photodiode

The *p-i-n* diode is a semiconductor junction detector formed from a sandwich of three layers of crystal, each layer with different band structures caused when impurities are added (doping) to the base material, usually silicon or germanium. The layers are doped in this arrangement: *p*-type on top, intrinsic, meaning undoped, in a thin middle layer, and *n*-type (or negative) on the bottom. For a silicon crystal a typical *p*-type impurity would be boron, and indium would be a *p*-type impurity for germanium. The change in potential at the interface has the effect of influencing the direction of current flow, creating a diode.

The structure of a typical *p-i-n* photodiode is shown in Fig. 24.7. The *p*-type and the *n*-type silicon layers form a potential difference at the intrinsic region; this potential gradient depletes the junction region of charge carriers, both electrons and holes, and results in the conduction band's bending. The junction drives holes into the *p*-type material and electrons into the *n*-type material.

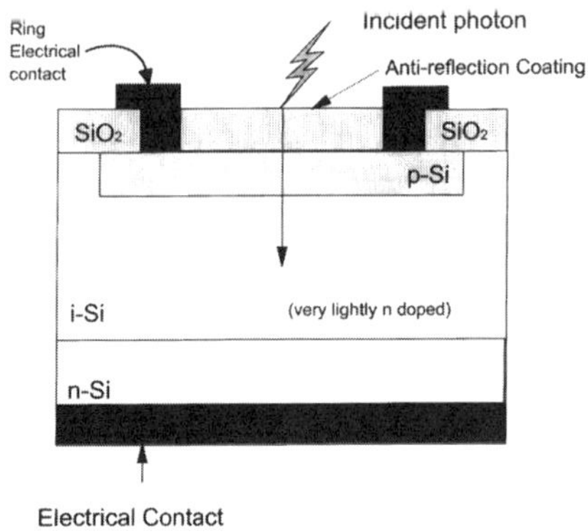

Fig. 24.7. Silicon *p-i-n* diode schematic.

The difference in potential of the two materials determines the energy an electron must have to flow through the junction. When photons fall on the active area of the device, they generate carriers near the junction, resulting in a voltage difference between the *p*-type and the *n*-type regions. If the diode is connected to external circuitry, a current will flow that is proportional to the illumination.

When there is no external bias, photodiodes are said to operate in the photovoltaic mode. When photodiodes are operated with a bias, reducing junction capacitance and improving speed of response, they operate in the photoconductive mode. Photodiodes operating in the photoconductive mode do not follow the bulk photoconductive detector equations!

The following equations apply to most linear-response photodiodes, such as *p-i-n* diodes, *p-n* diodes, and Schottky barrier diodes. (They do not apply to avalanche photodiodes, which are treated in the next section.)

To create an electron–hole pair, the photon must have enough energy to raise an electron across the bandgap. This is true when $h\nu \geq W_g$, where h is Plank's constant, ν is the frequency of the light, and W_g is the bandgap energy. The cutoff frequency ν_g and wavelength λ_g of a photodiodc is $\nu_g = W_g/h$, or $\lambda_g = hc/W_g = 1.234/W_g$ if the wavelength is in micrometers and the gap energy is in electron volts.

The photodiode current is given by

$$I = \frac{\eta e \Phi}{h\nu} + I_D = \frac{\eta e \Phi \lambda}{hc} + I_D, \tag{24.7}$$

where η is the quantum efficiency of the detector, Φ is the power of the incident optical signal, and I_D is the dark current of the detector.

The quantum efficiency η can be calculated by

$$\eta = \tau(1 - \rho)(1 - e^{-\alpha d}), \tag{24.8}$$

where ρ is the surface reflectivity, τ is the transmission of any lossy electrode layers (the *p* region of a *p-i-n* diode), *d* is the thickness of the absorbing layer, and α is the absorption coefficient.

The responsivity $\mathcal{R}$ is

$$\mathcal{R} = \frac{I}{\phi} = \frac{\eta e \lambda}{hc}. \tag{24.9}$$

The responsivity of an unfiltered silicon photodiode is shown in Fig. 24.8. The responsivity of an ultraviolet-enhanced silicon is also shown.

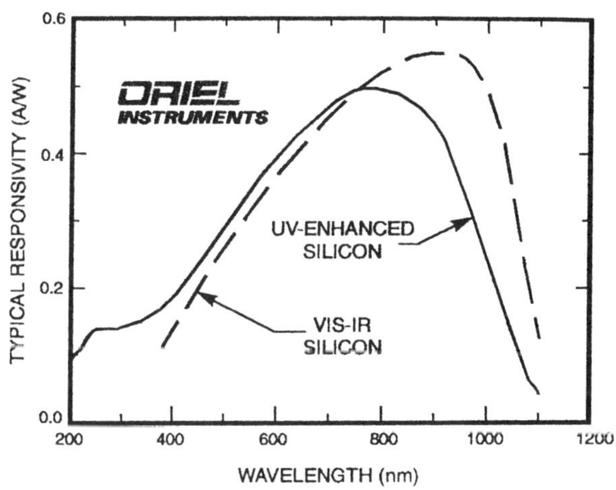

Fig. 24.8. Typical responsivity of silicon detectors. (Reprinted with permission from *The Book of Photon Tools,* Oriel Instruments, 2000.)

The temporal bandwidth Δf of a *p-i-n* diode is determined by the transport time τ of electrons across the intrinsic region, the capacitance of the device *C*, and the load resistance *R*, which can be calculated by

$$\omega = \frac{2.8}{\tau} + \frac{1}{RC}, \tag{24.10}$$

where ω is the upper (angular) frequency limit. The dark current is approximately

$$I_D \approx \frac{\beta kT}{eR_0}, \tag{24.11}$$

where β is the ideality factor, which varies from 1 to 2, depending on diode structure, and R_0 is the diode resistance. For long-wavelength detectors, β is closer to 2.

The shot noise (noise from the conversion of photons to free carriers) is

$$I_{\text{SH}} = \sqrt{2e\mathcal{R}\Phi\Delta f}, \tag{24.12}$$

where $\mathcal{R}$ is the responsivity of the detector, Φ is the incident optical power, and Δf is the temporal bandwidth.

Silicon photodiodes are inherently slow because the *p* and the *n* regions are absorbing and generate carriers as well as a thick intrinsic region for the electrons to traverse.

24.5. Avalanche Photodiode

The avalanche photodiode is a variant of the *p-i-n* diode described in the preceding section. For a *p-i-n* photodiode, each absorbed photon ideally creates one electron–hole pair, which, in turn, sets one electron flowing in the external circuit. If the bias voltage is increased significantly, the photogenerated carriers have enough energy to start an avalanche process, knocking more electrons free from the lattice, which contributes to amplification of the signal. This is known as an avalanche photodiode; it provides higher responsivity, especially in the near infrared, but also produces higher noise because of the electron-avalanche process.

The structure of the avalanche photodiode, shown in Fig. 24.9, creates a strong electrical field in a portion of the depletion region. Primary carriers—the free electrons and holes created by absorbed photons—within this field are accelerated by the field, thereby gaining several electron volts of kinetic energy. A collision of these fast carriers with neutral atoms causes the accelerated carrier to use some of its energy to raise a bound electron from the valence band to the conduction band. A free-electron–hole pair is thus created. Carriers created in this way, through collision with a primary carrier, are called secondary carriers. This process of creating secondary carriers is known as collision ionization. A primary carrier can create several new secondary carriers, and secondary carriers themselves can accelerate and create new carriers. The whole process is called photomultiplication. The number of electrons set flow-

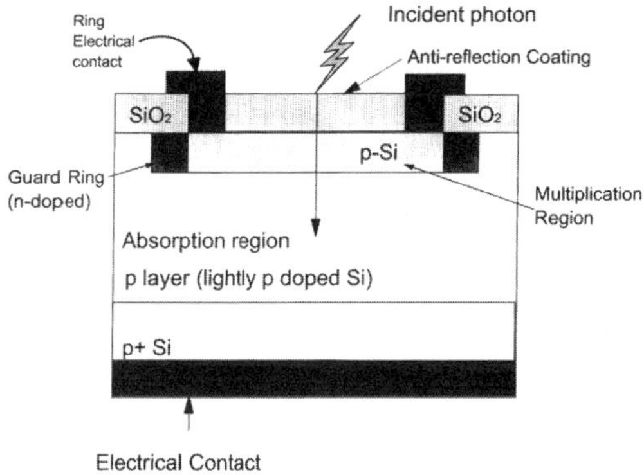

Fig. 24.9. Avalanche photodiode schematic.

ing in the external circuit by each absorbed photon depends on the avalanche photodiode's multiplication factor. Typical multiplication ranges in the tens and hundreds.

The low-frequency multiplication factor M_{DC} varies with the bias voltage as described by

$$M_{DC} = \frac{1}{\left(1 - \frac{V}{V_B}\right)^n}, \tag{24.13}$$

where V_B is the breakdown voltage and n varies between 3 and 6, depending on the semiconductor.

The current generated is

$$I = \frac{M\eta e\Phi}{h\nu} = \frac{M\eta e\Phi\lambda}{hc}, \tag{24.14}$$

and the responsivity $\mathcal{R}$ is

$$\mathcal{R} = \frac{M\eta e}{h\nu} = \frac{M\eta e\lambda}{hc}, \tag{24.15}$$

which is the same expression as Eqs. (24.2) and (24.4) for photomultiplier tubes.

The shot noise in an avalanche photodiode is that of a p-i-n diode multiplied by M times an excess noise factor $F(M)$:

$$I_{sh} = M\sqrt{2e\mathcal{R}\Phi\Delta f F(M)}, \tag{24.16}$$

where

$$F(M) = \beta M + (1 - \beta)(2 - 1/M). \tag{24.17}$$

In this case, β is the ionization coefficient of the holes divided by the ionization coefficient of the electrons. In III-V semiconductors, $F = M$. It is also important to remember that the dark current of avalanche photodiodes is also multiplied in the same manner as the shot noise.

24.6. Charge-Coupled Device Array

A charge coupled device (CCD) is a solid-state device composed of an array of metal–insulator–semiconductor capacitors. The major type of CCD available is composed of a silicon wafer (semiconductor) with an SiO_2 (insulator) surface grown on it and a matrix of metal squares deposited on top of the SiO_2 layer, known as metal–oxide semiconductor technology (Fig. 24.10). When a

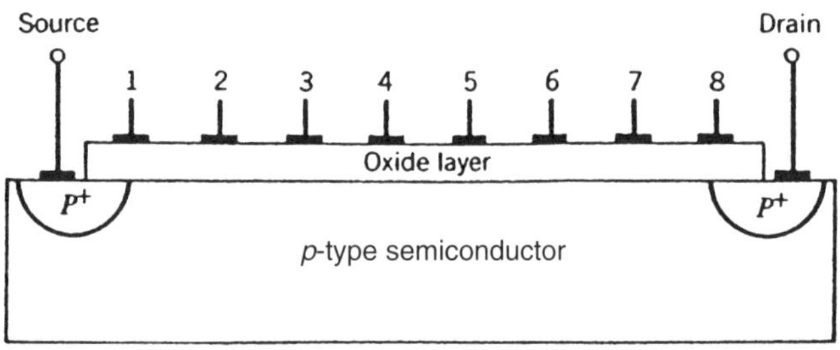

Fig. 24.10. CCD schematic: metal–oxide–semiconductor field-effect transistor with eight gates. (Adapted from E.L. Dereniak and D.G. Crowe, *Optical Radiation Detectors,* Wiley, 1984. Reprinted with permission of Wiley.)

photon of energy greater than that of the bandgap is absorbed into the metal surface, an electron–hole pair is formed. However, unlike *p-i-n* diodes, this array will store the charge, as it is composed of capacitors. Pulsed voltages will move the stored charges from their location to the next capacitor and eventually to the digital electronics where they can be processed. (The detector must be shuttered during this stage of the process.) Signal storage makes CCDs significantly different from the real-time operation of single-detector elements. Integration time is an important factor in setting up the experiment; with long integration times, charge from the dark current will build up.[4]

The number of electrons per pixel per second, $\dot{N}_e$, is related to the optical power Φ by

[4] *An Introduction to Scientific Imaging Charge-Coupled Devices,* Scientific Imaging Technologies, 1994.

$$\dot{N}_e = \frac{\eta\Phi}{h\nu} = \frac{\eta\Phi\lambda}{hc}, \tag{24.18}$$

where η is the quantum efficiency, h is Plank's constant, and ν is the frequency.

Note that this is not the photocurrent, as with previous detectors, because the charge is stored. The quantum efficiency of a typical CCD array used in spectroscopy is shown in Fig. 24.11.

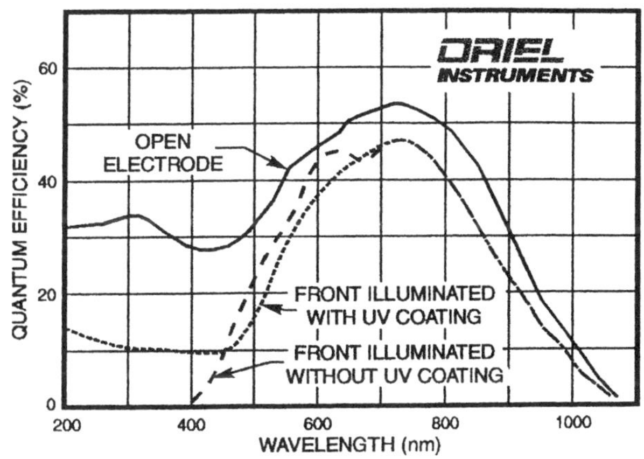

Fig. 24.11. Quantum efficiencies of various InstaSpec™ IV CCD sensors. (Reprinted with permission from *The Book of Photon Tools*, Oriel Instruments, 2000.)

24.7. Photodiode Array

This technology is in competition with CCD arrays. Silicon arrays saturate at 2 orders of magnitude higher light levels than CCD arrays, which has advantages for some applications. The individual diode elements respond to incident flux by producing photocurrents, which charge individual storage capacitors. The capacitors are read and discharged by circuitry switches, which are built onto the chip.

The quantum efficiency of a typical photodiode array is shown in Fig. 24.12.

24.8. Schottky-Barrier Photodiode

Schottky-barrier photodiodes are not *p-i-n* based. Instead, a thin metal layer replaces one half of the *p–n* junction (see Fig. 24.13). However, it does result in the same voltage characteristics. When an incident photon hits the metal layer, carriers are created in the depletion region, and their movement again sets cur-

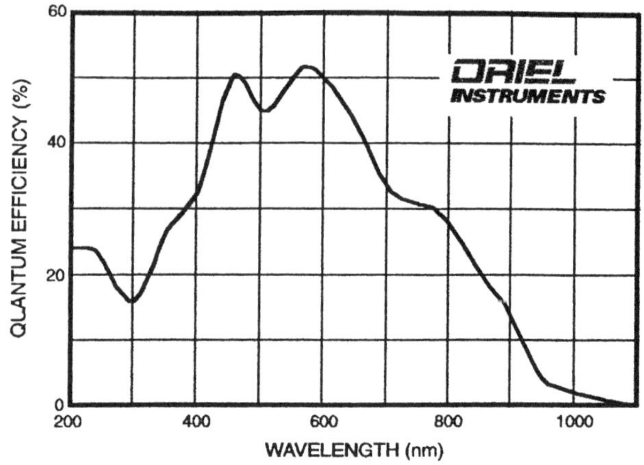

Fig. 24.12. Quantum efficiency of InstaSpec™ II photodiode array detectors. (Reprinted with permission from *The Book of Photon Tools,* Oriel Instruments, 2000.)

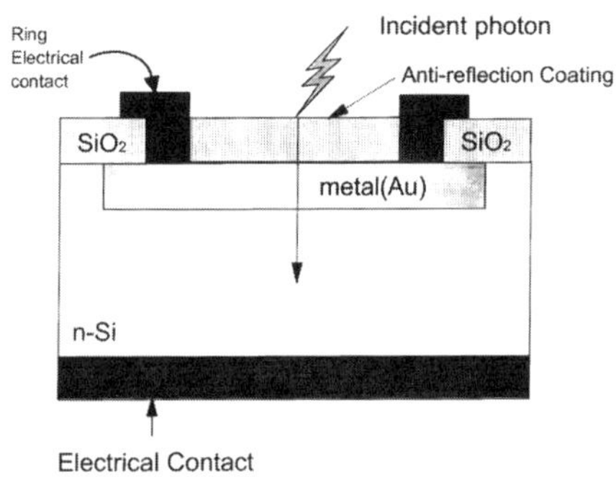

Fig. 24.13. Schottky-barrier photodiode schematic.

rent flowing. The voltage and field relations derived for *p-i-n* junctions can be applied to Schottky-barrier diodes if the metal layer is treated as if it were an extremely heavily doped semiconductor. The metal layer is a good conductor, and so electrons leave the junction immediately. This results in faster operation—up to 100-Ghz modulation has been reported.[5] It also lowers the chance of recombination, increasing efficiency, and increases the types of semiconductor that can be used, as it is less limited by lattice-matching fabrication constraints. However, the sensitivity is lower than that of *p-i-n* diodes with the usual-sized intrinsic layer. The photodiode equations given in Section 24.4 under *p-i-n* diodes also apply to Schottky diodes, with an ideality factor β of 1.

[5] E. Garmire, "Sources, modulators, and detectors for fiber-optic communication systems," in *Handbook of Optics,* McGraw-Hill, 2001, Vol. IV, Chap. 4, p. 4.78.

24.9. Metal–Semiconductor–Metal Detectors and Arrays

A metal–semiconductor–metal detector is made by the formation of two Schottky diodes on top of a semiconductor layer. The top metal layer has two contact pads and a series of "interdigitated fingers," which form the active area of the device. One diode gets forward biased, and the other diode gets reverse biased. When illuminated, these detectors create a time-varying electrical signal.

Like Schottky diodes, speed is an advantage of metal–semiconductor–metal detectors. The response speed of metal–semiconductor–metal detectors can be increased by a reduction in the finger spacing and thus in the carrier transit times. The other advantages of metal–semiconductor–metal detectors are their ease in fabrication and integration into electrical systems. Metal–semiconductor–metal detectors have the same disadvantages as other Schottky-based systems—weak signal and noise.

Most metal–semiconductor–metal detectors are photoconductive and follow the rules for photoconductive current and responsivity:

$$I = \frac{\eta e \phi}{h\nu}\frac{\tau_0}{\tau_d} = \frac{\eta e \phi \lambda}{hc}\frac{\tau_0}{\tau_d},$$

$$\mathcal{R} = \frac{\eta e}{h\nu}\frac{\tau_0}{\tau_d} = \frac{\eta e \lambda}{hc}\frac{\tau_0}{\tau_d},$$

where τ_0 is the recombination time and τ_d is the drift time for a carrier across the length d.

The photoinduced carrier density, which is the number of electrons per pixel per second $\dot{N}_e$, is related to the optical power Φ and is the same expression as that for CCD detectors:

$$\dot{N}_e - \frac{\eta\Phi}{h\nu} = \frac{\eta\Phi\lambda}{hc}.$$

The output time-varying resistance results from the time-varying photoinduced carrier densities of electrons, n, and holes, p, are

$$R(t) = \frac{L}{q(n\mu_e + p\mu_h)wd}, \tag{24.19}$$

where L is the distance between electrodes, q is the electronic charge, the μ's are the electron and hole mobilities, w is the length along the electrodes excited by light, and d is the absorption depth in the semiconductor.

At 1300–1550 nm, a typical noise-equivalent power is 0.14 mW rms and a sensitivity is –30 dBm. At 850 nm a typical noise-equivalent is 0.5 mW rms and sensitivity is –24.5 dBm.[6]

24.10. Resonant-Cavity Enhanced Photodetector

The principle of resonant-cavity enhanced photodetectors is that a fast photodetector is placed in a Fabry–Perot cavity to enhance the signal magnitude (see Fig. 24.14). In practice, the Fabry–Perot cavity's end mirrors are made of quarter-wave stacks of GaAs/AlAs, InGaAs/InAlAs, or any other semiconductor or insulator materials that have the desired refractive-index contrast. A typical example of a fast photodetector is a *p-i-n* photodetector with a thin *i*-layer. If the insulator layer is made thinner, detector speed increases, but the signal decreases. Schottky-barrier photodiodes and metal–semiconductor–metal detectors can also have their signal magnitude increased by use of resonant-cavity enhanced photodetector techniques.

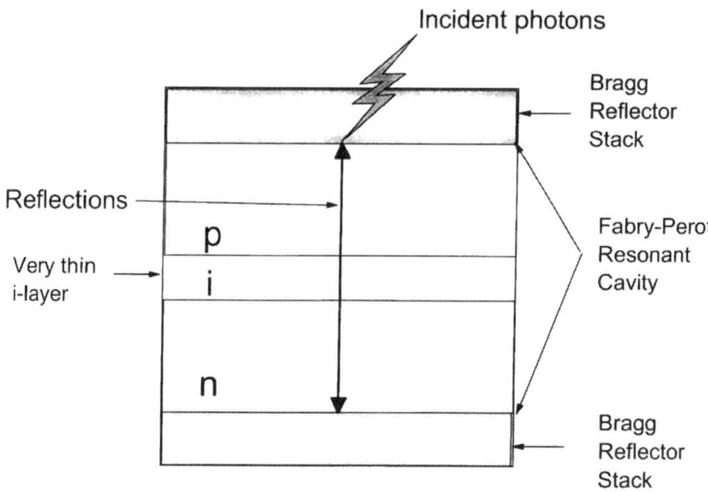

Fig. 24.14. Resonant-cavity detector schematic.

The light passes through a detector in a Fabry–Perot cavity several times, and this results in an increased signal. However, because of the resonance properties of the Fabry–Perot cavity, this detector is highly wavelength selective. It is also highly sensitive to the position of the detector in the Fabry–Perot cavity, as it is a resonance effect. The maximum quantum efficiency η_{max} of a resonant-cavity enhanced photodetector can be calculated by

[6] *Anadigics Datacom TIA Applications Note,* Anadigics, 2000, p. 3 (www.anadigics.com).

$$\eta_{max} = \frac{1 + \rho_2 e^{-\alpha d}}{(1 - \sqrt{\rho_1 \rho_2}\, e^{-\alpha d})^2}(1 - \rho_1)(1 - e^{-\alpha d}), \tag{24.20}$$

where ρ_1 and ρ_2 are the reflectivity of the top and the bottom mirrors, respectively, α is the absorption of the active area of the detector, and d is the thickness of the active area of the detector. If the reflection of the bottom mirror $\rho_2 = 0$, this reduces to Eq. (24.8), the quantum efficiency of a photodiode,

$$\eta = \tau(1 - \rho)(1 - e^{-\alpha d}),$$

when there is no loss in the p region, and thus the transmission through it, τ, is unity. The resonant-cavity enhancement can allow resonant-cavity enhanced detectors to have quantum efficiencies of nearly unity at their peak wavelength:

$$RE = \frac{1 + \rho_2 e^{-\alpha d}}{(1 - \sqrt{\rho_1 \rho_2}\, e^{-\alpha d})^2}. \tag{24.21}$$

The theoretical wavelength dependence of the quantum efficiency is shown in Fig. 24.15. Enhancements of the quantum efficiencies of 50%–80% of many types of detectors have been observed in the laboratory.[7]

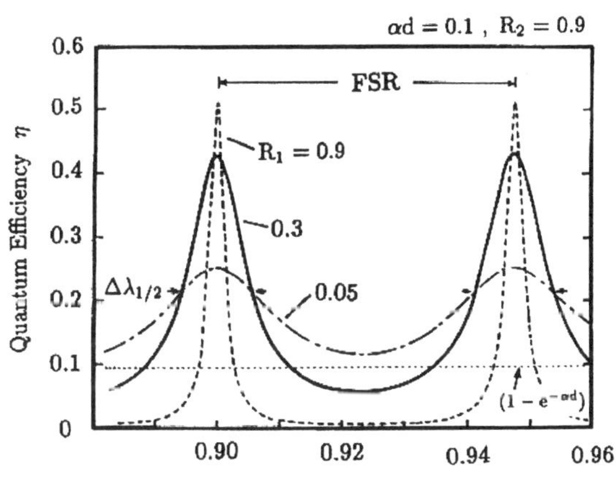

Fig. 24.15. Quantum efficiency of resonant-cavity detectors as a function of wavelength. FSR, free spectral range. (Reprinted by permission of Selim Ünlü and Samuel Strite, "Resonant cavity enhanced (RCE) photonic devices," *http://photon.bu.edu/selim/papers/apr-95/node 1.html*)

[7] M. Selim Ünlü and Samuel Strite, "Resonant cavity enhanced (RCE) photonic devices," *http://photon.bu.edu/selim/papers/apr-95/node 1.html*

Further Reading

On-Line Tutorials: These tutorials are available free of charge on the World Wide Web:

CORD Module 4-8 Optical Detectors: *http://utopia.cord.org/cm/leot/course04_mod08.htm*

An Introduction to Scientific Imaging Charge-Coupled Devices, Scientific Imaging: *http://www.site-inc.com/pdf/introdat.pdf*

Notes for B. Van Zeghbroeck's University of Colorado class ECEN6355, Heterojunction Devices: *http://ece-www.colorado.edu/~bart/ecen6355/notes.htm*

"Resonant cavity enhanced (RCE) photonic devices" by Selim Ünlü and Samuel Strite: *http://photon.bu.edu/selim/papers/apr-95/node 1.html*

Books

E. L. Dereniak and D. G. Crowe, *Optical Radiation Detectors,* Wiley, 1984.

H. Dutton, *Optical Communications,* Academic, 1999.

E. Garmire, "Sources, modulators, and detectors for fiber-optic communication systems," in *Handbook of Optics,* M. Bass, ed., McGraw-Hill, 2001, Vol. 4, Chap. 4.

C. J. Sher DeCusatis, "Fundamentals of detectors," in *Handbook of Applied Photometry,* C. M. DeCusatis, ed., American Institute of Physics, 1998, Chap. 4.

C. J. Sher DeCusatis and Ching-Long (John) Jiang, "Detectors for fiber optics," in *Handbook of Fiber Optic Data Communications,* 2nd ed., C. M. DeCusatis, ed., Academic, 2002, Chap. 4.

A. Yariv, "Detection of optical radiation," in *Optical Electronics,* Holt, Reinhart & Winston, 1985, Chap. 11.

25

Figures of Merit

William L. Wolfe

25.1. Introduction

There are many different ways to describe detectors and sensors. Some are specific, others are general. Most of these figures of merit apply both to a detector and to a sensor.

25.2. Signal

The output signal is usually a current or voltage, but may also be an accumulation of charge. The symbols are *V, I,* and *Q*—usually. The signal level of a detector is often of use in understanding the design of the subsequent electronics. It is conventional and convenient to describe these in terms of their spectra and the integrals of their spectra over some bandwidth.

25.3. Noise

The noise is a statistical variation around the average value of the signal and depends on the specific situation. Chapters 24, 25, and 29 describe some specific noises. The noises are also conveniently described by their spectra and the related integrals. The total rms noise is the square root of the sum of the squares of all the contributing noises:

$$v_{\mathrm{rms,total}} = \sqrt{\langle (v - \langle v \rangle)^2 \rangle} = \sqrt{\int \sum v_i^2 \mathrm{d}f},$$

where v_{rms} is the rms deviation from the mean and the v_i are the individual noise-voltage spectra.

25.4. Signal-to-Noise Ratio

The ratio of the signal to the total noise is an important parameter of a system, but it is a function of the conditions. The noise must be the rms value; the signal can be either the rms value or the peak-to-peak value. Usually the signal-to-noise ratio is the voltage or current ratio, but it can be the ratio of the signal power to the noise power.

25.5. Responsivity

Responsivity is the ratio of an output to the input. The input can be over a spectral band, or as a function of the spectral variable, or for a receiver with a special spectral response, and can be a photon rate or power. The output is usually either voltage or current.

25.5.1. *Spectral Responsivity*

Thus the spectral power—voltage spectrum responsivity is

$$\mathcal{R}_v(\omega, \lambda) = \frac{V(\omega)}{\Phi(\lambda)}.$$

The voltage can be a function of frequency; the flux can be a function of frequency and wavelength. In general the responsivity is a function of both variables.

25.5.2. *Band-Limited Responsivity*

The band-limited, current, photon responsivity is

$$\mathcal{R}_i(\Delta\lambda) = \frac{I(\omega)}{\Phi_q(\Delta\lambda)}.$$

Other combinations have similar definitions. The spectrum of the signal is given by

$$v(\omega) = \mathcal{R}_v(\omega, \lambda)\Phi(\omega, \lambda).$$

25.5.3. *Luminous Responsivity*

The luminous responsivity $\mathcal{R}_V$ is the output divided by the input that has been normalized to the luminous efficacy (Chaps. 21 and 30):

$$\mathcal{R}_V = \frac{v}{\int \Phi(\lambda)V(\lambda)\mathrm{d}\lambda},$$

where v is the output voltage, $V(\lambda)$ is the luminous efficacy, and $\Phi(\lambda)$ is the spectral distribution of the power.

The luminous responsivity is a function of the spectral distribution of the source used to measure it and may differ for the intended application. This difference may be taken into account. The true output current is given by

$$I = \mathcal{R}_V \Phi_V \frac{\int \phi_c v \mathrm{d}\lambda \int r\phi \mathrm{d}\lambda}{\int \phi v \mathrm{d}\lambda \int r\phi_c \mathrm{d}\lambda},$$

where ϕ is the normalized spectral distribution of the flux, v is the normalized luminous efficacy, and r is the normalized spectral responsivity. The subscript c indicates calibration conditions. All of them are functions of wavelength, although not written so explicitly.

25.5.4. Other Normalizations

Although the normalization to the luminous efficacy is the most prevalent, other normalizations exist. These include normalization to the lead sulfide detector response, to the sunburn response, the chlorophyll response, and more. The same principles apply.

25.6. Noise Equivalents

The noise-equivalent power is the power that yields a signal-to-noise ratio of 1:

$$\mathrm{NEP} = \frac{\Phi}{\mathrm{SNR}},$$

where the signal-to-noise ratio and the flux can be over the bands or spectra (in both senses). The noise-equivalent power may also be defined as the slope of the signal-to-noise ratio versus the Φ curve:

$$\mathrm{NEP} = \frac{\partial \Phi}{\partial \mathrm{SNR}}.$$

Note that this is the reciprocal of the change of the signal-to-noise ratio with respect to the flux. The noise-equivalent power is used for both detectors and systems. Other noise equivalents are the noise-equivalent flux density, spectral radiance, temperature difference, reflectivity difference, absorptivity, transmissivity, etc. These are usually used for systems. The noise-equivalent flux density is

$$\mathrm{NEFD} = \frac{\frac{\Phi}{A}}{\mathrm{SNR}}.$$

The noise-equivalent spectral radiance is

$$\mathrm{NESR} = \frac{L_\lambda}{\mathrm{SNR}}.$$

The noise-equivalent temperature difference is

$$\mathrm{NETD} = \frac{\Delta T}{\mathrm{SNR}}.$$

The expression for the signal-to-noise ratio can be evaluated for each situation.

25.7. Detectivity

Detectivity is a figure of merit for detectors. Detectivity is the reciprocal of the noise-equivalent power; it is the signal-to-noise ratio per input watt:

$$D = \frac{1}{\mathrm{NEP}} = \frac{\mathrm{SNR}}{\Phi}.$$

The specific detectivity is the detectivity normalized to a detector with an area of 1 cm^2 and has a 1-Hz noise bandwidth:

$$D^*(\lambda, f, B) = D\sqrt{A_d B} = \frac{\sqrt{A_d B}}{\Phi}\,\mathrm{SNR} = \frac{\sqrt{A_d B}}{\mathrm{NEP}},$$

where the parameters λ, f, and B indicate, respectively, that the detectivity is spectral, that the modulation frequency of the incident light is f, and the bandwidth is 1 Hz. (It is redundant to specify the bandwidth at 1 Hz, but that is the custom.) Detectivity can also be expressed with a normalization to the field of view, the solid angle; it is

$$D^{**}(\lambda, f, B) = D^*(\lambda, f, B)\sqrt{\frac{\Omega}{\pi}}.$$

The specific detectivity can be expressed in several ways. One is the spectral D^*, which is given by

$$D^*(\lambda) = \frac{\sqrt{A_d B}}{\Phi_\lambda(\lambda)} \text{SNR}.$$

Another is the blackbody D^*, which is

$$D^*(T) = \frac{\sqrt{A_d B}}{\Phi^{\text{BB}}(T)} \text{SNR},$$

where the flux is the blackbody flux for a source at temperature T.

The relationship between the spectral specific detectivity and the blackbody specific detectivity is

$$D^*(T, f, B) = \frac{\int_0^\infty \Phi_\lambda^{\text{BB}}(\lambda, T) D_\lambda^*(\lambda, f, B) \text{d}\lambda}{\int_0^\infty \Phi_\lambda^{\text{BB}}(\lambda, T) \text{d}\lambda}.$$

It is the weighted average spectral detectivity, where the weighting is the spectral flux on the detector. The blackbody detectivity at a second temperature, T_2, can be found from the known value at a first temperature, T_1, by

$$D^*_\lambda(T, f, B) = \frac{\int_0^\infty \Phi_\lambda^{\text{BB}}(\lambda, T_2) D_\lambda^*(\lambda, f, B) \text{d}\lambda}{\int_0^\infty \Phi_\lambda^{\text{BB}}(\lambda, T_1) D_\lambda^*(\lambda, f, B) \text{d}\lambda} \frac{\int_0^\infty \Phi_\lambda^{\text{BB}}(\lambda, T_1) \text{d}\lambda}{\int_0^\infty \Phi_\lambda^{\text{BB}}(\lambda, T_2) \text{d}\lambda}$$

$$= \frac{\int_0^\infty \Phi_\lambda^{\text{BB}}(\lambda, T_2) D_\lambda^*(\lambda, f, B) \text{d}\lambda}{\int_0^\infty \Phi_\lambda^{\text{BB}}(\lambda, T_2) \text{d}\lambda} D_\lambda^*(\lambda, T_1, B).$$

When a detector is limited only by the noise in the incident photon flux, it is considered to be background limited. The specific detectivity for a background-limited infrared detector is called the D-star BLIP, and it is given by

$$D^*_{\lambda_{\max},\text{BLIP}} = \frac{hc}{\lambda_{\max}} \sqrt{\frac{\eta}{gE_q}},$$

where this is the detectivity of a background-limited infrared detector at the wavelength of its maximum sensitivity, h is Planck's constant, c is the speed of light, $\lambda_{\max}$ is the wavelength of the maximum response of the detector, η is the quantum efficiency, g is the recombination factor (2 or 4), and E_q is the photon flux density on the detector. This can be evaluated for a family of maximum wavelengths and photon incidances.

The equivalent maximum specific detectivity for a thermal detector can be written as D^*_{BLIT}, the background-limited infrared thermal detector:

$$D^*_{\mathrm{BLIT}} = \frac{\eta}{\sqrt{8\sigma k(T_d + T_b)^4}}.$$

25.8. Quantum Efficiency

Quantum efficiency applies to photon detectors. It is the ratio of the output electron rate to the input photon rate. External quantum efficiency takes into account the surface reflectivity; internal quantum efficiency does not:

$$\eta_{\mathrm{internal}} = e^{-\alpha x},$$

$$\eta_{\mathrm{external}} = \frac{(1-\rho)^2\eta_{\mathrm{internal}}}{1-\rho^2\eta^2_{\mathrm{internal}}} = \frac{(1-\rho)^2 e^{-\alpha x}}{1-\rho^2 e^{-2\alpha x}}.$$

An equivalent concept for a thermal detector is the thermal efficiency. It is the ratio of the absorbed power to the incident power, which is equal to the surface absorptivity α, which is equal to the emissivity ε, and is sometimes specified as η.

25.9. Detective Quantum Efficiency

Detective quantum efficiency is the ratio of the signal to noise of a photon detector to an ideal detector under the same conditions. The ideal quantum detector has a quantum efficiency of 1 and is limited by photon noise. The ideal thermal detector has an absorptivity of 1 and is limited by thermal noise.

25.10. Responsive Quantum Efficiency

Responsive quantum efficiency is the ratio of the output signal current to the input flux of photons; it is equal to the quantum efficiency if expressed as the ratio of the flux of electrons to the flux of photons.

25.11. Time Constants

Time constants measure the speed with which a detector responds, both as a rise time and a decay time. For a standard, single time-constant detector (or circuit), the normalized response is described by

$$\mathcal{R}(t) = 1 - e^{-t/\tau}.$$

Therefore after one time constant the response is 63.2%; after two, it is 86.4; and after three, 95.0% of the final value. The *responsive time constant* is the time it takes for a detector or circuit *response* to reach 63.2% of its final value.

For some systems, both the signal and the noise fall off with increasing frequency. (This might be the case with recombination noise.) Then the detector is useful at time constants shorter than the responsive time constant. The *detective time constant* is the time it takes for the detectivity to rise or fall to 63.2% of its dc value.

25.12. Dynamic Range

Usually the dynamic range is a measure of the input range over which the responsivity is substantially constant. "Substantially" can be quantified in terms of a percentage deviation from linearity—the 10% dynamic range, 1% dynamic range, etc. The range itself is the ratio of the maximum value of input to the minimum value of input for which the constancy is maintained at the level specified.

25.13. Uniformity

This figure of merit is applied mostly to focal-plane arrays, but may also be applied to the overall image generated by a system. Ideally, every detector element in an array has exactly the same responsivity, no matter what the level of flux on the array. Uniformity is a measure of the degree to which this is attained. One specification is that 90% of the elements have a responsivity within ±1% of the average value. (Other numbers, of course, are used.) Sometimes dead elements are not counted. Uniformity can be specified either before or after any electronic processing. The uniformity, especially of infrared arrays, generally varies with the level of radiant incidance.

25.14. Contrast and Modulation Transfer

These are measures of the resolution performance of an optical system. Modulation is defined as

$$\frac{L_{\max} - L_{\min}}{L_{\max} + L_{\min}}.$$

The flux units could equally well be flux densities or flux. Contrast is usually defined as

$$\frac{L_{\max} - L_{\min}}{L_{\max}}.$$

Other definitions, which use the average radiance in the denominator, are also used. The transfer is the ratio of the output (in the imaging surface) to the input (in front of the optics). Thus the transfer will be the same, no matter what the definition.

25.15. Encircled Energy

This figure is based on the spot diagram and represents the size of an area into which most rays go. It can be encircled and be the radius of the circle that circumscribes all the rays or the circle that inscribes the rms number of them. For some applications the ensquared energy or the enrectangled energy is used, in which case care must be taken to understand the exact meaning.

25.16. Rayleigh, Sparrow, and Strehl Criteria

These are described in Chap. 19.

25.17. Optical Speed

This describes the collecting efficiency of an optical system for an extended object. One measure is the F number or focal ratio. It is the ratio of the effective focal length to entrance pupil diameter. Another is the numerical aperture, defined as $n \sin\theta$, where n is the refractive index of image space and θ is the angle of the outer ray of the pencil of light converging to the focus with the optical axis. For well-corrected systems and infinite object distances,

$$F = \frac{1}{2\mathrm{NA}}.$$

These measures of speed apply to circularly symmetric systems. They relate to the solid angle of incidence on the image surface in air as follows:

$$\Omega = \pi \sin^2\theta = \pi \mathrm{NA}^2 = \frac{\pi}{4F^2}.$$

25.18. Resolution, Resolving Power, Finesse, and Free Spectral Range

These are all discussed in Chap. 9.

Further Reading

R. Clark Jones, *Advances in Electronics,* Vol. 5, Academic, 1953.
W. L. Wolfe and G. J. Zissis, *The Infrared Handbook,* SPIE, 1989.
W. J. Smith, *Modern Optical Engineering,* 2nd ed., McGraw-Hill, 1990.

The first of these is a nice treatment of detector figures of merit. The second covers both detectors and systems. The third describes the optical figures.

26

Solid-State Cameras

Gerald C. Holst

26.1. Introduction

Boyle and Smith[1] and Amelio et al.[2] invented CCDs in 1970. Since then, considerable literature[3–7] has been written on CCD physics, fabrication, and operation. However, the array does not create an image by itself. The camera requires an optical system for imaging the scene onto the array's photosensitive area. The array requires bias and clock signals. Its output is a series of analog (or digital) pulses that represent the scene intensity at a series of discrete locations.

Cameras for the professional broadcast television and consumer camcorder markets are designed to operate in real time with an output that is consistent with a standard broadcast format. The resolution, in terms of array size, is matched to the bandwidth recommended in the standard. An array that provides an output of 768 horizontal by 484 vertical pixels creates a satisfactory image for conventional (U.S.) television. Eight bits (256 intensity levels or gray levels) provide an acceptable image in the broadcast and the camcorder industry.

Scientific applications generally require low noise, high responsivity, large dynamic range, and high resolution. To exploit a large dynamic range, scientific cameras may digitize the signal into 12, 14, or 16 bits. Theoretically, the array can be any size but manufacturing considerations may ultimately limit the array size. Footnote 8 provides details on scientific devices.

1 W. S. Boyle and G. E. Smith, "Charge coupled semiconductor devices," Bell Syst. Tech. J. **49,** 587–593, 1970.

2 G. F. Amelio, M. F. Tompsett, and G. E. Smith, "Experimental verification of the charge coupled concept," Bell Syst. Tech. J. **49,** 593–600, 1970.

3 M. J. Howes and D. V. Morgan, eds., *Charge-Coupled Devices and Systems,* Wiley, 1979.

4 C. H. Sequin and M. F. Tompsett, *Charge Transfer Devices,* Academic, 1975.

5 E. S. Yang, *Microelectronic Devices,* McGraw-Hill, 1988.

6 E. L. Dereniak and D. G. Crowe, *Optical Radiation Detectors,* Wiley, 1984.

7 A. J. P. Theuwissen, *Solid-State Imaging with Charge-Coupled Devices,* Kluwer Academic, 1995.

8 J. R. Janesick, *Scientific Charge-Coupled Devices,* Vol. PM83 of SPIE Press Monograph Series, Society of Photo-Optical Instrumentation Engineers, 2001.

Active pixel sensors are fabricated with complementary-metal-oxide-semiconductor technology. The advantage is that one or more active transistors can be integrated into the pixel. With the active pixel sensor approach, highly integrated image sensors are possible.[9] By placing processing on the chip, a complementary-metal-oxide-semiconductor camera is typically physically smaller than a CCD camera that requires clocks, image reformatting, and signal processing in separate hardware. A sophisticated active pixel sensor array could create[10] a "camera-on-a-chip." Complementary-metal-oxide-semiconductor and CCD arrays should not be compared directly. Rather, the complementary-metal-oxide-semiconductor camera should be compared with the CCD camera.

26.2. Charge-Coupled Devices

Devices may be described functionally according to their architecture (frame transfer, interline transfer, etc.) or by application. To minimize cost, array complexity, and electronic processing, the architecture is typically designed for a specific application. Full-frame and frame-transfer devices tend to be used for scientific applications. Interline-transfer devices are used in consumer camcorders and professional television systems. Linear arrays, progressive scan, and time delay and integration are used for industrial applications. Despite an ever-increasing demand for color cameras, black-and-white cameras are widely used for many scientific and industrial applications. The divisions among general imagery, machine-vision, scientific, and military devices become fuzzy as technology advances. Additional information appears in footnote 11.

26.2.1. Frame Arrays

Figure 26.1 illustrates a full-frame-transfer array. After integration, the image pixels are read out line by line through a serial register that then clocks its contents onto the output sense node. All charge must be clocked out of the serial register before the next line can be transferred. The number of pixels is often based on powers of 2 (e.g., 512×512, 1024×1024) to simplify memory mapping. Scientific arrays have square pixels, and this simplifies image-processing algorithms.

During readout, the photosites are continually illuminated, resulting in a

[9] S. K. Mendis, S. E. Kemeny, and E. R. Fossum, "A 128 × 128 CMOS active pixel sensor for highly integrated imaging systems," Tech. Dig. Int. Electron Devices Meet. 583–586, 1993.

[10] E. R. Fossum, "CMOS image sensors: Electronic camera-on-a-chip," IEEE Trans. Electron Devices **44,** 1689–1698, 1997.

[11] G. C. Holst, *CCD Arrays, Cameras, and Displays,* 2nd ed., JCD Publishing, 1998.

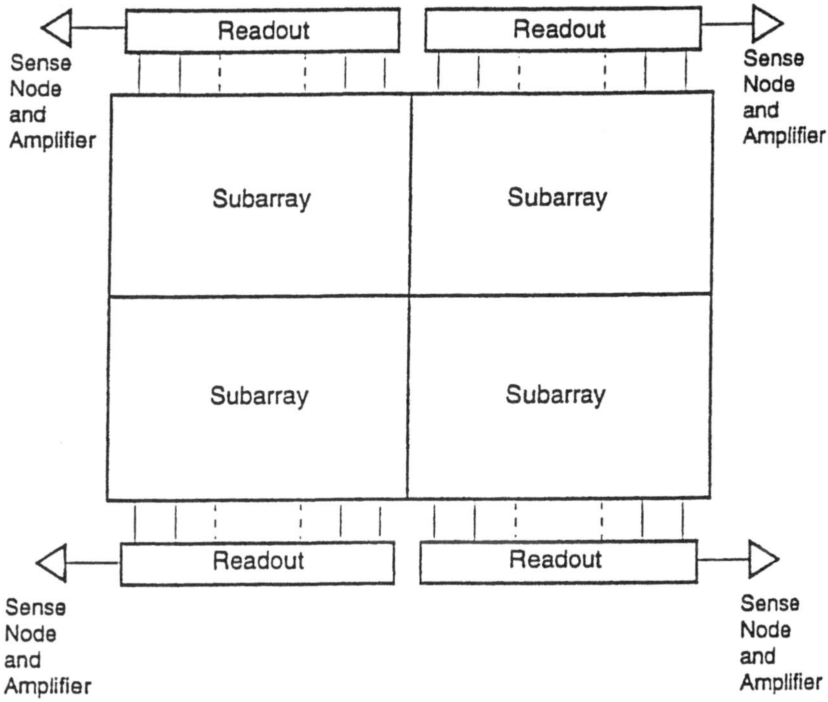

Fig. 26.1. Full-frame architecture consisting of four subarrays.

smeared image. An external electronic shutter can be used to shield the array during readout to avoid smear. If the image integration time is much longer than the readout time, then the smear may be considered insignificant. This situation often occurs with astronomical observations.

A frame-transfer imager consists of two almost identical arrays, one devoted to image pixels and one for storage. The storage cells are identical in structure to the light-sensitive cells but are covered with an opaque metal shield to prevent any light exposure. After the integration cycle, charge is transferred quickly from the light-sensitive pixels to the storage cells. Transfer time to the shielded area depends on the array size, but is typically less than 500 μs. Smear is limited to the time it takes to transfer the image to the storage area. This is less time than that required by a full-frame device.

26.2.2. *Interline Transfer*

The interline-transfer array consists of photodiodes separated by vertical transfer registers that are covered by an opaque metal shield (Fig. 26.2). After integration, the charge generated by the photodiodes is transferred to the vertical CCD registers in ~1 μs and smear is minimized. There is no need to shutter the

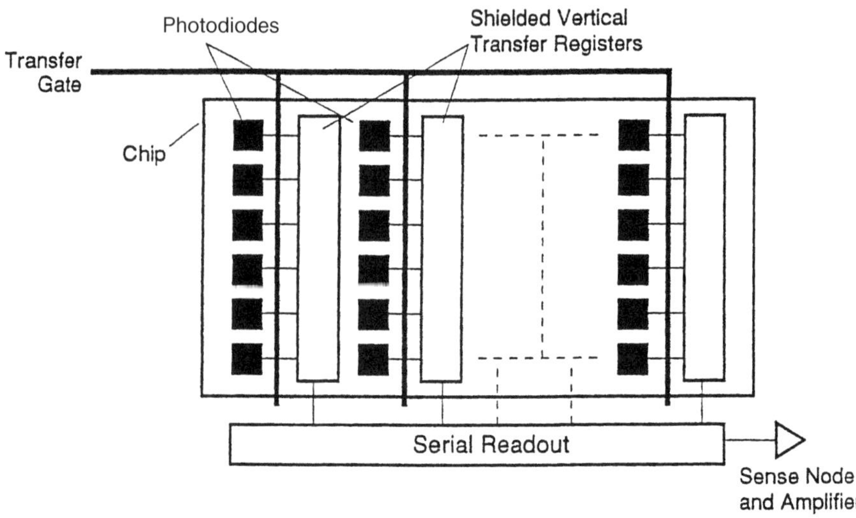

Fig. 26.2. Interline-transfer architecture. The charge is rapidly transferred to the interline-transfer registers by means of the transfer gate.

incoming light. The shields act as a Venetian blind that obscures half the information that is available in the scene. The area fill factor may be as low as 20%. Microlens assemblies (also called microlenticular arrays or lenslet arrays) increase the effective optical fill factor. However, it may not reach 100% because of slight misalignment of the microlens assembly, imperfections in the microlens itself, asymmetrical shielded areas, and transmission losses.

Because interline devices are most often found in general imagery products, most transfer register designs are based on standard video timing. Figure 26.3 illustrates a four-phase transfer register that stores charge under two gates.[12] With EIA-170, each field is read every 1/60 s. Because the fields alternate, the maximum integration time is 1/30 s for each field.

26.2.3. Progressive Scan

CCD arrays do not scan the image, but it is convenient to represent the output as if it were scanned. Progressive scan simply means the noninterlaced or sequential line-by-line scanning of the image. Any application that requires digitization and a computer interface will probably perform better with progressively scanned imagery. Although the image is captured in a progressive scan manner, the camera output is often processed into a standard video output so that it can be displayed on commercial monitors.

Because an entire frame is captured, progressive scan devices do not suffer

[12] A. J. P. Theuwissen, *Solid-State Imaging with Charge-Coupled Devices,* Kluwer Academic, 1995, pp. 161–165.

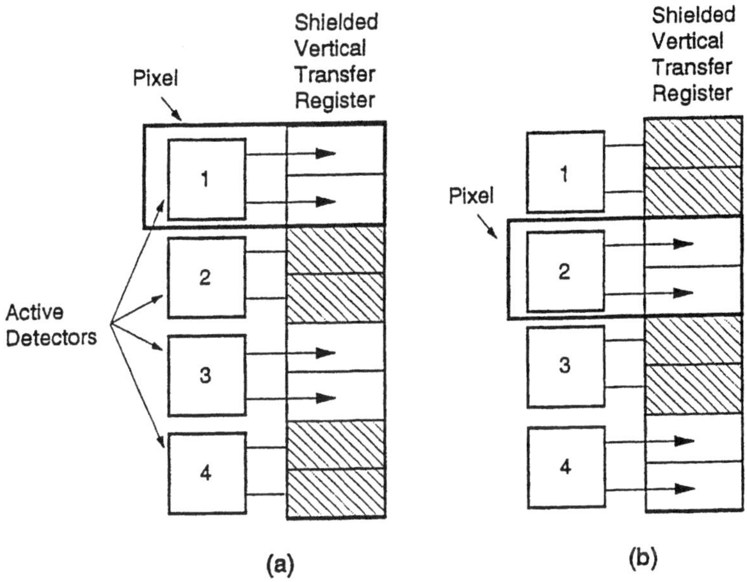

Fig. 26.3. Detailed layout of the 2:1 interlaced array: (a) The odd field is clocked into the vertical transfer register; (b) the even field is transferred. The vertical transfer register has four gates, and charge is stored under two wells. The pixel is defined by the detector center-to-center spacing, and it includes the shielded vertical register area. The transfer gate is not shown. (After footnote 12).

from same image motion effects seen with interlaced systems. Thus progressive scan cameras are said to have improved resolution over interlaced systems. Full-frame and frame-transfer devices inherently provide progressive scan readout. Although early interline-transfer architecture was designed for interlace scanning, some current devices also provide progressive scan output.

26.2.4. Active Pixel Sensor/Complementary-Metal-Oxide-Semiconductor Cameras

Complementary-metal-oxide-semiconductor cameras are either passive or active. In passive devices, the signal is read out without any further processing. With active devices, each cell may have buffers and amplifiers to decouple the signal from the photosite. These extra electronics reduce the fill factor. As shown in Fig. 26.4, when a row is activated, the data from the pixels in that row are simultaneously copied into the columns. Each column will have a load transistor, column-select switch, and a sampling switch. The chip may also have an analog-to-digital converter and provide correlated double sampling. The pixels are then reset and a new integration is started. The active pixel sensor does not rely on charge transfer; rather, the photodiode drives a capacitance line. The total capacitance limits the chip speed. Active pixel sensors can be fully addressable and subarrays can be read out at high frame rates.

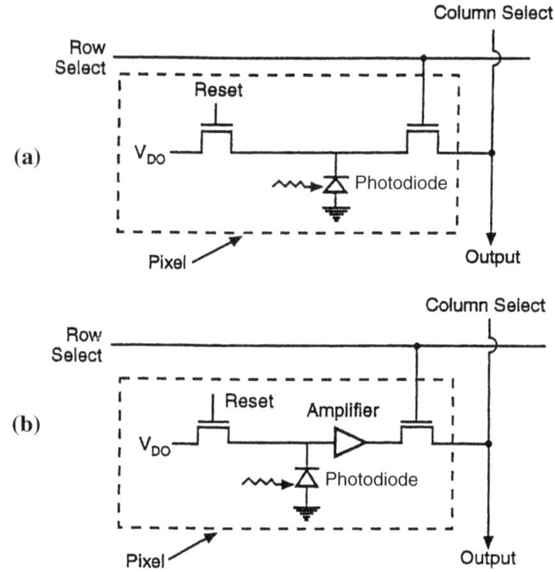

Fig. 26.4. (a) Passive pixel device; (b) active pixel sensor.

Complementary-metal-oxide-semiconductor devices tend to have higher dark currents because of the highly doped silicon used. As such, it does not appear that complementary-metal-oxide-semiconductors sensors will replace CCDs in low-noise scientific applications. Because the active devices often take up room, the area for the photosensor is reduced. This leads to reduced sensitivity that can be partially offset by a microlens. Because each pixel has its own amplifier, pattern noise is larger. However, more logic can be added to each pixel and it seems possible to use on-chip signal processing that suppresses fixed pattern noise.[13] With charge limited to a single pixel, it cannot overflow into neighboring pixels and create blooming, as seen with CCDs.

A further advantage is their low power consumption. Active pixel sensors operate from a 5-V supply or less (compared with 10–15 V for a CCD). This is an attractive feature for portable, battery-operated devices. It appears at this time that complementary-metal-oxide semiconductors will compete with CCDs when weight, size, and power consumption are factors.

26.3. Video Chip Size

Vidicon vacuum tubes were originally used for professional television applications. These were specified by the tube diameter. To minimize distortion and nonuniformities within the tube, the recommended[14] image size was consider-

[13] J. Nakamura, T. Nomoto, T. Nakamura, and E. R. Fossum, "On-focal-plane signal processing for current-mode active pixel sensors," IEEE Trans. Electron Devices **44,** 1747–1757, 1997.

[14] R. G. Neuhauser, "Photosensitive camera tubes and devices," in *Television Engineering Handbook,* K. B. Benson, ed., McGraw-Hill, 1986, p. 11.34.

ably less than the overall tube diameter. The CCD industry maintained the image diagonal but continued to use the tube format nomenclature (Table 26.1).

Table 26.1. Array Size for Standard Formats

Camera Format (in.)	Array Size (H × V) (mm)	Array Diagonal (mm)	Nominal Pixel Size (H × V) (μm)
1	12.8 × 9.6	16	20 × 20
2/3	8.8 × 6.6	11	13.75 × 13.75
1/2	6.4 × 4.8	8	10 × 10
1/3	4.8 × 3.6	6	7.5 × 7.5
1/4	3.2 × 2.4	4	5 × 5

26.4. Color Filter Arrays

Color cameras must produce three separate colors (R, G, and B). Single-chip cameras require a color filter array. An array that has 768 horizontal elements may devote 384 to green, 192 to red, and 192 to blue. This results in an unequal sampling of the colors. A birefringent crystal inserted between the lens and the array accommodates the different sampling rates. The video signal from the color filter array is embedded in the architecture of the color filter array. The color video from a single-chip color filter array must be decoded (unscrambled) to produce usable R, G, and B signals.

Higher-resolution cameras contain three separate chips—one devoted to each color. Beam splitters separate the colors. These cameras are often advertised as 3CCD cameras. Note that, if each signal is digitized into 8 bits, it is said that the camera produces $(2^8)^3$ or 16×10^6 "colors." Although these combinations exist digitally, it does not mean that the eye can discern all these colors.

26.5. Noise

Many books and articles (footnotes 15–18; see also footnotes 4–8 and 11) have been written on noise sources. The level of detail used in noise modeling depends on the application. Shot noise is due to the discrete nature of electrons.

[15] D. G. Crowe, P. R. Norton, T. Limperis, and J. Mudar, "Detectors," in *Electro-Optical Components,* W. D. Rogatto, ed., pp. 175–283. This is Vol. 3 of *The Infrared and Electro-Optical Systems Handbook,* J. S. Accetta and D. L. Shumaker, eds., copublished by Environmental Research Institute of Michigan and the Society of Photo-Optical Instrumentation Engineers, 1993.

[16] J. R. Janesick, T. Elliott, S. Collins, M. M. Blouke, and J. Freeman, "Scientific charge-coupled devices," Opt. Eng. **26,** 692–714, 1987.

[17] T. W. McCurnin, L. C. Schooley, and G. R. Sims, "Charge-coupled device signal processing models and comparisons," J. Electron. Imaging **2,** 100–107, 1994.

[18] M. D. Nelson, J. F. Johnson, and T. S. Lomheim, "General noise process in hybrid infrared focal plane arrays," Opt. Eng. **30,** 1682–1700, 1991.

It occurs when photoelectrons are created and when dark-current electrons are present. Additional noise is added when the charge (reset noise) is read and is introduced by the amplifier (1/*f* noise and white noise). The amplifier noise (provided by the array manufacturer) may be called readout noise, multiplexer noise, noise-equivalent electrons, or the noise floor. If the output is digitized, the inclusion of quantization noise may be necessary. Switching transients that are coupled through the clock signals also appear as noise. It can be minimized by good circuit design.

Pattern noise refers to any spatial pattern that does not change significantly from frame to frame. Dark current varies from pixel to pixel, and this variation is called fixed pattern noise. Fixed pattern noise is due to differences in detector size, doping density, and foreign matter that is trapped during fabrication. Photoresponse nonuniformity is the variation in pixel responsivity and is seen when the device is illuminated. This noise is due to differences in detector size, spectral response, and thickness in coatings. These "noises" are not noise in the usual sense. This variation appears as spatial noise to the observer. Although fixed pattern noise and photoresponse nonuniformity are different, they are sometimes collectively called scene noise, pixel noise, pixel nonuniformity, or simply pattern noise.

The total system noise can often be represented by three noise sources:

$$\langle n_{\mathrm{SYS}} \rangle = \sqrt{\langle n_{\mathrm{SHOT}}^2 \rangle + \langle n_{\mathrm{FLOOR}}^2 \rangle + \langle n_{\mathrm{PATTERN}}^2 \rangle}, \tag{26.1}$$

where $\langle n_i^2 \rangle$ is the noise variance for source i and $\sqrt{\langle n_i^2 \rangle} = \langle n_i \rangle$ is the standard deviation measured in rms units.

Both photoelectrons and dark current contribute to shot noise. In Poisson statistics, the variance is equal to the mean:

$$\langle n_{\mathrm{SHOT}}^2 \rangle = n_{\mathrm{PE}} + n_{\mathrm{DARK}}. \tag{26.2}$$

Although the dark-current average value can be subtracted from the output to provide only the signal that is due to photoelectrons, the dark-current shot noise cannot. Cooling the array can reduce the dark current to a negligible value and thereby reduce the dark-current shot noise. The dark current typically decreases by a factor of ~2 for every 8–9 °C drop in temperature.

As a multiplicative noise, photoresponse nonuniformity is traditionally expressed as a fraction of the total number of charge carriers. This approach assumes that the detectors are operating in a linear region with the only difference being responsivity differences. If U is the fixed pattern ratio or nonuniformity, then

$$\langle n_{\mathrm{PATTERN}} \rangle \approx \langle n_{\mathrm{PRNU}} \rangle = U n_{\mathrm{PE}}. \tag{26.3}$$

The manufacturer usually supplies photoresponse nonuniformity. Off-chip-gain/level-correction algorithms can minimize fixed pattern noise and photoresponse nonuniformity.

26.6. Signal

For a circular aperture ($A_0 = \pi D^2/4$) and F-number ($F = f/D$), the number of photoelectrons is

$$n_{\mathrm{PE}} = \int_{\lambda_1}^{\lambda_2} \mathcal{R}_q(\lambda) \frac{\pi}{4} \frac{L_q(\lambda) t_{\mathrm{INT}} A_D}{F^2} T_{\mathrm{OPTICS}}(\lambda) d\lambda, \tag{26.4}$$

where $\mathcal{R}_q(\lambda)$ has units of electrons per photon and is simply the detector's quantum efficiency, t_{INT} is the integration time, $L_q(\lambda)$ is the photon spectral-radiance sterance, and $T_{\mathrm{OPTICS}}(\lambda)$ is the optical transmittance. The variable A_D is the detector's effective sensitive area. This value depends on the microlens used. Without a microlens, it is simply the detector area. Scientific devices tend to have large areas to increase the sensitivity. In addition, the charge-well capacity is roughly proportional to cell size, and therefore these larger devices will have a larger dynamic range.

The camera output is

$$V_{\mathrm{CAMERA}} = n_{\mathrm{PE}} \frac{Gq}{C} G_1 G_2. \tag{26.5}$$

The gain G of the source-follower amplifier is approximately 1 and q is the electronic charge (1.6×10^{-19} C). The output gain conversion is Gq/C and typically ranges from 0.1 $\mu V/e^-$ to 10 $\mu V/e^-$. This signal is then amplified, G_1G_2, by electronics external to the CCD sensor.

The photon level that saturates the charge wells is the camera's maximum input signal, and $V_{\mathrm{WELL}} = G_1 G q n_{\mathrm{CHARGE\ WELL}}/C$, where $G_2 = 1$. However, there is no requirement that the camera manufacturer set the maximum output voltage equal to V_{WELL}. The gain G_2 is used to reduce the minimum detectable signal.

The signal-to-noise ratio is

$$\mathrm{SNR} = \frac{n_{\mathrm{PE}}}{\langle n_{\mathrm{SYS}} \rangle} = \frac{n_{\mathrm{PE}}}{\sqrt{\langle n^2_{\mathrm{SHOT}} \rangle + \langle n^2_{\mathrm{FLOOR}} \rangle + \langle n^2_{\mathrm{PATTERN}} \rangle}}. \tag{26.6}$$

The highest achievable signal-to-noise ratio occurs when the system is shot noise limited. Then $\mathrm{SNR} = \sqrt{n_{\mathrm{PE}}}$.

26.7. Dynamic Range

Dynamic range is

$$\mathrm{DR} = \frac{n_{\mathrm{CHARGE\ WELL}} - n_{\mathrm{DARK}}}{\sqrt{\langle n_{\mathrm{FLOOR}}^2 \rangle + \langle n_{\mathrm{DARK}}^2 \rangle}}. \tag{26.7}$$

With no dark current, Eq. (26.7) reduces to the more commonly used definition

$$\mathrm{DR} = \frac{n_{\mathrm{CHARGE\ WELL}}}{\langle n_{\mathrm{FLOOR}} \rangle}. \tag{26.8}$$

Dynamic range is often expressed in decibels. The dynamic range as defined is not directly related to the size of the analog-to-digital converter. Do not confuse dynamic range with the signal-to-noise ratio. The signal-to-noise ratio includes shot noise and can never be as high as the dynamic range.

26.8. Resolution

Many manufacturers specify resolution by the number of detector elements. This is not related to spatial resolution. Spatial resolution, a measure of the smallest detail that can be resolved, is given by the detector angular subtense

$$\alpha = d/f, \tag{26.9}$$

where d is the linear extent of the pixel. For higher resolution, use either a smaller detector or a longer focal-length optical system. For square detectors, the number of photoelectrons is proportional to $(\alpha)^2$. Hence there is a trade-off between resolution (smaller α wanted) and sensitivity (larger n_{PE} wanted).

26.9. Analog-to-Digital Converters

An analog-to-digital converter is an essential component of all digital cameras. For consumer applications, an 8-bit converter is used. Sometimes seemingly slight deviations from perfection in an analog-to-digital converter can modify image fidelity[19] when examined by a trained photointerpreter or by an automatic target-detection algorithm.

Analog-to-digital converter linearity is specified by differential nonlinearity and integral nonlinearity (Fig. 26.5). Differential nonlinearity is a test of missing codes or misplaced codes. Two adjacent codes should be exactly 1 least-significant bit apart. Any positive or negative deviation from this is the differ-

[19] C. Sabolis, "Seeing is believing," *Photonics Spectra*, 119–124, October 1993.

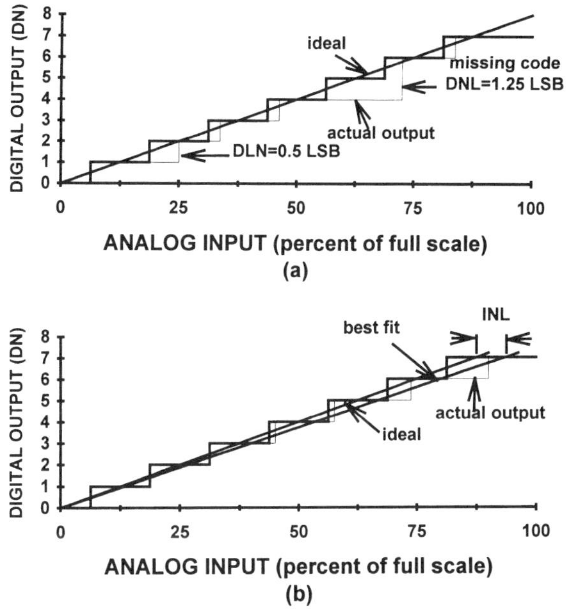

Fig. 26.5. Analog-to-digital converter nonlinearity for 3-bit analog-to-digital converter: (a) differential nonlinearity, (b) integral nonlinearity. Missing codes can be caused by the accumulation of differential nonlinearities. Accumulation can also cause integral nonlinearity.

ential nonlinearity. If the differential nonlinearity exceeds 1 least-significant bit, there is a missing code. Many nonlinearity specifications simply state "no missing codes." This means that the differential nonlinearity is less than 1 least-significant bit. However, 1.01 least-significant-bit differential nonlinearity (fail specification) and 0.99 least-significant-bit differential nonlinearity (pass specification) probably look the same visually. Poor differential nonlinearity can affect the cosmetic quality of images.

Integral nonlinearity is the deviation of the transfer function from an ideal straight line. The integral nonlinearity is usually specified in terms of the deviation from a straight line between the end points. Good integral nonlinearity is important for radiometric applications. An integral nonlinearity of less than 0.5 least-significant bit is considered excellent.

It would initially appear that the least-significant bit should be equal to the noise floor:

$$2^N = \frac{n_{\text{FULL WELL}}}{n_{\text{FLOOR}}}, \tag{26.10}$$

where N is the number of bits. However, the eye can integrate for ~0.2 s; therefore, if the frame rate is 30 Hz, the eye integrates over 6 frames. If the noise is random (temporally), the *perceived* signal-to-noise ratio improves by $\sqrt{6} \approx 2.5$. This suggests that 2 bits can be devoted approximately to noise and still provide a perceivable image. Under carefully controlled situations, the eye can detect (well-recognized) objects when the signal-to-noise ratio is below 0.1. That means that 3 bits can be devoted to noise. Note that, even though the target can be detected in these situations, the image will appear noisy—a condition typically avoided for cosmetic reasons.

26.10. Video Formats

There are three color-broadcast standards: the National TV System Committee, PAL, and SECAM. The standards specify the line frequency and blanking period. The blanking period has a tolerance of approximately ±0.25% so that the reported, active, time line varies slightly. EIA-170 is the U.S. standard for monochrome closed-circuit television. EIA-170 and the National TV System Committee standards have 525 lines/frame whereas PAL and SECAM standards have 625 lines/frame. The number of active (displayed) lines is approximately 480 and 575, respectively. These standards are incompatible for a variety of reasons.

A 4:3 aspect ratio means that there are 4/3 times more detectors in the horizontal direction than in the vertical direction. If the number of elements in the vertical direction is equal to the number of active scan lines, then the number of horizontal detectors should be ~640 for the National TV System Committee standard and 766 for PAL and SECAM standards. The number of pixels per line is often chosen so that the horizontal clock rate is an integer multiple of the color subcarrier. Therefore arrays that are compatible with the National TV System Committee standard often have 768 pixels in the horizontal direction.

For greater vertical resolution, more lines are required. EIA-343A was created for monochrome high-resolution closed-circuit television. Although the standard encompasses equipment that operates from 675 to 1023 lines, the recommended values are 675, 729, 875, 945, and 1023 lines per frame. The U.S. military primarily uses 875 lines/frame but it requires "standards conversion" boxes that can record and present imagery on EIA-170 compatible equipment.

With the advent of electronic still cameras, video conferencing, and high-speed industrial imaging, new digital standards are emerging that offer direct coupling to a computer. An imaging system is only as fast as its slowest link: camera, frame grabber, or computer bus. Many video-conferencing cameras have built-in image compression.

EIA-644 is similar to EIA-422 but operates at lower voltage differences and thereby provides higher transmission bandwidths. It is also known as low-voltage differential signaling. IEEE-1394 is marketed by Apple as FireWire and by

Sony as i.LINK. The maximum date rates drop as the cable length increases. Table 26.2 provides current digital standards.

Table 26.2. Digital Standards

Standard	Adapter	Max Date Rate	Cable Length (m)
EIA-422	Frame grabber	30 Mbit/s	5
EIA-644	Frame grabber	400 Mbit/s	20
Camera link	Frame grabber	1.923 Gbit/s	10
IEEE-1394	Standard adapter	400 Mbit/s	4.5
IEEE-1394B	Standard adapter	3.2 Gbit/s	100
USB-2	Standard adapter	480 Mbit/s	5

The common computer busses are industry standard architecture (ISA), peripheral component interconnect (PCI), and Versa Module Europa (VME). The industry standard architecture supports a 16-bit 8-MHz bus. Nearly all IBM-compatible personal computers before the Pentium were industry standard architecture based. Industry standard architecture cannot support real-time video and therefore frame grabbers for those machines require onboard memory. The peripheral component interconnect handles real-time video burst rates of 32-bit signals at 132 MHz (Table 26.3).

Table 26.3. Computer Busses

Bus	Industry Standard Architecture	Peripheral Component Interconnect	Versa Module Europa (2x)	Versa Module Europa (2x) 320
Speed	128 Mbits/s	4 Gbits/s	1.2 Gbits/s	2.5 Mbits/s

Composite video encodes luminance, color, and timing information onto a single channel. It can be transmitted on a single wire, such as cable television. National TV System Committee, PAL, and SECAM signals are composite video signals. The signal must be decoded to create an image.

Component analog video carries each part on a separate wire. RGB, or their complementary values, are carried on three separate cables. It is common for the green to be a composite line and to carry the sync signals. Component videos do not require encoding, and therefore there is less chance that there will be color or luminance cross coupling. This implies that component imagery will produce a higher-quality image.

Most standards deal with color encoding (chrominance encoding). The luminance channel is identical with the green channel and is labeled Y. The other signals are color, and the exact composition depends on the standard. The chrominance values are related through matrices, but the signals are not compatible because of different line frequencies and sync signals. Y/C consists of a Y (signal) channel that is identical to the monochrome EIA-170 and a chrominance channel that is identical to the chrominance subcarrier defined in the National TV System Committee standard. When two signals are used, higher-quality images are available compared with those given in the National TV System Committee standard. Y/C is also known as S-video, supervideo, and S-VHS.

Matrixing can convert the RGB into any color coordinate system. With the ease of digital implementation, multiple outputs are available from many cameras. That is, a camera may offer RGB, National TV System Committee, and Y/C outputs to provide flexibility to the user.

Further Reading

K. B. Benson, ed., *Television Engineering Handbook,* McGraw-Hill, 1986.

E. L. Dereniak and D. G. Crowe, *Optical Radiation Detectors,* Wiley, 1984.

G. C. Holst, *CCD Arrays, Cameras, and Displays,* 2nd ed., JCD Publishing, 1998.

J. R. Janesick, *Scientific Charge-Coupled Devices,* Vol. PM83 of SPIE Press Monograph Series, Society of Photo-Optical Instrumentation Engineers, 2001.

A. J. P. Theuwissen, *Solid-State Imaging with Charge-Coupled Devices,* Kluwer Academic, 1995.

E. S. Yang, *Microelectronic Devices,* McGraw-Hill, 1988.

27

Infrared Imagers

J. M. Mooney

27.1. Introduction

Infrared imagers cover a broad range of scanned or staring and cooled or uncooled sensor architectures. Here the focus is on staring photon detectors that image radiation in bands from 3 to 14 μm.

27.2. Sampled Imagery

The modern treatment of imaging has been heavily influenced by the introduction of focal-plane arrays and by the application of computers for processing, displaying, and archiving imagery. The current practice includes electron counting and sampling. Therefore a digital image is a two-dimensional mosaic of integers, the imager is a physical device that maps object space into an image, and the display converts the mosaic of integers to a form easily interpreted by eye.

In this model each integer in the mosaic is directly related to the number of photoelectrons collected at a spatial location on a focal-plane array:

$$N_{i,j} = G_c(Q_{ij}), \tag{27.1}$$

where $N_{i,j}$ is the integer signal captured by the analog-to-digital conversion, $G_c(\ )$ is the conversion function of the sensor, and $Q_{i,j}$ is the total number of carriers (e.g., electrons) collected by the (i, j) pixel. Ideally $G_c(\ )$ is a linear function such that

$$N_{i,j} = g_c Q_{ij} + N_0, \tag{27.2}$$

where $g_c = \{[\mathrm{d}G_c(Q)]/(\mathrm{d}Q)\}$ and is referred to as the conversion gain, N_0 is the conversion offset, and

$$Q_{ij} = S_{ij} + D_{ij}, \tag{27.3}$$

where S_{ij} is the signal gathered on the (i, j) pixel (e.g., electrons) and D_{ij} is the dark current collected by the pixel.

The electrons are integrated on a node of limited capacity, Q_{full}, on the focal-plane array; therefore

$$N_{\max} = G_c(Q_{\mathrm{full}}) \leq 2^{N_{\mathrm{bits}}}. \tag{27.4}$$

27.3. Sensor Signal

Each pixel of the imager is assumed to collect photons from a region of object space as mapped by the sensor. For imaging applications the target is treated as an extended source, and performance is characterized by the minimum resolvable temperature as a function of spatial frequency. For the detection of small sources at long range, the target is treated as a point source and performance is characterized by the minimum detectable temperature. These are also called the minimum resolvable temperature difference and the minimum detectable temperature difference.

27.3.1. Imaging Extended Scenes

Collected photons originate from emission by an object, reflection off of the object, and scattering into the optical path by diffraction or some other process. Usually the target is assumed to be a few degrees different from the rest of the scene, and the signal is characterized as a function of temperature. However, the target may differ from the scene in emissivity, the spectral character of its emissivity, or its polarization.

Because of the infinite possibilities of multiple reflections and various scattering processes, the exact expression of the signal becomes quite complicated. Only one reflection is treated here. The reader can incorporate additional reflections as necessary.

It is also assumed that the spectral and the spatial terms are separable. A complete expression involves a convolution of the wavelength-dependent point-spread function of the sensor with the wavelength-dependent description of the image. This can be done if necessary. It is assumed that each pixel in object space can be represented by a single temperature and a single wavelength-dependent emissivity.

The signal collected by a pixel is approximated by

$$\begin{aligned} S_{ij} = \int_{\lambda_1}^{\lambda_2} & \mathcal{R}_{ij}(\lambda)\tau_s(\lambda)\{\epsilon_{ij}(\lambda)L_q(\lambda, T_{ij}) \\ & + [1 - \epsilon_{ij}(\lambda)]\tau_B(\lambda)L_q(\lambda, T_B)\}\mathrm{d}\lambda, \end{aligned} \tag{27.5}$$

where $L_q(\lambda, T)$ is the photon radiance as a function of wavelength and temperature, T_{ij} is the scene temperature in the cell of object space corresponding to the (i, j) pixel in degrees Kelvin, T_B is the temperature of the background in degrees Kelvin, $\tau_s(\lambda)$ is the atmospheric transmission from the scene to the sensor, $\tau_B(\lambda)$ is the atmospheric transmission from the background to the scene where it is reflected, $\epsilon_{ij}(\lambda)$ is the emissivity of the (i, j) cell in object space, and $\mathcal{R}_{ij}(\lambda)$ is the responsivity as a function of wavelength for the (i, j) cell in image space (i.e., the i, j detector). The first term in the braces corresponds to the emitted signal from the scene, and the second term approximates the signal reflected off of the scene. (Our assumption that the background is at a uniform temperature is a coarse approximation.)

The effect of the second term is to reduce the contrast that is due to emissivity variations. When the thermal contrast between the scene and the background is small, the radiation reflected from a low-emissivity target is indistinguishable from that emitted from a high-emissivity target.

The responsivity is given by

$$\mathcal{R}_{ij}(\lambda) = \frac{\pi a_{ij} \tau_{\text{optics}}(\lambda) t_{\text{int}} \eta_{ij}(\lambda) \cos^4(\phi_{i,j})}{4F^2 + 1}, \tag{27.6}$$

where a_{ij} is the optically active area of the (i, j) pixel, $\tau_{\text{optics}}(\lambda)$ is the optical transmission, t_{int} is the integration time, $\eta_{ij}(\lambda)$ is the wavelength-dependent quantum efficiency, $\phi_{i,j}$ is the field angle of the pixel at index (i, j), and

$$F = \frac{f_{\text{effective}}}{d_{\text{optics}}}, \tag{27.7}$$

where d_{optics} is the diameter of the circular optical aperture,

$$\frac{1}{f_{\text{effective}}} = \frac{1}{f_{\text{optics}}} - \frac{1}{r_{\text{scene}}}, \tag{27.8}$$

f_{optics} is the focal length of the optics, and r_{scene} is the range to the scene. When r_{scene} is more than $10 f_{\text{optics}}$, the approximation $f_{\text{effective}} \approx f_{\text{optics}}$ is valid.

If the optical aperture is not circular, or has an obscuration, the expression for d_{optics} can be modified to reflect the effect on the sensor's collection efficiency:

$$d_{\text{optics}} = 2\sqrt{A_{\text{optics}}/\pi}, \tag{27.9}$$

where A_{optics} is the active area of the optical aperture. Equation (27.9) is applicable only to photon collection and should not be used in calculating the spatial resolution of the imager.

For a Lambertian, blackbody radiator (radiance is independent of angle),

$$L(\lambda, T) = M(\lambda, T)/\pi, \tag{27.10}$$

where the photon flux density is

$$M(\lambda, T) = \frac{2\pi c}{\lambda^4[\exp(hc/\lambda kT) - 1]}, \tag{27.11}$$

c is the speed of light, h is Plank's constant, and k is Boltzmann's constant. The field-dependent responsivity variations [e.g., $\cos^4(\phi)$ shading] are more apparent in infrared imagers than in visible imagers. In infrared images the scene contrast is a relatively small variation on top of a large uniform pedestal (the entire scene is within a few degrees of the nominally 300 K background). The product of the field-dependent responsivity variation with the background can be larger than the scene variations and leads to the need for responsivity nonuniformity correction.

27.3.2. Signal from a Point Source

When the sensor is used to detect a source that geometrically maps into a single pixel,

$$S_{ij} = \int_{\lambda_1}^{\lambda_2} \mathcal{R}_{ij}(\lambda)\tau_s(\lambda)I_q(\lambda)\mathrm{d}\lambda, \tag{27.12}$$

where $I_q(\lambda)$ is the radiant photon intensity from the point source, and the responsivity is

$$\mathcal{R}_{ij}(\lambda) = \frac{A_{\text{optics}}\tau_{\text{optics}}(\lambda)t_{\text{int}}\eta_{ij}(\lambda)\cos^4(\phi_{i,j})}{r^2}, \tag{27.13}$$

where r is the distance to the point source. Because the source is a point, it does not obscure the scene, and the signal in Eq. (27.12) is the additional signal that is due to the presence of the source.

Treatment of the transition between extended and point-source behavior is given in the discussion of minimum detectable temperature in Subsection 27.4.3.

27.3.3. Spatial Dependence

The spatial dependence of imager performance is given by the convolution of the imager point-spread function with the geometrical image or by taking the product of all the constituent optical transfer functions, as described in Chaps. 1 and 21.

The spatial resolving capability of an imager is often described by the magnitude of the optical transfer function, the modulation transfer function, which characterizes the modulation as a function of spatial frequency. The ability of the modulation transfer function to predict the performance of an imager is limited by the assumption that the imager is linear and shift invariant. Although the modulation transfer function is useful, sampled systems are inherently shift variant, so its utility is somewhat limited.

27.3.3.1. Modulation Transfer Function

For a circular aperture the modulation transfer function is given by[1]

$$\mathrm{MTF}_{(\mathrm{diffraction})}(\rho) = \frac{2}{\pi}\{\arccos(\lambda F\rho) - \lambda F\rho[1-(\lambda F\rho)^2]^{1/2}\}\mathrm{cyl}(\lambda F\rho/2), \qquad (27.14)$$

where $\rho = \sqrt{\xi^2+\zeta^2}$, ξ and ζ are the spatial frequencies, and $\mathrm{cyl}(x)$ is 1 for $x < 0$ and 0 otherwise. For optical systems that do not have an unobstructed circular aperture, a more complicated expression results.

For a rectangular pixel the active area of the pixel determines the pixel modulation transfer function, which is given by

$$\mathrm{MTF}_{(\mathrm{pixel})}(\xi,\zeta) = \left|\frac{\sin(\pi F_x p_x \xi)}{\pi F_x p_x \xi}\,\frac{\sin(\pi F_y p_y \zeta)}{\pi F_y p_y \zeta}\right|, \qquad (27.15)$$

where p_x and p_y are the pixel dimensions, F_x and F_y are the fraction of the pixel that is optically active, measured linearly in the x and y direction. The product $F_xF_yp_xp_y = \langle a_{ij}\rangle$ is the average pixel-active area. Note that focal-plane arrays with a small pixel active area have better modulation than those with a large active area; however, this is usually not desirable because the small active area has lower sensitivity and increased aliasing.

If photoelectrons diffuse into the depletion region of a neighboring pixel the modulation transfer function will be adversely affected. The resulting modulation transfer function is[2]

$$\mathrm{MTF}_{\mathrm{diffusion}}(\rho) = \frac{\cosh(d_{\mathrm{absorption}}/L_n)}{\cosh[d_{\mathrm{absorption}}(1/L_n^2 + 2\pi\rho^2)^{1/2}]}, \qquad (27.16)$$

[1] J. M. Lloyd, "Fundamentals of electro-optical imaging systems analysis," in M. C. Dudzik, ed., *Electro-Optical System Design, Analysis, and Testing,* Chap. 1, and in J. S. Accetta and D. L. Shumaker, eds., *The Infrared and Electro-Optical Systems Handbook,* Chap. 4, Society of Photo-Optical Instrumentation Engineers, 1993.

[2] D. F. Barbe, "Imaging devices using the charge-coupled concept," Proc. IEEE **63,** 38, 1975.

where the photon is absorbed a distance $d_{\text{absorption}}$ from the depletion region and L_n is the diffusion length of electrons in the semiconductor (L_p should be used when the signal depends on hole diffusion).

When the charge in a CCD is trapped in a well and picked up by a subsequent charge packet, the resulting modulation transfer function is given by

$$\text{MTF}_{\text{CTE}}(\xi) = \exp\{-n_{\text{tr}}\epsilon_{\text{tr}}[1 - \cos(\pi p_x \xi)]\}, \tag{27.17}$$

where n_{tr} is the number of transfers and ϵ_{tr} is the charge-transfer efficiency. Here we have assumed that the transfer direction corresponds to the ξ direction. There is no degradation in the orthogonal direction.

Other factors that affect the system modulation transfer function are optical cross talk, in which photons are scattered or diffracted in the focal-plane array and are collected by the wrong pixel, and blooming, in which the signal from a saturated pixel bleeds into neighboring pixels, which tends to affect pixels confined to a row or column. Although easily named, their effect on performance is difficult to quantify.

27.3.3.2. Field Of View and Instantaneous Field of View

The sensor field of view of an imager is approximated by

$$\text{FOV}_x = 2\arctan\left(\frac{N_x p_x}{2f_{\text{effective}}}\right), \tag{27.18}$$

where N_x is the number of pixels in a row. For nonsquare geometries the field of view will be orientation dependent.

The sensor instantaneous field of view is defined as the angle corresponding to the full width at half-maximum of the sensor impulse response (Rayleigh's criterion), which is given by

$$\text{IFOV} = \arctan(2r_{\text{blur}}/f_{\text{effective}}), \tag{27.19}$$

where r_{blur} is the blur-circle radius defined as half-width at half-maximum of the sensor impulse response. All of the factors discussed with regard to the modulation transfer function will affect the instantaneous field of view. The radius of the total system blur circle and the resulting system instantaneous field of view can be estimated by the quadrature sum of the constituents.

In many imaging systems (typically terrestrial), a common practice is to design the sensor so that the diameter of the first minimum of the diffraction-limited blur circle matches the pixel dimension,

$$d_{\text{pixel}} \approx d_{(\text{1st zero})} = 2.44\lambda F, \tag{27.20}$$

where the diffraction blur diameter and wavelength are in the same units. Spatial resolution for these systems is limited by the pixel size, and the instantaneous field of view is set by the imager effective focal length and the pixel size. In this case the instantaneous field of view is determined geometrically,

$$\mathrm{IFOV}_{\mathrm{geometrical}} = \arctan(d_{\mathrm{pixel}}/f_{\mathrm{optics}}) \simeq d_{\mathrm{pixel}}/f_{\mathrm{optics}}, \tag{27.21}$$

with the results of the approximation in radians. For nonsquare pixels the geometrically determined instantaneous field of view will depend on orientation.

For strategic and astronomical systems, the optics becomes the cost and performance driver. The blur circle can be usefully oversampled by the pixels up to the point that the aliasing in the image is eliminated. Greater sampling than that increases the noise relative to the signal without adding spatial resolution.

For a diffraction-limited optical system, the aliasing is eliminated when

$$d_{\mathrm{pixel}} = \lambda F/2.0. \tag{27.22}$$

In this case the instantaneous field of view is limited by the diffraction of the optics and is given by

$$\mathrm{IFOV}_{\mathrm{diffraction}} = 1.22\lambda/d_{\mathrm{optics}}, \tag{27.23}$$

where d_{optics} is the entrance pupil diameter and the units of λ and d_{optics} are the same.

27.3.4. Dark Current

Even when there is no optical signal, some current will flow in the pixel. The number of carriers collected is

$$D_{ij} = d_{ij} t_{\mathrm{int}} J_{\mathrm{dark}}, \tag{27.24}$$

where J_{dark} is the dark-current density and d_{ij} is the electrically active area. (When microlenses are used, the electrically active area of the pixel can be less than the optically active area.)

Local doping variations can lead to variations in J_{dark}, which can be identified by their temperature dependence. J_{dark} can also vary with distance from the output node (a heat source) or distance to thermal contact because of the corresponding variations in the focal-plane array temperature.

The detector dark current consumes well capacity and is a source of noise. Many infrared focal-plane arrays use diode detectors (those that do not are too complicated to be described here). The dark current as a function of voltage for a diode is obtained from the Shockley equation[3]:

$$J_{\text{dark}} = J_s\,(e^{qV/kT_{\text{diode}}} - 1), \tag{27.25}$$

where q is the charge on an electron, V is the voltage across the diode, k is Boltzmann's constant, T_{diode} is the diode temperature, and J_s is the saturation-current density. Usually the diodes are operated in reverse bias, so $J_{\text{dark}} = J_s$.

27.3.4.1. *The* p–n *Diode*

For a *p–n* diode,

$$J_s = \frac{qD_p p_{no}}{L_p} + \frac{qD_n n_{po}}{L_n}, \tag{27.26}$$

where

$$D_p = \mu_p kT_{\text{diode}}/q, \quad D_n = \mu_n kT_{\text{diode}}/q, \tag{27.27}$$

μ_n and μ_p are the mobilities of electrons and holes, respectively,

$$p_{no} = n_i^2/n_{no} \simeq n_i^2/N_D, \quad n_{po} = n_i^2/p_{po} \simeq n_i^2/N_A, \tag{27.28}$$

where N_D and N_A are the donor and the acceptor concentrations, respectively, on the two sides of the junction $N_D n$,

$$n_i^2 = N_C N_V \exp(E_g/kT_{\text{diode}}), \tag{27.29}$$

N_C and N_V are the effective densities of states in the conduction and the valance bands, respectively, E_g is the bandgap, and

$$L_p \equiv \sqrt{D_p \tau_p}, \quad L_n \equiv \sqrt{D_n \tau_n}, \tag{27.30}$$

where τ_p and τ_n are the minority carrier lifetimes. The temperature dependence of J_s for a *p–n* diode is approximately proportional to

$$J_{\text{dark}} \propto \exp(-E_g/kT_{\text{diode}}). \tag{27.31}$$

27.3.4.2. *Schottky Diode*

For a Schottky diode the saturation-current density is

$$J_s = A^{**}T^2 \exp(-q\psi_B/kT_{\text{diode}}), \tag{27.32}$$

[3] S. M. Sze, *Physics of Semiconductor Devices,* Wiley, 1981.

where A^{**} is the effective Richardson constant and ψ_B is the Schottky barrier height.

27.4. Imager Noise

In an analysis of modern thermal imager performance it is useful to separate the noise into a term that varies rapidly in time (temporal noise) and a term that varies slowly in time (spatial noise). For example, photoelectron noise (often referred to as shot noise) is strictly temporal in nature, focal-plane nonuniformities are spatial, and excess low-frequency noise is a temporal noise that appears as a time-varying spatial pattern on the imagery.

27.4.1. *Temporal Noise*

The carriers integrated by an ideal imager will have a Poisson probability density function. For this ideal imager the variance in the number of collected carriers for a given mean will equal the mean. Therefore the noise that is due to the variation in the number of collected carriers is

$$\sigma_p = \sqrt{\langle Q_{ij} \rangle}. \tag{27.33}$$

kTC noise is the variation in the number of electrons on a capacitor when it is disconnected from a circuit. It is relevant because the integration node of a focal-plane array can be modeled as a capacitor. The resulting noise is

$$\sigma_{kTC} = \sqrt{kT_c C/q}, \tag{27.34}$$

where k is Boltzmann's constant T_c is the temperature of the capacitor, C is its capacitance, and q is the charge on an electron. Anomalously low kTC noise can be measured when the transistor used to switch the capacitor is operated subthreshold (as can be the case when kTC noise is measured directly). In this case the transistor resistance is large and the RC time constant of the circuit is longer than the time between measurements. Under these conditions, there is little variation between measurements (as they are temporally correlated) and anomalously low noise is measured.

The process of digitizing a signal introduces noise. The resulting noise is

$$\sigma_{\mathrm{ADC}} = 1/\sqrt{12} = 0.2887 \tag{27.35}$$

(in units of least-significant bits). In a CCD, the fluctuations in a charge packet after n transfers is

$$\sigma_{\text{CTE}} = \sqrt{2\epsilon_{\text{tr}} n_{\text{tr}} S} \tag{27.36}$$

for $\epsilon_{\text{tr}} n_{\text{tr}} \ll 1$. When the product of $\epsilon_{\text{tr}} n_{\text{tr}}$ approaches 1, the number of carriers in adjacent charge packets becomes correlated, and the effect of charge transfer is to reduce the noise at the expense of severe image blur.

The total temporal noise is obtained when the various constituents are added in quadrature:

$$\sigma^2_{\text{temporal}} = \sigma^2_{kTC} + \sigma^2_{\text{cte}} + \sigma^2_p + \cdots +. \tag{27.37}$$

Many infrared sensors can adjust the integration time to keep the charge storage wells nearly full so that the total noise is dominated by the photoelectron noise (i.e., $\sigma^2_{\text{temporal}} \approx \sigma^2_p$). Because most vendors quote their performance at 80% of full well capacity,

$$\sigma_{\text{temporal}} \approx \sigma_p \approx \sqrt{0.8Q_{\text{full}}}. \tag{27.38}$$

The noise measured in digital units will be

$$\sigma_N = g_c\sqrt{0.8Q_{\text{full}}}. \tag{27.39}$$

27.4.2. Techniques for Reducing the Temporal Noise

Ideally, the noise that is due to the Poisson statistics of the photoelectrons will be much larger than the other noise terms, and at high illumination levels this is often true. However, at low illumination levels the other noise terms can be significant.

The noise that is due to the dark current is minimized by cooling of the focal plane. Correlated double sampling can be used to eliminate the *kTC* noise of the output node for many (but not all) focal-plane array architectures. With correlated double sampling the output node is reset, the output voltage is sampled, the signal is transferred to the node, the output voltage is sampled again, and, finally, the difference is taken between the two voltages.

For terrestrial applications, there is little to be done about read noise once the focal plane has been fabricated; however, for astronomical applications a widely used multiread clocking scheme all but eliminates it.[4] First, the reset and the measurement steps are separated. Immediately after the array is reset, it is read out repeatedly. The multiple reads are averaged and digitally stored. The array is then exposed and again repeatedly read. The difference between

[4] A. M. Fowler and I. Gatley, "Demonstration of an algorithm for read-noise reduction in infrared arrays," Astrophys. J. **353,** L33, 1990.

the multiple read averages is then subtracted on a pixel-by-pixel basis, and, ideally, the resulting read noise is reduced by the square root of the number of reads. This technique is referred to as the Fowler multiple-read technique. It is practical because of the long stare times common in astronomy.

27.4.3. Spatial Noise

If the response of the detectors on the focal plane varies from pixel to pixel, then the resulting imagery suffers from that nonuniformity. It is particularly significant in infrared imagery because of the high background and the low contrast that are characteristic of it. To a large extent, focal-plane array nonuniformities are deterministic and can be corrected; however, excess low-frequency noise, drift in the electronics, pixel-to-pixel variations in response nonlinearities, thermal drift in the optics, or any of a number of other factors lead to a residual spatial noise.[5]

27.4.3.1. Single-Point Nonuniformity Correction

A single-point nonuniformity correction uses a single uniform source as a reference to correct offset nonuniformities for each pixel. Optimally, the reference source is at the same temperature as that of the scene. A sequence of images of the uniform source is added together on a pixel-by-pixel basis to construct one time-averaged reference image. That reference image is then subtracted from subsequent imagery, and a constant is added (to avoid negative results). The reference image is given by

$$U_{ij} = \sum_{k \text{ frames}} N_{ij}(T_R)/k_U, \tag{27.40}$$

where the sum is over k frames and T_R is the temperature of the uniform reference. The number of frames in the average should be large enough to keep the temporal noise from introducing additional spatial noise into the corrected imagery. The minimum residual spatial noise that can be achieved from a k frame average is given by

$$\sigma_{\text{spatial}} \geq \sigma_{\text{temporal}}/\sqrt{k_U}. \tag{27.41}$$

The single-point corrected imagery is then

$$A_{ij} = N_{ij} \quad U_{ij} + U_0, \tag{27.42}$$

[5] J. M. Mooney, F. D. Shepherd, W. S. Ewing, J. E. Murguia, and J. Silverman, "Responsivity nonuniformity limited performance of infrared staring cameras," Opt. Eng. **28,** 1151, 1989.

where U_0 is the desired mean output level. The spatial noise after single-point correction has a minimum at T_R and increases for both lower and higher temperatures (as a function of temperature, it is shaped like a U).

Several tricks have been found for creating a reference image without using an artificial uniform source. We can obtain relatively uniform images by severely defocusing the imager, by pointing it at the sky or the ground, or by rapidly moving the field of view across the scene while acquiring the reference-image sequence. The utility of these tricks varies considerably among focal-plane array technologies, optical configuration, and applications.

27.4.3.2. Two-Point Nonuniformity Correction

A two-point nonuniformity correction corrects both offset and gain for each pixel. Two reference frames are used and are collected as for the single-point correction. Optimally the two temperatures will bracket the temperature range in the scene. The two-point corrected imagery is then

$$B_{ij} = \left(\frac{H_0 - L_0}{H_{ij} - L_{ij}} \right) (N_{ij} - L_{ij}) + L_0, \tag{27.43}$$

where H_{ij} is the high-temperature reference frame, L_{ij} is the low-temperature reference frame, and H_0 and L_0 are usually constants that are relevant to the high and the low frames, respectively (e.g., the average value in each frame or the value of a typical pixel). After two-point correction, the spatial noise will have two minima at the reference temperatures (as a function of temperature, it is shaped like a W). Again the minimum spatial noise is set by relation (27.41).

27.4.3.3. Stability of Nonuniformity Correction

If the focal-plane nonuniformities are stable and uniquely determined by the measured output, then they can, in principle, be fully corrected. However, detector excess low-frequency noise, spectral nonuniformities, cooler cycling, thermal and physical changes in the optics, and drift in clocks and biases combine to make complete correction of focal-plane array nonuniformities impossible. The net effect is a limitation on the length of time during which reference images remain valid.

Exotic nonuniformity correction techniques have been developed that utilize neural networks, temporal filters, and dithered images. The objective of each is to overcome the effects of excess low-frequency noise and other system drift on spatial noise. Each of these approaches attempts to update the nonuniformity correction in real time as the imager is being used.

27.4.3.4. Characterizing Spatial Noise

The effect of spatial noise on imager performance is dependent on the spatial-noise pattern, on the orientation of the pattern, and on the scene to be imaged.

For example, a pattern of vertical bars will degrade the sensor's performance at detecting vertical edges, but will have little effect on its performance on horizontal edges. A complete model for the impact of spatial noise on imager performance has yet to be developed.

27.4.4. Combining the Effects of Spatial and Temporal Noise

To an observer, the spatial-noise pattern is spatially fixed on the imagery that varies slowly in time. Because it is not correlated with the temporal noise, the two add in quadrature for a single frame. However, when multiple frames are averaged, the spatial noise will be correlated from frame to frame, whereas the temporal noise will not.[6] Therefore the variance after averaging is

$$\sigma_{\text{total}}^2 = \left(\frac{t_{\text{avg}}}{t_{\text{frame}}}\right)\sigma_{\text{temporal}}^2 + \left(\frac{t_{\text{avg}}}{t_{\text{frame}}}\right)^2 \sigma_{\text{spatial}}^2. \tag{27.44}$$

Although the variance increases with integration, the signal increases faster. The improvement in sensor performance (e.g., minimum resolvable temperature or minimum detectable temperature) after integration is

$$G_{\text{ST}} = \sqrt{\frac{(t_{\text{frame}}/t_{\text{avg}})\sigma_{\text{temporal}}^2 + \sigma_{\text{spatial}}^2}{\sigma_{\text{temporal}}^2}}, \tag{27.45}$$

where small values of G_{ST} indicate superior performance. In the limit of no spatial noise, $G_{\text{ST}} = (t_{\text{frame}}/t_{\text{avg}})^{1/2}$, and, when the spatial noise dominates, $G_{\text{ST}} = (\sigma_{\text{spatial}}/\sigma_{\text{temporal}})^{1/2}$, and increased frame averaging has no effect on performance.

Equation (27.44) can be applied to image-processing algorithms that use multiple frames of data to improve performance, but it is usually used when the imagery is observed by eye, where $t_{\text{avg}} = t_{\text{eye}}$ and t_{eye} is the integration time of the eye–brain system. Typically $t_{\text{eye}} = 0.1$ s; however, in bright conditions it can be much less.

27.4.5. Miscellaneous Noise Terms

The standard noise terms are usually well behaved, and the user can usually rely on the vendor to meet the quoted specifications. Unfortunately, additional undocumented features often create problems. Usually they are unexpected and their effect on performance is difficult to quantify.

[6] J. M. Mooney, "Effect of spatial noise on the minimum resolvable temperature of a staring sensor," Appl. Opt. **30,** 3324, 1991.

Stray reflections. In the focal plane stray reflections result in a cross pattern around bright point sources. In the optics stray reflections result in a (usually inverted and defocused) ghost image. The ghost image is usually more apparent in low-speed optical systems because of the increase in depth of focus with decrease in optical speed or focal ratio F.

Image retention. Usually caused by traps in an oxide on the focal-plane array, image retention results in a slowly decaying ghost image that does not track with sensor motion. Often the ghost image is contrast inverted. The decay time of the ghost is typically of the order of minutes.

Clock feedthrough. Usually caused by poor shielding or poor grounding in the camera electronics, clock feedthrough is apparent in the image as diagonal lines or a herringbone pattern that moves up or down the image.

Channel drift. Common on focal-array planes with more than one output, channel drift results in a distinct pattern on the imagery that is due to drift in one channel relative to the others. The pattern reflects the region of the focal-plane array read out by each channel (e.g., fields, rows, or quadrants). This effect can also result when the pixel layout is mirrored in the mask design.

Low signal threshold. This occurs when a multiplexer designed for high-flux applications is used in low-flux applications (for example, when a focal-plane array designed for terrestrial applications is used for astronomy). In effect, the low detector signal is never read out, and anomalously low signal and noise levels result.

Incomplete output reset. This is caused by a slow reset circuit or charge partitioning, and, when combined with correlated double sampling, an incomplete reset results in overshoots at rising edges and undershoots on falling edges in the readout direction.

Analog-to-digital converter nonlinearities. A very small amount of nonlinearity can be observed as a ring pattern correlated with the $\cos^4$ shading. The edges of the ring occur at higher-order bit-transition levels in the raw imagery (e.g., the bits corresponding to 32, 64, or 128 gray levels). After nonuniformity correction, low-level nonlinearities can be observed when a low-contrast scene is imaged.

27.4.6. Focal-Plane Nonuniformities and Dynamic Range

Ideally, the dynamic range is given by the largest signal that can be measured divided by the smallest observed noise. However, in the presence of focal-plane array nonuniformities, a large fraction of the potential dynamic range is consumed by the most extreme pixels, and the resulting pixel-averaged dynamic range is diminished. Some of the dynamic range can be recovered if extreme pixels are allowed to saturate, at the expense of a greater number of nonfunctioning pixels.

27.5. Characterizing Imager Performance

Most imager metrics characterize performance in terms of a minimum temperature contrast required for performing a particular function. It follows that low values (small temperature differences) indicate superior performance.

27.5.1. Noise-Equivalent Delta Temperature

The noise-equivalent temperature difference is the scene temperature difference required for yielding a signal equal to the noise at a given background level:

$$\mathrm{NE\Delta T} \equiv \sigma_{\mathrm{temporal}} \left[\frac{\mathrm{d}S(T)}{\mathrm{d}T} \right]^{-1}. \tag{27.46}$$

For photoelectron-noise-limited performance,

$$\mathrm{NE\Delta T}_{\mathrm{photon}} = \frac{\sqrt{S(T)}}{\mathrm{d}S(T)/\mathrm{d}T}. \tag{27.47}$$

When the noise-equivalent delta temperature is measured directly from the camera, it is best to limit the spatial extent of the thermally variable region in the scene. If the thermal object is too small, it will fail to fill the pixel and an anonymously large value of the noise-equivalent delta temperature will result. If the thermal object is too large (say, greater than 10% of the field of view), stray light, pixel cross talk, and any of a long list of undesirable artifacts can cause an anonymously small value of the noise-equivalent delta temperature to result. (Measurements of the noise-equivalent delta temperature made without a lens fall into this category, and should be avoided.) The temperature difference used in measurement should be large enough to mitigate noise in its measurement and small enough to avoid the effects of nonlinear response in temperature (e.g., 1 or 2 °C). In practice,

$$\mathrm{NE\Delta T}_{\mathrm{measured}} = \sigma_{\mathrm{temporal}} \frac{\Delta T}{\Delta S}, \tag{27.48}$$

where ΔT is the temperature difference and ΔS is the difference between the signals measured at high and low temperatures.

The noise-equivalent delta temperature should not be used to compare focal-plane arrays with different pixel sizes. The noise-equivalent delta temperature of a 256×256 focal-plane array will be half that of a 512×512 focal-plane array of the same physical size (assuming photoelectron-noise-limited operation and 100% fill factor). However, despite its larger noise-equivalent delta temperature, the 512×512 focal-plane array will have precisely the same image quality as that of the smaller array at low spatial frequencies and better performance at high spatial frequencies when characterized with image-based techniques (e.g., minimum resolvable temperature).

27.5.2. *Minimum Resolvable Temperature*

The minimum resolvable temperature is a subjective measurement of infrared imager performance. It is measured by a trained observer who declares when a thermal bar pattern displayed on a monitor is visible. The minimum resolvable temperature is the minimum temperature difference between the bars and the background such that all of the bars are just visible. By definition, the target has four bars, each with a 7×1 aspect, and the entire four-bar pattern is square. In practice, both positive and the negative contrast bars are measured and the minimum resolvable temperature is given by the average of the two values.

When the spatial resolution is set by the geometrical instantaneous field of view (sensor modulation at ξ_{max} is high), the spatial noise is negligible, and care has been taken to optimize the phasing of the bars relative to the imager sampling, the minimum resolvable temperature is accurately estimated by

$$\mathrm{MRT}(\xi) = \mathrm{SNR_{th}}\ \mathrm{NE\Delta T} \left(\frac{\tau_{\mathrm{frame}} \alpha}{\tau_{\mathrm{eye}} \beta} \right)^{1/2} \frac{\xi}{\xi_{\max}}, \tag{27.49}$$

where $\mathrm{SNR_{th}}$ is the signal-to-noise ratio required for unambiguously detecting a single pixel signal in noise ($\mathrm{SNR_{th}} \approx 3.0$), α is the aspect ratio of the pixels on the focal-plane array, β is the aspect ratio of the bars ($\beta = 7$), τ_{frame} is the frame time of the camera–monitor system (not necessarily the same as the pixel integration time), ξ is the frequency of the bar pattern, and ξ_{max} is half the sampling frequency of the imager ($\xi_{max} = 1/2p_x$).

A more general expression for the minimum resolvable temperature is

$$\mathrm{MRT}(\xi) = \mathrm{SNR_{th}} G_{\mathrm{ST}} \left[\frac{\mathrm{NE\Delta T}}{\mathrm{MTF}(\xi)} \right] \left(\frac{\alpha}{\beta} \right)^{1/2} \frac{\xi}{\xi_{\max}}, \tag{27.50}$$

where $\mathrm{MTF}(\xi)$ is the modulation transfer function of the entire imaging system. The use of $\mathrm{MTF}(\xi)$ in this case is not as well founded as one might think. Aliasing, which is common in staring infrared imagers, has a large systematic effect on the measured minimum resolvable temperature.[7]

At bar frequencies greater than $1/2\ \xi_{max}$, the modulation is strongly dependent on the relative phase. When $\xi = \xi_{max}$ the magnitude of the modulation will be zero when the bars are centered on the gap between active areas, but can be

[7] C. M. Webb, "Dynamic minimum resolvable temperature testing for staring array imagers," Opt. Eng. **38,** 845, 1999.

large when centered on the active areas. For imagers in which $\mathrm{MTF}_{\mathrm{pixel}}(\xi)$ dominates the system modulation transfer function, the modulation of a bar pattern at $\xi_{\max}$ can be nearly unity despite the fact that $\mathrm{MTF}_{\mathrm{pixel}}(\xi_{\max})$ is only 0.63.

The minimum resolvable temperature was originally created as a laboratory measure of system performance. The trained observer was allowed to vary the gain and the contrast of the monitor and move relative to it. In most applications the gain and the contrast are automatically set, there is glare on the monitor, the ambient-light level is high, and the observer is fixed relative to the monitor. These cases can be modeled if $\mathrm{SNR}_{\mathrm{th}}$ and $\mathrm{MTF}(\xi)$ are modified to include environmental effects. For example, if the observer is not free to move relative to the monitor, then $\mathrm{MTF}_{\mathrm{eye}}(\xi)$ should be included in the total system $\mathrm{MTF}(\xi)$, and if the ambient-light level is high, $\mathrm{SNR}_{\mathrm{th}}$ should be increased.

27.5.3. Minimum Detectable Temperature

The minimum detectable temperature[8] is figure a merit appropriate for characterizing a sensor's ability to detect unresolved targets. It is similar to the minimum resolvable temperature except that a uniform (usually square) target is used instead of a four-bar target. For the case of a square target, square pixels, and where the instantaneous field of view is limited by $\mathrm{IFOV}_{\mathrm{geometrical}}$,

$$\mathrm{MDT}(\omega) = \mathrm{NE\Delta T}\ \mathrm{SNR}_{\mathrm{th}} G_{\mathrm{ST}} \frac{\mathrm{IFOV}_{\mathrm{geometrical}}}{\omega} \sqrt{1 + \left(\frac{\mathrm{IFOV}}{\omega}\right)^2}, \qquad (27.51)$$

where ω is the angle subtended by the target (in one dimension).

At close range, where $\omega \gg \mathrm{IFOV}$ (i.e., the target subtends several pixels), the minimum detectable temperature varies as the square root of the number of subtended pixels (i.e., as $1/r$). At long range, the target subtends less than a pixel and the minimum detectable temperature varies as $1/r^2$.

For an asymetric target, ω in Eq. (27.50) should be replaced with $\sqrt{\omega_H \omega_V}$, and a similar substitution should be made for nonsymetric instantaneous fields of view.

27.5.4. Mean-Variance Technique

The mean-variance technique[9] is a useful tool for measuring the conversion gain g_c of a photoelectron-noise-limited imager. It can also be used to evaluate

8 M. J. Cantella, "Staring-sensor systems," in S. B. Campana, ed., *Passive Electro-Optical Systems,* Chap. 3, and in J. S. Accetta and D. L. Shumaker, eds., *The Infrared and Electro-Optical Systems Handbook,* Chap. 5, Society of Photo-Optical Instrumentation Engineers, 1993.

9 L. Mortara and A. Fowler, "Evaluations of charge-coupled device performance for astronomical use," in *Solid-State Imagers for Astronomy,* J. C. Geary and D. W. Latham, eds., Proc. SPIE **290,** 28, 1981.

sensor linearity and dynamic range and can be useful for determining whether an imager is operating in the photoelectron-noise limit.

The performance of thermal imagers is theoretically limited by fluctuations in the arriving photons or, more practically, by fluctuations in the collected photoelectrons. For most thermal imaging applications the probability density function of the photoelectron count will be Poisson. Therefore, when the signal is measured in electrons, its variance will be equal to its mean. In the absence of additional noise, the conversion of photoelectrons to voltage or current has no effect on the signal-to-noise ratio; therefore the noise is scaled by the same factor as the signal and the variance is scaled by the square of that factor. Then

$$\sigma_N^2 = g_c^2 \sigma_Q^2, \tag{27.52}$$

where σ_N is the noise in the digital data and σ_Q is the noise in the number of electrons. For a photoelectron-noise-limited process, $\sigma_Q^2 = \langle Q \rangle$. In the presence of additional noise terms, which we lump into a single term σ_{floor}, the image signal variance is

$$\sigma_N^2 = g_c(\langle N \rangle - N_0) + \sigma_{\text{floor}}^2, \tag{27.53}$$

and the slope of the variance that is plotted as a function of mean is equal to the conversion gain of the imager.

Sensor nonlinearities appear as a change in slope in the mean-variance plot. At high radiance levels the nonlinearities will compress the noise fluctuations and the slope of the mean-variance curve will decrease. At low radiance values poor conversion efficiency and poor transfer efficiency can introduce unexpected artifacts into the mean-variance plot.

The difficult part of the mean-variance technique is getting an accurate estimate of the variance. The precision of the measured variance is significantly less than that of the mean. If the ideal value of the variance is σ_c^2, the probability density function is Gaussian (a good approximation in the limit of a large number of carriers), and the process is stationary, then the relative error of the measured estimate will be

$$e \equiv \sigma_{\sigma_N^2}/\sigma_c^2 = \left(\frac{2}{k_{\text{mv}} - 1}\right)^{1/2}, \tag{27.54}$$

where k_{mv} is the number of samples used to estimate the variance. From Eq. (27.54) we see that 800 samples are required for obtaining a 5% estimate of the variance, whereas 20,000 are required for obtaining a 1% estimate. The slow improvement in precision with number of samples combined with the drift inherent in infrared measurements makes the use of spatial pixel averages attractive for estimating the variance.

27.6. Display and Interpolation

Even though the only aspect of an infrared imager that a pilot, navigator, driver, or analyst sees is the image on a monitor, the task of image display is often considered only as an afterthought. For a digital infrared image to be displayed, the image must be mapped spatially to the display pixels and mapped in intensity to the dynamic range of the display. The most common nontrivial interpolation and display algorithms are described in the next two subsections.

27.6.1. Parametric Cubic Interpolation

Infrared images are often displayed on monitors with higher spatial resolution than is present in the sampled image. To that end, the image is resampled to the display resolution by means of an interpolation technique. The simplest approach, nearest-neighbor replication, introduces high-spatial-frequency artifacts into the displayed image that are not present in the scene. A slightly more complicated approach, bilinear interpolation, eliminates the artifacts at the price of substantial blur. An even more computationally demanding approach, parametric cubic convolution, avoids the artifacts without undue blur.

In one dimension the parametric cubic convolution[10] assigns a value

$$V(x, y) = \sum_{i=0}^{N_x} \sum_{j=0}^{N_y} N_{ij} r(x - i) r(y - j) \tag{27.55}$$

to each resampled pixel, where

$$r(x) = \begin{cases} (\alpha + 2)|x|^3 - (\alpha + 3)|x|^2 + 1, & |x| < 1 \\ \alpha|x|^3 - 5\alpha|x|^2 + 8\alpha|x| - 4\alpha, & 1 \le |x| < 2 \\ 0 & \text{otherwise} \end{cases} \tag{27.56}$$

and α is a free parameter that is best set to -0.5.

27.6.2. Plateau Equalization

Digital infrared images typically have 12, 14, or 16 bits per pixel, although the human eye viewing a monitor can distinguish little more than 100 gray levels (i.e., fewer than 7 bits). Plateau equalization[11] is a way of mapping the infrared

[10] S. K. Park and R. A. Schowengerdt, "Image reconstruction by parametric cubic convolution," Comput. Vis. Graph. Image Process. **23,** 258, 1983.

[11] V. E. Vickers, "Plateau equalization algorithm for real-time display of high-quality infrared imagery," Opt. Eng. **35,** 1921, 1996.

image to the monitor in a way that maintains the high-contrast features without losing the low-contrast details.

Plateau equalization is a general class of display algorithms that map high dynamic range imagery to lower dynamic range displays. In the limit of large and small plateau values it includes both histogram equalization and histogram projection. Plateau equalization is defined in terms of the histogram of the raw image, in which C_b pixels have digital signal level b. From a user-defined plateau value P, a modified histogram is defined with c_b pixels in level b, where

$$c_b = \min(C_b, P), \tag{27.57}$$

and c_b is the lesser of C_b and the plateau value P. In the limit of $P = 1$, histogram projection will result; when P is greater than the number of pixels on the focal plane, histogram equalization will result. Typically, users find an intermediate plateau value most useful.

The output mapping is constructed from the partial and the complete sums. The partial sum is

$$s_m(P) \equiv \begin{cases} 0, & \text{for } m = 0 \\ \sum_{b=0}^{m-1} c_b, & \text{for } 1 \le m \le 2^{N_{\text{bits}}}, \end{cases} \tag{27.58}$$

where N is the number of digital bits used to represent the raw image. The complete sum is

$$S(P) \equiv s_{2^N}(P). \tag{27.59}$$

The display output d_m for input level m is

$$d_m = \text{floor}\left[\frac{Ds_m(P)}{S(P)}\right], \tag{27.60}$$

where floor [ ] denotes integer truncation and D is the maximum display level (for an 8-bit display, $D = 255$).

Further Reading

J. M. Lloyd, *Thermal Imaging Systems,* Plenum, 1975.

R. D. Fiete, "Image quality and $\lambda FN/p$ for remote sensing," Opt. Eng. **38,** 1229, 1999.

G. C. Holst, *Testing and Evaluation of Infrared Imaging Systems,* JCD Publishing, 1993.

J. D. Howe, "Electro-optical imaging system performance prediction," in M. C. Dudzik, ed., *Electro-Optical System Design, Analysis, and Testing,* Chap. 2, and in J. S. Accetta and D. L. Shumaker, eds., *The Infrared and Electro-Optical Systems Handbook,* Chap. 4, Society of Photo-Optical Instrumentation Engineers, 1993.

C. H. Séquin and M. F. Tompsett, *Charge Transfer Devices,* Academic, 1975.

J. Silverman, J. M. Mooney, and F. D. Shepherd, "Infrared video cameras," Sci. Am. **266,** 78, 1992.

W. L. Wolfe and G. J. Zissis, *The Infrared Handbook,* The Environmental Research Institute of Michigan International and Society of Photo-Optical Instrumentation Engineers, 1985.

Special Issue on Infrared Physics and Technology, Proc. IRE **47,** September 1959.

In the first four references, various aspects of imaging, both infrared and visual, are discussed. The *Handbook* by Accetta and Shumaker is an eight-volume treatise on infrared systems in general. The next two cover the operation of charge-transfer devices and infrared cameras. *The Infrared Handbook* has data and information on infrared components and systems. The last is a fascinating historical perspective.

28

Infrared Sensors

Bjørn F. Andresen and William L. Wolfe

28.1. Introduction

Infrared sensors, as used here, are not meant to supply an image. They may generate an image as an intermediate step, but the functions are detection, tracking, and location.

28.2. Signal-to-Noise Equations

System signal-to-noise ratios can be calculated in several different ways. They depend on whether the target is resolved (extended) or not (point), whether the detector is photon-noise limited, and whether one uses specific detectivity, responsivity, and noise or counts photons.

28.2.1. Point Source, Specific Detectivity

The equation for a point source in terms of specific detectivity is

$$\mathrm{SNR} = \frac{\int D^*(\lambda)\Phi_\lambda(\lambda)\mathrm{d}\lambda}{\sqrt{A_d B}} = \frac{\int D^*(\lambda)\tau_a(\lambda)\tau_o(\lambda)L_\lambda(\lambda)\mathrm{d}\lambda}{\sqrt{A_d B}}\frac{A_o A_s}{R^2}$$

$$= \frac{\int D^*(\lambda)\tau_a(\lambda)\tau_o(\lambda)L_\lambda(\lambda)\mathrm{d}\lambda}{\sqrt{A_d B}} A_o\Omega_{so} = \frac{\int D^*(\lambda)\tau_a(\lambda)\tau_o(\lambda)L_\lambda(\lambda)\mathrm{d}\lambda}{\sqrt{A_d B}} A_s\Omega_{os},$$

where SNR is the voltage or current signal-to-noise ratio, $D^*(\lambda)$ is the spectral specific detectivity, $\Phi_\lambda(\lambda)$ is the spectral power per unit wavelength on the detector, B is the effective noise bandwidth, the τ_o and τ_a are the optical and the atmospheric transmission, respectively, $L_\lambda(\lambda)$ is the source spectral radiance, the A's are areas, with the subscripts d for detector, o for optical aperture, and

s for source, R is the range from the aperture to the source, and Ω is the solid angle. The integration is over the spectral response of the system. In most cases the difference between the target and the background in radiance or power should be used rather than the radiance or power itself. Another useful form when a vertical detector array with one or more columns scans horizontally over a point source is

$$\mathrm{SNR} = \frac{\pi}{2\sqrt{2}} \tau_a \tau_o I D^* \frac{D_o}{R^2 F} \sqrt{\frac{m_{tdi} t_h}{\Theta_h \alpha_v}},$$

where m_{tdi} is the number of *tdi* steps, Θ_h is the horizontal field, α_v is the angular subtense of a detector element in the vertical direction, D_o is the entrance pupil diameter, F is the focal ratio, t_h is the time it takes for one horizontal scan, and I is the integral of the spectral intensity over the band; it is assumed that there is no horizontal oversampling.

28.2.2. Point Source, Photon Limited

If the detector is photon limited, the equations are

$$\mathrm{SNR} = \frac{A_s A_o}{\sqrt{A_d B}} \frac{1}{R^2} \int \tau_a(\lambda) \tau_o(\lambda) D^*(\lambda) L_\lambda(\lambda) \mathrm{d}\lambda$$

$$= \frac{A_s A_o}{\sqrt{A_d B}} \frac{1}{R^2} \frac{D_m^*}{\lambda_m} \int \tau_a(\lambda) \tau_o(\lambda) \lambda L_\lambda(\lambda) \mathrm{d}\lambda,$$

$$\mathrm{SNR} = \frac{A_s A_o}{\sqrt{A_d B}} \frac{1}{R^2} \frac{1}{hc} \sqrt{\frac{\eta}{g \int E_{q\lambda}(\lambda) \mathrm{d}\lambda}} \int \tau_a(\lambda) \tau_o(\lambda) \lambda L_\lambda(\lambda) \mathrm{d}\lambda,$$

$$\mathrm{SNR} = \frac{A_s A_o}{\sqrt{A_d B}} \frac{1}{R^2} \sqrt{\frac{\eta}{g}} \frac{\int \tau_a(\lambda) \tau_o(\lambda) L_{q\lambda}(\lambda) \mathrm{d}\lambda}{\sqrt{\int E_{q\lambda}(\lambda) \mathrm{d}\lambda}},$$

where $L_{q\lambda}(\lambda)$ is the spectral photon radiance, h is Planck's constant, c is the speed of light, η is the quantum efficiency (assumed to be constant over the spectral band), and λ_m is the wavelength for which D_λ^* has its maximum, D_m^*, g is 4 if there is recombination, 2 if not, and $E_{q\lambda}(\lambda)$ is the spectral photon incidance on the detector from all sources. All spectral integrations extend over the spectral band of the sensor.

28.2.3. Extended Source, Specific Detectivity

If the source is extended, the equation in terms of specific detectivity is

$$\text{SNR} = \frac{\pi}{4}\frac{D_o\sqrt{\alpha_h\alpha_v}}{F\sqrt{B}}\int D^*(\lambda)\tau_a(\lambda)\tau_o(\lambda)L_\lambda(\lambda)\mathrm{d}\lambda.$$

28.2.4. Extended Sources, Photon Limited

If the detector is photon limited, the equation is

$$\text{SNR} = \frac{\pi}{4}\frac{D_o}{F}\frac{\sqrt{\alpha_h\alpha_v}}{\sqrt{gB}}\frac{\sqrt{\eta}}{hc}\frac{\int \lambda\tau_a(\lambda)\tau_o(\lambda)L_\lambda(\lambda)\mathrm{d}\lambda}{\sqrt{\int E_{q\lambda}(\lambda)\mathrm{d}\lambda}}$$

$$= \frac{\sqrt{\pi}}{2}\frac{D_o\alpha}{\sqrt{gB}}\frac{\int \tau_a(\lambda)\tau_o(\lambda)L_{q\lambda}(\lambda)\mathrm{d}\lambda}{\sqrt{\int L^b_{q\lambda d}}},$$

where D_o is the aperture diameter and F is the focal ratio. If the system has a circularly symmetric aperture and is diffraction limited (the main lobe is on the detector) then

$$\text{SNR} = \lambda_m\sqrt{\frac{\eta}{gB}}\frac{\int \tau_a(\lambda)\tau_o(\lambda)L_{q\lambda}\mathrm{d}\lambda}{\sqrt{\int L^b_{q\lambda}\mathrm{d}\lambda}} \approx \frac{0.5\lambda_m}{\sqrt{B}}\frac{\int L_{q\lambda}\mathrm{d}\lambda}{\sqrt{\int L^b_{q\lambda}\mathrm{d}\lambda}},$$

where the approximation assumes that all efficiency factors are one and there is recombination.

28.2.5. Responsivity and Noise

The detailed equation with responsivity and noises is

$$\text{SNR} = \frac{\int \mathcal{R}_{v,\lambda}\Phi_\lambda\,\mathrm{d}\lambda}{\sqrt{\sum\int v_i^2(f)\mathrm{d}f}} = \frac{\int \mathcal{R}_{i,\lambda}\Phi_\lambda\mathrm{d}\lambda}{\sqrt{\sum\int i_i^2(f)\mathrm{d}f}},$$

where $\mathcal{R}_{v,\lambda}(\lambda)$ is the voltage responsivity, $\mathcal{R}_{i,\lambda}(\lambda)$ is the current responsivity, Φ_λ is the spectral power on the detector, and v_i (or i_i) is the noise voltage (or current) of the ith component of noise. The integration of the numerator is over the infrared spectral band; integration of the denominator is over the information bandwidth. The noises are described in Section 28.3.

28.2.6. *Counting Electrons*

This applies to photon detectors. The point-target signal equation is

$$N_s = \eta \tau_a \tau_o L_q \frac{A_s A_o}{R^2} t_i,$$

where N_s is the number of signal electrons in the integration time t_i from a source of in-band photon radiance L_q of area A_s, optics area A_o, and range R, with transmissions τ. The noise is given by

$$N_n = \sqrt{N_s + \sum N_i},$$

where N_i represents mean-square number of electrons. N_s are the mean-square signal–noise electrons and are equal to the signal electrons. The signal-to-noise ratio is the ratio of N_s to N_n.

28.3. Noises

The noises are Johnson, shot, generation-recombination, current, photon, temperature, switching, and sampling.

28.3.1. *Johnson Noise (Thermal or Schottky Noise)*

The mean-square Johnson noise current and noise voltage are

$$\bar{\imath}^2_{\text{Johnson}} = \frac{4kTB}{R}, \quad \bar{v}^2_{\text{Johnson}} = 4kTRB,$$

where k is Boltzmann's constant, T is the temperature, R is the resistance, and B is the bandwidth.

28.3.2. *Shot Noise*

The mean-square shot-noise current and voltage are

$$\bar{\imath}^2_{\text{shot}} = 2qIB, \quad \bar{v}^2_{\text{shot}} = 2qRIB.$$

28.3.3. *Generation-Recombination Noise*

The mean-square generation-recombination noise current in a near-intrinsic semiconductor is

$$i_{gr}^2(f) = 4I^2 \frac{(b+1)^2 N_e N_h}{(bN_e + N_h)^2 (N_e + N_h)} \frac{\tau}{1+\omega^2\tau^2},$$

where I is the dc bias current, b is the ratio of electron-to-hole mobility, and the N's are the total number of electrons and holes, respectively. For a true intrinsic device, this becomes

$$i_{gr}^2(f) = \frac{2I^2\tau}{N_e(1+\omega^2\tau^2)}.$$

28.3.4. *Temperature Noise*

The mean-square fluctuation in power is

$$\bar{\Phi}^2 = 4kGT^2.$$

The mean-square fluctuation in temperature is

$$\bar{T}^2 = \frac{4kT^2B}{G(1+\omega^2\tau^2)},$$

where the time constant τ is the time constant of the detector.

28.3.5. *Photon Noise*

The Planck expression for blackbody radiation provides an expression for the average arrival rate of photons:

$$\langle \dot{N} \rangle = \frac{2\pi c}{e^{\frac{hc}{\lambda kT}} - 1} = \frac{2\pi c}{e^{\frac{c_2}{\lambda T}} - 1} = \frac{2\pi c}{e^x - 1}.$$

The fluctuation about this mean is related to the photon noise. The mean-square deviation from the mean rate is equal to the mean rate for a Poisson process, and this is a good approximation for photon noise. A more accurate expression, with Bose–Einstein statistics taken into account, leads to a correction factor:

$$\sqrt{\langle(\langle\dot{N} - \langle\dot{N}\rangle\rangle)^2\rangle} = \dot{N}_{\mathrm{rms}} = \langle\dot{N}\rangle \frac{e^x}{e^x - 1}.$$

28.3.6. *Excess Noise (Current Noise)*

The mean-square excess-noise current is

$$\bar{i}^2_{\mathrm{excess}} = \frac{KI^{\alpha}B}{f^{\beta}},$$

where K, α, and β are constants.

28.3.7. *Quantization Noise*

If the signal represented by v has a maximum value $v_{\max}$, then the quantization noise is

$$v_{\mathrm{rms}} = \frac{v_{\max}}{2^{m+2}\sqrt{3}},$$

where m is the number of switches.

28.4. Sources of Photon Noise in Space Sensors

There are several sources of photon incidance that produce photon noise: the sun, the moon, the system, stars, zodiacal light, and intergalactic light.

28.4.1. *The Sun*

The sun may be modeled in the infrared as a 5900 K blackbody. Its spectral radiance is given by

$$L_{\lambda} = \frac{2c^2 h}{\lambda^5\left(e^{\frac{c_2}{5900\lambda}} - 1\right)} \text{ W sr}^{-1}\text{ cm}^{-2}.$$

Its photon radiance is

$$L_{q\lambda} = \frac{2c}{\lambda^4\left(e^{\frac{c_2}{5900\lambda}} - 1\right)} \text{ s}^{-1}\text{ sr}^{-1}\text{ cm}^{-2}.$$

The radiant incidance at the Earth, not accounting for the atmosphere, is obtained by multiplying by the solid-angle subtense of the sun, which is

$$\Omega = \frac{\pi}{4}\left(\frac{33}{60}\,\frac{\pi}{180}\right)^2 = 7.237 \times 10^{-5}\ \text{sr}.$$

The energetic incidance and the photon incidance are

$$E_{\Delta\lambda} = \Omega \int L_\lambda(\lambda)\mathrm{d}\lambda, \quad E_{q,\Delta\lambda} = \Omega \int L_{q,\lambda}(\lambda)\mathrm{d}\lambda,$$

where the integration is over the spectral band of the system.

Solar spectrum data are compiled in the *Handbook of Chemistry and Physics.*[1]

These equations provide the incidance levels at the front of the baffle tube. The flux on the detector is obtained by multiplying these values of incidance by the point-source transmittance of the system (see Chap. 22).

28.4.2. *The Earth*

The Earth can contribute to the background incidance, either directly into the sensor or by reflection off a baffle tube. The Earth's radiation is complicated and variable, based in part on cloud cover and the region of the Earth that is viewed as well as the spectral band that is used. In the transmission bands, a good approximation is the radiation from a 300 K blackbody. In regions of atmospheric absorption an equivalent approximation is the radiation from a 250 K blackbody. *The Infrared Handbook* (see the section on Further Reading) has data on the radiation from the Earth as a function of latitude and season.

28.4.3. *The Moon*

The moon contributes in two ways to the radiant incidance at the Earth, by reflecting sunlight and by emitting radiation. The reflected sunlight is obtained by calculating the radiant incidance at the moon from the sun and multiplying by the lunar diffuse reflectivity (0.04 sr^{-1}) and by the angular subtense of the irradiated moon: 30 arcmin.

One obtains the radiant incidance at the Earth from radiation emitted by the moon by multiplying the blackbody radiance of the moon at either its cool (120 K) or its hot (370 K) temperature, by its emissivity (0.88), and by the solid-angle subtense of the cool and hot parts of the moon, respectively.[2]

[1] D. R. Lide, ed., *Handbook of Chemistry and Physics,* 73rd ed., CRC, 1993, p. 14-8.
[2] Ibid., pp. 14-26–14-27.

28.4.4. The System

The system radiates to the detector according to its temperature and emissivity. If the system has n elements (mirrors, lenses, windows, filters), then the photon incidance E_{qd} on the detector may be calculated as

$$E_{qd} = \sum_{i=1}^{n} \varepsilon_i L(T_i)\Omega \sum_{j=1}^{n-1} \tau_{i-j}$$

where $\tau_m = 1$ for $m \leq 0$, the counting starts at the element closest to the detector, the τ's are the transmittances of the elements, $L_q^{BB}(T_i)$ are the blackbody photon radiances of the elements, Ω is the limiting solid angle, and the ε_i are the emissivities of the elements. All elements must be imaged into the optical space of the detector.

28.4.5. Stars

Stars are point sources and can be mistaken for targets. Either position information or two-color techniques need to be used for discrimination. Star flux tables are given in *The Infrared Handbook.*

28.4.6. Zodiacal Light

This is a diffuse radiation that is usually low enough to be ignored. Data are in *The Infrared Handbook.* The upper limits of the spectral radiance in the ecliptic plane at 160° from the sun have been measured as 3×10^{-11} W cm^{-2} sr^{-1} μm^{-1} (3–5 μm), 6×10^{-11} (12–14 μm), and 2.5×10^{-11} (16–23 μm). The radiance increases toward the sun and decreases away from the ecliptic plane.

28.4.7. Galactic Radiance

The radiance of the galaxy from its stars is given by

$$L_\lambda = \frac{1}{2} \sum n_i I_{\lambda,I}^{3/2} \int_0^\infty \frac{R_c^2}{R_c^2 + R^2} E_\lambda^{-3/2} \exp\left(\frac{I_{\lambda,I}}{E_\lambda} \frac{\sin^2 b}{\sigma_i^2}\right) dE_\lambda,$$

where

$$R^2 = \frac{I_{\lambda,i}}{E_\lambda} \cos^2 b + R_0^2 - 2R_0 \frac{I_{\lambda,i}^{1/2}}{E_\lambda} \cos b \cos l, \quad R_0 = 3.09 \times 10^{22} \text{ cm},$$

$$R_c = 9.27 \times 10^{21} \text{ cm}, \quad I_{\lambda,i} = a_i \lambda^{c_i},$$

l and b are the galactic longitude and latitude, respectively, n_i is the mean spatial density of the sources at the galactic center, and σ_i is the dispersion of the

ith class normal to the galactic plane. The necessary constants for the three different classes are given in Table 28.1.

Table 28.1. Constants for the Galactic Equation

Class	σ_i (cm)	n_i (cm^{-3})	a_i (W sr^{-1} μm^{-1})	c_i
Dwarfs	4.94×10^{20}	2.63×10^{-56}	2.77×10^{24}	−3.693
Giants	13.9×10^{20}	9.46×10^{-60}	1.07×10^{28}	−1.915
Supergiants	3.09×10^{20}	7.90×10^{-62}	1.09×10^{29}	−1.108

28.4.8. *Extragalactic, Diffuse Radiation*

The equation is

$$L_\lambda = \frac{n_g c}{4\pi H_0} \sum F_i \Phi_i,$$

where n_g is the number density of galaxies, c is the velocity of light, H_0 is the Hubble constant, and F_i is the fraction of galaxies with flux Φ_i. The results for some wavelengths are given in Table 28.2.

Table 28.2. Extragalactic Background Spectral Radiances

λ	3	5	10	15	20	25	30
L_λ	11.0	7.2	3.8	2.6	2.0	1.7	1.4

28.5. Range Equations

The range equation for a subresolution source in terms of specific detectivity is

$$R^2 = \int D^*(\lambda)\tau_a(\lambda)\tau_o(\lambda)I_\lambda(\lambda)d\lambda \frac{A_s A_o}{\sqrt{A_d B}\ \mathrm{SNR}}$$

$$= \int D^*(\lambda)\tau_a(\lambda)\tau_o(\lambda)I_\lambda(\lambda)d\lambda \frac{A_o}{\sqrt{A_d B}\ \mathrm{SNR}},$$

where $L_\lambda(\lambda)$ is the spectral radiance of the source and $I_\lambda(\lambda)$ is its spectral radiant intensity; if the detector is photon limited, the range is

$$R^2 = \frac{A_s A_o}{\sqrt{A_d B}} \sqrt{\frac{\eta}{g}} \frac{1}{\mathrm{SNR}} \frac{\int \tau_a(\lambda)\tau_o(\lambda)L_{q\lambda}(\lambda)d\lambda}{\sqrt{E_{q\lambda}d\lambda}}.$$

Note that the atmospheric transmission is a function of the range, which is a function of the atmospheric transmission.

28.6. Search Rate and Bandwidth

The search rate is most generally expressed in angular dimensions. It can be expressed as

$$\dot{\Omega} = \eta_{sc} \frac{\Theta_x \Theta_y}{t_f},$$

where $\dot{\Omega}$ is the solid angle that is scanned per unit time, η_{sc} is the scan efficiency, the Θ's are the full-field angles, and t_f is the frame time. For specific situations, there are specialized equations.

28.6.1. Look-Down Mode

This can be single-detector, pushbroom, or whiskbroom mode. For the single detector the rate is

$$\dot{\Omega} = \eta_{sc} \frac{n_x \alpha n_y \beta}{t_d},$$

where the n's are the number of angular pixels, α and β are the pixel angular subtenses, and t_d is the dwell time, which has to be larger than the time constant of the detector. For a pushbroom system, with a full array of detectors across the field, the rate is

$$\dot{\Omega} = \frac{n_x \alpha\beta}{t_d} = \frac{n_x \alpha\beta}{t_l},$$

where n_x is the number of across-track detectors and t_l is the line time, which is equal to the dwell time, which must be greater than the detector time constant. The line time is determined by the angular subtense at the nadir,

$$t_l = \frac{\alpha h}{v},$$

and the angular subtense is determined by the resolution requirement at the edge of the field,

$$\alpha = \frac{x}{h} \cos^2(\Theta_{1/2,x}),$$

where $\Theta_{1/2,x}$ is the half-field angle in the across-track direction. All detectors of the array must be read out in less than a line time.

The whiskbroom detector has an array of $m_x \times m_y$ detectors that is scanned in the across-track direction. The rate is

$$\dot{\Omega} = \eta_{\text{sc}} \frac{\alpha m_y \beta}{t_d}.$$

In this case, the line time is the pixel dwell time times the number of detectors in the in-track direction.

28.6.2. Look-Ahead Mode

This is a downward look ahead of the vehicle. The resolution is determined by the projected angular subtense on the ground at the maximum range, which is the position at the edge of the scan and at the look-ahead angle.

28.7. Equations in the Spatial Frequency Domain

For a point source the equations can be cast in the following form.

When all the modulation transfer functions are taken into account, the spatial spectrum must be multiplied by the transfer functions, and the result, contained in parentheses, inverted by a two-dimensional Fourier transform:

$$\text{FT}^{-1}\ \text{SNR} = (W_T H_a H_o H_{\text{ds}} H_{\text{dt}} H_e) \frac{\pi}{4F^2} \frac{\int D^* L_\lambda \tau_a \tau_o d\lambda}{\sqrt{A_d B |H_{\text{dt}}|^2 |H_E|^2}},$$

where W_T is the spatial frequency spectrum of the target and the H's are the modulation transfer functions of the atmosphere, optics, the spatial response of the detector, the temporal response of the detector, and the electronics, two dimensional where appropriate.

The target may often be represented as a true point, in which case the spectrum is flat. If it is rectangular, then it is a two-dimensional sinc function (see Chap. 2):

$$W_T = \text{sinc}(\pi\alpha\xi)\text{sinc}(\pi\beta\zeta).$$

If it is circular, then it is a sombrero function (see Chap. 2):

$$W_T = \text{jinc}(\pi\xi r) = \text{somb}(\pi\xi r).$$

It may also be represented as a Gaussian of the average periphery.

The atmosphere is usually too general for a single representation, but the turbulence modulation transfer function may be approximated by

$$H_a = \exp[-2\pi^2 2.91\lambda^{1/3}(f\xi)^{5/3}C_n^2 R],$$

where C_n is the atmospheric structure function and R is the range.

The optics are often diffraction limited and circular, in which case the modulation transfer function is

$$\begin{aligned}\mathrm{MTF} &= \frac{2}{\pi}\left[\arccos\frac{\xi}{\xi_c} - \frac{\xi}{\xi_c}\sin\left(\arccos\frac{\xi}{\xi_c}\right)\right] \\ &= \frac{2}{\pi}\left[\arccos\frac{\xi}{\xi_c} - \frac{\xi}{\xi_c}\left(1-\frac{\xi^2}{\xi_c^2}\right)^{1/2}\right],\end{aligned}$$

where ξ is the spatial frequency and ξ_c is the cutoff frequency:

$$\xi_c = \frac{2\mathrm{NA}}{\lambda} = \frac{1}{F\lambda}.$$

More generally, the modulation transfer function of the optics can be obtained from the design program that is used.

The spatial response of a rectangular detector of sides a and b is the two-dimensional sinc function (see Chap. 2):

$$H_d = \mathrm{sinc}(\pi a\xi)\mathrm{sinc}(\pi b\zeta).$$

The temporal transfer function of the detector is

$$H_{\mathrm{dt}} = \frac{1}{\sqrt{1+\omega^2\tau^2}} = \frac{1}{\sqrt{1+f^2/f_o^2}},$$

where ω is the radian frequency, τ is the detector time constant, and f_o is the 3-dB down frequency.

The transfer function of the electronics is a function of its design. Several include single-time-constant systems and doubly and triply cascaded ones. They are the temporal response of the detector raised to the power of the cascade:

$$H_e = \left(\frac{1}{\sqrt{1+\omega\tau}}\right)^n.$$

28.8. Frame-to-Frame Substraction

In some applications the target position is obtained by a form of moving-target indication that is accomplished by the subtraction of two successive image frames and the determination of the differences between them. There must be good registration and stability to be able to subtract out the backgrounds effectively.

The drift can be characterized by the rate of motion of the field, drift rate v_D compared with the speed of the target v_T:

$$\frac{v_D}{v_T} = \frac{\dot{\Theta} t_f}{\alpha} = \frac{v_D t_f}{s},$$

where $\dot{\Theta}$ is the angular rate of the field of view, α is the subtense of the target, t_f is the frame time, and s is the linear size of the target. Typical values are from 1% to 10%.

Frame-to-frame differencing can be one time, two time, or three time (and even more). The first three orders of differencing are

$$S_{d1} = S_i - S_{i-1}, \quad S_{d2} = S_i - 2S_{i-2}, \quad S_{d3} = S_i - 3S_{i-1} + 3S_{i-2} - S_{i-3},$$

where the S's indicate the output from each pixel, d_i indicates the level of difference, and the other subscripts indicate the current frame and the previous, preprevious, and third previous.

28.9. Background Clutter

This noise arises from the structure of the background. The rms clutter is given by

$$\sigma = x^2 \left[\int_{-\infty}^{\infty} |\mathrm{MTF}(\zeta)|^2 S(\zeta) \mathrm{d}\zeta \right],$$

where x is the linear subtense at the background. The signal-to-clutter ratio is

$$\mathrm{SCR} = \frac{\int_{-\infty}^{\infty} S_T(\zeta) \mathrm{d}\zeta}{x \left[\int_{-\infty}^{\infty} |\mathrm{MTF}(\zeta)|^2 S_B(\zeta) \mathrm{d}\zeta \right]^{1/2}}.$$

Various background spectra have been measured. In general they have an almost linear decay on a log–log plot, with a maximum of between 10^6 and 10^7 at zero spatial frequency to ~1 at ~100 cycles/km. Of course, these vary with the terrain and the clouds.

28.10. Probability of Detection and False-Alarm Rate

This is a complicated calculation that depends on the signal processing methods. Examples are given here, but the reader should consult the references in the Further Reading section.

28.11. Cruciform Detector

Four detectors are placed in the form of a cross, and the target image is scanned across it in a circular pattern. The time between detections on each of the four detectors determines the azimuth and the elevation. The figure shows the position of the center of the circular scan (x, y) on the cruciform detector and the times the target crosses each detector. Its coordinates are

$$x = r \sin\left(\frac{t_2 - t_3}{2}\right) = r \sin\left(\frac{t_1 - t_4}{2}\right),$$

$$y = r \sin\left(\frac{t_1 - t_2}{2}\right) = r \sin\left(\frac{t_3 - t_4}{2}\right).$$

The second pair of relations can be used for redundancy.

28.12. Quadrant Detector

The quadrant detector can be used in either of two modes. One is with the conical scan, in which the length of time on each quadrant determines the position; the other is with a defocused image of the target, wherein the relative signal levels on each quadrant are used to determine the position. For the conical scan, the relations are the same as those for the cruciform except that the time differences of the cruciform are replaced with the time durations in each quadrant. If the defocused beam is used, then the relative signals (fluxes) in each quadrant are used. If the spot is uniform in flux density and the detector has uniform sensitivity, the approximate equations for x and y are

$$x = \frac{A}{2D} - \frac{V_1 + V_4}{D}, \quad y = \frac{A}{2D} - \frac{V_2 + V_3}{D},$$

where the V's are the outputs from quadrants 1–4, A is the area of the spot, and D is its diameter. The approximation is good for the spot not more than ~20% off center. In practice, nonuniformities of the spot and the responsivity make more precise calculations ill advised.

28.13. Imaging Trackers

In a sense, imaging trackers are extensions of quadrant cells. The target may subtend one or more pixels in the array. One of many algorithms is then used to determine the position of the target in the array.

Further Reading

W. L. Wolfe and G. J. Zissis, *The Infrared Handbook,* ERIM and SPIE, 1989.

W. L. Wolfe, *Introduction to Infrared System Design,* SPIE, 1998.

W. L. Wolfe, *Infrared Design Examples,* SPIE, 1999.

J. S. Acetta and D. L. Shumacher, *The Infrared and Electro-Optics Handbook,* ERIM and SPIE, 1993.

I. J. Spiro and M. Schlesinger, *Infrared Technology Fundamentals,* Marcel Dekker, 1989.

J. M. Lloyd, *Thermal Imaging Systems,* Plenum, 1975.

L. M. Biberman, *Reticles in Electro-Optical Devices,* Pergamon, 1966.

The Infrared Handbook has extensive information about optics, the atmosphere, targets, backgrounds, detectors, and system design. *The Infrared and Electro-Optics Handbook* is an updated version with additional chapters on active systems, visible systems, and related topics. Chapter 4 of Vol. 4 is especially pertinent. *Introduction to Infrared System Design* is a beginning tutorial, and *Infrared Design Examples* treats six design approaches in considerable detail. *Thermal Imaging Systems* is the classic on that topic. *Infrared Technology Fundamentals* is an introductory text with emphasis on space applications.

29

Scanners

William L. Wolfe and Douglas W. Wolfe

29.1. Introduction

Scanners are used to move a light beam. The pattern may be a raster pattern, as with many imaging systems and laser printers, or in many directions, as with bar-code readers. Similar devices are optical addressing systems that cause the beam to be pointed in successive specific directions.

29.2. Significant Characteristics

There are several significant properties of scanners. Their values and calculation vary with type. They include the resolution, the field, scan efficiency, size, speed, reliability, and cost.

29.2.1. Resolution

Resolution is often specified in terms of the number of resolution elements in a scan line. Commercial TV has ~500; facsimile, ~1500; a good photograph, ~5000; a recon photograph, ~10,000; and a large map, ~100,000. If the resolution spot has an angle $\Delta\theta$ and the total field is θ, then the number of resolution elements is $N = \theta/\Delta\theta$. If the beam is diffraction limited then the full angle between the first zeros is for small angles $a\lambda/D$, where a is a function of the shape of the aperture, and the number of resolution elements is

$$N = \frac{\theta D}{a\lambda}.$$

29.2.2. Field

The field is usually specified linearly in degrees or radians. It is determined by the "throw" of the scanner—the angular extent of a mirror, the prism angle, the facet angle, etc.

29.2.3. Scan Efficiency

The efficiency is determined by the type of scanning, number of facets, turn-around time, linearity, and other factors.

29.2.4. Size

Size is determined by the required aperture size and field.

29.2.5. Reliability

Reliability can be measured in terms of the mean time before failure or in terms of long-term stability or equivalent measures. Galvanometers, taut-band scanners, and piezoelectric devices in general have better mean times before failure than the mechanical scanners. The latter vary, based on the size and the drive motors.

29.2.6. Cost

Although important for final engineering designs, cost is application dependent and is not treated here.

29.3. Scan Types and Definitions

Pushbroom—The pushbroom scanner is used in satellites and aircraft that provide the only scan. There are enough detectors to cover the entire width of the scanned swath.

Whiskbroom—The whiskbroom scanner uses an array of detectors aligned in the direction of travel of the aircraft or satellite and oscillates (or rotates) in the transverse direction.

Barrel Roll—Some use this appellation for a whiskbroom that rotates rather than oscillates.

Palmer—This is a translated circular scan. The optics scan a circle while the vehicle moves forward.

Preobjective—The image- or beam-forming element precedes the scanner in the optical train.

Postobjective—The image- or beam-forming element follows the scanner in the optical train.

Raster—This is a two-dimensional scan that usually covers a rectangular field in a series of parallel lines.

Rosette—This is sometimes used in trackers and has a series of loops that pass through the center.

Hypocycloid—This is similar to a rosette, but is generated by an eccentric circular motion.

29.4. Galvanometers

These devices typically scan at *rates* from 1 to 25 kHz over *fields* up to 30° for iron core and 5° for wound armature. Beam *sizes* can be up to 10 cm for iron core and lower frequencies, but a few millimeters for wound armatures and higher speeds. These are analog scanners; the rate and the amplitude can be varied by varying the applied voltage. If they are not operated at their natural frequencies, they require ~0.1 to 2 W. If they are operated resonantly they take 10 times less power, and if they stay within the field, the *efficiency* is $2/\pi$. They can also be used as sawtooth scanners with dead time.

29.5. Taut-Band Scanners

These devices typically scan at fixed *rates* from 5 to 1000 Hz over *fields* up to 30°, with *sizes* up to ~10 cm. A taut band (spring) is driven by an armature. The amplitude may be varied with drive power. They are difficult to phase lock. The *efficiencies* are the same as those of galvanometers.

29.6. Torsion-Rod Scanners

These devices typically scan at fixed rates from 0.5 to 200 Hz over *fields* up to 20°, with beam *sizes* up to ~25 mm. They are small mirrors on the end of a rod that is driven by an armature. The amplitude can be varied by varying the drive power, typically from 10 to 200 mW. They are easily phased locked to other scanners. The *efficiencies* are the same as those of galvanometers.

29.7. Piezoelectric Scanners

These devices scan at *rates* up to 45 kHz over *fields* up to 2°, with beam *sizes* as large as a few centimeters. A bimorph is bent by an electrostatic force. Usually operation is at a high frequency over *fields* of less than 1°.

29.8. Polygons

Polygons can be either reflecting or refracting, can have one, two, or many sides, and can be used either as prismatic polygons or as pyramids. They can be regular or irregular. They can be external or internal (inverted). The line rate is the rotational rate times the number of facets. Rotation rates are limited by the drive, the weight, and the bearings.

The maximum scan *rate,* based on physical integrity, is given by

$$\dot{\omega} = \frac{1}{2\pi r_0}\sqrt{\frac{8\mathrm{UTS}}{\rho(3+\eta)}},$$

where r_0 is the edge–center distance, UTS is the ultimate tensile strength, η is Poisson's ratio, and ρ is the volumetric density.

The system may also be limited by distortion or windage. Windage is determined by several different equations:

$$T = C_M \frac{\rho}{2}\omega^2 r^5,$$

$$T = \frac{A}{2}(\rho^{1-B})(\omega^{2-B})(r^{5-2B})\mu^B,$$

$$T = C_D \frac{\rho}{2} N\omega^2 r_e^3 A,$$

$$C_M = A\,\mathrm{Re}^{-B},$$

where T is the torque that is due to windage, ρ is the volume density, μ is the dynamic viscosity of the fluid in which the polygon resides, ω is the rotational rate (in radians per second), r is a representative dimension, like the tip radius, N is the number of facets, C_D is the drag coefficient, Re is the Reynolds number, r_e is the effective radius (the radius to the center of percussion, but about the mean or the inscribed and circumscribed radii), and A and B are coefficients that depend upon the configuration. Over a reasonable range, A divided by L/r is approximately 2.5 and B is 0.25. However, both A and B can be sensitive functions of the geometry—holes, vanes, bolt heads, etc. Smoother is better, and the higher the Reynolds number, the greater the effects of protrusions.

Distortion may be prevented, or at least minimized, if the polygon has at least 12 facets, is made of aluminum or steel, has a solid radial cross section at least as thick as the facet is wide, and has a central hole no larger than 3/8 the diameter of the polygon inscribed circle.

Size is determined by physical and mechanical considerations, with beams from 1 to 50 cm, typically. The *field* is determined by the number of facets, $2\pi/N$.

The *efficiency* is determined by several factors, including the relative size of the aperture and the polygon facet, their separation, and whether the interior of the scanner can be neglected as a source. Figure 29.1 illustrates this for a four-sided, reflective polygon. The upper part is assumed to be the interior of the scanner and radiates a uniform amount to the optics. In position 1, the quadrigon reflects light from straight down horizontally to the right to the optics. The beam is shown by the solid lines. As it moves to position 2, the polygon reflects light from an angle (from downward and right of vertical) horizontally to the optics, as shown by the dashed lines. The projected facet is no longer as large as the aperture, and some of the beam misses the optics; there is vignetting. This continues as shown by position 3 and the dashed–dotted lines. In this position the optics would obscure part of the beam unless they are moved far to the right. With a little more rotation, the facet would be parallel to the optics and not reflect any field to them. The efficiency is determined by the amount of vignetting that is acceptable. It is determined by the relative placing of the scanner and any stops in front of the optics. Note also that the stops can be used to reduce beam wander across the optics at the expense of aperture size. This applies in general to both reflective and refractive polygons used outward or inward.

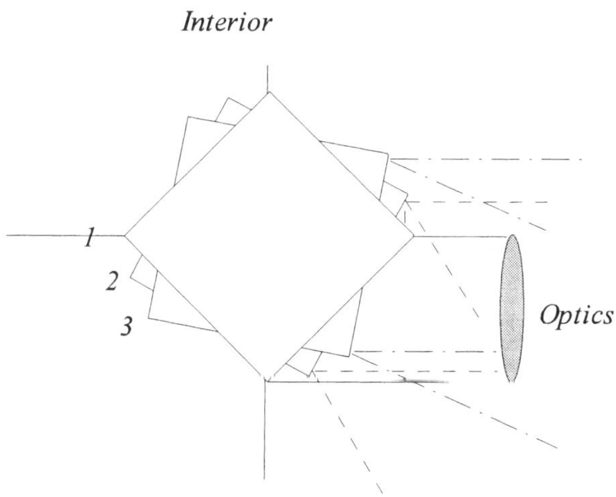

Fig. 29.1. Polygon scan efficiency.

29.8.1. Reflecting Prismatic Polygons

These can have one side, in which case they are simple plane mirrors. Both sides of the substrate may be reflectorized. When the mirror has three sides it is trigonal; other dimensions are pentagons, hexagons, and so on. When the face is underfilled the *efficiency* is

$$\eta = \frac{\tan\left(\frac{180^\circ}{\mathrm{N}}\right)}{\tan\left[\frac{1}{2}\cos^{-1}\left(\frac{r}{R}\right)\right]}.$$

The image does not rotate with prismatic reflecting polygons.

29.8.2. *Reflecting Pyramidal Polygons*

The Egyptian pyramid is the best example of the pyramidal polygon; it has four sides that taper from the bottom to the apex. The number of faces N is set by the required field. If the required field is θ, then N is $360°/\theta$ or $2\pi/N$. An integral number must be chosen, smaller than the calculation. For instance, an 80° field indicates 4.5 facets, and this must be reduced to 90°. For an underfilled facet the efficiency is given by

$$\eta_{sc} = \frac{1}{\pi}\arcsin\left(\frac{1}{\rho - 1}\right),$$

where ρ is the ratio of the facet radius to the beam radius. The image rotates at the same rate at which the pyramidal polygon rotates.

29.8.3. *The Plane-Parallel Plate*

The plane-parallel plate can be used in a converging beam as a scanner. As the plate is rotated through an angle θ, the image moves in the focal plane a distance given by

$$\delta y = t\left[\tan\theta - \sqrt{\left(\frac{n}{n\sin\theta}\right)^2 - 1}\right],$$

where t is the plate thickness, θ is the tip angle, and n is the refractive index of the plate material. The plane-parallel plate may be rotated or oscillated; it is hardly ever rotated.

The field is limited by the large size of the plate required for large angles. It is given by the y position in the focal plane divided by the image distance and the tilt angle:

$$l = \frac{i}{F\sin\theta}\left[1 - \frac{\left(\frac{s}{i} - 1\right)\tan\theta + \frac{1}{F}}{\frac{\tan\theta}{i} - \frac{1}{F}}\right].$$

29.8.4. Refracting Prismatic Polygons

These have the same displacement as the plane-parallel plates. They are, in fact, a set of plane-parallel plates.

29.8.5. Refracting Prisms

The best known of these are the rotating Risley prisms. The patterns that can be generated by them are dictated by the following equations:

$$\delta_x = \delta_1 \cos(\omega_1 t) + \delta_2 \cos(\omega_2 t - \phi),$$
$$\delta_y = \delta_1 \sin(\omega_1 t) + \delta_2 \sin(\omega_2 t - \phi),$$

where δ_x and δ_y represent the angular deviation in the x and the y directions, respectively, δ_1 and δ_2 are the respective prism deviations, the ω's represent the radian frequencies of rotation, t is time, and ϕ is the relative phase of the prisms, the angle between their respective apexes. The deviation by a prism is given by

$$\delta = n \sin \alpha,$$

where n is the refractive index and α is the prism angle.

The *field* is determined by the prism angle and the refractive index and can be up to almost 180°. It may be limited by total internal reflection. The *size* is determined by the material availability and is typically about 30 cm or less. The *efficiency* is that of a resonant scanner.

29.8.6. Axe Blades

These may be thought of as two mirrors attached to each other at an acute angle. They also come in fours, similar to a pyramid.

The *field* is determined by the number of facets: $2\pi/N$.

The *efficiency* is determined by the portion of the field that is covered without double fielding. It is

$$\eta_{sc} = \frac{1}{\pi} \arcsin\left(\frac{r}{R - r}\right).$$

29.8.7. Weiler Wheel

The Weiler wheel is a polygon scanner with many facets, each of which is tilted slightly with respect to its neighbors. Such polygons are also called irregular. A full rotation of the wheel provides a two-dimensional raster without auxiliary scanners. The *size* is determined by the facet size, but the wheel will be larger according to the number of lines in the frame. The frame *rate* is equal to the rotation rate of the wheel. The number of lines N is related to the required *field* Θ by $N = 2\pi/\Theta$.

29.8.8. Nipkow Disk

This is a flat wheel with a spiral of holes that is placed in the image of the fore-optics. The holes generate a raster with one full rotation of the wheel. The frame *rate* is equal to the rotation rate of the disk. A single detector is used, and optics with a demagnification of Θ/α is required, where Θ is the full field and α is the resolution. For a nominal TV raster, this is ~500! The number of lines N is related to the required *field* Θ by $N = 2\pi/\Theta$. The resolution is determined by the angular subtense of the holes at the detector. The *size* is determined by the focal length of the reimaging lens, which determines the size of the holes for a given angular subtense. The resolution is determined by the angular subtense of the detector.

29.9. Pupil Control

Many systems use a scanning flat mirror to generate the slow component of a two-dimensional scan and a polygon to generate the fast (line) scan. This normally generates beam wander. To minimize this, one can offset the scan axis of the flat mirror. The position of the axis can be found by layout techniques as illustrated in Figure 29.2. The lens or other optics is positioned as shown, behind a scanning mirror that pivots about an axis at P. A horizontal beam is directed to the lens when the mirror is at an angle of 45 degrees to the verti-

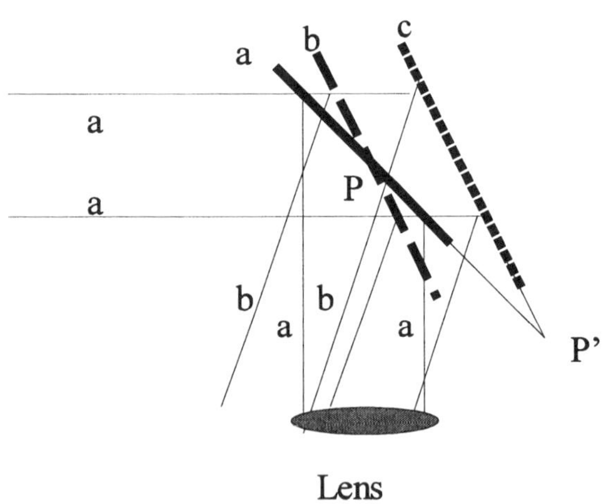

Fig. 29.2. Pupil and beam control.

cal—the center-field position. The rays and the mirror are in the position denoted by *a*. When the mirror is rotated to a different angle, the beam wanders off the lens, as denoted by the mirror position and ray positions *b*. If the mirror is translated to the *c* position, the beam moves onto the lens. This translation can be accomplished by rotating the mirror about the axis denoted by P', which is determined by the extension of the mirrors. Other positions are also available.

29.10. Holographic Scanners

See Chap. 11.

29.11. Electro-Optic and Acousto-Optic Scanners

See Chap. 32.

Further Reading

L. Beiser, "Laser scanning systems," in M. Ross, *Laser Applications,* Academic, 1974, Vol. 2.

G. Marshall, ed., *Optical Scanning,* Marcel Dekker, 1991.

G. Marshall, ed., *Laser Beam Scanning,* Marcel Dekker, 1985.

G. Marshall and L. Beiser, eds., *Beam Deflection and Scanning Techniques,* Proc. SPIE **1454,** 1991.

M. Gottlieb, C. Ireland, and J. M. Levy, *Electro-Optic and Acousto-Optic Scanning and Deflection,* Marcel Dekker, 1983.

W. L. Wolfe and G. Zissis, *The Infrared Handbook,* Society of Photo-Optical Instrumentation Engineers, 1989.

W. L. Wolfe, *Introduction to Infrared System Design,* Society of Photo-Optical Instrumentation Engineers, 1996.

30

Visual Considerations for Visual Displays

Vasudevan Lakshminarayanan and Aparna Raghuram

30.1. Introduction

In this chapter the various visual factors, perceptual and optical, that need to be considered in designing visual displays are described. These factors are accommodation, aging, color, contrast, exit pupil, eye relief distance, field of view, flicker, glare, interpupillary distance, presbyopia, ocularity, resolution, simulator, and motion sickness. These factors, listed in alphabetical order, are discussed in detail in this chapter. For further information, see the cited references.

Figure 30.1 illustrates the main features of the eye. Additional descriptions regarding the optical elements and functional properties of the eye can be found in the books by Grosvenor,[1] Oyster,[2] Hart,[3] and Atchison and Smith.[4]

The structures that optically aid in retinal image formation are the cornea and the lens. The cornea is a transparent layer, which accounts for nearly 42 diopters, D, of the power of the eye. (A diopter is the reciprocal of the focal length in meters). It is the first refractive element that light comes in contact with and has an anterior radius of curvature of ~7.8 mm. The thickness is ~0.5 mm at the center of the cornea and ~0.7 mm at the periphery. After undergoing refraction at the cornea, the light passes through the aqueous humor, which fills the anterior chamber. It is a transparent, clear fluid and does not scatter light to a significant extent. The light next passes through the circular opening in the iris, called the pupil, and then falls on the crystalline lens. The pupil plays the role of an aperture stop in the optical system of the eye. Change in pupil size controls the amount of light entering the eye and can affect the retinal image quality that is due to diffraction and depth of focus. The diameter of the pupil can change because of factors such as illumination, age, accommo-

1 T. P. Grosvenor, *Primary Care Optometry,* Professional Press Books/Fairchild Publications, 1989.

2 C. W. Oyster, *The Human Eye: Structure and Function,* Sinauer, 1999.

3 W. M. Hart Jr., *Adler's Physiology of the Eye: Clinical Application,* 9th ed., Mosby, 1992.

4 D. A. Atchison and G. Smith, *Optics of the Human Eye,* Butterworth and Heinmann, 2000.

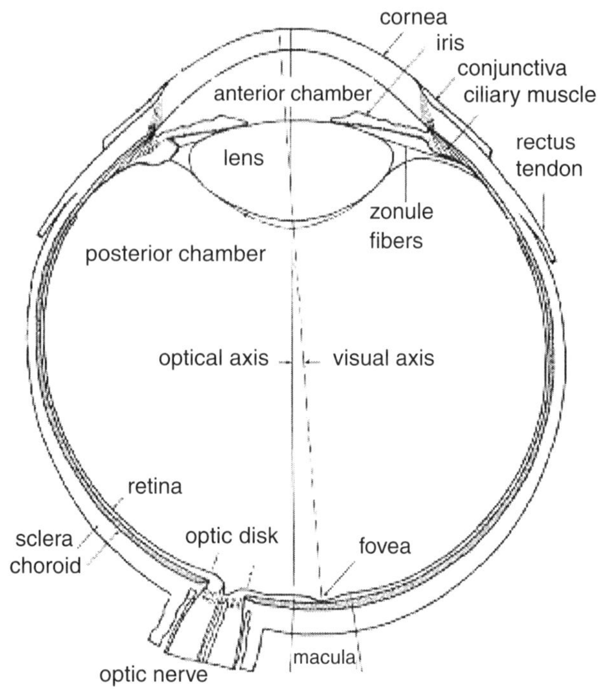

Fig. 30.1. Schematic section of the eye.

dation, and drugs. The crystalline lens is a structure that grows continuously throughout life. New fibers are added on to the center of the lens and the old fibers move peripherally toward the cortex.

Hence the lens does not have a uniform index. The highest refractive index is found in the center and the lowest in the cortex.

The overall power of the eye in the unaccommodated state is ~60 D (approximately 42 D from the cornea and 18 D from the lens). The lens is suspended by fibers, called the zonular fibers, that attach to the ciliary muscle. It is the contraction and the relaxation of the ciliary muscle that allow the lens to change its shape. After undergoing refraction from the lens the light passes through the vitreous humor. This is a transparent gellike structure that is relatively homogeneous in early life but changes in adults. Light is then focused on the retina. The retina contains millions of photoreceptors, rods, and cones. These receptors absorb visible light and convert the absorbed light into nerve impulses that are sent to the brain. The central retina contains the macula, which is a specialized area for perceiving fine details and color. The center of the macula is called the fovea. The fovea contains only cones and is completely devoid of rods. The concentration of cones decreases toward the retinal periphery whereas the concentration of rods increases. The peripheral retina with its higher concentration of rods helps us to see in the dark and aids in motion perception. All nerve impulses after passing through the retina meet at the optic nerve head/optic disk, a region that is devoid of photoreceptors. It is here that the axons of retinal cells collect to form the optic nerve. The region of the optic nerve head corresponds to the blind spot. The optic nerve is made of millions of nerve fibers that collect information from the eye and send it to different parts of the brain.

30.2. Accommodation

This is the process by which the power of the human crystalline lens is changed to maintain clear focus of targets at close distances. As an object comes close to the eye, the ciliary muscle in the eye changes the shape of the lens capsule, thereby increasing its power. This process is called accommodation. This enables adequate bending of the light rays by the lens to form a sharp focus on the retina. If the rays are focused in front of or behind the retina, it causes the image to blur. The brain immediately responds to the blur and signals the ciliary muscle to alter the shape of the lens, so that the ray can be focused on the retina. A good review of the mechanisms of accommodation can be found in the paper by Koretz.[5]

Accommodation has four components to its response. They are discussed in the following four subsections.

30.2.1. Reflex Accommodation

This occurs because of certain changes in the image being viewed. Stimulus blur and chromatic aberrations are considered to be the most crucial factors. The strongest stimulus to reflex accommodation is blur. If the eye is inappropriately accommodated to a particular object, the object appears blurred; the eye then attempts to focus the image by changing the accommodative state of the lens.

Chromatic aberration occurs as the human lens system focuses the different wavelengths of the light at different points. Chromatic aberration of the image also plays an important role in accommodation response.

30.2.2. Tonic Accommodation

This is also called the resting point of accommodation and refers to the position at which the eye is focused when there is no stimulus to accommodation.

[5] J. F. Koretz, "Development and aging of human visual focussing mechanisms," in *Vision Science and Its Applications*, V. Lakshminarayanan, ed., Vol, 35 of OSA Trends in Optics of Photonics Series, Optical Society of America, 2000, p. 246.

It is generally assumed that, in the absence of any visual stimulation, the eye will relax and focus at infinity. However, the eye tends to focus at a finite distance, making the observer short sighted (myopic). This phenomenon is found to occur in darkness, and the resulting state of accommodation is called dark-focus accommodation. The same thing can occur in a bright, unobstructed field, where it is called empty-field myopia. The resting point of accommodation differs among individuals; it usually varies from a distance of 0.5–2 m (0.5–2.0 D) and, on average, the resting point of accommodation is approximately 0.8 m (30.5 in.).[6]

30.2.3. Convergence Accommodation

When an observer changes from one fixation point to another in addition to accommodation there is also change in the *convergence* (see Subsection 30.2.5). For every dioptric change in accommodation there is a corresponding change in the convergence and vice versa.

30.2.4. Proximal and Psychic Accommodation

This refers to change in the accommodative state of the eye that is due to effects of the observer's knowledge of the proximity of the target.

30.2.5. Convergence

Convergence response has four components:

1. Proximal convergence: Convergence guided because of awareness of the position of the object.
2. Tonic convergence: The resting state of convergence. On average, the resting point of vergence is found to be ~1.14 m (45 in.).
3. Accommodative convergence
4. Fusional or reflex convergence: Aligns corresponding points on the retina.

As the observer's gaze moves among natural objects at different distances, the lens in the eye changes focus to obtain a clear image on the retina. The eyes converge on that target to facilitate fusion of the two images on the retina into a single percept. The distance of the target determines the demand on the focusing and accommodative capabilities of the eye. The distance of the target

[6] H. Kruger, "Visual function in office—including VDUs: Introductory paper," in *Ergonomics and Health in Modern Offices*, Taylor & Francis, 1985.

from the eye gives the accommodative demand, stated in meters. A target at a distance of 1 m has an accommodative demand of 1 D.

Convergence demand is determined in prism diopters, and it is dependent on both distance of the target and the interpupillary distance, IPD. Convergence demand equals accommodative demand times interpupillary distance, i.e., $D \times \text{IPD}$ (in centimeters).

30.2.6. Effects of Accommodative and Vergence Factors on Accommodative Response in Visual Displays

Reflex accommodation: The images on visual display units and head-mounted displays are not as sharp as natural objects. This can be due to the optical setup, the lens design, or the poor resolution of the display units. The resulting blur can cause a difference in the accommodative state of the eye when compared with that of real objects.

Tonic and proximal accommodation: In some displays, especially those that are monocular, a condition called instrument myopia may occur. Instrument myopia refers to improper accommodation of the eyes. It arises from many factors:

1. Proximal accommodation. This is a condition in which the observer tends to overaccommodate because of the awareness of the nearness of the target. This causes the observer to be near sighted and results in transient myopia. It disappears once the display use is completed. It does tend to cause accommodative spasm that might have adverse effects on children. Many virtual reality games utilize head-mounted displays to project the image.[7] It has been found that children who use these devices for more than two hours have a myopic shift. This might be critical in the biological development process because the critical period is up until the child is 9 years old. Hence these devices have restricted use among children. Long-term use of the display can also lead to a myopic refractive error.
2. The second effect of instrument myopia relates to resolution and image clarity. If the display does not provide adequate blur cues for the correct level of accommodation, then the accommodative state of the eye lapses toward the resting point of accommodation. For example, in displays in which the requirement is to make the individual pixel less visible, a filter is placed in front of the display. This reduces the resolution

[7] S. K. Rushton and P. M. Riddell, "Developing visual systems and exposure to virtual reality and stereo displays: Some concerns and speculation about the demands on accommodation and convergence," Appl. Ergonom. **30,** 69, 1999.

of the display, causing the image to blur. Although the accommodative status of the eye might be ideal for the distance, the eye tends to focus more for the induced blur.

Kotulak and Morse[8] found the resolution ability of the eye to be optimal at the resting focus of the eye. The average distance was found to be ~50 cm. Displays with image distances close to the range of the tonic accommodation seemed to be the best focus for the display. Jaschinski-Kruza[9] measured eyestrain for monitor distances of 50–100 cm and found that the more distal the dark vergence of the individual, the greater was the eyestrain for a distance of 50 cm. Therefore it has been suggested that the focal distance of the head-mounted display should be set to between approximately 0.60 and 1 m to obtain best resolution results.

Studies have shown that overaccommodation occurs when the exit pupil is small.[10] With natural pupils, observers noticed the blur resulting from improper accommodation and its correction. Displays with small exit pupils and therefore large depths of focus have problems with instrument myopia. Using devices with small depths of field (large exit pupils) provides visual feedback of the blur because of improper accommodation and may be the best method for controlling instrument myopia.

30.2.7. Mandelbaum Effect

Observers who try to read a distant sign through a mesh screen tend to focus on the screen rather than on the sign. This is called the Mandelbaum effect. Owens[11] has shown that the Mandelbaum effect occurs when the interposing screen is at or near the resting point of accommodation of the eye. This indicated that it is the tendency of the accommodative mechanism to revert to the resting point of accommodation, thereby rendering the eye temporarily myopic. This kind of effect was found to affect piloting of aircraft. When the cockpit window is dirty and the resting point of accommodation is near the position of the window, visual performance is degraded.

In an optical system this effect can be avoided if collimation is used, i.e., the image is viewed at optical infinity. The other interfering factor, even with

[8] J. C. Kotulak and S. E. Morse, "Relationship among accommodation, focus, and resolution with optical instruments," J. Opt. Soc. Am. A **11,** 71, 1994.

[9] W. Jaschinski-Kruza, "Eyestrain in VDU users: Viewing distance and the resting position of ocular muscles," Hum. Factors **33,** 69–83, 1991.

[10] R. Jones, "Proximal accommodation in virtual displays," Soc. Inf. Display Dig. 195–197, 1993.

[11] D. A. Owens, "The Mandelbaum effect: Evidence for accommodation bias towards intermediate viewing distance," J. Opt. Soc. Am. **69,** 649, 1979.

collimated images, is the psychic awareness of the proximity of the displayed image.

When a collimated image is viewed, it is assumed that the eye focuses at optical infinity and hence requires zero vergence. Studies conducted by Hull et al.[12] show that this is not necessarily the case. The tendency of the eye to return to its resting point of accommodation is opposed by the acuity demand of the visual task. The degree of accommodation is determined by the spatial frequencies that must be resolved to perform the task as well as spatial frequencies of visible textural gradients surrounding task-related objects. If either the foveal or the surrounding texture is obscured by reduced illumination, contrast, and/or optical defocusing, stimulus adequacy is degraded and focus tends toward the resting point of accommodation.

Adding pictorial cues to the display will provide a rich textural content and can make distance and size judgments more accurate. The better the display simulates the real world, the more accurate will be the distance and size judgments.[13,14]

Jones[10] has shown that there is greater accommodation for a near target when viewed through small artificial pupils (3 and 0.5 mm), but no significant accommodative change was found for normal pupils. Accommodative accuracy for targets in the virtual image was not affected by a mismatch between the perceived target and the optical distance unless the exit pupil was very small. A larger exit pupil can augment possible deterioration effects that are due to proximal accommodation in a display. Proximal accommodation was smaller with binocular viewing than with monocular viewing.

30.2.8. Accuracy of Accommodation and Convergence Response

Accommodation and convergence are synergistically interconnected by nerves in natural vision. This should be duplicated in the display. If the accommodative demand is not matched with the convergence demand, the situation is equivalent to having a prism in front of the eyes in the real world (Fig. 30.2). If such a condition prevails and if the misalignment is large enough, vision may appear to be blurred or even doubled. For small misalignment, though, the image remains clear and single; longer use of the display results in severe eyestrain.

[12] J. C. Hull, R. T. Gill, and S. N. Roscoe, "Locus of the stimulus to visual accommodation where in the world or where in the eye?" Hum. Factors **30,** 689, 1982.

[13] J. W. Meehan and T. J. Triggs, "Apparent size and distance in an imaging display," Hum. Factors **34,** 303, 1992.

[14] D. Drasic and P. Milgram, "Perceptual issues in augmented reality," in *Stereoscopic Displays and Virtual Reality Systems III,* S.S. Fisher, J.O. Merritt, and M.T. Bolas, eds., Proc. SPIE **2653,** 126, 1996.

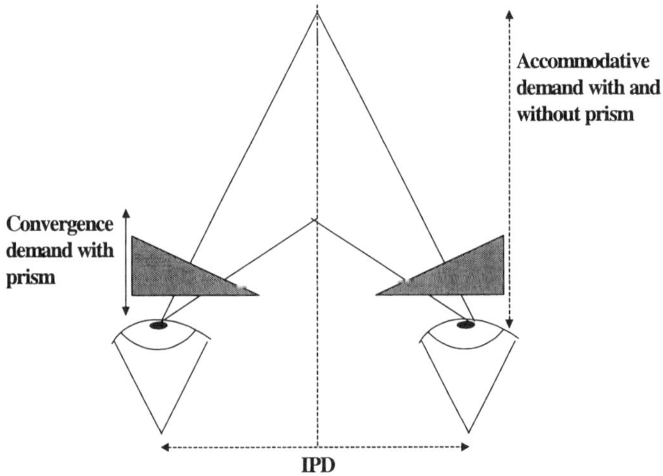

Fig. 30.2. Decoupling of convergence and accommodation demands the use of prisms, similar to the effect caused by a stereo display. IPD, interpupillary distance.

In biocular displays a satisfactory compromise can be achieved for matching accommodative and convergence demands. In displays with stereo systems it is much more complicated. In a binocular stereo display, such as a head-mounted display, the images are always displaced laterally on a screen at a fixed distance. The accommodative demand is kept constant to the distance of the screen, but convergence should change toward the simulated distance of the target either in front of or behind the screen. This results in the misalignment between convergence and accommodation.

Although initial use of a stereo device causes considerable eyestrain, longer duration or a longer period of use may allow adaptation and an increase in the fusional reserves of the observer (similar to the principle of visual training).

Although a considerable amount of discomfort is caused because of the decoupling, much stronger evidence and research are required for reaching firm conclusions.

Studies have been conducted by Kajiki et al.,[15] Omura et al.,[16] and Shiwa et al.[17] on designing a head-mounted display that provides a hardware system

[15] Y. Kajiki, H. Yoshikawa, and T. Honda, "Ocular accommodation by super multi-view stereogram and 45-view stereoscopic display," in *Proceedings of the Third International Display Workshop,* Society of Information Display, 1996, pp. 302–303.

[16] K. Omura, S. Shiwa, and F. Kishini, "3-D display with accommodative compensation (3DDAC) employing real time gaze detection," in *SID International Symposium, Digest of Technical Papers,* Society of Information Display, 1996, p. 889.

[17] S. Shiwa, K. Omura, and T. Miyasato, "Development of three-dimensional display with accommodative compensation," in *Proceedings of the Third International Display Workshop,* Society of Information Display, 1996, pp. 497–500.

that more closely simulates the real-world situation of coupling convergence and accommodation demands. Their approaches were based on monitoring eye movements to determine which target the user is fixating on at any particular time. The stereoscopic depth of the target was known. A mechanical optical system was used to modify the distance of the virtual display so as to match the accommodative demand with the convergence demand.

30.2.9. Accommodation Effects on See-Through Devices

See-through devices are ones in which the virtual image is superimposed on the real world, such as head-up displays. As a result, real objects will appear at varying distances whereas the display images will be at a fixed distance. As there is a difference in the accommodative states, one of them tends to be blurred.

The distances at which the *virtual* images appear affect the perception (detection and recognition) of the *real* objects.[18] The closer the apparent distance of the virtual image, the greater the difficulty in perceiving and detecting real objects.

Therefore the rules are as follows:

1. Make judicious use of the depth of focus,[19] perhaps by placing the virtual image one depth of focus away from the real object.
2. Use pinhole optics to increase the depth of focus. However, this reduces the light reaching the retina. Maxwellian view[20] has to be used to increase brightness.

The other change caused in displays when focus shifts from a far point inward is the reduction in the size of the image, which causes accommodative micropsia (optical effect by which an object appears smaller than the real size). A reduction in the apparent size of an object at a given retinal angle causes objects to appear smaller and farther away. The micropsia induced by these virtual images causes poor judgment of the apparent distance and size. Accommodative micropsia can be due to accommodative convergence.[21] Errors in perceived distances can occur when vergence distance is not matched to the image size

[18] J. Norman and S. Ehrlich, "Visual accommodation and virtual image displays: Target detection and recognition," Hum. Factors **28,** 135, 1996.

[19] C. Schor, "Simultaneous focus mechanism for head mounted displays," Opt. Vis. Sci. (suppl.) **72,** 169, 1995.

[20] G. Westheimer, "The Maxwellian view," Vis. Res. **6,** 669, 1966.

[21] G. Smith, J. W. Meehan, and R. H. Day, "The effect of accommodation and retinal image size," Hum. Factors **34,** 289, 1992.

of a perceptual task. These errors can result in perceptual image expansion (macropsia) or constriction (micropsia), particularly in binocular displays that stimulate a range of convergence values. These effects can be minimized when the range of binocular disparity is limited to a few degrees for near objects and biocular displays are used when views of distant objects are presented.

30.3. Aging

Light is attenuated by absorption, scatter, and reflection in the ocular media. The attenuation increases with age. Thus the effects of normal changes that are due to aging should be considered in the design of a display. The following features need to be taken into account:

1. The pupil constricts with age; this is called senile miosis. Both screen and room illuminances need to be higher for older people.
2. The ocular media become more attenuating with age, and the lens gets thicker. The attenuation in a normal 60-year-old person is approximately three times that of a normal 20-year-old person.[22] With age there is accumulation of insoluble proteins that are the by-products of oxidative mechanisms, thereby causing yellowing of the lens. UV rays are also found to play a major role in the yellowing of the lens. Said and Weale[23] attributed yellowing of the lens partly to pigment absorption and partly to scattering. This results in difficulty in perceiving short-wavelength light, which causes a shift in the appearance of white to yellow. Difficulty in discriminating pastel violets and yellow–green also occurs. Thus color-matching and color-discrimination abilities are also found to decrease with age because of the increase in ocular media absorption. Some 40% or more of the age-related loss in sensitivity to light of short wavelengths is due to light loss in the ocular media and the rest to changes in the receptors.[24,25]
3. Reduced retinal illuminance can be attributed to senile miosis, yellowing of the lens, or an increase in lens scatter.
4 Sensitivity to glare increases with age, and this can be attributed to increased lenticular light scatter.

[22] R. A. Weale, "Aging and vision," Vis. Res. **26,** 1507–1512, 1986.

[23] F. S. Said and R. A. Weale, "The variation with age of the spectral transmissivity of the living human crystalline lens," Gerontologica **3,** 213, 1959.

[24] J. S. Werner and V. G. Steele, "Sensitivity of human foveal color mechanisms throughout the life span," J. Opt. Soc. Am. A **5,** 2122, 1988.

[25] C. A. Johnson, A. J. Adams, J. D. Twelker, and J. M. Quigg, "Age-related changes of the central visual field for short wavelength sensitive pathways," J. Opt. Soc. Am. A **5,** 2131, 1988.

5. Contrast sensitivity also decreases with age. Sensitivity losses occur predominantly in the intermediate and the high spatial frequencies. This reduces the peak of the contrast sensitivity function toward lower frequencies. This reduction in contrast is due to a decrease in the retinal illuminance produced by senile miosis. Absorption of short-wavelength light results in fluorescence of the lens, which reduces contrast. The other factors that are considered for reduction in the contrast sensitivity function with age are increased light scatter, which is due to cataractous changes in the lens, and aging of the neural elements.
6. A usable functional field of view is found to decrease with age.[26] Acuity for moving targets also decreases with age.[27]
7. The human eye is considered to be able to adjust to wide differences in light intensity. This process is known as adaptation. Dark adaptation occurs when the eye is exposed from a light to a dark environment. There is a significant decrease in the adaptation abilities with age.[28]
8. There is a decline in critical flicker frequency threshold with age.[29,30] This decrease in critical flicker frequency, to a certain extent, is attributed to the reduction in retinal illuminance.
9. With increase in age there is a gradual shift from "with-the-rule astigmatism" (in which the vertical corneal meridian is steeper than the horizontal) to "against-the-rule astigmatism." Lyle[31] showed that the prevalence of against-the-rule astigmatism increased from 6% at ages 41–50 to 27% at age 61. There is also a slight increase in hyperopic refractive error with age.

These factors need to be considered when a display for the elderly is designed. It would be advisable to increase the intensity of the screen. Creating strong contrast by use of brightness or color differences can enhance visibility of the objects. Adequate measures should be adopted to avoid effects of glare. Color displays should use colors that do not hinder perception or performance. Adequate ambient light should be provided.

[26] R. Sekuler and K. Ball, "Visual localization with age and practice," J. Opt. Soc. Am. A **3,** 864, 1986.

[27] A. Burg, "Visual acuity as measured by dynamic and static tests: A comprehensive evaluation," J. Appl. Psychol. **50,** 460, 1966.

[28] R. A. McFarland, B. Warren, and C. Karis, "Alterations in critical flicker frequency as a function of age and light: Dark ratio," J. Exp. Psychol. **56,** 529, 1958.

[29] R. A. Weale, "On the eye," in A. T. Welford and J. E. Brien, eds., *Behavior, Aging and the Nervous System; Biological Determinants of Speed of Behavior and Its Changes with Age,* American Lecture Series, Thomas, 1965.

[30] A. A. Rosenbloom Jr. and M. W. Morgan, *Vision and Aging,* Butterworth-Heinemann, 1993.

[31] W. M. Lyle, "Changes in corneal astigmatism with age," Am. J. Optom. Arch. Am. Acad. Optom. **48,** 467, 1971.

30.4. Color

Color is usually described by the terms hue, brightness, and saturation. Good expositions of color vision can be found in the books by Kaiser and Boynton[32] or Wyszecki and Stiles.[33]

30.4.1. Definitions

Hue is an attribute of visual sensation according to which an area appears to be similar to one of the perceived colors, red, yellow–green, or blue (i.e., the primary hues) or to a combination of the two (i.e., the intermediate hues). It is determined by the dominant wavelength.

Saturation is an attribute of visual sensation that refers to the degree to which a given monochromatic stimulus differs from achromatic light. It describes the amount of color in a stimulus as compared with the amount of white, black, or gray. Saturation basically represents purity of the color. Hue and saturation together make the chromaticity of a given color.

Brightness is an attribute of a visual sensation according to which an area appears to exhibit more or less light.

Colorfulness is an attribute of the visual system according to which an area appears to exhibit more or less hue.

Chroma is defined as the colorfulness of objects in proportion to the brightness of a similarly illuminated area that appears to be white or highly transmitting.

Intensity is the flux density of light; it is a physical quantity, whereas brightness is determined by the perception of color and hence is physiological. Intensity has other meanings (see Chap. 22).

Illuminance is another term for intensity, the flux density of light.

Luminance is the flux density per unit solid angle.

Color constancy is a psychophysical or perceptual phenomenon that accounts for the ability of humans to perceive nearly the same color under different levels of illumination. The spectral content of the sunlight varies continuously throughout the day, yet the colors that we perceive remain the same.

CIE[34]: The Commission Internationale de l'Eclairage (The International Commission on Illumination) is the primary international organization concerned with color, color measurement, and other aspects of light.

The CIE chromaticity diagram is an international standard for primary colors. It allows all other colors to be defined as a weighted sum of three primary

[32] P. K. Kaiser and R. M. Boynton, *Human Color Vision,* 2nd ed., Optical Society of America, 1996.

[33] G. Wyszecki and W. S. Stiles, *Color Science: Concepts and Methods, Quantitative Data and Formulae,* Wiley, 1992.

[34] *Colorimetry,* CIE Pub. 15.2, 2nd ed., 1986.

colors. In reality, no three colors can be combined to give all possible colors. Therefore the standard colors established by CIE do not correspond to real colors. These primary colors, X,Y, and Z, are the virtual colors corresponding to red, green, and blue. For a given real color, its components with respect to the primaries are given by

$$x = \mathrm{X}/(\mathrm{X} + \mathrm{Y} + \mathrm{Z}),$$
$$y = \mathrm{Y}/(\mathrm{X} + \mathrm{Y} + \mathrm{Z}),$$
$$z = \mathrm{Z}/(\mathrm{X} + \mathrm{Y} + \mathrm{Z}),$$

where $x + y + z = 1$. Therefore, if x and y are known, z can be determined.

The CIE chromaticity diagram (Fig. 30.3) is a two-dimensional graph of the chromaticity coordinates (x as the abscissa and y as the ordinate), which shows the spectrum locus. The chromatic coordinates of monochromatic light fall between 380 and 770 nm.

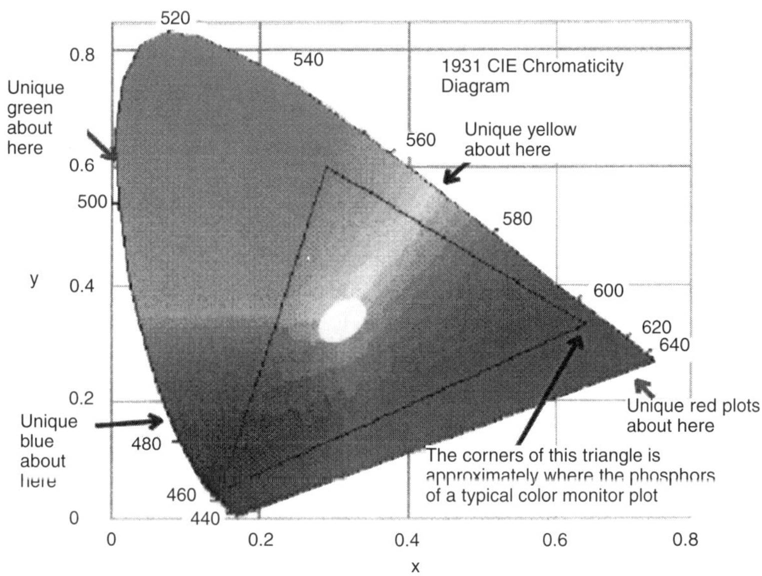

Fig. 30.3. CIE chromaticity diagram.

Color space is a three-dimensional solid enclosing all possible colors. These dimensions may be described in various geometries, giving rise to various spacing within the solid.

The RGB color space or color model is a model that consists of three independent image planes, one in each of the primary colors, red, green, and blue (Fig. 30.4). A particular color can be specified by the amount of each of the three primaries needed to form the color. The model is built on a cube with Cartesian coordinates. Each dimension of the cube represents a primary color.

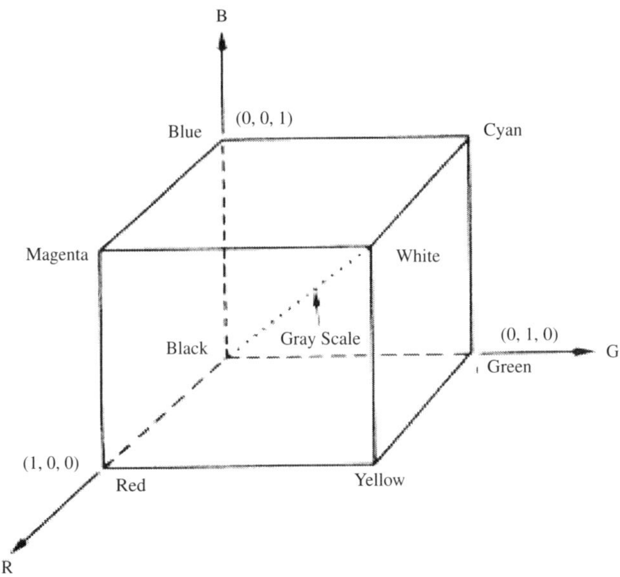

Fig. 30.4. RGB color model.

Each point within the cube represents a particular hue. The coordinates of the point give the contribution of each primary toward the color.

*CIELAB (CIE L*a*b*)* is the color space in which values L^*, a^*, and b^* are plotted at right angles to one another to form a three-dimensional coordinate system. Value L^* represents lightness, value a^* represents the redness–greenness axis, and value b^* represents the yellowness–blueness axis. This color space is popular for use in measuring reflective and transmissive objects and in representing subtractive systems.

*CIELUV (CIE L*u*v*)* is the color space in which values L^*, u^*, and v^* are plotted at right angles to one another to form a three-dimensional coordinate system. Value L^* represents lightness, value u^* represents the redness–greenness axis, and value v^* represents the yellowness–blueness axis. This color space is a popular color space for use in measuring emissive objects, e.g., CRT displays. They represent the additive color system.

The color that is seen on a computer display is obtained by the addition of the light of the wavelengths that are emitted by the three-phosphor dots. They are based on the additive theory model with RGB color space.

30.4.2. Effective Use of Color in Displays

1. Colors need to be used efficiently. They add realism to the display. The overuse of colors tends to make tasks more complex, increase er-

rors, and reduce user productivity. It is generally said that colors are effective maximally when used minimally.[35]

2. Background colors affect the effectiveness of color highlights. One can understand the effect of surround on color by being aware of simultaneous contrast. The appearance of color is dependent not only on inherent properties such as hue and intensity but also on the color of the surrounding regions.
3. Simultaneous contrast can also affect the colorfulness of the image. Colorfulness increases against a low-colored background and decreases against a high-colored surround. For instance, greens appear more yellow on blue and more blue on yellow.
4. To avoid adverse effects of simultaneous contrast it is advised that a neutral surround like midgray be used. In a display both foreground and background should not be colored. Either colored text on a neutral background or black or white text on a colored background needs to be used.
5. Ambient light and reflected light can also affect the perceived color. Colors tend to appear different in different lighting environments. Hence colors that are distinct enough to be visible under a wide range of lighting conditions must be used.
6. If colors used are of the same intensity then the color in the middle of the visible spectrum is perceived as brighter than the others. Therefore colors like yellow would appear brighter. Using opponent colors such as blue on yellow and red on green would give effective results.
7. Saturated colors should be used restrictedly, as they tend to produce visual fatigue. Because of chromatic aberration, different wavelengths are imaged at different depths behind the lens. The lens therefore has to adjust its focus for the different wavelengths. Saturated colors are found to cause more refocusing.[36] This constant refocusing tends to fatigue the muscles controlling accommodation.
8. Saturated colors can also produce a sensation of false depth. The eye normally focuses on yellow–green colors. Red tends to fall beyond the retina, and blue converges in front of the retina. This frequently causes red and blue to appear in different planes. This effect is called *chromostereopsis.* Saturated colors that are at the same distance are perceived to be at different distances. For instance, when viewed against a black background, red appears to advance and blue appears to recede. This is noticed in ~60% of the population. In these cases

[35] L. J. Najjar, "Using color effectively (or peacocks cannot fly)," IBM, 1990.

[36] G. M. Murch, "Physiological principles for the effective use of color," IEEE Comput. Graph. Appl. **4**(11), 49, 1984.

blue can be used effectively as a background. Some people perceive the reverse; in such cases red would be used as a preferred background. Unintended depth effects can occur if saturated red and blue are placed next to each other.

9. It is better to avoid using blue for text. Saturated blue should never used for depicting data with high resolution. The reasons are threefold. First, because the center of the fovea is devoid of blue cones, the sharpness of the letters in blue is reduced[37]; the lens absorbs twice as much in the blue region of the spectrum compared with yellow or red, and this tends to increase with age. Second, in the relaxed state the eye focuses best on green light. Saturated blue falls in front of the retina and the lens might not be able to change its shape to focus on the blue accurately. This causes multicolored text to appear blurred. Third, according to the opponent-color theory (red–green and blue–yellow), brightness information is transmitted by use of inputs from the red and green cone photoreceptors. The blue photoreceptors are not involved in transmitting brightness information. This implies that the perception of brightness of blue objects is reduced and that blue text objects will tend to appear blurred. Boundaries and fine spatial details should be depicted by colors that activate the red–green opponent process or the achromatic system.[38,39]
10. For dark-on-light polarity displays with high-luminance contrast, any text color is satisfactory if the background is chosen appropriately.[40] Light-on-dark polarity combinations involving saturated colors tend to be unsatisfactory, and less saturated combinations are generally preferred. Cool or short-wavelength background colors like cyan and blue may be preferable for displays with light text and dark background. Unsaturated color combinations tend to yield the best results. However, for a dark-on-light polarity, saturation of the text or foreground color affects performance and ratings only marginally. Red and green should not be used for background, and blue is preferred because the retinal periphery is relatively insensitive to reds and greens.
11. Colors should not be alternated too quickly, as they will be perceived as a fused, single color. Changing an object quickly between red and green, e.g., will cause it to be perceived as yellow.

[37] D. R. Williams, D. I. MacLeod, and M. M. Hayhoe, "Foveal tritanopia," Vis. Res. **21,** 1341, 1981.

[38] R. M. Boynton, *Human Color Vision,* Holt, Rinehart & Winston, 1979.

[39] L. Thorell, "Using color on displays: A biological perspective," in Vol. 2 of *SID Seminar Lecture Notes,* Society for Information Display, 1983.

[40] S. Pastoor, "Legibility and subjective preference of color combination in text," Hum. Factors **32,** 157, 1995.

12. Color vision tends to change with age. It is therefore important to use colors that are not affected by age.[41]

30.4.3. Color Blindness

Color blindness can be classified into different types:

1. Subjects with complete color blindness: A rare condition in which the subject does not perceive color at all. They are monochromatic or achromatic to light.
2. Subjects with partial color blindness come under the categories called *dichromatic* and *anomalous trichromat.* Dichromats are subjects who miss one of the color-sensitive photopigments in retinal cones. Anomalous trichromats are subjects who have all three-cone pigments but one or more are found to be abnormal.

30.4.4. Incidence of Color Blindness

Approximately 8% of men and fewer than 1% of women have some form of color blindness. Red–green forms of color blindness are found to be the most common, each affecting ~1% of men and 0.02% of women. Blue–yellow color blindness is quite rare and occurs in only 0.02% of men and 0.01% of women.

If a display is to be used by those with color blindness, it is essential to use color that effectively conveys information to them. This can be achieved if the color is varied in terms of intensity and saturation and not in terms of hue. If colors in the display vary only with respect to hue it is difficult for a color-blind subject to distinguish them. Single-color distinction should be avoided. Color contrast can be achieved by varying both chromaticity and luminance. Varying the luminance alone can benefit the color-blind subject. Kovacs et al.[42] designed a filter that can be used along with a CRT display and can compensate for the color deficiency. To utilize this technique, a thorough knowledge of the exact shape of the response function of abnormal and anomalous cones is required.

[41] J. M. Enoch, J. S. Werner, G. Haegerstrom-Portnoy, V. Lakshminarayanan, and M. Rynders, "Forever young: Visual functions not affected or minimally affected by aging: A review," J. Gerontol. A **54,** B336–B351, 1999.

[42] G. Kovacs, G. Abraham, I. Kucsera, and K. Wenzel, "Improving color vision for color deficient patients on video displays," in *Vision Science and its Applications,* V. Lakshminarayanan, ed., Vol. 35 of OSA Trends in Optics and Photonics Series, Optical Society of America, 2000.

30.4.5. Accommodative and Convergence Demands for Multichromatic Displays

Differences in accommodation and convergence response are expected for images of various colors on a display on the basis of chromatic aberration.[43] To bring targets of varying colors into focus, there is an accompanying change in accommodation and convergence according to the wavelength of the target.

Bedford and Wyszecki[44] calculated the refractive state of the eye for different monochromatic lights. They found that a 644-nm red target located at a distance of 50 cm (2.0 D) required a refractive state of 2.30 D whereas a blue target (450 nm) required a refractive state of 1.00 D. A green of 545 nm required a refractive state of 1.80 D. Therefore simultaneous viewing of equidistant red and blue monochromatic light requires refocusing of the lens and reconvergence of the two eyes each time the point is moved from one color to the other. Although the colors produced on a display are not monochromatic, they are highly saturated and are subject to wavelength-dependent chromatic aberration. Shorter wavelengths are overaccommodated, and longer wavelengths are underaccommodated.[45] Single-gun colors show the effect of chromatic aberration on accommodative and convergence responses. Red is focused behind the retina, requiring an increase in accommodative response in order to bring the image to focus on the retina, whereas blue requires a reduction in the accommodative response. Similarly, the convergence demand is more for red and less for blue. Accommodative distance for multiple-gun colors seems to vary at ~2.0 D, indicating that the wave band of the color mix influences accommodation. Hence broadband or unsaturated colors are affected less by chromatic aberration. The absolute difference in the accommodative shift required for focusing on displayed colored targets is much less than that observed for monochromatic lights. Hence multichromatic displays are not subject to a great change in the accommodative and convergence response.

30.4.6. Flicker

Flicker perception depends on stimulus size, luminance, retinal location, and temporal modulation. Chromaticity has little or no effect on critical flicker frequency if the luminance is held constant.[46]

[43] P. A. Howarth and A. Bradley, "The longitudinal chromatic aberration of the human eye, and its correction," Vis. Res. **26,** 361, 1986.

[44] R. E. Bedford and G. Wyszecki, "Axial chromatic aberration of the human eye," J. Opt. Soc. Am. **47,** 564, 1947.

[45] G. M. Murch, "Visual accommodation and convergence to multichromatic visual display terminals," Proc. Soc. Inf. Disp. **24,** 67, 1983.

[46] H. Delange, "Research into the dynamic nature of the fovea–cortex systems with intermittent and modulated light. II. Phase shift in brightness and delay in color perception," J. Opt. Soc. Am. **48,** 784, 1958.

30.5. Contrast

This is a measure of how much darker or brighter an object is compared with the mean luminance of the background. There are several definitions of contrast. The simplest is

$$C_1 = L_c/L_b,$$

where L_c is the character luminance and L_b is the average background luminance. The optimum is 1:6 to 1:8.

In visual display monitors the contrast of the characters does not remain constant. The luminance changes with the quality of the phosphor, the number of electrons striking the phosphor, and the reflection of the ambient light by the screen. Hence contrast increases with brightness of the display and decreases with increase in the ambient illumination:

$$C_1 = L_c/(L_b + L_d + L_s),$$

where L_c is the character luminance, L_b is the screen background luminance, L_d is the diffuse-reflection luminance, and L_s is the specular-reflection luminance.

Contrast is also defined as

$$C_2 = (L_{\text{object}} - L_{\text{background}})/L_{\text{background}}.$$

This is sometimes called the aperiodic form.

The luminance of the object can be less than that of the background. Contrast measured in the above-mentioned way can take either a positive or a negative value. This might be used to indicate the polarity of the object with respect to the background.

Another method used to determine contrast is called the Michaelson contrast (periodic form) and is commonly used.

$$C_3 = (L_{\max} - L_{\min})/(L_{\max} + L_{\min}).$$

In this case the contrast varies between 0 to 1.

The optimal level of contrast for displays such as a head-mounted display depends on the type of the display, i.e., whether it is an immersive display or a see-through display. In the latter case the virtual image is superimposed on the real-world scene. The real world provides a background over which the virtual image is observed. At nighttime the background luminance provided by the real environment can be quite low, thereby increasing the contrast and visibility of the symbols if the characters are bright. If the symbols are dark this would reduce the overall contrast. The same display on a bright sunny day causes a decrease in contrast and reduces the visibility of the image displayed. Maintaining the overall luminance of the background at a constant level and changing the luminance of the viewed display image accordingly can offset this problem. Using a filter or a visor around the head-mounted display can also prove to be effective.

For immersive displays a contrast value of 9 to 24 is generally desired. In such a display the luminance of the real environment does not affect the overall contrast of the display. If bright characters are used in a dark background then the maximum contrast should be maintained at 1:12 or 1:18.[47] At much higher contrast a considerable amount of glare that would be due to diffuse or specular reflection would result. There is no perceptual or comfort limit on contrast in the presentation of dark symbols on a bright background. The eye adapts to the bright background. The background luminance should be maintained at a level that keeps reflected screen light to an acceptable level.

30.6. Contrast Sensitivity Function and Modulation Transfer Function

The *contrast sensitivity function* is a subjective measurement of a person's ability to see targets of varying spatial frequencies as functions of contrast.

Figure 30.5 shows the normal human contrast sensitivity function.[48] The eye is most sensitive to the intermediate spatial frequencies, whereas sensitivity drops at low and high spatial frequencies. At high luminance (photopic vision) the human contrast sensitivity function peaks at ~5–6 cycles/deg (0.286–0.344 cycles/mrad) and the high-frequency cutoff is ~50–60 cycles/degree (2.86–3.44 cycles/mrad). With a decrease in luminance, the peak of the sensitivity function shifts toward smaller spatial frequencies.

The contrast sensitivity function describes the visual performance by taking into account both the optical image formation and neural processing of the eye. Lakshminarayanan et al.[49] have deduced a convenient analytical function to describe the human function. The function has three parameters: w, which influences the low-frequency behavior of the function, bl, which is the portion of the contrast sensitivity peak, and c, which is the high-frequency behavior of the function. The expression is

$$\mathrm{CS}(\xi) = a\xi \exp(-b\xi)[1 + c \exp(b\xi)]^{1/2},$$

[47] S. Kokoschha and P. Haubner, "Luminance ratios at visual display workstation and visual performance," Light. Res. Technol. **17**(3), 138, 1984.

[48] S. H. Schwartz, *Visual Perception: A Clinical Orientation,* Appleton & Lange, 1994.

[49] V. Lakshminarayanan, A. J. Lang, and V. Portney, "The 'expected visual outcome' (EVO) model: Methodology and clinical validation," Vis. Res. **72,** 511, 1995.

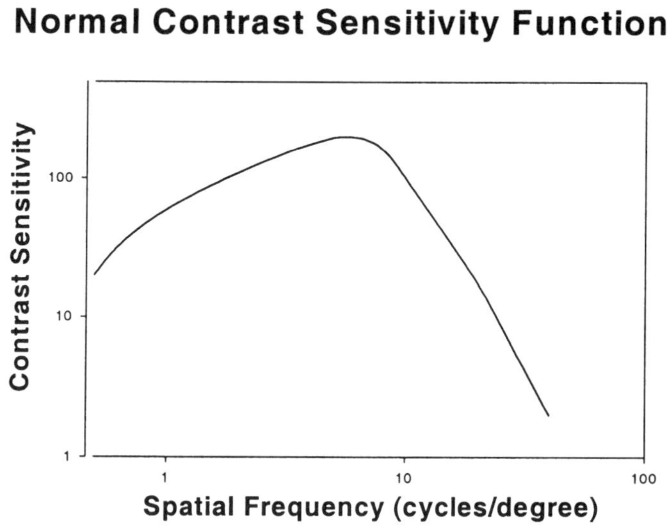

Fig. 30.5. Adapted from Devalois and Devalois, *Spatial Vision,* Oxford University Press, 1988.

where ξ is the spatial frequency and the parameters a and b are defined as

$$a = \frac{540(1 + 0.7/L)^{-0.2}}{1 + \{12/[w(1 + \xi/3)^2]\}},$$

$$b = \mathrm{b}_1(1 + 100/L)^{0.15}.$$

Here L is the display luminance (in candelas per square meter) and w is the square root of the angular area of the display.

The contrast on the retina can then be evaluated by

$$\mathrm{CS}_{\mathrm{ret}}(k) = \mathrm{MTF}(k)C_{\mathrm{obj}}(k).$$

The *modulation transfer function.* describes the optical performance of an optical system. In the case of the human eye, the modulation transfer function describes how well the cornea and lens form an image on the retina. It is defined as the modulation of the image divided by the modulation of the object, where modulation is the periodic contrast, C_3, previously defined.

The contrast sensitivity function is limited by the modulation transfer function at high frequencies. High-frequency cutoff for the human eye is ~60 cycles/deg (1.047 cycles/rad). High-frequency cutoff that is due to receptor sampling (neural processing) parallels the optics of the eye. The contrast sensitivity function has a low-frequency falloff that is considered to be due to neural factors.

The modulation transfer function as a function of pupil size: Optical performance of the human eye is near the diffraction limit for a 2.0-mm-diameter pupil. As the pupil size increases, aberration causes performance to decrease. Campbell and Green[50] found that the modulation transfer function was of optimum quality for 2.0–2.8-mm pupil diameters (Fig. 30.6).

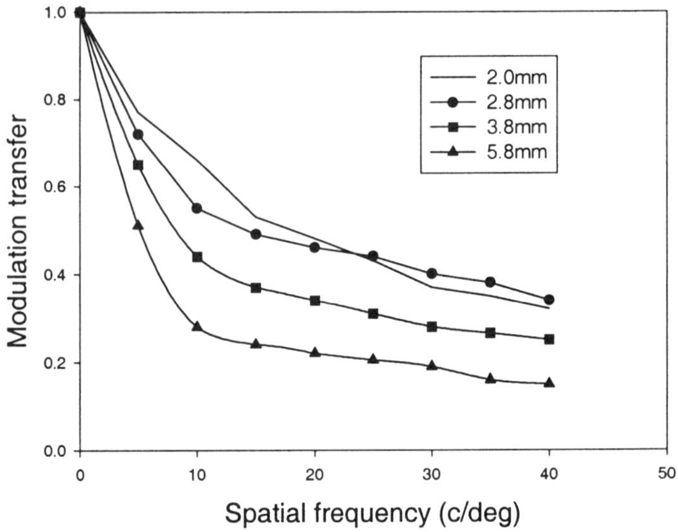

Fig. 30.6. Modulation transfer function from a range of pupil sizes. Adapted from Campbell and Green, "Optical and retinal factors affecting visual resolution," J. Physiol. **181,** 576, 1965.

The modulation transfer function as a function of retinal eccentricity: Retinal image quality declines steadily with retinal eccentricity.[51,52] This agrees well with the falloff in visual acuity with eccentricity, as the optics of the peripheral retina is poor because of focusing errors in the form of oblique astigmatism and field curvature (Fig. 30.7).

For experimental determination of the modulation transfer function both for center and periphery, Navarro et al.[52] and Williams et al.[53] used the fit

$$\mathrm{MTF}(\sigma) = (1 - C)\exp(-A\sigma) + C\exp(-B\sigma),$$

where σ is the spatial frequency in cycles per degree and *A, B,* and *C* are constants for a particular object angle.

[50] F. W. Campbell and D. G. Green, "Optical and retinal factors affecting resolution," J. Physiol. **181,** 576, 1965.

[51] J. A. M. Jennings and W. N. Charman, "Optical image quality in the peripheral retina," Am. J. Optom. Physiol. Opt. **55,** 582, 1978.

[52] R. Navarro, P. Artal, and D. R. Williams, "Modulation transfer of the human eye as a function of retinal eccentricity," J. Opt. Soc. Am. A **14,** 2884, 1993.

[53] D. R. Williams, P. Artal, and R. Navarro, "Off axis optical quality and retinal sampling in the human eye," Vis. Res. **36,** 1103, 1996.

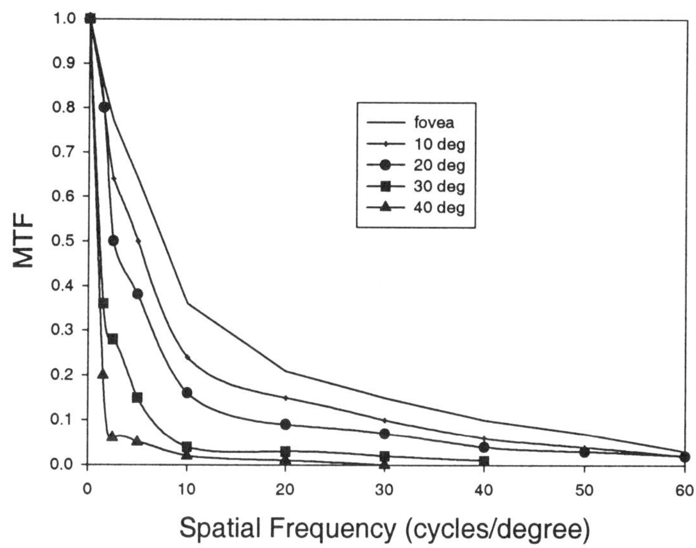

Fig. 30.7. Redrawn from R. Navarro, P. Artal, and D. R. Williams, "Modulation transfer of the human eye as a function of retinal eccentricity," J. Opt. Soc. Am. **10,** 201, 1993.

30.7. Exit Pupil

A head-mounted display can incorporate a one-lens system with a simple microscope type of display. Sometimes it may utilize a two-lens system. In such cases the lens closest to the display is called the objective lens and the other is called the eyepiece. The major difference between the single-lens and the double-lens system is that in the latter a real exit pupil is formed. The exit pupil is the image of the objective lens as formed by the eyepiece (Fig. 30.8). As long as the eye is within the exit pupil, the entire field of view of the display is visible.

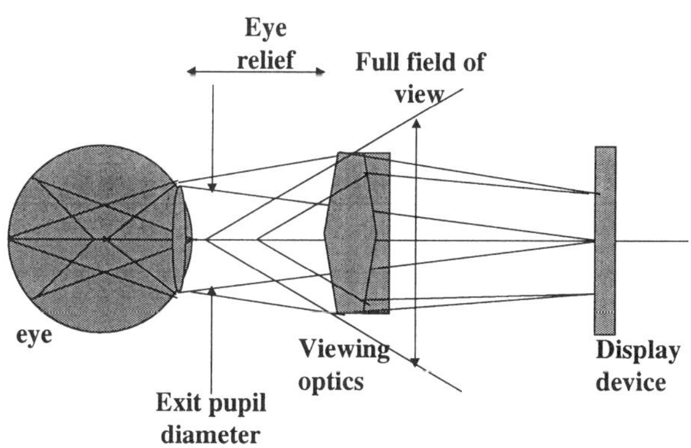

Fig. 30.8. Basic parameters that are needed for head-mounted display optics.

A single-lens head-mounted display does not form a real exit pupil. Instead, there is a volume of space within which the observer should place the eye in order to capture the entire field of view of the display. The size of this space is determined by

1. lens diameter
2. lens focal length
3. size of the display.

The larger exit pupil diameter gives space for more eye movement without any compromise of the field of view or the performance. A large exit pupil can accommodate a wider range of the image pupil distance without requiring an adjustment of the display.

30.8. Eye-Relief Distance

This is the distance between the last component of the viewing optics and the cornea of the eye. In general the eye-relief distance is ~25 mm. This distance is of importance to observers using a fixed-focus display who require eyeglass correction when viewing with a head-mounted display. For a given field of view, the larger the eye-relief distance, the larger and heavier the lens needs to be. Eye-relief distance often limits field of view (Fig. 30.9).

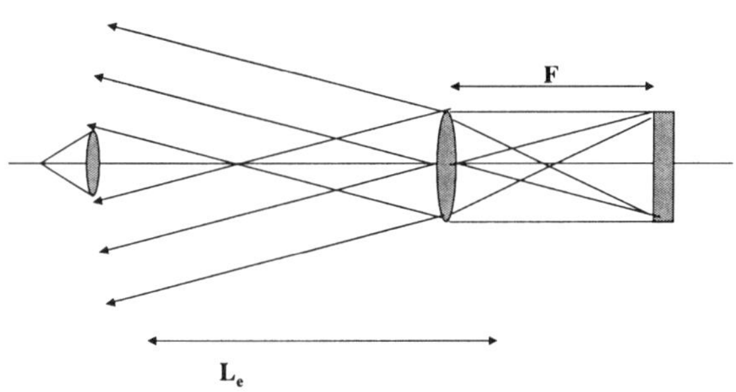

Fig. 30.9. Eye-relief distance L_e positioned far back so that the entire field of view is not visible and is limited by the diameter of the lens.

30.9. Field of View

This is one of the most important parameters to be considered when a display is designed. The field of view is the angle subtended by the virtual images as viewed by the observer. It is determined for a monocular display by the size of the display screen (image size) and the focal length of the lens that is used to magnify the image:

$$\text{field of view} = 2\tan^{-1}(x/2f),$$

where x is either the horizontal–vertical or diametral length of the display and f is the focal length of the lens. A complete field of view can be obtained only if the eye-relief distance (eye-to-lens distance) is minimal. If the distance d is large and the diameter of the lens D is small such that $D/d < x/f$, then the field of view is limited by the diameter of the lens:

$$\text{field of view} = 2\tan^{-1}(D/2d).$$

The field of view is also defined as the sensor's field of view without any eye movements. The normal values for a monocular field of view is 160° horizontal and 135° vertical and for a binocular field they are approximately 200° horizontal and 135° vertical. The field of view determined in such a manner is called the instantaneous field of view; it is always less than the optical field of view. In a well-designed head-mounted design the eye-relief distance and the diameter of the lens will be controlled in such a way that the instantaneous and the optical fields of view nearly coincide. This will make the full display visible.

If the display is rectangular, the field of view of the display can be expressed in terms of horizontal and vertical angles. It can also be expressed as the diagonal of the rectangle. The field of view of a display can be increased by use of a larger screen or a shorter focal length [Figs. 30.10(a) and 30.10(b)].

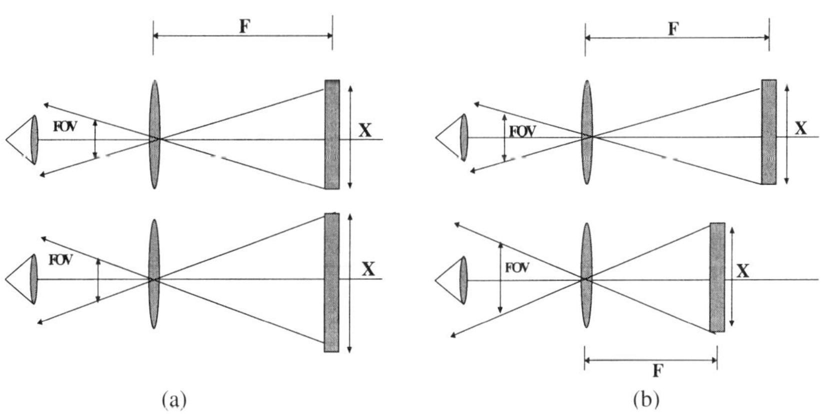

Fig. 30.10. Increasing field of view (FOV): (a) by increasing display size (X), (b) by decreasing focal length of the lens.

30.9.1. Trade-Off between Visual Field and Resolution

Visual field and resolution are inversely proportional. Hence a trade-off exists between these two parameters. The angular resolution of a display with a fixed number of pixels is the full field divided by the number of pixels. The trade-off needs to be made depending on the functional requirements and expectation of the user. If a display requires a wide field of view, it causes a decrease in the resolution and an increase in the weight of the display. Both these factors affect the performance of the user.

Some of the following techniques are used to compensate for this:

1. Use of multiresolution features such as high resolution in the center and low resolution at the periphery.
2. Binocular head-mounted display with partial overlap of the field.

30.9.2. Effects of Small Field of View on Performance

A small field of view has the following effects on performance:

1. More head movements required by the user for determination and location of targets.
2. Bodily discomfort, unsteadiness, and disorientation can occur.

Larger fields of view permit greater peripheral vision and a resultant better ego orientation and understanding of the events that occur in the panoramic world. The performance of the user of a head-mounted display is decreased with a decreased field of view, but the degree depends on the complexity of the task. A field of view of 20° is good for a simple task whereas a field of view of 60° is found to be optimum for more complicated tasks.[54]

The optimum field of view also depends on the type of head-mounted display, task requirements, and environmental conditions. In see-through head-mounted displays, the intradisplay field of view (which is due to the head-mounted display) must be well meshed with the paradisplay (real environment). If the intradisplay environment requires a higher resolution, a smaller field of view is desired. In general the intradisplay field of view is approximately 20°–60° and the paradisplay field of view is wider as the observer needs to move around the real environment.

For an opaque immersive display, a larger field of view is desired. Studies have suggested that a horizontal field of view wider than 60°–70° would be

[54] M. J. Wells and M. Venturino, "Performance and head movements using helmet mounted display with different fields of view," Opt. Eng. **29,** 870, 1990.

better but it is independent of task requirements. Arthur[55] has shown that even a slight decrease in the field of view from 176° to 112° decreases visual performance by ~23%.

One method of addressing the challenge of providing a wide field of view without considerable loss in resolution is to optically tile several smaller field-of-view displays to form a single, wide-field-of-view high-resolution image.[56]

30.10. Flicker

The images on a monitor are not continuous in time. The image on a CRT display is formed by an electron beam, which sweeps across from the top to the bottom of the screen, causing phosphors to glow. The beam has to go back to the top of the screen to refresh the phosphors. Hence the image on the screen is continuously refreshed. The rate at which the screen is refreshed plays a crucial role in making the image appear continuous, although the actual luminance of a point on the image is intermittent. The visual system is sensitive to temporal changes in the stimuli. The image appears to be continuous as the visual system integrates the responses with respect to time.

Flicker arises when the display images are not repeated quickly enough. Various factors such as luminance, refresh rate, phosphor persistence, background lighting, screen size, eccentricity, and adaptation tend to affect flicker perception.

Temporal vision with respect to a CRT is frequently studied with stimuli whose luminance varies sinusoidally over time. The visibility of a temporally modulated stimulus is related to its depth of modulation.

The depth of modulation is the same as the contrast, C_3, previously defined.

30.10.1. Critical Flicker Frequency

Critical flicker frequency refers to the temporal frequency at which flicker is not noticed. This implies that the stimulus at that particular temporal frequency cannot be resolved and appears to be steady.

31.10.2. Effect of Luminance and Illuminance on Flicker

Display luminance is the average luminance generated by the display.

Total luminance is the sum of the display luminance and the luminance caused by reflected illuminance from the ambient.

[55] W. K. Arthur, "Effects of field of view on performance with head mounted display," Ph.D. dissertation, University of North Carolina, 2000.

[56] M. J. Hoppe and J. E. Melezer, "Optical tiling for wide-field-of-view head mounted displays," in *Current Developments in Optical Design and Optical Engineering III*, R. E. Fischer and W. J. Smith, eds., Proc. SPIE **3779**, 146, 1999.

Illuminance is the light flux that is incident on the display surface per unit area.

Effect of ambient illumination on flicker: Isensee and Bennett[57] found that, at higher levels of display luminance, ambient illumination has less effect on the perception of flicker (Fig. 30.11).

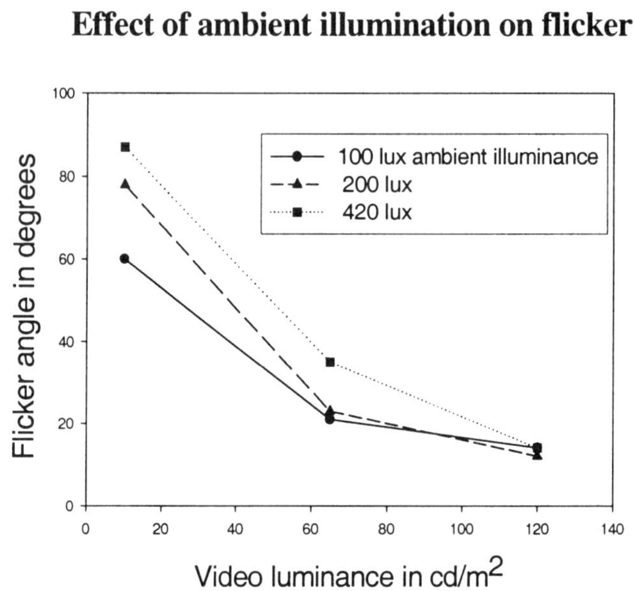

Fig. 30.11. Angle at which flicker is noticed. For high video luminance, ambient illumination has less effect on flicker perception. Adapted from S. H. Isensee and C. A. Bennett, "The perception of flicker and glare on computer CRT displays," Hum. Factors **30,** 689–702, 1983.

The recommended American National Standards Institute's standards for room illumination are ~200–500 lx. Illuminance affects flicker perception on a CRT screen only to a small extent under these conditions.

Luminance: As the luminance of the CRT increases so does perceived flicker. Bauer[58] has shown that for luminance from 80 to 320 cd/m^2 with refresh rates between 50 and 95 Hz, the critical flicker frequency increases linearly with log luminance.

The critical flicker frequency depends on the time-averaged luminance and the amplitude of modulation of the stimuli. Hecht and Verrijp[59] show that the

[57] S. H. Isensee and C. A. Bennett, "The perception of flicker and glare on computer CRT displays," Hum. Factors **30,** 689, 1983.

[58] D. Bauer, "Use of slow phosphors to eliminate flicker in VDU's with bright background," Displays **8,** 29, 1987.

[59] S. Hecht and C. D. Verrijp, "Intermittent stimulation by light. The relation between intensity and critical flicker frequency for different retinal locations," J. Gen. Physiol. **17,** 251, 1933.

critical flicker frequency increases as luminance is increased. DeLange[60] demonstrated that, for a particular temporal frequency of a stimulus, sensitivity to flicker increases as the amplitude of the luminance modulation is increased. Time-averaged luminance as well as modulation amplitude should therefore be considered for the detection of flicker.

30.10.3. Ferry Porter Law

This law states that the critical flicker frequency increases approximately linearly with the log of the retinal illuminance (Fig. 30.12). Retinal illuminance is given by the product of luminance and pupil area. Its unit is in trolands.

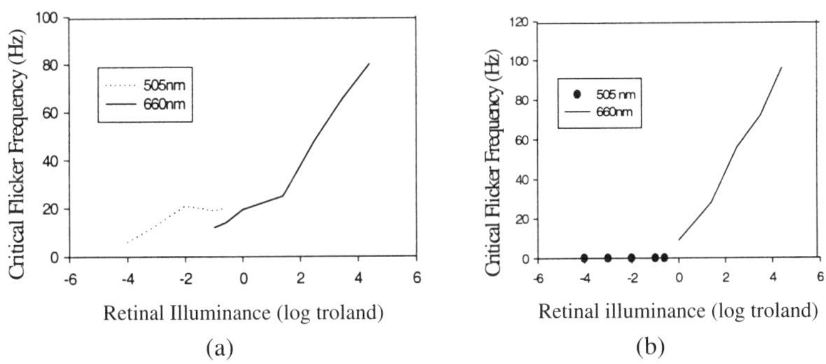

Fig. 30.12. As the critical flicker frequency (CFF) increases linearly, the log of retinal illuminance increases. Graph (a) shows a shift in the rod-mediated limb of the psychophysical critical flicker frequency intensity function with a test field wavelength. At low retinal illuminance, a slight increase in the critical flicker frequency for a wavelength of 505 nm is attributed to the scotopic luminance (action of rods). All data are collected by the forced choice technique for a 5.5° field at 35° eccentricity. Adapted from C. W. Tyler and R. D. Hammer, "Analysis of visual modulation sensitivity. IV. Validity of the Ferry–Porter Law," J. Opt. Soc. Am. A **7,** 743, 1990.

At high frequencies it is the modulation amplitude that has an effect on the flicker threshold and not the time-averaged luminance. At low frequencies the eye is sensitive to the percent change in the luminance, and therefore the critical flicker frequency depends on the ratio of the amplitude to the time-averaged luminance.[61]

30.10.4. Talbot–Plateau Law

This states that the brightness of a temporally modulated stimulus when fused (that is, when presented above the critical flicker frequency) is equal to the

[60] H. DeLange, "Research into the dynamic nature of the human fovea—cortex systems with intermittent and modulated light. I. Attention characteristics with white and colored light," J. Opt. Soc. Am. **48,** 777–784, 1958.

[61] D. H. Kelly, "Visual responses to time-dependent stimuli. I. Amplitude sensitivity measurements," J. Opt. Soc. Am. **59,** 422, 1961.

brightness of a steady light with the same time-averaged luminance. This law basically shows that the visual system integrates temporal information.

30.10.5. Bloch's Law

Studies on absolute threshold for detection of stimuli have shown that brief flashes require proportionately more luminance or light quanta to be detected than longer flashes. This is known as temporal summation. This summation is linear, provided the stimulus duration does not exceed a critical period. This is described as Bloch's law, which states that

$$Lt = c(0 < t < t_c).$$

The critical duration, on average, has been found to be ~100 ms. It depends on many factors, the most important being retinal eccentricity. It is important to consider this factor and the Talbot–Plateau law when a display is designed because it will affect the perception of the brightness of the stimuli.

30.10.6. Stimulus Size

The Granit–Harper law states that the critical flicker frequency increases with the log of the stimulus area. For a given percentage of modulation, flicker is more likely to be noticed if the stimulus size is large (Fig. 30.13). These results are consistent with the observation that the peripheral retina is more sensitive to flicker than the central retina is. A large stimulus falls more in the peripheral retina rather than a smaller stimulus. The peripheral retina is more sensitive to flicker because of the greater density of rods, which are very sensitive to changes in luminance.

30.10.7. Effect of Retinal Eccentricity on the Critical Flicker Frequency

The critical flicker frequency as a function of retinal eccentricity is dependent on the size of the stimuli. For a two-degree test field, the high-luminance critical flicker frequency decreases from the fovea to the periphery, whereas the opposite is true for the low-luminance critical flicker frequency (Fig. 30.14). For a larger stimulus size, the critical flicker frequency is found to be higher in the periphery than in the fovea as luminance is increased.[62]

[62] D. Bauer, M. Bonacker, and C. R. Cavonius, "Frame repetition rate for flicker-free viewing of bright VDU screens," Displays **4,** 31, 1983.

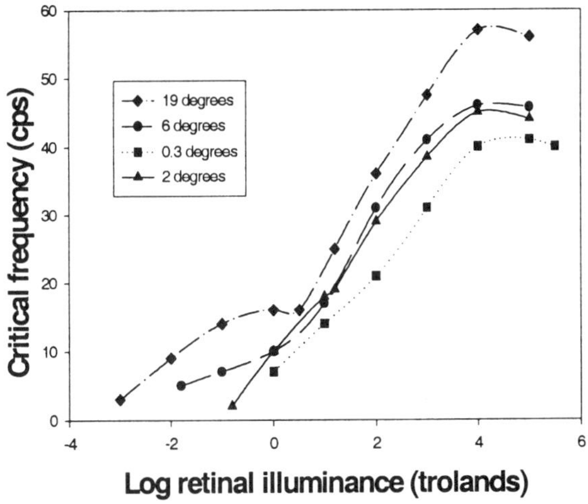

Fig. 30.13. Critical flicker frequency as a function of retinal illuminace for centrally fixated light of different areas. As size of the stimulus increases, the critical flicker frequency increases. Adapted from Hecht and Smith, J. Gen. Physiol. **10,** 979–989, 1936.

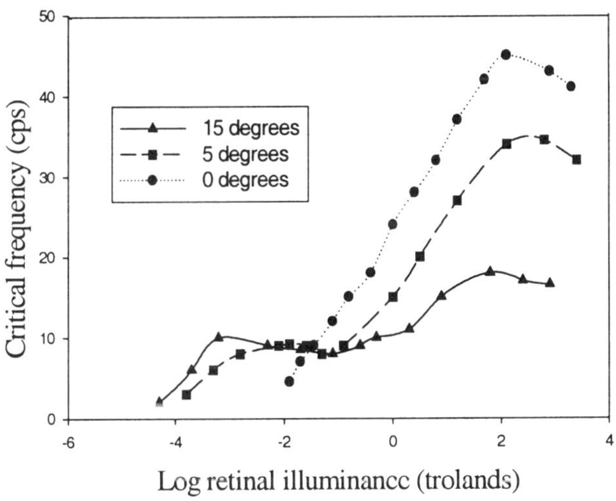

Fig. 30.14. Relation between critical frequency and log retinal illuminance for white light for three different retinal locations: at the fovea and 5° and 15° above the fovea. Adapted from Hecht and Verrjip, "Intermittent stimulation by light III. The relation between intensity and critical frequency for different retinal locations," J. Gen. Physiol. **17,** 251–265, 1933.

30.10.8. Display Polarity

Displays with dark characters on a bright background are called displays with positive contrast, and displays with bright characters on a dark background are displays with negative contrast. When an image is displayed with positive contrast, more that 85% of the pixels are activated. Such displays will tend to flicker more because increased luminance increases the perception of flicker. However, there are advantages to such a setup. Displays with positive contrast simulate the normal paper-based environment. The reading speed on a CRT with dark characters on a bright background is more or less similar to the reading speed on paper.[63]

Displays with reverse polarity are affected by reflections from ambient illumination. This can cause veiling glare, which decreases the contrast and makes the display less readable. Display polarity can be chosen depending on the viewing and working condition and whether it is flicker or glare that is going to be more troublesome.

30.10.9. Refresh Rate and Flicker

Refresh rate is also termed regeneration rate. It is one of the most important factors in the perception of flicker on a display. If the rate at which the screen is refreshed is greater than the critical flicker frequency of the observer then the observer will not perceive flicker. Refresh rates need to be higher for displays with higher luminance and a wider field of view in order to make the display flicker free.

Higher refresh rates generally cost more and should be minimized. Refresh rates can be decreased by an increase in the phosphor persistence. The normal refresh rate for a flicker-free response is ~72–80 Hz.

30.10.10. Phosphor Persistence

Along with refresh rate and screen luminance, phosphor persistence plays a vital role in flicker perception. The characteristics of phosphor persistence have been separated as medium short, medium, or long. This is dependent on how long it takes for the intensity of the light emitted from the phosphor to decay to ~10% of the peak intensity after the source of excitation has been removed.

The medium-short phosphors, P20 and P4, caused maximum flicker.[64,65] Most of the data on phosphor characteristics are obtained with the assumption of an exponential decay persistence. However, many phosphors have nonexponential persistence characteristics, especially modern CRT phosphors that contain a blend of two or more phosphors. Each component is known to have its own distribution persistence characteristics. In such cases, it would not be ac-

[63] W. H. Cushmann, "Reading from microfiche, a VDT, and the printed page: Subjective fatigue and performance," Hum. Factors **28,** 63, 1986.

[64] R. E. Turnage, "The perception of flicker in cathode ray tube displays," Inf. Disp. **3,** 38, 1966.

[65] J. E. Bryden, "Some notes on measuring performance of phosphors used on CRT displays," in *Proceedings of the Seventh Annual SID Symposium,* Society for Information Display, 1966.

curate to describe the phosphor persistence by a single-word description such as long or medium or short.

The ripple ratio is considered a better index to describe the persistence of the phosphor in a CRT. The ripple ratio is defined as the ratio of the amplitude of the fundamental frequency of the Fourier series to the mean value of the periodic form. The modulation index of the fundamental frequency gives the ripple ratio. Ripple ratios are determined for various vertical refresh frequencies and are tabulated elsewhere.[66]

The smaller the ripple ratio, the less susceptible the monitor is to flicker. For refresh rates between 50 and 80 Hz, the medium-short P4 phosphor has a ripple ratio of 1.97, one of the highest for any phosphor. PC214 has a ripple ratio of 0.4. It corresponds to a long-persistence phosphor. Ripple ratios are useful in calculating the absolute energy that comes from a CRT. The ripple ratio alone does not add much information to the overall perception of flicker on a CRT. Other factors, such as luminance and angle of viewing, also play major roles. Pearson obtained a series of calibration curves by performing a test on varying clinical population groups and for different levels of visual display tube screen luminance. For each screen luminance, the effect of different visual angles was plotted. The curves obtained were a combination of flicker frequency and ripple ratio at which 85% of the population does not perceive flicker. In his study he concluded that for any particular CRT phosphor along with the general calibration curve the probability of perceiving flicker could be found as a function of frequency and ripple ratio. This is considered to be a better index for the perception of flicker.

Selecting a phosphor with an adequate ripple ratio is limited to the relatively few available. The other factor that can be changed is the visual display tube refresh rate, adjustment of which is an effective route to flicker-free design.

The effect of the phosphor persistence on the amplitude of the refresh frequency, assuming the phosphor has a single-time constant decay, is given by

$$A(\omega) = 1/[1 + (\omega\tau)^2]^{1/2},$$

where A is the amplitude, τ is the time constant of the exponential describing the phosphor persistence, and ω is the refresh frequency (in radians) of the display, which is equal to $2\pi f$.

A display with fast phosphor will have a relatively large amplitude at a specific refresh rate. A faster phosphor needs a higher refresh rate than a slower one.

[66] R. A. Pearson, "Predicting VDT flicker," Inf. Disp. **7&8,** 22, 1991.

Farrell et al.[67] have shown that the longer the persistence of the phosphor the lower the amplitude so the display will require a lower refresh frequency to be flicker free. If two displays have the same refresh frequency but different phosphor characteristics, the one with longer persistence will flicker less.

30.10.11. Jitter

When a display uses an interlaced scan, characters sometimes seem to move or jitter. This occurs more often when the pixels are relatively very large. The effect can be largely eliminated by use of high-resolution displays. In high-resolution displays the user will not notice that only alternate lines are updated at that refresh frequency.

30.10.12. Method to Predict Monitor Flicker

Various empirical methods[68] have been proposed to see whether a particular visual display terminal will appear to flicker in a given environmental condition. However, these methods are cumbersome and time consuming, although they do provide a thorough analysis. Farrell[69,70] has developed an analytical method for predicting whether a visual display tube will flicker given the screen phosphor persistence, refresh frequency, and other factors like distance between the user and the display.

30.11. Glare

Glare can be of two types, direct or reflected glare. Direct glare is from high levels of screen luminance. Reflected glare, also called veiling glare, is caused by the reflection of light from external sources on the CRT (such as luminaires, windows etc.)

Reflected glare affects legibility by decreasing the contrast between background and image. Low to moderate levels of ambient illumination of 100 to 260 lx and moderate levels of video luminance of ~65 cd/m^2 produce the least amount of discomfort that is due to glare and flicker.[57]

[67] J. E. Farrell, E. J. Casson, C. R. Haynie, and B. L. Benson, "Designing flicker free video display terminals," Displays **9**(7), 115, 1988.

[68] B. E. Rogowitz, "A practical guide to flicker measurement: Using the flicker-matching technique," Behav. Inf. Technol. **5,** 359–378, 1986.

[69] J. E. Farrell, "An analytical method for predicting perceived flicker," Behav. Inf. Technol. **5,** 349–358, 1986.

[70] J. E. Farrell, "Predicting flicker thresholds for video display terminals," Proc. Soc. Inf. Disp. **28**(4), 18, 1987.

Diffusers and up-lighting should be used. One can minimize glare from the sun by positioning the monitor perpendicular to the windows and by having adjustable blinds. Neutral-density filters, polarizing materials, and antireflection coatings are some of the options to reduce glare that is due to increased screen luminance.

30.12. Interpupillary Distance

The distance between the optical centers of the two eyes is called the interpupillary distance. An interpupillary distance range of 50–74 mm has been observed among adult male and female populations.

Binocular or biocular displays require the alignment of the optical centers of the optics (intraocular distance) with the observer's interpupillary distance. For a wide range of observer interpupillary distances the devices usually have movable lenses that can be so aligned. There are displays that do not use this feature. Such fixed-focus displays must have a large exit pupil to avoid the effects of mismatch in interpupillary distance. Using large exit pupils requires a larger optical system with a larger lens diameter. This makes the system heavier and can affect the overall performance, other things being equal. For real-world targets, a line of demand can be plotted that represents the accommodation and convergence needs for all observers with a given interpupillary distance (Fig. 30.15).

Convergence demand is dependent on the interpupillary distance of the observer. From the graph, it is seen that for different interpupillary distances the demand line converges for distances beyond 25 cm. In general, if the accommodative demand is set less than 1 D (i.e., farther than 1 m), then only one demand line need be considered for all users. For displays at a distance farther

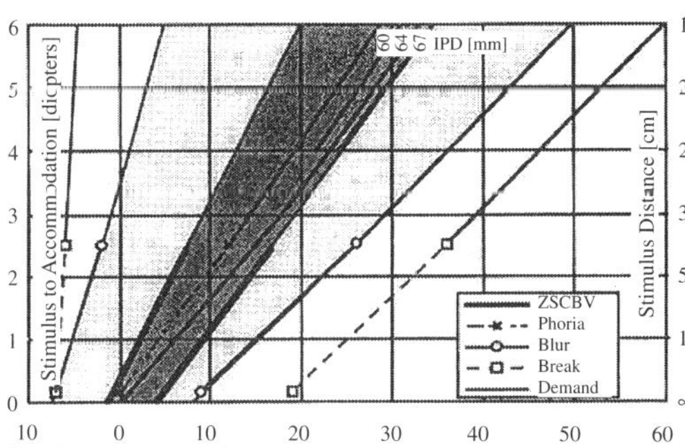

Fig. 30.15. Graphic representation of accommodation and convergence demands and the zones of single clear binocular vision (ZSCBV) for an average person. A demand line is associated with the person's interpupillary distance, but the various lines converge at greater distances. Adapted from P. Mouroulis, *Visual Instrumentation: Optical Design and Engineering Principles*, McGraw-Hill, 1999.

than 1 m, the demand on convergence and accommodation is more or less the same for observers with different interpupillary distances. The effect of different interpupillary distances is minimized in such cases.

Prismatic effect induced by the mismatches in the interpupillary distance with the intraocular distance: If the observer's interpupillary distance is smaller than the intraocular distance of the display, then the observer will be looking through the inner elements of the lens system. This results in a prismatic effect induced by the lens, and in this case it results in a *base-out* prismatic effect. The images in both the eyes are shifted outward, and the eye therefore needs to converge in order to place the images back on the central retina to attain fusion and perception of a single image. Similarly, if the interpupillary distance is larger than the intraocular distance a *base-in* prismatic effect results. The images are shifted inward, and the eye needs to diverge to bring the image to focus. Because it is easier to converge than to diverge the eyes, smaller interpupillary distances are preferred in fixed-focus head-mounted displays.

Prismatic effect (prism diopters) is given by the product of the decentration c and lens power $P = c\Phi$, where P is the prismatic power in diopters, c is the decentration in centimeters, and Φ is the power of the lens (this is called a Prentice rule).

For example, when an observer with an interpupillary distance smaller than the intraocular distance looks through the display, the eye must converge more than normal (Fig. 30.16). This results in an increase in the esophoria (a latent deviation of the visual axes of the two eyes in which the visual axes of one eye deviated inward with respect to the other, when all stimuli to fusion are eliminated) of the observer, with consequent eyestrain. However, if the observer has a strong binocular system, this is compensated for by the use of fusional vergence (convergence or divergence stimulated by retinal disparity, resulting in the avoidance of diplopia). Once the display use is discontinued it has been

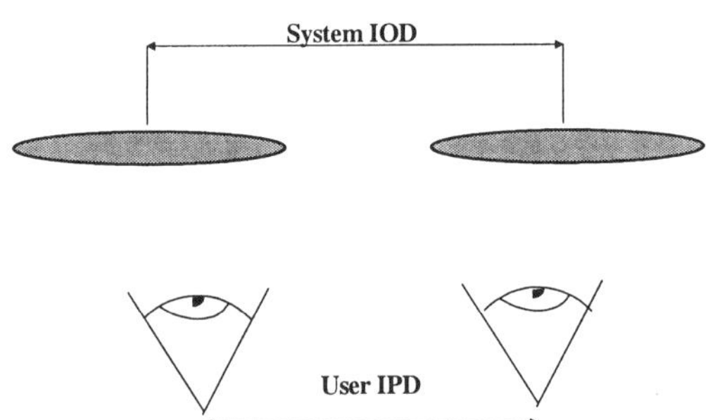

Fig. 30.16. Base-out prismatic effect induced when user interpupillary distance (IPD) is smaller than a system's intraocular distance (IOD) for a positive power lens.

found that small amounts of esophoria can be measured in the observer, but it is a transient change. The eye returns to the normal position after some time.

Head-mounted displays generally use high-powered lenses, so decentration effects can result in large prismatic effects. This induced prismatic effect is small if the optical image distance is sufficiently large. Only the power of the lens in excess of that required for bringing the image on the virtual screen participates in the decentration effects.

For a virtual screen placed at 2 m and a 10-mm difference between the observer interpupillary distance and the intraocular distance, the prismatic effect induced by the system is ~0.5^{Δ}, which is within the tolerable limit. (Δ refers to a prism diopter unit and is defined as a light beam deviation of 1 cm at 1-m distance. It is conventionally used to denote the power of an ophthalmic thin prism). If a large screen is placed at 2 m or beyond, the effect of mismatch between the interpupillary distance and the intraocular distance is within acceptable limits.[71]

30.13. Presbyopia

Presby means old, and presbyopia is an age-related condition of the eye in which the lens gradually loses its accommodative ability and the subjects have difficulty seeing near targets. Correcting for this refractive error seems to play a crucial role in presbyopic subjects working with a visual display. The distance between the display and the subject falls in the intermediate range of vision. While being corrected for this condition, the subject needs to be given proper correction for distant, intermediate, and near vision. Trifocal lenses and progressive lenses are given for these subjects so that they can enjoy clear vision for all three distances. A spectacle lens suitable for presbyopic users would be a single-vision lens with correction for the intermediate range.[72] If clear distant vision is also required, then a bifocal lens with a larger near portion to aid for the screen distance needs to be prescribed. (Editor's note: I do quite well with full magnifiers.)

30.14. Ocularity

Displays for virtual environments that use a head-mounted display for the presentation of images can be monocular, binocular, or biocular. The decision on the type of head-mounted display depends on the visual needs of the observer. Each of these has advantages and disadvantages.

[71] E. Peli, "Real vision and virtual reality," Opt. Photon. News **6**(7), 28–34, 1995.

[72] D. Burns, H. Obslfeld, and J. Saunders, "Prescribing for presbyopes who use VDU's," Ophthalmic Physiol. Opt. **13**, 409, 1993.

30.14.1. Monocular Head-Mounted Display

As the name suggests, the image formed by the monocular head-mounted display is typically presented to one eye, usually the dominant eye, the right eye for 90% of the population. It has only one set of optics, there by making it lighter, cheaper, and simpler than binocular systems. These monocular head-mounted displays can be one of two kinds: see-through or immersive full-image-filling.

In the first device the unaided eye has an unrestricted field of view. In the second, the unaided eye may produce binocular rivalry with the other. In the immersive full-image-filling display, one eye views the entire field of view (in the real world) while the other sees the display. Hence binocular rivalry might result. Binocular rivalry is the superposition of different images to the two eyes; this results in alternating times of monocular dominance. Some users have problems in sensing a shifting view from one eye to the other.

The monocular display has a smaller field of view than that of binocular and biocular displays. Monocular cues tend to aid the displays in creating depth perception.

In head-mounted displays used outdoors on a bright sunny day, the ambient light surrounding the head-mounted display tends to reduce the display contrast. If the head-mounted displays are used by children younger than 10 years of age, they can have an adverse effect on the binocular function of the eye.

30.14.2. Binocular Head-Mounted Displays

In these, the same image is presented to both eyes with a little retinal disparity. They provide a wider field of view than monocular devices (~120° with ~60° overlap). This is an increased burden on the optical system, requiring two lenses and accurate coalignment. Some head-mounted displays provide partial overlap of the images, providing a binocular view of the central 100° of the visual field. Approximately 20°–30° of field on either side of the image is left for monocular viewing. This reduces the burden on the optics of the head-mounted display, making it lighter.

Sometimes distortions in the (combined) image may be caused by retinal disparities that result from a magnification difference between the two monoculars or misalignment between them.

30.14.3. Biocular Head-Mounted Displays

In these, the same image is presented to both eyes. They contain two sets of optics. They are costlier and heavier compared with monocular displays. Both eyes view exactly the same image, and this eliminates binocular rivalry or monocular suppression seen in monocular displays. These devices need accurate optical alignment for adequate performance.

30.15. Radiation Hazards to the Eye

Optical radiation hazards to the eye are found to be dependent on two factors: wavelength of light and ocular exposure duration.

30.15.1. Damage Mechanisms

The principal mechanisms by which optical radiation damages the eye are photochemical and thermal reactions. Photochemical damage to tissue structures occurs in the UV end of the spectrum, and thermal damage occurs in the IR and the visible spectral regions. Photochemical damage is caused by chemical changes from brief but intense exposure to irradiance or from lengthy, weak radiance, that is, total exposure. It is highly wavelength dependent. Thermal damage occurs when the temperature rise in irradiated tissue is high enough to cause local coagulation and scarring.

The sensitivity of the eye is not constant over the whole electromagnetic spectrum. The human eye is sensitive to wavelengths between 380 and 780 nm, which we call the visible spectrum or light. Even over the visible range, the eye reacts differently to different wavelengths. Its sensitivity is zero at either end of the visible spectrum and peaks at 555 nm for photopic (daylight) vision and 507 nm for scotopic (night) vision. While the luminous intensity of a light source is being measured, information regarding the relative sensitivity of the eye to different wavelengths is required. The luminous intensity of a white-light source is defined by multiplying the watts emitted at each wavelength by the efficiency of the wavelength in exciting the eye, relative to the efficiency at 555 nm. This efficiency factor is referred to as the $V(\lambda)$ curve. The photopic $V(\lambda)$ curve can be seen in Fig. 30.17. This curve is obtained by the averaging of the experimental results obtained from many observers and has long been standardized as an essential component in the quantitative description of light. This is known as the CIE $V(\lambda)$,[73] and separate curves exist for the photopic and scotopic vision.

Radiant exposure H is defined as the time-integrated irradiance falling on a surface. It is also referred to as the dose, and is measured in joules per square meter:

[73] L. S. Pedrotti and F. L. Pedrotti, *Optics and Vision,* Prentice-Hall, 1998.

$$H = Et,$$

where E is the irradiance measured in watts per square meter and t is the exposure duration in seconds.

Both E and H are defined over the entire optical spectrum, but to obtain the photochemical effects it is necessary to use the action spectrum. The action spectrum describes the relative effectiveness of different wavelengths in causing photobiological effects. The photopic $V(\lambda)$, UV hazard $S(\lambda)$, and blue-light hazard $B(\lambda)$ curves are examples of action spectra (Fig. 30.17). These action spectra are used to spectrally weight the incident light:

$$E = \sum EA(\lambda)\Delta(\lambda),$$

where $A(\lambda)$ is the action spectrum.

To calculate the amount of light falling on the retina, the retinal irradiance is given by

$$E_{\text{ret}} = \pi L_s \tau d_e^2 / 4f^2,$$

where L_s is the source radiation in watts times inverse square centimeters times inverse steradians, f is the effective focal length of the eye in centimeters, d_e is the pupil diameter in centimeters, and τ is the transmittance of the ocular media.

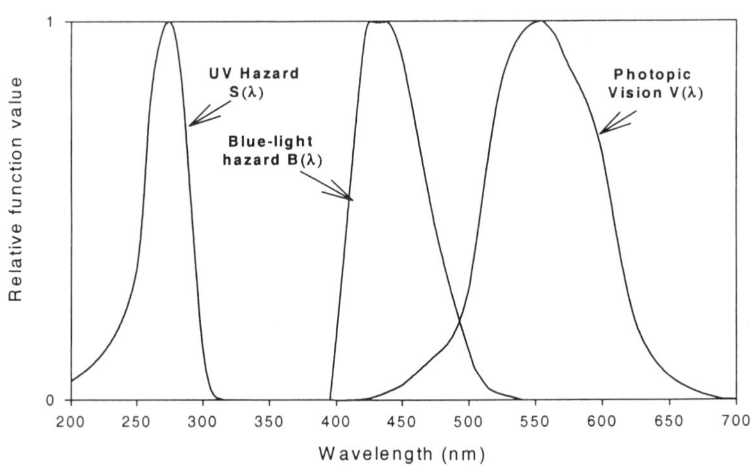

Fig. 30.17. Action spectra. The ACGIH UV hazard function $S(\lambda)$, the blue-light hazard function $B(\lambda)$, and the CIE spectral sensitivity function $V(\lambda)$ for the human eye.

30.15.2. Effect of Transmission and Absorption of Optical Radiation by Ocular Tissues as a Function of Wavelength

The CIE has defined different spectral regions as follows:

UVC: 100–280 nm
UVB: 280–315 nm
UVA: 315–380 nm
Light: 380–400 to 760–780 nm
IRA: 760–1400 nm
IRB: 1.4–3 μm
IRC: 3.0 μm–1 mm

These wavelength divisions have a considerable overlap. The human eye is quite well adapted to protect against the potential hazards from optical radiation. The cornea, lens, and aqueous and vitreous humors all help in absorbing certain parts of the electromagnetic spectrum before light reaches the retina.

The tear layer absorbs small amounts of radiation, absorbing UV radiation below 290 nm and IR radiation above 3000 nm. It transmits radiation from 290 to 3000 nm. The cornea, like the tear layer, absorbs UV below 290 nm and IR above 3000 nm. It also partially transmits UV in 290–315 nm and IR in 1000–3000 nm. Hence corneal transmission is from 315 to 1000 nm, which includes longer UV wavelengths, all the visible spectrum, and shorter IR wavelengths. Transmission of the cornea, particularly for shorter wavelengths, decreases with age. The aqueous humor protects the lens by partially absorbing IRB and IRC. Its absorption characteristics are similar to those of the cornea. Radiation that is transmitted by the cornea is also transmitted by the aqueous humor to the iris and lens.

The absorption of the lens and the cornea varies with age. Lenses of children absorb UV below 310 nm and IR beyond 2500 nm. Adult lenses absorb all radiation below 375 nm, but there is no change in the absorption of IR with increasing age. The vitreous humor largely absorbs UV below 290 nm and IR above 1600 nm. Nearly 94% of the IR radiation of 770 nm reaches the retina, falling off to 90% at 900 nm and to a very low level beyond 1500 nm.

30.15.3. Effects of Radiant Energy on the Eye

The region of the spectrum that affects the eye most is from the near-UV to the near-IR region. The effects are manifested mostly at the cornea and the lens, where the greater portion of absorption takes place. However, the retina can be affected significantly even if only a small amount of radiation is absorbed. The effect is dependent on

1. size and intensity of the source
2. size of the pupil
3. size of the image.

Long-IR rays cause thermal injuries to the cornea, resulting in conjunctival irritation and keratitis. Short- or near-IR radiation can cause thermal hazards to the lens, producing clouding of the lens, which leads to cataracts. Burnlike lesions can also occur in the retina and the choroid. Lasers with wavelengths between 400 and 1400 nm can cause thermal injury (photocoagulation) of the retina. These injuries occur even with a short exposure duration. Permanent changes in the eye occur mainly because of local heating of the tissue. When the rate of heat dissipated is slower than the rate of heat inflow, protein coagulation occurs. Secondary heating can also be produced, like heating of the lens by heating of the iris.

Short-UV rays cause inflammation of conjunctiva and cornea. Photochemical injury to the cornea is called photokeratitis. It generally occurs at approximately 180–400 nm. Exposure for long intervals causes changes in cell structures of the lens and the retina.

Long-UV rays produce lens damage other than causing fluorescence in the lens. Such photochemical injury to the crystalline lens occurs generally at 295–325 nm. Blue-light photochemical injury to the retina occurs generally at 400–550 nm. In aphakic eyes (a condition in which the lens is not present in the eye) photochemical injury occurs at 310–550 nm. Intense or excessive *visible light* produces discomfort and asthenopia (eyestrain).

30.15.4. Protection of Eyes Against Excessive Radiation

The following conditions may protect the eyes against radiation:

1. nervous reflexes that result in aversion of the head and eyes from bright-light sources
2. blink reflex
3. pupillary constriction.

30.15.5. Exposure Guidelines

UVC and UVB (Wavelengths below 315 nm)

The American Conference of Governmental Industrial Hygienists[74] has formulated the exposure limits for UV radiation, which are recommended by the Na-

[74] American Conference of Governmental Industrial Hygienists, "TLVs, threshold limit values and biological indices for 1999," American Conference of Governmental Industrial Hygienists, 1999.

tional Institute of Occupational Safety and Health Administration and are widely used in analysis.

The effective UV irradiance is

$$E_{\mathrm{eff}} = \sum E_\lambda S_\lambda \Delta\lambda,$$

where E_{eff} is the spectrally weighted irradiance and $\Delta\lambda$ is the wavelength.

The permissible exposure duration $t_{\max}$ for any 24-h period is given by

$$t_{\max} = 0.003(\mathrm{J/cm^2})/E_{\mathrm{eff}}(W/\mathrm{cm^2}),$$

where $t_{\max}$ is given in seconds.

UVA Sources (315–400 nm)

The biological effectiveness for UVA is far less than that of UVB and UVC. The total irradiance E in the UVA spectral region is given by

$$E_{\mathrm{eff}} = \sum E_\lambda \Delta\lambda,$$

where E_λ is the spectral irradiance in watts per square centimeter.

The permissible exposure duration $t_{\max}$, in seconds, is given by

$$t_{\max} = 1(\mathrm{J/cm^2})/E_{\mathrm{eff}}(\mathrm{W/cm^2}).$$

In general, by the standards given by the American Conference of Governmental Industrial Hygienists, the total irradiance on the unprotected eyes should not exceed 10 W m^{-2} for periods $>10^3$ s and, for periods $<10^3$ s, irradiance should not exceed 10^4 J m^{-2}.

Visible Light Sources (400–760 nm) and Near-IR (760–1400 nm)

Retinal lesions are due to many factors:

1. pupil size
2. spectral transmittance of the ocular media
3. spectral absorption of retina and choroid
4. optical quality of retinal image
5. size of source
6. location and size of retinal image
7. exposure duration
8. type of source and its spectral characteristic.

Retinal injury that is due to viewing broadband light sources can be either thermal or photochemical. Thermal injury occurs for a short duration (<10 s) and photochemical (blue-light) injury predominates for long exposures.

The American Conference of Governmental Industrial Hygienists criterion for evaluating the two potential hazards requires two spectral weighting functions, R_λ for thermal injury and B_λ for photochemical injuries.

The exposure limits are given by the following equations.

1. For photochemical (blue light):

$$\sum_{400}^{1400} L_\lambda t B_\lambda < 10^6 \text{ J m}^{-2} \text{ sr}^{-1} \; t < 10^4 \text{ s},$$

$$\sum_{400}^{1400} L_\lambda B_\lambda < 10^2 \text{ W m}^{-2} \text{ sr}^{-1} \; t > 10^4 \text{ s},$$

   where L_λ is the spectral radiance of the source at wavelength λ (in nanometers) and t is the exposure time (in seconds).
2. For thermal injury:
   The burn hazard function R_λ is limited to an exposure duration of <10 s as pupillary constriction and eye movements limit the added risk from greater exposures. The exposure limit is given by

$$\sum_{400}^{1400} L_\lambda R_\lambda \Delta\lambda \leq 5/\alpha t^{0.25},$$

   where α is the angle (in radians) subtended at the eye by the largest dimension of the source and t is the exposure duration.

The American Conference of Governmental and Industrial Hygienists gives us a more specific exposure limit for the near-IR radiance, which is

$$\sum_{770}^{1400} L_\lambda < 6 \times 10^3/\alpha$$

for a 7-mm-diameter pupil and a 100-W/m^2 limit on irradiance, summed for all wavelengths between 770 and 1400 nm.

IR Limits

The exposure limits applicable to IRA (760–1400 nm) and IRB (1400–3000 nm) specify a maximal irradiance of continued exposure of 10 W/cm^2 over any 1000-s period, but not to exceed an irradiance of $1.8t^{-0.75}$ W/cm^2.

A detailed description for UV, IR, and visible radiation exposure can be found in footnotes 75 and 76.

Effect of Lasers on the Eye

Significant damage to the ocular structures occurs because of the high radiance of laser light.

Exposure guidelines and standards: Laser products are classified according to their output power and potential hazard during normal use of the product. This is a numerical system with four broad classes, 1–4.

The classification can be obtained from organizations such as the Center for Devices and Radiological Health of the U.S. Food and Drug Administration,[77] the American National Standards Institute,[78] and the International Electrotechnical Commission,[79] to name a few.

Exposure Guidelines can be obtained from published data by the American National Standards Institute, the American Conference of Governmental and Industrial Hygienists,[80] the International Commission on Non-Ionizing Radiation Protection,[81] and the International Electrotechnical Commission. The guidelines include exposure limits for the eye and the skin. The American National Standards Institute Z136.1 standard applies to a single pulse of 100 fs up to 8 h of continuous exposure. The exposure limit is called the maximum permissible exposure. The guidelines include a number of wavelength-dependent correction factors. Whenever exposure limit is determined, it will include a reference to any correction factors that may be used.

Exposure limits for lasers with visible emissions where the duration of exposure is the human aversion response time (0.25 s):

$$\text{MPE} = 1.8t^{3/4} \times 10^{-3}\ \text{J/cm}^{-2}.$$

For these lasers no correction factor is applicable. Hence, for $t = 0.25$ s,

[75] M. Bass, ed., *Handbook of Optics,* Optical Society of America and McGraw-Hill, 2001, Vol. III.

[76] J. Marshall, ed., "The susceptible visual apparatus," in *Vision and Visual Dysfunction,* CRC, 1991, Vol. 16.

[77] "FDA performance standards for light-emitting products," in *Regulation for the Administration and Enforcement of the Radiation Control for Health and Safety Act of 1968,* HHS Publication FDA 88-8035, Center for Devices and Radiological Health, 1988, paragraph 1040.30.

[78] American National Standards Institute, "Safe use of lasers," ANSI Z136.1-2000, Laser Institute of America, 2000.

[79] IEC, "Radiation safety of laser products, equipment classification, requirements and users guide," IEC 825, International Electrotechnical Commission, 1984.

[80] *Documentation of Threshold Limit Values,* 4th ed., American Conference of Governmental Industrialist Hygienists, 1992.

[81] International Commission on Non-Ionizing Radiation Protection, "Guidelines on limits for laser radiation of wavelengths between 180nm and 1000μm," Health Phys. **71,** 804, 1996.

$$\text{MPE} = 6.4 \times 10^{-3}\text{J/cm}^2.$$

For an IR laser beam the exposure limit is

$$1.8C_A t^{3/4} \times 10^{-3}\ \text{J/cm}^2,$$

where C_A is the frequency-dependent correction factor.

These correction factors are predetermined for different lasers, depending on their applicable wavelength.

30.16. Resolution

Resolution can be specified either as the total number of pixels or the angular subtense of a single pixel. Angular measure is usually a better method as it can be compared directly with the visual capability limit of the eye. The visual limit of the human eye is ~1 arcmin (mrad).

Another method used to determine the resolution of a display is to estimate the highest visible spatial frequency. One way to calculate the spatial frequency is

$$\xi = 1/[2\ \arctan(h/dN_l)] = 1/[2\ \arctan(a_{\text{px}}/d)] = 1/(2\alpha),$$

where h is the height of the monitor, d is the distance of the observer to the monitor, N_l is the number of (vertical) lines, a_{px} is the (vertical) size of a pixel, and α is the angular subtense of a pixel.

The resolution limit of the eye is ~30 cycles/deg (0.5 cycles/arcmin). (Note that this is one over twice the angular resolution of 1 arcmin.) The monitors currently available do not match the visual limit of the eye. Sometimes a compromise on the resolution is made, depending on the functional requirement of the user. Complete characterization of the display requires the use of both the resolution (pixel size) and addressability (center-to-center distance of the pixels).[82]

Addressability is controlled by the hardware driving the display, whereas resolution is determined by the design of the display. The vertical resolution can be estimated by the number of scan lines, but the horizontal resolution is more difficult to assess because the scan line is continuous. If the electron-beam spot is smaller than the addressable cells, then the cells can be used as the resolution size. The electron-beam spot can often be characterized as a Gaussian shape, and its modulation transfer function can be used.

[82] G. M. Murch and R. J. Beaton, "Matching display resolution and addressability to human capacity," Displays **9**(1), 23, 1988.

Murch et al.[83] investigated the trade-off between display resolution and display addressability. They proposed a method to determine the best pixel width for any particular spacing. It is based on the ratio of display resolution to display addressability (RAR):

$$\mathrm{RAR} = W/S,$$

where W is the resolution (spot full-width half-maximum) and S is the display addressability. Resolution of the display was taken to be the spot width at half of the maximum luminance. Display addressability is the peak-to-peak separation between the adjacent pixels.

A suitable ratio of display resolution to display addressability that would make the raster invisible is sought. That is, the modulation of the grating pattern produced, when all display lines are illuminated, should be below the threshold of visibility. An optimum ratio of display resolution to display addressability depends on the type of application. A ratio of display resolution to display addressability of 1.2 to 1.3 is considered to produce a smooth appearance of the raster.

Murch et al. also published a useful empirical formula to predict the modulation (M) of a spatial frequency pattern created by displays with ratio values within 0.35 and 2.4:

$$M = (2/\pi)\exp(3.6\ \mathrm{RAR} - 7.0\ \mathrm{RAR}^2 + \mathrm{RAR}^3).$$

They suggested that display engineers use this formula along with published human contrast sensitivity functions to determine whether the raster pattern is invisible.

Resolution can also be described by the modulation transfer function.[84] Of the many methods that estimate the modulation transfer function, the discrete frequency technique is found to be the most dependable, although it is time consuming.[85] A fractional Fourier transform of the line spread function is a common method used by automated systems.

The system's final modulation transfer function should then be multiplied by the contrast sensitivity function of the human eye. Bringing in the eye's response is important in assesing the "information transfer" an observer can

[83] G. M. Murch, L. Virgin, and R. J. Beaton, "Resolution and addressability: How much is enough," J. Soc. Inf. Disp. **26**(4), 305–308, 1985.

[84] C. Infante, "On the resolution of raster-scanned CRT displays," J. Soc. Inf. Disp. **26**(1), 23, 1985.

[85] R. W. Verona, "Comparisons of CRT display measurements techniques," in *Helmet Mounted Displays III,* T. M. Lippert, ed., Proc. SPIE **1695,** 117, 1992.

achieve. The modulation transfer function area, developed by Charman and Olin,[86] is one such procedure that takes into account the response of the human visual system; it is the area bounded by the display system's modulation transfer function and the detection threshold curve of the human eye. The greater the modulation transfer function area, the greater information perceived by the human eye. Crossing between the system modulation transfer function and the detection threshold curve defines the highest spatial frequency that can be detected.[87,88]

A new method in head-mounted displays is to use lasers to form an image on the retina directly; this is called retinal scanning display.[89] Unlike a CRT or a LCD, which builds images on the screen, a retinal scanning display builds images on the retina and the point source of light is moved like a raster on the retina.

The advantage of this procedure is that high resolution is possible. In a CRT or an LCD, resolution is limited by the pixel size. In a retinal scanning display it is the radiating area of the source that limits the resolution. Color pixels are spatially superimposed, which increases the resolution by a factor of 3. It has other advantages, such as an increase in the brightness, a larger color range, good image quality, and a wider field of view. Although lasers have the potential to be dangerous to the eye, instruments such as the retinal scanning display have adequate neutral-density filters, which attenuate the intensity of the source.

30.17. Simulator and Motion Sickness

Motion sickness is the sickness that people experience in simulators and displays that incorporate motion. Vestibular (physical motion) simulation alone is found to be sufficient to cause motion sickness. Simulator sickness tends to be caused by the compounding effect of vision and motion, but is generally found in situations in which visual motion is present without actual motion of the subject (vection). The parameters can be categorized as individual factors, simulator factors, and simulated task factors.[90]

[86] W. N. Charman and A. Olin, "Tutorial: Image quality criteria for aerial camera systems," Photograph. Sci. Eng. **9,** 385, 1965.

[87] P. G. Barten, "Effects of quantization and pixel structure on the image quality of color matrix displays," J. Soc. Inf. Disp. **1**(2), 147, 1993.

[88] H. L. Snyder, "Image quality: Measures and visual performance," in *Flat-Panel Displays and CRTs,* Van Nostrand Reinhold, 1985, p. 70.

[89] G. C. de Wit, "Retinal scanning display: Light sources moving over the retina," Sci. Prog. **82,** 135–149, 1999.

[90] E. M. Kolasinski, S. A. Jones, R. S. Kennedy, and R. D. Gilsone, "Postural stability and its relation to simulator sickness," in *Proceedings of the 38th Annual Meeting of the Human Factors and Ergonomics Society,* 1994, Vol. 2, p. 980.

Motion and simulator sickness tend to present more or less the same symptoms: ataxia, nausea, dizziness, eyestrain, headache, drowsiness, fatigue, and vomiting. Although symptoms are similar, the causative factors might not be the same.

Simulator sickness is generally found in flight simulation in which all the cues for visual motion are present but a corresponding vestibular component is not. Another kind of sickness is experienced by some backseat car riders. All cues to the actual motion of the car are present but visual cues are absent. Such visual-vestibular conflicts are found to arise with use of displays like the head-mounted display.

The retinal image is stabilized even during head movements. This stabilization occurs because of the vestibulo-ocular reflex. This reflex is controlled by the semicircular canals in both the ears. As the head moves, the acceleration is sensed by the semicircular canals. These send the input to the vestibular nuclei through the VIII cranial nerve. From here the information is passed to the oculomotor system to produce compensatory eye movements that stabilize the image on the retina (Fig. 30.18).

Cue conflict is one of the major causes of simulator sickness. It refers to the discrepancy between the senses concerning information about body orientation and motion. A primary conflict that tends to cause simulator sickness is one that occurs between vision and vestibular senses. In the case of fixed-base simulators, the visual system senses motion whereas the vestibular system does not. In a moving-base simulator the visual stimuli may not coincide with the corresponding motion that the vestibular stimuli sense.

Head-mounted displays create the sensory disparities given in the subsequent subsections.

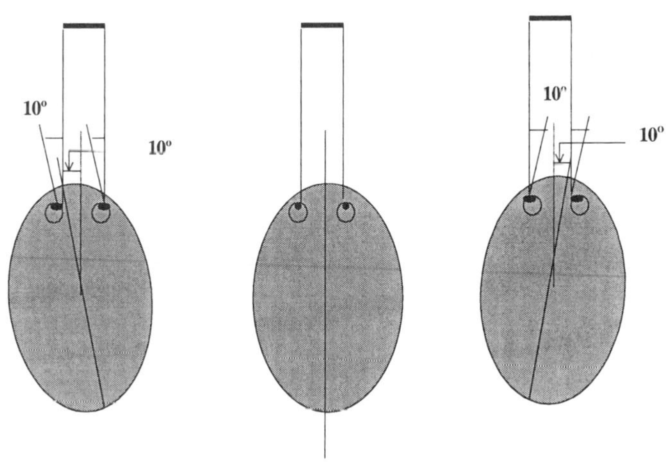

Fig. 30.18. Graphical representation of the vestibular ocular reflex.

30.17.1. Time Lag

This occurs when a delay exists between the movements made by the user and the response of the display. Both vision lag and motion lag have been observed. Vision lag occurs when the subject moves the head but the visual scene is updated at a slower rate. The inverse is motion lag. Visual lag is the more disruptive, and tends to reduce performance more.[91]

30.17.2. Update Rate (Frame Rate)

Update rate refers to the speed of simulation. This is the rate at which the subsequent frames of the moving screen can be generated and rendered into the frame buffer of the display. In many displays a mismatch occurs between the refresh and the screen update rate. A screen would generally be refreshed at 50–60 Hz whereas the update rate might be ~20 Hz. Presenting the same screen more than once can avoid such a mismatch. This would be suitable for a static image display, but for a display with moving images it causes an increase in simulator sickness. The more complex the image, the more it would affect the update rate. The speed at which the new image is rendered decreases. These factors tend to cause a visual lag.

30.17.3. Time Delay

Time delay between the head movement and the image update affects the head tracking performance of a head-mounted display. Many studies have been conducted to find the acceptable amount for time delay, as zero delay is not possible with the current technology. Kalawksy[92] suggested a 5-ms delay as satisfactory.

30.17.4. Image Stabilization

To stabilize images on the retina with a head-mounted display, the head-mounted display is usually accompanied by a tracking device, which provides the computer with the information about the position of the user's head. This aids in providing the adequate amount of eye movements to stabilize the image motion. If there is an error in the tracking there would be a discrepancy be-

[91] L. H. Frank, J. G. Casali, and W. W. Wierwille, "Effects of visual display and motion system delays on operator performance and uneasiness in a driving simulator," Hum. Factors **30,** 201, 1988.

[92] R. S. Kalwsky, *The Science of Virtual Reality and Virtual Environments,* Addison-Wesley, 1993.

tween where the image is located in the display and where the user thinks the image should be.

30.17.5. Flicker

Another feature that has been found to increase simulator sickness is flicker. Pausch et al.[93] and Rinalducci and Mac Arthur[94] have found that flicker is associated with simulator sickness. Flicker is affected by many factors such as luminance, stimulus size, refresh rates, etc. When a display is selected to be flicker free, several trade-offs are necessary.

Display field of view: It has been found by a study conducted by Regan and Price[95] and Kennedy et al.[96] that wider fields of view cause more simulator sickness than narrow fields of view. It was thought that this might be due to increased vection arising from increased stimulation of the peripheral retina. A wider field of view tends to increase the likelihood to flicker, due to the peripheral retina being more sensitive to flicker. This is due to rods being more sensitive to low luminance than cones.

Here are some possible solutions to decrease head-mounted display simulator sickness:

1. Increase the update rate.
2. Include a highly efficient head-tracking system.
3. Decrease the head-mounted display's head-supported weight, as added weight causes a problem between the motor commands and head movements.
4. Avoid using a wide field of view.
5. Postural instability should be decreased (occurs more in immersive displays with wide field of view, in which observers tend to have localization problems).

A see-through and look-around head-mounted display has been found to have a decrease in the perception of simulator sickness.[97] This is due to the stability provided by a view of the outside world as well as position of one's own body.

[93] R. Pausch, T. Crea, and M. Conway, "A literature survey for virtual environments: Military flight simulator visual systems and simulator sickness," Presence **1,** 344, 1992.

[94] E. Rinalducci and M. Mac Arthur, "Annotated bibliography on the effects of flicker on simulator sickness," Institute for Simulation and Training, University of Central Florida, Division of Sponsored Research, Orlando, Florida, 1990.

[95] E. C. Regan and E. R. Price, "The frequency of occurrence and severity of side-effects of immersion virtual reality," Aviat. Space Environ. Med. **65,** 527, 1994.

[96] R. S. Kennedy, M. G. Lilienthal, K. S. Berbaum, D. R. Baltzley, and M. E. McCauley, "Simulator sickness in U.S. Navy flight simulators," Aviat. Space Environ. Med. **60,** 10, 1989.

[97] A. Buckert-Donelson, "Interview with Dean Inman," VR World **3,** 22, 1995.

This provides the user with a better visual reference to prevent simulator sickness.

30.18. Visual Complaints with Visual Display Terminal Use

Studies have shown that use of a visual display terminal is associated with an increased occurrence of certain eye discomfort symptoms.[98–100] The most common complaints are eyestrain, such as red, itchy and watery eyes, or dry eye; fatigue, including heaviness of eyelids and forehead; headaches; and difficulty in focusing the eyes. Other nonvisual symptoms are neckache, backache, and muscle spasms. These symptoms experienced by visual display terminal users are caused either by visual conditions or visual environmental factors. Most of the complaints arise because of the prolonged use of computers.

These steps can be taken to reduce visual complaints:

1. It is important to correct even marginal refractive errors.[101] Even half a diopter or less of uncorrected refractive error is found to be uncomfortable for visual display terminal users.
2. Presbyopic visual display terminal users need to be given different glasses suitable for computer use. In the normal presbyopic correction, bifocals are generally prescribed for reading at ~16 in. and for a downward viewing angle of 25°. The visual display terminal is generally viewed from a distance of 20–24 in. (50–60 cm) from the eye and the viewing angle is ~10–15°. A standard bifocal would not suit for this environment. The user would have to tilt the head back to properly use the glasses to view the monitor. The visual display terminal is used at an intermediate viewing distance. Occupational glasses specifically designed for the visual display terminal are preferred.
3. Eye irritation is one of the most common complaints of visual display terminal users. This occurs as the observer needs to stare at fixed objects for long periods of time. This reduces the blink rate and results in dry eye. Artificial tears need to be used, and users should be advised to take rest breaks every 20 min and to blink frequently.
4. Another important way to eliminate visual display terminal problems is to increase the quality of the image on the screen.[102] The main factors

[98] U. O. Bergqvist and B. G. Knave, "Eye discomfort and work with visual display terminals," Scand. J. Work Environ. Health **20,** 27, 1994.

[99] L. Ivarm, "VDT work, oculomotor strain, and subjective complaints: An experimental and clinical study," Ergonomics **37,** 1419, 1994.

[100] J. E. Sheedy, "Vision problems at video display terminals: A survey of optometrists," J. Optom. Assoc. **63,** 687, 1992.

[101] J. E. Sheedy, "The bottom line on fixing computer-related vision and eye problems," J. Am. Optom. Assoc. **67,** 512, 1996.

[102] J. W. Eichenbaum, "Computers and eyestrain," J. Ophthal. Nurs. Technol. **15,** 23, 1996.

in image quality include image stability, spacing resolution, luminance, and chromaticity. Screen flicker has also been identified as a significant factor in producing eyestrain. The characters on the screen are not as sharp as seen in print. This causes problems with accommodation.

5. Users who are in front of the monitor for long periods of time strain their eyes and often have ocular fatigue or tired eyes. These complaints arise because of binocular vision problems.[103] The most frequent binocular vision problems experienced by the visual display terminal user are convergence insufficiency, esophoria, and exophoria. [Esophoria is a state in which the eyes tend to wander inward (toward the nose) when both eyes are in a straight-ahead direction. With exophoria, the eyes (one or both) tend to wander outward (away from the nose) when both eyes are in a straight-ahead direction.] The options available to treat these problems are correction of any significant ametropia present, the use of added lens power, prism, and vision therapy. Taking a break every 15–20 min and focusing on a distant object would help in relaxing eye muscles.
6. Improving the arrangement of the work environment can eliminate most of the complaints of a visual display terminal user.[104] A comfortable workstation is crucial for reducing fatigue and eyestrain. Light focused directly on textural material is preferred to overhead fluorescent lighting, which produces annoying monitor reflection. The overall illumination of the room should not be three times brighter than the screen. A desk lamp can be used instead of overhead lights. Glare should be controlled from over headlights and windows. Glare filters on the computer and antireflection coatings can be used to limit glare. Light sources should be arranged in such a way that they minimize glare and reflection on the screen. Reference material should be kept close to the video display terminal. This will reduce the amount of unnecessary eye and head movements and possible neckache and backache.

Further Reading

J. F. Amos, *Diagnosis and Management in Vision Care,* Butterworth, 1987.

R. J. Beaton, "Linear system metrics of image quality for flat-paned displays," in *Three-Dimensional Imaging and Remote Sensing Imaging,* W. E. Robbins, ed., Proc. SPIE **902,** 125, 1988.

I. M. Borish, *Clinical Refraction,* 3rd ed., Professional Press, 1975, Vol. I.

J. L. Brown, "Flicker and intermittent stimulation," in *Vision and Visual Perception,* C. H. Graham, ed., Wiley, 1965.

[103] M. Scheiman, "Accommodative and binocular vision disorders associated with video display terminals: Diagnosis and management issues," J. Am. Optom. Assoc. **67,** 531, 1996.

[104] D. Koh, C. N. Ong, and J. Jeyaratnam, "The safe use of visual display units," Sing. Med. J. **35,** 381–385, 1994.

S. P. Butzon and S. R. Eagels, "Prescribing for the moderate-to-advanced ametropic presbyopic VDT user," J. Optom. Assoc. **68,** 495, 1997.

F. W. Campbell, "The human eye as an optical filter," Proc. IEEE **56,** 1009, 1968.

F. W. Campbell and D. G. Green, "Optical and retinal factors affecting visual resolution," J. Physiol. **181,** 576, 1965.

F. W. Campbell and R. W. Gubisch, "Optical quality of the human eye," J. Physiol. **186,** 558, 1966.

K. Carr and R. England, *Simulated and Virtual Reality,* Taylor & Francis, 1995.

C. Carterette and P. F. Friedman, eds., *Seeing,* Vol. V of Handbook of Perception Series, Academic, 1975.

R. Chaplin and R. A. Fremantle, "Measurement of perceived flicker in VDT products," Displays **8,** 22, 1987.

W. C. Charman and G. Walsh, "The optical transfer function of the eye and perception of spatial phase," Vis. Res. **25,** 619, 1985.

CIE, "International lighting vocabulary," CIE Pub. 17.4, 1987.

T. N. Cornsweet, *Visual Perception,* Academic, 1970.

W. Cowan, "Displays for vision research," in *Handbook of Optics,* M. Bass, ed., McGraw-Hill, 1995, Vol. I.

B. H. Crawford, "The dependence of pupil size upon external light stimuli under static and variable conditions," Proc. R. Soc. London B **121,** 373, 1936.

J. Cronly-Dillon, E. S. Rosen, and J. Marshall, eds., "Hazards of light: Myths and realities: Eye and skin," in *Proceedings of the First International Symposium of the Northern Eye Institute,* University of Manchester, 1985.

H. DeLange, "Eye's response at flicker fusion to square wave modulation of a test field surrounded by a large steady field of equal mean luminance," J. Opt. Soc. Am. **51,** 415, 1961.

G. K. Edgar, J. C. D. Pope, and I. Craig, "Visual accommodation problem with head-up and helmet-mounted displays?," Displays **65,** 68, 1994.

S. Eriksson and L. Backstrom, "Temporal and spatial stability in visual displays," in *Work with Display Units,* B. Knave and P. G. Wideback, eds., North-Holland, 1987, pp. 461–473.

T. E. Fannin and T. Grosvenor, *Clinical Optics,* Butterworth, 1987.

J. E. Farrell, "Objective methods for evaluating screen flicker," in B. Knave and P. G. Wideback, eds., *Work with Display Units,* North-Holland, 1987, p. 449.

J. C. Feltz, "Development of modular transfer function and contrast transfer function for discrete systems, particularly charge-coupled devices," Opt. Eng. **29,** 893, 1990.

R. E. Fisher, "Optics for head mounted displays," Inf. Disp. **10,** 112, 1994.

R. E. Fisher, *Fundamental's of HMD Optics: Head Mounted Displays: Designing for the User,* McGraw-Hill, 1997.

J. D. Gould, "Visual factors in the design of computer controlled CRT displays," Hum. Factors **40,** 554, 1968.

S. Gur, S. Ron, and A. Heicklen-Klein, "Objective evaluation of visual fatigue in VDU workers," Occupat. Med. **44,** 201, 1994.

G. Haegerstrom-Portnoy, M. E. Schneck, and J. A. Brabyn, "Seeing into old age: Vision function beyond acuity," Optom. Vis. Sci. **76,** 141, 1999.

W. T. Ham Jr., H. A. Muller, and D. H. Sliney, "Retinal sensitivity to damage from short wavelength light," Nature (London) **260,** 153, 1976.

K. Harwood and O. Foley, "Temporal resolution: an insight into the video display terminal (VDT) problem," Hum. Factors **29,** 447, 1987.

P. A. Howarth, "Visual issues with head mounted displays," Optom. Vis. Sci. (suppl.) **72,** 169, 1995.

P. A. Howarth, "Empirical studies of accommodation, convergence and HMD use," in *Proceedings of the Hoso-Bunka Foundation Symposium: The Human Factors in 3D Imaging,* Hosu-Bunka Foundation, 1996.

R. W. G. Hunt, *Measuring Color,* Wiley, 1987.

J. E. Kaufmann and H. Haynes, eds., *IES Lighting Handbook (Reference Volume),* Illuminating Engineering Society of North America, 1984.

C. Infante, "On the modulation transfer function of matrix displays," J. Soc. Inf. Disp. **26,** 449, 1993.

J. Ivaecchia, H. Ivaecchia, and S. N. Roscoe, "Eye accommodation to head up virtual images," Hum. Factors **30,** 689–702, 1988.

J. A. M. Jennings and W. N. Charman, "Analytic approximation of the off-axis modulation transfer function of the eye," Vis. Res. **37,** 697, 1997.

R. Jones, "Binocular fusion factors in visual displays," in *SID 92 Digest,* Society of Information Display, 1992, p. 267.

D. H. Kelly, "Sine waves and flicker fusion," Doc. Ophthalmol. **18,** 16, 1964.

D. H. Kelly, "Diffusion model of linear flicker responses," J. Opt. Soc. Am. **59,** 1665, 1669.

D. H. Kelly, "Flicker," in *Visual Psychophysics,* Vol. VIII/4 of Handbook of Sensory Physiology Series, D. Jameson and L. M. Hurvich, eds., Springer-Verlag, 1972.

P. B. Kruger and J. Polla, "Stimuli for accommodation: Blur, chromatic aberration and size," Vis. Res. **26,** 951, 1986.

J. J. Kulikowski, W. Walsh, and I. J. Murray, eds., "Limits of vision," in *Vision and Visual Dysfunction,* CRC, 1991, Vol. 5.

T. M. Lippert, "Fundamental monocular/binocular HMD factors," in *Helmet-Mounted Displays II,* R. J. Lewandowski, ed., Proc. SPIE **1290,** 220–223, 1990.

M. A. Losado, R. Navarro, and J. Santamaria, "Relative contributions of optical and neural limitations to human contrast sensitivity at different luminance levels," Vis. Res. **33,** 2321, 1993.

N. P. Lyons and J. E. Farrell, "Linear system analysis of CRT displays," in *SID 89 Digest,* Society of Information Display, 1989, pp. 220–223.

J. Mandelbaum, "An accommodation phenomenon," Arch. Ophthalmol. **63,** 923, 1960.

C. A. Maxwell, "Flicker science and the consumer," Inf. Disp. **11,** 7, 1992.

E. J. Melzer and K. Moffitt, eds., In *Head Mounted Displays: Designing for the User,* McGraw-Hill, 1997.

M. Mon-Williams, J. P. Wann, and S. Rushton, "Binocular vision in a virtual world: Visual deficits following the wear of a head mounted display," Ophthalm. Physio. Opt. **13,** 387, 1993.

M. W. Morgan, *The Optics of Ophthalmic Lenses,* Professional Press, 1978.

P. Mouroulis, *Visual Instrumentation: Optical Design and Engineering Principles,* McGraw-Hill, 1999.

C. Neveu, T. Blackmon, and L. Stark, "Evaluation of the effects of a head-mounted display on ocular accommodation," Presence-Teleoperat. Virtual Environ. **7,** 278, 1998.

A. V. Oppenheim and A. S. Willsky, *Signals and Systems,* Prentice-Hall, 1983.

S. E. Palmer, *Vision Science: Photons to Phenomenology,* MIT Press, 1999.

M. Pantazis, *Visual Instrumentation and Engineering Principles,* McGraw-Hill, 1999.

E. Peli, "Visual issues in the use of head mounted monocular display," Opt. Eng. **29,** 883, 1990.

E. Peli, "The visual effects of HMD are not distinguishable from those of desk top computer display," Vis. Res. **38,** 2053, 1998.

T. Piantanida, "Another look at HMD safety," Cyber Edge J. **9–10,** 12, 1993.

T. Piantanida, D. Boman, and J. Gille, "Human perceptual issues and virtual reality," Virtual Real. Syst. **1,** 43, 1993.

T. Piantanida, D. Bowman, J. Larimer, J. Gille, and C. Reed, "Studies of the field of view/resolution tradeoff in virtual reality systems," in *Proceedings of the SPIE/IS&T Symposium on Electronic Imaging, Science and Technology,* 1992.

D. G. Pitts, A. P. Cullen, and P. D. Hacker, "Ocular effects of ultraviolet radiation from 295nm to 365nm," Invest. Ophthalmol. Vis. Sci. **16,** 932, 1977.

D. G. Pitts, "The human ultraviolet action spectrum," Am. J. Optom. Physiol. Opt. **51,** 946, 1974.

D. G. Pitts and T. J. Tredici, "The effects of ultraviolet on the eye," Am. Ind. Hyg. Assoc. J. **32,** 235, 1971.

S. N. Roscoe, "Judgements of size and distance with imaging displays," Hum. Factors **26,** 617, 1984.

J. A. Roufs, ed., "The man machine interface," in *Vision and Visual Dysfunction,* CRC, 1991, Vol. 15.

S. Sanders and E. J. McCormick, *Human Factors in Engineering and Design,* McGraw-Hill, 1993.

R. Sekuler, D. Kline, and K. Dismukes, *Aging and Human Visual Function,* Liss, 1982.

S. Sherr, *Electronic Displays,* 2nd ed., Wiley, 1993.

S. Sherr, *Video and Digital Electronic Displays: A User's Guide,* Wiley, 1982.

S. Shiwa, K. Omura, and T. Miyasato, "Development of three-dimensional display with accommodative compensation," in *Proceedings of the Third International Display Workshop,* Society for Informational Display, 1996.

J. G. Sivak and G. C. Woo, "Color of visual display terminals and the eye: Green VDTs provide the optimal stimulus to accommodation," Am. J. Optom. Physiol. Opt. **60,** 678–687, 1983.

D. H. Sliney, "Laser effects on vision and ocular exposure limits," Appl. Occupat. Environ. Hyg. **11,** 313, 1996.

D. H. Sliney, "Eye protection techniques for bright light," Ophthalmology **90,** 937, 1983.

D. H. Sliney, "Ocular injuries from laser accidents," in *Laser Inflicted Eye Injuries: Epidemiology, Prevention and Treatment,* Proc. SPIE **2674,** 25–33, 1996.

D. H. Sliney and B. C. Frasier, "The evaluation of optical radiation hazards," Appl. Opt. **12,** 1, 1973.

R. Steen, D. Whitaker, D. B. Elliot, and J. M. Wild, "Age-related effects of glare on luminance and color contrast sensitivity," Optom. Vis. Sci. **71,** 792, 1994.

L. N. Thibos and A. Bradley, "New methods of discriminating neural and optical losses in vision," Optom. Vis. Sci. **70,** 279, 1993.

L. N. Thibos, F. E. Cheney, and D. Walsh, "Retinal limits to the detection and resolution of gratings," J. Opt. Soc. Am. A **4,** 1524, 1987.

C. W. Tyler and R. D. Hamer, "Eccentricity and the Ferry–Porter law," J. Opt. Soc. Am. A **10,** 2084, 1993.

J. P. Wann, S. Rushton, and M. Mon-Williams, "Natural problems for stereoscopic depth perception in virtual environments," Vis. Res. **35,** 2731, 1995.

B. A. Watson, "Temporal sensitivity," in *Handbook of Perception and Human Performance,* L. Kaufmann, R. K. Boff, and P. J. Thomas, eds., Wiley, 1986, Vol. I, Chap. 6.

N. P. Wiggins and K. M. Daum, "Visual discomfort and astigmatic refractive errors in VDT use," J. Am. Optom. Assoc. **62,** 680, 1991.

D. R. Williams, D. H. Brainard, M. J. McMahon, and R. Navarro, "Double pass and interferometric measures of the optical quality of the eye," J. Opt. Soc. Am. A **11,** 3123, 1994.

World Health Organization, Environmental Health Criteria 23, *Lasers and Optical Radiation.*

M. Ziefle, "Effects of display resolution on visual performance," Hum. Factors **40,** 554, 1998.

J. A. Zuelich, "Ultraviolet induced photochemical damage in ocular tissues," Health Phys. **56,** 671, 1989.

31

Electro-Optics and Magneto-Optics

William L. Wolfe

31.1. Introduction

The application of electric or magnetic fields can polarize the light in them, which is then useful in such things as spectral line splitting and shutters.

31.2. Zeeman Effect[1]

In the presence of a magnetic field spectral lines of a gas are split into one at a higher frequency and one lower, when viewed parallel to the field (longitudinal Zeeman effect), separated by $2\Delta\nu$, where

$$\Delta\sigma = m' \frac{gqH}{4\pi mc} = 4.67 \times 10^{-5} gH,$$

where $m' = \pm 1/2, \pm 3/2 \cdots \pm j$, j is the angular quantum number, g is the Landé g factor, q is the electronic charge, H is the magnetic-field strength, m is the mass of the electron, and c is the speed of light; H must be in gauss for the number to be correct. The two lines are oppositely, circularly polarized.

If the lines are viewed perpendicular to the field (lateral effect), the separation is the same, but the original, center line remains. The two outside lines are linearly polarized in one direction; the center one in the other.

31.3. Anomalous Zeeman Effect

There are actually more than two or three lines generated by the magnetic field. (Like anomalous dispersion, this is the general case and not anomalous, but was discovered laer).

[1] P. Zeeman, *Researches in Magneto Optics,* Macmillan, 1913.

31.4. Stark Effect

This is an analogy to the anomalous Zeeman effect. Intense electric fields introduce the splitting of spectral lines. When viewed perpendicular to the electric field, the lines are linearly polarized, with both polarizations; when viewed parallel, only the perpendicular components appear.

The change in wave number in a hydrogenlike atom is given by

$$\Delta\sigma = AE + BE^2 + CE^3 + \cdots,$$

where

$$A = \frac{3h}{8\pi^2 mqc} n(n_2 - n_1) = 6.42 \times 10^{-5} n(n_2 - n_1),$$

$$B = \frac{h^5}{2^{10}\pi^6 m^3 q^6 c} n^4[17n^2 - 3(n_2 - n_1) - 9m_l^2 + 19]$$
$$= 5.22 \times 10^{-16} n^4[17n^2 - 3(n_2 - n_1) - 9m_l^2 + 19],$$

$$C = \frac{3h^9}{2^{15}\pi^{10} m^5 q^{11} c} n^4[23n^2 - (n_2 - n_1)^2 + 11m_l^2 + 39]$$
$$= 1.53 \times 10^{-25} n^4[23n^2 - (n_2 - n_1)^2 + 11m_l^2 + 39],$$

where the field strengths must be in volts per centimeter for the numbers to be accurate, $m_l = n - n_2 - n_1 - 1$, n is the total quantum number, and the others are electric quantum numbers. For n and $n_2 - n_1 = 1$, the change in wave numbers for the primary Stark effect is only 6.42×10^{-5} reciprocal centimeters. Strong fields are necessary to split the spectral lines.

31.5. Inverse Zeeman Effect

This is the Zeeman effect in absorption. The frequency shifts and lines are the same, but the polarizations are reversed.

31.6. Inverse Stark Effect

This is the Stark effect in absorption. The frequency shifts and lines are the same, but the polarizations are reversed.

31.7. Paschen-Back Effect

In the presence of strong magnetic fields, all the anomalous Zeeman patterns become normal triplets. The spectral shift is slightly different:

$$\Delta\sigma = \sigma_0 - (m_l + 2m_s)\frac{Hq}{4\pi mc^2} - am_l m_s,$$

where the m's are the magnetic quantum numbers, $m = \pm 1/2$, $m_l = 0, \pm 1$.

31.8. Faraday Effect

The application of a strong magnetic field to a material can make it optically active. The rotation angle θ is given by

$$\theta = VHd,$$

where V is the Verdet constant, H is the magnetic field, and d is the thickness of the material.

31.9. Voigt Effect[2]

When light is passed through a gas perpendicular to a magnetic field, the light is doubly refracted.[2] This effect is approximately proportional to the magnetic-field strength and depends on the gas and its density.

31.10. Cotton–Mouton Effect

This is the Voigt effect in a liquid. The effect is much stronger than the Voigt effect (sometimes 1000 times) and is proportional to the square of the magnetic-field strength.

31.11. Kerr Electro-Optic Effect

When a glass or liquid is subjected to an electric field, it becomes doubly refracting, like a uniaxial crystal with the optical axis parallel to the field. The optical path difference between the two components is given by

$$\text{OPD} = K\frac{lV^2\lambda}{d^2},$$

where K is the Kerr constant, l is the length of the cell, d is the plate separation, λ is the optical wavelength in the medium, and V is the voltage between the plates.

This effect is used in the Kerr cell, an electro-optical switch.

[2] W. Voigt, *Magneto-und Elektro-optik,* Teubner, 1908.

31.12. Kerr Magneto-Optic Effect[3]

Linearly polarized light incident on a magnetic material becomes elliptically polarized with the major axis rotated with respect to the incident polarization direction. There are three orientations (conditions) for the effect, called polar, longitudinal, and transverse; these arrangements are shown in Figs. 31.1, 31.2, and 31.3, respectively.

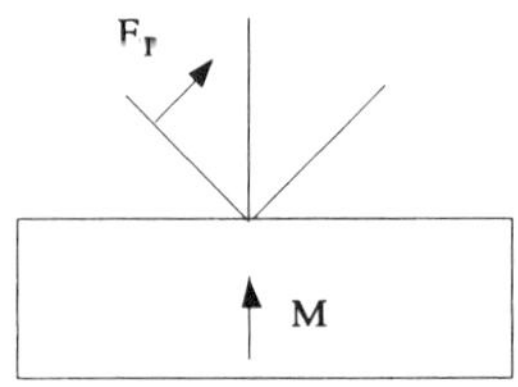

Fig. 31.1. Polar.

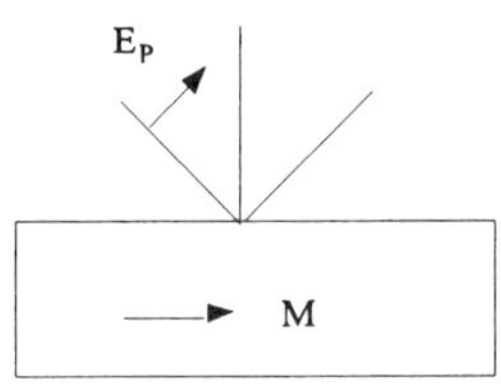

Fig. 31.2. Longitudinal.

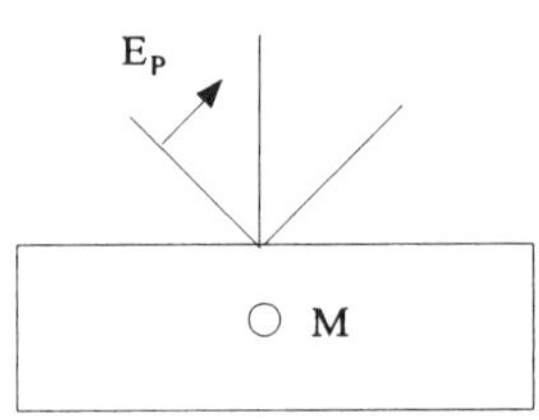

Fig. 31.3. Transverse.

In all three, the parallel component of the electric field is shown and the perpendicular one is out of the paper. Similarly, in the transverse case, the magnetic moment is out of the plane of the paper.

Each of these has a different degree of induced ellipticity and rotation as a function of angle of incidence and material properties. The effect can be used to generate various degrees of extinction when used with polarized input light and analyzers.

[3] M. Mansuripur, *The Physical Principles of Magneto-Optical Recording,* Cambridge Univ. Press, 1995.

31.13. Pockels Effect

This is also called the linear electro-optical effect. The index ellipsoid in an anisotropic crystal is described by

$$\frac{x_1^2}{n_1^2} + \frac{x_2^2}{n_2^2} + \frac{x_3^2}{n_3^2} + \frac{x_2x_3}{n_4^2} + \frac{x_1x_3}{n_5^2} + \frac{x_1x_2}{n_6^2} = 1.$$

When an electric field is applied, the refractive indices change according to

$$\Delta\left(\frac{1}{n^2}\right)_i = r_{ij}E_j$$

and summation is over repeated indices. The index ellipsoid written in terms of the electro-optic coefficients is

$$\left(\frac{1}{n_1^2} + r_{1j}E_j\right)x_1^2 + \left(\frac{1}{n_2^2} + r_{2j}E_j\right)x_1^2 + \left(\frac{1}{n_3^2} + r_{3j}E_j\right)x_1^2 + 2\left(\frac{1}{n_4^2} + r_{4j}E_j\right)x_2x_3$$

$$+ 2\left(\frac{1}{n_5^2} + r_{5j}E_j\right)x_1x_3 + 2\left(\frac{1}{n_6^2} + r_{6j}E_j\right)x_1x_2 = 1.$$

Use of symmetry can reduce the complexity. If the beam is directed along x_2 and the field along x_3, then the ellipse perpendicular to the beam is given by

$$\left(\frac{1}{n_1^2} + r_{13}E_3\right)x_1^2 + \left(\frac{1}{n_3^2} + r_{33}E_3\right)x_3^2 + 2\left(\frac{1}{n_5^2} + r_{53}E_3\right)x_1x_3 = 1.$$

If x_1, x_2, and x_3 are aligned with the crystal axes, a further simplification is obtained:

$$\left(\frac{1}{n_x^2} + r_{13}E_z\right)x^2 + \left(\frac{1}{n_z^2} + r_{33}E_z\right)x^2 + 2r_{53}E_zxz = 1.$$

When r_{53} is negligible, and it often is, then the changes is refractive index can be written as

$$\Delta n_x = -\frac{1}{2}n_x^3r_{13}E_z, \quad \Delta n_z = -\frac{1}{2}n_z^3r_{33}E_z.$$

31.14. Digital Light Deflectors

These are two-position beam devices. They consist of a Pockels cell that changes the refractive properties of a medium and an analyzer that transmits one or another of the polarized beams. Several geometries have been proposed: simple birefringent plate, total internal reflection polarizer, Wallaston prism, and birefringent wedge, as described in the subsequent subsections.

31.14.1. Simple Plate

This plate (see Fig. 31.4) is arranged such that the *o* ray passes through undeviated, while the *e* ray is refracted through an angle δ, the maximum value of which is given by

$$\delta_{\max} = \arctan \frac{n_o^2 - n_e^2}{2n_o n_e};$$

the two beams exit parallel, a distance d apart, given by

$$d = t \tan \delta,$$

where n_o and n_e are the ordinary and the extraordinary refractive indices, respectively, and t is the thickness of the cell. The *e* ray is polarized vertically, the *o* ray out of the plane of the paper. The angle of the optical axis is given by $\arctan(n_e/n_o)$.

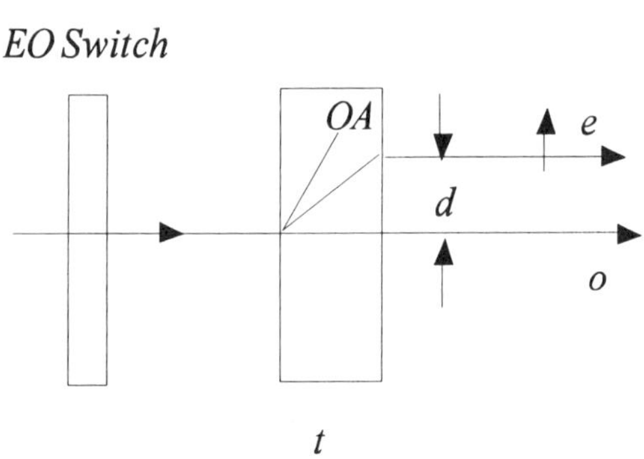

Fig. 31.4. Simple plate. OA, optical axis.

31.14.2. Total Internal Reflection Plate

The birefringent plate is placed in a fluid that matches the higher of the two refractive indices, typically the *e* ray (see Fig. 31.5). Then the *o* ray passes through, while the *e* ray is reflected up and then forward by the mirror. The separation d is established by the position of the mirror. The optical axis is out of the plane of the paper.

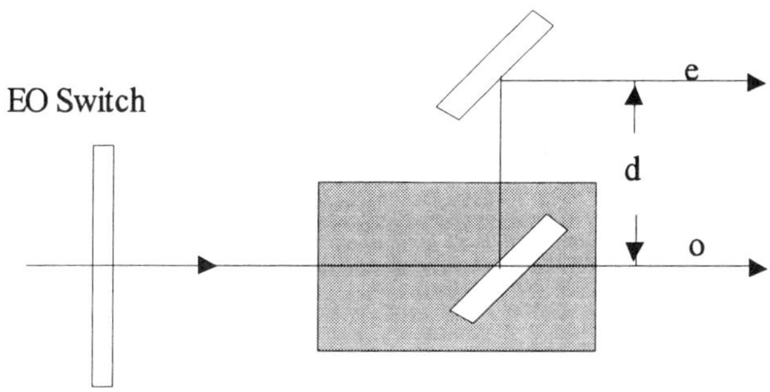

Fig. 31.5. Total internal reflection plate.

31.14.3. Wallaston Prism

The Wallaston prism (see Fig. 31.6) consists of two birefringent plates with their optical axes orthogonal, so that the *o* ray of the first prism becomes the *e* ray of the second. The angular separation between the two beams is given by

$$\delta = 2(n_e - n_o)\tan \gamma$$

for normal incidence (the most common angle) and is about the same for small incidence angles, but the angles are different.

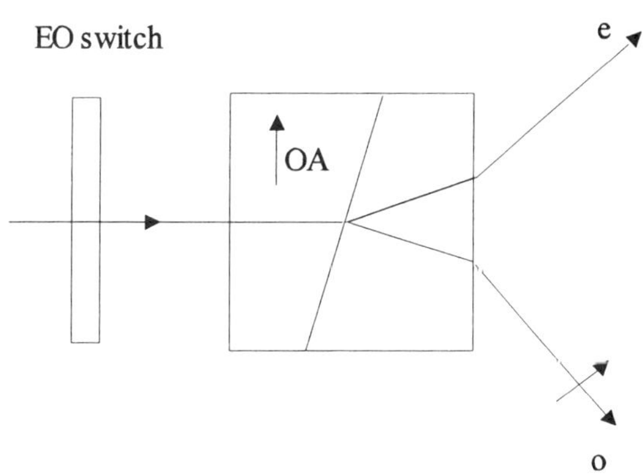

Fig. 31.6. Wallaston prism.

31.14.4. Simple Birefringent Wedge

For a prism of angle γ, the beam separation δ is given by

$$\delta \approx (n_o - n_e)\sin \gamma.$$

The polarization of the o ray is shown in Fig. 31.7; that of the e ray is out of the plane of the paper, as is the optical axis.

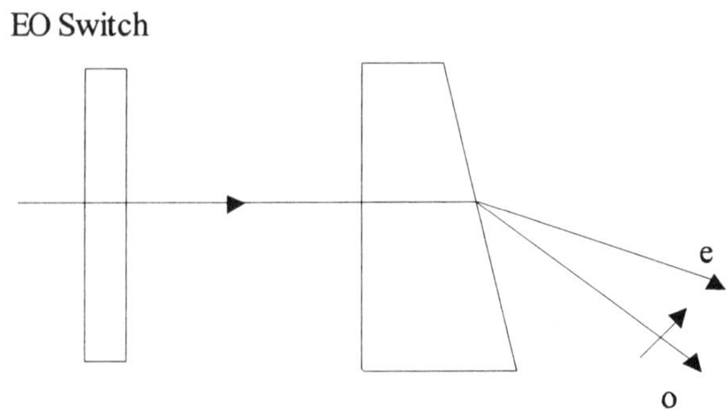

Fig. 31.7. Birefringent wedge.

31.15. Analog Beam Deflectors

Most analog deflectors consist of one or more prisms, the refractive index of which is changed electro-optically. The angle of deflection is given by

$$\delta = \theta_1 + \arcsin\left\{n \sin\left[\alpha - \arcsin\left(\frac{\sin\theta}{n}\right)\right]\right\} - \alpha.$$

Its change is due to a change in refractive index, given by

$$\Delta\delta = \frac{\sin\alpha + \dfrac{\dfrac{\sin\theta}{n^2}}{\sqrt{1-\left[\arcsin\left(\dfrac{\sin\theta}{n}\right)\right]^2}}}{\sqrt{1-\left[n\sin\alpha - \arcsin\left(\dfrac{\sin\theta}{n}\right)\right]^2}}\,\Delta n,$$

where θ is the angle of incidence on the front face and n is the refractive index relative to the medium in which the prism is placed. The change in index is given by

$$\Delta n_x = -\frac{1}{2}\,n_x^3 r_{13} E_z, \quad \Delta n_z = -\frac{1}{2}\,n_z^3 r_{33} E_z.$$

Further Reading

F. A. Jenkins and H. E. White, *Fundamentals of Optics,* 3rd ed., McGraw-Hill, 1957.

H. E. White, *Introduction to Atomic Spectra,* McGraw-Hill, 1934.

M. Gottlieb, C.L.M. Ireland, and J.M Ley, *Electro-Optic and Acousto-Optic Scanning and Deflection,* Marcel Dekker, 1983.

32

Acousto-Optics

Neelam Gupta

32.1. Introduction

Acousto-optics deals with the interaction of acoustic (usually ultrasonic) and light waves in a liquid, solid, or gaseous medium. Ultrasound waves can modulate the amplitude and phase of light beams: they can deflect light beams; they can Doppler shift the frequency of light; and they can control optical polarization and filter the white light into different colors. In general, there are three main classes of acousto-optic devices that are currently in use: deflectors, modulators, and tunable filters. Any such device—RF spectrum analyzer, spatial-frequency filter, correlator, convolver, or processor—belongs to one of these three basic classes of instruments. There are two types of acousto-optic interactions: isotropic and anisotropic. In the isotropic interaction (the refractive index has the same value for both the incident and the diffracted light beams) the polarization of the diffracted light is usually the same as that of the incident light, whereas in the anisotropic interaction (the refractive index for the diffracted beam is different than that for the incident beam) the polarization of the diffracted beam is orthogonal to the linearly polarized incident beam; i.e., if the incident beam is an ordinary wave, it becomes an extraordinary wave after diffraction or vice versa. There are two main regimes of these diffraction processes: Bragg diffraction, in which the incident light is diffracted into a single order, and Raman–Nath diffraction, in which light gets diffracted into multiple orders. In birefringent media there also exist Bragg regimes of interaction with more than one maximum (known as the degenerate case). In Bragg diffraction, close to 100% of the incident light is generally diffracted in +1 or −1 order of diffraction, whereas the diffracted light intensity in any of the Raman–Nath maxima is always less than 34%. Therefore the Bragg diffraction regime is usually used in designing practical devices. Most modulators or Bragg cells are based on the isotropic interaction whereas filters and high-performance deflectors are based on the anisotropic interaction. In optical information processing and laser technology, deflectors and modulators are used to electronically manipulate laser beams. Some compact spectrometers and multi- or hyperspectral imaging systems use acousto-optic tunable filters.

32.2. Acousto-Optic Interactions

All acousto-optic interactions are based on the photoelastic or elasto-optic effect. An RF electric signal is applied to a piezoelectric transducer that generates ultrasound in a medium. The acoustic signal produces a strain that changes the optical properties of the medium. As this acoustic wave propagates in the medium, it gets absorbed at the other end by an acoustic absorber, as shown in Fig. 32.1. An expanded cross-sectional view of the transducer is shown in Fig. 32.2. (The thickness of the transducer is usually equal to about

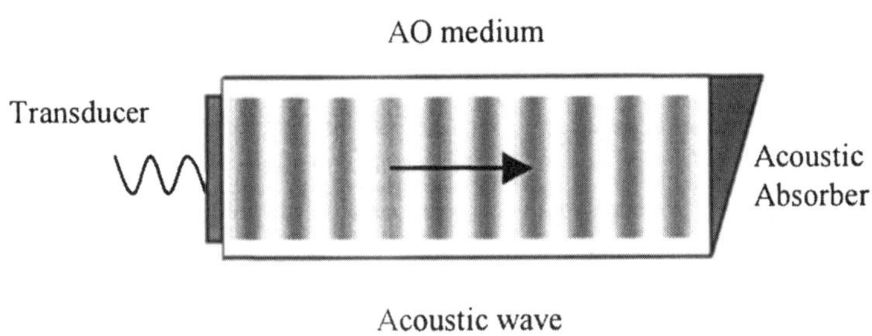

Fig. 32.1. Schematic drawing of an acousto-optic (AO) cell.

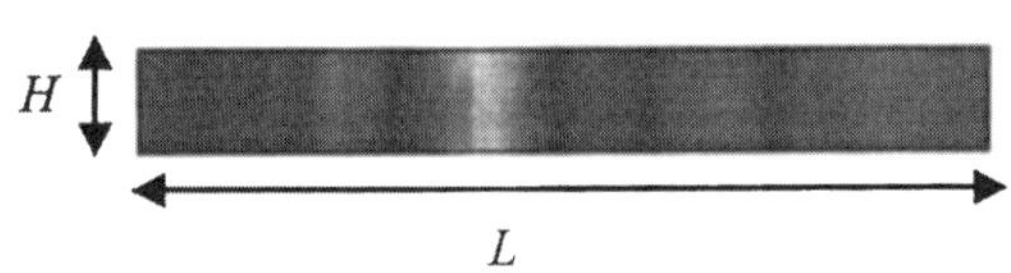

Fig. 32.2 Cross-sectional view of a crystal facet with transducer of length L and width (or height) H.

one half-wavelength of ultrasound in the piezoelectric material). The traveling sound wave produces regions that are more or less dense; that is, a volume optical phase grating with its period equal to the wavelength of sound is set up in the medium. When a light beam passes through the cell it may get deflected or modulated, and its frequency is shifted either up or down by the frequency of the sound wave (Doppler shift). The periodic regions of density variations correspond to a periodic variation in refractive index n of the medium. The strain and stress in the medium are related by a generalized Hooke's law. The resulting change in refractive index is[1]

[1] J. Xu and R. Stroud, *Acousto-optic Devices: Principles, Design, and Applications,* Wiley, 1992.

$$\Delta\left(\frac{1}{n^2}\right)_i \equiv \Delta B_i = p_{ij}S_j, \quad i, j = 1,2, \ldots ,6, \tag{32.1}$$

where ΔB_i are the changes in components of relative impermeability tensor, p_{ij} are the components of the photoelastic tensor, and S_j are the strain components. In general, there are 36 components of the photoelastic tensor but for the most common crystals of higher symmetry only a few are nonzero. The three strain components S_j with $j = 1, 2$, and 3 represent longitudinal waves along the x, y, or z axis, while S_j with $j = 4, 5$, and 6 represent three shear waves. For anisotropic materials, the strain-induced changes in the material not only modify the index of refraction but also may cause polarization rotation.

32.3. Isotropic Interactions

A good acousto-optic material should have low optical and acoustic attenuation, good optical quality, and large size needed for device fabrication. Such materials are useful only in their spectral transparency region. A single parameter known as the acousto-optic figure of merit, M_2, that depends only on the intrinsic properties of a material, is defined to compare the relative performance of various acousto-optical materials. This figure of merit is useful in expressing the bandwidth and diffraction efficiency of an acousto-optic device, i.e., a deflector, a modulator, etc. It is defined by the following expression.

$$M_2 = \frac{n^7p^2}{\rho V}, \tag{32.2}$$

where ρ is the density, V is the acoustic phase velocity in the material, and p is the relevant photoelastic tensor component. Materials that have large refractive indices, large photoelastic constants, low density, and slow acoustic velocity have high value of M_2. Some of the commonly used acousto-optic materials, their physical properties, and highest values of relative M_2 with respect to fused silica (absolute value of $M_2 = 1.5 \times 10^{-18}$ s^3/g) are listed in Table 32.1.

To derive the characteristics of the diffracted light beams for the isotropic interaction, one needs to solve the wave equation in the acousto-optic medium after Raman and Nath.[2] When an acoustic wave propagates along the x axis and a plane-wave optical beam propagates in the x–z plane at an angle θ_0 from the z axis, the wave equation can be written as

$$\nabla^2\mathbf{E} = \left[\frac{n(x, t)}{c}\right]^2 \frac{\partial^2\mathbf{E}}{\partial t^2}, \tag{32.3}$$

[2] C. V. Raman and N. S. Nath, "The diffraction of light by high-frequency sound waves: Part IV," Proc. Indian Acad. Sci. **3**, 459–465, 1936.

Table 32.1. Selected Acousto-Optic Materials

Material	Transparency Window (μm)	Refractive Indices	Acoustic Velocity ($\times 10^5$ cm/s)	Density (g/cm^3)	M_2
Crystal quartz (SiO_2)	0.2–4.5	$n_o = 1.54$ $n_e = 1.55$	6.32	2.2	1.0
$LiNbO_3$	0.4–4.5	$n_o = 2.20$ $n_e = 2.29$	7.33	4.64	4.6
GaP	0.54–10.5	$n = 3.31$	6.3	4.13	30
Ge	2.0–20	$n = 4.0$	5.5	5.33	170
$CaMoO_4$	0.4–4.5	$n_o = 1.95$ $n_e = 1.96$	6.0	4.25	14
$PbMoO_4$	0.42–5.5	$n_o = 2.39$ $n_e = 2.26$	3.63	6.95	24
TeO_2	0.35–4.5	$n_o = 2.26$ $n_e = 2.41$	0.62	6.0	800
Tl_3AsSe_3	1.1–17	$n_o = 3.38$ $n_e = 3.19$	1.0	7.7	2900
Hg_2Cl_2	0.35–20	$n_o = 1.96$ $n_e = 2.62$	0.35	7.18	700
Hg_2Br_2	0.40–30	$n_o = 2.12$ $n_e = 2.98$	0.27	7.31	2600

where $n(x, t)$ is the refractive index in the acousto-optic interaction region, c is the speed of light, and **E** is the electric field. For a planar, sinusoidal traveling acoustic wave, $n(x, t)$ can be written as

$$n(x, t) = n + \delta n \sin(2\pi f_a t - K_a x), \qquad (32.4)$$

where δn is the change in refractive index that is due to the acoustic wave, and $K_a = 2\pi/\Lambda$ and f_a are the wave number and the frequency of the acoustic wave of wavelength Λ, respectively.

The following coupled-wave equation, known as the Raman–Nath equation, can be derived from Eq. (32.3).

$$\frac{dE_m}{dz} - j\Delta k_m E_m = \frac{\nu}{2L}(E_{m-1} - E_{m+1}), \qquad (32.5)$$

where

$$\nu = \frac{2\pi\delta nL}{\lambda_0 \cos\theta_0}, \tag{32.6}$$

$$\Delta k_m = \frac{mK_a}{\cos\theta_0}\left[\frac{mK_a}{2k_i} - \sin\theta_0\right], \tag{32.7}$$

$$\Delta k_m = k_0 + mK_a - k_m, \tag{32.8}$$

where E_m and $k_m = 2\pi n_m/\lambda_0$ are the amplitude and wave number of the mth diffracted light beam, L is the length of the acoustic column, λ_0 is the wavelength of the incident optical beam in free space, and $k_0 = 2\pi/\lambda_0$ is the corresponding wave number, respectively. The coupling coefficient ν is known as the Raman–Nath parameter, and it represents the optical phase shift induced by the interaction as well as the strength of the acousto-optic interaction. The parameter Δk_m is the momentum mismatch, and $k_i = 2\pi n/\lambda_0$ is the wave number of the incident light. An examination of Eq. (32.5) shows that different optical waves propagate in the material and for each optical wave; energy is exchanged only between adjacent waves, and ν acts as the coupling factor between the adjacent waves. The amount of energy transferred depends on the coupling constant and the factor $m^2K_a^2/2k_i \cos\theta_0$, which indicates the degree of synchronization or phase matching. The phase mismatch is given by $\Delta k_m L/2$.

To evaluate the amount of light transfer, another parameter Q, called the Klein–Cook parameter, is defined as

$$Q = \frac{K_a^2 L}{k_i \cos\theta_0} = \frac{2\pi\lambda_0 L}{n\Lambda^2 \cos\theta_0}. \tag{32.9}$$

With the use of parameter Q, Eq. (32.5) can be rewritten as

$$\frac{\mathrm{d}E_m}{\mathrm{d}z} + \frac{\nu}{2L}(E_{m+1} - E_{m-1}) = j\left[\frac{m^2Q}{2L} - mK_a \tan\theta_0\right]E_m. \tag{32.10}$$

It is clear from Eq. (32.10) that an appreciable amount of light is transferred from the zeroth order to the $m = \pm 1$ order, only when either or both coefficients on the right-hand side of the equation are small. This can happen if Q is small and θ_0 is close to 0°, or if Q is large and the terms inside the square brackets are equal to each other. Also, Q is a measure of the momentum mismatch. Based on the value of Q, the acousto-optic interaction is divided into three regimes: (1) $Q \ll 1$, Raman–Nath; (2) $Q \gg 1$, Bragg; and (3) intermediate for $Q \sim 1$.

32.4. Raman–Nath Regime

When $Q \leq 0.3$, the acousto-optic interaction is called Raman–Nath and the multiple-order diffraction pattern produced is similar to those produced by a thin diffraction grating. In this regime the diffraction can be described as a pure phase modulation similar to a holographic volume grating moving at the speed of sound. Figure 32.3 shows the general scheme of the Raman–Nath diffraction. In this case light is transferred from the zeroth to the first order, from

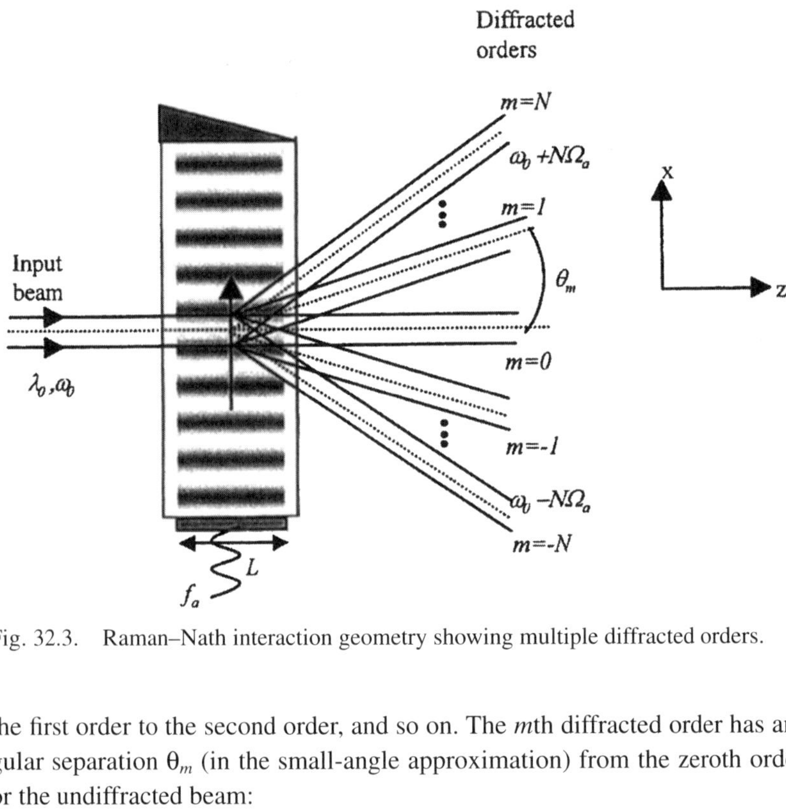

Fig. 32.3. Raman–Nath interaction geometry showing multiple diffracted orders.

the first order to the second order, and so on. The mth diffracted order has angular separation θ_m (in the small-angle approximation) from the zeroth order or the undiffracted beam:

$$\theta_m \approx \pm \frac{m\lambda_0}{n\Lambda}. \tag{32.11}$$

Because $Q \leq 0.3$, the first term on the right-hand side of Eq. (32.10) can be neglected, and by using the boundary conditions at the input of the cell $E_0(0) = E_0$ and $E_m(0) = 0$, the solution is

$$E_m(z) = E_0 e^{-jmX} J_m\left(\frac{2\nu \sin X}{K_a L \tan \theta_0}\right), \tag{32.12}$$

where $X = 0.5(K_a z \tan \theta_0)$, and J_m is the Bessel function of the mth order. The normalized intensity for the mth order at $z = L$, the length of the transducer that is approximately the same as the interaction length, is

$$I_d^m = J_m^2 \left\{ \nu \frac{\sin[(K_a L \tan \theta_0/2)]}{(K_a L \tan \theta_0)/2} \right\}. \tag{32.13}$$

32.5. Bragg Regime

For the values of Klein–Cook parameter $Q > 7$, the acoustic grating is no longer thin, the interaction is called Bragg diffraction, and the diffracted light appears in a single order, as shown in Fig. 32.4. The diffraction condition is given by

$$\sin \theta_0 = \sin \theta_B = \frac{K_a}{2k_i} = \frac{\lambda_0}{2n\Lambda} = \frac{\lambda_0 f_a}{2nV}, \tag{32.14}$$

where θ_B is called the Bragg angle. In general, θ_B is small, no larger than a few

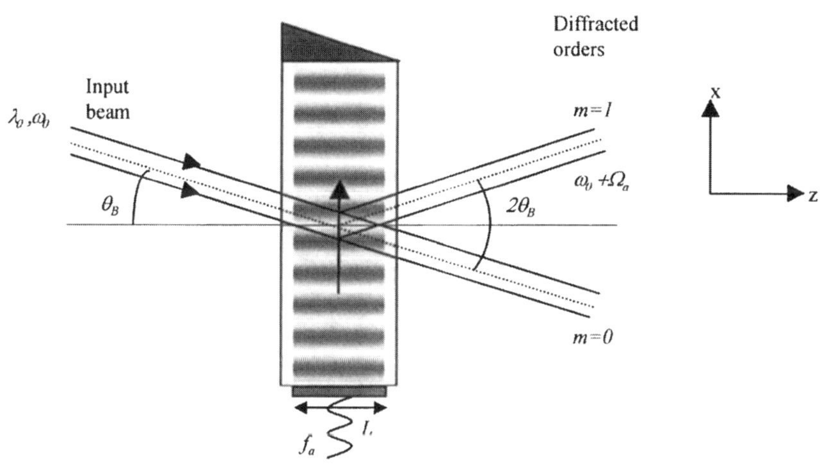

Fig. 32.4. Bragg interaction geometry showing a single diffracted order.

degrees. The wave-vector diagram showing the momentum conservation for the Bragg diffraction given by the following equation is shown in Fig. 32.5 for the +1 order:

$$\mathbf{k}_i + \mathbf{K}_a = \mathbf{k}_d. \tag{32.15}$$

In Bragg diffraction the energy is transferred from the incident optical wave E_0 to the diffracted optical wave E_1. These two waves are given by

$$\frac{dE_0}{dz} + \frac{\nu}{2L} E_1 = 0, \tag{32.16}$$

$$\frac{dE_1}{dz} - \frac{\nu}{2L} E_0 = j\frac{2q}{L} E_1, \tag{32.17}$$

where $q = 0.5K_aL(\sin\theta_B/\cos\theta_0 - \tan\theta_0)$ and it is equal to zero when $\theta_0 = \theta_B$. Solutions of Eqs. (32.16) and (32.17) are

$$E_1(z) = \sin\left(\frac{z\nu}{2L}\right), \tag{32.18}$$

$$E_0(z) = \cos\left(\frac{z\nu}{2L}\right). \tag{32.19}$$

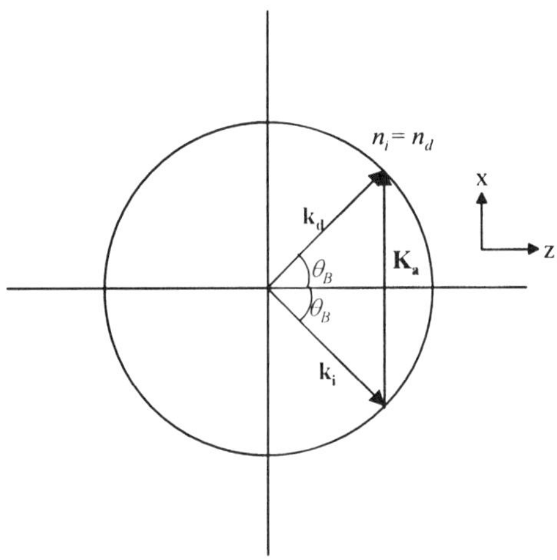

Fig. 32.5. Wave-vector diagram for the isotropic Bragg diffraction.

The Raman–Nath parameter ν is related to the acoustic power P_a by

$$\nu = \frac{2\pi}{\lambda_0 \cos\theta_0} M_2^{1/2} \left(\frac{P_aL}{2H}\right)^{1/2}, \tag{32.20}$$

where H is either width (or height) of the acoustic beam defined by the width (or height) of the transducer, as shown in Fig. 32.2.

The intensity of the first-order diffracted beam in the Bragg regime of operation and during wave-vector phase matching is given by

$$I_{d1} = \sin^2\left(\frac{\nu}{2}\right). \tag{32.21}$$

In general, the ratio I_{d1}/I_0 is called the diffraction efficiency η, and for small ν it can be approximated as

$$\eta \approx \frac{\nu^2}{4} = \frac{\pi^2}{2\lambda_0^2 \cos^2\theta_0} M_2 \left(\frac{P_a L}{H}\right). \tag{32.22}$$

This shows that, for a given acoustic power and device geometry, the materials that have larger values of M_2 will have higher diffraction efficiency. Also, for small ν the efficiency is linearly proportional to the applied acoustic power and the acoustic column length and inversely proportional to the transducer width (or height) and the square of the optical wavelength.

For the Bragg interaction, the 3-dB acoustic full width at half-maximum (FWHM) bandwidth Δf, around the center frequency f_a, for small values of $\eta \ll 1$ is given by

$$\Delta f = 1.8 n V^2 \frac{\cos\theta_0}{\lambda L f_a}. \tag{32.23}$$

Note that Δf is inversely proportional to the acoustic column length L and the acoustic frequency f_a. This is because the acoustic diffraction spread decreases with higher frequency and larger interaction length.

32.6. Intermediate Regime

For intermediate values of Q, Eq. (32.10) does not have simple analytical solutions and needs to be solved numerically. The acousto-optic interaction in this transition regime is a combination of the Raman–Nath and the Bragg regimes and appears as a few dominant diffraction orders. For smaller values of Q there are multiple diffraction orders similar to the Raman–Nath, and as Q increases the number of multiple orders decreases until the Bragg limit is reached and there is only one diffracted beam. In 1967, Klein and Cook carried out a detailed numerical analysis of this region.[3]

32.7. Anisotropic Acousto-Optic Interactions

In 1967, Dixon[4] investigated the acousto-optic interaction in optically anisotropic crystals. He found that a special type of Bragg diffraction can take place that rotates the polarization of the diffracted beam by 90° with respect to the incident beam. He also proved that Bragg matching conditions in a birefringent medium are different from the corresponding phase-matching conditions

[3] W. R. Klein and B. D. Cook, "Unified approach to ultrasonic diffraction," IEEE Trans. Sonics Ultrason. **SU-14,** 123–134, 1967.

[4] R. W. Dixon, "Acoustic diffraction of light in anisotropic media," IEEE J. Quantum Electron. **QE-3,** 85–93, 1967.

valid in an optically isotropic medium. If the incident beam is ordinary polarized then the diffracted beam is extraordinary polarized and vice versa. This is a useful feature of the anisotropic acousto-optic interaction because it can be used to carry out polarization filtering of light. In a birefringent material, the ordinary and extraordinary beams propagate with different speeds and have refractive indices n_o and n_e, respectively, associated with them. For a negative uniaxial crystal $n_o > n_e$ and for a positive uniaxial crystal $n_o < n_e$. To cover the anisotropic case, an equation corresponding to Eq. (32.10) is[5]

$$\frac{\mathrm{d}E_m}{\mathrm{d}z} - j\Delta K_m E_m = \frac{\omega_m^2}{4jk_i c_m c^2}(X_m S E_{m-1} + X_{m+1} S^* E_{m+1}), \qquad (32.24)$$

where ΔK_m is the magnitude of the momentum mismatch given by

$$\Delta K_m = \frac{k_i}{2c_m}\left[1 - \frac{n_m^2}{n_0^2} + 2m\frac{K_a}{k_i}\cos(\theta_a + \theta_0) + \left(\frac{mK_a}{k_i}\right)^2\right], \qquad (32.25)$$

ω_m is the frequency of the mth order, $c_m = \cos\theta_0 + m(K_a/k_i)\cos\theta_a$, n_m is the refractive index of the mth order, X_m is the appropriate component of the acoustically induced susceptibility, X is the susceptibility tensor that describes the photoelastic effect, and S is the strain of the acoustic wave. The coupled-wave equation for zero order can be obtained by observing that $\Delta K_a = 0$, $c_0 = \cos\theta_0$, $E_{-1} = 0$, and $X_1 = -n_d^2 n_i^2 p$, where p is the appropriate photoelastic coefficient, and $n_e = n_d$ and $n_o = n_i$ are the refractive indices for the diffracted and incident beams, respectively. It is given by

$$\frac{\mathrm{d}E_0}{\mathrm{d}z} = j\frac{k_i n_d^2 p S^*}{4\cos\theta_0}E_1. \qquad (32.26)$$

Similarly,

$$\frac{\mathrm{d}E_1}{\mathrm{d}z} - j\Delta K_1 E_1 = j\frac{k_1^2 n_d^2 p S^*}{4k_i c_1}E_0. \qquad (32.27)$$

The figure of merit M_2 needs to be redefined as

$$M_2^e = \frac{n_i^3 n_d^3 p^2}{\rho V^3}. \qquad (32.28)$$

The diffraction efficiency η and 3-dB bandwidth Δf can be approximated by

[5] A. P. Goutzoulis and D. R. Pape, eds., *Design and Fabrication of Acousto-Optic Devices,* Marcel Dekker, 1994.

$$\eta \approx \frac{\pi^2}{2\lambda_0^2} M_2^e \left(\frac{P_a L}{H}\right), \tag{32.29}$$

$$(\Delta f)^2 = \frac{3.6 n_i V^2 \cos\theta_0}{L\lambda_0}, \tag{32.30}$$

where L is the anisotropic interaction length, same as in the isotropic case. The anistropic 3-dB bandwidth is related to the isotropic 3-dB bandwidth by

$$(\Delta f_a)^2 = 2 f_a \Delta f_i. \tag{32.31}$$

This means that by using anisotropic interaction there is a significant increase in the bandwidth.

The frequency dependence of the angles of the incident and diffracted optical beams as measured with respect to the direction perpendicular to the acoustic wave vector $\mathbf{K}_a$ are shown in Fig. 32.6, where wave vectors are shown for an interaction plane perpendicular to the optic axis, and in Fig. 32.7 for a special case when they are parallel. The second case corresponds to a collinear acousto-optic filter design in which both the incident and diffracted optical beams travel in the same direction, and the acoustic wave vector could be either in the same or opposite direction. There is a special case when $\mathbf{K}_a$ is tangential to the inner circle (or ellipse), and the phase matching takes place over

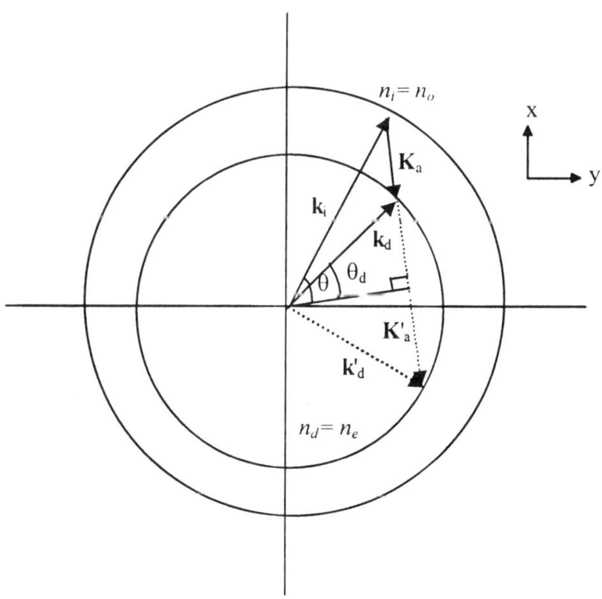

Fig. 32.6. Wave-vector diagram for the anisotropic interaction in the *xy* plane of a crystal.

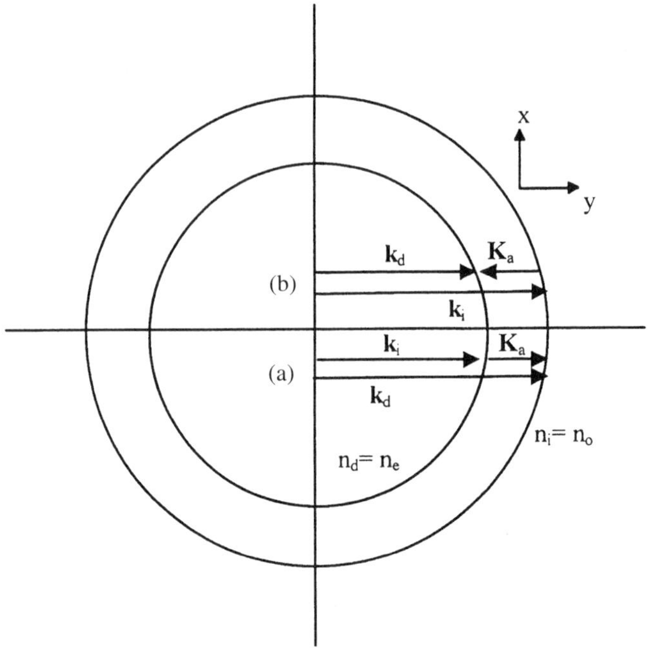

Fig. 32.7. Wave-vector diagram for collinear anisotropic interaction when both incident optical and acoustic beams are in (a) the same and (b) opposite directions.

a wide range of acoustic frequencies. Two types of tangential birefringent band shapes are possible: single and double peak. In general, the double-peak approach gives maximum bandwidth and is used most often in designing wideband deflectors. For most birefringent materials the tangential phase-matching case for the visible-light wavelength region corresponds to gigahertz frequencies. For certain types of crystal, i.e., tellurium dioxide, mercury halides, etc., the frequency is much lower. However, the acoustic frequency of the tangential matching of the $\mathbf{K}_a$ vector in all crystals can be easily tuned by tilting the acousto-optic interaction plane with respect to the optic axis. By changing the direction of light or ultrasound propagation one can also obtain the interaction with the tangential phase matching of optical wave vectors.[6] In this case the tangential planes to $\mathbf{k}_i$ and $\mathbf{k}_d$ remain parallel, as shown in Fig. 32.8. This corresponds to a wide-angle noncollinear acousto-optic tunable filter, in which the incident, diffracted, and acoustic wave fronts all travel in different directions. However, the direction of energy flow of the diffracted beam in the crystal is parallel to the energy flow of the incident beam.

The frequency dependence of the Bragg angles of the incident and diffract-

[6] I. C. Chang, "Acousto-optic filters," Opt. Eng. **20,** 824–829, 1981.

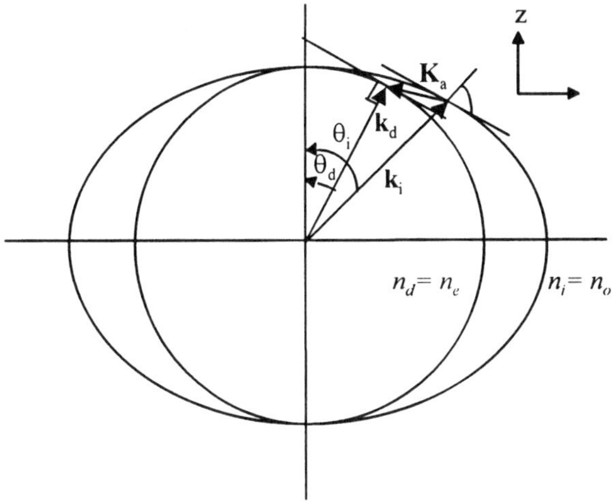

Fig. 32.8. Wave-vector diagram for parallel condition of the tangential planes in a noncollinear acousto-optic tunable filter.

ed optical beams evaluated relatively to the acoustic wave front is given by the following two Dixon equations:

$$\sin \theta_i = \frac{f_a \lambda_0}{2 n_i V} \left[1 + \left(\frac{V}{f_a \lambda_0} \right)^2 (n_i^2 - n_d^2) \right], \tag{32.32}$$

$$\sin \theta_d = \frac{f_a \lambda_0}{2 n_d V} \left[1 - \left(\frac{V}{f_a \lambda_0} \right)^2 (n_i^2 - n_d^2) \right]. \tag{32.33}$$

An examination of Eqs. (32.34) and (32.35) makes it clear that their first term is similar to the isotropic Bragg diffraction, whereas the second term is due to the anisotropic diffraction.

The angles of incidence and diffraction as a function of frequency are plotted in Fig. 32.9, which corresponds to the vector diagram in Fig. 32.6. Two regions in this graph are of great interest: (1) low frequency and (2) middle frequency. The low-frequency region is known as the collinear interaction region, where $\partial\theta/\partial f \rightarrow \infty$ and the angular frequency selectivity is maximum. In this region, phase velocities of the incident, diffracted, and acoustic waves are all collinear to each other. For such an interaction

$$f_a = (n_i - n_d) \frac{V}{\lambda_0}, \tag{32.34}$$

where $n_i = n_o$ and $n_d = n_e$. In the middle-frequency region, $\partial\theta/\partial f \rightarrow 0$ and the angular frequency selectivity is minimum. This region is useful for designing deflectors. Here the center frequency, f_{ac}, is given by

$$f_{ac} = \frac{V}{\lambda_0} \sqrt{|n_i^2 - n_d^2|}. \qquad (32.35)$$

The relationship among the acoustic frequency, the diffracted wavelength, and the angle of incidence for the tangential phase-matching case corresponding to the noncollinear acousto-optic tunable filter shown in Fig. 32.8 and for $\Delta n/n \ll 1$ is approximately given by

$$f_a = \frac{\Delta n V}{\lambda_0} [\sin^2 2\theta_i + \sin^4 \theta_i]^{1/2}, \qquad (32.36)$$

where θ_i is defined relative to the optical axis of the crystal and Δn is the difference in the two refractive indices for a birefringent medium. (It is important to distinguish this Δn from the strain-induced refractive-index change δn.) This expression reduces to Eq. (32.34) for a collinear case when $\theta_i = 90°$.

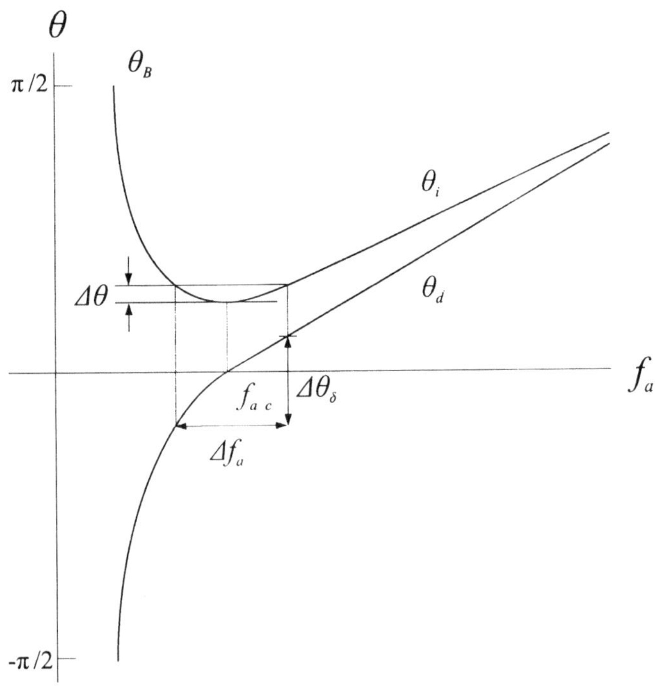

Fig. 32.9. Plot of angles of incidence and diffraction for anisotropic interaction as a function of acoustic frequency.

32.8. Design and Fabrication of an Acousto-Optic Device

The design of an acousto-optic device depends on its application, i.e., whether it is to be used as a light-beam deflector, a modulator, or a tunable filter. The material selection is governed by the transparency of the material in the spectral region of interest, low acoustic attenuation, and strong interaction. A list of selected materials is provided in Table 32.1.

One single parameter can be used to classify whether a device is an acousto-optic deflector, modulator, or filter. This parameter is the ratio of the divergence angle $\Delta\phi$ of the light beam and the divergence angle $\Delta\psi$ of the acoustic beam within the device. We can write this as

$$a \equiv \frac{\Delta\phi}{\Delta\psi}. \tag{32.37}$$

The value of a is -1 for deflectors and Bragg cells, ≈ 1 for modulators, and $+1$ for tunable filters.

To fabricate an acousto-optic cell after the appropriate material is selected, a crystal grown in the proper orientation is obtained and characterized for the orientation of its axes. After that it is cut according to the device design. The next step is to polish it and apply antireflection coatings to the input and output facets. Next a piezoelectric transducer is bonded to it if a plate transducer is used or a thin-film transducer is grown on it. An acoustic absorber is coated on the opposite side. A typical transducer structure consists of a metal top or driving electrode, a piezoelectric crystal, and one or more metal or organic bonding layers that attach the piezoelectric crystal to the acousto-optic medium. Another metal electrode serves as a ground electrode. In general, a single transducer can be used but, depending on the application requirements, more than one transducer can be used. Because the transducer converts an electrical signal into an acoustic signal, both electrical and acoustic losses need to be minimized. For reducing the electrical loss, a 50-Ω impedance matching circuit is included because most transducers have a low value of impedance (a few ohms). The acoustic loss is minimized by the proper choice of the bonding-layer materials and thickness to optimize the acoustic impedance matching. Most commonly used transducers are made of a thin plate of $LiNbO_3$, which has a low dissipation loss and efficiently couples both shear and/or longitudinal acoustic waves into the acousto-optic medium. Another way of fabricating a transducer is by the direct deposition of a ZnO or CdS thin film onto the crystal. The transducer thickness is typically chosen to be close to half of the center acoustic wavelength in the piezoelectric material. The layout of a device with both the driving and ground electrodes is shown in Fig. 32.10.

32.9. Acousto-Optic Deflectors

The acousto-optic effect can be used to electronically manipulate the direction of propagation of light by changing the applied acoustic frequency. A deflector

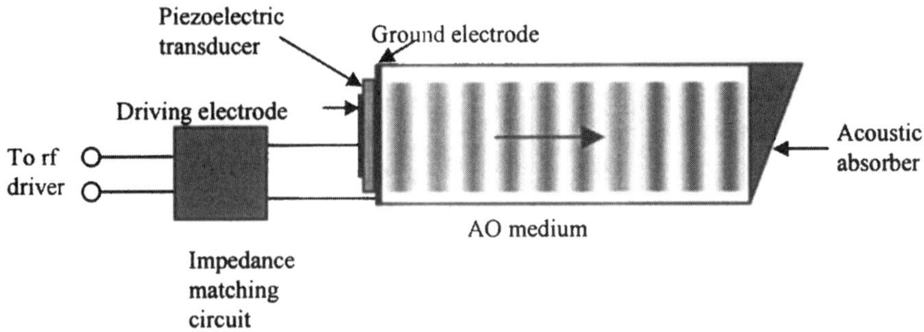

Fig. 32.10. Piezoelectric transducer bonded on an acousto-optic device.

can be designed by using the Bragg interaction such that most of the diffracted light from a collimated incident quasi-monochromatic light beam is deflected in a single order, as shown in Fig. 32.11. The deflection angle is linearly proportional to the applied RF and inversely proportional to the acoustic velocity as given by Eq. (32.14) and for small angles can be written as

$$2\theta_B = \frac{\lambda_0 f_a}{nV}. \tag{32.38}$$

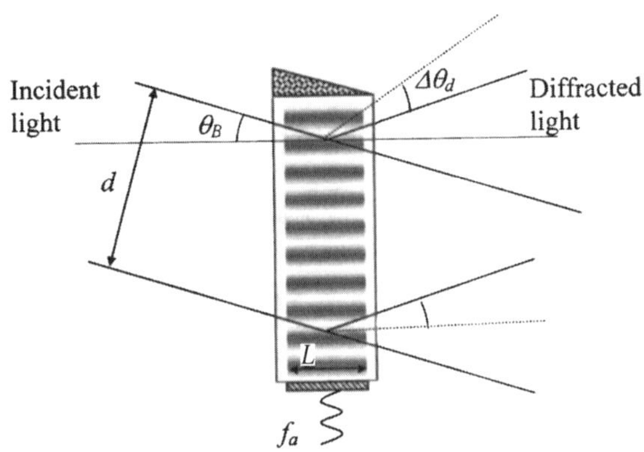

Fig. 32.11. Operation of a deflector.

The most commonly used materials to design the deflectors are $LiNbO_3$, GaP, Ge, TeO_2, $PbMoO_4$, and quartz. The time it takes for the sound to pass across the light beam of width d in the overlap region of length L is known as the time aperture of the device. The main operational parameters that are considered in the design of an acousto-optic deflector are the 3-dB bandwidth, the diffraction efficiency, and the time aperture τ. The time aperture for an incident light beam of width d is given by

$$\tau = \frac{d}{V}. \tag{32.39}$$

The total angular width scanned through for an input RF signal containing a range of acoustic frequencies Δf_a in air is given by

$$\Delta\theta_d = \left(\frac{\lambda}{v}\right)\Delta f_a. \tag{32.40}$$

The optical beam divergence in air is

$$\Delta\phi = \frac{\lambda}{d}. \tag{32.41}$$

To obtain the number of resolvable spots N, $\Delta\theta_d$ is divided by the divergence $\Delta\phi$ of the light beam:

$$N = \tau\Delta f_a. \tag{32.42}$$

This is also known as the time–bandwidth product (TB) of the deflector.

A deflector is designed by optimizing one or more of these parameters for the desired performance requirements and material selection. Four separate Bragg interaction geometries can be used in designing a deflector: isotropic with a single transducer, isotropic with phased-array transducers, anisotropic with a single transducer, and anisotropic with phased-array transducers. If the bandwidth is less than 500 MHz the first interaction geometry is used; otherwise, for higher bandwidths, the other three geometries are preferable. The diffraction efficiency will be given by either Eq. (32.22) or Eq. (32.29), depending on whether the interaction is isotropic or anistropic. A two-dimensional deflector can be designed by use of two identical acoustic beams, one on the horizontal plane and the other on the vertical plane, to produce an *XY* deflection.

32.10. Acousto-Optic Modulators

Modulators are used widely in a large variety of practical systems including laser printers, television displays, *Q*-switching of lasers, optical heterodyning, etc. Numerous optical processing architectures have also been developed that use acousto-optic modulators. The main idea behind these modulators is that, when the amplitude of the electrical signal that drives the transducer is varied, it modulates the amplitude of the diffracted light beam. Also, the diffracted beam is Doppler shifted, thus providing frequency modulation. The most important consideration for optical processing is that the sound cell, used with a modulated carrier, produces an optical replica of an electronic signal that is available simultaneously for parallel processing of information. Figure 32.12

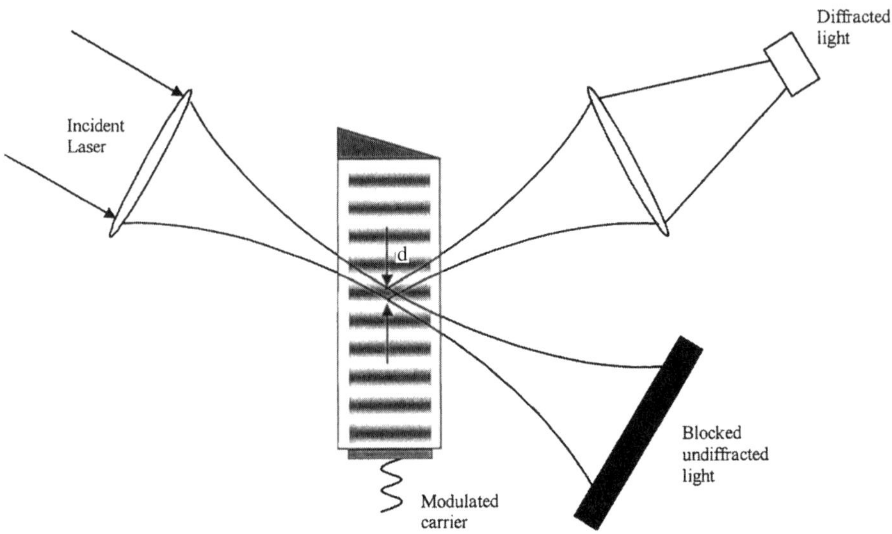

Fig. 32.12. Operation of a modulator.

shows the basic operation of an acousto-optic modulator. Most modulators operate in the Bragg regime. The important operational parameters for a modulator are optical wavelength, 3-dB modulation bandwidth, rise time (determined by transit time $\tau = d/V$ of the sound wave across an optical beam), diffraction efficiency, and dynamic range.

The divergence of the optical beam should be approximately equal to that of the acoustic beam as discussed earlier. The actual value of the divergence ratio of these two beams depends on the trade-off between the desired efficiency and the modulation bandwidth.

Focusing optics can be used to get a divergent optical beam, as shown in Fig. 32.12. The divergence angle of the incident and diffracted focused Gaussian beams with a beam waist of diameter d is given by

$$\Delta\phi = \frac{4\lambda}{\pi n d}. \tag{32.43}$$

The divergence $\Delta\psi$ of an acoustic beam produced by a planar transducer, shown in Fig. 32.2, in the acoustic interaction plane is given by

$$\Delta\psi = \frac{\Lambda}{L}. \tag{32.44}$$

The design of a modulator depends on the choice of suitable parameters to make the two divergences approximately equal. It should be noted here that this depends directly on L and the applied frequency. The transit time τ for a modulator is defined by Eq. (32.39). For pulse modulation, the modulation speed is given by the rise time. This is the time duration needed for the intensity of the diffracted light to increase from 10% to 90% of its steady-state value. This rise time increases with the acoustic transit time τ, and its value is the same order of magnitude.

32.11. Acousto-Optic Tunable Filters

An acousto-optic cell based on the anisotropic Bragg interaction geometry can be designed to filter white light into narrowband single-color light beams by changing the applied RF. Such a filter, called an acousto-optic tunable filter, can be designed with either a collinear or a noncollinear geometry based on the crystal symmetry. These filters can be used to design a no-moving-parts spectrometer with a point detector or a multispectral or hyperspectral imager by use of a two-dimensional detector array. Such filters can also be used to tune the wavelengths of lasers either in an internal or external cavity. These electronically tunable filters are agile, provide random-wavelength access and are fast. Because the anisotropic diffraction involves rotation of the plane of polarization, the diffracted beam can be used to extract the polarization state of the incident beam. In the collinear case, because both the diffracted and the undiffracted zero-order beams travel in the same direction as the incident beam, a set of crossed polarizers is used to separate the diffracted beam from the undiffracted beam. The first polarizer is used to linearly polarize the incident beam, and the analyzer is used to separate the orthogonally polarized diffracted beam from the undiffracted zero-order beam, as shown in Fig. 32.13. For

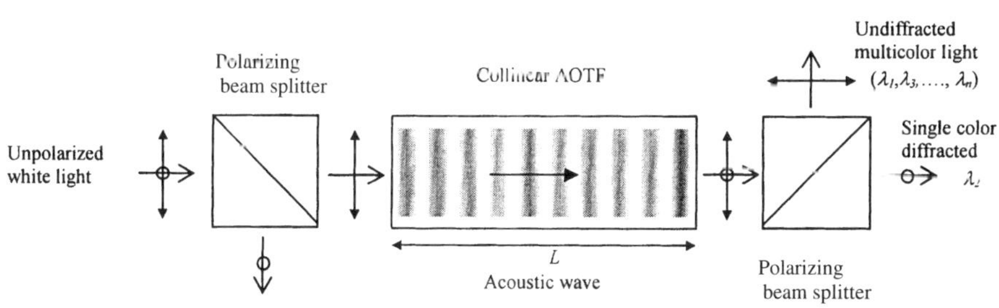

Fig. 32.13. Filtering operation of a collinear acousto-optic tunable filter.

the noncollinear case, polarizers are not needed because the two diffracted beams are spatially separated from the undiffracted beam, as shown in Fig. 32.14. The polarization dependence is exploited to design spectropolarimetric imagers that can provide both spectral and polarization signatures of objects. Lithium niobate ($LiNbO_3$) based integrated optics and bulk paratellurite

(TeO_2) devices are being used in the optical communication field for wavelength-division-multiplexing applications.

Most important parameters in the collinear design are the center wavelength of the passband, resolving power ($R = \lambda_o/\Delta\lambda$), and full angular aperture outside the crystal. Approximations to these (neglecting the spectral dependence of Δn) for an interaction length L are given for FWHM quantities as follows:

$$\lambda_0 = \frac{V\Delta n}{f_a}, \tag{32.45}$$

$$\frac{\Delta\lambda}{\lambda_0} = \frac{0.9\lambda_0}{\Delta nL} \equiv \frac{1}{R}, \tag{32.46}$$

$$\delta\theta \cong \left(\frac{0.9\lambda_0}{\Delta nL}\right)^{1/2}. \tag{32.47}$$

Operation of a noncollinear filter is shown in Fig. 32.14. For such a design, the

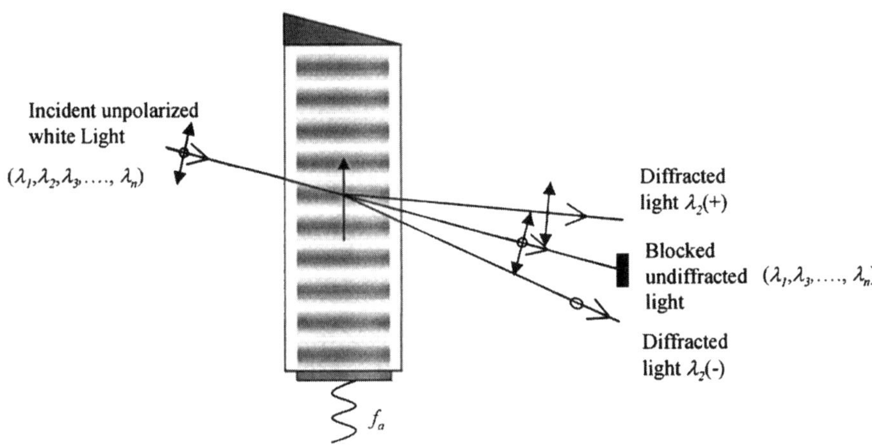

Fig. 32.14. Filtering operation of a noncollinear acousto-optic tunable filter.

most important parameters are the center wavelength of the passband, resolution, angle of incidence relative to the z axis, diffraction efficiency, and angular aperture. The relevant expressions (neglecting the spectral dependence of Δn) are given by the following[7]:

[7] I. C. Chang, "Tunable acousto-optic filters: An overview," Opt. Eng. **16,** 455–460, 1977; J. Xu and R. Stroud, *Acousto-Optic Devices: Principles, Design, and Applications,* Wiley, 1992.

$$\lambda_0 = \frac{\Delta n V}{f_a}[\sin^2 2\theta_i + \sin^4 \theta_i]^{1/2}, \tag{32.48}$$

$$\frac{\Delta\lambda}{\lambda_0} = \frac{0.9\lambda_0^2}{\Delta n L \sin^2 \theta_i} \equiv \frac{1}{R}, \tag{32.49}$$

$$\delta\theta = 2n\left[\frac{0.9\lambda_0}{L\Delta n|(3\cos^2\theta_i - 1)|}\right]^{1/2}, \tag{32.50}$$

and

$$\delta\phi = 2n\left[\frac{0.9\lambda_0}{L\Delta n(\cos^2\theta_i + 1)}\right]^{1/2}, \tag{32.51}$$

where $\delta\theta$ and $\delta\varphi$ are the FWHM polar and azimuth angles, respectively, outside the crystal. For certain geometries $\delta\theta$ and $\delta\varphi$ may be nearly equal to each other, depending on the acoustic propagation direction.

32.12. Guided-Wave Acousto-Optic Devices

So far all our discussion has been focused on bulk acousto-optic interaction and devices. However, planar integrated-optics acousto-optic devices are also widely used, especially in optical communication. In such devices both light and sound are confined to a thin layer on a substrate. The most commonly chosen material for such devices is $LiNbO_3$ in which a waveguiding layer is formed by the creation of a graded-index region using diffusion techniques. Surface acoustic waves are excited by using an interdigital transducer fabricated on top of the optical guiding layer. Propagation of these waves creates a moving index grating in the optical waveguide that diffracts an incident guided light wave. Moving corrugation gratings are also formed at both the air–waveguide and the waveguide–substrate interfaces. The dominant mechanism for diffraction of light is the coplanar interaction of the guided acoustic and optical waves, although the contributions from the corrugated gratings can be significant in some cases. The guided-wave interaction can be Raman–Nath with several diffraction orders or the isotropic or anisotropic Bragg type, depending on the device design. Guided-wave devices are easy to combine with laser diodes and fiber optics. High-performance guided-wave Bragg modulators, deflectors with gigahertz bandwidth, and acousto-optic tunable filters are commonly used in wide-bandwidth optical networks and optical communication. Apodization, acoustic focusing, and beam steering are widely used in the integrated optic devices to improve operating parameters of these compact instruments. An integrated-optic modulator is shown in Fig. 32.15.

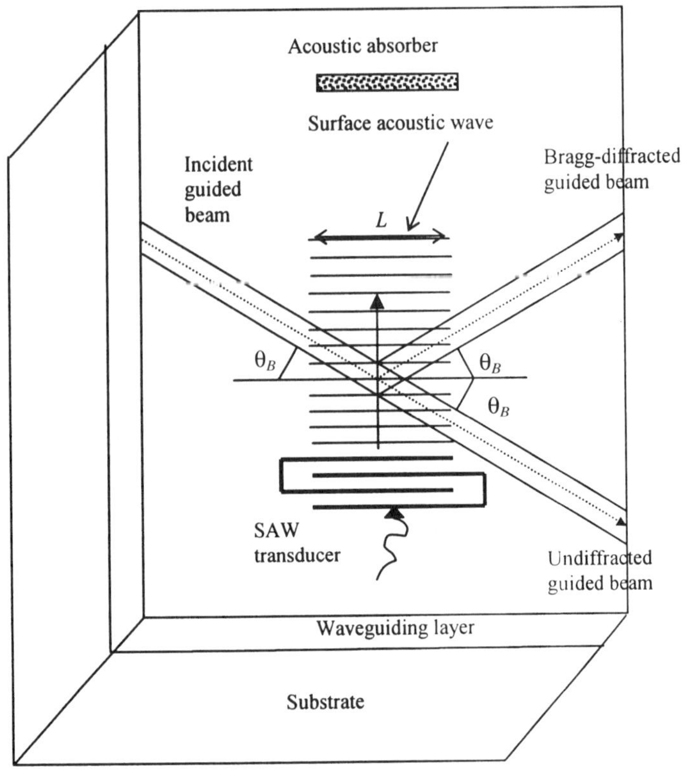

Fig. 32.15. An integrated-optic Bragg modulator.

Further Reading

Acousto-Optics

L. Brillouin, "Diffusion de la lumiere et des rayons x par un corps transparent homogene: Influence de l'agitation thermique," Ann. Phys. (Paris) **17,** 88–122, 1921.

R. Lucas and P. Bquard, "Proprietes optiques des milleux solides et liquids soumis aux vibration elastiques ultr sonores," J. Phys. Radium **3,** 464–477, 1932.

P. Debye and F. W. Sears, "On the scattering of light by supersonic waves," Proc. Natl. Acad. Sci. USA **18,** 409–414, 1932.

C. V. Raman and N. S. Nath, "The diffraction of light by high frequency sound waves: Part I," Proc. Indian Acad. Sci. **2,** 406–412, 1935.

S. M. Rytov, "Light diffraction on ultrasonic waves," USSR Phys. Acad. Sci. Bull. **2,** 223–259, 1937.

A. B. Bhatia and W. J. Noble, "Diffraction of light by ultrasonic waves. I. General theory," Proc. R. Soc. London Ser. A **220,** 356–368, 1953.

R. Adler, "Interaction between light and sound," IEEE Spectrum, 42–54, May 1967.

J. Sapriel, *Acousto-Optics,* S. Francis and B. Kelly, trans., Wiley, 1979.

A. Korpel, *Acousto-Optics,* Marcel Dekker, 1988.

A. Yariv and P. Yeh, *Optical Waves in Crystals,* Wiley, 1984.

I. C. Chang, "Acousto-optic devices and applications," in *Handbook of Optics II,* M. Bass, E. W. Van Stryland, D. R. Williams, and W. L. Wolfe, eds., McGraw-Hill, 1995, pp. 12.1–12.54.

N. J. Berg and J. M. Pellegrino, eds., *Acousto-Optic Signal Processing Theory and Implementation,* 2nd ed. Marcel Dekker, 1996.

Materials

Y. Ochida and Y. Ohmachi, "Elastic and photoelastic properties of TeO_2 single crystal," J. Appl. Phys. **40,** 4692–4695, 1969.

D. A. Pinnow, "Guide lines for the selection of acoustooptic materials," IEEE J. Quantum Electron. **QE-6,** 223–238, 1970.

N. Uchida, "Optical properties of single-crystal paratellurite (TeO_2)," Phys. Rev. B **4,** 3739–3745, 1971.

G. A. Coquin, D. A. Pinnow, and A. W. Warner, "Physical properties of lead molybdate relevant to acousto-optic device applications," J. of Appl. Phys. **42,** 2162–2168, 1971.

N. Uchida and N. Niizeki, "Acoustooptic deflection materials and techniques," Proc. IEEE **61,** 1073–1092, 1973.

N. B. Singh and D. J. Todd, eds., *Growth and Characterization of AO Materials,* Materials Science Forum Vol. 61, Trans Tech Publications, 1990.

N. B. Singh and R. Mazelsky, "Development of mecurous halides for acousto-optic devices," Curr. Top. Crystal Growth Res. **2,** 435–444, 1995.

Acousto-Optic Deflectors

A. Korpel, A. Adler, D. Desmares, and T. M. Smith, "An ultrasonic light deflection system," IEEE J. Quantum Electron. **QE-1,** 60–61, 1965.

E. I. Gordon, "A review of acoustooptical deflection and modulation devices," Proc. IEEE **54,** 1391–1400, 1966.

A. Korpel, A. Adler, D. Desmares, and W. Watson, "A television display using acoustic deflection and modulation of coherent light," Appl. Opt. **5,** 1667–1675, 1966.

E. G. H. Lean, C. F. Quale, and H. J. Shaw, "Continuous deflection of laser beams," Appl. Phys. Lett. **10,** 48–51, 1967.

A. W. Warner, D. L. White, and W. A. Bonner, "Acousto-optic light deflectors using optical activity in paratellurite," Appl. Phys. **43,** 4489–4495, 1972.

I. C. Chang and D. L. Hecht, "Doubling acousto-optic light deflector resolution utilizing second-order birefringent diffraction," Appl. Phys. Lett. **27,** 517–518, 1975.

G. A. Coquin, J. P. Griffin, and L. K. Anderson, "Wide-band acoustooptic deflectors using acoustic beam steering," IEEE Trans. Sonics Ultrason. **SU-17,** 34–140, 1970.

Acousto-Optic Modulators

A. H. Rosenthal, "Application of ultrasonic light modulation to signal recording, display, analysis and communication, IRE Trans. Ultrason. Eng., 1–5, 1961.

H. V. Hance and J. K. Parks, "Wide-band modulation of a laser beam, using Bragg-angle diffraction by amplitude-modulated ultrasonic waves," J. Acoust. Soc. Am. **38,** 14–23, 1964.

A. Korpel and R. L. Whitman, "Visualization of a coherent light field by heterodyning with scanning laser beam," Appl. Opt. **8,** 1577–1580, 1969.

D. Maydan, "Acousto-optic pulse modulators," IEEE J. Quantum Electron. **QE-6,** 15–24, 1970.

D. Maydan, "Fast modulator for extraction of internal laser power," J. Appl. Phys. **41,** 1552–1559, 1970.

H. Carleton and R. Soref, "Modulation of 10.6 μm laser radiation by ultrasonic diffraction," Appl. Phys. Lett. **9,** 110–112, 1966.

T. M. Turpin, "Time integrating optical processors," Real-Time Signal Processing Proc. **154,** 196–203, 1978.

L. Bademian, "Parallel-channel acousto-optic modulation," Opt. Eng. **25,** 303–308, 1986.

W. T. Rhodes, "Acousto-optic signal processing: Convolution and correlation," Proc. IEEE **69,** 65–79, 1981.

V. Voloshinov and V. Molchanov, "Acousto-optic modulation of radiation with arbitrary polarization direction," Opt. Laser Technol. **27,** 307–313, 1995.

Integrated-Optic Acousto-Optic Devices

Y. Ohmachi and J. Noda, "$LiNbO_3$ TE-TM mode converter collinear acousto-optic interaction," IEEE J. Quantum Electron. **QE-13,** 43–46, 1977.

C. S. Tsai, "Guided-wave acoustooptic Bragg modulators for wide-band integrated optic communications and signal processing," IEEE Trans. Circuits Syst. **CAS-26,** 1072–1098, 1979.

K. W. Cheung, D. A. Smith, J. E. Baran, and J. J. Johnson, "Wavelength-selective circuit and packet switching using acousto-optic tunable filters," Proc. IEEE Global Telecommunications Conf. **3,** 1541–1547, 1990.

D. A. Smith and Z. Bao, "Technology and applications of the integrated acousto-optic filter," in *Proceedings of MELECON 96,* IEEE, 1997, pp. 100–107.

Acousto-Optic Tunable Filters

S. E. Harris and R. W. Wallace, "Acousto-optic tunable filter," J. Opt. Soc. Am. **59,** 744–747, 1969.

S. E. Harris, S. T. K. Smith, and D. K. Winslow, "Electronically tunable acousto-optic filter," Appl. Phys. Lett. **15,** 325–326, 1969.

S. E. Harris, S. T. K. Smith, and R. S. Feigelson, "$CaMoO_4$ electronically tunable optical filter," Appl. Phys. Lett. **17,** 223–225, 1970.

J. A. Kusters, D. A. Wilson, and D. L. Hammond, "Optimum crystal orientation for acoustically tuned optic filter," J. Opt. Soc. Am. **64,** 434–440, 1974.

R. B. Watson, S. A. Rappaport, and E. E. Frederick, "Imaging spectrometer study of Jupiter and Saturn," Icarus **27,** 417–423, 1976.

I. B. Belikov, G. Ya. Buimistryuk, V. B. Voloshinov, L. N. Magdich, M. I. Mit'kin, and V. N. Parygin, Sov. Tech. Phys. Lett. **10,** 517–519, 1985.

V. M. Epikhin, F. L. Vizen, and L. L. Pal'tsev, "Acoustooptic filtering of arbitrarily polarized radiation," Sov. Phys. Tech. Phys. **32,** 1149–1153, 1987.

W. H. Smith and K. M. Smith, "A polarimetric spectral imager using acousto-optic tunable filters," Exp. Astron. **1,** 329–343, 1991.

D. R. Suhre, M. Gottlieb, L. H. Taylor, and N. T. Melamed, "Spatial resolution of imaging noncollinear acousto-optic tunable filters," Opt. Eng. **31,** 2118–2121, 1992.

V. Voloshinov, "Close to collinear acousto-optic interaction in paratellurite," Opt. Eng. **31,** 2089–2094, 1992.

I. C. Chang, "Collinear beam acousto-optic tunable filters," Electron. Lett. **28,** 1255–1256, 1992.

V. Voloshinov, V. Molchanov, and J. Mosquera, "Spectral and polarization analysis of optics images by means of acousto-optics," Opt. Laser Technol. **28,** 119–127, 1996.

E. N. Lewis, P. J. Treado, and I. W. Levin, "A miniaturized, no-moving-parts Raman spectrometer," Appl. Spectrosc. **47,** 539–543, 1993.

V. I. Pustovoit and V. E. Pozhar, "Collinear diffraction of light by sound waves in crystals: Devices, applications, new ideas," Photon. Optoelectron. **2,** 53–69, 1994.

D. R. Suhre and J. G. Theodore, "White-light imaging by use of a multiple passband acousto-optic tunable filter," Appl. Opt. **35,** 4494–4501, 1996.

N. Gupta and N. F. Fell, Jr., "A compact collinear Raman spectrometer," Talanta **45,** 279–284, 1997.

N. Gupta, "Acousto-optic tunable filters," Opt. Photon. News **8**(11), 23–27, 1997.

N. Gupta, ed., *Proceedings of the First Army Research Laboratory Acousto-Optic Tunable Filter Workshop,* ARL-SR-54, Army Research Laboratory, 1997.

D. R. Suhre and E. Villa, "Imaging spectroradiometer for 8–12 μm region with a 3-cm^{-1} passband acousto-optic tunable filter," Appl. Opt. **37,** 2340–2345, 1998.

L. J. Denes, M. S. Gottlieb, and B. Kaminsky, "Acousto-optic tunable filters in imaging applications," Opt. Eng. **37,** 1262–1267, 1998.

N. Gupta and R. Dahmani, "AOTF Raman spectrometer for remote detection of explosive," Spectrochem. Acta Part A **58,** 1453–1456, 2000.

N. B. Singh, D. Suhre, N. Gupta, W. Rosch, and M. Gottlieb, "Performance of TAS crystal for AOTF imaging," J. Cryst. Growth **225,** 124–128, 2001.

N. Gupta, L. Denes, M. Gottlieb, D. Suhre, B. Kaminsky, and P. Metes, "Object detection using a field-portable spectropolarimetric imager," Appl. Opt. **40,** 6626–6632, 2001.

G. Georgiev, D. A. Glenar, and J. J. Hillman, "Spectral characterization of acousto-optic filters used in imaging spectroscopy," Appl. Opt. **41,** 209–217, 2002.

N. Gupta, R. Dahmani, and S. Choy, "Acousto-optic tunable filter based visible-to near-infrared spectropolarimetric imager," Opt. Eng. **41,** 1033–1038, 2002.

J. Sapriel, D. Charissoux, V. Voloshinov, and V. Molchanov, "Tunable acousto-optic filters and equalizers for WDM applications," J. Lightwave Technol. **20,** 864–871, 2002.

33

Miscellany

William L. Wolfe

33.1. Introduction

This chapter contains information that may be of use to the optical engineer that just does not fit anywhere else.

33.2. Maxwell's Fish Eye

This is one example of an absolute instrument for which the optical length of any curve in object space is equal to the optical length of its image. The Maxwell fish eye has a medium of refractive index $n(r)$:

$$n(r) = \frac{1}{1 + (r/a)^2} n_0.$$

33.3. Weierstrass Sphere

In a sphere of relative refractive index n and radius r, the aplanatic points are at r/n and nr. This concept has been used for immersion lenses. See also Chap. 13.

33.4. Lagrange Invariant

The Lagrange invariant is

$$ny \sin \theta = n'y' \sin \theta',$$

where n is the refractive index, y is the ray height, and θ is the angle the ray makes with the optical axis. The primes indicate a different part of the system, perhaps after a lens.

33.5. Throughput Invariance

Throughput is invariant throughout an optical system. This is related to the Lagrange invariant in the following way. Let L be the Lagrange invariant and T be the throughput; then

$$L_1 = L_2, \quad L_1^2 = L_2^2, \quad T_1 = T_2.$$

33.6. Radiance Invariance

The radiance divided by the square of the refractive index is invariant in an optical system.

33.7. Sine Condition

If the Lagrange invariant is obeyed, then the sine condition is obeyed.

33.8. Fresnel Plates

Fresnel zone plates are usually circles of alternating transparent and opaque annuli, as shown in Fig. 33.1. Arranged in the proper way, they can be used as

Fig. 33.1. Fresnel zone plate.

lenses. If the distances are successive half-wave increments, then there is a bright spot on axis. The imaging formula is

$$\frac{1}{i} + \frac{1}{o} = \frac{m\lambda}{s_m^2} = \frac{1}{f},$$

where the radii of the zones are given by

$$r_m = \sqrt{m\lambda \frac{ab}{a+b}}.$$

The areas of the zones are given by

$$A_m = \frac{\pi ab\lambda}{a+b}.$$

They all have the same area.

There are also fainter spots on axis at $f/3$, $f/5$, etc.

Wood[1] made phase-reversal zone plates; that is, instead of alternate transparent and opaque zones, the zones have opposite phases. This is accomplished by a variation in the thickness of the transparent zones. This provides a more intense image.

33.9. Radiometric Temperatures

There are several defined temperatures that relate to true temperatures that can be measured radiometrically.

33.9.1. Radiation Temperature

The radiation temperature is the temperature of a blackbody that has the same total radiation as the real body. The relationship is

$$T_{\text{rad}} = \varepsilon^{1/4} T,$$

where ε is the total emissivity of the real body and T is its temperature.

33.9.2. Brightness Temperature

The brightness temperature, also called the radiance temperature, is the temperature of a blackbody that has the same radiance at a particular wavelength as the real body. The relationship is

$$T_b = \frac{T}{1 - \frac{1}{x} \ln \varepsilon}$$

if the sources can be approximated by the Wien expressions, and

[1] R. W. Wood, *Physical Optics,* Macmillan, 1921.

$$T_b = \frac{x}{\ln\left(1 + \frac{e^x - 1}{\varepsilon}\right)}$$

if the approximation cannot be made;

$$x = \frac{c_2}{\lambda T}.$$

33.9.3. Ratio Temperature

The ratio temperature is the temperature of a blackbody that has the same ratio of radiances at two wavelengths as the real body. The relationship is

$$T_{\text{ratio}} = \frac{T}{1 - \frac{1}{X}\ln \varepsilon_r}$$

if the Wien approximation may be used, where

$$X = \frac{c_2}{\Lambda T},$$

$$\Lambda = \frac{\lambda_1 \lambda_2}{\lambda_1 + \lambda_2},$$

$$\varepsilon_r = \frac{\varepsilon_2}{\varepsilon_1}.$$

If the Wien approximation cannot be made, the resulting expression is very complicated:

$$\frac{e^{c_2/\lambda_1 T_r} - 1}{e^{c_2/\lambda_2 T_r} - 1} = \frac{e^{c_2/\lambda_1 T} - 1}{e^{c_2/\lambda_2 T} - 1}.$$

33.9.4. Distribution Temperature

One could also use three wavelengths, or four, but these are not defined. The natural extension is the distribution temperature. The distribution temperature is the temperature of a blackbody that has the closest relative spectrum to that of the real body. The relationship is the same as that for the ratio temperature except that is is summed over all wavelengths and one calculates the best fit on a mean-square basis.

33.9.5. Color Temperature

Color temperature is the temperature of a blackbody that has the same color coordinates as the real body.

34

Conversions and Standards

William L. Wolfe and Kiebong Nahm

34.1. Introduction

The tables presented in this chapter should be used in the following way. The unit in the left column equals the number in the cell times the unit on the top row. For example, in Table 34.1, the example table, 1 in. = 2.54 cm; also, there are 2.54 cm/in. Thus, multiply the unit in the left column by the number to get the unit at the top. For some values, only one direction of conversion is given. Using the reciprocal will give the reverse information. Standard exponential notation is sometimes used; that is, 1.5E-6 = 1.5×10^{-6} and 1.5E6 = 1.5×10^{6}.

Not all tables have entries in all cells. Some conversions just seemed very unlikely. Most of the conversions from other units to the SI unit are from the *Handbook of Chemistry and Physics.*[1] The conversions in other columns were obtained by calculation from these values.

Table 34.1. Examples

	Inch	Centimeter
1 in. =	1	2.54
1 cm =	0.3937	1

34.2. Length

The standard of length is the meter (m); it is defined as the distance traveled by light in a vacuum during a time interval of 1/299792458 s. The SI base unit is the meter.

Table 34.2 provides metric length conversions and is almost a definition of the numerical prefixes. The entries in the top half of the table are the reciprocals of the corresponding lower-half entries.

[1] D. R. Lide, ed., *Handbook of Chemistry and Physics,* 73rd ed., CRC, 1993.

Table 34.2. Metric Length Conversions

	Å	nm	μm	mm	cm	m	km	Mm
1 Å =	1							
1 nm =	1E1	1						
1 μm =	1E4	1E3	1					
1 mm =	1E7	1E6	1E3	1				
1 cm =	1E8	1E7	1E4	1E1	1			
1 m =	1E10	1E9	1E6	1E3	1E2	1		
1 km =	1E13	1E12	1E9	1E6	1E5	1E3	1	
1 Mm =	1E16	1E15	1E12	1E9	1E8	1E6	1E3	1

Table 34.3 gives some common English and metric length conversions, and Table 34.4 gives some astronomical length conversions.

Some infrequently used units are described here. The *cubit* is an old measure based on the length of a person's forearm—from the elbow to the tip of the middle finger. It ranges from 17 to 21 in. A *furlong* (fur.) is the length of a furrow = 220 yd = 201 m = 0.125 mi = 0.2 km. A *rod* (rd) = 5.5 yd = 16.5 ft = 5.029 m. A *span* is the distance between the tips of the thumb and the little finger of a fully extended hand = 9 in. = 23 cm. A *hand* is the width of a closed, flattened hand = 4 in. A *chain* has 100 links. The *surveyor's chain* = 66 ft = 20 m; an *engineer's chain* = 100 ft = 30 m; a *football chain* = 10 yd. A *fathom* (fath) is generally used to measure ocean depths = 6 ft = 1.8288 m. A *barleycorn* is one-third of an inch. A *mil* is often used to describe wire gauge; it is 0.001 in. or 2.54×10^{-5} m. The *pica* is a unit used in printing. It is 4.217518E-3 m = 6 picas ≈ 1 in. An *em* is the length of an em dash. It is based on the type size; an em dash in 12-point type = 12 points = 1 pica. An *en* is half the length of an *em*.

34.3. Area

There is no standard of area; it is based on the length standard. For the most part, area conversions are just the square of the length conversions, but there are a few measures that are not. Conversions are given in Table 34.5. A *barn*, a unit of nuclear cross section, is 10^{-24} cm^2; an *acre* is 43,560 ft^2, 1/640 mi^2, and 4047 m^2.

34.4. Volume

As with area, there is no standard for volume. The derived SI unit is the cubic meter (m^3). There exist both liquid and dry measures in the English system. A *U.S. gallon* (gal, U.S.) is 3.785412E-3 m^3. A *stere* is equal to 1 m^3. Table 34.6

Table 34.3. Common English and Metric Length Conversions

	Centimeter	Meter	Kilometer	Inch[a]	Foot	Statute Mile	Nautical Mile
1 cm =	1	1E-2	1E-5	3.937007E-1	3.28083E-2	6.21371E-6	5.39595E-6
1 m =	1E2	1	1E-3	3.937007E1	3.28084	6.21371E-4	5.39595E-4
1 km =	1E5	1E3	1	3.9370E4	3.28084E-3	6.21371E-1	5.39595E-1
1 in. =	2.54	2.54E-2	2.54E-5	1	8.3333E-2	1.57828E-5	1.3705E-5
1 ft =	3.048E1	3.0480E-1	3.048E-4	12	1	1.8939E-4	1.64579E-4
1 statute mi =	1.6093E5	1.6093E3	1.6093	6.3360E4	5.280E3	1	8.68952E-1
1 naut. mi =	1.85200E5	1.852E3	1.85200	72913.4	6.07611E3	1.15081	1
1 intl. mi =	1.85205E5	1.609344E3	1.609344	72915.394	6.076276E3	1.0000273	1.150841

[a] Note that the abbreviation for inches, in., always takes a period.

Table 34.4. Astronomical Length Conversions

	Meter	Kilometer	Statute Mile	Astronomical Unit[a]	Light Year	Parsec
1 m =	1					
1 km =	1E3	1				
1 st mi =	1.6093E3	1.6093	1			
1 AU =	1.495979E11	1.495979E8	9.2975699E7	1		
1 light year =	9.46073E15	9.46073E12	3.6528926E9	6.32411E4	1	
1 pc =	3.085678E16	3.08568E13	1.91735E13	2.06265E5	3.2615644	1

[a] The astronomical unit, AU, should not be confused with the atomic unit, a.u.

Table 34.5. Area Conversions

	Square Centimeter	Square Meter	Square Inch	Square Foot	Acre	Square Mile
1 cm^2 =	1					
1 m^2 =	1E4	1				
1 in.^{2} =	6.45160	6.4516E-4	1			
1 ft^2 =	9.290304E2	9.29030E-2	144	1		
1 acre =	4.04687E7	4.046873E3	6.2726688E6	43560	1	
1 mi^2 =	2.5899E10	2.58999E6	4.0144896E9	2.7878400E7	6.3999785E2	1

provides conversions for different measures of dry volume. One cubic centimeter (cc or cm^3) equals one milliliter (ml). A *dry barrel* is 105 dry quarts (115 liters). Dry-volume conversions are given in Table 34.6; wet-volume conversions are in Table 34.7. Table 34.8 gives relations among U.S. and Imperial units. In both systems a pint is equal to four gills and 1 qt = 2 pt. In the U. S. system 1 pt = 16 oz, but 1 Imperial pint = 20 oz.

Liquid-volume measures and their equivalents are given in Table 34.7.

One *shot* is usually 1 oz, and one *jigger* is 1.5 oz or 45 ml. A *pinch* and a *dash* are both small quantities, the first being approximately what you can hold between the thumb and forefinger, the second approximately equal to a *squirt*. Two unusual measures of volume are the butt and the hogshead (hhd). The butt is 2 hhd, and 1 hhd is 63 gal or 238 liters. A barrel (bbl) is generally 31.5 U.S. gal (11 liters), but a barrel of oil is 42 U.S. gal (159 liters). A keg is 5–10 gal (19–38 liters), or 100 lb of nails.

34.5. Angle Conversions

There is no standard for an angle; it is the ratio of two lengths. The two basic systems are degrees (° or deg), arcminutes (arc min), arcseconds (arc sec), and radians (rad), milliradians (mrad), and microradians (μrad). The SI-derived unit is the radian. Because a full circle is 360° and 2π rad, the relationship is that 1 rad = π/180° and 1° = 180/π rad. The military mil is defined as 1/6400 of a circle, so there are 6400 mils in 2π rad and 360°. The angular mil is 1 part in 1000, and is approximately 1 mrad. Table 34.9 includes the military mil, written as mil.

34.6. Mass

The standard of mass is the prototype international kilogram (kg). The SI base unit is the *mole* (mol), which is the amount of substance that contains as many elementary entities as there are atoms in 0.012 kg of carbon 12. For example,

Table 34.6. Dry-Volume Conversions

	Cubic Meter	Cubic Inch	Cubic Foot	Cubic Yard	Acre Foot	Board Foot	Cord	Stere
1 m^3 =	1							
1 cu in. =	1.63871E-5	1						
1 cu ft =	2.831685E-2	1.728E3	1					
1 cu yd	7.64555E-1	1.5552E4	27	1				
1 acre-ft =	1.23350E3	7.52720E7	4.35602E4	1.61334E3	1			
1 fbm =	2.35974E-3	144	8.3333E-2	3.08642E-3	1.9130418E-6	1		
1 cd =	3.62456	2.7648E4	16	1.77777	2.938435E-3	192	1	
1 s =	1	6.102361E4	35.3145	3.92384	1.23350E3	4.2377604E2	2.758955E-1	1

Table 34.7. Liquid U.S. Volume Conversions

	Fluid Ounce	Milliliter	Drop	Teaspoon	Tablespoon
1 fl oz =	1	2.957353E1	360	6	2
1 ml =	3.3814E-2	1	12.17656	0.20289	6.763E-2
1 drop =	2.777E-3	8.2125E-2	1	1.666E-2	5.5555E-3
1 tsp =	1.6667E-1	4.9287	60	1	0.33
1 tbs =	0.5	1.47867E1	180	3	1
1 fifth =	25.6	7.57082E2	9.216E3	153.6	51.2
1 gill =	4	1.18294E2	1.440E3	24	8
1 cup =	8	2.36588E2	2.880E3	48	16
1 pt =	16	4.73176E2	5.760E3	96	32
1 qt =	32	9.4635E2	1.1520E4	192	64
1 gal	128	3.78541E3	4.6080E4	768	256

Table 34.8. U.S. and Imperial Conversions

	Cubic Meter	U.S. Ounce	U.S. Pints	U.S. Quart	U.S. Gallons	Imperial Ounce	Imperial Pint	Imperial Quart	Imperial Gallon
1 m^3 =	1								
1 U.S. oz =	2.957353E-5	1							
1 U.S. pt =	4.731765E-4	8	1						
1 U.S. qt =	9.463529E-4	16	2	1					
1 U.S. gal =	3.785412E-3	64	8	4	1				
1 Imp. oz =	2.841307E-5	0.9607601	6.00475E-2	3.00237E-2	7.5059386E-3	1			
1 Imp. pt =	5.682614E-4	1.9215204E1	1.2009493	0.6004751	0.1501187	20	1		
1 Imp. qt =	1.136522E-3	3.843038E1	2.4018987	1.2009495	0.3002373	40	2	1	
1 Imp. gal =	4.546090E -3	1.5372159E2	9.638898	4.8194495	1.2009499	160	8	4	1

Table 34.9. Angle Conversions

	Degree	Arcminute	Arcsecond	Radian	Milliradian	Microradian	Mil
1° =	1	60	3600	1.74532E-2	1.7453E1	1.7453E4	1.777777E1
1 arc min =	1.66666E-2	1	60	2.90882E-4	2.90888E-1	2.90888E2	
1 arc sec =	2.777E-4	1.66666E-2	1	4.848136E-6	4.848136E-3	4.8481336	
1 rad =	57.2957795	3.43774677E3	2.06265E5	1	1E3	1E6	1.018592E3
1 mrad =	5.72957795E-2	3.43774677	2.06265E2	1E-3	1	1E3	1.018592
1 μrad =	5.72957795E-5	3.437746E-3	2.06265E-1	1E-6	1E-3	1	1.01859E-2
1 mil =	5.625E-2			9.81747E-4	0.981747	9.81747E2	1

1 mol of H_2 contains ~6.022×10^{23} such molecules and 1 mol of photons at 10^{14} Hz has an energy of ~39.9 kJ. The *short ton* or *avoirdupois ton* is 2000 lb. The *long ton* is 12% larger and is also called the *freight ton* or *displacement ton.* The *metric ton,* also called *tonne,* is 1000 kg. *Shipping ton* and *measurement ton* are also used as volume measures, typically 40 ft^3.

A *dram* (dr) is an apothecary measure equal to 60 grains, 1/8 oz, 3.89 g, and sometimes 1/16 oz, 27.34 grains, or 1.77 g. The *atomic mass unit* (amu), also called the dalton, is 1/12 the mass of the carbon 12 atom. The *avoirdupois pound* (lb avdp) is 7000 grains. The *troy pound* (lb t) is 5760 grains. Table 34.10 contains mass conversions.

34.7. Density

Density is the ratio of the mass of an object to its volume. Thus all the mass and volume conversions apply. The most common English unit is pounds per cubic foot; the metric equivalents are kilograms per cubic meter and grams per cubic centimeter. Specific gravity is the ratio of the density of an object to the density of water. Because water has a density of 1 g/cm^3 in the metric system (0.99984 at 0 °C and 1 atm, exactly), the specific gravity and the density are (almost) numerically equal. The density of water in the English system is 62.5 lb/ft, so the density is 62.5 times the specific gravity in these units. Table 34.11 shows some of the most common conversions.

34.8. Time

The standard of time is the duration of 9,192,631,770 periods of radiation corresponding to the transition between two hyperfine levels of the ground state of cesium 133. The SI base unit is the second. Table 34.12 provides some useful time conversions based on the common year and common divisions of time. A *common* year (yr) is 365 days. A *lunar* year, an *equinoctial* year, a *solar* year, and a *tropical* year are synonyms for a year of 12 lunar months, 12 lunar orbits, 365 days, 5 hours (h), 48 minutes (min), and 46 s = 365.2418 days. A *siderial* year is 20 min longer; 365.256 mean solar days = 8766.144 mean solar h. A *shake* is 1E-8 s. A jiff or a jiffy is a very short time.

34.9. Speed Conversions

Speed is the time rate of change of position; velocity is a vector in which speed is the magnitude. The main speed conversions are presented in Table 34.13. A knot is 1 nautical mile per hour (mph). The miles in the table are statute miles.

Table 34.10. Mass Conversions

	Milligram	Gram	Kilogram	Tonne	Ounce	Pound	Ton
1 mg =	1						
1 g	1E3	1					
1 kg =	1E6	1E3	1				
1 tonne =	1E9	1E6	1E3	1			
1 oz =	2.834952E4	2.834952E1	2.834952E-2	2.834952E-5	1		
1 lb =	453592E5	453.5924E2	4.539524E-1	4.539524E-4	32	1	
1 ton =	9.071847E8	9.071847E5	9.071847E2	0.9071847	6.4E4	2000	1
1 long ton =	1.016047E9	1.016047E6	1.016047E3	1.016047	71680	2240	1.112

Table 34.11. Density Conversions

	Grams per Cu-bic Centimeter	Kilograms per Cubic Meter	Grams per Liter	Milligrams per Milliliter	Pounds per Cubic Inch	Pounds per Cubic Foot	Ounces per Cubic Inch
g cm^{-3} =	1						
kg m^{-3} =	1E-3	1					
g l^{-1} =	1E-3	1E3	1	1			
mg ml^{-1} =	1E-3	1E3	1	1			
lb $in.^{-3}$ =	2.767990E1	2.767990E4	2.767990E-2	2.767990E-2	1	1728	32
lb ft^{-3} =	1.601846E-2	1.601846E1	1.601846E-5	2.19732E-5	5.787037E-4	1	1.85185E-2
oz $in.^{-3}$ =	1.729994	1.729994E3	1.729994E-3	1.729994	3.125E2	54	1

Table 34.12. Time Conversions

	Second	Minute	Hour	Day	Week	Year
1 s =	1					
1 min =	60	1				
1 h =	3600	60	1			
1 day =	8.6400E4	1.440E3	24	1		
1 week =	6.04800E5	1.0080E4	168	7	1	
1 yr =	3.1536000E7	5.24160E5	8760	365	52	1
1 siderial yr =	3.1558118E7	5.2596863E5	8.7661439E3	365.255	52.179	1.003455

34.10. Force

Force is a vector quantity that is defined either as mass times acceleration or the time rate of change of momentum. The units ounce (oz) and pound (lb) in the table are avoirdupois or force. A unit called the *kip* is 1000 lb. Table 34.14 provides the conversions.

34.11. Torque

Torque about a point acting on a body is the product of the linear distance from the point and the force acting perpendicular to that line. The SI unit is the Newton-meter (Nm). Table 34.15 gives conversions among other units.

34.12. Pressure

Pressure is force perpendicular to a surface per unit area. The SI-derived unit is the torr (abbreviated Torr), which is the pressure of 1 mm Hg at 0 °C and standard gravity. Table 34.16 provides a number of pressure conversions. The Pascal (Pa) is the same as the Newton per square meter (N/m^2). The heights of water are at 6 °C and 4 °C for inches and feet, respectively.

34.13. Work and Energy

The SI-derived unit for work and energy is the joule (J). Table 34.17 provides conversions among the different units for work. There are several different types of calorie (cal). The one in the table is the 15° calorie. The mean calorie is 4.19002 J; the thermochemical calorie is 4.184; the 15° calorie, 4.18580; the 20° calorie, 4.18190. The kilogram-calories (kg-cal), the ones that measure food intake, are 1000 times these values, kilocalories. There are also several different British thermal unit (Btus). The one in the table is the current mean value. Before 1956 the International Steam Table value was 1055.04. The 39 °F value is 1059.67 J; the 60 °F value, 1054.68 J; the thermochemical,

Table 34.13. Speed Conversions

	Meters per Second	Kilometers per Second	Kilometers per Hour	Inches per Second	Feet per Second	Miles per Hour	Knots
1 m s^{-1} =	1						
1 km s^{-1} =	1E3	1					
1 km h^{-1} =	2.777778E-1	2.777778E-4	1				
1 in. s^{-1} =	2.54E-3	2.54E-6	7.055555E-10	1			
1 ft s^{-1} =	3.048E-1	3.048E-4	8.466666E-8	12	1		
1 mph =	4.470319E-1	4.470319E-4	1.6093151	17.6	1.466666	1	
1 knot =	5.144444E-1	5.144444E-4	1.8519998	20.25408	1.68784	1.1508	1

Table 34.14. Force Conversions

	Newton	Dyne	Kilogram	Ounce	Pound	Poundal
1 N =	1					
1 dyn =	1E-5	1				
1 kg =	9.806650	9.806650E5	1			
1 oz =	2.780139E-1	2.780139E4	2.8349528E-7	1		
1 lb =	4.448222	4.44822E5	0.4535924	32	1	
1 poundal =	1.382550E-1	1.382550E4	1.4098E-2	9.994591E-1	0.0310809	1

Table 34.15. Torque Conversions

	Newton-Meter	Dyne-Centimeter	Kilogram-Meter	Ounce-Inch	Pound-Inch	Pound-Foot
1 Nm =	1					
1 dyne-cm =	E-7	1				
1 kg-m =	9.806650	9.806650E-7	1			
1 oz-in. =	7.061552E-3	7.06155E-10	7.200779E-4	1		
1 lb-in. =	1.129848E-1	1.129848E-8	1.15212E-2	32	1	
1 lb-ft =	1.355818	1.355818E-7	1.382563E-1	12	12	1

1054.35 J. The ton of energy is a ton of TNT, trinitrotoluene, as used in measuring the energy of nuclear weapons.

34.14. Temperature

Temperature scales are set by transition points. The International Practical Temperature Scale is based on various transition points.[2] The most recent definition is the 1990 scale. It is described in Table 34.18. The standard is 1/273.16 of the thermodynamic temperature of the triple point of water.

To convert from the Kelvin scale to the Celsius or centigrade scale, add 273.15 to the Celsius temperature to get the temperature in degrees kelvin:

$$T(\mathrm{K}) = T(\mathrm{C}) + 273.15.$$

For instance, the freezing point of water is 0 °C and 273.15 K.

To convert a temperature in Fahrenheit to one in Rankine, add 459.67 to the Fahrenheit temperature:

$$T(\mathrm{R}) = T(\mathrm{F}) + 459.67.$$

For instance, the freezing point of water is 32 F and 491.67 R.

A Fahrenheit or Rankine degree is 9/5 or 1.8 times larger than a Celsius or Kelvin degree. To convert from Rankine to Kelvin, multiply by 5/9. To convert from Kelvin to Rankine, multiply by 9/5. For instance, the freezing point of water is 491.67 R and 273.15 K.

The equations from Celsius and Fahrenheit and vice versa are

$$T(\mathrm{F}) = 1.8T(\mathrm{C}) + 32,$$

$$T(\mathrm{C}) = [T(\mathrm{F}) - 32]9/5.$$

Specific temperatures are listed in Table 34.19.

[2] H. Preston-Thomas, "The International Temperature Scale of 1990," Metrologia **27,** 3, 1990.

Table 34.16. Pressure Conversions

	Pascal	Bar	Atmosphere	Torr	μm Hg	Pounds per Square Inch
1 Pa =	1	1E-5	9.8692E-6	7.5006E-3	7.5006	1.450377E-4
1 bar =	1E5	1	9.8692E-1	7.5006E-3	7.5006E5	1.450377E1
1 atm =	1.01325E5	1.01325	1	760	760000	14.69594
1 Torr =	133.322	1.33322E-3	1.31579E-3	1	1000	1.933672E-2
1 μm Hg	0.133322	1.33322E-6	1.31579E-6	1E-3	1	1.933672E-5
1 psi	6.894757E3	6.894757E-2	6.8046E-2	5.17151E1	5.17151E4	1

Table 34.17. Work and Energy Conversions

	Joule	Erg	Calorie	Electron Volt	British Thermal Unit	Watt-Hour
1 J =	1					
1 erg =	1E-7	1				
1 cal =	4.18580	4.18580E7	1			
1 eV =	1.60219E-19	1.60219E-12	3.82679E-20	1		
1 Btu =	105587E3	1.05587E10	252.2504	6.59016E21	1	
1 Whr =	3.60000E3	3.60000E10	860.0506	2.24692E22	3.4095	1
1 ton =	4.2E9	4.2E16	1.00339E9	2.621412E28	3.97776E6	1.166666E6
1 ft lb =	1.355818	1.35582E7	0.323908	8.46228E18	1.28407E-3	3.76616E-4
1 ft poundal =	4.214011E-2	4.214011E5	0.010067	2.630157E17	3.99103E-5	1.170558E-5

Table 34.18. Fixed Points of the 1990 International Temperature Scale

Material	State[a]	Temperature (deg K)
He	VP	3–5
eq-H_2	TP	13.8033
eq-He	VP	≈17
He	VP	≈20.3
Ne	TP	24.5561
O_2	TP	54.3584
Ar	TP	83.8058
Hg	TP	234.3156
H_2O	TP	273.16
Ga	MP	302.9146
In	FP	429.7485
Sn	FP	505.078
Zn	FP	692.677
Al	FP	933.473
Ag	FP	1234.93
Au	FP	1337.33
Cu	FP	1357.77

[a] TP = triple point, MP = melting point, FP = freezing point, eq-H_2 = equilibrium between the ortho and para states.

Table 34.19. Specific Temperatures

Rankine	Kelvin	Fahrenheit	Celsius	Point
0	0	–459.6	–273.15	Absolute zero
491.67	273.15	32	0	Freezing point of H_2O
491.68	273.16	32.01	0.01	Triple point of H_2O
671.67	373.15	212	100	Boiling point of H_2O
558.27	310.15	98.6	37	Body temperature

34.15. Power Conversions

Power is the time rate of energy change. The watt, equal to 1 J s^{-1}, is the SI-derived unit of power. There are five different types of horsepower: boiler, electric, metric, water, and imperial. The relations are given in Table 34.20. Table 34.21 contains conversions for other units.

34.16. Spectroscopic Conversions

Various units are used in spectroscopy to specify a wavelength or frequency. Table 34.22 gives some of these. The electron volt is an energy unit and is related to frequency and wavelength by way of the energy of a photon, hc/λ or

Table 34.20. Horsepower

	Watt	Boiler	Electric	Metric	Water	Imperial
W	1					
boiler	9.80950E3	1	13.149469	13.337204	13.149469	13.154754
electric	7.46000E2	7.60487E-2	1	13.148706	0.9999464	1.0004023
metric	7.35499E2	7.49782E-2	1.0142774	1	0.98587	0.98632
water	7.46043E2	7.60531E-2	1.0000536	1.01433	1	1.00046
imperial	7.4570E2	7.60181E-2	0.99959	1.013869	0.99954	1

$h\nu$. The Kayser is a unit of wave number, the number of waves in a centimeter (usually), and is related to energy by the speed of light.

34.17. Photometric Conversions

The photometric SI standard is the candela (cd). It is the intensity in a given direction of a source that emits monochromatic radiation of frequency 5.40×10^{14} Hz and has a radiant intensity in that direction of 1/683 W sr^{-1}. In use with other units, the candela is usually written as the candle. The lumen (lm) is an SI-derived quantity for luminous flux.

Luminous energy is usually given as lumen hour, lumen minute, and lumen second; the last is also called a talbot. Luminous energy density is given in lumen seconds per cubic meter and related units.

Luminous flux (areal) density is another derived unit, lumens per unit area. The SI-derived unit is the lux (lx), one lumen per square meter. Table 34.23 provides some conversions. The phot is 1 lm cm^{-2}. The foot-candle is 1 lm ft^{-2}. The lux is 1 lm m^{-2}.

The SI-derived unit for luminance is the candela per square meter, sometimes called the nit. Table 34.24 provides some luminance conversions. One stilb is a candela per square centimeter; one foot-candle is a candela per square foot.

34.18. Nail Sizes

Information about nail sizes is given in Table 34.25.

34.19. Wire Gauge

Wire gauges start at 0, and this gauge is 0.060 in. Each increase in gauge by 1 corresponds to an increase in diameter of 0.013 in.

34.20. Hardness

The four popular measures of hardness are Moh, Knoop, Vickers, and Brinell. Moh values are determined by which material scratches which and are

Table 34.21. Power Conversions

	Watt	British Thermal Unit per Minute	Foot Pound per Second	Kilogram Calorie per Minute	Erg per Second	Horsepower
1 Watt =	1					
1 Btu/min =	1.75725E1	1				
1 ft lb s^{-1} =	1.355818	7.71556E-2	1			
1 kg cal/min =	6.973333E-2	3.96832E-3	5.14326E-2	1		
1 erg s^{-1} =	1E-7	5.69071E-9	7.37562E-8	1.434034E-6	1	
1 hp =	7.456999E2	42.43556	550	1.069359E4	7.456999E9	1
1 kg m s^{-1} =	9.806650	0.558068	7.2330136	1.406307E2	9.806650E7	1.31509E-2

Table 34.22. Spectroscopic Conversions

	Erg	Electron Volt	Hertz	Kayser
1 erg =	1	6.241924E11	1.509189E26	5.0341119E15
1 eV =	1.60207E-12	1	2.417826E14	8.06555E3
1 Hz =	h = 6.6260755E-27	4.135946E-15	1	3.33564E-11
1 K = cm^{-1} =	hc = 1.9864477E-16	1.239842E-4	c = 2.997925E10	1

Table 34.23. Photometric Flux Density Conversions

	Lux	Phot	Milliphot	Foot-Candle
1 lux =	1	1E-4	0.1	9.29E-2
1 phot =	1E4	1	1E3	9.29E2
1 milliphot =	10	1E-3	1	9.29E-1
1 foot-candle =	1.0764E1	1.0764E-3	1.0764	1

Table 34.24. Luminance Conversions

	Candela per Square Meter	Stilb	Candela per Square Foot	Apostilb	Lambert	Foot-Lambert
1 cdm^{-2}	1					
1 stilb =	1E4	1				
1 cd ft^{-2} =	10.764	1.0764E-3	1			
1 apostilb =	1/π = 3.183098E-1	1/πE4 = 3.183098E3	(2.9/π)E-2 = 9.230986E-3	1		
1 L =	(1/π)E4 = 3.183098E3	1/π = 3.183098E-1	929/π = 2.9570988E2	1E4	1	
1 ft-L =	10.76/π = 3.42501	1/(929π) = 3.42637E-4	1/π = 3.183098E-1	10.764	1.076E4	1
1 skot =	(1/π)E-3 = 3.183098E-4	(1/π)E-7 = 3.183098E-8	2.95717E-5	1E-3	1E-7	9.293666E-5

Table 34.25. Nail Sizes

	Common Nails			Other Nails	
Length (in.)	Penny	Gauge	Nails per Pound	Gauge	Nails per Pound
1	2	15	875	15.5	1010
1.25	3	14	585	14.5	635
1.5	4	12.5	315	14	470
1.75	5	12.5	270	14	405
2	6	11.5	180	12.5	235
2.25	7	11.5	160	12.5	210
2.5	8	10.25	105	11.5	145
2.75	9	10.25	95	11.5	130
3	10	9	70	10.5	95
3.25	12	9	62	10.5	86
3.5	16	8	50	10	70
4	20	6	30	9	50

nonlinear. Knoop values are obtained by pressing a pyramidal indenter into the material and measuring the area of the indentation. Vickers hardness is measured the same way, but expressed as the ratio of the indented area to the load (in square millimeters per kilogram). The Brinell indenter is a sphere. The Moh values are listed in Table 34.26. A good relationship between the Moh values and the Knoop values is

$$\text{Knoop} = \text{Moh}^2 + 4.5\ \text{Moh}^3.$$

Table 34.26. Hardness Values

Material	Moh	New Moh	Knoop
Talc	1	Talc	
Gypsum	2	Gypsum	32
Calcite	3	Calcite	135
Fluorite	4	Fluorite	163
Apatite	5	Apatite	430
Orthoclase	6	Orthoclase	560
Quartz	7	Vitreous silica	820
Topaz	8	Quartz, Stellite	1340
Titanium nitride	9	Topaz	1800
Diamond	10	Garnet	7000
	11	Fused zirconia	
	12	Fused alumina	
	13	Silicon carbide	
	14	Boron carbide	
	15	Diamond	

Further Reading

C. D. Hodgman, comp., *Mathematical Tables from the Handbook of Chemistry and Physics,* CRC, 1947.

R. S. Burrington, *Mathematical Tables,* Handbook Publishers, 1948.

D. R. Lide, ed., *Handbook of Chemistry and Physics,* 73rd ed., CRC, 1992.

M. Bass, J. Enoch, E. Van Stryland, and W. L. Wolfe, *Handbook of Optics,* McGraw-Hill, 2001, Vol. III.

Index

C